Springer Series on
SIGNALS AND COMMUNICATION TECHNOLOGY

Signals and Communication Technology

Circuits and Systems Based on Delta Modulation
Linear, Nonlinear and Mixed Mode Processing
D.G. Zrilic ISBN 3-540-23751-8

Functional Structures in Networks
AMLn – A Language for Model Driven Development of Telecom Systems
T. Muth ISBN 3-540-22545-5

RadioWave Propagation for Telecommunication Applications
H. Sizun ISBN 3-540-40758-8

Electronic Noise and Interfering Signals
Principles and Applications
G. Vasilescu ISBN 3-540-40741-3

DVB
The Family of International Standards for Digital Video Broadcasting, 2nd ed.
U. Reimers ISBN 3-540-43545-X

Digital Interactive TV and Metadata
Future Broadcast Multimedia
A. Lugmayr, S. Niiranen, and S. Kalli
ISBN 3-387-20843-7

Adaptive Antenna Arrays
Trends and Applications
S. Chandran (Ed.) ISBN 3-540-20199-8

Digital Signal Processing with Field Programmable Gate Arrays
U. Meyer-Baese ISBN 3-540-21119-5

Neuro-Fuzzy and Fuzzy Neural Applications in Telecommunications
P. Stavroulakis (Ed.) ISBN 3-540-40759-6

SDMA for Multipath Wireless Channels
Limiting Characteristics and Stochastic Models
I.P. Kovalyov ISBN 3-540-40225-X

Digital Television
A Practical Guide for Engineers
W. Fischer ISBN 3-540-01155-2

Multimedia Communication Technology
Representation, Transmission and Identification of Multimedia Signals
J.R. Ohm ISBN 3-540-01249-4

Information Measures
Information and its Description in Science and Engineering
C. Arndt ISBN 3-540-40855-X

Processing of SAR Data
Fundamentals, Signal Processing, Interferometry
A. Hein ISBN 3-540-05043-4

Chaos-Based Digital Communication Systems
Operating Principles, Analysis Methods, and Performance Evalutation
F.C.M. Lau and C.K. Tse ISBN 3-540-00602-8

Adaptive Signal Processing
Application to Real-World Problems
J. Benesty and Y. Huang (Eds.)
ISBN 3-540-00051-8

Multimedia Information Retrieval and Management
Technological Fundamentals and Applications
D. Feng, W.C. Siu, and H.J. Zhang (Eds.)
ISBN 3-540-00244-8

Structured Cable Systems
A.B. Semenov, S.K. Strizhakov, and I.R. Suncheley ISBN 3-540-43000-8

UMTS
The Physical Layer of the Universal Mobile Telecommunications System
A. Springer and R. Weigel
ISBN 3-540-42162-9

Advanced Theory of Signal Detection
Weak Signal Detection in Generalized Obeservations
I. Song, J. Bae, and S.Y. Kim
ISBN 3-540-43064-4

Wireless Internet Access over GSM and UMTS
M. Taferner and E. Bonek
ISBN 3-540-42551-9

The Variational Bayes Method in Signal Processing
V. Šmídl and A. Quinn
ISBN 3-540-28819-8

Topics in Acoustic Echo and Noise Control
Selected Methods for the Cancellation of Acoustical Echoes, the Reduction of Background Noise, and Speech Processing
E. Hänsler and G. Schmidt (Eds.)
ISBN 3-540-33212-x

Eberhard Hänsler · Gerhard Schmidt
(Eds.)

Topics in Acoustic Echo and Noise Control

Selected Methods for the Cancellation of Acoustical Echoes, the Reduction of Background Noise, and Speech Processing

With 316 Figures and 32 Tables

Springer

Editors

Professor (em.) Dr.-Ing. Eberhard Hänsler
Technische Universität Darmstadt
Institute of Telecommunications
Merckstrasse 25
D-64283 Darmstadt
Germany

Dr.-Ing. Gerhard Schmidt
Harman/Becker Automotive Systems
Acoustic Signal Processing
Soeflinger Strasse 100
D-89077 Ulm
Germany

Library of Congress Control Number: 2006923555

ISBN-10 3-540-33212-x Springer Berlin Heidelberg New York

ISBN-13 978-3-540-33212-1 Springer Berlin Heidelberg New York

Springer is a part of Springer Science+Business Media.

springer.com

Printed in The Netherlands

Typesetting by the authors and SPI Publisher Services

Cover design: Design & Production, Heidelberg

Printed on acid-free paper SPIN: 11412748 62/3100/SPI- 5 4 3 2 1 0

Preface

The demand for devices that utilize digital speech processing is constantly growing. The desire to carry out tasks "hands-free" is very often the motivation. Examples are voice controlled technical devices, speech or speaker recognition systems, hands-free communication with remote partners, systems to ease communication in noisy environments without using close-talking microphones, and to improve the hearing ability of impaired persons, to name only a few. In the majority of these applications, the existence of acoustical echoes and background noise lead to considerable performance degradations. Methods for the cancellation of echoes and the suppression of background noise, therefore, are of high interest to designers of speech processing systems.

The idea for this book arose immediately after the editors had finished their book on "Acoustic Echo and Noise Control"[1] since a number of subjects could not be treated in sufficient detail and some important topics had to be completely omitted. The editors also came to the conclusion that the value of an additional book would be considerably increased if scientists that are internationally recognized for their work on the related topics would report on the state of the art and on their findings.

The editors approached most of the authors at EUSIPCO '2004 and on the spot they agreed to contribute to this project. The editors feel bound to express their sincere thanks to all of them. Not only did they finish their contributions in good time, they also accepted the proposals of the editors with respect to notation of variables and references. Thus, it should be easier for the reader to jump between different chapters.

The book is organized in five parts. Part I just contains a brief introduction into acoustic echo and noise control, a few remarks on current research topics, and a description of the contents of the book. Part II deals with multi-microphone processing. Having the outputs of more than one microphone

[1] Eberhard Hänsler and Gerhard Schmidt: *Acoustic Echo and Noise Control,* New York, NY: Wiley, 2004

available opens an additional degree of freedom to the designer of speech processing systems.

In Part III advanced methods for echo cancellation such as the identification of sparse impulse responses, selective-tap update, and the application of nonlinear echo paths models are presented. Attempts for an intelligent control of hands-free telephones are introduced. Part IV is devoted to noise reduction procedures. An in-depth treatment of conventional and of advanced time- and frequency-domain methods is given, followed by a model-based approach using Kalman filters.

Selected applications of acoustic echo and noise control systems are outlined in Part V. Auditory scene analysis, spatial sound reproduction by using wave field synthesis, in-car communication, and adaptive signal processing in high-end hearing aids are the topics of this part of the book.

All the authors and the editors hope that this book will become a useful resource for researchers and developers, as well as for doctoral students, who design new advanced procedures or who are on the "rocky road from algorithms to systems".

It is much more than a pure matter of duty that the editors wish to thank all who helped during the preparation of this book. The dedication of the authors has already been mentioned. Further thanks go to members of the Signal Processing and Signal Theory Group at Darmstadt University of Technology and the Acoustic Signal Processing Group at Harman/Becker Automotive Systems at Ulm (Germany) for proof reading and various valuable hints.

Finally, the editors have to thank the Springer Publishing Company, especially Dr. Dietrich Merkle and his colleagues, for their encouragement and their help.

Darmstadt and Ulm, Germany

Eberhard Hänsler
Gerhard Schmidt

Contents

3 Blind Source Separation of Convolutive Mixtures of Audio Signals in Frequency Domain

S. Makino, H. Sawada, R. Mukai, S. Araki

4 Localization and Tracking of Acoustical Sources

G. Doblinger

Part III Echo Cancellation

---

5 Adaptive Algorithms for the Identification of Sparse Impulse Responses

J. Benesty, Y. Huang, J. Chen, P. A. Naylor

List of Contributors

Shoko Araki
NTT Communication Science Laboratories
Kyoto, Japan

Jacob Benesty
Université du Québec
Montreal, Canada

Christina Breining
Siemens AG
Ulm, Germany

Jingdong Chen
Bell Labs, Lucent Technologies
Murray Hill, NJ, USA

Gerhard Doblinger
Vienna University of Technology
Vienna, Austria

Pia Dreiseitel
Smiths-Heimann
Wiesbaden, Germany

Eghart Fischer
Siemens Audiological Engineering Group
Erlangen, Germany

Volkmar Hamacher
Siemens Audiological Engineering Group
Erlangen, Germany

Eberhard Hänsler
Darmstadt University of Technology
Darmstadt, Germany

Tim Haulick
Harman/Becker Automotive Systems
Ulm, Germany

Wolfgang Herbordt
ATR – Spoken Language Communication Research Laboratories
Kyoto, Japan

Ulrich Heute
University of Kiel
Kiel, Germany

Guoning Hu
Ohio State University
Columbus, OH, USA

Yiteng (Arden) Huang
Bell Labs, Lucent Technologies
Murray Hill, NJ, USA

Walter Kellermann
University Erlangen-Nuremberg
Erlangen, Germany

Andy H. W. Khong
Imperial College
London, UK

Ulrich Kornagel
Siemens Audiological Engineering Group
Erlangen, Germany

Fabian Küch
University Erlangen-Nuremberg
Erlangen, Germany

Andreas Mader
Smiths-Heimann
Wiesbaden, Germany

Shoji Makino
NTT Communication Science Laboratories
Kyoto, Japan

Ryo Mukai
NTT Communication Science Laboratories
Kyoto, Japan

Satoshi Nakamura
ATR – Spoken Language Communication Research Laboratories
Kyoto, Japan

Patrick A. Naylor
Imperial College
London, UK

Henning Puder
Siemens Audiological Engineering Group
Erlangen, Germany

Rudolf Rabenstein
University Erlangen-Nuremberg
Erlangen, Germany

Hiroshi Sawada
NTT Communication Science Laboratories
Kyoto, Japan

Gerhard Schmidt
Harman/Becker Automotive Systems
Ulm, Germany

Sascha Spors
University Erlangen-Nuremberg
Erlangen, Germany

Peter Steffen
University Erlangen-Nuremberg
Erlangen, Germany

DeLiang Wang
Ohio State University
Columbus, OH, USA

Abbreviations and Acronyms

1D	One-dimensional
2D	Two-dimensional
3D	Three-dimensional
A/D	Analog-to-digital (also AD)
ABF	Adaptive beamformer
AD	Analog-to-digital (also A/D)
AEC	Acoustic echo canceller
AEVD	Adaptive eigenvalue decomposition
AGC	Automatic gain control
AGC-i	Input controlled automatic gain control
AGC-o	Output controlled automatic gain control
AI	Articulation index
AI-DI	Articulation index - directivity index (weighted-average directivity index)
AM	Amplitude modulation
ANSI	American National Standards Association
AP	Affine projection
APA	Affine projection algorithm
AR	Auto-recursive
ASA	Auditory scene analysis
ASR	Automatic speaker- or speech-recognition
AVC	Automatic volume control
BSS	Blind source separation
BTE	Behind-the-ear
CA	Coefficient adjustment
CAN	Controller area network
CASA	Computational auditory scene analysis
CB	Codebook
CCR	Cartesian coordinate representation
CD	Compact Disc
CMOS	Comparison mean opinion score

CPSD	Cross-power spectral density
CPU	Central processing unit
CU	Categorical loudness unit
CWT	Continuous wavelet transform
D/A	Digital-to-analog (also DA)
DA	Digital-to-analog (also D/A)
DACF	Differential autocorrelation function
DCR	Diagonal coordinate representation
DCT	Discrete cosine transform
DFT	Discrete Fourier transform
DI	Directivity index
DOA	Direction-of-arrival
DRT	Diagnostic rhyme test
DSL i/o	Desired sensation level
DSP	Digital signal processor
DVD	Digital video disc
DWPA	Discrete wavelet-packet analysis
DWT	Discrete wavelet transform
ECF	Echo cancelling filter
EG	Exponentiated gradient
EG±	Exponentiated gradient with positive and negative weights
EIC	Echo and interference canceller
EIR	Echo-to-interference ratio
EM	Estimate maximize
EOS	Equivalent orthogonalized structure
ERLE	Echo-return loss enhancement
ERLS	Exponentiated RLS
F0	Fundamental frequency
FERLS	Fast exponentiate RLS
FFT	Fast Fourier transform
FIR	Finite impulse response
FRLS	Fast RLS
GCC	Generalized cross-correlation
GDCT	Generalized discrete cosine transform
GDFT	Generalized DFT
GEIC	Generalized echo and interference canceller
GMM	Gaussian mixture model
GSAEC	Generalized sidelobe acoustic echo canceller
GSC	Generalized sidelobe canceller
GSM	Global System for Mobile Communications
GSVD	Generalized singular value decomposition
HMM	Hidden Markov Model
IC	Integrated circuit
ICA	Independent component analysis
IDFT	Inverse discrete Fourier transform

IFFT	Inverse fast Fourier transform
IID	Independent, identically distributed
IIR	Infinite impulse response
InfoMax	Information maximization approach
IP	Internet Protocol
IPAPA	Improved proportionate APA
IPNLMS	Improved proportionate NLMS
IR	Interference suppression
ISDN	Integrated Services Digital Network
ITD	Interaural time difference
ITE	In-the-ear
ITU	International Telecommunication Union
JADE	Joint approximate diagonalization of eigenmatrices
KEMAR	Knowles electronic manikin for acoustic research
LBG	Linde-Gray-Buzo
LCLSE	Linearly-constrained LSE
LCMV	Linearly-constrained minimum variance
LEM	Loudspeaker-enclosure-microphone
LMS	Least mean square
LPF	Low-pass filter
LS	Least squares
LSE	Least-squares error
LTI	Linear time invariant
LVQ	Learning vector quantization
LWG	Lambert based gradient
MAC	Multiply-accumulate
MAP	Maximum-a-posteriori
Max-LMS	LMS updating the coefficient with largest amplitude
MC-FDAF	Multi channel frequency-domain adaptive filter
MDF	Multidelay filter
MDVF	Multidelay Volterra filter
MIMO	Multi-input multi-output
MIP	Millions of instructions per second
MISO	Multi-input single-output
MLP	Multilayer perceptron
MMax-AP	AP updating M coefficients with largest input amplitudes
MMax-LMS	LMS updating M coefficients with largest input amplitudes
MMax-RLS	RLS updating M coefficients with largest input amplitudes
MMSE	Minimum mean-square error
MOS	Mean opinion score
MPEG	Motion Picture Expert Group
MRP	Mouth reference point
MRT	Modified rhyme test
MSE	Mean square error
MUSIC	Multiple signal classification

NAL-NL1	National Acoustic Laboratories of Australia - nonlinear, version 1
NAPA	Natural APA
NIIR	Nonlinear infinite impulse response
NL	Nonlinear
NLMS	Normalized LMS
NN	Neural network
NPR	Near-perfect reconstruction
NRM	Real-valued multiplications per output sample
PAMS	Perceptual analysis/measurement system
PAPA	Proportionate APA
PC	Personal Computer
PCM	Pulse code modulation
PDA	Personal digital assistent
PDF	Probability density function
PDS	Power density spectrum
PESQ	Perceptual evaluation of speech quality
PHAT	Phase transform
PNLMS	Proportionate NLMS
PPN	Polyphase network
PPN-FFT	Polyphase analysis system based on FFT
PR	Perfect reconstruction
PSD	(Auto)-Power spectral density
RBF	Radial basis function
RGSC	Robust generalized sidelobe canceller
RLS	Recursive least squares
SAB	Self-adjusting back-propagation
SAEC	Stereophonic AEC
SAM	Short-term average magnitude
SCOT	Smoothed coherence transform
SER	Signal-to-echo ratio
SFG	Signal-flow graph
SIR	Signal-to-interference ratio
SIRP	Sperically invariant random process
SM-NLMS	Set-membership-NLMS
SNR	Signal-to-noise ratio
SOM	Self-organizing map
SPL	Sound pressure level
SPNLMS	Sparse patial update NLMS
SPU-NLMS	Selective-partial-update-NLMS
SRT	Speech reception threshold
STFT	Short-time Fourier transform
SVD	Singular value decomposition
SVF	Second-order Volterra filter
TDD	Time delay difference
T-F	Time-frequency

TOSQA	Telecommunication objective speech quality assessment
UMTS	Universal Mobile Telephone System
VAD	Voice activity detector
VF	Volterra filter
WEVN	Weight error vector norm
WFS	Wave field synthesis
WGN	White Gaussian noise
XM	Exclusive maximum
XMNL	Combination of XM and NL

Part I

Introduction

1

Acoustic Echo and Noise Control – Where did we come from and where are we going?

Eberhard Hänsler[1] and Gerhard Schmidt[2]

[1] Darmstadt University of Technology, Darmstadt, Germany
[2] Harman/Becker Automotive Systems, Ulm, Germany

The invention of the telephone about 150 years ago extended the range of verbal communication between humans beyond the bounds given by the power of their voices. Using this technology, however, was – and still is – inherently connected with some inconveniences. The talkers have to hold a handset such that the loudspeaker is close to their ear and the microphone is adjacent to their mouth. Even then, speech quality is reduced and ambient noise may be picked up. Replacing the handset by a microphone and a loudspeaker now positioned a – short – distance from the talker increases the loss of the transmission loop by say 20 dB [6]. Furthermore, the level of ambient noise collected by the microphone is increased and the echo from the loudspeaker signal is picked up.

Methods of acoustic echo and noise control aimed to remedy these disadvantages exhibit a long history. Originally, efforts were focussed on the development of hands-free telephone systems. In the following section we will highlight some of the important steps toward today's systems. We will also point out that the developments were always linked to the technology available at the time of their proposal.

This introductory chapter will close with an overview of important current developments reported in the following chapters of this book in detail by international experts in the field of acoustic echo and noise control.

1.1 The Journey to Maturity

1.1.1 The Problems to be Solved

To restore the comfort of a face-to-face conversation over a hands-free telephone connection three major problems have to be solved:

- Comfortable volumes of the speech signals have to be provided for both partners without destabilizing the electro-acoustic loop.

- The echoes of the loudspeaker signal(s) picked up by the microphone(s) have to be reduced to an acceptable level without affecting double talk performance.[3]
- Ambient noise has to be removed from the microphone output signal(s) to below a level that might be tolerable in case of binaural listening.

If there is only one hands-free telephone (locally) and the remote talker uses a handset it is this party who suffers mostly from inappropriate solutions to the problems mentioned above. The local talker may move the loudspeaker closer to his ear or increase the volume – increasing the risk that the loop starts howling. His echo and noise problem is minor at most since his partner holds the microphone close to his mouth. Consequently, when algorithms for echo and noise control are designed, most attention has to be paid to the situation of the remote communication partner.

Historically, when efforts to stabilize the electro-acoustic loop started, only classical acoustic means were available. Loudspeakers and microphones in separate units or combined in one housing were put in favorable positions. Furthermore, the walls, floor, and ceiling of the enclosing room had to be treated with absorptive materials [4].

It was not until the 1950s that signal processing means could be considered. Voice controlled switching of the receiving and sending circuit, center clipping and frequency shifting were employed at that time.

1.1.1.1 Voice Controlled Switch

Voice controlled switching (see Fig. 1.1) means that either the receiving or the sending line is interrupted [3]. Thus, only half-duplex communication is provided, double talk is impossible. Proper control of the switching is difficult. It is based on the *estimated* activities of the incoming and the outgoing line. Noise and echoes can cause malfunctions. The beginning and the end of utterances may be "chopped off". For the "inactive" partner of a conversation it is not possible to break in.

A considerable number of modifications to the loss control circuit have been proposed over the years. Instead of switching lines completely on and off, a finite attenuation is inserted and is distributed on the incoming and the outgoing circuit according to the estimated activities. Short-term power estimations – utilized for speech activity detection – with different time constants can improve the performance at the starts and at the ends of words.

All these modifications can reduce, but not completely remove, the problems described above. Nevertheless, voice controlled switching is still used in modern echo control systems. In an environment with adaptive circuits, minimum levels of echo attenuation as required by international standards can only be guarantied by such circuits. However, only the difference of the

[3] The term *double talk* describes periods in which both – the local and the remote – communication partners speak at the same time.

attenuation already provided by echo cancellation and/or echo suppression (see below) and the one called for by the standards has to be inserted. Thus, in these cases, the impact of loss control on speech quality may be hardly noticeable, if at all.

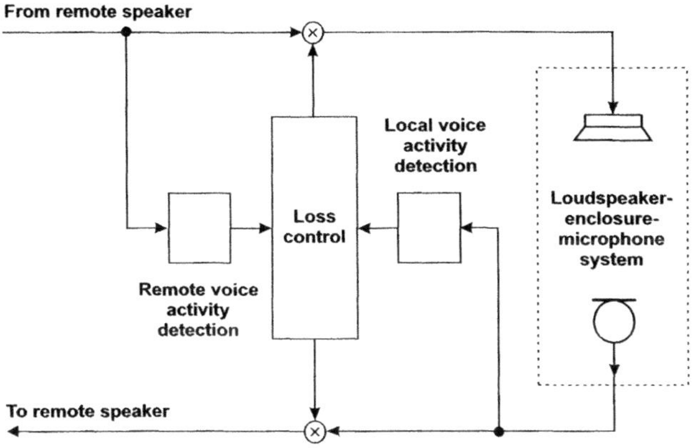

Fig. 1.1. Principle of a loss control circuit.

1.1.1.2 Center Clipper

A center clipper (see Fig. 1.2) inserted into the transmission circuit suppresses small output signals [2]. If these signals contain only the acoustical echo – plus some small ambient noise – the echo is removed completely. However, if the echo is superimposed onto a local speech signal, the center clipper proves to be ineffective and only distorts the speech signal. Again, a large number of modifications – including adaptive thresholds and adaptive slopes – have been proposed over the years. Nevertheless, the use of center clippers in acoustic echo and noise control seems to remain a makeshift solution.

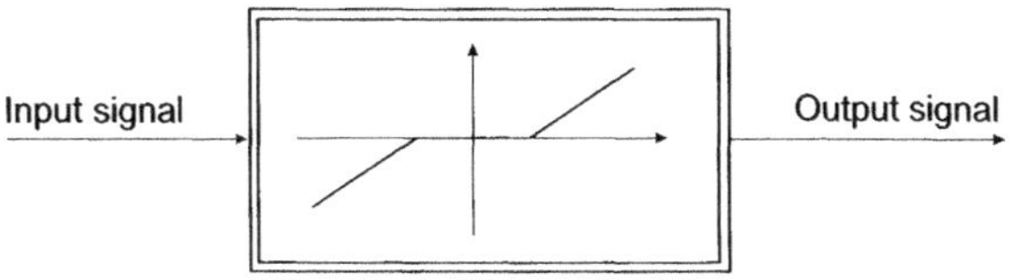

Fig. 1.2. Center clipper.

1.1.1.3 Frequency Shift

The magnitude of the transfer function of a typical loudspeaker-enclosure-microphone (LEM) system exhibits a sequence of maxima and minima with a separation of 5 to 10 Hz (see Fig. 1.3). Peaks and valleys are, respectively, about 10 dB above and below the average magnitude. Based on this observation a *frequency shift* of the loop signal can increase the stability margin [16]. This method was proposed especially for systems like public–address systems where the loudspeaker output signal feeds back directly into the talker's microphone. It can be used in hands-free telephone applications as well. Its primary component is a single–sideband modulator that performs a shift of of the loop signal by a few Hertz. Thus, stationary howling can not build up. It is moved to higher or lower frequencies – depending on whether the modulation frequency is positive or negative – until it "falls" into a minimum of the transfer function of the LEM system.

In speech communication systems frequency shifts of about 3 to 5 Hz are scarcely noticeable. The stability gain achievable with this method depends on the signal and the acoustical properties of the enclosure. For speech signals and rooms with short reverberation times, the stability gain is of the order of 3 to 5 dB; for rooms with long reverberation times it can go up to about 10 dB [15].

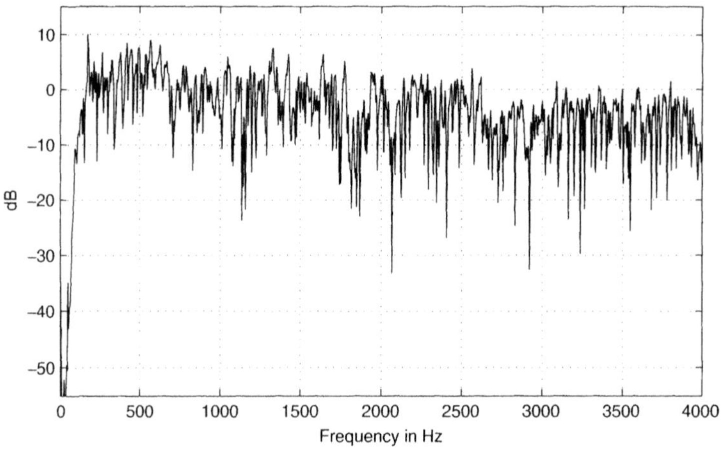

Fig. 1.3. Absolute value of a transfer function measured in a small lecture room.

1.1.1.4 Echo Cancellation and Echo Suppression

The invention of the least mean square (LMS) algorithm in 1960 [19] can be considered as the most important development for adaptive filtering. This procedure became the "work horse" for today's existing enormous variety of algorithms for filter adaptation. Its numerical complexity is proportional to $2N$, where N is the number of filter coefficients. Given a proper step size, it does not cause stability problems. However, its speed of convergence is low especially in case of correlated inputs like speech signals.

The potential of the LMS algorithm for echo cancellation or suppression was recognized soon after its publication. The first application was the cancellation of *electrical* echoes on long distance transmission lines [11, 17]. Compared to *acoustic* echoes, line echoes are considerably shorter. Thus, they require less complex filters. In contrast, the processing of acoustical echoes necessitates adaptive filters that are extremely demanding with respect to signal processing power. It is, therefore, not astonishing that the application of adaptive filters to acoustic echo and noise control was not considered before the late 1970's [14]. Even at that time the signal processing technology to implement those filters could only be seen on the distant horizon.

Simulations and laboratory experiments in the 1980's affirmed the weakness of the LMS algorithm with respect to correlated – e.g. speech – signals. These results started a strong effort by researchers to utilize the recursive least squares (RLS) algorithm for acoustic echo processing. In contrast to the LMS algorithm, the complexity of this procedure grows quadratically with the number N of filter coefficients that have to be adapted. It can handle correlated signals very well since it has a "built in" decorrelation facility. This, however, needs the inversion of the short-term $N \times N$ correlation matrix of the input signal. In the applications considered here, N may range up to the order of several thousands. This matrix can become singular by the nature of the input signal or the estimation procedure. As a consequence, the RLS algorithm frequently becomes instable for echo processing. The stabilization and reduction of the complexity to a linear dependency on the number of filter coefficients was one of the major topics at the first *International Workshop on Acoustic Echo Control* held in 1989 in Berlin, Germany. Despite all the efforts at that time and in the following years, the problems of applying the RLS algorithm to acoustic echo cancellation still seem unsolved. The situation can be highlighted by a cartoon (see Fig. 1.4).

The LMS and the RLS algorithms may be considered extremes in the world of adaptive algorithms. This holds with respect to complexity and numerical problems, but also with respect to their dependence on past signals and settings of the filter coefficients. The LMS algorithm uses only current inputs whereas the RLS procedure looks back on past inputs according to a forgetting factor. In order to stabilize the RLS algorithm it may be necessary to furnish the algorithm with a long memory. This turns out to be a handicap when changes to the LEM system have to be tracked.

The affine projection (AP) algorithm [13], and especially its fast version [9], provides a good compromise between the LMS and RLS algorithms. Compared to the LMS algorithm, numerical complexity is modestly increased. The speed of convergence for speech inputs nearly reaches that of the RLS procedure.

Fig. 1.4. "Better to have a sparrow in the hand than a pigeon on the roof" (by Prof. Helmut Lortz).

These properties are achieved by optimizing the filter coefficients not just with respect to the current input signals – as the LMS algorithm does – but also optimizing for $M-1$ preceding inputs. M is called the order of the algorithm. For $M = 1$ it is equal to the LMS procedure. Like the RLS algorithm, the AP method needs the inversion of a matrix. This, however, is of size $M \times M$ only. For speech inputs M can be chosen in the order of 2 to 5. By comparison, in this situation the RLS algorithm would require inversion of an $N \times N$ matrix with N in the order of 1000.

1.1.1.5 Echo Cancellation

Echo cancellation is achieved by using the output of a filter that attempts to match the LEM system (see Fig. 1.5). Since the latter is changing constantly, the filter has to be adaptive.

During the development of echo cancellation filters (ECFs), a long-winded discussion took place whether a transversal (FIR) or a recursive (IIR) filter is better suited to model the LEM system. Since a long impulse response has to be modelled by the ECF (see Fig. 1.6), an IIR filter seems best suited at first glance. However, upon further inspection, the impulse response exhibits a highly detailed and irregular shape. To achieve a sufficiently good match, the replica must offer a large number of adjustable parameters. Several studies have shown that an IIR filter does not provide a sufficiently large advantage over an FIR filter to justify the enormous cost of controlling its stability [10, 12, 20]. The even more important argument in favor of an FIR filter is that adaptation algorithms for FIR filter are available and that the stability of these filters need no extra control.

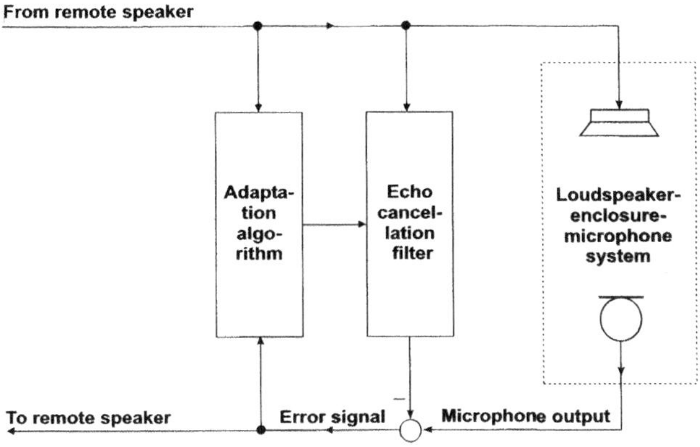

Fig. 1.5. Principle of echo cancellation.

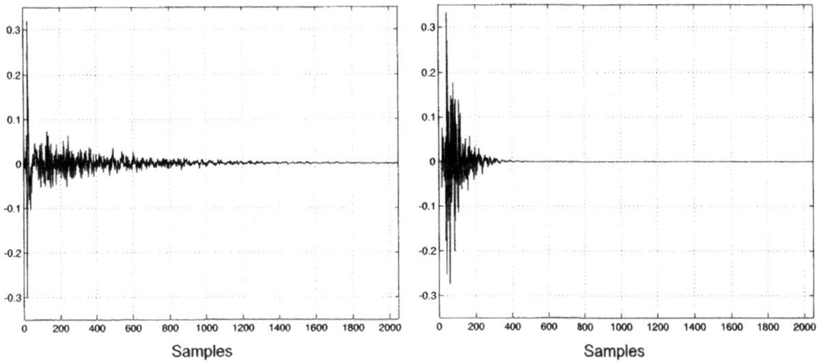

Fig. 1.6. Impulse response of LEMs measured in an office (left) and in a car (right). The sampling rate is 8 kHz.

1.1.1.6 Control of the Filter Adaptation

From a control engineering point of view, the adaptation of the echo cancellation filter is equivalent to the identification of a highly complex system. To make things even more difficult, the adaptation has to be performed in an environment where the signal-to-noise ratio often falls below 0 dB. A short example may help illustrate the complexity of the task: Assume, that the error signal (see Fig. 1.5) suddenly rises. This can have two reasons:

- The local speaker started talking or a local noise started.
- The local speaker changed his position and thus changed the impulse response of the LEM system.

The control of the adaptive filter that can only rely on the output signals of the microphone and the ECF cannot distinguish between the two cases. The reactions, however, have to be diametrical: Adaptation has to be frozen in the first case whereas it has to be opened as much as possible in the second instance. No algorithm for adaptive filters can handle this situation without additional information.

This information has to come from estimates of various quantities. Most of them are not directly measurable. Independent of the currently available processing power and processing speed, the reliability of these estimates depends critically on the length of the signal segment the estimation can be based on. This simply means that it may be necessary to delay control actions until dependable estimates are available. In cases where erroneous control signals lead to a rapid divergence of the filter coefficients – as in the first case of the example given above – rapid actions based on temporary estimates are necessary to prevent "dangerous" situations.

In this respect, the question of applying adaptation algorithms that result in a high speed of convergence of the filter coefficients becomes an additional consideration: Fast adaptation requires a reliable and fast acting control structure. The reaction time of the latter, however, is limited by the time necessary to acquire a sufficiently long signal segment. If this condition is not fulfilled, an algorithm not reacting "nervous like a race horse" may lead to better results.

1.1.1.7 Echo and Noise Suppression

With echo cancellation, the achievable echo attenuation is limited to at most 30 dB in an ordinary office. This is due to thermal fluctuations [5], nonlinearities within the A/D and D/A converters, within the electro-acoustic converters [18], and, lastly, the insufficient length of the echo cancellation filter. To improve echo attenuation, a filter in the transmission circuit is necessary (see Fig. 1.7). The transfer function of this filter is adapted according to the spectrum of the speech signal. A similar filter can be used to suppress ambient noise picked up by the microphone. The coefficients of both filters, however, have to be adapted according to the different properties of the residual echo and the noise.

In contrast to echo cancellation by a filter parallel to the LEM system, echo and noise suppression affects the quality of the transmitted speech signal. Therefore a compromise between speech quality and echo and noise suppression is always necessary.

1.2 State of the Art

Acoustic echo and noise control are among to the most challenging problems in digital signal processing. Many authors confirm this statement. The effort in research and development over the last three decades has been overwhelming. As a result, the problems around "classical" single channel hands-free systems are very well understood and are basically solved. There are systems available

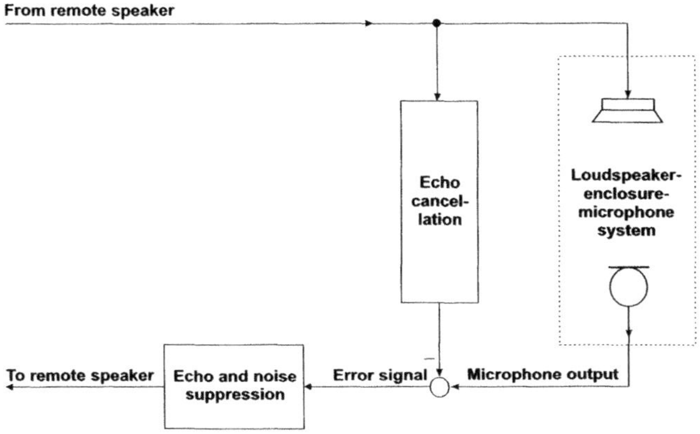

Fig. 1.7. Principle of echo and/or noise suppression.

that function satisfactorily. The fact that systems with poor performance are still in use seems to depend on two reasons: The sales price of consumer products is calculated by the cent. The hands-free functionality has to be implemented with absolutely minimum cost. Furthermore, the benefit of a high-quality echo and noise control system is with the remote communication partner; and he is not the one who pays for the system.

Stereophonic systems still offer open questions. Due to the fact that both signals may be fully dependent on each, the optimal settings of the coefficients of the ECFs are not unique. A remedy is found by artificially distorting one or both signals [1]. Fortunately, real systems behave well and only small distortions are necessary. From a conceptual point of view, a more "elegant" solution seems desirable. Promising approaches towards this goal are the subject of current research.

Parallel to the growth of processing power, new applications that require more and more sophisticated systems move into the field of vision of researchers. Advanced methods are based not only on measured and estimated signals but also on expert knowledge of the underlying processes.

New algorithms for filter adaptation are proposed that are tailored to the specific properties of the echo and noise control process.

Multi microphone and/or multi loudspeaker configurations offer additional degrees of freedom to the echo and noise control problem. Microphone arrays allow the speaker to be located and tracked. The same holds for noise sources. All these methods aim to improve the signal-to-noise ratio of the audio signal. With loudspeaker arrays, radiation patterns are generated such that microphones are located within the minima of sound intensity. Both array approaches are able to reduce the echo problem. Since the electro-acoustic properties of inexpensive microphones exhibit considerable variances, automatic scaling improves the performance of microphone arrays.

Methods of sound source separation isolate individual speakers from mixtures of speech signals and noise. Blind methods for this task are under investigation.

The application of the Kalman filter offers considerably better results in noise reduction than "classical" procedures like spectral subtraction. Processing in subbands overcomes the complexity problem of the filter.

Based on models for speech production and on the properties of human sound perception, methods for enhancing speech signals beyond the quality of the still widely used telephone speech are proposed. Making use of code books for narrow and wide band speech signals and of the masking properties, it is possible to fool the human ear such that the impression of listening to wide band speech is generated from processing narrow band speech signals without using any side information.

The deverberation of speech signals and the inclusion of dictionaries improves the reliability of speech recognition systems.

The development of new methods is – at least partly – related to new application areas. The technology for implementing demanding algorithms in hearing aids has become available only in the last few years. Digital implementations allow for more adjustable parameters. Thus, a better match to hearing impairments is possible.

Demands for greater passenger comfort initiated the search for solutions for in-car passenger communication.

This list is far from complete. With a look into the proceedings of recent signal processing conferences and into signal processing journals it can easily be extended.

The quality of early real-time implementations of echo and noise control systems was bounded by the capacity of signal processors available at that time. This limitation no longer exists. Forgetting for a moment the above remarks on the allowed costs of consumer products, algorithms implemented now or in the near future can perform at their theoretical limits.

The availability of powerful general purpose computers and high level simulation tools allow the simulation of algorithms for acoustic echo and noise control with low effort. Reality, however, turns out to be much more complex than even sophisticated models. Results based only on simulations should be handled with extreme care.

1.3 Outline of this Book

It is the purpose of this book to describe a number of highly important developments in acoustic echo and noise control in more detail. Distinguished authors present overviews and results of their research. Their contributions are organized in four Parts focusing on multi-microphone processing (Part II), echo cancellation (Part III), noise reduction (Part IV), and selected applications (Part V).

In Part II, Chapter 2 addresses the problem of time-varying echo paths, high level background noise, and frequent double-talk. A new joint acoustic echo canceller and beamformer is derived and evaluated. The advantages of the joint system are shown by a realization that integrates a stereophonic echo canceller and a generalized sidelobe canceller. The described solution requires only one echo canceller for an arbitrary number of microphones and no separate adaptation control.

Chapter 3 treats the problem of separating multiple sources of audio signals – as occurs during teleconferences or in hearing aids, to name only two application areas. The input signals to several microphones are convolutive mixtures of speech signals and ambient noise. The solution uses blind independent component analysis in the frequency domain. The phenomena of permutation and circularity are addressed and successfully solved. The authors present a complete solution for source separation. Experimental results are included.

In Chapter 4 techniques for the localization of acoustic sources are presented. The methods are based on only two microphones and perform a precise time-delay estimation. Using these techniques, a moving speaker can be tracked and the direction of high sensitivity of a microphone array can be steered such that it points towards this speaker.

Part III starts with adaptation algorithms for filters that have to model systems with sparse impulse responses (Chapter 5). Sparse in this context means that only a small percentage of the sample values of the impulse response of the original system exhibits values significantly larger than zero. Based on this prior knowledge, general procedures for filter adaptation like the LMS or the RLS algorithm can be considerably improved. It is shown how refined algorithms can be derived and how known procedures and the algorithms developed in this chapter are related to each other. Further, the distinction between algorithms with linear and with nonlinear updates is made. In the case of acoustic echo cancellation, procedures with nonlinear updates can be advantageous.

Cancellation of acoustical echoes needs the update of filters with up to several thousands of coefficients. In Chapter 6 it is shown that the computational complexity associated with this task is reduced by updating only a fraction of the coefficients at a time. Through proper selection the performance of the filter degrades only by a small degree. The sorting of the filter coefficients, however, may lead to a considerable overhead. Fast sorting algorithms described in this chapter overcome this problem. It is also shown that selective update methods may be a remedy against the misalignment of the ECFs due to their non-uniqueness in stereophonic systems.

Realistic electro-acoustic echo paths may contain a number of elements with non-negligible nonlinearities, such as low cost loudspeakers, overloaded amplifiers and non-ideal converters. In Chapter 7 a nonlinear model of the echo path is formulated consisting of a cascade of linear and nonlinear filters. It is explained that second order Volterra filters are suited to model loudspeaker nonlinearities, whereas power filters are proper models of memoryless nonlinearities as they occur, for instance, with overloaded amplifiers or low cost converters. Adaptation algorithms for both filter types are developed and applications to real systems are discussed.

Systems for acoustic echo and noise control require sophisticated procedures to supervise all subsystems in order to avoid performance degradations in case of "dangerous" communication events. Incidents like these are, for example, the sudden onset of double talk or changes in the echo path by movements of the local speaker. To arrive at a robust control structure, the outputs of detectors and estimators have to be combined in an intelligent way. Chapter 8 presents several systematic approaches for this combination. Their additional computational effort is made up for by an improved overall system performance.

Chapter 9 in Part IV gives an in-depth treatment of noise reduction algorithms. The emphasis is put on single microphone solutions based on Wiener filtering and spectral subtraction. The design of Wiener filters is described in the time and in the frequency domain; filtering effects and realisations are explained. A second focal point is spectral subtraction methods and their relationship to Wiener filters. Central to both methods is the estimation of the noise power spectral density that is discussed in the following sections. Finally, techniques for the design of uniform and non-uniform filter banks – including wavelets – are described.

In Chapter 10 a Kalman filter based single channel noise reduction method is presented. It starts with an analysis of speech signals and car noise and the formulation of parametrical models needed for the Kalman filter. A special procedure for the estimation of speech parameters from noisy signals is developed. The complexity problem of the Kalman filter is overcome by a subband approach. Methods are discussed to enhance the noise reduction performance of the filter. A comparison with more conventional noise reduction methods closes this chapter.

Part V opens with considerations about the assessment of the quality of acoustic echo and noise control systems. Subjective listening tests are the most reliable means. To perform such tests, however, is time consuming and expensive. Therefore, especially during the algorithm development phase, the availability of objective tests is desirable. In Chapter 11 both classes of tests are discussed. It is described how they can be performed and how their results have to be evaluated.

Chapter 12 is concerned with auditory scene analysis. This techniques is inspired by the ability of humans to segregate a sound source from a mixture of multiple sources even from only a monophonic signal. A system for computational acoustic scene analysis (CASA) performs successively four subfunctions. Firstly, a peripheral analysis is performed where the auditory scene is decomposed into a time-frequency representation. A feature extraction follows. These features provide the basis of a segmentation and, finally, of a grouping. Here, segments for the sound source of interest – the target – and the interferers are created. Finally, the waveform of the target is synthesized from the related segments. The approach described here is primarily feature-based. Except for unvoiced grouping, no prior knowledge is assumed.

Chapter 13 deals with the synthesis of wave fields, a novel method for spatial sound reproduction. It applies arrays of large numbers of loudspeakers to recreate a sound field in a listening area. Even if the main applications of this techniques are in the areas of entertainment and performing arts, it may also be used to recreate sound fields for human communications. The technique is based on the physical properties of wave propagation. It applies the solution of the acoustic wave equation by Green's function. Signal processing methods to derive the input signals for the loudspeakers of the wave field synthesis system are reported. Exemplary implementations close this contribution.

Chapter 14 deals with so called in-car communication systems. They help to ease communication between passengers in a car. Such a system is especially helpful for passengers seated in the back of the car to understand those seated in the front. The problem that has to be solved is comparable to the one present with public address systems where the electro-acoustic loop is closed within the enclosure and where only very short processing delays are tolerable. In contrast to e.g. hands-free systems, the (local) speech signal and the echo signal are highly correlated. Therefore, new control structures have to be developed. Since standardized quality measures for in-car communication systems do not yet exist, measurements and subjective test are also reported.

The continuous improvements in semi-conductor technology allowed the changeover from analog to digital technology. Chapter 15 describes algorithms implemented in high-end hearing aids to improve the hearing ability and the hearing comfort of impaired people. The procedures have to take into account the special requirements of these devices. For example, the loudspeaker and the microphones are very close together and a high amplification is inevitable. Furthermore, different listening situations call for their automatic classification, enabling the selection of different parameter sets.

References

[1] J. Benesty, D. R. Morgan, M. M. Sondhi: A Better Understanding and an Improved Solution to the Specific Problems of Stereophonic Acoustic echo Cancellation, *IEEE Trans. Speech Audio Process.*, **T-SA-6**(2), 156–165, 1998.

[2] D. A. Berkley, O. M. M. Mitchell: Seeking the ideal in "hands-free" telephony, *Bell Lab. Rec.*, **52**, 318–325, 1974.

[3] A. Busala: Fundamental considerations in the design of a voice-swiched speakerphone, *B.S.T.J.*, **39**, 265–294, 1960.

[4] W. F. Clemency, W. D. Goodale Jr.: Functional design of a voice–switched Speakerphone, *B.S.T.J.*, **40**, 649–668, 1961.

[5] G. W. Elko, E. Diethorn, T. Gänsler: Room impulse response variation due to thermal fluctuation and its impact on acoustic echo cancellation, *Proc. IWAENC '03,* 67–70, Kyoto, Japan, 2003.
[6] J. W. Emling: General aspects of hands-free telephony, *Comm. and Electronics,* **76**(5), 201–205, 1957.
[7] Y. Epharaim, D. Malah: Speech enhancement using a minimum mean-square error short-time spectral amplitude estimator, *IEEE Trans. Acoust. Speech Signal Process.,* **32**(6), 1109–1121, 1984.
[8] Y. Epharaim, D. Malah: Speech enhancement using a minimum mean-square error log-spectral amplitude estimator, *IEEE Trans. Acoust. Speech Signal Process.,* **33**(2), 443–445, 1985.
[9] S. Gay, S. Travathia: The fast affine projection algorithm, *Proc. ICASSP '95,* **3**, 3023–3027, Detroit, MI, USA, 1995.
[10] A. P. Liavas, P. A. Regalia: Acoustic echo cancellation: do IIR filters offer better modelling capabilities than their FIR counterparts? *IEEE Trans. Signal Process.,* **46**(9), 2499–2504, 1998.
[11] R. W. Lucky, H. R. Rudin: Generalized automatic equalization for communication channels, *Proc. IEEE,* **54**(3), 439–440, 1966.
[12] M. Mboup, M. Bonnet: On the adequatness of IIR adaptive filtering for acoustic echo cancellation, *Proc. EUSIPCO '92,* **1**, 111–114, Bruxells, Belgium, 1995.
[13] K. Ozeki, T. Umeda: An adaptive filtering algorithm using an orthogonal projection to an affine subspace and its properties, *Electron. Commun. Jpn.,* **67-A**(5), 19–27, 1984.
[14] G. Pays, J. M. Person: Modèle de laboratoire d'un poste téléphonique à haut-parleur, *FASE,* **75**, 88–102, Paris, France, 1975 (in French).
[15] T. Schertler, P. Heitkämper: Erhöhung der Stabilitätsgrenze von Saalbeschallungsanlagen, *Proc. DAGA '95,* **1**, 331–334, Saarbrücken, Germany, 1995 (in German).
[16] M. R. Schroeder: Improvement of acustic-feedback stability by frequency shifting, *J. Acoust. Soc. Am.,* **36**(9), 1718–1724, 1964.
[17] M. M. Sondhi: An adaptive echo canceller, *B.S.T.J.,* **46**, 497–511, 1967.
[18] A. Stenger, W. Kellermann, R. Rabenstein: Adaptation of acoustic echo cancellers incorporating a memoryless nonlinearity, *Proc. IWAENC '99,* 168–171, Pocono Manor, NJ, USA, 1999.
[19] B. Widrow, M. E. Hoff Jr.: Adaptive switching circuits, *IRE WESCON Conv. Rec.* **IV**, 96–104, 1960.
[20] A. von Zitzewitz: Considerations on acoustic echo cancelling based on realtime experiments, *Proc. EUSIPCO '90,* **3**, 1987–1990, Barcelona, Spain, 1990.

Part II

Multi-Microphone Processing

2

Joint Optimization of Acoustic Echo Cancellation and Adaptive Beamforming

Wolfgang Herbordt[1], Walter Kellermann[2], and Satoshi Nakamura[1]

[1] ATR – Spoken Language Communication Research Laboratories, Kyoto, Japan
[2] Telecommunications Laboratory, University Erlangen-Nuremberg, Germany

For full-duplex hands-free acoustic human/machine interfaces, often a combination of acoustic echo cancellation and speech enhancement is required to suppress acoustic echoes, local interference, and noise. To optimally exploit positive synergies between acoustic echo cancellation and speech enhancement, various approaches were presented in the literature. However, efficient solutions for situations with high levels of background noise, with time-varying echo paths, and frequent double talk are still a challenging research topic. In this contribution, we address this problem by a joint least-squares (LS) optimization criterion for integrating acoustic echo cancellation and adaptive linearly-constrained minimum variance (LCMV) beamforming. After summarizing the state-of-the-art of this field, we derive the joint acoustic echo cancellation and beamforming system and show its relation to existing approaches. A realization of the joint system integrating a stereophonic acoustic echo canceller (AEC) and a robust generalized sidelobe canceller (RGSC) shows the advantages of the proposed system for high levels of background noise, time-varying echo paths, and frequent double talk. The proposed solution requires only one AEC for an arbitrary number of microphones. A separate adaptation control for the AEC is not necessary. Moreover, for AECs for multiple reproduction channels, the problem of slow convergence due to cross-correlated loudspeaker signals is avoided.

2.1 Introduction

For audio signal acquisition in hands-free human/machine interfaces, adaptive beamforming microphone arrays can be efficiently employed for enhancing a desired signal while suppressing interference and noise [11].

For full-duplex communication systems, not only interference and noise corrupt the desired signal, but also acoustic echoes originating from loudspeakers. For suppressing acoustic echoes, acoustic echo cancellers (AECs) using adaptive filters are the optimum choice since they exploit the reference information provided by the loudspeaker signals [12,32,41,42].

To simultaneously suppress interferers and acoustic echoes, it is thus desirable to combine acoustic echo cancellation with adaptive beamforming in the acoustic human/machine interface. To achieve optimum performance, synergies between the AECs and the beamformer should be maximally exploited while the computational complexity should be kept moderate. When designing such a joint acoustic echo cancellation and beamforming system, it proves necessary to consider especially the time-variance of the acoustic echo path, the background noise level, and the reverberation time of the acoustic environment.

To combine acoustic echo cancellation with beamforming, various strategies were studied in the literature [4,21,44,47,49,54,55,57,63,66,67], reaching from cascades of AECs and beamformers to integrated solutions. These combinations address aspects such as maximization of the echo and noise suppression for slowly time-varying echo paths and high echo-to-interference ratios (EIRs) [55, 57, 66, 67], strongly time-varying echo paths, and low EIRs [21,47,49,63], or minimization of the computational complexity [4,44]. Overviews and comparisons of these methods can be found in [48,58].

In this chapter, we review the state-of-the-art of joint acoustic echo cancellation and beamforming and compare the various approaches. Especially, we analyze the joint acoustic echo cancellation and beamforming system after [47,49] in more detail. We show that this method, which is based on a joint linearly-constrained minimum variance (LCMV) optimization criterion, is especially efficient for low numbers of microphones ($M = 4 \dots 8$), low and moderate reverberation times in the range of $T_{60} = 50\,\text{ms}$ and $400\,\text{ms}$, low EIRs, and/or strong time-variance of the echo path. A separate adaptation control for the AEC is not required so that the difficult task of designing a robust adaptation control for the AEC is avoided. For multichannel reproduction systems such as, for example, stereophonic or 5.1-channel systems, the commonly known problem of slow convergence due to highly cross-correlated loudspeaker signals [5,84] is avoided since the system identification problem is reduced to an interference cancellation problem [48].

Our proposed approach is based on the robust generalized sidelobe canceller (RGSC) after [48]. The RGSC provides high suppression of both strongly time-varying interference such as competing speakers and slowly time-varying diffuse noise, (as typical for, e.g., the interior of cars,) while preserving signal integrity of the desired speech, even for relatively small array apertures and limited numbers of microphones, and even in reverberant environments or for a moving desired speaker.

This chapter is organized as follows: In Sec. 2.2, we introduce the concepts of acoustic echo cancellation and of adaptive beamforming and discuss the previously presented combinations of acoustic echo cancellation and beamforming. In Sec. 2.3, the joint LCMV approach to acoustic echo cancellation and beamforming and its realization as a generalized sidelobe canceller (GSC) are presented. Sec. 2.4 outlines a practical realization based on the RGSC. Sec. 2.5 gives experimental results.

2.2 Concepts for Joint Acoustic Echo Cancellation and Adaptive Beamforming

We consider the scenario of an acoustic human/machine front-end with Q loudspeakers and a microphone array with M microphones. The microphones capture the desired speech signal of the user, interference from other sound sources, such as speech of other human talkers, and ambient noise, such as noise from air conditioning or from computer fans. We can thus identify two problems for the acoustic human/machine interface: acoustic echo cancellation for multiple reproduction channels and noise and interference suppression by microphone arrays.

In the following, we provide a concise overview of the problems and solutions for the individual tasks – acoustic echo cancellation (Sec. 2.2.1) and adaptive beamforming (Sec. 2.2.2). The integration into a joint system is discussed in Sec. 2.2.3.

2.2.1 Acoustic Echo Cancellation

With acoustic echo cancellation being considered from several points of view in this book (Chapters 5, 6, 7, and 8), we only review the main aspects of the general multichannel concept. The principle of multichannel acoustic echo cancellation is illustrated in Fig. 2.1. For simplicity, the multichannel AEC is shown only for a single recording channel.

The signals $x_q(n)$, $q = 0, 1, \ldots Q-1$, are played back by the Q loudspeakers and fed back to the microphones, where the signals $x_q(n)$ appear as acoustic echoes $d_q(n)$. With the assumption that the amplifiers and the transducers are linear, a linear model is commonly used for the echo paths between the loudspeaker signals $x_q(n)$ and the microphone signals $y(n)$. See, e.g., [38, 59, 87] and Chapter 7 of this book for the case where nonlinearities of the transducers and of the amplifiers cannot be neglected. To cancel the acoustic echoes in the microphone channel, adaptive filters $\hat{\boldsymbol{h}}_q(n)$, $q = 0, 1, \ldots Q-1$, are placed in parallel to the echo paths between the loudspeakers and the microphones with the loudspeaker signals $x_q(n)$ as references. The adaptive filters form replicas of the echo paths such that the output signals $\hat{d}_q(n)$ of the adaptive filters are replicas of the acoustic echoes. Subtracting the output signals of the adaptive filters from the microphone signal thus suppresses the acoustic echoes. Acoustic echo cancellation is thus a system identification problem, where the echo paths are usually identified by adaptive linear filtering. The design of the adaptation algorithm requires consideration of the nature of the echo paths and of the echo signals:

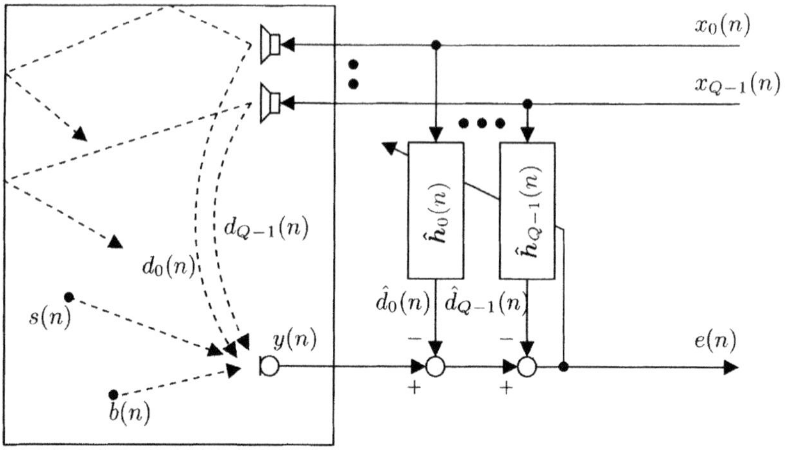

Fig. 2.1. Principle of multichannel acoustic echo cancellation with Q loudspeakers and a single microphone.

Time-variance of acoustic echo paths. The acoustic echo paths may vary strongly over time due to moving sources or changes in the acoustic environment requiring a good tracking performance of the adaptation algorithm [12].

Reverberation time of the acoustic environment. The reverberation time of the acoustic environment typically ranges from, e.g., $T_{60} \approx 50\,\mathrm{ms}$ in passenger cabins of vehicles to $T_{60} > 1\,\mathrm{s}$ in public halls. With

$$N_{\hat{h}} \approx \frac{ERLE}{60} f_{\mathrm{s}} T_{60}, \tag{2.1}$$

where $ERLE$ is the desired echo suppression of the AEC in dB [12], as a rule of thumb it becomes obvious that with many realistic acoustic environments and sampling rates $f_{\mathrm{s}} = 8 - 48\,\mathrm{kHz}$, FIR filters with several thousands coefficients are needed to achieve $ERLE \approx 20\,\mathrm{dB}$. For environments with long reverberation times, this means that the time for convergence – even for fast converging adaptation algorithms – cannot be neglected and that, after a change of the echo paths, noticeable residual echoes may be present until the adaptation algorithm has re-converged.

Auto- and cross-correlation of loudspeaker signals. The spatial sound impression in multi-loudspeaker systems is often artificially generated by weighting and delaying the spectrally colored source signals according to the position of the sources. This leads to a high auto- and cross-correlation of the loudspeaker signals [5, 14, 15, 30, 84]. With increasing auto- and cross-correlation and with an increasing number of reproduction channels, the condition number of the loudspeaker signals' correlation matrix increases, which reduces the rate of convergence of many adaptation algorithms in turn [43]. To reduce the auto-correlation of the loudspeaker signals, prewhitening filters can be applied [28,95,96]. Furthermore, to reduce the cross-correlation of the loudspeaker signals, inaudible nonlinearities [5,30,84] or inaudible time-varying filters can be introduced into the loudspeaker channels [2,53] or inaudible noise with no correlation between the channels can be added to the loudspeaker signals [29,34].

Double talk. The presence of disturbing sources such as desired speech, interference, or ambient noise may lead to instability and divergence of the adaptive filters. To prevent these instabilities, adaptation control mechanisms are required which adjust the step size of the adaptation algorithm to the present acoustic conditions [12,42,64]. With a decrease in the power ratio of acoustic echoes and disturbance a smaller step size becomes mandatory, which increases the time until the adaptive filters have converged to efficient echo path models.

As the discussion about adaptive filtering for acoustic echo cancellation shows, the convergence time of the adaptive filters is a crucial factor in acoustic echo cancellation and limits the performance of AECs in realistic acoustic environments. With the aim of reducing the convergence time while assuring robustness against instabilities and divergence even during double talk, various adaptation algorithms, such as the normalized least mean-squares (NLMS) algorithm, the affine projection algorithm, or the recursive least-squares (RLS) algorithm have been studied for realizations in the time-domain, in the DFT-domain, or in frequency subbands using filterbanks [8,12,32,42,56,81,83]. Acoustic echo cancellation in the DFT-domain or in frequency subbands has the advantage that sparseness of desired speech, interference, and noise can be exploited for selecting the step size of the adaptation algorithm differently for different frequencies as a function of the disturbance level to obtain faster convergence.

Even with fast converging adaptation algorithms, there are typically residual echoes present at the output of the AEC. Furthermore, it is desirable to combine the echo cancellation with noise reduction. Therefore, single-channel echo and noise reduction is often cascaded with the AEC to suppress residual echoes and noise at the AEC output [10,26,39,40,68,69]. These methods are typically based on spectral subtraction or Wiener filtering [9,61] so that estimates of the noise spectrum and of the spectrum of the acoustic echoes at the AEC output are required. These are often difficult to obtain in single-microphone systems for time-varying noise spectra and frequently changing echo paths.

2.2.2 Adaptive Beamforming

To overcome the limitations of single-channel noise reduction especially for interference and noise with time-varying spectra, beamforming with microphone arrays is promising for many applications as, thereby, the spatial domain supports separation of desired and undesired signals. In practical situations, source positions and signal characteristics change over time so that adaptive, data-dependent beamforming algorithms are preferable over fixed data-independent beamformers [92].

For speech and audio signal processing, adaptive data-dependent beamforming can be classified into LCMV beamforming, minimum mean-squared error (MMSE) beamforming, and maximum-a-posteriori (MAP) beamforming, disregarding special combinations with automatic speech recognition [78, 80].

LCMV beamforming [31,48,51,75]. In LCMV beamforming, with the GSC as one implementation, the variance of the output signal of the beamformer is minimized subject to constraints which prevent distortion of the desired signal. Estimates of the auto-power spectral densities (PSDs) and of the cross-power spectral densities (CPSDs) of interference and noise at the sensors are not required so that the efficient suppression of signals with highly time-varying spectra, such as speech signals, becomes possible. Adaptive differential microphone arrays [25, 89] are a special case of the LCMV beamformer.

However, reverberation of the acoustic environment w.r.t. the desired signal [18, 77, 94], moving desired sources, or array imperfections, such as position errors or gain and phase mismatch of the microphones [16, 35, 52, 97], may lead to distortion of the desired signal by the adaptive LCMV beamformer due to 'leakage' of the desired signal. To resolve this problem, the filter coefficients can be updated only when interference and noise are present [51, 74, 90, 93], quadratic [20, 33, 48, 50–52, 75, 88] or adaptive spatio-temporal constraints [31, 48, 51] can be used, or the speech distortion can be controlled directly [23, 86].

The suppression of ambient noise and 'cocktail-party' noise is limited due to the limited number of spatial degrees of freedom of the microphone array. To overcome this limitation for such noise scenarios, two methods have been proposed: First, LCMV beamformers can be combined with single-channel noise reduction ('post-filtering') [19, 65, 70, 71, 79, 82]. This leads to a structure that is basically equivalent to the MMSE beamformer [24, 82], but which exploits the advantages of the LCMV beamformer. Second, a spatial pre-processor in the structure of the GSC can be combined with single-channel noise reduction [72, 76] or with a so-called 'speech distortion weighted multichannel Wiener filter' [23, 86].

MMSE beamforming [1, 22, 23, 27, 79, 86]. MMSE beamforming is an extension of single-channel noise reduction to the multichannel case. In contrast to adaptive LCMV beamforming, multichannel MMSE beamformers are inherently robust against array imperfections and reverberation of the acoustic environment, so that the problem of cancellation of the desired signal due to signal leakage is avoided. However, the minimization of the mean squared error inherently allows for desire signal distortion which may not be acceptable for applications where high speech quality is required. Moreover, MMSE beamformers require estimates of the CPSDs of interference and noise at the sensors so that – at least from today's point of view– there is limited suppression of noise and interfering signals with highly time-varying PSDs.

MAP beamforming [62]. While the derivation of multichannel MMSE estimators is often difficult, multichannel MAP estimators often provide simpler mathematical descriptions. Thereby, multichannel MAP estimation allows, for example, the use of general statistical models, such as super-Gaussian probability density functions for speech and noise.

Adaptive data-dependent beamformers are generally realized using time-averaging over a finite temporal aperture to estimate the relevant statistics of the sensor data. For directly considering this temporal averaging in the optimization criterion of the MMSE beamformer, the term least-squares error (LSE) beamformer is used in [91]. Following [46, 48], we use in this work the term linearly-constrained least-squares error (LCLSE) beamformer for including this temporal averaging into the optimization criterion of the LCMV beamformer.

2.2.3 Joint Acoustic Echo Cancellation and Adaptive Beamforming

In this section, we briefly discuss solutions to the problem of joint acoustic echo cancellation and adaptive beamforming which were presented previously in the literature, namely 'AEC first', 'beamformer first', AEC integrated into the GSC ('GSAEC'), and a joint system of 'AEC first' and 'beamformer first'.

'AEC first' [13, 21, 44, 48, 54, 57, 58, 66]. The AECs can be captured by a matrix of time-variant impulse responses $\hat{\boldsymbol{H}}(n)$ in the sensor channels. This matrix $\hat{\boldsymbol{H}}(n)$ directly models the echo path between all loudspeakers and all microphones, without interaction with the beamforming (Fig. 2.2). For the adaptive beamformer described by a vector of time-variant impulse responses $\boldsymbol{w}(n)$, positive synergies can be exploited after convergence of the AECs: The acoustic echoes are efficiently suppressed by the AECs, and the adaptive beamformer $\boldsymbol{w}(n)$ does not depend on the echo signals. Thus, all degrees of freedom of the beamformer are available for the suppression of interference and noise. Obviously, one AEC is necessary for each sensor channel so that an M-fold complexity, where M is the number of microphones, is required at least for the filtering and for the filter update in comparison to AEC for a single microphone [57]. Even with a moderate number of microphones ($4 \leq M \leq 8$), this is a limiting factor for the use of 'AEC first' in cost-sensitive systems. Moreover, in the presence of strong interference and noise, the adaptation of the AECs must be slowed down or even stopped in order to avoid instabilities of the adaptive filters $\hat{\boldsymbol{H}}(n)$. This reduces the tracking capability and, consequently, the efficiency of the AECs for frequently changing echo paths. Limited echo suppression of the AECs, however, limits the positive synergies with the adaptive beamformer so that the performance improvement of 'AEC first' relative to an adaptive beamformer alone strongly depends on the acoustic environment.

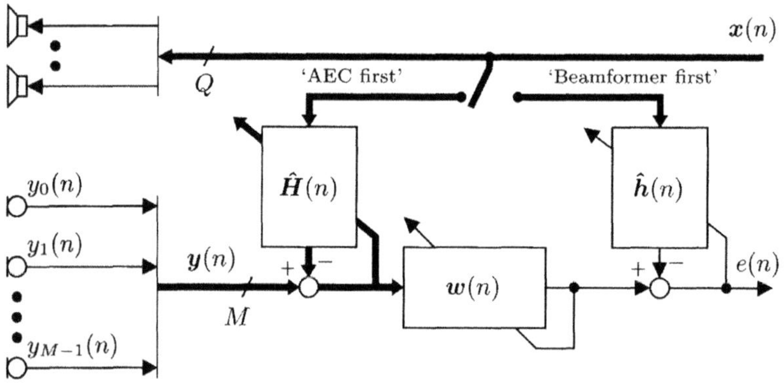

Fig. 2.2. Combinations of AEC and beamforming [58, 66].

'Beamformer first' [4, 44, 48, 54, 57, 58, 66]. Alternatively, the AEC can be placed behind the adaptive beamformer (Fig. 2.2). Obviously, the complexity is reduced to that of AEC for a single microphone. However, positive synergies cannot be exploited for the adaptive beamformer, since the beamformer always 'sees' not only interference but also acoustic echoes. On the other hand, the AEC captured in a vector of time-variant impulse responses $\hat{\boldsymbol{h}}(n)$ generally cannot track the relatively fast time-variance of $\boldsymbol{w}(n)$, which results from the dependency of $\boldsymbol{w}(n)$ on the time-varying spectra of the sensor signals and the generally smaller number of filter taps of $\boldsymbol{w}(n)$ relative to $\hat{\boldsymbol{h}}(n)$ [48].

AEC integrated into the GSC (GSAEC). Another solution would be to integrate acoustic echo cancellation and adaptive beamforming so that the AEC does not depend on the time-variance of the adaptive beamformer [58]. One option, which is based on the structure of the GSC [37] (see Sec. 2.3.2), was proposed in [44]. For this so-called GSAEC, the AEC is placed in the reference path behind the quiescent weight vector $\boldsymbol{w}_c$ of the GSC so that the AEC is independent of the time-varying sidelobe-cancelling path (Fig. 2.3), which consists of the blocking matrix $\boldsymbol{B}(n)$ and the interference canceller $\hat{\boldsymbol{h}}(n)$.

However, acoustic echoes may leak through the sidelobe-cancelling path although they may be efficiently suppressed by the AEC in the reference path, so that the overall performance of 'AEC first' cannot be expected. Moreover, analogously to 'AEC first', the performance of this integrated system is limited for strong interference and noise or for frequently changing echo paths.

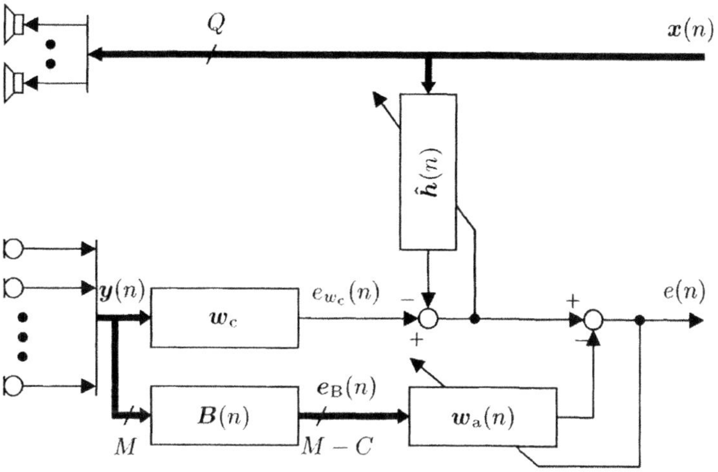

Fig. 2.3. AEC integrated into the GSC (GSAEC) [44].

To overcome the problems of these structures in environments with frequently changing echo paths, frequent double talk, interference, and background noise, we study here the joint optimization of adaptive beamforming and acoustic echo cancellation. We focus on an LCLSE optimization criterion to derive the beamformer weight vector. MMSE/LSE and MAP criteria are not considered since they require estimates of the interference spectra at the microphones, which are difficult to obtain for mixtures of non-stationary signals.

2.3 Joint Optimization of Acoustic Echo Cancellation and Adaptive Beamforming

In contrast to 'beamformer first' in Fig. 2.2, where different signals are used to optimize $\boldsymbol{w}(n)$ and the AEC $\hat{\boldsymbol{h}}(n)$, we propose to use the output signal $e(n)$ to optimize both AEC and the adaptive beamformer as shown in Fig. 2.4. The reference loudspeaker signals $\boldsymbol{x}(n)$ can thus be interpreted as additional input signals for the adaptive beamformer. This idea was first used in [21] for a combination of acoustic echo cancellation and multichannel noise-reduction based on the generalized singular value decomposition (GSVD). In [63], a similar approach is used for a combination of blind source separation with acoustic echo cancellation.

We assume that the sensor signals $\boldsymbol{y}(n)$ are given by the superposition of the desired signal $\boldsymbol{s}(n)$, interference and noise $\boldsymbol{b}(n)$, and acoustic echoes $\boldsymbol{d}(n)$,

$$\boldsymbol{y}(n) = \boldsymbol{s}(n) + \boldsymbol{b}(n) + \boldsymbol{d}(n)\,, \tag{2.2}$$

where $\boldsymbol{s}(n)$, $\boldsymbol{b}(n)$, and $\boldsymbol{d}(n)$ are zero-mean and mutually uncorrelated. The output signal $e(n)$ of the combined system can be written as a function of

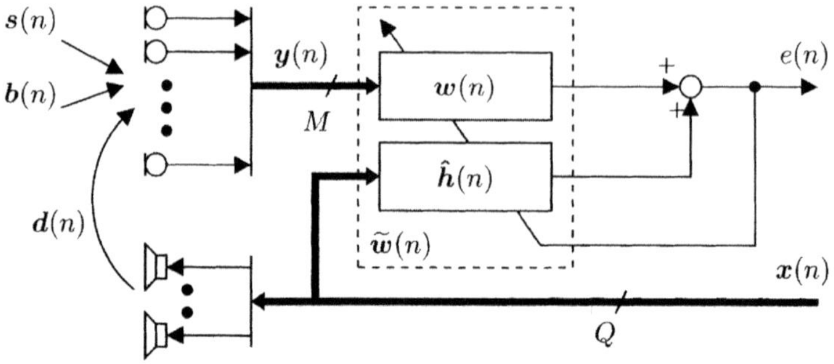

Fig. 2.4. Joint optimization of adaptive beamforming and acoustic echo cancellation.

the sensor signals $\boldsymbol{y}(n)$, the loudspeaker signals $\boldsymbol{x}(n)$, the stacked beamformer weight vector $\boldsymbol{w}(n)$, and the stacked AEC weight vector $\hat{\boldsymbol{h}}(n)$ as

$$e(n) = \boldsymbol{w}^{\mathrm{T}}(n)\boldsymbol{y}(n) + \hat{\boldsymbol{h}}^{\mathrm{T}}(n)\boldsymbol{x}(n)\,, \tag{2.3}$$

where

$$\boldsymbol{y}(n) = \left[\boldsymbol{y}_0^{\mathrm{T}}(n),\, \boldsymbol{y}_1^{\mathrm{T}}(n),\, \ldots,\, \boldsymbol{y}_{M-1}^{\mathrm{T}}(n)\right]^{\mathrm{T}}, \tag{2.4}$$

$$\boldsymbol{y}_m(n) = \left[y_m(n),\, y_m(n-1),\, \ldots,\, y_m(n-N_w+1)\right]^{\mathrm{T}}, \tag{2.5}$$

$$\boldsymbol{x}(n) = \left[\boldsymbol{x}_0^{\mathrm{T}}(n),\, \boldsymbol{x}_1^{\mathrm{T}}(n),\, \ldots,\, \boldsymbol{x}_{Q-1}^{\mathrm{T}}(n)\right]^{\mathrm{T}}, \tag{2.6}$$

$$\boldsymbol{x}_q(n) = \left[x_q(n),\, x_q(n-1),\, \ldots,\, x_q(n-N_{\hat{h}}+1)\right]^{\mathrm{T}}, \tag{2.7}$$

$$\boldsymbol{w}(n) = \left[\boldsymbol{w}_0^{\mathrm{T}}(n),\, \boldsymbol{w}_1^{\mathrm{T}}(n),\, \ldots,\, \boldsymbol{w}_{M-1}^{\mathrm{T}}(n)\right]^{\mathrm{T}}, \tag{2.8}$$

$$\boldsymbol{w}_m(n) = \left[w_{0,m}(n),\, w_{1,m}(n),\, \ldots,\, w_{N_w-1,m}(n)\right]^{\mathrm{T}}, \tag{2.9}$$

$$\hat{\boldsymbol{h}}(n) = \left[\hat{\boldsymbol{h}}_0^{\mathrm{T}}(n),\, \hat{\boldsymbol{h}}_1^{\mathrm{T}}(n),\, \ldots,\, \hat{\boldsymbol{h}}_{Q-1}^{\mathrm{T}}(n)\right]^{\mathrm{T}}, \tag{2.10}$$

$$\hat{\boldsymbol{h}}_q(n) = \left[\hat{h}_{0,q}(n),\, \hat{h}_{1,q}(n),\, \ldots,\, \hat{h}_{N_{\hat{h}}-1,q}(n)\right]^{\mathrm{T}}. \tag{2.11}$$

N_w and $N_{\hat{h}}$ are the number of filter coefficients of the beamformer weight vectors $\boldsymbol{w}_m(n)$ and of the AEC filters $\hat{\boldsymbol{h}}_q(n)$, respectively. With stacked vectors

$$\tilde{\boldsymbol{w}}(n) = \left[\boldsymbol{w}^{\mathrm{T}}(n),\, \hat{\boldsymbol{h}}^{\mathrm{T}}(n)\right]^{\mathrm{T}}, \tag{2.12}$$

$$\tilde{\boldsymbol{x}}(n) = \left[\boldsymbol{y}^{\mathrm{T}}(n),\, \boldsymbol{x}^{\mathrm{T}}(n)\right]^{\mathrm{T}}, \tag{2.13}$$

we can write $e(n)$ as

$$e(n) = \widetilde{\boldsymbol{w}}^{\mathrm{T}}(n)\widetilde{\boldsymbol{x}}(n) . \tag{2.14}$$

which reflects that the AEC input signals $\boldsymbol{x}(n)$ and the AEC filters $\hat{\boldsymbol{h}}(n)$ can be interpreted as additional channels of a beamformer $\widetilde{\boldsymbol{w}}(n)$.

2.3.1 Linearly-Constrained Least-Squares Error (LCLSE) Minimization

An LCLSE optimization criterion is obtained when we aim at minimizing the windowed sum of squared output signal samples $e^2(n)$ subject to constraints which assure that the desired signal is not distorted by $\widetilde{\boldsymbol{w}}(n)$. That is,

$$\min_{\widetilde{\boldsymbol{w}}(n)} \sum_{i=0}^{n} g_i(n)\, e^2(i) \quad \text{subject to} \quad \widetilde{\boldsymbol{C}}^{\mathrm{T}}(n)\widetilde{\boldsymbol{w}}(n) = \boldsymbol{c}(n) . \tag{2.15}$$

The windowing function $g_i(n)$ extracts desired samples from the output signal $y(n)$ which should be included into the optimization.[3] For example, infinite memory with exponential decay is obtained with $g_i(n) = \lambda^{n-i}$ [43]. The constraint matrix $\widetilde{\boldsymbol{C}}(n)$ of size $(MN_w + QN_{\hat{h}}) \times C$ and the constraint column vector $\boldsymbol{c}(n)$ of length C put C spatial constraints onto $\widetilde{\boldsymbol{w}}(n)$ in order to assure unity beamformer response for the direction-of-arrival of the desired signal [91]. Since the Q loudspeaker signals $\boldsymbol{x}(n)$ can safely be assumed to be orthogonal to the desired signal, the constraints are only required for the microphone signals, just as for conventional LCMV beamformers [91]. We can thus write $\widetilde{\boldsymbol{C}}(n)$ as

$$\widetilde{\boldsymbol{C}}(n) = \left[\boldsymbol{C}^{\mathrm{T}}(n),\, \boldsymbol{0}_{C \times QN_{\hat{h}}}\right]^{\mathrm{T}} , \tag{2.16}$$

where $\boldsymbol{C}(n)$ of size $MN_w \times C$ is a conventional constraint matrix known from LCMV beamforming [91]. We thus obtain with Eq. 2.15 a formally simple optimization criterion, where only one single error signal needs to be minimized for an arbitrary number of microphones. This combined optimization allows us to update the beamformer and the AEC simultaneously without reducing the step size for the AEC – in contrast to the previously discussed combinations, where the adaptation of the AEC at least has to be slowed down if interference, noise, or the desired signal are active. Thereby, the structural problems for tracking in 'AEC first' and the leakage in GSAEC can be avoided. The number of spatial degrees of freedom for interference suppression and for echo cancellation are increased by the number of loudspeakers Q relative to a beamformer alone. Due to the correlation of $\boldsymbol{y}(n)$ and $\boldsymbol{x}(n)$, however, it must be expected that the conditioning of the optimization problem is worsened relative to the individual optimization problems.

---

[3] The corresponding LCMV optimization criterion is obtained by replacing the windowed sum of squared output signal samples $e^2(n)$ by the expected value of $e^2(n)$. The solution of the LCMV optimization criterion is analogous to that of the LCLSE criterion shown here.

2.3.2 Realization as a Generalized Sidelobe Canceller (GSC)

A direct solution of Eq. 2.15 can be determined using Lagrange multipliers [91]. However, with regard to an efficient realization of this combined system, we transform the constrained optimization problem into an unconstrained one using the structure of the GSC [17,37].

To obtain the GSC, the stacked weight vector $\widetilde{\boldsymbol{w}}(n)$ is projected onto two orthogonal subspaces,

$$\widetilde{\boldsymbol{w}}(n) = \left[\boldsymbol{P}_{\mathrm{c}}(n) + \boldsymbol{P}_{\mathrm{a}}(n)\right] \widetilde{\boldsymbol{w}}(n) . \tag{2.17}$$

The first subspace $\widetilde{\boldsymbol{w}}_{\mathrm{c}}(n) := \boldsymbol{P}_{\mathrm{c}}(n)\widetilde{\boldsymbol{w}}(n)$ (constrained subspace) fulfills the constraint equation. That is,

$$\widetilde{\boldsymbol{C}}^{\mathrm{T}}(n)\widetilde{\boldsymbol{w}}_{\mathrm{c}}(n) \stackrel{!}{=} \boldsymbol{c}(n) . \tag{2.18}$$

From (2.16), it follows that $\widetilde{\boldsymbol{w}}_{\mathrm{c}}(n)$ can be chosen as

$$\widetilde{\boldsymbol{w}}_{\mathrm{c}}(n) = \left[\boldsymbol{w}_{\mathrm{c}}^{\mathrm{T}}(n),\, \boldsymbol{0}_{1\times QN_{\hat{h}}}\right]^{\mathrm{T}} \tag{2.19}$$

in order to fulfill Eq. 2.18. The weight vector $\boldsymbol{w}_{\mathrm{c}}(n)$ of size $MN_w \times 1$ is known as quiescent weight vector [91]. The quiescent weight vector $\boldsymbol{w}_{\mathrm{c}}(n)$ steers the sensor array to the position of the desired source and enhances the desired signal relative to interference and noise (Fig. 2.5).[4]

The second (orthogonal) subspace is chosen as

$$\boldsymbol{P}_{\mathrm{a}}(n)\, \widetilde{\boldsymbol{w}}(n) := -\widetilde{\boldsymbol{B}}(n)\, \widetilde{\boldsymbol{w}}_{\mathrm{a}}(n) , \tag{2.20}$$

where the columns of the matrix $\widetilde{\boldsymbol{B}}(n)$ are orthogonal to the columns of the constraint matrix $\widetilde{\boldsymbol{C}}(n)$, i.e.,

$$\widetilde{\boldsymbol{C}}^{\mathrm{T}}(n)\, \widetilde{\boldsymbol{B}}(n) \stackrel{!}{=} \boldsymbol{0} . \tag{2.21}$$

The cascade of $\widetilde{\boldsymbol{B}}(n)$ and $\widetilde{\boldsymbol{w}}_{\mathrm{a}}(n)$ is termed the sidelobe-cancelling path [37]. From Eq. 2.16, it may be seen that Eq. 2.21 is met for

$$\widetilde{\boldsymbol{B}}(n) = \begin{bmatrix} \boldsymbol{B}(n) & \boldsymbol{0}_{MN_w \times QN_{\hat{h}}} \\ \boldsymbol{0}_{QN_{\hat{h}} \times (M-C)N_{\mathrm{wa}}} & \boldsymbol{I}_{QN_{\hat{h}} \times QN_{\hat{h}}} \end{bmatrix} , \tag{2.22}$$

where $\boldsymbol{I}_{QN_{\hat{h}} \times QN_{\hat{h}}}$ is the identity matrix of size $QN_{\hat{h}} \times QN_{\hat{h}}$ and where $\boldsymbol{B}(n)$ meets $\boldsymbol{C}^{\mathrm{T}}(n)\boldsymbol{B}(n) = \boldsymbol{0}$. Since the constrained subspace generally contains the desired signal, the matrix $\boldsymbol{B}(n)$, which fulfills the requirement that the second subspace is orthogonal to the constrained subspace, suppresses desired signal

[4] Note that we used in Fig. 2.3 for the GSAEC structure a fixed quiescent weight vector. This assumption is relaxed here for generality of the derivation.

components. Therefore, the matrix $\boldsymbol{B}(n)$ is generally referred to as a blocking matrix [91]. The identity matrix assures that acoustic echoes are not cancelled by $\widetilde{\boldsymbol{B}}(n)$. As a consequence, ideally only acoustic echoes, interference, and noise are present at the output of $\widetilde{\boldsymbol{B}}(n)$, so that the weight vector $\widetilde{\boldsymbol{w}}_{\mathrm{a}}(n)$ can be determined by unconstrained LS minimization of $e(n)$,

$$\min_{\widetilde{\boldsymbol{w}}_{\mathrm{a}}(n)} \sum_{i=0}^{n} g_i(n) \left[\left(\widetilde{\boldsymbol{w}}_{\mathrm{c}}(n) - \widetilde{\boldsymbol{B}}(n) \widetilde{\boldsymbol{w}}_{\mathrm{a}}(n) \right)^{\mathrm{T}} \widetilde{\boldsymbol{x}}(i) \right]^2 . \tag{2.23}$$

Introducing Eqs. 2.19 and 2.22 into Eq. 2.23 and identifying the result with Eq. 2.3, it may be seen that $\widetilde{\boldsymbol{w}}_{\mathrm{a}}(n)$ is equivalent to a stacked weight vector consisting of a weight vector $\boldsymbol{w}_{\mathrm{a}}(n)$ and of the AEC $\hat{\boldsymbol{h}}(n)$,

$$\widetilde{\boldsymbol{w}}_{\mathrm{a}}(n) := \left[\boldsymbol{w}_{\mathrm{a}}^{\mathrm{T}}(n), \hat{\boldsymbol{h}}^{\mathrm{T}}(n) \right]^{\mathrm{T}} . \tag{2.24}$$

We obtain for the output signal $e(n)$ the expression

$$e(n) = \left[\boldsymbol{w}_{\mathrm{c}}(n) - \boldsymbol{B}(n)\, \boldsymbol{w}_{\mathrm{a}}(n) \right]^{\mathrm{T}} \boldsymbol{y}(n) - \hat{\boldsymbol{h}}^{\mathrm{T}}(n)\, \boldsymbol{x}(n) , \tag{2.25}$$

which can be put into the structure depicted in Fig. 2.5. The combined system thus corresponds to the GSC, where $\boldsymbol{w}_{\mathrm{a}}(n)$ is combined with the AEC $\hat{\boldsymbol{h}}(n)$, and where the loudspeaker signals $\boldsymbol{x}(n)$ are used as additional channels of the sidelobe-cancelling path. $\boldsymbol{w}_{\mathrm{a}}(n)$ is generally called an interference canceller since $\boldsymbol{w}_{\mathrm{a}}(n)$ is optimized to cancel interference and noise at the output of the GSC. Analogously, we refer to $\widetilde{\boldsymbol{w}}_{\mathrm{a}}(n)$ as the 'echo and interference canceller' (EIC) and to the combined system of AEC and GSC as the 'generalized echo and interference canceller' (GEIC).

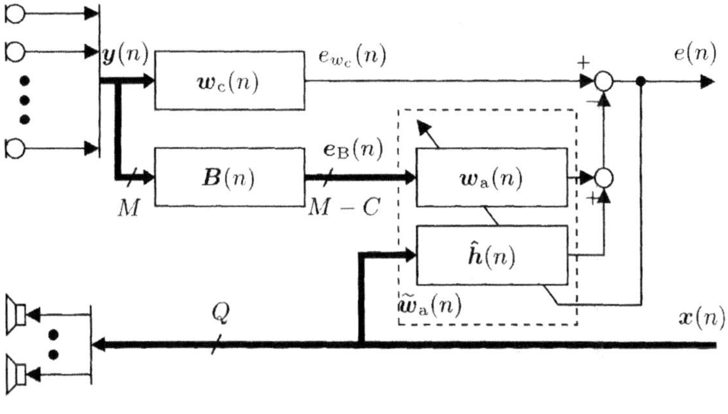

Fig. 2.5. Generalized echo and interference canceller (GEIC).

The optimum weight vector $\widetilde{\boldsymbol{w}}_{\mathrm{a}}(n)$ is now obtained by setting the derivative of Eq. 2.23 w.r.t. $\widetilde{\boldsymbol{w}}_{\mathrm{a}}(n)$ equal to zero and by solving the obtained system of linear equations for $\widetilde{\boldsymbol{w}}_{\mathrm{a}}(n)$:

$$\widetilde{\boldsymbol{w}}_{\mathrm{a,opt}}(n) = \left[\widetilde{\boldsymbol{B}}^{\mathrm{T}}(n)\,\widetilde{\boldsymbol{\Phi}}(n)\,\widetilde{\boldsymbol{B}}(n)\right]^{+}\widetilde{\boldsymbol{B}}^{\mathrm{T}}(n)\,\widetilde{\boldsymbol{\Phi}}(n)\,\widetilde{\boldsymbol{w}}_{\mathrm{c}}(n)\,, \tag{2.26}$$

$$\widetilde{\boldsymbol{\Phi}}(n) = \sum_{i=0}^{n} g_i(n)\,\widetilde{\boldsymbol{x}}(i)\,\widetilde{\boldsymbol{x}}^{\mathrm{T}}(i) = \begin{bmatrix} \boldsymbol{\Phi}_{yy}(n) & \boldsymbol{\Phi}_{yx}(n) \\ \boldsymbol{\Phi}_{xy}(n) & \boldsymbol{\Phi}_{xx}(n) \end{bmatrix}. \tag{2.27}$$

The $(\cdot)^{+}$ is the pseudoinverse of a matrix, and $\widetilde{\boldsymbol{\Phi}}(n)$ is the sample correlation matrix of the stacked data vector $\widetilde{\boldsymbol{x}}(n)$ [43] for a given windowing function $g_i(n)$. As shown in (2.27), $\widetilde{\boldsymbol{\Phi}}(n)$ can be decomposed into the submatrices

$$\boldsymbol{\Phi}_{yy}(n) = \sum_{i=0}^{n} g_i(n)\,\boldsymbol{y}(i)\,\boldsymbol{y}^{\mathrm{T}}(i)\,, \tag{2.28}$$

$$\boldsymbol{\Phi}_{xx}(n) = \sum_{i=0}^{n} g_i(n)\,\boldsymbol{x}(i)\,\boldsymbol{x}^{\mathrm{T}}(i)\,, \tag{2.29}$$

$$\boldsymbol{\Phi}_{yx}(n) = \sum_{i=0}^{n} g_i(n)\,\boldsymbol{y}(i)\,\boldsymbol{x}^{\mathrm{T}}(i)\,, \tag{2.30}$$

$$\boldsymbol{\Phi}_{xy}(n) = \boldsymbol{\Phi}_{yx}^{\mathrm{T}}(n)\,, \tag{2.31}$$

with the sample correlation matrix of the sensor signals $\boldsymbol{\Phi}_{yy}(n)$, the sample correlation matrix of the loudspeaker signals $\boldsymbol{\Phi}_{xx}(n)$, and the sample cross-correlation matrices between the sensor signals and the loudspeaker signals $\boldsymbol{\Phi}_{xy}(n)$ and $\boldsymbol{\Phi}_{yx}(n)$, respectively. The solution of the optimum weight vector $\widetilde{\boldsymbol{w}}_{\mathrm{a,opt}}(n)$ is formally equivalent to the optimum weight vector of the GSC [17]. Finally introducing Eqs. 2.19, 2.22, and 2.27 into Eqs. 2.26, 2.26 can be written as

$$\begin{bmatrix} \boldsymbol{w}_{\mathrm{a,opt}}(n) \\ \hat{\boldsymbol{h}}_{\mathrm{opt}}(n) \end{bmatrix} = \begin{bmatrix} \boldsymbol{B}^{\mathrm{T}}(n)\,\boldsymbol{\Phi}_{yy}(n)\,\boldsymbol{B}(n) & \boldsymbol{B}^{\mathrm{T}}(n)\,\boldsymbol{\Phi}_{yx}(n) \\ \boldsymbol{\Phi}_{xy}(n)\,\boldsymbol{B}(n) & \boldsymbol{\Phi}_{xx}(n) \end{bmatrix}^{+} \times \begin{bmatrix} \boldsymbol{B}^{\mathrm{T}}(n)\,\boldsymbol{\Phi}_{yy}(n)\,\boldsymbol{w}_{\mathrm{c}}(n) \\ \boldsymbol{\Phi}_{yx}(n)\,\boldsymbol{w}_{\mathrm{c}}(n) \end{bmatrix}. \tag{2.32}$$

Because of the structural equivalence of the GEIC to the GSC, any implementation of the GSC can be used to realize the GEIC. Especially, any linear constraints can be used for designing the quiescent weight vector and the blocking matrix. Furthermore, the echo and interference canceller can be calculated directly employing Eq. 2.32 or iteratively using recursive adaptation algorithms [48,91]. With regard to practical realizations, the matrix inversion in Eq. 2.32 can be avoided by using recursive adaptation algorithms, and, thus, the computational complexity can be reduced.

For the GSC, the number of filter taps $N_{\mathrm{w_a}}$ is generally chosen such that fast convergence of $\boldsymbol{w}_{\mathrm{a}}(n)$ is assured. Typically, $N_{\mathrm{w_a}} = 64 \ldots 512$ for an $f_s = 8\,\text{kHz}$ sampling rate independently of the reverberation time T_{60} of the acoustic environment [48]. The number of filter taps $N_{\hat{h}}$ of the AEC $\hat{\boldsymbol{h}}(n)$, however, is typically chosen as a function of the reverberation time T_{60}, and is typically $N_{\hat{h}} = 256\ldots 2048$ for $T_{60} = 0.05\ldots 0.5\,\text{s}$ (see Eq. 2.1). In most cases, the number of filter taps $N_{\hat{h}}$ should thus be greater than N_{w_a} depending on the reverberation time of the acoustic environment (typically $T_{60} \geq 100\,\text{ms}$) in order to assure optimum performance of $\boldsymbol{w}_\text{a}(n)$ and $\hat{\boldsymbol{h}}(n)$. However, different numbers of filter taps are problematic for the convergence behavior of $\widetilde{\boldsymbol{w}}_\text{a}(n)$ for a time-varying sample correlation matrix $\widetilde{\boldsymbol{\Phi}}(n)$, since the convergence speed of $\boldsymbol{w}_\text{a}(n)$ differs from that of $\hat{\boldsymbol{h}}(n)$. Consider as an extreme case $N_{\hat{h}} \to \infty$: Then, the convergence speed of $\hat{\boldsymbol{h}}(n)$ tends to zero, which yields inefficiency of the AEC. It is thus necessary to limit $N_{\hat{h}}$ to N_{w_a}. This may reduce the performance of GEIC relative to 'AEC first' in situations where 'AEC first' does not exhibit tracking or adaptation problems as, for example, for presence of weak interference and noise and/or for slowly time-varying acoustic echo paths. The influence of the acoustic environment on the performance of GEIC will be investigated experimentally in Sec. 2.5.

2.3.3 Simplification to General Sidelope Acoustic Echo Canceller (GSAEC)

The joint optimization of $\hat{\boldsymbol{h}}(n)$ and $\boldsymbol{w}_\text{a}(n)$ introduces the off-diagonal matrices into the first correlation matrix on the right side of (2.32). Setting the off-diagonal matrices equal to zero corresponds to separate optimization of $\hat{\boldsymbol{h}}(n)$ and $\boldsymbol{w}_\text{a}(n)$, which yields for the optimum weight vector:

$$\boldsymbol{w}_{\text{a,opt}}(n) = \left[\boldsymbol{B}^\text{T}(n)\,\boldsymbol{\Phi}_{yy}(n)\,\boldsymbol{B}(n)\right]^+ \boldsymbol{B}^\text{T}(n)\,\boldsymbol{\Phi}_{yy}(n)\,\boldsymbol{w}_\text{c}(n)\,, \tag{2.33}$$

$$\hat{\boldsymbol{h}}_{\text{opt}}(n) = \boldsymbol{\Phi}_{xx}^+(n)\,\boldsymbol{\Phi}_{xy}(n)\,\boldsymbol{w}_\text{c}(n)\,. \tag{2.34}$$

It may be noticed that Eq. 2.33 corresponds to the LS solution of a GSC interference canceller [17] and that Eq. 2.34 is equivalent to the LS solution of an AEC which is located after the quiescent weight vector. Eqs. 2.33 and 2.34 can thus be described by the system depicted in Fig. 2.3, which is recognized as the structure of the GSAEC [44].

Independent optimization of the GSC and of the AEC after the quiescent weight vector allows to choose the number of filter taps of the interference canceller, N_{w_a}, and of the AEC, $N_{\hat{h}}$, independently so that the coupling problems of the echo and interference canceller of GEIC can be avoided. However, [48] describes in detail in that efficient cancellation of the acoustic echoes in the reference path of the GSC leads to leakage of acoustic echoes through the sidelobe-cancelling path of the GSC so that the performance of GSAEC is reduced relative to 'AEC first'. Moreover, for the presence of strong interference and noise and/or time-varying echo paths, GSAEC exhibits the same convergence problems as 'AEC first'. Experimental results can be found in Sec. 2.5.

2.4 Implementation

In this section, we describe the practical implementation of the joint acoustic echo cancellation and adaptive beamforming systems examined experimentally in Sec. 2.5, namely GEIC, 'AEC first', GSAEC, and GSC. For all joint acoustic echo cancellation and adaptive beamforming systems, the beamformer is realized as a GSC with an adaptive blocking matrix (RGSC, Sec. 2.4.1). The AEC is implemented as a stereophonic AEC (Sec. 2.4.2). A detailed description including parameter setting can be found in [48].

2.4.1 Robust Generalized Sidelobe Canceller (RGSC)

To realize the adaptive beamformer, it is crucial to obtain (a) tracking of moving sources with time-varying spectra and (b) robustness against cancellation of the desired signal due to reverberation, source movements, and array imperfections. To solve these problems, we choose the RGSC in the discrete Fourier transform (DFT) domain [48] with an adaptive blocking matrix [51] as the adaptive beamformer, as depicted in Fig. 2.6.

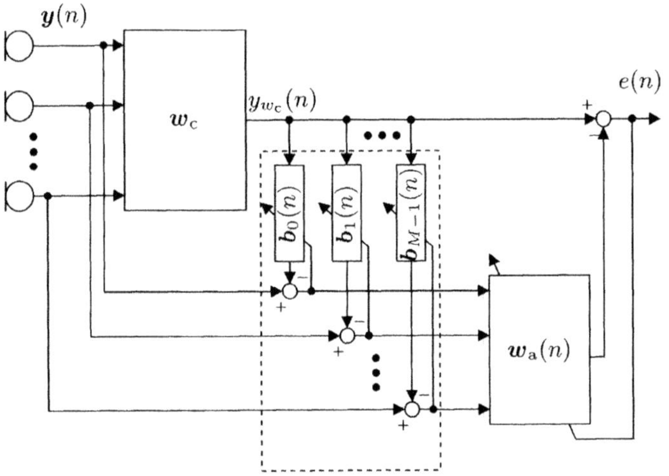

Fig. 2.6. GSC with an adaptive blocking matrix after [51].

For adaptation of the blocking matrix and of the interference canceller, we use computationally efficient multichannel DFT-domain adaptive filters (MC-FDAFs) [6, 7, 15]. Their RLS-like convergence behavior leads to fast convergence and they allow for a frequency-selective adaptation to exploit sparseness of the sensor signals.

2.4.1.1 Quiescent weight vector

The quiescent weight vector is realized as a fixed beamformer $\boldsymbol{w}_{\mathrm{c}}(n) := \boldsymbol{w}_{\mathrm{c}}$. We thus assume that the position of the desired speaker is roughly known, as it can be safely assumed for, for example, laptop PCs or personal digital assistants (PDAs). The width of the mainlobe of the quiescent weight vector needs to be adjusted to the expected variations of the source position.

2.4.1.2 Blocking Matrix

The blocking matrix is realized by adaptive filters $\boldsymbol{b}_m(n)$ between the output of the time-invariant quiescent beamformer $\boldsymbol{w}_{\mathrm{c}}$ and each of the inputs of the interference canceller $\boldsymbol{w}_{\mathrm{a}}(n)$. The adaptive filters $\boldsymbol{b}_m(n)$ use the output of $\boldsymbol{w}_{\mathrm{c}}$ as a reference for the desired signal and subtract the desired signal from the sidelobe-cancelling path. Orthogonality of the reference path and of the sidelobe-cancelling path is thus assured for the desired signal. Since the quiescent beamformer cannot produce an estimate of the desired signal that is free of interference, the filters $\boldsymbol{b}_m(n)$ should only be adapted when the signal-to-interference ratio (SIR) is high in order to prevent suppression of the interference by the blocking matrix [51,90].

In [48], the adaptive blocking matrix is formally linked to LCLSE beamforming and to the derivation of the GSC in Sec. 2.3.2.

Realization of the blocking matrix by adaptive filters yields greater robustness against distortion of the desired signal than fixed realizations [31, 48,51,90]: For the GSC, the distortion results from the interference canceller, which cancels desired signal components leaking through the blocking matrix due to inherent mismatched constraints. The inherent mismatch results from possible array imperfections and especially from the fact that the required exact spatio-temporal information for the desired signal is not perfectly given. Adaptive filters, however, allow tracking of time-varying propagation for the desired source and time-varying array imperfections so that the desired signal is efficiently cancelled by the blocking matrix.

2.4.1.3 Interference Canceller

The interference canceller $\boldsymbol{w}_{\mathrm{a}}(n)$ adaptively subtracts the signal components from the reference path, which are correlated with the output signals of the blocking matrix. However, the blocking matrix – due to limited convergence speed, limited tracking capability, and limited number of filter coefficients – generally does not produce an estimate of the interference which is perfectly free of the desired signal. Therefore, the interference canceller (1) is realized using a (usually quadratic) norm constraint [20,33,48,50–52,75,88] and (2) is only adapted when the SIR is low in order to maximally prevent distortion of the desired signal [51,74,90,93].

2.4.1.4 Adaptation Control

The blocking matrix and the interference canceller cannot be adapted simultaneously but should only be adapted when the SIR is high and low, respectively. By exploiting sparseness in the spectra of desired speech and interference, i.e., by considering individual frequency components separately, the blocking matrix and the interference canceller can be adapted more often than adaptation in the fullband enabling a better tracking capability and a better convergence speed to be obtained. Experiments show that the exploitation of sparseness is also necessary for the interference canceller of the RGSC to track the time variance of the adaptive blocking matrix and to efficiently suppress non-stationary interference [45, 48].

Obviously, to exploit the sparseness, an activity detector is required, which detects 'desired signal only' (adaptation of the blocking matrix), 'interference only' (adaptation of the interference canceller), and 'double talk' (no adaptation) in discrete frequency bins [48].

2.4.2 Acoustic Echo Canceller

The design of the AEC – as long as it is realized independently of the beamformer – requires consideration of the tracking performance, of the convergence speed, and of the robustness against double talk. For the joint adaptation of the AEC and beamformer, acoustic echoes can simply be interpreted as additional interference, and these aspects do not need to be explicitly taken into account. However, at the EIC input, the variance of the output signals of the blocking matrix needs to be adjusted to the variance of the loudspeaker signals by an automatic gain control to have similar signal levels.

Especially because of the high convergence speed with moderate computational complexity, we employ MC-FDAFs to realize the AECs. Adaptation of the AECs of 'AEC first' and of GSAEC is controlled by a double talk detector based on a shadow filter [85] with a constant frequency-independent step size during adaptation. The GEIC is realized as an RGSC with additional channels of the interference canceller for the AECs. For the experiments described below, the time-averaged variance of the loudspeaker signals is manually adjusted to the time-averaged variance of the blocking matrix output signals. For all structures, the loudspeaker signals are de-cross-correlated by a simple time-invariant nonlinearity to increase the convergence speed of the adaptive filters [5].

2.4.3 Computational Complexity

The computational complexity of GEIC is compared to that of 'AEC first', GEIC, GSAEC, and RGSC in Fig. 2.7 as a function of the filter length $N_{\hat{h}}$ of the AEC for $M = 4$ microphones (Fig. 2.7a) and $M = 8$ microphones

(Fig. 2.7b) for a stereophonic AEC. The filter length of the interference canceller is $N_{w_a} = 256$ for all systems. For GEIC, the filter length of the AEC is adjusted to the filter length of the interference canceller, i.e., $N_{w_a} = N_{\hat{h}} = 256$. The adaptation control of the AEC and of the RGSC is not taken into account. Furthermore, the filter length N_{w_a} is not changed since experimental results in environments with various reverberation times show that the optimum filter length does not change with the reverberation time in our implementations of the RGSC and the GEIC. The computational complexity is measured as 'real-valued multiplications per output sample' (NRM)[5]. Comparing Fig. 2.7a with Fig. 2.7b, roughly speaking, it may be noticed that doubling the number of sensors doubles NRM. The relative complexity reduction from 'AEC first' to GEIC rises with increasing $N_{\hat{h}}$: For $N_{w_a} = N_{\hat{h}} = 256$, the relative complexity reduction from 'AEC first' to GEIC is 21 % for $M = 4$ and 25 % for $M = 8$, while 59 % ($M = 4$) and 37 % ($M = 8$) for $N_a = 2048$. Obviously, the complexity of the RGSC dominates the complexity of the additional AECs for 'AEC first'.

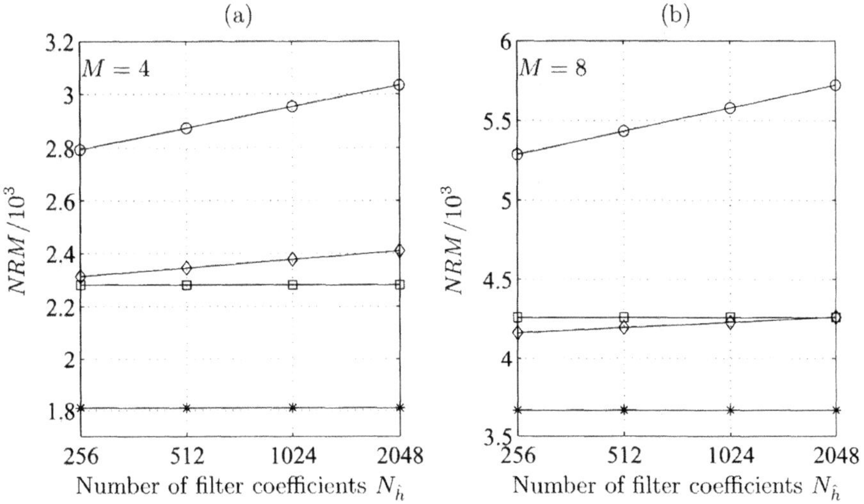

Fig. 2.7. Comparison of the number of real multiplications per sample (NRM) of '○' 'AEC first', '□' GEIC, '◇' GSAEC, and '∗' RGSC for (a) $M = 4$ and for (b) $M = 8$ ($N_{w_a} = 256$ and for GEIC $N_{\hat{h}} = 256$).

[5] The results differ from the results in [48], since, here, NRM includes the inversion of the CPSD matrix of the input signals of the interference canceller and of the EIC. The matrix inversion is assumed to be carried out using the matrix inversion lemma [36].

2.5 Experimental Results

We illustrate the performance of the joint acoustic echo cancellation and adaptive beamforming systems by experiments in the passenger cabin of a car and in an office room. In Sec. 2.5.1, we analyze the performance for time-invariant echo paths, for a fixed position of the desired source, and for variable noise level. In Sec. 2.5.2, we examine the influence of time-varying echo paths and of a time-varying position of the desired source on the performance of joint AEC-beamforming systems. Section 2.5.3 illustrates the influence of the reverberation time on the performance of GEIC.

2.5.1 Time-Invariant Echo Paths and Time-Invariant Source Position

In this section, we study the performance of GEIC for variable SIR and time-invariant echo paths in the passenger cabin of a car relative to the other concepts presented in Sec. 2.2. The interference is slowly time-varying car noise recorded with a microphone array setup inside of the car's passenger cabin (Fig. 2.8).

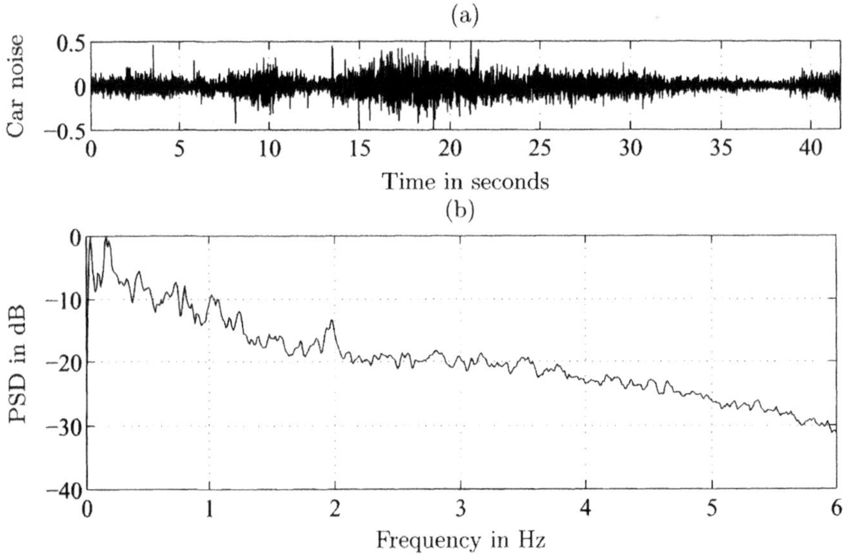

Fig. 2.8. Temporal signal (a) and power spectral density (PSD) of the car noise (b) measured at one of the microphones (before highpass filtering).

The desired source and two loudspeakers are located in broadside direction ($\theta = 90°$) and in the two endfire directions ($\theta = 0°$, $180°$), respectively, at a distance of 60 cm from the array center. The room impulse responses between the two loudspeakers and the microphones and between the desired source position and the microphones are simulated using the image method [3] with a simulated reverberation time $T_{60} = 50\,\text{ms}$. The desired source signal is a subset of 50 utterances of the TIDigits database [60], while the loudspeaker signals are stereophonic pop music. The microphone signals are obtained by convolving the clean source signal with the room impulse responses followed by superposing noise with variable SIR and a fixed signal-to-echo ratio $SER = 7\,\text{dB}$. The microphone array consists of $M = 4$ sensors or $M = 8$ sensors with sensor spacing $d = 4\,\text{cm}$. The frequency range is 200 Hz–4 kHz. The echo suppression $ERLE$ and the interference suppression IR averaged over the whole test data are given in Fig. 2.9 ($M = 4$) and in Fig. 2.10 ($M = 8$). The filter lengths are chosen as follows: 'AEC first', GSAEC: $N_{\hat{h}} = 512$, $N_w = 256$; GEIC, RGSC: $N_{\hat{h}} = N_w = 256$).

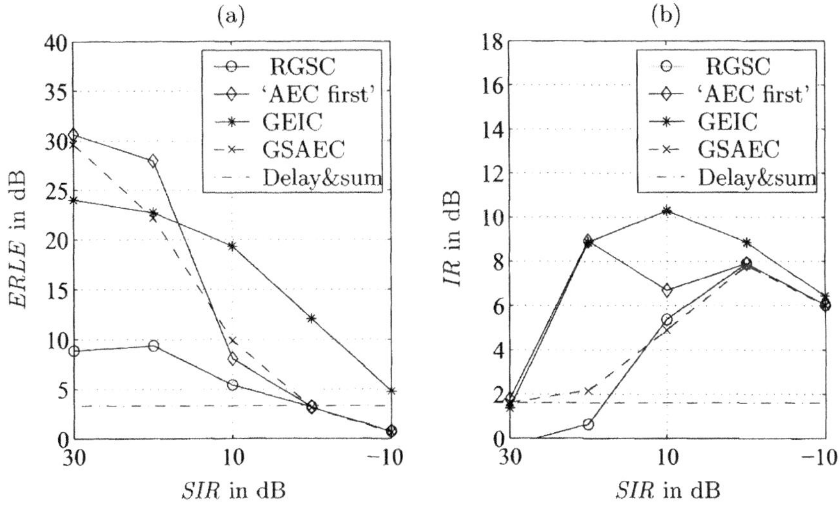

Fig. 2.9. Interference suppression IR and echo suppression $ERLE$ for RGSC alone, 'AEC first', GSAEC, and GEIC for fixed echo paths and fixed source position in the car environment for $M = 4$ (Signal-to-echo ratio $SER = 7\,\text{dB}$).

For high SIR (equivalent to high EIR, since $SER = 7\,\text{dB}$), the AECs of 'AEC first' converge in pauses of the desired speaker and provide high echo suppression, which translates to a greater $ERLE$ and IR of 'AEC first' relative to GSC and GEIC. With decreasing EIR, the echo suppression of the AECs of 'AEC first' decreases until the AECs are inefficient and $ERLE$ and IR of 'AEC first' are equivalent to the RGSC. Here, the GEIC outperforms 'AEC first', since the number of degrees of freedom does not depend on the EIR. Nevertheless, $ERLE$ of GEIC falls with decreasing EIR, since the system concentrates on the suppression of the stronger car noise. For $M = 4$ (Fig. 2.9), it can be noticed that the improvement of $ERLE$ and IR relative to RGSC

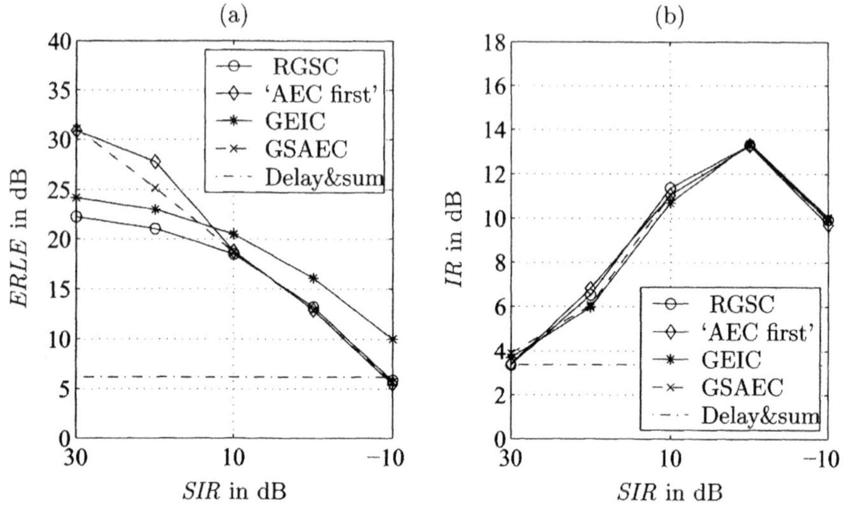

Fig. 2.10. Interference suppression *IR* and echo suppression *ERLE* for RGSC alone, 'AEC first', GSAEC, and GEIC for fixed echo paths and fixed source position in the car environment for $M = 8$ (Signal-to-echo ratio $SER = 7\,\mathrm{dB}$).

is larger than for $M = 8$ (Fig. 2.10). This is due to the RGSC's greater number of degrees of freedom, where the additional degrees of freedom of GEIC due to the AEC yield a relative lower performance improvement. In fact, the improvement of *IR* can even be neglected in this scenario. The performance of GSAEC decreases relative to 'AEC first', since, after convergence of the AEC in the reference path of the GSC (Fig. 2.3), acoustic echoes leak through the sidelobe-cancelling path of the GSC, which leads to reduced echo and interference suppression relative to 'AEC first' [44,48].

2.5.2 Time-Varying Echo Path and Time-Varying Source Position

In this section, we compare the performance of joint acoustic echo cancellation and adaptive beamforming for a time-varying echo path and a moving desired source. Because of the better tracking during double talk, we expect that the performance gap between echo and noise suppression of GEIC and that of 'AEC first' and GSAEC increases. The position of the desired source is switched randomly for each file of the TIDigits database in the interval $\theta = 80^\circ \ldots 100^\circ$ in steps of 2° with equal probability for all directions. This range corresponds to the 5 dB width of the mainlobe of the uniformly weighted delay&sum beamformer at 4 kHz. The desired signal is thus attenuated by less than 5 dB at 4 kHz. While one of the loudspeakers is located at $\theta = 180^\circ$, the position of the second loudspeaker is switched every 20000 samples between $\theta = 0^\circ$ and $\theta = 60^\circ$. The distance between the sources and the array center is fixed at 60 cm. The interference suppression *IR* and the echo suppression *ERLE* are averaged over the entire data set across all variations of the loudspeaker position.

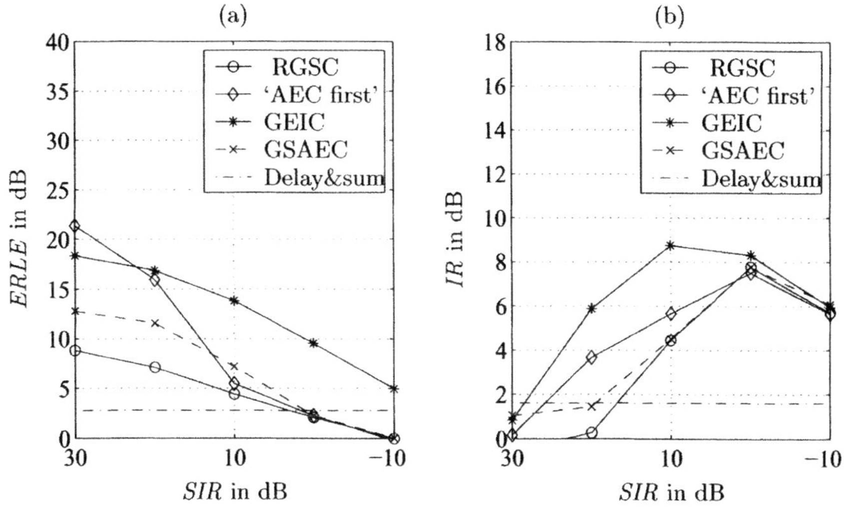

Fig. 2.11. Interference suppression *IR* and echo suppression *ERLE* for RGSC alone, 'AEC first', GSAEC, and GEIC for time-varying echo paths and fixed source position in the car environment for $M = 4$ (Signal-to-echo ratio $SER = 7\,\text{dB}$).

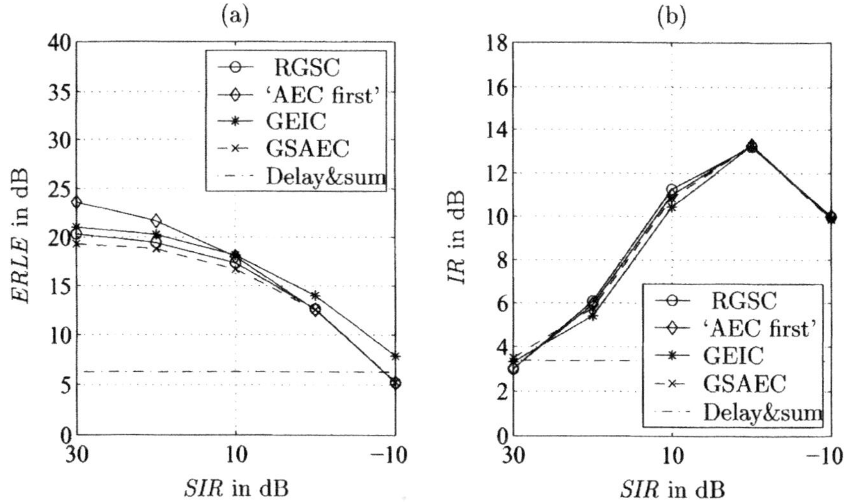

Fig. 2.12. Interference suppression *IR* and echo suppression *ERLE* for RGSC alone, 'AEC first', GSAEC, and GEIC for time-varying echo paths and fixed source position in the car environment for $M = 8$ (Signal-to-echo ratio $SER = 7\,\text{dB}$).

Figs. 2.11 and 2.12 depict the results for $M = 4$ and $M = 8$, respectively. The difference in performance between $M = 4$ and $M = 8$ can be explained similarly as in Figs. 2.3 and 2.9 by the greater number of degrees of freedom of the beamformer for $M = 8$. It may be further noticed that the performance of 'AEC first' and GSAEC is considerably reduced for $SIR \geq 20$ dB relative to fixed echo paths (see Figs. 2.9 and 2.10). This effect can be explained by the reduced efficiency of the AECs of 'AEC first' and of GSAEC due to the missing capability to adapt the AECs while desired speech and acoustic echoes are simultaneously active. The performance loss is mainly related to the time-variance of the echo paths: Experiments showed that the performance for fixed echo paths and the time-varying position of the desired source ($\theta = 80° \ldots 100°$) can almost not be distinguished from the results in Figs. 2.9 and 2.10. As for fixed echo paths and the fixed position of the desired source, the echo suppression and the interference suppression converge with an increasing number of microphones.

Note that the AECs of 'AEC first' and of GSAEC are realized using a frequency-independent double talk detector with a constant step size during adaptation. When using a DFT bin-wise step-size control with variable frequency-dependent step size as, for example, proposed in [26, 73], it is possible to exploit sparseness of desired speech and of interference. The AECs can therefore be adapted more frequently, which improves the performance of 'AEC first' and of GSAEC for time-varying acoustic conditions and high EIRs.

2.5.3 Reverberation Time

In this section, we study the dependency of the echo and noise suppression of GEIC on the reverberation time T_{60}. Because of the limited number of filter taps of the EIC, we expect the performance of the GEIC to decrease compared to 'AEC first' and to GSAEC with increasing reverberation time. The experimental setup is the same as in Sec. 2.5.1, except for the fact that the impulse responses between the loudspeakers and the microphones are taken from three different acoustic environments: the environment with $T_{60} = 50$ ms as above, and measured impulse responses from office rooms with $T_{60} = 250$ ms and with $T_{60} = 400$ ms. The microphone array with $M = 4$ sensors is used, $SER = 7$ dB, and $SIR = 10$ dB. The results are depicted in Fig. 2.13.

It can be seen that the average echo suppression $ERLE$ (Fig. 2.13a) decreases with increasing reverberation time from 19.5 dB for $T_{60} = 50$ ms to 15.5 dB for $T_{60} = 400$ ms. The interference suppression IR decreases from 10 dB for $T_{60} = 50$ ms to 9.5 dB for $T_{60} = 400$ ms. Considering that the number of filter taps of the AEC is only $N_{\hat{h}} = 256$ for a reverberation time $T_{60} = 400$ ms, where, according to (2.1), $N_{\hat{h}} = 1240$ is required for $ERLE = 15.5$ dB, these results reflect that the AECs within GEIC are better interpreted as interference cancellers than as system identifiers. The usage of GEIC –despite the limitation on the number of filter taps– is thus not

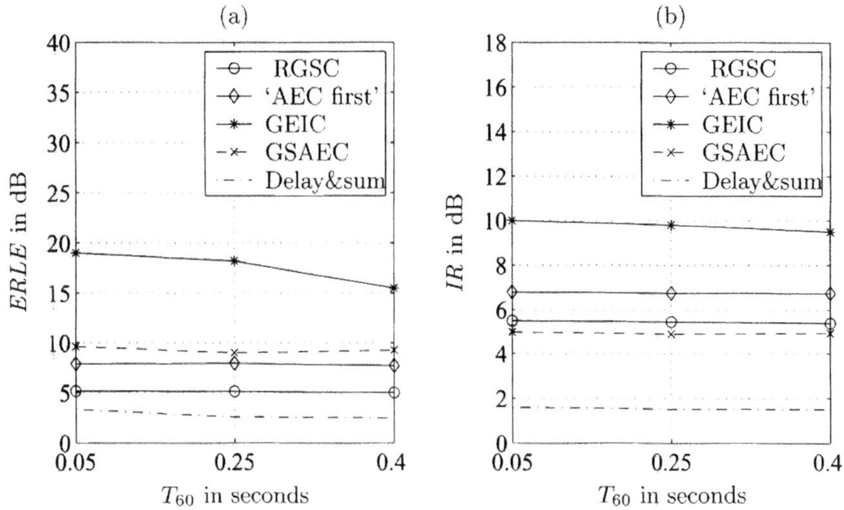

Fig. 2.13. Interference suppression *IR* and echo suppression *ERLE* for RGSC alone, 'AEC first', GSAEC, and GEIC as a function of the reverberation time T_{60} for $M = 4$ (Signal-to-echo ratio $SER = 7\,\text{dB}$, Signal-to-interference ratio $SIR = 10\,\text{dB}$).

restricted to environments with low reverberation times but still gives acceptable echo suppression in environments with longer reverberation time such as office or home environments, at least for slowly time-varying conditions. Note, however, that the performance of joint acoustic echo cancellation and adaptive beamforming systems based on the GSC depends on the robustness of the GSC against distortion of the desired signal in reverberant environments. It is thus not assured that all GSC realizations yield an undistorted desired signal.

2.6 Conclusion

We presented a technique for joint optimization of acoustic echo cancellation and adaptive LCMV beamforming. The derivation of the system shows that it can be interpreted as a straightforward extension of the GSC with additional input channels of the interference canceller (GEIC). With a realization example based on the RGSC and a stereophonic AEC, we showed that the GEIC is especially efficient for (a) transient echo paths if frequent double talk between acoustic echoes, local interference, and desired speakers is to be expected and (b) high levels of background noise. For stationary conditions and low levels of background noise, the performance of GEIC is reduced relative to 'AEC first' due to a constraint on the number of filter taps of the weight vector of the AEC. However, the proposed solution requires only one AEC for an arbitrary number of microphones and no separate adaptation control for the

AEC. For acoustic echo cancellation with multiple reproduction channels, the problem of slow convergence due to cross-correlated loudspeaker signals can be avoided, since the system identification problem is reduced to an interference cancellation problem.

References

[1] R. Aichner, W. Herbordt, H. Buchner, W. Kellermann: Least-squares error beamforming using minimum statistics and multichannel frequency-domain adaptive filtering, *Proc. IWAENC '03*, Kyoto, Japan, 223–226, September 2003.

[2] M. Ali: Stereophonic acoustic echo cancellation system using time-varying all-pass filtering for signal decorrelation,*Proc. ICASSP '98*, **6**, 3689–3692, Washington, DC, USA, May 1998.

[3] J.B. Allen: Multimicrophone signal-processing technique to remove room reverberation from speech signals, *Journal of the Acoustical Society of America*, **62**(4), 912–915, October 1977.

[4] H.J.W. Belt, C.P. Janse: Method and device for acoustic echo cancellation combined with adaptive beamforming, *United States Patent*, US 2002/0015500 A1, 2002.

[5] J. Benesty, D.R. Morgan, M.M. Sondhi: A better understanding and an improved solution to the specific problems of stereophonic acoustic echo cancellation, *IEEE Trans. on Speech and Audio Processing*, **6**(2), 156–165, 1998.

[6] J. Benesty, D.R. Morgan: Frequency-domain adaptive filtering revisited, generalization to the multi-channel case, and application to acoustic echo cancellation, *Proc. ICASSP '00*, 261-264, Istanbul, Turkey, June 2000.

[7] J. Benesty, D.R. Morgan: Multi-channel frequency-domain adaptive filtering, in S.L. Gay, J. Benesty, (eds.), *Acoustic Signal Processing for Telecommunication*, Chapter 7, 121–133, Boston, MA, USA: Kluwer Academic Publishers, 2000.

[8] J. Benesty, Y. Huang (eds): *Adaptive Signal Processing: Applications to Real-World Problems*, Berlin, Germany: Springer, 2003.

[9] J. Benesty, S. Makino, J. Chen (eds.): *Speech Enhancement*, Berlin, Germany: Springer, 2005.

[10] R. Le Bouquin, P. Scalart, G. Faucon, C. Beaugeant: Combined noise and echo reduction in hands-free systems: A survey, *IEEE Trans. on Speech and Audio Processing*, **9**(8), 808–820, November 2001.

[11] M.S. Brandstein, D.B. Ward (eds.): *Microphone Arrays: Signal Processing Techniques and Applications*, Berlin, Germany: Springer, 2001.

[12] C. Breining, P. Dreiseitel, E. Hänsler, A. Mader, B. Nitsch, H. Puder, T. Schertler, G. Schmidt, J. Tilp: Acoustic echo control - an application

of very-high-order adaptive filters, *IEEE Signal Processing Magazine*, **16**(4), 42–69, July 1999.

[13] H. Buchner, W. Herbordt, W. Kellermann: An efficient combination of multi-channel acoustic echo cancellation with a beamforming microphone array, *Proc. Int. Workshop on Hands-Free Speech Communication*, 55–58, Kyoto, Japan, April 2001.

[14] H. Buchner, W. Kellermann: Acoustic echo cancellation for two or more reproduction channels, *Proc. IWAENC '01*, 99–102, Darmstadt, Germany, September 2001.

[15] H. Buchner, J. Benesty, W. Kellermann: Multichannel frequency-domain adaptive filtering with application to multichannel acoustic echo cancellation, in J. Benesty, Y. Huang (eds.), *Adaptive Signal Processing: Applications to Real-World Problems*, Berlin, Germany: Springer, 2003.

[16] M. Buck: Aspects of first order differential microphone arrays in the presence of sensor imperfections, *European Trans. on Telecommunications*, **13**(2), 115–122, March/April 2002.

[17] K.M. Buckley: Broad-band beamforming and the generalized sidelobe canceller, *IEEE Trans. on Acoustics, Speech, and Signal Processing*, **34**(5), 1322–1323, October 1986.

[18] A. Cantoni, L.C. Godara: Resolving the direction of sources in a correlated field incident on an array, *Journal of the Acoustical Society of America*, **67**(4), 1247–1255, April 1980.

[19] I. Cohen, B. Berdugo: Microphone array post-filtering for non-stationary noise suppression, *Proc. ICASSP '02*, **1**, 901–904, Orlando, FL, USA, May 2002.

[20] H. Cox, R.M. Zeskind, T. Kooij: Robust adaptive beamforming, *IEEE Trans. on Acoustics, Speech, and Signal Processing*, **35**(10), 1365–1376, October 1987.

[21] S. Doclo, M. Moonen, E. De Clippel: Combined acoustic echo and noise reduction using GSVD-based optimal filtering, *Proc. ICASSP '00*, **2**, 1051–1054, Istanbul, Turkey, June 2000.

[22] S. Doclo, M. Moonen: GSVD-based optimal filtering for single and multi-microphone speech enhancement, *IEEE Trans. on Signal Processing*, **50**(9), 2230–2244, September 2002.

[23] S. Doclo, A. Spriet, J. Wouters, M. Moonen: Speech distortion weighted multichannel Wiener filtering techniques for noise reduction, in J. Benesty, S. Makino, J. Chen (eds.), *Speech Enhancement*, Chapter 9, 199–228. Berlin, Germany, Springer, 2005.

[24] D.J. Edelblute, J.M. Fisk, G.L. Kinneson: Criteria for optimum-signal-detection theory for arrays, *Journal of the Acoustical Society of America*, **41**(1), 199–205, January 1967.

[25] G.W. Elko: Microphone array systems for hands-free telecommunication, *Speech Communication*, **20**, 229–240, 1996.

[26] G. Enzner, P. Vary: Robust and elegant, purely statistical adaptation of acoustic echo canceler and postfilter, *Proc. IWAENC '03*, 43–46, Kyoto, Japan, September 2003.

[27] D.A. Florencio, H.S. Malvar: Multichannel filtering for optimum noise reduction in microphone arrays, *Proc. ICASSP '01*, **1**, 197–200, Salt Lake City, UT, USA, May 2001.

[28] R. Frenzel, M. Hennecke: Using prewhitening and step-size control to improve the performance of the LMS algorithm for acoustic echo cancellation, *Proc. ISCAS '92,* **4**, 1930–1932, San Diego, CA, USA, 1992.

[29] T. Gänsler, P. Eneroth: Influence of audio coding on stereophonic acoustic echo cancellation, *Proc. ICASSP '98*, **6**, Washington, DC, USA, 3649–3652, May 1998.

[30] T. Gänsler, J. Benesty: New insight into the stereophonic acoustic echo cancellation problem and an adaptive nonlinearity solution, *IEEE Trans. on Speech and Audio Processing*, **10**(5), 257–267, July 2002.

[31] S. Gannot, D. Burshtein, E. Weinstein: Signal enhancement using beamforming and nonstationarity with applications to speech, *IEEE Trans. on Signal Processing*, **49**(8), 1614–1626, August 2001.

[32] S.L. Gay, J. Benesty (eds.): *Acoustic Signal Processing for Telecommunications,* Boston, MA, USA: Kluwer Academic Publishers, 2000.

[33] E.N. Gilbert, S.P. Morgan: Optimum design of directive antenna arrays subject to random variables, *Bell Systems Technical Journal*, **34**, 637–663, May 1955.

[34] A. Gilloire, V. Turbin: Using auditory properties to improve the behavior of stereophonic acoustic echo cancellers, *Proc. ICASSP '98*, **6**, 3681–3684, Washington, DC, USA, May 1998.

[35] L.C. Godara: Error analysis of the optimal antenna array processors, *IEEE Trans. on Aerospace and Electronic Systems*, **22**(4), 395–409, July 1986.

[36] G.H. Golub, C.F. van Loan: *Matrix Computations,* 2-nd edition, Baltimore, MD, USA: John Hopkins University Press, 1989.

[37] L.J. Griffiths, C.W. Jim: An alternative approach to linearly constrained adaptive beamforming, *IEEE Trans. on Antennas and Propagation,* **30**(1), 27–34, January 1982.

[38] A. Guerin, G. Faucon, R. Le Bouquin-Jeannes: Nonlinear acoustic echo cancellation based on Volterra filters, *IEEE Trans. on Speech and Audio Processing*, **11**(6), 672–683, November 2003.

[39] S. Gustafsson, R. Martin, P. Vary: Combined acoustic echo control and noise reduction for hands-free telephony, *Signal Processing*, **64**(1), 21–32, January 1998.

[40] S. Gustafsson, R. Martin, P. Jax, P. Vary: A psychoacoustic approach to combined acoustic echo cancellation and noise reduction, *IEEE Trans. on Speech and Audio Processing*, **10**(5), 245–256, July 2002.

[41] E. Hänsler: The hands-free telephone problem – an annotated bibliography, *Signal Processing*, **27**(3), 259–271, March 1992.

[42] E. Hänsler, G. Schmidt: *Acoustic Echo and Noise Control,* Hoboken, NJ, USA: John Wiley & Sons, 2004.
[43] S. Haykin: *Adaptive Filter Theory,* 3rd edition, Englewood Cliffs, NJ, USA: Prentice Hall, 1996.
[44] W. Herbordt, W. Kellermann: Acoustic echo cancellation embedded into the generalized sidelobe canceller, *Proc. EUSIPCO '00*, **3**, 1843–1846, Tampere, Finland, September 2000.
[45] W. Herbordt, H. Buchner, W. Kellermann: An acoustic human-machine front-end for multimedia applications, *EURASIP Journal on Applied Signal Processing*, **1**, 1–11, January 2003.
[46] W. Herbordt, W. Kellermann: Adaptive beamforming for audio signal acquisition, in J. Benesty, Y. Huang, (eds.), *Adaptive Signal Processing: Applications to Real-World Problems*, 155–194, Berlin, Germany: Springer, 2003.
[47] W. Herbordt, W. Kellermann, S. Nakamura: Combined optimization of LCMV beamforming and accoustic echo cancellation, *Proc. EUSIPCO '04*, **3**, , 2003–2006, Vienna, Austria, 2004.
[48] W. Herbordt: *Sound Capture for Human/Machine Interfaces: Practical Aspects of Microphone Array Signal Processing*, Berlin, Germany: Springer, 2005.
[49] W. Herbordt, S. Nakamura, W. Kellermann: Joint optimization of LCMV beamforming and acoustic echo cancellation for automatic speech recognition, *Proc. ICASSP '05*, **3**, 77–80, Philadelphia, PA, USA, March 2005.
[50] M.W. Hoffman, K.M. Buckley: Robust time-domain processing of braodband microphone array data, *IEEE Trans. on Speech and Audio Processing*, **3**(3), 193–203, May 1995.
[51] O. Hoshuyama, A. Sugiyama, A. Hirano: A robust adaptive beamformer for microphone arrays with a blocking matrix using constrained adaptive filters, *IEEE Trans. on Signal Processing*, **47**(10), 2677–2684, October 1999.
[52] N.K. Jablon: Adaptive beamforming with the generalized sidelobe canceller in the presence of array imperfections, *IEEE Trans. on Antennas and Propagation*, **34**(8), 996–1012, August 1986.
[53] Y. Joncour, A. Sugyiama: A stereo echo canceler with pre-processing for correct echo path identification, *Proc. IEEE ICASSP '98*, **6**, 3677–3680, Washington, DC, USA, May 1998.
[54] M. Kallinger, J. Bitzer, K.D. Kammeyer: Study on combining multichannel echo cancellers with beamformers, *Proc. IEEE ICASSP '00*, **2**, 797–800, Istanbul, Turkey, June 2000.
[55] K.D. Kammeyer, M. Kallinger, A. Mertins: New aspects of combining echo cancellers with beamformers, *Proc. ICASSP '05*, **3**, 137–140, Philadelphia, PA, USA, March 2005.

[56] W. Kellermann: Analysis and design of multirate systems for cancellation of acoustical echoes, *Proc. ICASSP '88*, **5**, New York, NY, USA, 2570–2573, April 1988.
[57] W. Kellermann: Strategies for combining acoustic echo cancellation and adaptive beamforming microphone arrays, *Proc. ICASSP '97*, **1**, 219–222, Munich, Germany, April 1997.
[58] W. Kellermann: Acoustic echo cancellation for beamforming microphone arrays, in M.S. Brandstein, D.B. Ward (eds.), *Microphone Arrays: Signal Processing Techniques and Applications*, Chapter 13, 281–306, Berlin, Germany: Springer, 2001.
[59] F. Küch, W. Kellermann: Partitioned block frequency-domain adaptive second-order Volterra filter, *IEEE Trans. on Signal Processing*, **53**(2), 564–575, February 2005.
[60] R.G. Leonard: A database for speaker independent digit recognition, *Proc. ICASSP '84*, **3**, 42.11.1–42.11.4, San Diego, CA, USA, March 1984.
[61] J.S. Lim: *Speech Enhancement,* Englewood Cliffs, NJ, USA: Prentice Hall 1983.
[62] T. Lotter: Single- and multi-microphone spectral amplitude estimation using super-Gaussian speech models, in J. Benesty, S. Makino, J. Chen (eds.), *Speech Enhancement,* Chapter 4, 67–95, Berlin, Germany: Springer, 2005.
[63] S.Y. Low, S. Nordholm: A blind approach to joint noise and acoustic echo cancellation, *Proc. ICASSP '05*, **3**, 69–72, Philadelphia, PA, USA, March 2005.
[64] A. Mader, H. Puder, G.U. Schmidt: Step-size controls for acoustic echo cancellation filters - an overview, *Signal Processing*, **80**(9), 1697–1719, September 2000.
[65] C. Marro, Y. Mahieux, K.U. Simmer: Analysis of noise reduction and dereverberation techniques based on microphone arrays with postfiltering, *IEEE Trans. on Speech and Audio Processing*, **6**(3), 240–259, May 1998.
[66] R. Martin: *Freisprecheinrichtungen mit mehrkanaliger Echokompensation und Störgeräuschreduktion,* PhD thesis, Aachener Institut für Nachrichtengeräte und Datenkommunikation, 1995 (in German).
[67] R. Martin, P. Vary: Combined acoustic echo cancellation, dereverberation, and noise-reduction: a two microphone approach, *Proc. Int. Workshop on Acoustic Echo Control*, 125–132, Lannion, France, September 1993.
[68] R. Martin, P. Vary: Combined acoustic echo and noise reduction for hands-free telephony – state of the art and perspectives, *Proc. EUSIPCO '96*, 1107–1100, Trieste, Italy, September 1996.
[69] R. Martin, P. Vary: The echo shaping approach to acoustic echo control, *Speech Communication*, **20**(3/4), 181–190, December 1996.

[70] R. Martin: Small microphone arrays with postfilters for noise and acoustic echo cancellation, in M.S. Brandstein, D.B. Ward, (eds.), *Microphone Arrays: Signal Processing Techniques and Applications,* 255–279, Berlin, Germany: Springer, 2001.

[71] I. McCowan, H. Bourlard: Microphone array post-filter for diffuse noise field, *Proc. ICASSP '02*, **1**, 905–908, Orlando, FL, USA, May 2002.

[72] M. Mizumachi, M. Akagi: Noise reduction by paired-microphones using spectral subtraction, *Proc. ICASSP '98*, **2**, 1001–1004, Washington, DC, USA, May 1998.

[73] B. H. Nitsch: A frequency-selective stepfactor control for an adaptive filter algorithm working in the frequency domain, *Signal Processing*, **80**(9), 1733–1745, September 2000.

[74] S. Nordholm, I. Claesson, B. Bengtsson: Adaptive array noise suppression of handsfree speaker input in cars, *IEEE Trans. on Vehicular Technology*, **42**(4), 514–518, November 1993.

[75] S. Nordholm, H.Q. Dam, N. Grbic, S.Y. Low: Adaptive microphone arrays employing spatial quadratic soft constraints and spatial filtering, in J. Benesty, S. Makino, J. Chen (eds.), *Speech Enhancement,* Chapter 9, 229–246, Berlin, Germany: Springer, 2005.

[76] Y. Ohashi, T. Nishikawa, H. Saruwatari, A. Lee, K. Shikano: Noise robust speech recognition based on spatial subtraction array, *Int. Workshop on Nonlinear Signal and Image Processing*, 324–327, Sapporo, Japan, May 2005.

[77] N.L. Owsley: An overview of optimum-adaptive control in sonar array processing, in K.S. Narendra, R.V. Monopoli (eds.), *Applications of Adaptive Control*, 131–164, New York, NY, USA: Academic Press, 1980.

[78] D. Raub, J. McDonough, M. Wölfel: A cepstral domain maximum likelihood beamformer for speech recognition, *Proc. Int. Conf. on Spoken Language Processing*, **2**, 817-820, Jeju Island, Korea, October 2004.

[79] J. Rosca, R. Balan, C. Beaugeant: Multi-channel psychoacoustically motivated speech enhancement, *Proc. ICASSP '03*, **1**, 84–87, April 2003.

[80] M.L. Seltzer: *Microphone array processing for robust speech recognition*, PhD thesis, Carnegie Mellon University, Pittsburgh, PA, USA, July 2003.

[81] J.J. Shynk: Frequency-domain and multirate adaptive filtering, *IEEE Signal Processing Magazine*, **9**(1), 14–37, January 1992.

[82] K.U. Simmer, J. Bitzer, C. Marro: Post-filtering techniques, in M.S. Brandstein, D.B. Ward (eds.), *Microphone Arrays: Signal Processing Techniques and Applications*, Chapter 3, 39–60, Berlin, Germany: Springer, 2001.

[83] M.M. Sondhi, W. Kellermann: Echo cancellation for speech signals, in S. Furui, M.M. Sondhi (eds.), *Advances in Speech Signal Processing*, 327–356, New York, NY, USA: Marcel Dekker, 1991.

[84] M.M. Sondhi, D.R. Morgan, J.L. Hall: Stereophonic acoustic echo cancellation – an overview of the fundamental problem, *IEEE Signal Processing Letters*, **2**(8), 148–151, August 1995.
[85] W.-J. Song, M.-S. Park: A complementary pair LMS algorithm for adaptive filtering, *Proc. ICASSP '97*, **3**, 2261–2264, Munich, Germany, 1997.
[86] A. Spriet, M. Moonen, J. Wouters: Spatially pre-processed speech distortion weighted multi-channel Wiener filtering for noise reduction in hearing aids, *Signal Processing,* **84**(12), 2367–2387, December 2004.
[87] A. Stenger: *Kompensation akustischer Echos unter Einfluß von nichtlinearen Audiokomponenten*, PhD thesis, University Erlangen-Nuremberg, Erlangen, Germany, 2000 (in German).
[88] Z. Tian, K.L. Bell, H.L. Van Trees: A recursive least squares implementation for LCMP beamforming under quadratic constraint, *IEEE Trans. on Signal Processing*, **49**(6), 1138–1145, June 2001.
[89] H. Teutsch, G.W. Elko: An adaptive close-talking microphone array, *Proc. WASPAA '01*, 163–166, New Paltz, NY, USA, October 2001.
[90] D. Van Compernolle: Switching adaptive filters for enhancing noisy and reverberant speech from microphone array recordings, *Proc. ICASSP '90*, **2**, Albuquerque, NM, USA, 833–836, 1990.
[91] H.L. Van Trees: *Optimum Array Processing, Part IV of Detection, Estimation, and Modulation Theory*, New York, NY, USA: John Wiley & Sons, Inc., 2002.
[92] B.D. Van Veen, K.M. Buckley: Beamforming: A versatile approach to spatial filtering, *IEEE ASSP Magazine*, **5**(2), 4–24, April 1988.
[93] J. Vanden Berghe, J. Wouters: An adaptive noise canceller for hearing aids using two nearby microphones, *Journal of the Acoustical Society of America*, **103**(6), 3621–3626, June 1998.
[94] B. Widrow, K.M. Duvall, R.P. Gooch, W.C. Newman: Signal cancellation phenomena in adaptive antennas: Causes and cures, *IEEE Trans. on Antennas and Propagation*, **30**(3), 469–478, May 1982.
[95] S. Yamamoto, S. Kitayama: An adaptive echo canceller with linear predictor, *IEICE Trans. on Information and Systems*, **62**(12), 851–857, December 1979.
[96] H. Yasukawa, S. Shimada: An acoustic echo canceller using subband sampling and decorrelation methods, *IEEE Trans. on Signal Processing*, **41**(2), 926–930, February 1993.
[97] W.S. Youn, C.K. Un: Robust adaptive beamforming based on the eigenstructure method, *IEEE Trans. on Signal Processing*, **42**(6), 1543–1547, June 1994.

3

Blind Source Separation of Convolutive Mixtures of Audio Signals in Frequency Domain

Shoji Makino, Hiroshi Sawada, Ryo Mukai, and Shoko Araki

NTT Communication Science Laboratories, Japan

This chapter overviews a total solution for frequency-domain blind source separation (BSS) of convolutive mixtures of audio signals, especially speech. Frequency-domain BSS performs independent component analysis (ICA) in each frequency bin, and this is more efficient than time-domain BSS. We describe a sophisticated total solution for frequency-domain BSS, including permutation, scaling, circularity, and complex activation function solutions. Experimental results of separating speech signals for the cases of 2×2, 3×3, 4×4, 6×8, and 2×2 moving sources (#sources $\times$ #microphones) in a room are promising.

3.1 Introduction

Blind source separation (BSS) [14, 19, 28] is an approach to estimating source signals by using only the information of mixed signals observed at each input channel. The estimation is performed blindly, i.e., without possessing information on each source such as its location and active time. Typical examples of such source signals include mixtures of simultaneous speech signals that have been picked up by several microphones. Potential audio signal applications of BSS include speech enhancement for speech recognition, teleconferences, and hearing aids. In such applications, signals are mixed in a convolutive manner with reverberations. This makes the BSS problem difficult. We need very long finite impulse response (FIR) filters (e.g., around a thousand taps for 8-kHz sampling) to separate the acoustic signals mixed under such conditions.

Independent component analysis (ICA) [18, 27] is a major statistical tool for dealing with the BSS problem. If signals are mixed instantaneously, we can directly employ an instantaneous ICA algorithm to separate them. However, signals are mixed in a convolutive manner in the applications mentioned above. Therefore, we need to extend the ICA/BSS technique so that it can be used for convolutive mixtures.

The first approach is time-domain BSS, where ICA is directly extended to the convolutive mixture model [1,11,15,22,30,49]. This approach is theoretically sound and achieves good separation once an algorithm converges, since the algorithm correctly evaluates the independence of separated signals. However, an ICA algorithm for convolutive mixtures is not as simple as an ICA algorithm for instantaneous mixtures, and it is computationally expensive for long FIR filters because it includes convolution operations.

The second approach is frequency-domain BSS, where complex-valued ICA for instantaneous mixtures is employed in each frequency bin [4–7,20,24,31,33,35–38,40,41,43,45,48,50]. The merit of this approach is that the ICA algorithm remains simple and can be performed separately at each frequency. Also, any complex-valued instantaneous ICA algorithm can be employed with this approach. The computational time for BSS can be reduced by employing a fast algorithm such as FastICA [10,17] and/or by performing parallel computation for multiple frequency bins. However, the permutation ambiguity of the ICA solution becomes a serious problem. We need to align the permutation in each frequency bin so that a separated signal in the time domain contains frequency components from the same source. This problem is well known as the permutation problem of frequency-domain BSS [4,7,20,24,31,33,35–37,41,43,45,48], which is the main focus of this chapter. Another problem relates to the circularity effect of discrete frequency representation. Frequency responses calculated in the frequency domain assume a periodic time-domain filter for their implementation. However, such a periodic filter is unrealistic, and we usually use its one-period operation for the separation filter. Therefore, the frequency responses should be smoothed so that the one-period operation does not rely on the circularity effect [7,40]. This chapter also discusses this problem.

The third approach uses both the time and frequency domains. In some time-domain BSS methods, convolutions in the time domain are speeded up by the overlap-save method in the frequency domain [11,21]. Furthermore, in some methods [8, 25, 26], filter coefficients are updated in the frequency domain while nonlinear functions for evaluating independence are applied in the time domain. The permutation problem does not occur in either case since the independence of separated signals is evaluated in the time domain. Nor does the circularity problem occur when there is an appropriate constraint for filter coefficients [46] by such means as rectangular windowing. However, the algorithm moves back and forth between the two domains at every iteration, spending non-negligible time on discrete Fourier transforms (DFTs) and inverse DFTs. Therefore, dealing with the permutation and circularity problems seems to be inevitable if we hope to benefit from the merits of frequency-domain BSS.

This chapter deals with the second approach, i.e., frequency-domain BSS. We begin by formulating the BSS problem for convolutive mixtures in Sec. 3.2. Sec. 3.3 provides an overview of frequency-domain BSS. We then present several important techniques that enable this approach to achieve effective separation of many sources mixed in a reverberant environment. Sec. 3.4 discusses

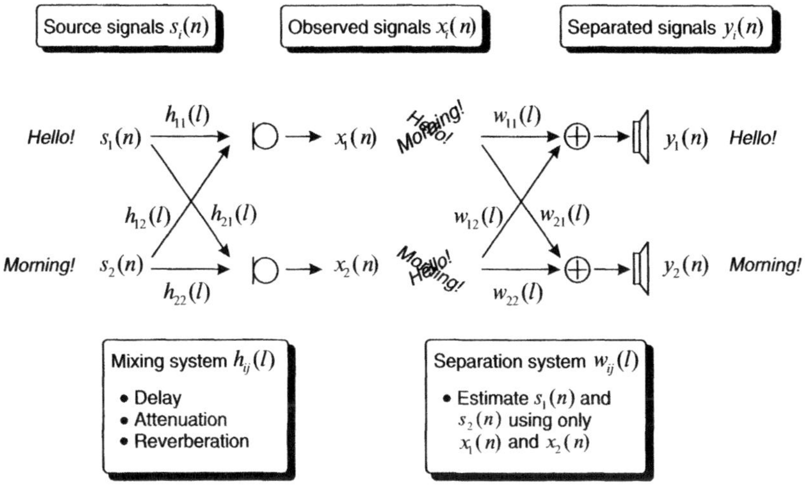

Fig. 3.1. BSS system configuration.

complex-valued ICA for instantaneous mixtures. Understanding the separation mechanism of BSS in Sec. 3.5 greatly helps us to cope with the problem. Sec. 3.7 presents a method for solving the permutation problem, which is the most important task in frequency-domain BSS. To solve this problem, information on source location is very useful. This can be estimated from ICA solutions as shown in Sec. 3.6. The key point with respect to source localization is that the estimation of the mixing system is easily obtained. This is because the ICA algorithm is just for instantaneous mixtures, and thus it is straightforward to calculate the (pseudo)-inverse of a separation matrix, which corresponds to the mixing system. This fact also makes it easy to solve the scaling ambiguity as shown in Sec. 3.8. Sec. 3.9 discusses a spectral smoothing technique designed to solve the circularity problem. The experimental results shown in Sec. 3.10 are very promising. Sec. 3.11 concludes this chapter.

3.2 Blind Source Separation for Convolutive Mixtures

In the case of audio source separation, several sensor microphones are placed in different positions so that each records a mixture of the original source signals at a slightly different time and level. In the real world, where the source signals are speech and the mixing system is a room, the signals that are picked up by the microphones are affected by reverberation. Suppose that N source signals $s_i(n)$ are mixed and observed at M sensors:

$$x_j(n) = \sum_{i=1}^{N} \sum_{l} h_{ji}(l)\, s_i(n-l), \quad j = 1, \ldots, M, \tag{3.1}$$

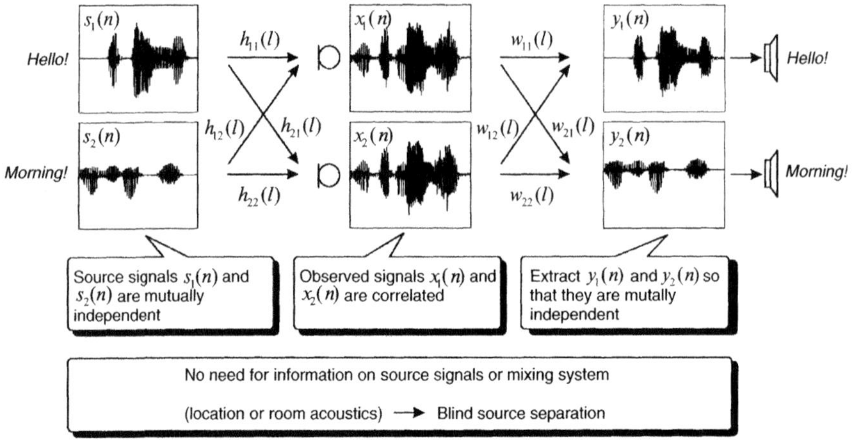

Fig. 3.2. Task of blind source separation of speech signals.

where $h_{ji}(l)$ represents the impulse response from source i to sensor j. We assume that the number of sources N is known or can be estimated in some way (e.g., by [42]) and that the number of sensors M is more than or equal to N ($N \leq M$).

The separation system typically consists of a set of FIR filters $w_{ij}(l)$ of length L to produce N separated signals at the outputs:

$$y_i(n) = \sum_{j=1}^{M} \sum_{l=0}^{L-1} w_{ij}(l)\, x_j(n-l), \quad i = 1, \ldots, N. \tag{3.2}$$

The separation filters are estimated so that the separated signals become mutually independent. The separation filters $w_{ij}(l)$ should be obtained blindly, i.e., without knowing $s_i(n)$ or $h_{ji}(l)$.

A two-input, two-output convolutive BSS problem, i.e., $N = M = 2$, is shown in Figs. 3.1 and 3.2. It is assumed that the source signals $s_1(n)$ and $s_2(n)$ are mutually independent. This assumption usually holds for sounds in the real world. There are two microphones that pick up the mixed speech. Only the observed signals $x_1(n)$ and $x_2(n)$ are available, and they are correlated. The goal is to adapt the separation systems $w_{ij}(l)$ and to extract $y_1(n)$ and $y_2(n)$ so that they are mutually independent. With this operation, we can obtain $s_1(n)$ and $s_2(n)$ in the output $y_1(n)$ and $y_2(n)$. No information is needed on the source positions or period of source existence/absence. Nor is any information on the mixing systems $h_{ji}(l)$ required. Thus, this task is called *blind* source separation.

Fig. 3.3 shows a block diagram of BSS. The ideal goal of BSS is to separate and deconvolve the mixtures $x_j(n)$ and to obtain a delayed version of source $s_i(n)$ at each output i. However, this is very difficult if $s_i(n)$ is a colored signal, which is the case when separating natural sounds such as speech [15]. A practical alternative goal [30, 49] is to obtain the convolved version of a

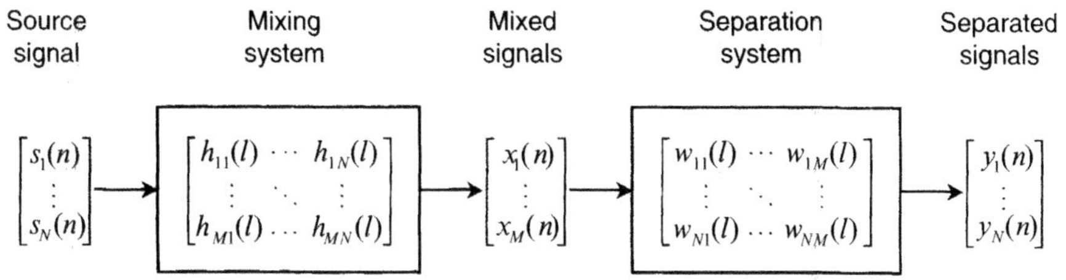

Fig. 3.3. BSS for convolutive mixtures.

source $s_i(n)$ measured at a sensor J_i:

$$y_i(n) \stackrel{!}{=} \sum_l h_{J_i i}(l)\, s_i\Big(n - \frac{L}{2} - l\Big), \tag{3.3}$$

where the sensor index J_i can be selected according to each output i. The way used to attain this goal will be discussed in Sec. 3.8.

3.3 Overview of Frequency-Domain Approach

Fig. 3.4 and more visually Fig. 3.5 show the flow of frequency-domain BSS. Time-domain signals $x_j(t)$ sampled at frequency f_s are converted into frequency-domain time-series signals $X_j(e^{j\Omega}, n)$ with an L-point short-time Fourier transform (STFT):

$$X_j\left(e^{j\Omega}, n\right) = \sum_{r=-\frac{L}{2}}^{\frac{L}{2}-1} x_j(n+r)\, win(r)\, e^{-j\Omega r}, \tag{3.4}$$

where $\Omega \in \{0, \frac{1}{L}2\pi, \ldots, \frac{L-1}{L}2\pi\}$ is a normalized frequency, $win(r)$ is a window that tapers smoothly to zero at each end, such as a Hanning window $\frac{1}{2}(1 + \cos\frac{2\pi r}{L})$, and n is an index representing time.

The remaining operations are performed in the frequency domain. The advantage is that the convolutive mixtures in Eq. 3.1 can be approximated as instantaneous mixtures in each frequency bin:

$$X_j\left(e^{j\Omega}, n\right) = \sum_{i=1}^{N} H_{ji}\left(e^{j\Omega}\right) S_i\left(e^{j\Omega}, n\right), \tag{3.5}$$

where $H_{ji}(e^{j\Omega})$ is the frequency response from source i to sensor j, and $S_i(e^{j\Omega}, n)$ is a frequency-domain time-series signal of $s_i(n)$ obtained by the same operation as Eq. 3.4. The vector notation of the mixing model (Eq. 3.5) is

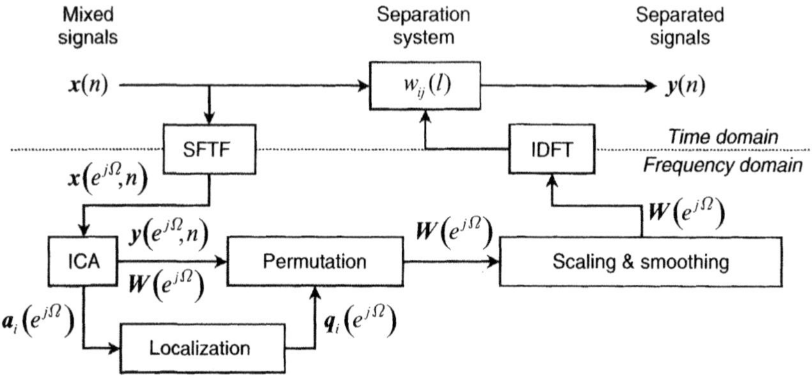

Fig. 3.4. Flow of frequency-domain BSS.

$$\boldsymbol{x}\left(e^{j\Omega},n\right)=\sum_{i=1}^{N}\boldsymbol{h}_i\left(e^{j\Omega}\right)S_i\left(e^{j\Omega},n\right), \tag{3.6}$$

where

$$\boldsymbol{x}\left(e^{j\Omega},n\right)=\left[X_1\left(e^{j\Omega},n\right),\ \ldots,\ X_M\left(e^{j\Omega},n\right)\right]^{\mathrm{T}} \tag{3.7}$$

is a sensor sample vector and

$$\boldsymbol{h}_i\left(e^{j\Omega}\right)=\left[H_{1i}\left(e^{j\Omega}\right),\ \ldots,\ H_{Mi}\left(e^{j\Omega}\right)\right]^{\mathrm{T}} \tag{3.8}$$

is the vector of the frequency responses from source $s_i(n)$ to all M sensors.

To obtain the frequency responses $W_{ij}(e^{j\Omega})$ of separation filters $w_{ij}(l)$ in Eq. 3.2, complex-valued ICA

$$\boldsymbol{y}\left(e^{j\Omega},n\right)=\boldsymbol{W}\left(e^{j\Omega}\right)\boldsymbol{x}\left(e^{j\Omega},n\right) \tag{3.9}$$

is solved, where

$$\boldsymbol{y}\left(e^{j\Omega},n\right)=\left[Y_1\left(e^{j\Omega},n\right),\ \ldots,\ Y_N\left(e^{j\Omega},n\right)\right]^{\mathrm{T}} \tag{3.10}$$

is a vector of separated signals,

$$\boldsymbol{W}\left(e^{j\Omega}\right)=\left[\boldsymbol{w}_1\left(e^{j\Omega}\right),\ \ldots,\ \boldsymbol{w}_N\left(e^{j\Omega}\right)\right]^{\mathrm{H}} \tag{3.11}$$

is an $N\times M$ separation matrix,

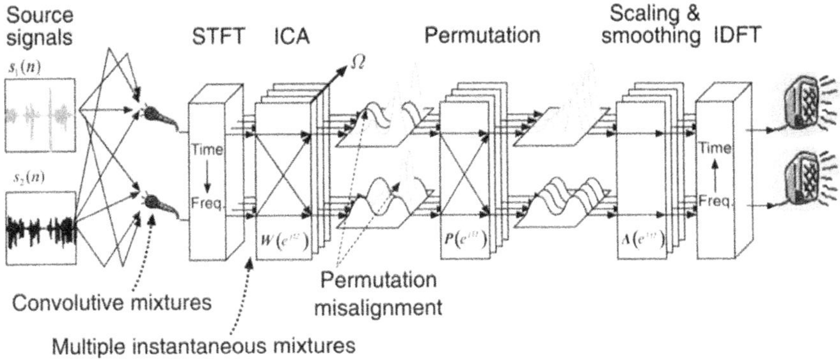

Fig. 3.5. Frequency-domain BSS overview ($N = M = 2$).

$$\boldsymbol{w}_i\left(e^{j\Omega}\right) = \left[W_{i1}\left(e^{j\Omega}\right), \ldots, W_{iM}\left(e^{j\Omega}\right)\right]^{\mathrm{H}} \tag{3.12}$$

and

$$W_{ij}\left(e^{j\Omega}\right) = \left[\boldsymbol{W}\left(e^{j\Omega}\right)\right]_{ij}. \tag{3.13}$$

The details of the ICA algorithm are discussed in Sec. 3.4.

Calculating the Moore-Penrose pseudoinverse $\boldsymbol{W}^{+}(e^{j\Omega})$ (reduced to the inverse $\boldsymbol{W}^{-1}\left(e^{j\Omega}\right)$ if $N = M$) of $\boldsymbol{W}(e^{j\Omega})$ as

$$\left[\boldsymbol{a}_1\left(e^{j\Omega}\right), \cdots, \mathbf{a}_N\left(e^{j\Omega}\right)\right] = \boldsymbol{W}^{+}\left(e^{j\Omega}\right), \tag{3.14}$$

$$\boldsymbol{a}_i\left(e^{j\Omega}\right) = \left[A_{1i}\left(e^{j\Omega}\right), \ldots, A_{Mi}\left(e^{j\Omega}\right)\right]^{\mathrm{T}} \tag{3.15}$$

is very useful for source localization and scaling alignment, as described in Sec. 3.6 and Sec. 3.8, respectively. It should be noted that it is not difficult to make $\boldsymbol{W}(e^{j\Omega})$ invertible by using an appropriate ICA procedure (for an example, see Sec. 3.4). By multiplying both sides of Eq. 3.9 by $\boldsymbol{W}^{+}(e^{j\Omega})$, the sensor sample vector $\boldsymbol{x}(n)$ is represented by a linear combination of basis vectors $\boldsymbol{a}_1(e^{j\Omega}), \ldots, \boldsymbol{a}_N(e^{j\Omega})$:

$$\boldsymbol{x}\left(e^{j\Omega}, n\right) = \sum_{i=1}^{N} \boldsymbol{a}_i\left(e^{j\Omega}\right) Y_i\left(e^{j\Omega}, n\right). \tag{3.16}$$

It is well-known that an ICA solution (Eq. 3.9) has permutation and scaling ambiguities: even if we permute the rows of $\boldsymbol{W}(e^{j\Omega})$ or multiply a row by a constant, it is still an ICA solution. In matrix notation,

$$\boldsymbol{W}\left(e^{j\Omega}\right) \leftarrow \boldsymbol{\Lambda}\left(e^{j\Omega}\right) \boldsymbol{P}\left(e^{j\Omega}\right) \boldsymbol{W}\left(e^{j\Omega}\right) \tag{3.17}$$

is also an ICA solution for any permutation $\boldsymbol{P}(e^{j\Omega})$ and diagonal $\boldsymbol{\Lambda}(e^{j\Omega})$ matrix. Permutation alignment is to decide $\boldsymbol{P}(e^{j\Omega})$ so that a time-domain separated signal contains frequency components from the same source. Sec. 3.7 presents a method for solving this problem. Scaling alignment is to decide $\boldsymbol{\Lambda}(e^{j\Omega})$ so that a time-domain separated signal satisfies the goal (Eq. 3.3), as discussed in Sec. 3.8.

Then, we perform spectral smoothing so that a time-domain separation filter tapers smoothly to zero at each end. This is typically achieved by multiplying the time-domain filter by a Hanning window, which is equivalent to smoothing the frequency-domain separation matrices as

$$\boldsymbol{W}\left(e^{j\Omega}\right) \leftarrow \frac{1}{4}\left[\boldsymbol{W}\left(e^{j\Omega-\Delta\Omega}\right) + 2\boldsymbol{W}\left(e^{j\Omega}\right) + \boldsymbol{W}\left(e^{j\Omega+\Delta\Omega}\right)\right],$$

where $\Delta\Omega = \frac{2\pi}{L}$ is the difference from the adjacent frequency. However, this smoothing changes the ICA solution and causes an error. Sec. 3.9 discusses the error and how to minimize it.

Finally, separation filters $w_{ij}(l)$ are obtained by applying inverse DFT to $W_{ij}(e^{j\Omega}) = [\boldsymbol{W}(e^{j\Omega})]_{ij}$:

$$w_{ij}(l) = \sum_{\Omega \in \{0,\, \frac{1}{L}2\pi,\, \dots,\, \frac{L-1}{L}2\pi\}} W_{ij}\left(e^{j\Omega}\right) e^{j\Omega(l-\frac{L}{2})},$$

where $l = 0, \dots, L-1$. The reason for using $e^{j\Omega(l-\frac{L}{2})}$ instead of $e^{j\Omega l}$ is to make the separation filter $w_{ij}(l)$ causal. Then, the separated signals $y_i(n)$ are produced by Eq. 3.2.

3.4 Complex-Valued Independent Component Analysis

This section discusses how to solve the ICA equation 3.9. One of the advantages of frequency-domain BSS is that we can employ any ICA algorithm for instantaneous mixtures, such as the information maximization approach (InfoMax) [9] combined with the natural gradient [2], FastICA [17], JADE [13], or an algorithm based on the non-stationarity of signals [29]. Here, we explain a procedure that was shown to be efficient by the experiments described in Sec. 3.10. The procedure consists of the following three steps:

1. Dimension reduction and whitening by eigenvalue decomposition,
2. ICA by a unitary matrix (FastICA),
3. ICA by InfoMax combined with the natural gradient.

The first step performs a linear transformation

$$\boldsymbol{z}\left(e^{j\Omega}, n\right) = \boldsymbol{V}\left(e^{j\Omega}\right) \boldsymbol{x}\left(e^{j\Omega}, n\right)$$

for M-dimensional sensor observations $\boldsymbol{x}(e^{j\Omega}, n)$ such that the dimension of $\boldsymbol{z}(e^{j\Omega}, n)$ is reduced (if necessary) to the number of sources N and $\boldsymbol{z}(e^{j\Omega}, n)$ is spatially whitened (sphered), i.e., $\mathrm{E}\left\{ \boldsymbol{z}(e^{j\Omega},n)\,\boldsymbol{z}^{\mathrm{H}}(e^{j\Omega},n) \right\} = \boldsymbol{I}$, where $\boldsymbol{I}$ is the $N \times N$ identity matrix. The linear transformation $\boldsymbol{V}(e^{j\Omega})$ is typically obtained by eigenvalue decomposition. Let $\lambda_1(e^{j\Omega}) \geq \cdots \geq \lambda_M(e^{j\Omega})$ be sorted eigenvalues of the spatial correlation matrix $\boldsymbol{R}(e^{j\Omega}) = \mathrm{E}\left\{ \boldsymbol{x}(e^{j\Omega},n)\,\boldsymbol{x}^{H}(e^{j\Omega},n) \right\}$ and $\boldsymbol{e}_1(e^{j\Omega}), \ldots, \boldsymbol{e}_M(e^{j\Omega})$ be their corresponding eigenvectors. Then, the linear transformation is

$$\boldsymbol{V}\left(e^{j\Omega}\right) = \boldsymbol{D}^{-1/2}\left(e^{j\Omega}\right)\boldsymbol{E}^{\mathrm{H}}\left(e^{j\Omega}\right),$$

where

$$\boldsymbol{D}\left(e^{j\Omega}\right) = \mathrm{diag}\left[\lambda_1\left(e^{j\Omega}\right), \ldots, \lambda_N\left(e^{j\Omega}\right)\right]$$

is the diagonal matrix of the N largest eigenvalues,

$$\boldsymbol{E}\left(e^{j\Omega}\right) = \left[\boldsymbol{e}_1\left(e^{j\Omega}\right), \ldots, \boldsymbol{e}_N\left(e^{j\Omega}\right)\right]$$

is the matrix of their corresponding eigenvectors, and

$$\boldsymbol{e}_i\left(e^{j\Omega}\right) = \left[e_{1i}\left(e^{j\Omega}\right), \ldots, e_{Mi}\left(e^{j\Omega}\right)\right]^{\mathrm{T}}.$$

This step has practical importance for the following two reasons. First, the outputs $\boldsymbol{y}\left(e^{j\Omega},n\right)$ of ICA (Eq. 3.9) adhere to the signal subspace that is identified by the N eigenvectors $\boldsymbol{e}_1(e^{j\Omega}), \ldots, \boldsymbol{e}_N(e^{j\Omega})$. This means that the following ICA algorithm does not pursue its solution in the noise subspace, which consequently stabilizes the algorithm and also has a noise/reverberation reduction effect [7]. A geometrical interpretation of the dimension reduction is given in [50]. Second, the whitening $\mathrm{E}\left\{ \boldsymbol{z}(e^{j\Omega},n)\,\boldsymbol{z}^{\mathrm{H}}(e^{j\Omega},n) \right\} = \boldsymbol{I}$ is necessary for FastICA, and it also provides an efficient convergence for InfoMax even if the step size is constant over all frequency bins.

The second step performs ICA in a constrained form:

$$\boldsymbol{y}\left(e^{j\Omega},n\right) = \boldsymbol{B}\left(e^{j\Omega}\right)\boldsymbol{z}\left(e^{j\Omega},n\right),$$

where $\boldsymbol{B}(e^{j\Omega})$ is an $N \times N$ unitary matrix: $\boldsymbol{B}(e^{j\Omega})\,\boldsymbol{B}^{\mathrm{H}}(e^{j\Omega}) = \boldsymbol{I}$. This is performed by a complex-valued version of FastICA [10,17]. It is very efficient because a fairly good solution can be obtained with only several iterations. The efficiency comes from the fact that $\boldsymbol{z}(e^{j\Omega},n)$ is whitened and $\boldsymbol{B}(e^{j\Omega})$ is unitary. However, there remains room for improving the solution by using another ICA algorithm. One of the reasons is that the output $\boldsymbol{y}(e^{j\Omega},n)$ of FastICA is whitened $\mathrm{E}\left\{ \boldsymbol{y}(e^{j\Omega},n)\,\boldsymbol{y}^{\mathrm{H}}(e^{j\Omega},n) \right\} = \boldsymbol{I}$ and thus uncorrelated, whereas original sources $S_1(e^{j\Omega},n), \ldots, S_N(e^{j\Omega},n)$ are not always completely uncorrelated with a limited number of samples.

The third step improves the ICA solution obtained so far as an initial value

$$\boldsymbol{y}\left(e^{j\Omega},n\right) = \boldsymbol{W}\left(e^{j\Omega}\right)\boldsymbol{x}\left(e^{j\Omega},n\right) = \boldsymbol{B}\left(e^{j\Omega}\right)\boldsymbol{V}\left(e^{j\Omega}\right)\boldsymbol{x}\left(e^{j\Omega},n\right)$$

by employing another ICA algorithm that does not have the unitary constraint. Based on the use of InfoMax combined with the natural gradient, a separation matrix $\boldsymbol{W}(e^{j\Omega})$ is gradually improved by the learning rule:

$$\boldsymbol{W}\left(e^{j\Omega}\right) \leftarrow \boldsymbol{W}\left(e^{j\Omega}\right) + \mu\left[\boldsymbol{I} - \mathrm{E}\left\{\boldsymbol{\Phi}\Big(\boldsymbol{y}\left(e^{j\Omega}, n\right)\Big)\boldsymbol{y}^{\mathrm{H}}\left(e^{j\Omega}, n\right)\right\}\right]\boldsymbol{W}\left(e^{j\Omega}\right), \tag{3.18}$$

where μ is a step-size parameter. $\boldsymbol{\Phi}(\boldsymbol{y}) = [\Phi(y_1), \ldots, \Phi(y_N)]^{\mathrm{T}}$ is an element-wise nonlinear function defined by

$$\Phi(y_i) = -\frac{\partial}{\partial y_i}\log p(y_i), \tag{3.19}$$

where $p(y_i)$ is the probability density function (PDF) of a complex-valued signal $y_i = |y_i|\, e^{j\cdot\arg(y_i)}$. Since y_i is a frequency-domain signal whose phase can be shifted arbitrarily by shifting the STFT window position (Eq. 3.4), a feasible assumption is that the PDF is independent of the phase $p(y_i) = \beta \cdot p(|y_i|)$, where β is a constant. This assumption reduces Eq. 3.19 to

$$\Phi(y_i) = \varphi(|y_i|)\, e^{j\arg(y_i)}, \tag{3.20}$$

$$\varphi(|y_i|) = -\frac{\partial}{\partial |y_i|}\log p(|y_i|). \tag{3.21}$$

If we assume the Laplacian distribution $p(|y_i|) = \frac{1}{2}e^{-|y_i|}$, which is typical for speech modeling, we have $\varphi(|y_i|) = 1$ and thus a simple nonlinear function

$$\Phi(y_i) = e^{j\arg(y_i)}.$$

A nonlinear function of the form of Eq. 3.20 has a better convergence property [38] than one where the nonlinearity is applied separately to the real and imaginary parts of a complex-valued signal y_i.

3.5 Separation Mechanism of Blind Source Separation

The mechanism of BSS based on ICA has been shown to be equivalent to that of an adaptive microphone array system, i.e., N sets of adaptive beamformers (ABFs) with an adaptive null directivity aimed in the direction of unnecessary sounds [5, 6]. From the equivalence between BSS and ABF, it becomes clear that the physical behavior of BSS reduces the jammer signal by making a spatial null toward the jammer and extracts the target.

The separation performance of BSS is compared with that of ABF. Fig. 3.6 shows the directivity patterns obtained by BSS and ABF. In Fig. 3.6, (a) and (b) show directivity patterns by $\mathbf{W}$ obtained by BSS, and (c) and (d) show directivity patterns by $\mathbf{W}$ obtained by ABF. When $T_{\mathrm{R}} = 0$[1], a sharp spatial null is obtained by both BSS and ABF [see Figs. 3.6(a) and (c)]. When $T_{\mathrm{R}} = 300$ ms, the directivity pattern becomes duller for both BSS and ABF [see Figs. 3.6(b) and (d)].

[1] T_{R} abbreviates the reverberation time.

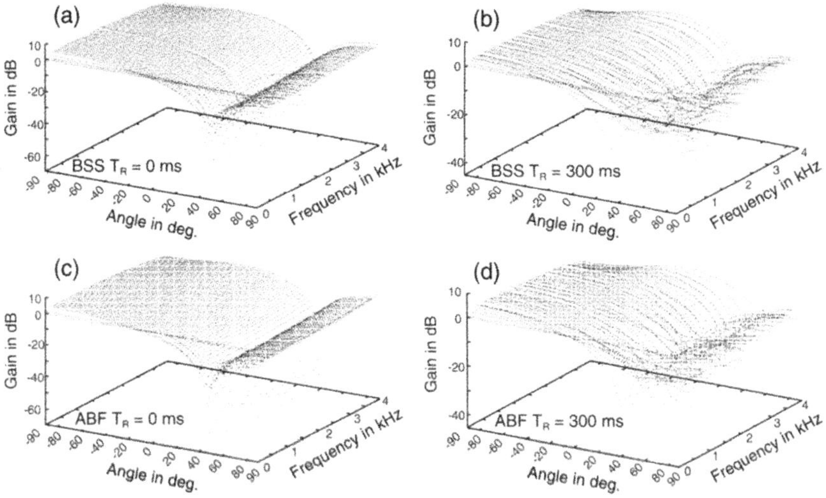

Fig. 3.6. Directivity patterns (a) obtained by BSS ($T_{\mathrm{R}} = 0$ ms), (b) obtained by BSS ($T_{\mathrm{R}} = 300$ ms), (c) obtained by ABF ($T_{\mathrm{R}} = 0$ ms), and (d) obtained by ABF ($T_{\mathrm{R}} = 300$ ms).

BSS can be regarded as an intelligent version of ABF in the sense that it can adapt without any information on the source positions or period of source existence/absence [28].

3.6 Source Localization

This section presents a source localization method by analyzing the ICA solution (Eq. 3.9 or equivalently Eq. 3.16). The information on source locations can be used to solve the permutation problem, as described in the next section. Many source localization methods have been proposed. A widely used method is MUSIC (MUltiple SIgnal Classification) [44], which employs subspace analysis with second-order statistics. The ICA-based method, on the other hand, employs higher-order statistics (or multiple second-order statistics based on non-stationarity). In this sense, the ICA-based method has certain advantages over the subspace-based method [39].

The source localization technique that employs ICA is a by-product of research on frequency-domain BSS. Direction-of-arrival (DOA) estimation methods [20, 24, 37] have been proposed based on beamforming theory [51]. They calculate directivity patterns as shown in Fig. 3.6 from the separation matrix $\boldsymbol{W}(e^{j\Omega})$ and then search the null directions, which correspond to the directions of sources [6]. However, it is simpler and more effective to estimate the directions directly from the basis vectors $\boldsymbol{a}_i(e^{j\Omega})$, which are given by the pseudoinverse of $\boldsymbol{W}(e^{j\Omega})$. The source localization method [31, 33, 39, 41]

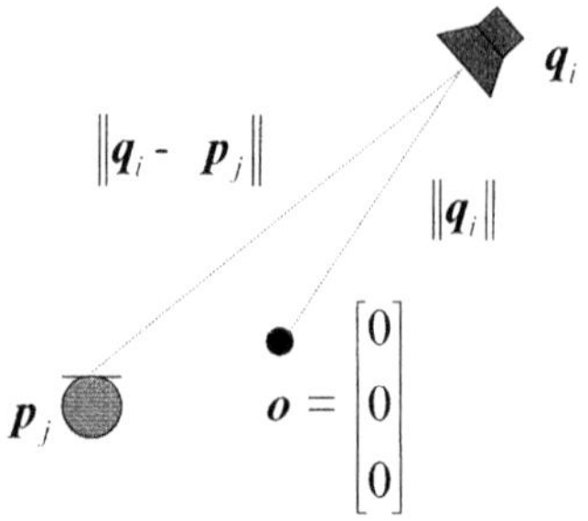

Fig. 3.7. Nearfield (direct-path) model.

presented in this section is based on this idea. Such an idea was taken for granted in research on blind identification [12, 47], where the mixing system is estimated directly.

3.6.1 Basic Theory of Nearfield Model

Let us assume a mixing model that is suitable for source localization. Although the mixing model (Eq. 3.1) in the time domain is a multi-path mixing model, we approximate the frequency response $H_{ji}(e^{j\Omega})$ in Eq. 3.5 with a nearfield (direct-path) model (Fig. 3.7):

$$H_{ji}\left(e^{j\Omega}\right) \approx \frac{1}{\|\boldsymbol{q}_i - \boldsymbol{p}_j\|} e^{j\frac{\Omega f_s}{c}\left(\|\boldsymbol{q}_i - \boldsymbol{p}_j\| - \|\boldsymbol{q}_i\|\right)}, \tag{3.22}$$

where $\boldsymbol{p}_j$ and $\boldsymbol{q}_i$ are three-dimensional vectors representing the locations of sensor j and source i, respectively, and c is the propagation velocity of the signals. We assume that the amplitude is attenuated based on the distance $\|\boldsymbol{q}_i - \boldsymbol{p}_j\|$. We also assume that the phase depends on the difference between the distances $\|\boldsymbol{q}_i - \boldsymbol{p}_j\| - \|\boldsymbol{q}_i\|$ from the source to the sensor and to the origin $\boldsymbol{o} = [0, 0, 0]^{\mathrm{T}}$. This makes the phase zero at the origin. If the phase $2\frac{\Omega f_s}{c}\left(\|\boldsymbol{q}_i - \boldsymbol{p}_j\| - \|\boldsymbol{q}_i\|\right)$ is outside the range $(-\pi, \pi)$, this model suffers from spatial aliasing. Therefore, the model is feasible as long as the condition

$$f = \frac{\Omega}{2\pi} f_s < \left| \frac{c}{2\left(\|\boldsymbol{q}_i - \boldsymbol{p}_j\| - \|\boldsymbol{q}_i\|\right)} \right|$$

is satisfied.

The ICA-based source localization discussed in this section estimates the location $\boldsymbol{q}_i$ of source i from information on sensor locations $\boldsymbol{p}_j$ and the separation matrix $\boldsymbol{W}(e^{j\Omega})$ obtained by ICA (Eq. 3.9). Let us assume here that the decomposition (Eq. 3.16) of observations $\boldsymbol{x}(e^{j\Omega}, n)$ has been obtained in each frequency bin by the pseudoinverse of $\boldsymbol{W}(e^{j\Omega})$. By comparing Eq. 3.6 and

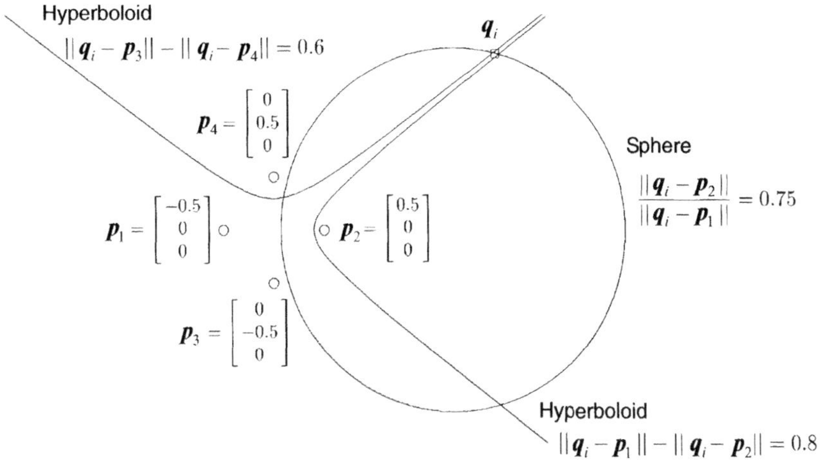

Fig. 3.8. Source localization by intersection of two hyperboloids and a sphere.

Eq. 3.16, we observe the following fact. If the ICA algorithm works well and the outputs $y_1(n), \ldots, y_N(n)$ are the estimation of the sources $s_1(n), \ldots, s_N(n)$, then the basis vectors $\boldsymbol{a}_1(e^{j\Omega}), \ldots, \boldsymbol{a}_N(e^{j\Omega})$ are also estimations of the mixing vectors $\boldsymbol{h}_1(e^{j\Omega}), \ldots, \boldsymbol{h}_N(e^{j\Omega})$ up to the permutation and scaling ambiguity.

Following the model (Eq. 3.22), the ratio between two elements $a_{ji}(e^{j\Omega})$ and $a_{j'i}(e^{j\Omega})$ of the same basis vector $\boldsymbol{a}_i(e^{j\Omega})$ provides the key equation for source localization:

$$\begin{aligned}\frac{a_{ji}\left(e^{j\Omega}\right)}{a_{j'i}\left(e^{j\Omega}\right)} &= \frac{\alpha_i H_{ji}\left(e^{j\Omega}\right)}{\alpha_i H_{j'i}\left(e^{j\Omega}\right)} \\ &= \frac{\|\boldsymbol{q}_i - \boldsymbol{p}_{j'}\|}{\|\boldsymbol{q}_i - \boldsymbol{p}_j\|} e^{j\frac{\Omega f_s}{c}\left(\|\boldsymbol{q}_i - \boldsymbol{p}_j\| - \|\boldsymbol{q}_i - \boldsymbol{p}_{j'}\|\right)}, \qquad (3.23)\end{aligned}$$

where the scaling ambiguity α_i is cancelled out by calculating the ratio. The permutation ambiguity still remains. However, if we estimate the location $\boldsymbol{q}_i$ for all $i = 1, \ldots, N$, the set of all estimated locations does not depend on the permutation.

With respect to the phase differences, the set of vectors $\boldsymbol{q}_i$ in the argument of Eq. 3.23,

$$\|\boldsymbol{q}_i - \boldsymbol{p}_j\| - \|\boldsymbol{q}_i - \boldsymbol{p}_{j'}\| = \frac{\arg\left(a_{ji}\left(e^{j\Omega}\right)/a_{j'i}\left(e^{j\Omega}\right)\right)}{(\Omega f_s)/c}, \qquad (3.24)$$

defines a surface where the difference between the distances from $\boldsymbol{p}_j$ and $\boldsymbol{p}_{j'}$ is constant. The surface is one sheet of a two-sheet hyperboloid.

Alternatively, with respect to the level differences, the set of vectors $\boldsymbol{q}_i$ in the modulus of Eq. 3.23,

$$\frac{\|\boldsymbol{q}_i - \boldsymbol{p}_{j'}\|}{\|\boldsymbol{q}_i - \boldsymbol{p}_j\|} = \left| \frac{a_{ji}\left(e^{j\Omega}\right)}{a_{j'i}\left(e^{j\Omega}\right)} \right|, \tag{3.25}$$

defines a sphere where the ratio of the distances from $\boldsymbol{p}_j$ and $\boldsymbol{p}_{j'}$ is constant. Therefore, with these two equations 3.24 and 3.25, we can estimate the possible location $\boldsymbol{q}_i$ of source i. Such hyperboloid and sphere are defined by a pair of sensors j and j'. If we select another pair of sensors, a different hyperboloid and sphere are obtained. In this way, the location $\boldsymbol{q}_i$ is estimated as the intersection of several hyperboloids and spheres. An example is shown in Fig. 3.8.

3.6.2 Direction of Arrival Estimation with Far-field Model

Although it is useful to estimate a three-dimensional location, calculating the intersections of hyperboloids and spheres is computationally demanding. In many cases, it is sufficient to estimate only the direction-of-arrival (DOA) of source signal $s_i(n)$. If we assume the source location $\boldsymbol{q}_i$ is far from sensors $\boldsymbol{p}_j$ and $\boldsymbol{p}_{j'}$, Eq. 3.24 can be approximated as a far-field model (Fig. 3.9):

$$\left[\boldsymbol{p}_j - \boldsymbol{p}_{j'}\right]^{\mathrm{T}} \frac{\boldsymbol{q}_i}{\|\boldsymbol{q}_i\|} = \frac{\arg\Big(a_{ji}\left(e^{j\Omega}\right) / a_{j'i}\left(e^{j\Omega}\right)\Big)}{(\Omega\, f_{\mathrm{s}})/c}, \tag{3.26}$$

and the cosine of angle $\theta_i^{jj'}$ between the two vectors $\boldsymbol{q}_i$ and $\boldsymbol{p}_j - \boldsymbol{p}_{j'}$ can be calculated as

$$\begin{aligned} \cos\theta_i^{jj'} &= \frac{\left[\boldsymbol{p}_j - \boldsymbol{p}_{j'}\right]^{\mathrm{T}} \boldsymbol{q}_i}{\|\boldsymbol{p}_j - \boldsymbol{p}_{j'}\| \, \|\boldsymbol{q}_i\|} \\ &= \frac{\arg\Big(a_{ji}\left(e^{j\Omega}\right) / a_{j'i}\left(e^{j\Omega}\right)\Big)}{\dfrac{\Omega\, f_{\mathrm{s}}}{c} \|\boldsymbol{p}_j - \boldsymbol{p}_{j'}\|}. \end{aligned} \tag{3.27}$$

The set of vectors $\boldsymbol{q}_i$ that satisfy Eq. 3.26 represents a cone [31], which is the asymptotic surface of the corresponding hyperboloid (Eq. 3.24). To estimate the DOA of a source, the intersections of several cones should be obtained. Let us assume that we select u cones whose corresponding sensor pairs are (j_1, j_1'), ..., (j_u, j_u'). The set of equations 3.26 for u sensor pairs is represented as

$$\boldsymbol{D} \frac{\boldsymbol{q}_i}{\|\boldsymbol{q}_i\|} = \frac{\boldsymbol{r}_i\left(e^{j\Omega}\right)}{(\Omega\, f_{\mathrm{s}})/c}, \tag{3.28}$$

where

$$\boldsymbol{D} = \left[\boldsymbol{p}_{j_1} - \boldsymbol{p}_{j_1'}, \ldots, \boldsymbol{p}_{j_u} - \boldsymbol{p}_{j_u'}\right]^{\mathrm{T}},$$

$$\boldsymbol{r}_i\left(e^{j\Omega}\right) = \left[\arg\left(\frac{a_{j_1 i}\left(e^{j\Omega}\right)}{a_{j_1' i}\left(e^{j\Omega}\right)}\right), \ldots, \arg\left(\frac{a_{j_u i}\left(e^{j\Omega}\right)}{a_{j_u' i}\left(e^{j\Omega}\right)}\right)\right]^{\mathrm{T}}.$$

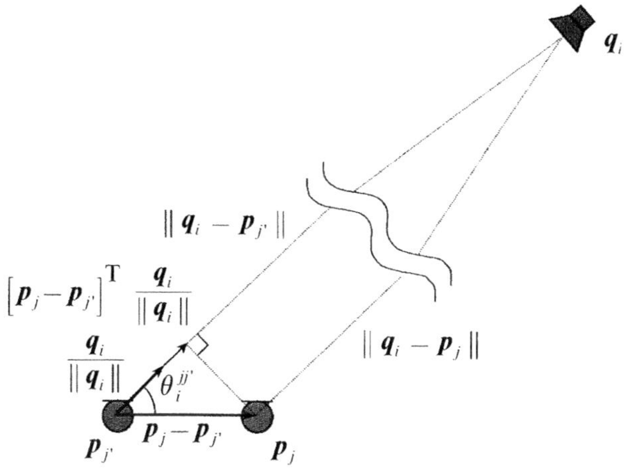

Fig. 3.9. Far-fieldmodel.

In practical situations, there is no exact solution for Eq. 3.28 because the u conditions do not coincide exactly. Therefore, we typically solve it with the least-square approach by using the Moore-Penrose pseudoinverse [33]:

$$\frac{\boldsymbol{q}_i}{\|\boldsymbol{q}_i\|} = \frac{\boldsymbol{D}^{+}\boldsymbol{r}_i\left(e^{j\Omega}\right)}{(\Omega\, f_{\mathrm{s}})/c}. \tag{3.29}$$

If $\mathrm{rank}(\boldsymbol{D}) \geq 3$, the set of vectors $\boldsymbol{q}_i$ that satisfy Eq. 3.29 represents a line in three-dimensional space, which represents the DOA of a source i.

The upper photo in Fig. 3.10 shows the case where eight microphones and three loudspeakers are arranged three-dimensionally, and the lower plot shows the DOA estimation results for this case. Each point shows a location vector $\bar{\boldsymbol{q}}_i(\Omega)$ that is normalized to unit norm

$$\bar{\boldsymbol{q}}_i(\Omega) = \frac{\boldsymbol{q}_i(\Omega)}{\|\boldsymbol{q}_i(\Omega)\|}.$$

The estimations are obtained for all frequencies Ω and all output indexes i. As shown in the plot, they form clusters, each of which corresponds to the location of each source.

If the sensor and source locations are limited to a two-dimensional plane, the dimensionality of location vectors, such as $\boldsymbol{p}_i$ and $\boldsymbol{q}_i$, can be reduced to two. In this case, $\mathrm{rank}(\boldsymbol{D}) \geq 2$ is sufficient to reach a solution in (3.29). Moreover, the DOA of source i can be represented simply by the angle θ_i that satisfies

$$\bar{\boldsymbol{q}}_i = \left[\cos(\theta_i),\, \sin(\theta_i)\right]^{\mathrm{T}}, \quad -180^\circ < \theta_i \leq 180^\circ. \tag{3.30}$$

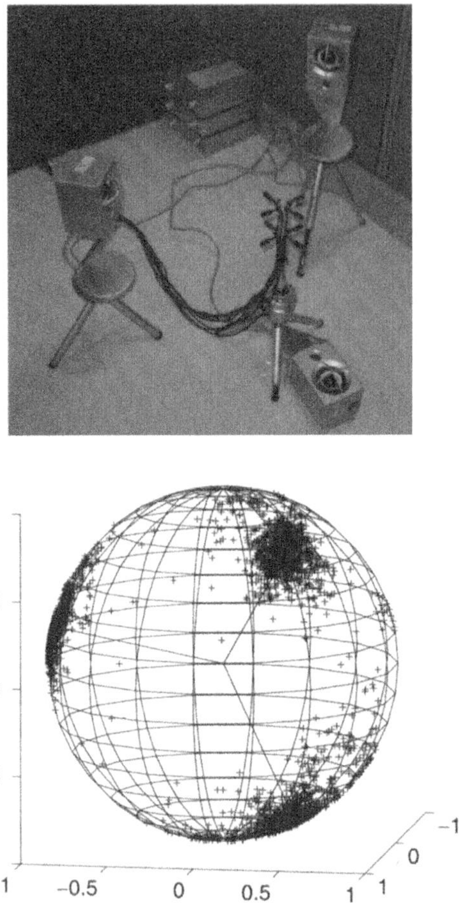

Fig. 3.10. Three-dimensional arrangement of eight microphones and three loudspeakers (upper picture) and DOA estimation results for this case (lower picture).

Fig. 3.19 shows the case where the sensor and source locations are limited to two dimensions. The DOA estimations in this case are shown in Figs. 3.20 and 3.21.

If the sensors are arranged linearly and the potential source location is in a two-dimensional half-plane, which is to one side of the sensor arrangement line, the angle $\theta_i^{jj'}$ $(0° \leq \theta_i^{jj'} \leq 180°)$ by Eq. 3.27 provides sufficient information on the source location. For example, Fig. 3.14 shows DOA estimation results for such a case with the conditions shown in Fig. 3.13.

3.7 Permutation Alignment

This section discusses how to solve the permutation problem. Various methods have already been proposed. With reference to the ICA Eq. 3.9 as well as to the decomposition (Eq. 3.16) of observations $\boldsymbol{x}(e^{j\Omega}, n)$, we classify these methods into four categories based on the following strategies:

1. Applying an operation to the separation matrix $\boldsymbol{W}(e^{j\Omega})$,
2. Utilizing the information on the separation matrix $\boldsymbol{W}(e^{j\Omega})$ itself,
3. Utilizing the information on the basis vectors $\boldsymbol{a}_1(e^{j\Omega}), ..., \boldsymbol{a}_N(e^{j\Omega})$,
4. Utilizing the information on the separated signals $Y_1(e^{j\Omega}, n), ..., Y_N(e^{j\Omega}, n)$.

The operation of the first strategy basically involves smoothing the separation matrices in the frequency domain. This has been realized by reducing the filter length through rectangular windowing in the time domain [11,36,45,48] or by averaging the separation matrices with adjacent frequencies [48]. However, this operation makes the separation matrix $\boldsymbol{W}(e^{j\Omega})$ different from the ICA solution (Eq. 3.9), which may have a detrimental effect on the separation performance. A possible way to solve this problem is to interleave the ICA update, e.g., Eq. 3.18, and this operation until convergence. In this sense, this strategy is related to the third approach to BSS discussed in the introduction.

The second category includes the beamforming approach [20,24,37], where the directivity patterns formed by the separation matrix are analyzed to identify the DOA of each source. The third category includes an approach that utilizes the results of source localization with the basis vectors [31,33,41,47]. The theory and operation for source localization were discussed in Sec. 3.6. These two approaches from the second and the third categories utilize basically the same information because the separation matrix $\boldsymbol{W}(e^{j\Omega})$ and the basis vectors $\boldsymbol{a}_1(e^{j\Omega}), \ldots, \boldsymbol{a}_N(e^{j\Omega})$ are directly connected by the pseudoinverse operation (Eq. 3.14). However, the information used in the third category is easier to handle since it directly represents the mixing system (Eq. 3.6). The last category includes an approach that employs the inter-frequency correlations of output signal envelopes [4, 35]. This is particularly effective for a non-stationary signal such as speech.

In the next two subsections, we explain the approaches of the third and fourth categories, respectively. Since these two approaches have different but complementary characteristics, integrating them is a good way to find a better solution to the permutation problem [41]. Subsection 3.7.3 presents a method that effectively integrates the two approaches to solve the permutation problem in a better way. In the following subsections, let Π_Ω be a permutation corresponding to the inverse $\boldsymbol{P}^{-1}(e^{j\Omega})$ of the permutation matrix of Eq. 3.17. The permutation problem can be formulated to obtain Π_Ω for every frequency Ω, which is a mapping from source index k to output index i:

$$i = \Pi_\Omega(k).$$

3.7.1 Localization Approach

The basic idea of this approach is to estimate the locations of sources and then cluster them to decide the permutation. ICA-based source localization (Sec. 3.6) estimates the location $\boldsymbol{q}_i(\Omega)$ of a source that corresponds to the i-th basis vector $\boldsymbol{a}_i(e^{j\Omega})$ for each frequency Ω. Let the following function "localize" estimate the location in this way:

$$\boldsymbol{q}_i(\Omega) = \text{localize}\Big(\Omega, \boldsymbol{a}_i\left(e^{j\Omega}\right)\Big)$$

If the DOA estimation alone is adequate, the location vector $\boldsymbol{q}_i(\Omega)$ should be normalized to the unit norm (see Eq. 3.30). If the locations of sensors and sources are limited to a two-dimensional plane, we simply obtain $\theta_i(\Omega)$ that satisfies Eq. 3.30 as a DOA estimation.

Then, we employ a clustering algorithm to find N clusters $C_1, \ldots, C_N$ formed by estimated locations $\bar{\boldsymbol{q}}_i(\Omega)$ or $\theta_i(\Omega)$. Each C_k corresponds to the location of source k. Let the following function "clustering" perform clustering for all of the estimated locations $\bar{\boldsymbol{q}}_i(\Omega)$ and return the centroid $\boldsymbol{c}_k$ and the variance σ_k^2 of each cluster C_k:

$$\begin{aligned}
\Big[\boldsymbol{c}_1, \sigma_1, \ldots, \boldsymbol{c}_N, \sigma_N\Big] &= \text{clustering}\Big(\forall\Omega, \bar{\boldsymbol{q}}_1(\Omega), \ldots, \bar{\boldsymbol{q}}_N(\Omega)\Big), \\
\boldsymbol{c}_k &= \sum_{\bar{\boldsymbol{q}} \in C_k} \frac{\bar{\boldsymbol{q}}}{|C_k|}, \\
\sigma_k^2 &= \sum_{\bar{\boldsymbol{q}} \in C_k} \frac{\|\boldsymbol{c}_k - \bar{\boldsymbol{q}}\|^2}{|C_k|},
\end{aligned}$$

where $|C_k|$ is the number of vectors in the cluster. The optimization criterion for clustering is to minimize the total sum $\sum_{k=1}^{N} \sigma_k^2$ of the variances. This optimization is efficiently performed with the k-means clustering algorithm [16]. Once we have N clusters, permutations for all frequencies Ω can be decided by

$$\Pi_\Omega = \text{argmin}_{\Pi_\Omega(k)} \sum_{k=1}^{N} \left\| \boldsymbol{c}_k - \bar{\boldsymbol{q}}_{\Pi_\Omega(k)}(\Omega) \right\|^2. \tag{3.31}$$

The advantage of this source localization approach is that it is very simple to decide the permutation Π_Ω for each frequency once the centroids of N clusters are obtained. However, the disadvantage of this approach is that the estimated locations or DOAs, and thus the permutations Π_Ω, are not accurate for some frequencies. Such situations typically arise at low frequencies, where the phase difference caused by the sensor spacing is very small, as shown in Fig. 3.14.

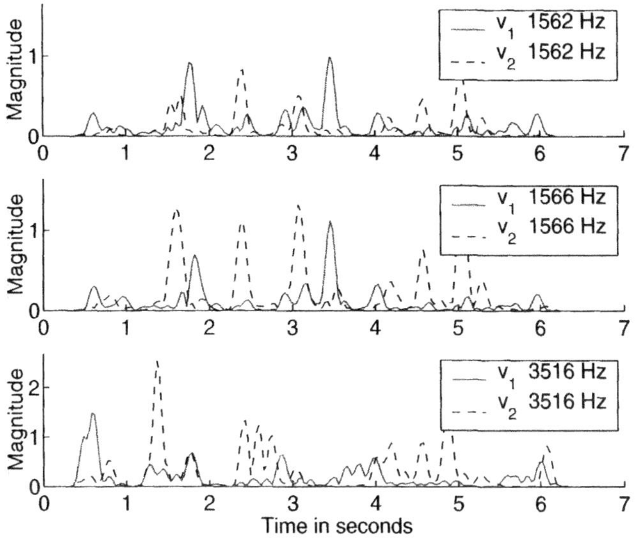

Fig. 3.11. Envelopes of two output signals at different frequencies.

3.7.2 Correlation Approach

This subsection presents an approach to permutation alignment based on the inter-frequency correlation of separated signals. The correlation should be calculated for the amplitude $|y_i(e^{j\Omega}, n)|$ or (log-scaled) power $|y_i(e^{j\Omega}, n)|^2$ of separated signals. The correlation of raw complex-valued signals $y_i(e^{j\Omega}, n)$ would be very low due to the STFT property. Here, we use the amplitude (so-called envelope)

$$v_i^{\Omega}(n) = \left| y_i \left(e^{j\Omega}, n \right) \right|$$

of a separated signal $y_i(e^{j\Omega}, n)$. The correlation of two sequences $x(n)$ and $y(n)$ is usually calculated by the correlation coefficient

$$\text{cor}(x, y) = (\mu_{xy} - \mu_x \, \mu_y)/(\sigma_x \, \sigma_y),$$

where μ_x is the mean and σ_x is the standard deviation of $x(n)$. Based on this definition, $\text{cor}(x, x) = 1$, and $\text{cor}(x, y) = 0$ if $x(n)$ and $y(n)$ are uncorrelated.

Envelopes have high correlations at neighboring frequencies if separated signals correspond to the same source signal. Fig. 3.11 shows an example. Two envelopes $v_1^{\Omega_1}$ and $v_1^{\Omega_2}$, as well as $v_2^{\Omega_1}$ and $v_2^{\Omega_2}$, are highly correlated. Ω_1 represents the frequency 1562 Hz $= \frac{\Omega_1}{2\pi} f_s$, Ω_2 was set according to 1566 Hz $= \frac{\Omega_2}{2\pi} f_s$. Thus, calculating such correlations helps us to align permutations.

A simple criterion for deciding Π_Ω is to maximize the sum of the correlations between neighboring frequencies within distance δ:

$$\Pi_{\Omega} = \mathrm{argmax}_{\Pi} \sum_{|\widetilde{\Omega}-\Omega|\leq\delta} \sum_{i=1}^{N} \mathrm{cor}\left(v_{\Pi_{\Omega}(i)}^{\Omega}, v_{\Pi_{\widetilde{\Omega}}(i)}^{\widetilde{\Omega}}\right), \tag{3.32}$$

where $\Pi_{\widetilde{\Omega}}$ is the permutation at frequency $\widetilde{\Omega}$. This criterion is based on local information and has a drawback in that mistakes in a narrow range of frequencies may lead to the complete misalignment of the frequencies beyond that range.

To avoid this problem, the method in [35] does not limit the frequency range in which correlations are calculated. It decides permutations one by one based on the criterion

$$\Pi_{\Omega} = \mathrm{argmax}_{\Pi} \sum_{i=1}^{N} \mathrm{cor}\left(v_{\Pi_{\Omega}(i)}^{\Omega}, \sum_{\widetilde{\Omega}\in\mathcal{F}} v_{\Pi_{\widetilde{\Omega}}(i)}^{\widetilde{\Omega}}\right), \tag{3.33}$$

where $\mathcal{F}$ is a set of frequencies in which the permutation is decided. This method assumes high correlations of envelopes even between frequencies that are not close neighbors. This assumption is not satisfied for all pairs of frequencies, e.g., $v_i^{\Omega_2}$ and $v_i^{\Omega_3}$ in Fig. 3.11 do not have a high correlation (Ω_3 corresponds to the Frequency 3516 Hz $= \frac{\Omega_3}{2\pi} f_s$). Therefore, this method still has the drawback of permutations possibly being misaligned at many frequencies.

If a source signal has a harmonic structure, as in the case of speech, there are strong correlations between the envelopes of a fundamental frequency f_0 and its harmonics $2f_0$, $3f_0$, ... Therefore, maximizing the correlation among harmonics is another idea for permutation alignment [41]:

$$\Pi_{\Omega} = \mathrm{argmax}_{\Pi} \sum_{\widetilde{\Omega}\in\mathcal{H}(\Omega)} \sum_{i=1}^{N} \mathrm{cor}\left(v_{\Pi_{\Omega}(i)}^{\Omega}, v_{\Pi_{\widetilde{\Omega}}(i)}^{\widetilde{\Omega}}\right), \tag{3.34}$$

where $\mathcal{H}(\Omega)$ provides a set of harmonic frequencies of Ω. The permutation accuracy improves if we take the harmonic structure of the signal into consideration. However, maximizing Eq. 3.32 and Eq. 3.34 simultaneously is not very straightforward and is computationally expensive.

3.7.3 Integrated Method

This subsection presents a method that integrates the two approaches discussed in the last two subsections. The intention behind this integration is to solve the permutation problem robustly and precisely. Let us review the characteristics of the above two approaches.

- **Robustness**: The localization approach is robust since a misalignment at one frequency does not affect other frequencies. The correlation approach is not robust since a misalignment at one frequency affects the results of other frequencies and may cause consecutive misalignments.

- **Preciseness**: The localization approach is not precise since the evaluation is based on a direct-path approximation (3.22) of the mixing system. The correlation approach is precise as long as signals are well separated by ICA, since the measurement is based on the separated signals themselves.

To benefit from both advantages, namely the robustness of the localization approach and the preciseness of the correlation approach, the integrated method first decides permutations with the localization approach and then refines the solution with the correlation approach. Implementation of the integrated method consists of the following four steps [41]:

1. Decide the permutations by the localization approach (3.31) at certain frequencies where the confidence of source localization is sufficiently high,
2. Decide the permutations based on neighboring correlations (3.32) as long as the criterion gives a clear-cut decision,
3. Decide the permutations at certain frequencies where the correlation among harmonics (3.34) is sufficiently high,
4. Decide the permutations for the remaining frequencies based on neighboring correlations (3.32).

The key to the first step is fixing a permutation only if the confidence of source localization is sufficiently high. We assume that the confidence is high if the squared distance between an estimated location and its corresponding centroid is smaller than the variance, i.e., $\|\boldsymbol{c}_k - \bar{\boldsymbol{q}}_{\Pi_\Omega(k)}(\Omega)\|^2 < \sigma_k^2$. In the second step, permutations are decided one by one for the frequency Ω where the sum of the correlations with fixed frequencies $\widetilde{\Omega} \in \mathcal{F}$ within distance $|\widetilde{\Omega} - \Omega| \leq \delta$ is the maximum. This is repeated as long as the maximum correlation sum is larger than a threshold th_{cor}. In the third step, the permutations are decided for frequencies Ω where the sum of the correlations among harmonics is larger than a threshold th_{ha}. The last step decides the permutations for the remaining frequencies with the same criterion as the second step.

Let us discuss the advantages of the integrated method. The main advantage is that it does not cause a large misalignment as long as the permutations fixed by the localization approach are correct. Moreover, the correlation part compensates for the lack of preciseness of the localization approach. The correlation part consists of three steps (steps 2, 3, 4) for two reasons. First, the harmonics part works well if most of the other permutations are fixed. Second, the method becomes more robust by quitting step 2 if there is no clear-cut decision. With this structure, we can avoid fixing the permutations for consecutive frequencies without high confidence. As shown in the experimental results (Sec. 3.10), this integrated method is effective in separating many sources.

3.8 Scaling Alignment

The scaling ambiguity $\boldsymbol{\Lambda}(e^{j\Omega})$ in Eq. 3.17 is easily solved by calculating the (pseudo)-inverse of a separation matrix $\boldsymbol{W}(e^{j\Omega})$ [30, 35]. The frequency-domain counterpart of the BSS goal (Eq. 3.3) is

$$y_i\left(e^{j\Omega}, n\right) \stackrel{!}{=} H_{J_i i}\left(e^{j\Omega}\right) s_i\left(e^{j\Omega}, n\right), \tag{3.35}$$

where J_i can be selected according to each output i but should be the same for all frequencies Ω. Let us assume that the ICA and the permutation problem have been solved. Then the $\boldsymbol{a}_i(e^{j\Omega})$ term in Eq. 3.16 is close to the $\boldsymbol{h}_i(e^{j\Omega})$ term in Eq. 3.6:

$$\boldsymbol{h}_i\left(e^{j\Omega}\right) s_i\left(e^{j\Omega}, n\right) \approx \boldsymbol{a}_i\left(e^{j\Omega}\right) y_i\left(e^{j\Omega}, n\right). \tag{3.36}$$

By substituting Eq. 3.35 into Eq. 3.36, we have the condition for scaling alignment:

$$\boldsymbol{h}_i\left(e^{j\Omega}\right) \approx \boldsymbol{a}_i\left(e^{j\Omega}\right) h_{J_i i}\left(e^{j\Omega}\right) \Leftrightarrow a_{J_i i}\left(e^{j\Omega}\right) \approx 1.$$

This condition, i.e., $a_{J_i i}(e^{j\Omega}) = 1$, is attained by

$$\begin{aligned} \boldsymbol{W}\left(e^{j\Omega}\right) &\leftarrow \boldsymbol{\Lambda}\left(e^{j\Omega}\right) \boldsymbol{W}\left(e^{j\Omega}\right), \\ \boldsymbol{\Lambda}\left(e^{j\Omega}\right) &= \operatorname{diag}\left[a_{J_1 1}\left(e^{j\Omega}\right), \ldots, a_{J_N N}\left(e^{j\Omega}\right)\right], \end{aligned}$$

where $a_{ji}\left(e^{j\Omega}\right) = [\boldsymbol{W}^{+}(e^{j\Omega})]_{ji}$ is an element of the pseudoinverse of $\boldsymbol{W}(e^{j\Omega})$.

3.9 Spectral Smoothing

The frequency-domain BSS described in this chapter is influenced by the circularity of discrete frequency representation. The circularity refers to the fact that frequency responses sampled at L points with an interval f_s/L (f_s: sampling frequency) represent a periodic time-domain signal whose period is L/f_s. Since this filter is unrealistic, we usually use its one-period operation. However, such one-period filters may cause a problem. Fig. 3.12 shows impulse responses from a source $s_k(n)$ to an output $y_i(n)$ defined by Eq. 3.47. Responses on the left $u_{11}(l)$ correspond to the extraction of a target signal, and those on the right $u_{14}(l)$ correspond to the suppression of an interference signal. The upper responses are obtained with infinite-length filters, and the lower ones with one-period filters. We can see that the one-period filters create spikes, which distort the target signal and degrade the separation performance. Note that these spikes are inevitable in the frequency-domain BSS, since we have an ICA solution in the frequency domain.

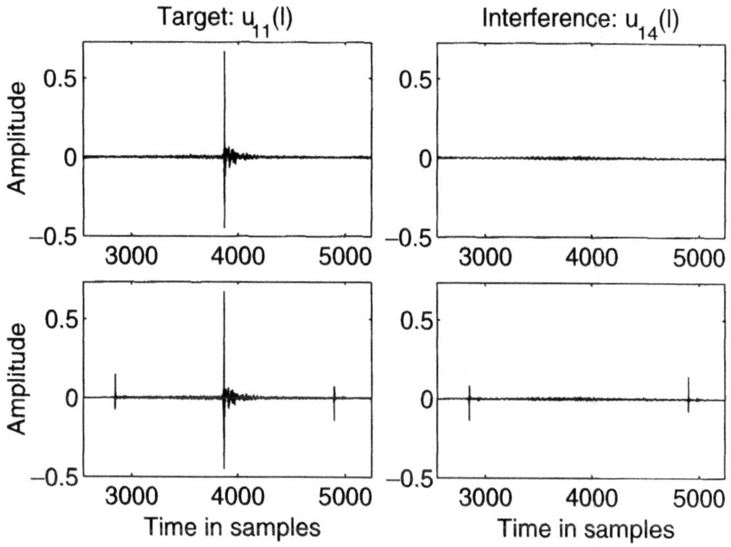

Fig. 3.12. Impulse responses $u_{ik}(l)$ obtained with periodic filters (above) and with their one-period operation (below).

3.9.1 Windowing

To solve this problem, we need to control the frequency responses $W_{ij}(e^{j\Omega})$ so that the corresponding time-domain filter $w_{ij}(l)$ does not rely on the circularity effect whereby adjacent periods work together to perform some filtering. The most widely used approach is spectral smoothing, which is realized by multiplying a window $g(l)$ that tapers smoothly to zero at each end, such as a Hanning window $g(l) = \frac{1}{2}(1 + \cos\frac{2\pi l}{L})$. This makes the resulting time-domain filter $w_{ij}(l)\, g(l)$ fit length L and have small amplitude around the ends [7]. As a result, the frequency responses $W_{ij}(e^{j\Omega})$ are smoothed as

$$\tilde{W}_{ij}\left(e^{j\Omega}\right) = \frac{1}{2\pi} \int_{\phi=0}^{2\pi} G\left(e^{j\phi}\right) W_{ij}\left(e^{j(\Omega-\phi)}\right) d\phi,$$

where $G(e^{j\Omega})$ is the frequency response of g(l). If a Hanning window is used, the frequency responses are smoothed as

$$\tilde{W}_{ij}\left(e^{j\Omega}\right) = \frac{1}{4}\left[W_{ij}\left(e^{j(\Omega-\Delta\Omega)}\right) + 2W_{ij}\left(e^{j\Omega}\right) + W_{ij}\left(e^{j(\Omega+\Delta\Omega)}\right)\right], \quad (3.37)$$

since the frequency responses $G(e^{j\Omega})$ of the Hanning window are $G(e^{j\,0}) = \frac{1}{2}$, $G(e^{j\Delta\Omega}) = G(e^{j(2\pi-\Delta\Omega)}) = \frac{1}{4}$, and zero for the other frequency bins. $\Delta\Omega$ is specified as $\Delta\Omega = \frac{2\pi}{L}$.

The windowing successfully eliminates the spikes. However, it changes the frequency response from $W_{ij}(e^{j\Omega})$ to $\tilde{W}_{ij}(e^{j\Omega})$ and causes an error. Let us evaluate the error for each row $\boldsymbol{w}_i(e^{j\Omega}) = [W_{i1}(e^{j\Omega}), \ldots, W_{iM}(e^{j\Omega})]^{\mathrm{T}}$ of the ICA solution $\boldsymbol{W}(e^{j\Omega})$. The error is

$$\begin{aligned} \boldsymbol{e}_i\left(e^{j\Omega}\right) &= \min_{\alpha_i}\left[\tilde{\boldsymbol{w}}_i\left(e^{j\Omega}\right) - \alpha_i\, \boldsymbol{w}_i\left(e^{j\Omega}\right)\right] \\ &= \tilde{\boldsymbol{w}}_i\left(e^{j\Omega}\right) - \frac{\tilde{\boldsymbol{w}}_i^{\mathrm{H}}\left(e^{j\Omega}\right)\boldsymbol{w}_i\left(e^{j\Omega}\right)}{\left\|\boldsymbol{w}_i\left(e^{j\Omega}\right)\right\|^2}\boldsymbol{w}_i\left(e^{j\Omega}\right), \end{aligned} \tag{3.38}$$

where $\tilde{\boldsymbol{w}}_i(e^{j\Omega}) = [\tilde{W}_{i1}(e^{j\Omega}), \ldots, \tilde{W}_{iM}(e^{j\Omega})]^{\mathrm{T}}$ and α_i is a complex-valued scalar representing the scaling ambiguity of the ICA solution. The minimization $\min_{\alpha_i}$ is based on least-squares, and it can be represented by the projection of $\tilde{\boldsymbol{w}}_i(e^{j\Omega})$ to $\boldsymbol{w}_i(e^{j\Omega})$. We can evaluate the error for the Hanning window case by substituting Eq. 3.37 for $\tilde{\boldsymbol{w}}_i(e^{j\Omega})$ of Eq. 3.38:

$$\boldsymbol{e}_i\left(e^{j\Omega}\right) = \frac{1}{4}\left[\boldsymbol{e}_i^-\left(e^{j\Omega}\right) + \boldsymbol{e}_i^+\left(e^{j\Omega}\right)\right], \tag{3.39}$$

where

$$\boldsymbol{e}_i^-\left(e^{j\Omega}\right) = \boldsymbol{w}_i\left(e^{j(\Omega-\Delta\Omega)}\right) - \frac{\boldsymbol{w}_i^{\mathrm{H}}\left(e^{j(\Omega-\Delta\Omega)}\right)\boldsymbol{w}_i\left(e^{j\Omega}\right)}{\left\|\boldsymbol{w}_i\left(e^{j\Omega}\right)\right\|^2}\boldsymbol{w}_i\left(e^{j\Omega}\right), \tag{3.40}$$

$$\boldsymbol{e}_i^+\left(e^{j\Omega}\right) = \boldsymbol{w}_i\left(e^{j(\Omega+\Delta\Omega)}\right) - \frac{\boldsymbol{w}_i^{\mathrm{H}}\left(e^{j(\Omega+\Delta\Omega)}\right)\boldsymbol{w}_i\left(e^{j\Omega}\right)}{\left\|\boldsymbol{w}_i\left(e^{j\Omega}\right)\right\|^2}\boldsymbol{w}_i\left(e^{j\Omega}\right). \tag{3.41}$$

This $\boldsymbol{e}_i^-(e^{j\Omega})$ [or $\boldsymbol{e}_i^+(e^{j\Omega})$] represents the difference between two vectors $\boldsymbol{w}_i(e^{j\Omega})$ and $\boldsymbol{w}_i(e^{j(\Omega-\Delta\Omega)})$ [or $\boldsymbol{w}_i(e^{j(\Omega+\Delta\Omega)})$]. Since these differences are usually not very large, the error $\mathbf{e}_i$ does not seriously affect the separation if we use a Hanning window for spectral smoothing.

3.9.2 Minimizing Error by Adjusting Scaling Ambiguity

Even if the error caused by the windowing is not very large, the separation performance is improved by minimizing the error [40]. The minimization is performed by adjusting the scaling ambiguity of the ICA solution before the windowing. Let $D_i(e^{j\Omega})$ be a complex-valued scalar for the scaling adjustment:

$$\boldsymbol{w}_i\left(e^{j\Omega}\right) \leftarrow D_i\left(e^{j\Omega}\right)\boldsymbol{w}_i\left(e^{j\Omega}\right). \tag{3.42}$$

We want to find $D_i(e^{j\Omega})$ such that the error (Eq. 3.38) is minimized. The scalar $D_i(e^{j\Omega})$ should be close to 1 to avoid any great change in the predetermined scaling. Thus, an appropriate total cost to be minimized is

$$\mathcal{J} = \sum_{\Omega} J_i(\Omega), \quad J_i(\Omega) = \frac{\left\|\boldsymbol{e}_i\left(e^{j\Omega}\right)\right\|^2}{\left\|\boldsymbol{w}_i\left(e^{j\Omega}\right)\right\|^2} + \beta\left|D_i\left(e^{j\Omega}\right) - 1\right|^2, \tag{3.43}$$

where β is a parameter indicating the importance of maintaining the predetermined scaling. With the Hanning window, the error after the scaling adjustment is easily calculated by substituting Eq. 3.42 for Eq. 3.39:

$$\boldsymbol{e}_i\left(e^{j\Omega}\right) = \frac{1}{4}\left[D_i\left(e^{j(\Omega-\Delta\Omega)}\right)\boldsymbol{e}_i^-\left(e^{j\Omega}\right) + D_i\left(e^{j(\Omega+\Delta\Omega)}\right)\boldsymbol{e}_i^+\left(e^{j\Omega}\right)\right], \quad (3.44)$$

where $\boldsymbol{e}_i^-(e^{j\Omega})$ and $\boldsymbol{e}_i^+(e^{j\Omega})$ are defined in Eq. 3.40 and Eq. 3.41, respectively.

The minimization of the total cost can be performed iteratively by

$$D_i\left(e^{j\Omega}\right) \leftarrow D_i\left(e^{j\Omega}\right) - \mu\frac{\partial\mathcal{J}}{\partial D_i\left(e^{j\Omega}\right)} \quad (3.45)$$

with a small step size μ. With the Hanning window, the gradient is

$$\begin{aligned}\frac{\partial\mathcal{J}}{\partial D_i\left(e^{j\Omega}\right)} &= \frac{\partial J_i(\Omega-\Delta\Omega)}{\partial D_i\left(e^{j\Omega}\right)} + \frac{\partial J_i(\Omega+\Delta\Omega)}{\partial D_i\left(e^{j\Omega}\right)} + \frac{\partial J_i(\Omega)}{\partial D_i\left(e^{j\Omega}\right)} \\ &= \frac{\boldsymbol{e}_i^{\mathrm{H}}\left(e^{j(\Omega-\Delta\Omega)}\right)\boldsymbol{e}_i^+\left(e^{j(\Omega-\Delta\Omega)}\right) + \boldsymbol{e}_i^{\mathrm{H}}\left(e^{j(\Omega+\Delta\Omega)}\right)\boldsymbol{e}_i^-\left(e^{j(\Omega+\Delta\Omega)}\right)}{8\left\|\boldsymbol{w}_i\left(e^{j\Omega}\right)\right\|^2} \\ &\quad +2\beta\left[D_i\left(e^{j\Omega}\right) - 1\right].\end{aligned} \quad (3.46)$$

With Eqs. 3.44 to 3.46, we can optimize the scalar $D_i(e^{j\Omega})$ for the scaling adjustment and minimize the error caused by the spectral smoothing (Eq. 3.37) with the Hanning window.

3.10 Experimental Results

The performance of BSS is evaluated by a signal-to-interference ratio (SIR), which is the power ratio between the target component and the interference components. Let $u_{ik}(l)$ be the impulse responses from source $s_k(n)$ to separated signal $y_i(n)$:

$$u_{ik}(l) = \sum_{j=1}^{M}\sum_{\tau=0}^{L-1} w_{ij}(\tau)h_{jk}(l-\tau). \quad (3.47)$$

Then, the SIR of output i is calculated as

$$\mathrm{SIR}_i = 10\log_{10}\frac{\left\langle\left|\sum_l u_{ii}(l)s_i(n-l)\right|^2\right\rangle_n}{\left\langle\left|\sum_{k\neq i}\sum_l u_{ik}(l)s_k(n-l)\right|^2\right\rangle_n}\ (\mathrm{dB}), \quad (3.48)$$

where $\langle\cdot\rangle_n$ denotes the averaging operator over time n.

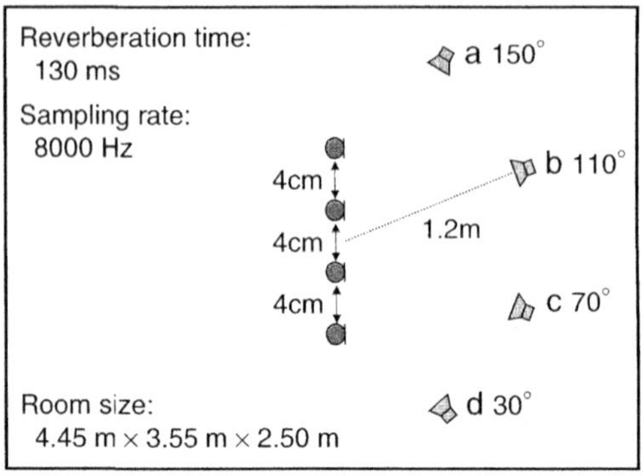

Fig. 3.13. Experimental conditions with linear array.

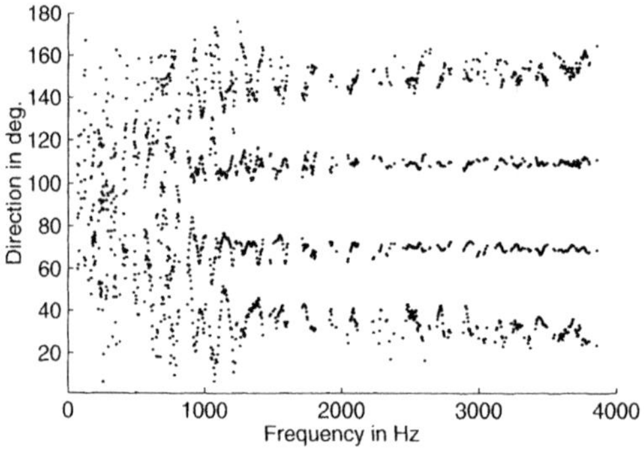

Fig. 3.14. DOA estimations by (3.27) with four sources.

Table 3.1. Separation performance with linear array.

#sources / position	2 / a c		3 / a b d		4 / a b c d	
Spectral smoothing	no	yes	no	yes	no	yes
Average SIR at microphones (dB)	0.1		-2.9		-4.6	
Average SIR of output (dB)	20.1	22.3	14.7	17.0	9.3	11.5
Execution time (s)	5.2	5.2	8.0	8.1	12.3	12.4

3.10.1 2×2, 3×3, and 4×4 with Linear Array

We performed experiments to separate speech signals in an environment whose conditions are summarized in Fig. 3.13. Our experiments involved two, three and four sources whose locations are indicated in Fig. 3.13. The individual selections are indicated in Tab. 3.1. The sensors were arranged linearly, and

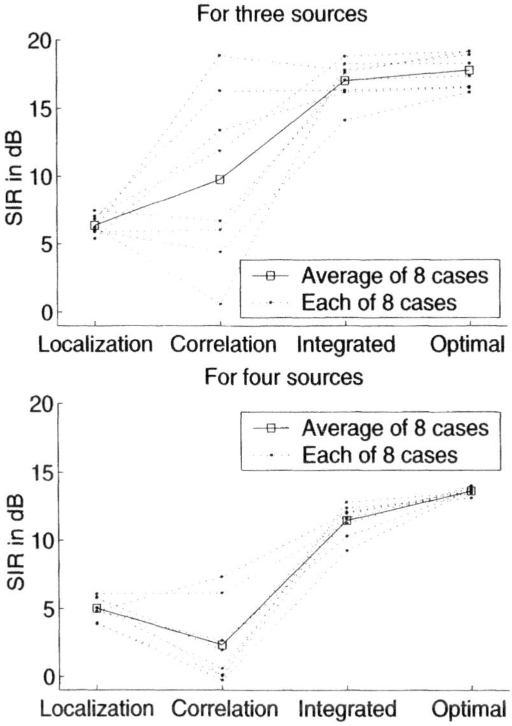

Fig. 3.15. Comparison of different methods for solving permutation problem.

the number of sensors used was the same as the number of sources. We used filters of length $L = 2048$ because this length provided the best performance under the conditions. The BSS program was coded in Matlab® and run on Athlon XP 3200+.

The results shown in Tab. 3.1 are the average SIRs of output for eight combinations of 7-second speeches. We can see that the spectral smoothing discussed in Sec. 3.9 improves the average SIR for every setup. The short execution time, as shown in Table 3.1, enables the BSS system to perform in real time if the number of source signals is not very large.

Fig. 3.14 shows DOA estimations for mixtures of four sources obtained with Eq. 3.27. Fig. 3.15 shows SIRs for three and for four sources with the different methods for solving the permutation problem discussed in Sec. 3.7. Here, "Localization" is the localization (DOA) approach (Eq. 3.31) alone, "Correlation" is the correlation approach (Eq. 3.32) alone, "Integrated" is the integrated method, and "Optimal" is the optimal solution obtained by utilizing the $s_i(n)$ and $h_{ji}(l)$ information. The performance of "Localization" was stable but insufficient. The performance of "Correlation" was unstable and very

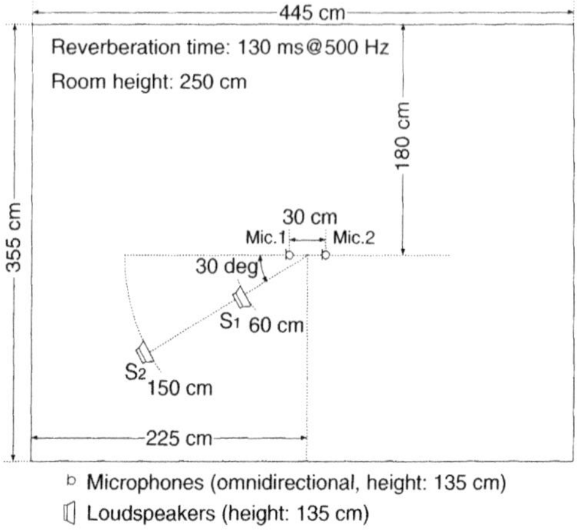

Fig. 3.16. Room layout.

Table 3.2. Experimental conditions.

Sampling rate	8 kHz
Data length	2 s
Window	Hanning
Frame length	1024 points (128 ms)
Frame shift	256 points (32 ms)
ICA algorithm	Infomax (complex valued)

poor in the four-source cases. The "Integrated" method performed very well, achieving nearly the same results as those by "Optimal".

We carried out experiments with two sources arriving from the same direction and two microphones using speech signals convolved with impulse responses measured in a room [32]. The room layout is shown in Fig. 3.16. The sources are located in the same direction from the microphone pair. The reverberation time of the room was 130 ms at 500 Hz. Other conditions are summarized in Tab. 3.2. The experimental procedure is as follows.

First, we apply ICA to observed signals $x_j(n)\,(j = 1, 2)$ and calculate separation matrix $\boldsymbol{W}(e^{j\Omega})$ for each frequency bin. Then we estimate radii $\widehat{R}_1$ and $\widehat{R}_2$ of two spheres on which each source signal exists by using $\boldsymbol{W}^{-1}(e^{j\Omega})$ and Eq. 3.25, and the permutation is aligned so that $\widehat{R}_2 \geq \widehat{R}_1$. In order to evaluate the reliability of the solution provided by the estimated spheres, we introduce a threshold parameter $\alpha \geq 1$, and we accept solutions only for frequency bins that satisfy the condition $\widehat{R}_2/\widehat{R}_1 \geq \alpha$. We then apply the correlation-based method to the remaining frequency bins. The permutation problem is solved simply by using the geometric information when $\alpha = 1$ and simply by using the correlation when $\alpha = \infty$.

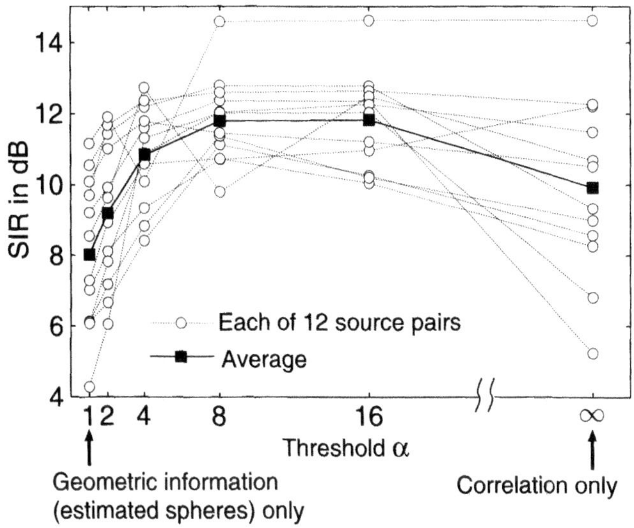

Fig. 3.17. Experimental results. SIRs are evaluated for 12 combinations of source signals with various values for threshold parameter α.

We define SIR as the average of SIR_1 and SIR_2 in order to cancel out the effect of the input SIR. We measured SIRs for 12 combinations of source signals using two male and two female speakers and varying the threshold parameter α.

Fig. 3.17 shows the experimental results. When we solve the permutation problem using only the estimated spheres ($\alpha = 1$), the performance is insufficient. In contrast, the performance we obtain using only the correlation ($\alpha = \infty$) is unstable. The combination of both methods yields good and stable performance. These tendencies are similar to the results we obtain when we use DOAs as geometric information [41].

We obtained good performance when the threshold parameter α was relatively large. When α was 8 to 16, the permutation of about 1/5 to 1/10 of the frequency bins was determined by the geometric information. This result suggests that we should use this geometric information for frequency bins where the estimation is highly reliable.

Fig. 3.18 shows the spatial gain patterns of the separation filters in one frequency bin (1000 Hz) drawn with the near-field model. The gain of the observed signal at microphone 1 is defined as 0 dB. We can see that the separation filter forms a spot null beam focusing on the interference signal. When source signals are located in different directions, a separation filter utilizes the phase difference of the input signals and makes a directive null toward the interference signal [6], whereas both the phase and level differences are utilized to make a regional null when signals come from the same direction.

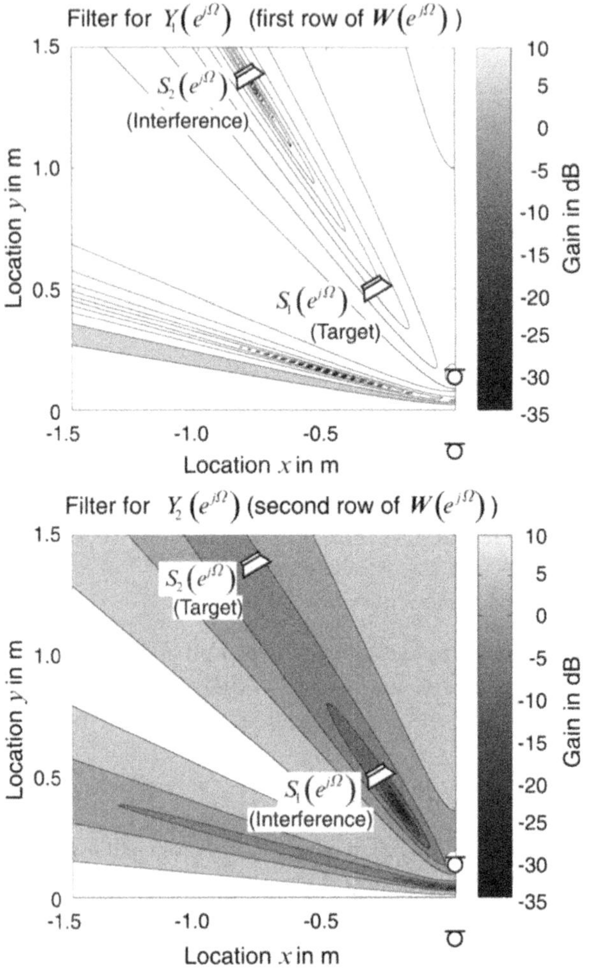

Fig. 3.18. Example spatial gain patterns of separation filters (around 1000 Hz).

3.10.2 6 × 8 with Planar Array

Next, we carried out experiments on separating six sources with a planar array of eight microphones. The room layout and other experimental conditions are shown in Fig. 3.19. All six sources produce 6-second speech signals, and two came from the same direction. The filter length was again $L = 2048$ for an 8-kHz sampling rate.

Let us explain the method for solving the permutation problem in this situation. First, the source directions were estimated with small-spacing microphone pairs (1-3, 2-4, 1-2 and 2-3 shown in the right-top corner of Fig. 3.19). This was performed based on Eqs. 3.26, 3.28 and 3.29. Fig. 3.20 shows a his-

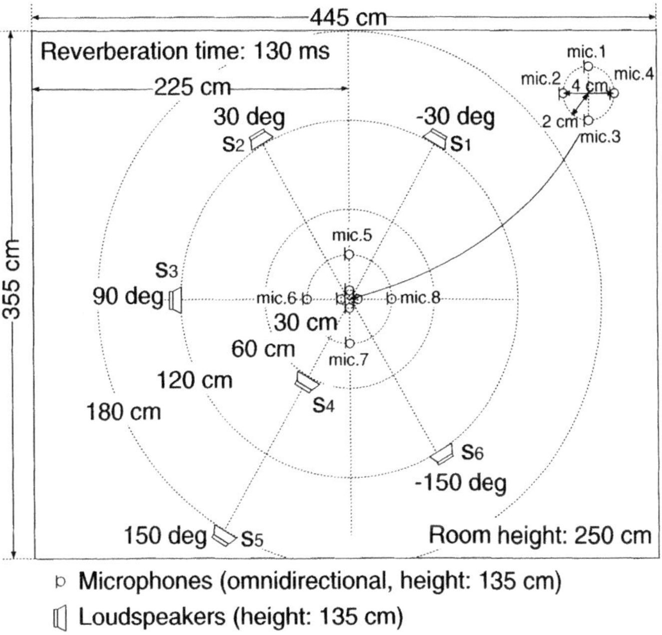

Fig. 3.19. Experimental conditions for planar array case.

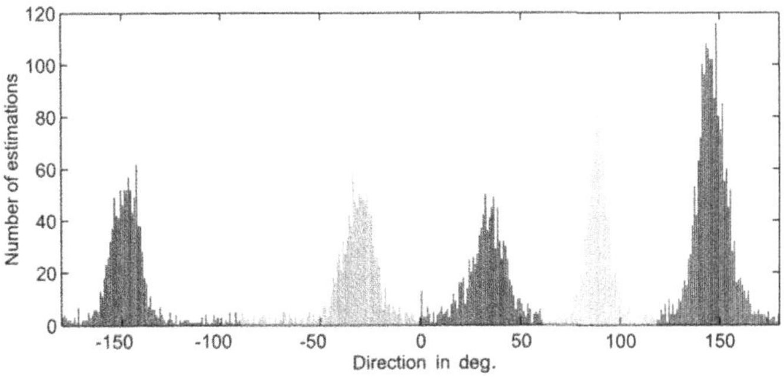

Fig. 3.20. Histogram of DOAs estimated with small spacing microphone pairs.

togram of the estimated DOAs. There are five clusters in this histogram, and one cluster is twice the size of the others. This implies that two sources came from the same direction (about 150°). We solved the permutation problem for the other four sources by using this DOA information as shown in the upper plot of Fig. 3.21.

Then, to distinguish between the two sources that came from the same direction, the spheres of these sources were estimated with large-spacing microphone pairs (7-5, 7-8, 6-5 and 6-8 shown in the center of Fig. 3.19). This was performed based on Eq. 3.25. The lower plot of Fig. 3.21 shows the radii of the

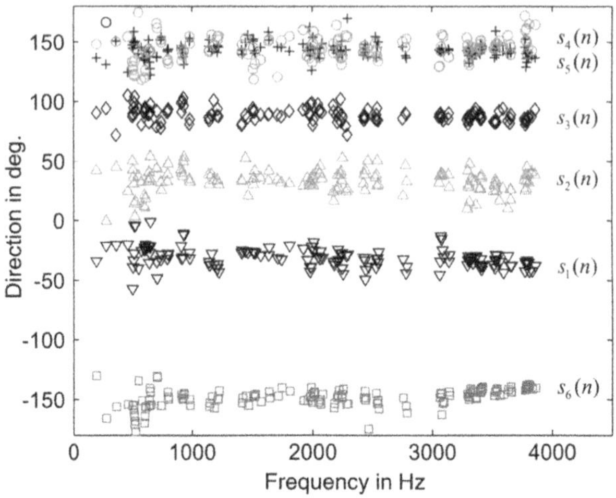

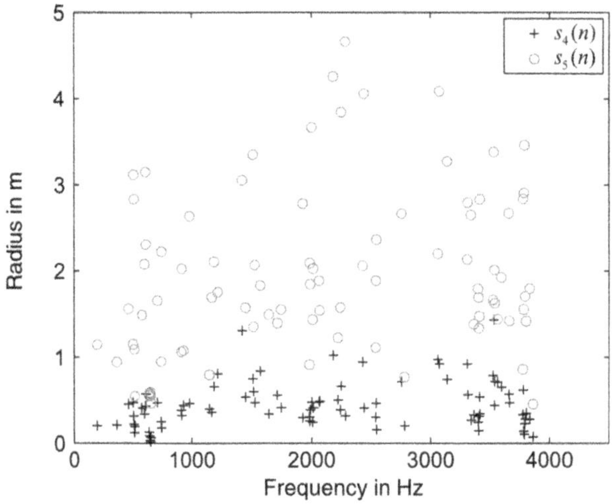

Fig. 3.21. Permutation solved by using estimated DOAs (upper) and spheres (lower).

spheres estimated with microphone pair 7-5. Although the radius estimations had large variances, it provided sufficient information to distinguish between the two sources. Consequently, the signal components of all frequencies were classified into six clusters. We decided the permutation only for frequency bins where the classification was reliable, as discussed in Sec. 3.7.3.

To show the effectiveness of this method, we compared SIRs by three different methods for the permutation problem. Tab. 3.3 shows the results. The last row, "DOA + Sphere + Correlation", shows the results obtained with the integrated method. The two methods for comparison were "Correlation" where only the correlations (Eq. 3.32) were maximized, and "DOA + Correlation"

Table 3.3. Separation performance with planar array measured by SIR (dB).

	SIR_1	SIR_2	SIR_3	SIR_4	SIR_5	SIR_6	Average
SIR at microphone 1	-8.3	-6.8	-7.8	-7.7	-6.7	-5.2	-7.1
Correlation	4.4	2.6	4.0	9.2	3.6	-2.0	3.7
DOA + Correlation	9.6	9.3	14.7	2.7	6.5	14.0	9.4
DOA + Sphere + Correlation	10.8	10.4	14.5	7.0	11.0	12.2	11.0

where only the DOA information was used for the source localization step in the integrated method. To see how much the SIRs improved, we also measured the SIR of the mixture observed at microphone 1 ("SIR at microphone 1"). The effectiveness of the two integrated methods can again be observed. If we compare the results of "DOA + Correlation" with "DOA + Sphere + Correlation", the improvement of the latter over the former is apparent for sources 4 and 5, which came from the same direction. This means that the sphere information was effective again in distinguishing between sources coming from the same direction (Fig. 3.18). The BSS program was again coded in Matlab® and run on Athlon XP 3200+. The computational time for separating six speeches of 6 seconds was around one minute.

3.10.3 2 × 2 Moving Sources

In most realistic applications, the source location may change. A mixing system is time-varying when source signals move. A naive approach for tracking a time-varying system is an online algorithm that updates the separation system sample by sample [3, 23].

Indeed, an online algorithm can track a time-varying system; however, its performance is generally worse than a batch algorithm, which can employ a number of samples, when the system is stationary. Although we are dealing with moving sources, we do not want to degrade the performance for fixed sources.

In this section, we describe a real-time BSS method [34] that employs frequency-domain ICA with a blockwise batch algorithm. This algorithm achieves better separation performance than an online algorithm for fixed source signals.

We measured the BSS performance using ICA. Fig. 3.22 shows the average and standard deviation of SIR for fixed sources (the target is at A and the interference at C in Fig. 3.23). This indicates that the blockwise batch algorithm outperforms the online algorithm (step size μ is tuned to optimize the performance) when we use the update equation 3.18. In addition, the deviation of the batch algorithm is smaller than that of the online algorithm, which is why we adopt the blockwise batch algorithm. We used block size T_b = 1.0 s in the experiments.

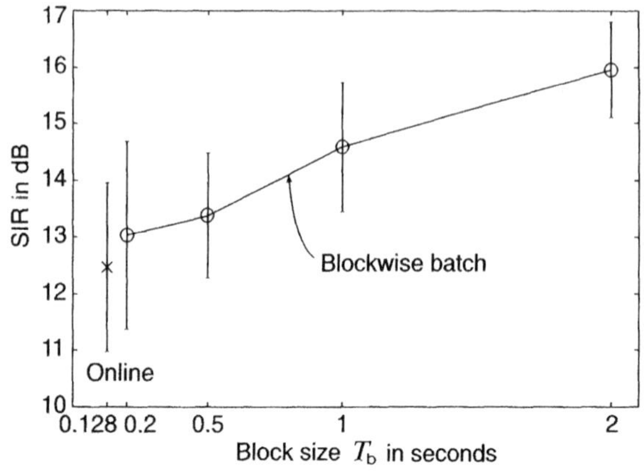

Fig. 3.22. Average and standard deviation of SIR for fixed sources.

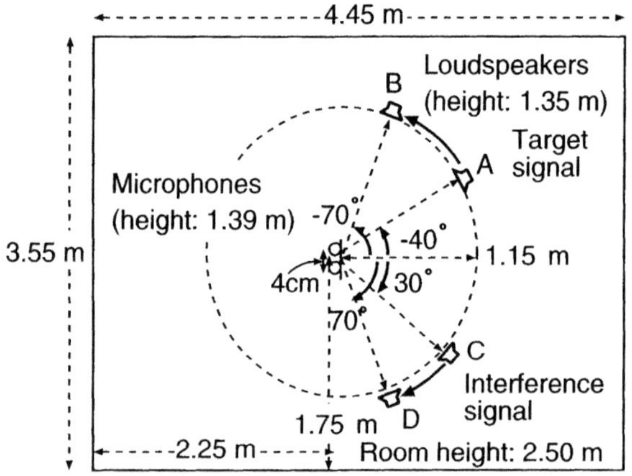

Fig. 3.23. Layout of room used in experiments ($T_R = 130$ ms).

We carried out experiments using speech signals recorded in a room. The reverberation time of the room was 130 ms. We used two omni-directional microphones with an inter-element spacing of 4 cm. The layout of the room is shown in Fig 3.23. The target source signal was first located at A and then moved to B at a speed of 30 deg/s. The interference signal was located at C and moved to D at a speed of 40 deg/s.

The step-size parameter μ in Eq. 3.18 affects the separation performance of BSS when the block size changes. We carried out preliminary experiments and chose μ to optimize the performance for each block size. The other conditions are summarized in Tab. 3.4. We measured SIRs with 30 combinations of source signals, using three male and three female speakers, and then averaged them.

We investigated the BSS performance for moving sources using the blockwise batch algorithm. Fig. 3.24 shows the SIR for a moving target (solid line) and that for a moving interference (dotted line). We can see that the SIR is

Table 3.4. Experimental conditions.

Common	Sampling rate $f_s = 8$ kHz Window = Hanning Reverberation time $T_R = 130$ ms
ICA part	Frame length $T_{ICA} = 1024$ points (128 ms) Frame shift = 256 points (32 ms) $g = 100.0$ μ = optimized for block size T_b Number of iterations $N_I = 100$

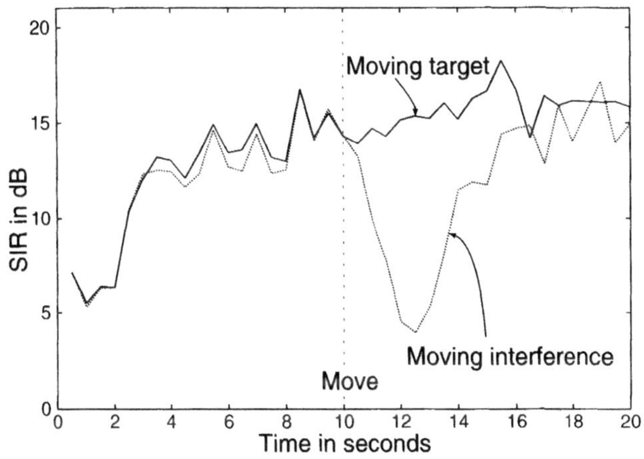

Fig. 3.24. SIR of blockwise batch algorithm without postprocessing. Target and interference signals moved at 10 s ($T_b = 1.0$ s)

not degraded even when the target moves. By contrast, interference movement causes a decline in the SIR.

This can be explained by the directivity pattern of the separation system obtained by ICA. The solution of frequency-domain BSS works in the same way as an adaptive beamformer that forms a spatial null toward an interference signal (Fig. 3.6). Because of this characteristic, BSS using ICA is robust with a moving target signal but fragile with a moving interference signal. Taking advantage of this nature, we can estimate residual crosstalk components, even when the interference signal moves, by employing postprocessing in the second stage [34].

3.11 Conclusion

This chapter presented a comprehensive description of frequency-domain BSS as well as various techniques that enable frequency-domain BSS to be used for separating many speech signals mixed in a real-room environment. The permutation problem has been a major concern with the frequency-domain approach. However, with the methods described in Sec. 3.7, this problem can be solved even in a practical situation. Moreover, the locations of sources can be estimated by the method described in Sec. 3.6. This ability is unique to the frequency-domain approach and cannot be seen in time-domain BSS. Our experimental results show that the separation performance was fairly good and the computational cost was feasible. These results demonstrate the effectiveness of frequency-domain BSS.

References

[1] S. Amari, S. Douglas, A. Cichocki, H. Yang: multichannel blind deconvolution and equalization using the natural gradient, *Proc. IEEE Workshop on Signal Processing Advances in Wireless Communications '97*, 101–104, 1997.

[2] S. Amari: Natural gradient works efficiently in learning, *Neural Computation*, **10**(2), 251–276, 1998.

[3] J. Anemüller, T. Gramss: On-line blind separation of moving sound sources, *Proc. ICA '99*, 331–334, Ottawa, Canada, 1999.

[4] J. Anemüller, B. Kollmeier: Amplitude modulation decorrelation for convolutive blind source separation, *Proc. ICA '00*, 215–220, Helsinki, Finland, 2000.

[5] S. Araki, R. Mukai, S. Makino, T. Nishikawa, H. Saruwatari: The fundamental limitation of frequency domain blind source separation for convolutive mixtures of speech, *IEEE Trans. Speech Audio Processing*, **11**(2), 109–116, March 2003.

[6] S. Araki, S. Makino, Y. Hinamoto, R. Mukai, T. Nishikawa, H. Saruwatari: Equivalence between frequency domain blind source separation and frequency domain adaptive beamforming for convolutive mixtures, *EURASIP Journal on Applied Signal Processing*, **2003**(11), 1157–1166, 2003.

[7] F. Asano, S. Ikeda, M. Ogawa, H. Asoh, N. Kitawaki: Combined approach of array processing and independent component analysis for blind separation of acoustic signals, *IEEE Trans. Speech Audio Processing*, **11**(3), 204–215, May 2003.

[8] A. D. Back, A. C. Tsoi: Blind deconvolution of signals using a complex recurrent network, *Proc. Neural Networks for Signal Processing*, 565–574, 1994.

[9] A. Bell, T. Sejnowski: An information-maximization approach to blind separation and blind deconvolution, *Neural Computation*, 7(6), 1129–1159, 1995.

[10] E. Bingham, A. Hyvärinen: A fast fixed-point algorithm for independent component analysis of complex valued signals, *International Journal of Neural Systems*, **10**(1), 1–8, February 2000.

[11] H. Buchner, R. Aichner, W. Kellermann: Blind source separation for convolutive mixtures: A unified treatment, in Y. Huang, J. Benesty (eds.), *Audio Signal Processing for Next-Generation Multimedia Communication Systems,* Boston, MA, USA: Kluwer Academic Publishers, 2004, 255–293.

[12] J.-F. Cardoso: Source separation using higher order moments, *Proc. ICASSP '89*, **4**, 2109–2112, Glasgow, Scotland, 1989.

[13] J. F. Cardoso: Blind beamforming for non-Gaussian signals, *IEE Proceedings-F*, 362–370, December 1993.

[14] A. Cichocki, S. Amari: *Adaptive Blind Signal and Image Processing,* Hoboken, NJ, USA: John Wiley & Sons, 2002.

[15] S. C. Douglas, X. Sun: Convolutive blind separation of speech mixtures using the natural gradient, *Speech Communication*, **39**, 65–78, 2003.

[16] R. O. Duda, P. E. Hart, D. G. Stork: *Pattern Classification*, 2nd ed. New York, NY, USA: Wiley Interscience, 2000.

[17] A. Hyvärinen: Fast and robust fixed-point algorithm for independent component analysis, *IEEE Trans. Neural Networks*, **10**(3), 626–634, 1999.

[18] A. Hyvärinen, J. Karhunen, E. Oja: *Independent Component Analysis,* Hoboken, NJ, USA: John Wiley & Sons, 2001.

[19] S. Haykin (ed.): *Unsupervised Adaptive Filtering (Volume I: Blind Source Separation),* Hoboken, NJ, USA: John Wiley & Sons, 2000.

[20] M. Z. Ikram, D. R. Morgan: A beamforming approach to permutation alignment for multichannel frequency-domain blind speech separation, *Proc. ICASSP '02*, 881–884, Orlando, FL, USA, 2002.

[21] M. Joho, P. Schniter: Frequency domain realization of a multichannel blind deconvolution algorithm based on the natural gradient, *Proc. ICA '03*, 543–548, Nava, Japan, 2003.

[22] M. Kawamoto, K. Matsuoka, N. Ohnishi: A method of blind separation for convolved non-stationary signals, *Neurocomputing*, **22**, 157–171, 1998.

[23] A. Koutras, E. Dermatas, G. Kokkinakis: Blind speech separation of moving speakers in real reverberant environment, *Proc. ICASSP '00,* Istanbul, Turkey, 1133–1136, 2000.

[24] S. Kurita, H. Saruwatari, S. Kajita, K. Takeda, F. Itakura: Evaluation of blind signal separation method using directivity pattern under reverberant conditions, *Proc. ICASSP '00*, 3140–3143, Istanbul, Turkey, 2000.

[25] R. H. Lambert, A. J. Bell: Blind separation of multiple speakers a multipath environment, *Proc. ICASSP '97*, 423–426, Munich, Germany, 1997.

[26] T. W. Lee, A. J. Bell, R. Orglmeister: Blind source separation of real world signals, *Proc. ICNN '97*, 2129–2135, Houston, TX, USA, 1997.

[27] T. W. Lee: *Independent Component Analysis - Theory and Applications,* Boston, MA, USA: Kluwer Academic Publishers, 1998.

[28] S. Makino: Blind source separation of convolutive mixtures of speech, in J. Benesty and Y. Huang (eds.), *Adaptive Signal Processing: Applications to Real-World Problems,* Berlin, Germany: Springer, 2003.
[29] K. Matsuoka, M. Ohya, M. Kawamoto: A neural net for blind separation of nonstationary signals, *Neural Networks*, **8**(3), 411–419, 1995.
[30] K. Matsuoka, S. Nakashima: Minimal distortion principle for blind source separation, *Proc. ICA '01*, 722–727, San Diego, CA, USA, 2001.
[31] R. Mukai, H. Sawada, S. Araki, S. Makino: Frequency domain blind source separation using small and large spacing sensor pairs, *Proc. ISCAS '04*, **5**, 1–4, Vancouver, Canada, 2004.
[32] R. Mukai, H. Sawada, S. Araki, S. Makino: Near-field frequency domain blind source separation for convolutive mixtures, *Proc. ICASSP '04*, **4**, 49–52, Montreal, Canada, 2004.
[33] R. Mukai, H. Sawada, S. Araki, S. Makino: Frequency domain blind source separation for many speech signals, *Proc. ICA '04 (LNCS 3195)*, 461–469, Granada, Spain, 2004.
[34] R. Mukai, H. Sawada, S. Araki, S. Makino: Blind source separation for moving speech signals using blockwise ICA and residual crosstalk subtraction, *IEICE Trans. Fundamentals*, **E87-A**(8), 1941–1948, 2004.
[35] N. Murata, S. Ikeda, A. Ziehe: An approach to blind source separation based on temporal structure of speech signals, *Neurocomputing*, **41**(1-4), 1–24, October 2001.
[36] L. Parra, C. Spence: Convolutive blind separation of non-stationary sources, *IEEE Trans. Speech Audio Processing*, **8**(3), 320–327, May 2000.
[37] H. Saruwatari, S. Kurita, K. Takeda, F. Itakura, T. Nishikawa, K. Shikano: Blind source separation combining independent component analysis and beamforming, *EURASIP Journal on Applied Signal Processing*, **11**, 1135–1146, 2003.
[38] H. Sawada, R. Mukai, S. Araki, S. Makino: Polar coordinate based nonlinear function for frequency domain blind source separation, *IEICE Trans. Fundamentals*, **E86-A**(3), 590–596, March 2003.
[39] H. Sawada, R. Mukai, S. Makino: Direction of arrival estimation for multiple source signals using independent component analysis, *Proc. ISSPA '03*, 411–414, Paris, France, 2003.
[40] H. Sawada, R. Mukai, S. de la Kethulle, S. Araki, S. Makino: Spectral smoothing for frequency-domain blind source separation, *Proc. IWAENC '03*, Kyoto, Japan, 311–314, 2003.
[41] H. Sawada, R. Mukai, S. Araki, S. Makino: A robust and precise method for solving the permutation problem of frequency-domain blind source separation, *IEEE Trans. Speech Audio Processing*, **12**, 530–538, September 2004.
[42] H. Sawada, S. Winter, R. Mukai, S. Araki, S. Makino: Estimating the number of sources for frequency-domain blind source separation, *Proc. ICA '04 (LNCS 3195)*, 610–617, Granada, Spain, 2004.

[43] H. Sawada, R. Mukai, S. Araki, S. Makino: Frequency-domain blind source separation, in J. Benesty, S. Makino, J. Chen (eds.), *Speech Enhancement,* Berlin, Germany: Springer, 2005.
[44] R. O. Schmidt: Multiple emitter location and signal parameter estimation, *IEEE Trans. Antennas and Propagation*, **34**, 276–280, March 1986.
[45] L. Schobben, W. Sommen: A frequency domain blind signal separation method based on decorrelation, *IEEE Trans. Signal Processing*, **50**(8), 1855–1865, August 2002.
[46] J. J. Shynk: Frequency-domain and multirate adaptive filtering, *IEEE Signal Processing Magazine*, **9**(1), 14–37, January 1992.
[47] V. C. Soon, L. Tong, Y. F. Huang, R. Liu: A robust method for wideband signal separation, *Proc. ISCAS '93*, **1**, 703–706, Chicago, IL, USA, 1993.
[48] P. Smaragdis: Blind separation of convolved mixtures in the frequency domain, *Neurocomputing*, **22**, 21–34, 1998.
[49] T. Takatani, T. Nishikawa, H. Saruwatari, K. Shikano: High-fidelity blind separation of acoustic signals using SIMO-model-based independent component analysis, *IEICE Trans. Fundamentals*, **E87-A**(8), 2063–2072, August 2004.
[50] S. Winter, H. Sawada, S. Makino: Geometrical understanding of the PCA subspace method for overdetermined blind source separation, *Proc. ICASSP '03*, 769–772, 2003.
[51] B. D. Van Veen, K. M. Buckley: Beamforming: a versatile approach to spatial filtering, *IEEE ASSP Magazine*, **5**, 4–24, April 1988.

4

Localization and Tracking of Acoustical Sources

Gerhard Doblinger

The Institute of Communications and Radio-Frequency Engineering, Vienna University of Technology

Speaker localization and automatic tracking in a reverberant environment are challenging and often needed tasks in many audio-based applications including hands-free mobile phones, speech recognition, and teleconferencing. In this chapter, we present signal processing algorithms for reliable location estimation of audio sources. We discuss high-quality techniques based on time-delay estimation using only two microphones. These algorithms can be used to estimate directions of sound waves travelling to a one-dimensional microphone array. We focus on this basic situation because it frequently occurs in practice. Furthermore, a precise and robust algorithm for time-delay estimation is fundamental to multi-dimensional source localization tasks as well. We present an automatically steered microphone array for speaker tracking using an adaptive beamformer in connection with a direction estimation subsystem. This array is very well suited to adjust the main lobe of the beam pattern to the direction of a moving speaker while suppressing sounds from other directions. In addition, the system is capable to track speaker movements or to switch among speakers in rooms with modest reverberation. The automatically steered microphone array uses a computationally efficient multi-input FFT filterbank. MATLAB® programs are available to facilitate algorithm implementation and testing by interested readers.

4.1 Introduction

Acoustical source localization is a well developed feature of the human auditory system. Using only two sensors, this biological system has a remarkable precision in resolving the position of speakers and other acoustical sources. The human ears in conjunction with the brain can accurately localize and track sources in a sound field around the head except two small ambiguity regions (cones of confusion) [1]. In addition, noise and reverberation do not greatly influence the precision of source localization. Achieving such a performance using two microphones and digital signal processing is a rather challenging task. In this chapter, our primary goal is the presentation of robust acoustical source localization algorithms which can be used to steer adaptive microphone arrays. Multiple microphones in array configurations offer many advantages over systems with a single microphone. Due to miniature piezo-electric sensors and powerful digital signal processors, microphone arrays can now be built in a compact and inconspicuous design. This leads to a number of applications of automatically steered microphone arrays like voice communications in cars, hands-free mobile phones, speech recognition, and teleconferencing. With these applications in mind, we focus on one-dimensional source localization since knowledge of the angle of arrival (azimuth in the xy-plane of a Cartesian coordinate system) is sufficient to adjust one-dimensional microphone arrays. To determine the position of a speaker in a room, we can use a multi-dimensional array or separate one-dimensional arrays.

The two-microphone technique of delay estimation is fundamental to all multi-dimensional source localization algorithms because different delay measurements can be combined by refined procedures to estimate a speaker's position and movement. However, extensions to multiple microphones and localization of multiple sources will not be treated in this chapter. Further readings on multi-microphone techniques for multi-source localization can be found in recent books [2–4].

The basic setup using two microphones is sketched in Fig. 4.1. If we assume far-field conditions (plane wave propagation), the estimation of azimuth Φ can easily be carried out by measuring the Time Delay Difference (TDD) between the two microphone signals.

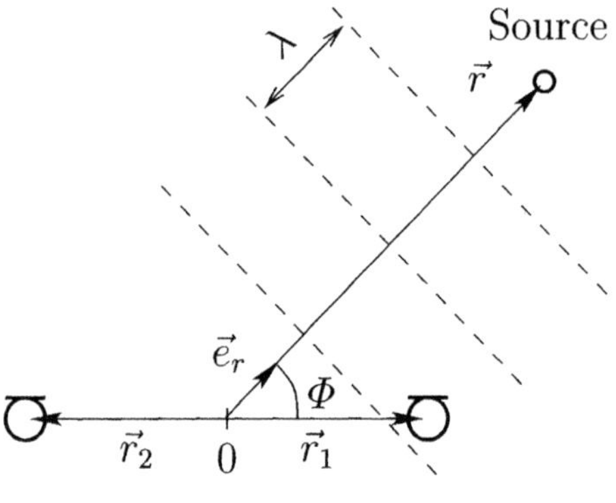

Fig. 4.1. Basic two-microphone layout for source localization (azimuth Φ of arrival direction, single frequency plane wave with wavelength λ).

Denoting microphone distance $d = \|\vec{r}_2 - \vec{r}_1\|$, sound velocity v_s, and TDD Δt, we get

$$\Phi = \arccos \frac{v_s \Delta t}{d} . \tag{4.1}$$

Due to the nonlinear relationship, accuracy is poor for Φ near 0° and 180°. In addition, discrete-time processing of the microphone signals results in quantized TDD estimates. If we estimate azimuth Φ from TDDs with accuracy $\pm\frac{T}{2}$ (sampling interval $T = 1/f_s$), we can expect an error behavior as shown in Fig. 4.2. Curves plotted in Fig. 4.2 obey the relationship

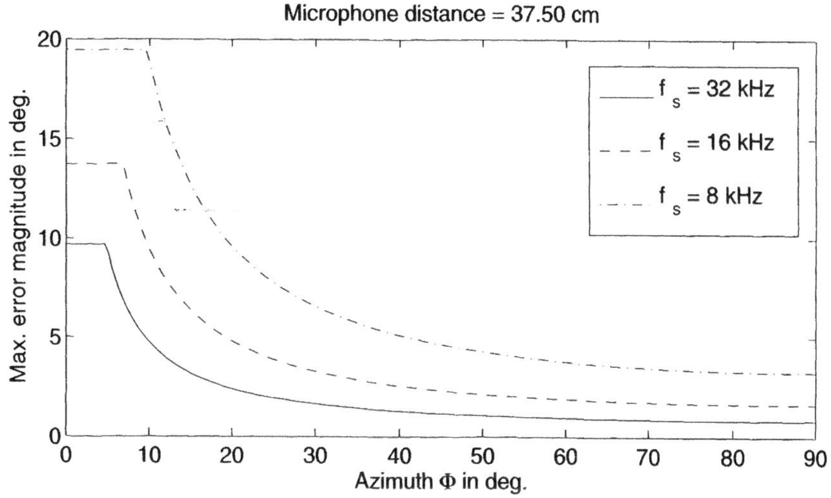

Fig. 4.2. Maximum azimuth error magnitude as a function of azimuth Φ and sampling frequency f_s.

$$\delta\Phi_{\max}(\Phi) \approx \min\left(\delta\Phi_0, \frac{\beta}{|\sin\Phi|}\right), \tag{4.2}$$

with $\beta = \frac{v_s}{2df_s} < 1$, and $\delta\Phi_0 = \arccos(1-\beta)$. As a consequence, we must use oversampling or a two-dimensional array (e.g. a quadratic array layout with 4 microphones) to reduce errors at $\Phi \approx 0°$ and $\Phi \approx 180°$. Later in this chapter, we will present an algorithm which exhibits an improved performance. It should be noted that Fig. 4.2 only shows the influence of delay quantization. In addition, errors resulting from TDD estimation must also be taken into account.

According to (4.2), the azimuth error at a given sampling frequency f_s can be reduced by increasing microphone distance d. For practical reasons, however, array size is limited in most situations like car cockpits. Additional problems affecting the performance of source localization algorithms are introduced by the specific nature of speech signals exhibiting speech pauses and segments with different spectral contents, and by noise and reverberation.

In the next sections, we will discuss algorithms which are rather robust in regard to these obstacles. We begin with a classical method using the Generalized Cross-Correlation (GCC) function [5]. The GCC method can efficiently be implemented using the Fast Fourier Transform (FFT). Motivated by binaural signal processing, an algorithm based on Interaural Time Differences (ITD) is presented next. This method offers an azimuth estimation with high accuracy but requires more computational load [6]. Afterwards, two source localization algorithms involving adaptive filters are described. One technique uses an adaptive eigenvalue decomposition to estimate TDDs [7]. This promising technique employes a normalized Least Mean-Square (LMS) adaptive algorithm suitable for implementation using the FFT. We conclude with a presentation of an adaptive microphone array comprised of an FFT filterbank beamformer and a source localization subsystem to automatically steer the beam pattern towards a moving speaker.

In order to facilitate implementation, algorithm variables and equations are formulated in a discrete-time framework. We do not use continuous-time variables, as sometimes found in the literature on TDD estimation. In addition, MATLAB® programs and test data for all algorithms presented in this chapter are available at `www.nt.tuwien.ac.at/dspgroup/gdobling.html`. Testing and comparison of the algorithms can thus be carried out with minimal effort.

4.2 Source Localization Using the Generalized Cross-Correlation Function

If we assume an ideal wave propagation model and an array with two microphones (see Fig. 4.1), then the analog (continuous-time) sensor signals are given by

$$x_{a1}(t) = s_a(t) + v_{a1}(t) \tag{4.3}$$
$$x_{a2}(t) = s_a(t - \tau_0) + v_{a2}(t), \tag{4.4}$$

with source signal $s_a(t)$ and noise disturbances $v_{a1,2}(t)$. In (4.3), (4.4), we have neglected any signal attenuation and spreading (caused by room acoustics). The discrete-time representations of the bandlimited sensor signals are

$$x_1(n) = s(n) + v_1(n) \tag{4.5}$$
$$x_2(n) = \underbrace{s_a(nT - \tau_0)}_{s_{\tau_0}(n)} + v_2(n), \tag{4.6}$$

with sampling interval T. In general, signal delay τ_0 is not an integer multiple of T. Therefore, $s_{\tau_0}(n)$ is not simply a delayed version of $s(n)$. Only if $\tau_0 = n_0 T$, then $s_{\tau_0}(n) = s(n - n_0)$. However, using the reconstruction property of a bandlimited analog signal

$$s_a(t) = \sum_{k=-\infty}^{\infty} s(k) \frac{\sin \frac{\pi}{T}(t - kT)}{\frac{\pi}{T}(t - kT)}, \tag{4.7}$$

we obtain

$$s_{\tau_0}(n) = s_a(nT - \tau_0) = \sum_{k=-\infty}^{\infty} s(k)\, \underbrace{h_a\big((n-k)T - \tau_0\big)}_{h_{\tau_0}(n-k)}, \tag{4.8}$$

with $h_{\tau_0}(n) = \frac{\sin \pi(n-\tau_0/T)}{\pi(n-\tau_0/T)}$. Thus, the discrete-time representation of the delayed microphone signal is an interpolated version of the non-delayed signal. An ideal lowpass interpolation function with parameter τ_0/T is used. If we determine signal delays in time domain, we have to use a sufficiently high sampling frequency or some kind of signal interpolation. As an alternative, signal delays can be obtained in the frequency domain from the phase spectrum. Application of the Fourier Transform to (4.5), (4.6) results in

$$X_1\left(e^{j\Omega}\right) = S\left(e^{j\Omega}\right) + V_1\left(e^{j\Omega}\right) \tag{4.9}$$

$$X_2\left(e^{j\Omega}\right) = S\left(e^{j\Omega}\right) e^{-j\Omega\frac{\tau_0}{T}} + V_2\left(e^{j\Omega}\right). \tag{4.10}$$

Assuming zero-mean uncorrelated noise disturbances, the cross-power spectrum is

$$S_{x_1x_2}(\Omega) = E\{X_1\left(e^{j\Omega}\right)X_2^*\left(e^{j\Omega}\right)\} = S_{ss}(\Omega)\, e^{j\Omega\frac{\tau_0}{T}}, \tag{4.11}$$

where $E\{\cdot\}$ means expectation and $*$ denotes complex conjugate operation. A computation of signal delays τ_0 from (4.11) requires a robust phase unwrapping algorithm and a least-squares procedure involving phase measurements at a set of different frequencies. In the context of microphone arrays, robust phase unwrapping has been proposed in [8,9]. However, these methods pose less robustness regarding room reverberation.

An alternative to phase unwrapping is delay estimation from the generalized cross-correlation (GCC) $R_{x_1x_2}(n)$:

$$\frac{\tau_0}{T} \approx n_0 = \arg\max_n R_{x_1x_2}(n), \tag{4.12}$$

with

$$R_{x_1x_2}(n) = \frac{1}{2\pi}\int_{-\pi}^{\pi} \psi_{12}\left(e^{j\Omega}\right) S_{x_1x_2}(\Omega)\, e^{j\Omega n} d\Omega. \tag{4.13}$$

Non-integer delays τ_0/T can only be approximately obtained from (4.12). To increase accuracy of delay estimation, an interpolation must be applied to $R_{x_1x_2}(n)$ prior to maximum detection. If we omit the weighting function $\psi_{12}\left(e^{j\Omega}\right)$ in (4.13), we obtain the classical cross-correlation between the sensor signals as the inverse Fourier Transform of the cross-power spectrum.

The benefits of using a weighting function $\psi_{12}\left(e^{j\Omega}\right) \neq 1$ are discussed in detail in [5]. The main idea is to create a sharp dominant peak and to reduce spurious peaks in $R_{x_1x_2}(n)$ caused by room reverberation and colored source signal spectra. A single sharp peak in the GCC function requires a flat cross-power spectrum magnitude. As a result, the weighting function must act as a pre-whitening filter. This leads to the SCOT (Smoothed Coherence Transform) algorithm with a weighting function

$$\psi_{12}\left(e^{j\Omega}\right) = \psi_S\left(e^{j\Omega}\right) = \frac{1}{\sqrt{S_{x_1x_1}(\Omega)\, S_{x_2x_2}(\Omega)}} \,. \tag{4.14}$$

Alternatively, we obtain the PHAT (Phase Transform) algorithm with the weighting function

$$\psi_{12}\left(e^{j\Omega}\right) = \psi_P\left(e^{j\Omega}\right) = \frac{1}{\left|S_{x_1x_2}(\Omega)\right|} \,. \tag{4.15}$$

Under ideal conditions as given in (4.11), the PHAT weighting function delivers an ideal GCC

$$R_{x_1x_2}(n) = \frac{1}{2\pi}\int_{-\pi}^{\pi} e^{j\Omega\frac{\tau_0}{T}}\, e^{j\Omega n} d\Omega = \frac{\sin\pi(n+\frac{\tau_0}{T})}{\pi(n+\frac{\tau_0}{T})} \,. \tag{4.16}$$

The PHAT weighting function has the computational advantage that only the cross-power spectrum is needed. Both the SCOT-GCC and the PHAT-GCC algorithm perform very well in practical situations with modest room reverberation, like medium-size office rooms, and car cabins. Furthermore, these GCC algorithms are robust against environmental noise and the specific nature of speech spectra. As shown by a comprehensive statistical analysis in [10], the PHAT-GCC is optimal among the class of GCC functions when used in reverberant environments. The GCC principle can be extended to more than one microphone pair, yielding better precision of source position estimates, especially in larger rooms [11].

Speech signals require an estimation of power spectra on a short-time basis. Therefore, the expectation operator in (4.11) will be replaced by a suitable time-average. Power spectra can be estimated from windowed signal frames of N samples (e.g. $N = 512$ at $f_s = 16$ kHz). Frames may overlap by some extend (typically $N/2$ to $3N/4$ samples). We use an exponential weighting of past frames resulting in the following cross-power spectrum estimate:

$$\widehat{S}_{x_1x_2}(m,k) = \alpha\widehat{S}_{x_1x_2}(m-1,k) + (1-\alpha)X_1(m,k)X_2^*(m,k), \tag{4.17}$$

with $\alpha = 0.7 \ldots 0.8$ to accommodate for the short-time stationarity of speech signals (m is the frame index, k the index of the discrete frequency axis, respectively). The Discrete Fourier Transforms (DFTs) of the windowed microphone signal frames are

$$X_i(m,k) = \sum_{n=0}^{N-1} x_i(mM+n)w(n)e^{-j\frac{2\pi}{N}nk}, \quad i = 1,2 \tag{4.18}$$

(frame index $m = 0, 1, 2, \ldots$, frequency index $k = 0, 1, \ldots, N-1$). The frame hop size M determines frame overlapping (no overlapping if $M \geq N$). A bell-shaped function $w(n)$ like Hann or Hamming windows may be used for time-windowing.

By means of the inverse DFT (IDFT), the cross-power spectrum estimate (4.17) can now be used to estimate the PHAT-GCC of the mth signal frame:

$$\widehat{R}_{x_1 x_2}(m,n) = \frac{1}{N} \sum_{k=0}^{N-1} \frac{\widehat{S}_{x_1 x_2}(m,k)}{\left|\widehat{S}_{x_1 x_2}(m,k)\right|} e^{j\frac{2\pi}{N}nk}, \quad n = 0, 1, \ldots, N-1. \quad (4.19)$$

Finding the maximum location of $\widehat{R}_{x_1 x_2}(m,n)$ in order to determine the TDD must be done with care. First, TDDs may be positive or negative depending on the azimuth of the sound wave (see Fig. 4.1). Therefore, indices $N-n$ must be used instead of $-n$ according to the periodicy of the DFT. Secondly, we do not need to carry out maximum search over the whole interval $n \in [0, N-1]$ because the maximum delay $\tau_{0\max}$ is limited by the microphone distance d ($\tau_{0\max} = d/v_s$). Third, and most important: In order to resolve fractional signal delays, we must use an interpolation of $\widehat{R}_{x_1 x_2}(m,n)$ before finding the maximum location. This can conveniently be done in the frequency domain by increasing the length (e.g. $N' = 4N$) of the IDFT in combination with proper zero-padding. Alternatively, GCC interpolation can efficiently be carried out in the time domain since the relevant GCC length is rather short.

Fig. 4.3 and Fig. 4.4 show a typical example of a PHAT-GCC azimuth estimation using a 50 seconds speech record of a moving speaker in a room with modest reverberation and noise. The initial speaker position is at azimuth 90°. After 16 seconds, the speaker moves towards 0°, and finally to 180°. Azimuth estimates are held constant during speech pauses detected by comparing the maximum of $\widehat{R}_{x_1 x_2}(m,n)$ with a threshold value. This speech activity detection is very robust at virtually no additional cost. Frame size is set to $N = 512$ samples with a frame hop size $M = 128$. FFT length is increased by a factor of 4 when calculating $\widehat{R}_{x_1 x_2}(m,n)$ in (4.19). In (4.18), however, an $N = 512$ point FFT is applied to compute the DFTs of the two microphone signals.

4.3 Source Localization Based on Interaural Time Differences

As briefly discussed in the introduction, human beings have an astonishing precise sound localization ability based on interaural differences in time delay and intensity between sound pressure signals at the two ears. Processing of these interaural differences is carried out to a great extend in the human brain. Several binaural models exist to describe numerous experimental data (see [12] for a detailed review). One of these models is the basis of the source localization algorithm presented in this section [6]. Basically, we create a set

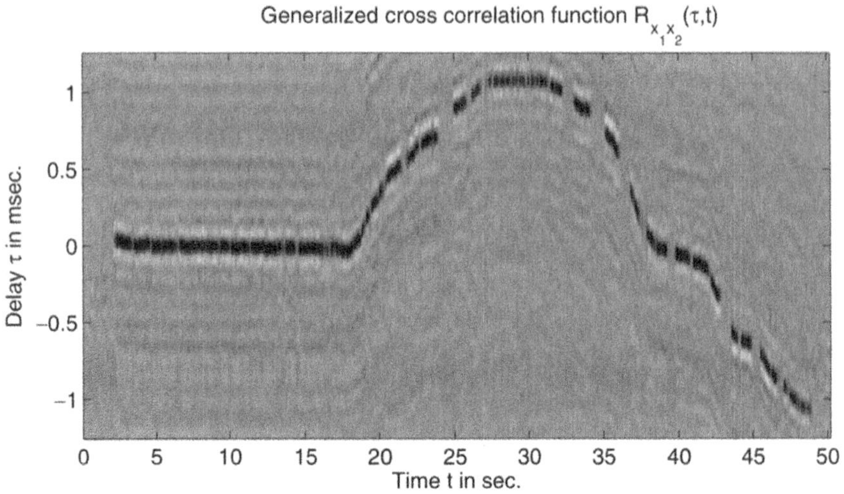

Fig. 4.3. PHAT-GCC map of a speaker movement in a medium-size office environment (Speech pauses are clearly visible as discontinuities).

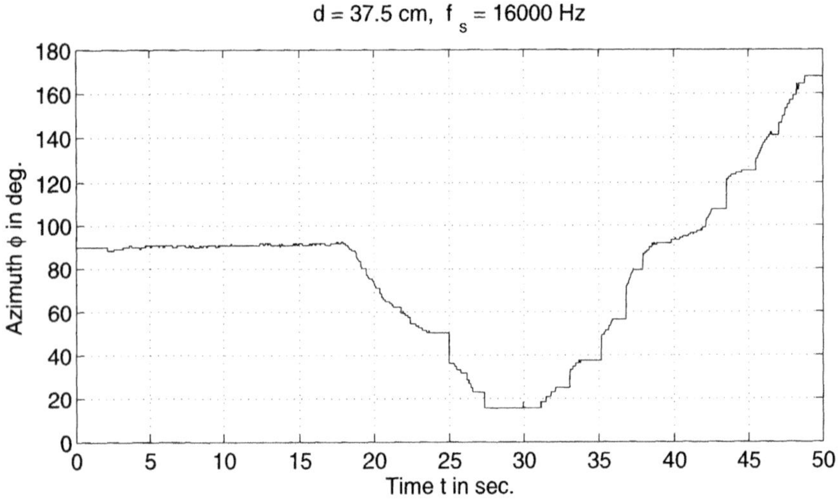

Fig. 4.4. Azimuth estimation using maximum search on the PHAT-GCC of Fig. 4.3 (Estimates are held constant during speech pauses).

of all relevant delays between the two microphone signals needed to estimate azimuth Φ to a given resolution. This set is searched for the optimum delay value resulting in the best coincidence of the two microphone signals. However, the matching procedure is implemented in the frequency domain to obtain fractional delays in an easy way.

The whole azimuth range $\Phi \in [0, \pi]$ is subdivided into an odd number I of equally spaced sectors. Using the array geometry of Fig. 4.1, each sector corresponds to a TDD[1]

$$\tau_i = \frac{d}{2v_s} \sin\left(\frac{i-1}{I-1}\pi - \frac{\pi}{2}\right), \quad i = 1, 2, \ldots, I, \tag{4.20}$$

with microphone distance d and sound velocity v_s. As an example, we need a set of $I = 73$ values τ_i to obtain an azimuth resolution of 2.5°. If we use an N-point DFT to represent the microphone signals in the frequency domain, this set of delays corresponds to phase factors

$$p_k(i) = e^{-j\frac{2\pi}{N}kf_S\tau_i}, \quad k = 0, 1, \ldots, \frac{N}{2}, \quad i = 1, 2, \ldots, I, \tag{4.21}$$

with sampling frequency f_s and τ_i from (4.20). The N-point DFTs $X_{1,2}(m,k)$ of the microphone signals are computed on a frame by frame basis as in (4.18). To find the optimum delay for each frequency index k, we can use the system shown in Fig. 4.5. The DFTs $X_{1,2}(m,k)$ are multiplied by phase factors

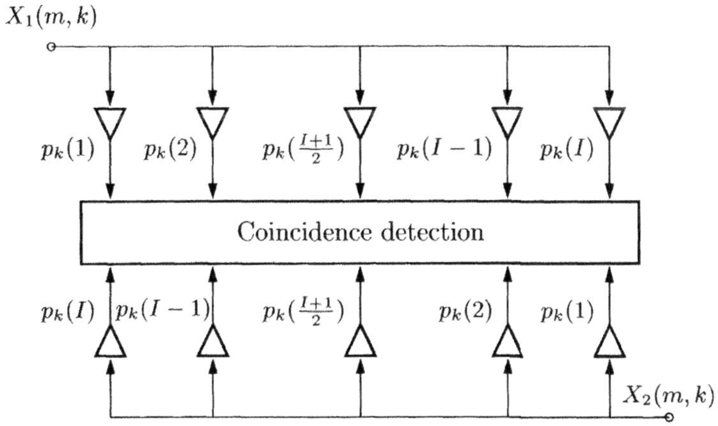

Fig. 4.5. Delay (phase) matching in frequency domain for each frequency index k (frame index m).

from (4.21) and compared in the coincidence detection box. Comparison is performed on each vertically aligned pair only, since the delays of the two microphone signals are coupled due to the array geometry and phase factors are properly arranged in Fig. 4.5. The coincidence detection is carried out according to the simple matching rule

[1] Delays τ_i are measured here with respect to the origin in Fig. 4.1.

$$i_{\text{opt}}(m,k) = \arg\min_{i} \Delta_i(m,k), \quad k = 0,1,\ldots,\frac{N}{2} \tag{4.22}$$

$$\Delta_i(m,k) = \left| p_k(i)X_1(m,k) - p_k(I-i+1)X_2(m,k) \right|^2, \quad i = 1,2,\ldots,I \tag{4.23}$$

(frame index $m = 0,1,2,\ldots$). With optimum delay indices from (4.22), optimum delays τ_i can be found for each frequency point k and frame m according to (4.20). To obtain the TDD, and thus the azimuth of the sound source from this set of data, we first build a histogram map $P_k(\tau_i, m)$ by counting τ_i values for each frequency point in several consecutive signal frames. τ_i values will gather around the actual delay corresponding to the azimuth of the signal source. In a similar manner as in [6], we use the following histogram averaging procedure in case of speech signals:

$$\begin{aligned} P_k(\tau_i, m) = \alpha P_k(\tau_i, m-1) + \delta\big(i - i_{\text{opt}}(m,k)\big), \\ i = 1,2,\ldots,I \\ k = 0,1,\ldots,\frac{N}{2} \\ m = 0,1,2,\ldots, \end{aligned} \tag{4.24}$$

where $\delta(\cdot)$ is the unit impulse and τ_i is the set of delays in (4.20). Forgetting factor α is chosen between 0.85 and 0.95.

An illustrative example of a histogram map is shown in Fig. 4.6 wherein delay values τ_i are replaced by corresponding azimuth values. A stationary broadband noise source emitting from azimuth direction 60° is used. In the frequency range below 2 kHz, a prominent population of azimuth values along a vertical line is observed. An additional curved pattern stems from phase ambiguity. Spatial aliasing occurs for signals with frequency contents above $f_{\max} = \frac{v_S}{2d}$ due to $\frac{\lambda}{2} < d$. With a microphone distance $d = 37.5$ cm, we get $f_{\max} \approx 450$ Hz.

To reduce the influence of phase ambiguity, we sum up histogram data over all frequency indices k for each azimuth (or τ_i, respectively). The optimum delay is then obtained by searching for the maximal sum. As a result, the azimuth of the source location is given by

$$\tau_{\text{opt}}(m) = \arg\max_{\tau_i} \sum_{k=0}^{\frac{N}{2}} P_k(\tau_i, m), \tag{4.25}$$

for each signal frame m. Despite the presence of phase ambiguity, the maximum in (4.25) is rather sharp. Further improvements, especially in case of multiple sources, are discussed in [6]. However, a high computationally effort is needed which is not justified in case of a single speaker or even for multiple speakers not talking at the same time. Our investigations show that no significant improvements by the refinements proposed in [6] are obtained in real acoustic environments.

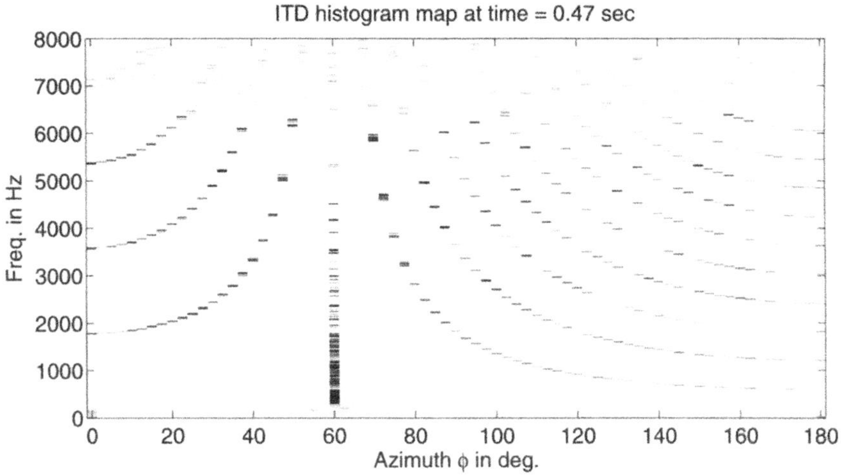

Fig. 4.6. Histogram map of a stationary white noise source, bandlimited between 300 Hz and 6400 Hz, emitting from azimuth direction $\Phi = 60°$, (FFT length $N = 512$, $\alpha = 0.9$, azimuth resolution 2.5°).

Using the same source signal as in Fig. 4.3, a representative example of a histogram map summed up over frequency is shown in Fig. 4.7. The result of azimuth estimation by searching for maxima locations in the ITD histogram map of Fig. 4.7 is presented in Fig. 4.8. Performance differences between the PHAT-GCC and ITD algorithm can barely be derived from these example figures. However, they can be better detected by using artificial broadband noise from known directions as test signals. The ITD method offers the advantage that the angular resolution can be selected by choosing the size I of the delay set in (4.20). In comparison with the PHAT-GCC algorithm, the accuracy is better for azimuths near 0° and 180°. Obviously, this is an advantage if two microphone pairs are used to find a speaker's position by calculating the cross point of the two azimuth estimates. Furthermore, there is no need for signal oversampling or increasing the FFT size because phase matching is done in the frequency domain. On the other hand, substantially more search algorithms are required for minima and maxima detections.

Our experiments with speech signals indicate less robustness against environmental noise and reverberation as compared to the PHAT-GCC method. The increased sensitivity with respect to room acoustics is due to the influence of sound reflections that smear maxima locations in the ITD histogram map. In [6] the authors suggest to set $P_k(\tau_i, m)$ to zero for values below a certain threshold. According to our experience, however, this does not improve the performance in reverberant rooms. Therefore, application of the ITD algorithm is limited to situations where accurate source localization under moderate environmental noise is needed. For automatic steering of microphone

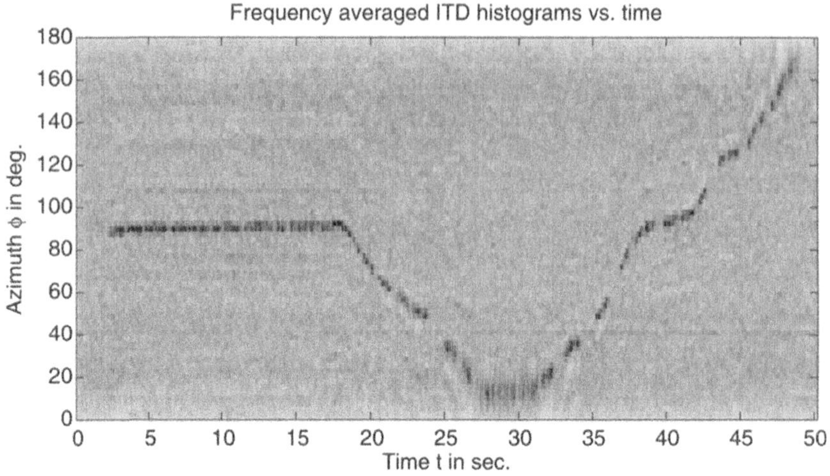

Fig. 4.7. ITD histogram map summed over frequency of a moving speaker (same acoustical environment as in Fig. 4.3, azimuth instead of delay values on vertical axis).

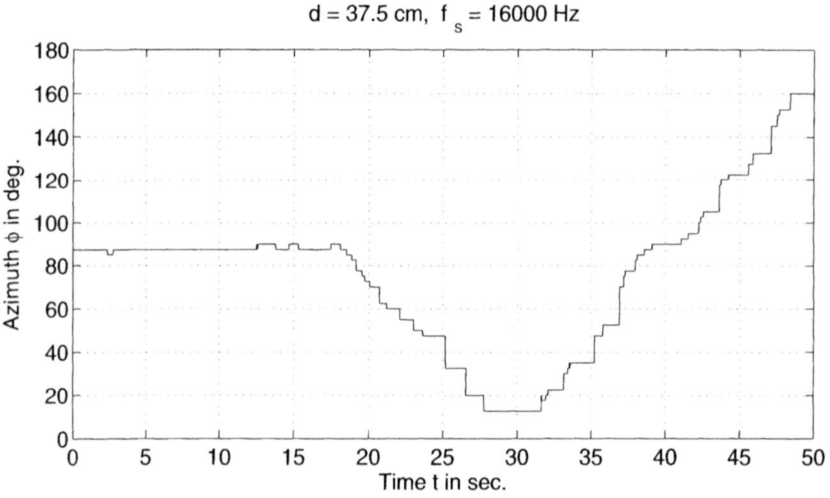

Fig. 4.8. Azimuth estimation by maximum search on the ITD histogram map from Fig. 4.7 (Estimates are held constant during speech pauses).

arrays, we prefer to use the PHAT-GCC method because of its robustness. Arrays of up to 8 microphones exhibit relatively broad main lobes in their array patterns. As a consequence, there is no need for an azimuth estimation accuracy less then $3^\circ \ldots 5^\circ$.

4.4 Source Localization Using Adaptive Filters

In the derivations of source localization algorithms, we have assumed an ideal wave propagation model so far. In such an environment with no sound reflections, the two microphone signals in Fig. 4.1 are simply delayed versions of the source signal. Although this model works remarkably well in real acoustic environments too, a more realistic approach is to find the signal delay from the actual impulse responses between source and microphones. In this section, two different adaptive systems for delay estimation are presented. The first system models the time delay between the two microphones. It is assumed that the direct path of sound propagation dominates. In the second method, we estimate the impulse responses by an adaptive eigenvalue decomposition. This method is more robust if strong reverberation is present. Both algorithms can efficiently be implemented by frequency-domain adaptive filters.

The first adaptive filtering technique is straight forward and shown in Fig. 4.9.[2] A detailed performance analysis can be found in [13]. We denote

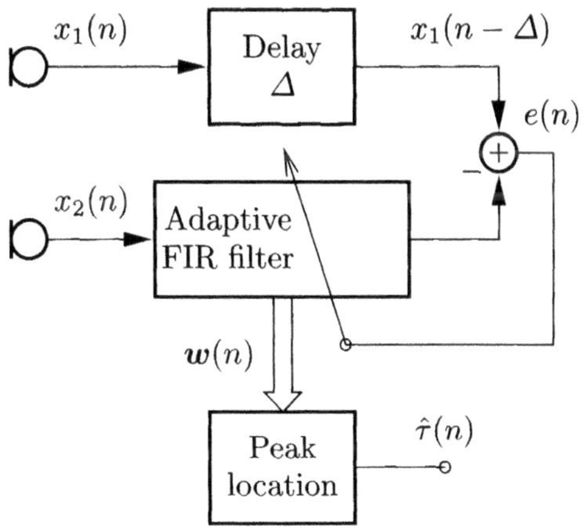

Fig. 4.9. Time delay estimation using an adaptive FIR filter (length L, coefficient vector $\boldsymbol{w}(n)$, delay $\Delta = \lfloor \frac{L-1}{2} \rfloor$).

FIR filter state as vector $\boldsymbol{x}_2(n)$ and coefficients as vector $\boldsymbol{w}(n)$ according to

$$\boldsymbol{x}_2(n) = \left[x_2(n)\; x_2(n-1)\; \cdots\; x_2(n-L+1)\right]^{\mathrm{T}} \tag{4.26}$$

$$\boldsymbol{w}(n) = \left[w_0(n)\; w_1(n)\; \cdots\; w_{L-1}(n)\right]^{\mathrm{T}}, \tag{4.27}$$

("T" denotes vector transpose). The error signal $e(n)$ is then given by

$$e(n) = x_1(n-\Delta) - \boldsymbol{w}^{\mathrm{T}}(n)\boldsymbol{x}_2(n), \tag{4.28}$$

[2] FIR = Finite Impulse Response Duration

($\Delta = \lfloor \frac{L-1}{2} \rfloor$). The Least Mean-Square (LMS) algorithm can be used to update the weight vector:

$$\boldsymbol{w}(n+1) = \boldsymbol{w}(n) + \mu_{\text{LMS}}\, e(n)\boldsymbol{x}_2(n). \tag{4.29}$$

In general, however, a better performance is achieved with the normalized LMS algorithm

$$\boldsymbol{w}(n+1) = \boldsymbol{w}(n) + \frac{\mu_{\text{NLMS}}}{\|\boldsymbol{x}_2(n)\|^2}\, e(n)\boldsymbol{x}_2(n), \tag{4.30}$$

with $\|\boldsymbol{x}_2(n)\|^2 = \boldsymbol{x}_2^{\mathrm{T}}(n)\boldsymbol{x}_2(n)$. In order to improve the convergence behavior, a pre-emphasis filter with impulse response $h_{pre}(n) = \delta(n) - 0.9\delta(n-1)$ (unit impulse $\delta(n)$) can be used for simple pre-whitening of speech signals. Such a pre-filter is not required if we use the following frequency-domain adaptive filter. Only three FFTs per frame plus one FFT every M samples ($M = 2000$, typically) are needed. In addition, convergence is superior in case of speech signals due to a frequency dependent adaptive filter step size. The algorithm is based on the fast block LMS adaptive filter as proposed in [14], combined with a frequency dependent step size as suggested in [15]. To implement the LMS adaptive filter in the frequency domain by means of the FFT, samples are grouped into frames and coefficients are held constant till the next frame is processed. The update of the adaptive filter coefficients in frequency-domain at each frame index m can be summarized as follows:

$$X_2(m,k) = \sum_{n=0}^{N-1} x_2(mL+n)e^{-j\frac{2\pi}{N}nk}, \quad k = 0,1,\ldots,N-1 \tag{4.31}$$

$$y(m,n) = \frac{1}{N}\sum_{k=0}^{N-1} W(m,k)X_2(m,k)e^{j\frac{2\pi}{N}nk}, \quad n = 0,1,\ldots,N-1 \tag{4.32}$$

$$\tilde{e}(m,n) = \begin{cases} 0 & n = 0,1,\ldots,L-1 \\ x_1(mL+n-\Delta) - y(m,n) & n = L, L+1,\ldots,N-1 \end{cases} \tag{4.33}$$

$$E(m,k) = \sum_{n=0}^{N-1} \tilde{e}(m,n)e^{-j\frac{2\pi}{N}nk}, \quad k = 0,1,\ldots,N-1 \tag{4.34}$$

$$S_{x_2x_2}(m,k) = \alpha S_{x_2x_2}(m-1,k) + (1-\alpha)|X_2(m,k)|^2, \quad k = 0,1,\ldots,N-1 \tag{4.35}$$

$$W(m+1,k) = W(m,k) + \frac{\mu}{S_{x_2x_2}(m,k)+\varepsilon} X_2^*(m,k)E(m,k)$$
$$k = 0,1,\ldots,N-1. \tag{4.36}$$

The frame length is $N = 2L$, with a frame hop size equal to the adaptive filter length L. An overlap-save method with an N point DFT/IDFT is used to perform the linear convolution needed in (4.28). Note that the step size of the weight update (4.36) is normalized by an estimate of the spectral power at each frequency point.[3] As a consequence, the convergence behavior of the adaptive algorithm is nearly independent on the signal spectrum.

Delay estimates are computed every M' frames (i.e. every $M = M'L$ samples) by finding peak locations of the adaptive filter coefficients

$$w(m', n) = \frac{1}{N} \sum_{k=0}^{N-1} W(m', k) e^{j\frac{2\pi}{N}nk}, \quad n = 0, 1, \ldots, N-1. \tag{4.37}$$

Due to the overlap-save method, the last L values of $w(m', n)$ are the valid filter coefficients to be searched to find the peak location. In addition, the search range can be further reduced because peak positions are limited to $[\Delta - N_d, \Delta + N_d]$, where $N_d = \lceil \frac{d}{v_S} f_s \rceil$ is the maximum delay between the microphone signals.

A typical example using the same microphone signals as before is shown in Fig. 4.10 and Fig. 4.11. The proposed frequency-domain adaptive filter is

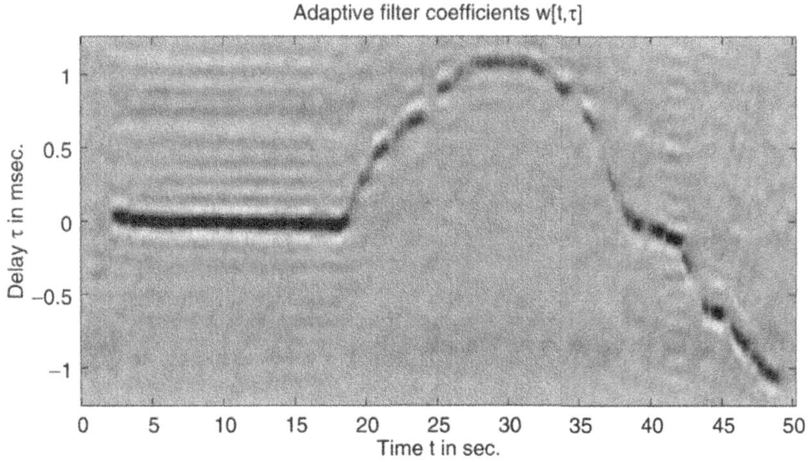

Fig. 4.10. Adaptive filter coefficient map of a moving speaker (same acoustical environment as in Fig. 4.3).

applied with length $L = 512$, step size $\mu = 0.2$, and $\alpha = 0.2$. The coefficient map is updated every $M = 2048$ samples to allow for sufficient convergence of the adaptive filter. Delay estimation is performed every M samples too by maximum detection using the coefficient map. Coefficients are oversampled by a factor of 4 to determine the peak location with sufficient accuracy.

[3] ε avoids division by zero during speech pauses.

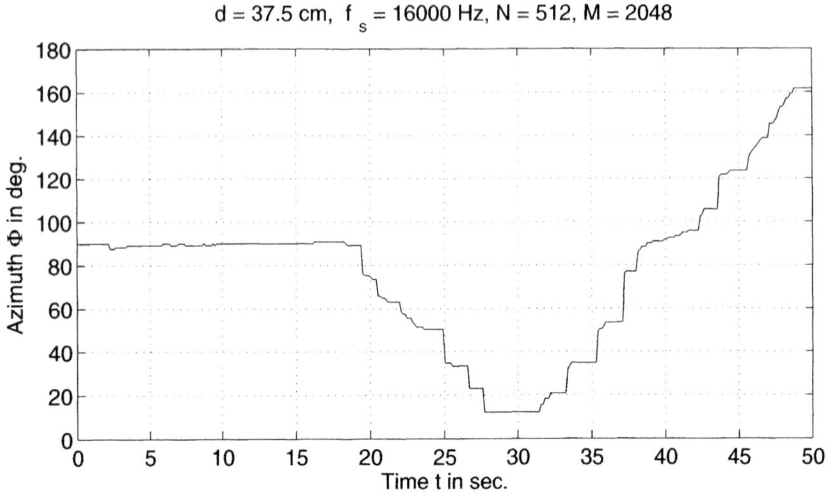

Fig. 4.11. Azimuth estimation by maximum search on the coefficient map from Fig. 4.10 (Estimates are held constant during speech pauses).

A different adaptive system showing a better performance in environments with strong reverberation is proposed in [7]. In principle, the impulse responses between source and microphones are estimated by means of an eigenvalue decomposition. Denoting $h_1(n)$ and $h_2(n)$ as impulse response from source to microphone 1, and microphone 2, respectively, we get the following discrete-time model:

$$x_1(n) = \sum_{k=-\infty}^{\infty} h_1(k)s(n-k) + v_1(n) \tag{4.38}$$

$$x_2(n) = \sum_{k=-\infty}^{\infty} h_2(k)s(n-k) + v_2(n), \tag{4.39}$$

(source signal $s(n)$, noise disturbances $v_{1,2}(n)$). At the moment, we assume a linear environment with time-invariant impulse responses. Later on, we will relax the time-invariance property by estimating $h_{1,2}(n)$ on a frame by frame basis. This allows for adaptation to sufficiently slow changes in the room acoustics, and for speaker movements. For the estimation of the impulse responses, we further assume that $h_{1,2}(n)$ can be approximated by filters with finite impulse response length L. Additionally, the noise signals $v_{1,2}(n)$ are neglected at first. This leads to the relation

$$(x_1 * h_2)(n) = (s * h_1 * h_2)(n) = (x_2 * h_1)(n) \tag{4.40}$$

between the convolutions since the order in which two stable sequences are convolved is unimportant (see Fig. 4.12). Equation (4.40) is the basis of an adaptive algorithm to estimate the impulse responses.

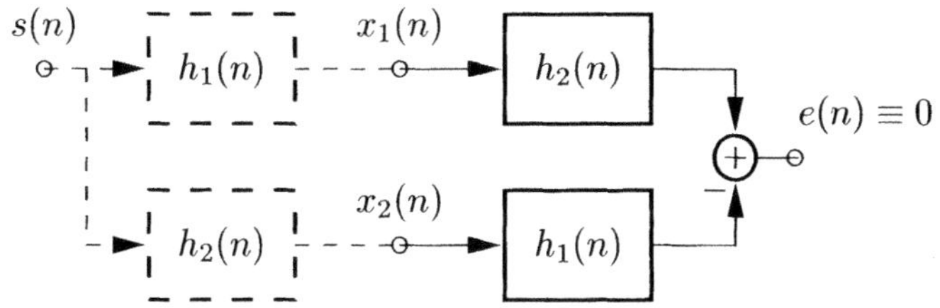

Fig. 4.12. Relationship between impulse responses according to (4.40) (signal model left, perfect estimation of impulse responses right).

If the impulse responses are approximated by length L filters, all data can be grouped in $L \times 1$ vectors

$$\boldsymbol{x}_i(n) = \left[x_i(n)\; x_i(n-1)\; \cdots\; x_i(n-L+1)\right]^{\mathrm{T}}, \quad i = 1, 2 \tag{4.41}$$

$$\boldsymbol{h}_i = \left[h_i(0)\; h_i(1)\; \cdots\; h_i(L-1)\right]^{\mathrm{T}}. \tag{4.42}$$

Equation (4.40) can now be rewritten as

$$\boldsymbol{x}_1^{\mathrm{T}}(n)\boldsymbol{h}_2 = \boldsymbol{x}_2^{\mathrm{T}}(n)\boldsymbol{h}_1. \tag{4.43}$$

Following the derivation outlined in [7], we introduce $2L \times 1$ vectors

$$\boldsymbol{x}(n) = \left[\boldsymbol{x}_1^{\mathrm{T}}(n)\;\; \boldsymbol{x}_2^{\mathrm{T}}(n)\right]^{\mathrm{T}} \tag{4.44}$$

$$\boldsymbol{u} = \left[\boldsymbol{h}_2^T\;\; -\boldsymbol{h}_1^T\right]^T \tag{4.45}$$

to rewrite (4.40):

$$\boldsymbol{x}^{\mathrm{T}}(n)\boldsymbol{u} = \boldsymbol{x}_1^{\mathrm{T}}(n)\boldsymbol{h}_2 - \boldsymbol{x}_2^{\mathrm{T}}(n)\boldsymbol{h}_1 = 0. \tag{4.46}$$

Left multiplying (4.46) by $\boldsymbol{x}(n)$ and taking expectation yields

$$\boldsymbol{R}_{\boldsymbol{xx}}(n)\boldsymbol{u} = \boldsymbol{0}. \tag{4.47}$$

$\boldsymbol{R}_{\boldsymbol{xx}}(n) = E\{\boldsymbol{x}(n)\boldsymbol{x}^{\mathrm{T}}(n)\}$ is the $2L \times 2L$ covariance matrix of the two microphone signals. Note that $\boldsymbol{R}_{\boldsymbol{xx}}(n)$ contains both temporal and spatial correlations of the microphone signals. Equation (4.47) indicates that $\boldsymbol{u}$ is the eigenvector of $\boldsymbol{R}_{\boldsymbol{xx}}(n)$ corresponding to eigenvalue 0. Therefore, both impulse responses can be found by determining this eigenvector.

If noise signals $v_{1,2}(n)$ are present, $\boldsymbol{u}$ may be estimated by minimizing $\boldsymbol{u}^{\mathrm{T}}\boldsymbol{R}_{\boldsymbol{xx}}(n)\boldsymbol{u}$ with constraint $\boldsymbol{u}^{\mathrm{T}}\boldsymbol{u} = 1$ [7]. Consequently, we get $\boldsymbol{u}$ by computing the normalized eigenvector of $\boldsymbol{R}_{\boldsymbol{xx}}(n)$ corresponding to the smallest eigenvalue. There exist several efficient algorithms to find the smallest eigenvalue and the associated eigenvector of a correlation matrix. Since the dimension of matrix $\boldsymbol{R}_{\boldsymbol{xx}}(n)$ is quite large, an adaptive algorithm will be used. As a main advantage, we need only a few iterations because the TDD between the microphone signals is of interest. There is no need to estimate the actual shapes of the impulse responses. According to (4.46) and Fig. 4.12, the error signal

$$e(n) = \boldsymbol{u}^{\mathrm{T}}(n)\boldsymbol{x}(n) \tag{4.48}$$

should be zero under ideal conditions. Actually, the cost function

$$J(n) = \frac{1}{2}E\left\{e^2(n)\right\} = \frac{1}{2}\boldsymbol{u}^{\mathrm{T}}(n)\boldsymbol{R}_{\boldsymbol{xx}}(n)\boldsymbol{u}(n) \tag{4.49}$$

can be minimized with the gradient-based adaptive algorithm

$$\boldsymbol{u}(n+1) = \boldsymbol{u}(n) - \mu_{\mathrm{LMS}}\,\boldsymbol{\nabla}_{\boldsymbol{u}}J(n) = \boldsymbol{u}(n) - \mu_{\mathrm{LMS}}\,\boldsymbol{R}_{\boldsymbol{xx}}(n)\boldsymbol{u}(n). \tag{4.50}$$

$\boldsymbol{\nabla}_{\boldsymbol{u}}J(n)$ is the cost function gradient with respect to vector $\boldsymbol{u}$. With the approximation $\boldsymbol{R}_{\boldsymbol{xx}}(n) = E\{\boldsymbol{x}(n)\boldsymbol{x}^{\mathrm{T}}(n)\} \approx \boldsymbol{x}(n)\boldsymbol{x}^{\mathrm{T}}(n)$, we get the LMS algorithm

$$\boldsymbol{u}(n+1) = \boldsymbol{u}(n) - \mu_{\mathrm{LMS}}\,e(n)\boldsymbol{x}(n). \tag{4.51}$$

The constraint $\boldsymbol{u}^{\mathrm{T}}\boldsymbol{u} = 1$ can be taken into account by normalization [7]:

$$\boldsymbol{v}(n) = \boldsymbol{u}(n) - \mu_{\mathrm{NLMS}}\,e(n)\boldsymbol{x}(n) \tag{4.52}$$

$$\boldsymbol{u}(n+1) = \frac{\boldsymbol{v}(n)}{\sqrt{\boldsymbol{v}^{\mathrm{T}}(n)\boldsymbol{v}(n)}}\,. \tag{4.53}$$

As mentioned above, only the delay between the two microphone signals is of interest. If we initialize the elements $u_i(n)$ of vector $\boldsymbol{u}(n)$ at $n = 0$ by

$$u_i(0) = \begin{cases} 0 & 0 \le i \le \lfloor \frac{L}{2} \rfloor - 1 \\ 1 & i = \lfloor \frac{L}{2} \rfloor \\ 0 & \lfloor \frac{L}{2} \rfloor + 1 \le i \le 2L - 1 \end{cases}, \tag{4.54}$$

then a negative peak will evolve in $\boldsymbol{u}(n)$ during adaptation. This peak corresponds to the direct path in the impulse response $\boldsymbol{h}_1$ (see (4.45)). The positive peak will remain at the initial position $i = \lfloor \frac{L}{2} \rfloor$. The index difference of these two peaks in $\boldsymbol{u}(n)$ determines the delay between the microphone signals. Since the position of the positive peak is fixed, we need to find the index of the negative peak only by searching vector elements $u_i(n)$, $\lfloor \frac{L}{2} \rfloor + 1 \le i \le 2L-1$. In a practical implementation, we will interpolate $\boldsymbol{u}(n)$ before peak position finding. Additionally, in case of a moving speaker we have to reset the adaptive algorithm periodically to allow tracking. Otherwise, peaks will stick at the first estimated positions, particularly for small step size values μ_{LMS}. Setting $u_i(nK) = u_i(0)$ for some period K removes all old negative peaks and allows the adaptive algorithm to adjust to the new delay position. Period K determines the tracking speed and is set to some 1000 samples, typically. During this period, the adaptive algorithm has plenty of time to converge.

The adaptive source localization algorithm can easily be implemented in the time domain. However, a significantly greater computational efficiency can be achieved by using a frequency-domain adaptive filter. As opposed to [7], an FFT based algorithm can be devised requiring only 4 FFTs per frame (plus one FFT at every initialization period) instead of 7 FFTs. This saving is obtained by eliminating the normalization of vector $\boldsymbol{u}$ in (4.53). It is argued in [7] that the normalization may avoid an error propagation in (4.51) if the algorithm runs over a long period of time. However, in order to ensure tracking, we have to periodically reset the adaptive algorithm. Thus, an eventual error propagation will efficiently be eliminated too.

The algorithm has a similar structure as the fast LMS algorithm (4.31) - (4.36):

$$X_1(m,k) = \sum_{n=0}^{N-1} x_1(mL+n)e^{-j\frac{2\pi}{N}nk}, \quad k = 0,1,\ldots,N-1 \tag{4.55}$$

$$X_2(m,k) = \sum_{n=0}^{N-1} x_2(mL+n)e^{-j\frac{2\pi}{N}nk}, \quad k = 0,1,\ldots,N-1 \tag{4.56}$$

$$e(m,n) = \frac{1}{N}\sum_{k=0}^{N-1}\Big[U_1(m,k)X_1(m,k) + U_2(m,k)X_2(m,k)\Big]\, e^{j\frac{2\pi}{N}nk}, \quad n = 0,1,\ldots,N-1 \tag{4.57}$$

$$\tilde{e}(m,n) = \begin{cases} 0 & n = 0,1,\ldots,L-1 \\ e(m,n) & n = L, L+1,\ldots,N-1 \end{cases} \tag{4.58}$$

$$E(m,k) = \sum_{n=0}^{N-1} \tilde{e}(m,n)e^{-j\frac{2\pi}{N}nk}, \quad k = 0,1,\ldots,N-1 \tag{4.59}$$

$$S_{x_1x_1}(m,k) = \alpha S_{x_1x_1}(m-1,k) + (1-\alpha)|X_1(m,k)|^2, \quad k = 0,1,\ldots,N-1 \tag{4.60}$$

$$S_{x_2x_2}(m,k) = \alpha S_{x_2x_2}(m-1,k) + (1-\alpha)|X_2(m,k)|^2, \quad k = 0,1,\ldots,N-1 \tag{4.61}$$

$$U_1(m+1,k) = U_1(m,k) - \frac{\mu}{S_{x_1x_1}(m,k)+\varepsilon}X_1^*(m,k)E(m,k), \quad k = 0,1,\ldots,N-1 \tag{4.62}$$

$$U_2(m+1,k) = U_2(m,k) - \frac{\mu}{S_{x_2x_2}(m,k)+\varepsilon}X_2^*(m,k)E(m,k), \quad k = 0,1,\ldots,N-1. \tag{4.63}$$

Similarly to the fast LMS algorithm, the DFT length is set to $N = 2L$, with impulse response length L. Vector $\boldsymbol{u}$ (see (4.45)) is split into two length L sub-vectors, i.e. $\boldsymbol{u} = [\boldsymbol{u}_1^{\mathrm{T}}\ \boldsymbol{u}_2^{\mathrm{T}}]^{\mathrm{T}}$. The updates of these sub-vectors are performed in the frequency domain. Delay estimates are computed every M' frames (i.e. every $M = M'L$ samples) by finding the dominant negative peak in $\boldsymbol{u}_2$. Likewise to (4.37), the elements of $\boldsymbol{u}_2$ are obtained by the IDFT

$$u_2(m', n) = \frac{1}{N} \sum_{k=0}^{N-1} U_2(m', k) e^{j \frac{2\pi}{N} nk}, \quad n = 0, 1, \ldots, N-1. \tag{4.64}$$

Nearly the same results as in Fig. 4.10, 4.11 are obtained if we use the same microphone signals and algorithm parameters $L = 512$, $\alpha = 0.2$, and $\mu = 0.2$.

4.5 Some Remarks on Algorithm Selection

Deciding which algorithm to choose depends on the specific area of application. In a car cabin, with no speaker movement, little reverberation, and heavy disturbing noise, the PHAT-GCC and the frequency-domain adaptive filter perform best. Both algorithms also exhibit the lowest computational demand. In situations with modest reverberation, the two adaptive source localization algorithms show the same performance. However, according to a detailed experimental comparison of algorithms in [7], the adaptive eigenvalue decomposition offers a better performance in rooms with strong reverberation and moderate noise. The best accuracy in azimuth estimation can be expected by the ITD based algorithm if nearly ideal sound propagation is present. However, the prize to be payed is the relatively high computational cost and memory demand.

If we compare the arithmetic operations per frame interval required by each algorithm, we get the coarse result listed in Tab. 4.1. The FFT length is equal to frame length N in case of PHAT-GCC and ITD algorithm. All FFTs use real-valued input data. The fast LMS algorithm (FLMS) and the adaptive eigenvalue decomposition (AEVD) require length $N = 2L$ FFTs (impulse response length L). One FFT is needed every M' frames only. Oversampling is not considered in Tab. 4.1. If we apply e.g. an oversampling (factor R) to find the GCC peak, one FFT must have a length RN. The IDT-algorithm requires only 2 real-input FFTs and no oversampling. However, the numbers of additions and multiplications depend on the azimuth resolution $\Delta\Phi \approx 180°/I$. In addition, $\frac{N}{2} + 1$ maximum/minimum search operations are needed.

Table 4.1. Comparison of computational requirements per frame of length N

Algorithm	FFT	Add.	Mult.	Div.	Sqrt.	Search
PHAT	3	$\frac{5}{2}N$	$8N$	$\frac{N}{2}$	$\frac{N}{2}$	1
ITD	2	$\left(4I + \frac{1}{2}\right)N$	$\left(\frac{11}{2}I + 2\right)N$	-	-	$\frac{N}{2} + 1$
FLMS	4	$\frac{9}{2}N$	$7N$	$\frac{N}{2}$	-	1
AEVD	5	$9N$	$14N$	N	-	1

4.6 Frequency-Domain Adaptive Beamformer with Speaker Tracking

In this section, we present an adaptive beamformer combined with source localization. The system automatically adjusts the main lobe of the array pattern to a speaker and suppresses sounds from all other directions. This behavior is preserved if the speaker moves. Applications include teleconferencing, hands-free telecommunications in cars, etc. The adaptive beamformer is based on the Frost constrained LMS algorithm [16]. However, as opposed to the original Frost beamformer, the adaptive algorithm is formulated in the frequency domain.

The main advantages of this approach are the possibility to use an efficient multi-input overlap-add FFT filterbank, the avoidance of variable fractional delay filters, and the inclusion of more constraints like nulls in the array pattern. In addition, the FFT filterbank beamformer can easily be combined with an adaptive post-filter for speech enhancement purposes [17–19]. Disadvantages are the signal delay introduced by the FFT block processing and a higher storage demand as compared with the time domain approach. However, signal delays are within usual tolerance limits if the frame size is properly chosen (e.g. 512 at 16 kHz sampling frequency). Additionally, memory requirements are no limiting factors with modern hardware.

The basic structure of the adaptive beamformer is shown in Fig. 4.13. Single channel overlap-add FFT filterbanks are used in many audio-based ap-

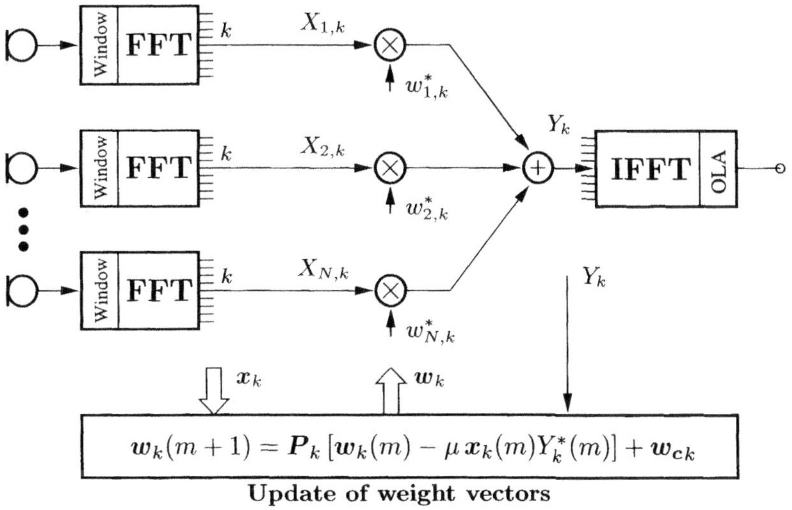

Fig. 4.13. Adaptive beamformer with N-channel overlap-add FFT filterbank and constrained LMS algorithm to compute weights $\boldsymbol{w}_k(m)$ (frequency index k, frame index m).

plications [20]. Such a multirate filterbank structure is highly efficient and offers a nearly perfect signal reconstruction property. In our extended filterbank system with multiple input channels, FFT spectra are modified by complex-valued weights on a frame by frame basis. For each frequency index k and frame index m the N-dimensional weight vectors $\boldsymbol{w}_k(m)$ are updated according to a constrained LMS algorithm. This algorithm will be derived in the sequel. Algorithm parameters $\boldsymbol{P}_k$ and $\boldsymbol{w}_{\boldsymbol{c}k}$ depend on the desired direction which is supplied by a source localization algorithm. The source localization algorithm makes use of the already available FFTs of the out-most two microphone signals of the array.

In the following derivation of the frequency-domain adaptive algorithm, the frame index m is omitted for clarity. The beamformer optimization problem to be solved by means of an adaptive algorithm may be defined by the minimization of a quadratic cost function under linear constraints:

$$\boldsymbol{w}_k = \arg\min_{\boldsymbol{w}_k} \boldsymbol{w}_k^H \boldsymbol{S}_{\boldsymbol{x}_k \boldsymbol{x}_k} \boldsymbol{w}_k, \quad \boldsymbol{C}_k^H \boldsymbol{w}_k = \boldsymbol{f} \tag{4.65}$$

(for each frequency index k). Superscript H denotes Hermitian transposition, i. e. transposition combined with complex conjugation. The minimization of the quadratic form stems from the desired minimization of the power of Y_k given by

$$E\{Y_k^2\} = \boldsymbol{w}_k^{\mathrm{H}} E\{\boldsymbol{x}_k \boldsymbol{x}_k^{\mathrm{H}}\} \boldsymbol{w}_k = \boldsymbol{w}_k^{\mathrm{H}} \boldsymbol{S}_{\boldsymbol{x}_k \boldsymbol{x}_k} \boldsymbol{w}_k, \tag{4.66}$$

with $\boldsymbol{w}_k = [w_{1,k}\, w_{2,k} \cdots w_{N,k}]^{\mathrm{T}}$, $\boldsymbol{x}_k = [X_{1,k}\, X_{2,k} \cdots X_{N,k}]^{\mathrm{T}}$. Matrix $\boldsymbol{S}_{\boldsymbol{x}_k \boldsymbol{x}_k}$ is the $N \times N$ spatio-spectral correlation matrix at frequency index k. This matrix depends on array geometry and sound field, and will be estimated by the adaptive algorithm. Minimization of $E\{Y_k^2\}$ has to be done with constraints. At least, signals from the desired direction must not be attenuated. In addition, signals from certain other directions may be suppressed by imposing nulls in the array pattern. These constraints are collected in (4.65) as a set of equations with matrix $\boldsymbol{C}_k$. The structure of this matrix is determined by the wave propagation model. If we assume plane waves and far field conditions, $\boldsymbol{C}_k$ is composed of steering vectors of the form

$$\boldsymbol{d}_k(\Phi) = \left[e^{j\Omega_k \tau_1(\Phi)}\, e^{j\Omega_k \tau_2(\Phi)} \cdots e^{j\Omega_k \tau_N(\Phi)} \right]^{\mathrm{T}}, \tag{4.67}$$

with $\Omega_k = 2\pi f_s \frac{k}{N_f}$ (sampling frequency f_s, FFT lengths N_f). Microphone signal delays τ_i depend on the direction (azimuth Φ) of the impinging wave. For simplicity, we are using a one-dimensional array with a coordinate system as shown in Fig. 4.1. This is not a restriction in general because delays τ_i can easily be calculated in a 3-dimensional coordinate system. In addition, more complicated steering vectors can be used if we apply other wave propagation models like those covering near field conditions. The structure of the optimization problem remains the same. We have to use different steering vectors only. Actually, knowledge of the sound propagation is very incomplete. Therefore, the simple steering vectors offer a convenient way to overcome this lag of information. However, a better beamformer performance can be achieved with more realistic steering vectors.

Suppose that the desired speaker direction has azimuth Φ_d and we want an array pattern null at azimuth Φ_s. Then $\boldsymbol{d}_k(\Phi_d)^{\mathrm{H}}\boldsymbol{w}_k = 1$ is the beamformer response in desired direction and $\boldsymbol{d}_k(\Phi_s)^{\mathrm{H}}\boldsymbol{w}_k = 0$ is the response in the unwanted direction. Therefore, matrix $\boldsymbol{C}_k$ is given by $\boldsymbol{C}_k = [\boldsymbol{d}_k(\Phi_d)\, \boldsymbol{d}_k(\Phi_s)]$ and vector $\boldsymbol{f}$ must be set to $\boldsymbol{f} = [1\,0]^{\mathrm{T}}$ in order to get the constraints in (4.65). We can include more array pattern nulls and extend the row dimension of matrix $\boldsymbol{C}_k$. To avoid an over-determined set of equations, the number of constraints must be less than the number N of microphones. In practice, only a few constraints should be used to obtain a good beamforming pattern with a strong main lobe and small side lobes.

We can solve the constrained optimization problem (4.65) with Lagrange multipliers by defining the cost function

$$L(\boldsymbol{w}_k, \boldsymbol{\lambda}) = \frac{1}{2}\boldsymbol{w}_k^{\mathrm{H}}\boldsymbol{S}_{\boldsymbol{x}_k\boldsymbol{x}_k}\boldsymbol{w}_k + \boldsymbol{\lambda}^{\mathrm{H}}\left(\boldsymbol{C}_k^{\mathrm{H}}\boldsymbol{w}_k - \boldsymbol{f}\right). \tag{4.68}$$

Evaluation of the gradient of this cost function yields

$$\nabla_{\boldsymbol{w}_k} L(\boldsymbol{w}_k, \boldsymbol{\lambda}) = \boldsymbol{S}_{\boldsymbol{x}_k\boldsymbol{x}_k}\boldsymbol{w}_k + \boldsymbol{C}_k\boldsymbol{\lambda}. \tag{4.69}$$

Using the gradient relationship, an iterative solution of the optimization problem on a frame by frame basis is given by

$$\boldsymbol{w}_k(m+1) = \boldsymbol{w}_k(m) - \mu_{\mathrm{LMS}}\, \nabla_{\boldsymbol{w}_k} L(\boldsymbol{w}_k, \boldsymbol{\lambda}). \tag{4.70}$$

Lagrange multiplier $\boldsymbol{\lambda}$ is obtained from (4.69) and (4.70) combined with the constraints $\boldsymbol{C}_k^{\mathrm{H}}\boldsymbol{w}_k(m+1) = \boldsymbol{f}$ (see (4.65)) according to

$$\begin{aligned}\boldsymbol{\lambda} = {} & \frac{1}{\mu_{\mathrm{LMS}}}\left(\boldsymbol{C}_k^{\mathrm{H}}\boldsymbol{C}_k\right)^{-1}\boldsymbol{C}_k^{\mathrm{H}}\boldsymbol{w}_k(m) - \left(\boldsymbol{C}_k^{\mathrm{H}}\boldsymbol{C}_k\right)^{-1}\boldsymbol{C}_k^{\mathrm{H}}\boldsymbol{S}_{\boldsymbol{x}_k\boldsymbol{x}_k}\boldsymbol{w}_k(m) \\ & - \frac{1}{\mu_{\mathrm{LMS}}}\left(\boldsymbol{C}_k^{\mathrm{H}}\boldsymbol{C}_k\right)^{-1}\boldsymbol{f}.\end{aligned} \tag{4.71}$$

Using this relationship in (4.69), we get from (4.70)

$$\boldsymbol{w}_k(m+1) = \boldsymbol{P}_k\Big[\boldsymbol{w}_k(m) - \mu_{\mathrm{LMS}}\,\boldsymbol{S}_{\boldsymbol{x}_k\boldsymbol{x}_k}\boldsymbol{w}_k(m)\Big] + \boldsymbol{w}_{\boldsymbol{c}k}, \tag{4.72}$$

with $N \times N$ matrix

$$\boldsymbol{P}_k = \boldsymbol{I} - \boldsymbol{C}_k\left(\boldsymbol{C}_k^{\mathrm{H}}\boldsymbol{C}_k\right)^{-1}\boldsymbol{C}_k^{\mathrm{H}}, \tag{4.73}$$

and $N \times 1$ vector

$$\boldsymbol{w}_{ck} = \boldsymbol{C}_k\left(\boldsymbol{C}_k^{\mathrm{H}}\boldsymbol{C}_k\right)^{-1}\boldsymbol{f}. \tag{4.74}$$

We finally arrive at the constrained LMS algorithm by replacing the unknown spatio-spectral correlation matrix by the basic estimate $\widetilde{\boldsymbol{S}}_{\boldsymbol{x}_k\boldsymbol{x}_k} = \boldsymbol{x}_k\boldsymbol{x}_k^{\mathrm{H}}$ and applying $Y_k(m) = \boldsymbol{w}_k^{\mathrm{H}}(m)\boldsymbol{x}_k(m)$ (see Fig. 4.13):

$$\boldsymbol{w}_k(m+1) = \boldsymbol{P}_k\Big[\boldsymbol{w}_k(m) - \mu_{\mathrm{LMS}}\,\boldsymbol{x}_k(m)Y_k^*(m)\Big] + \boldsymbol{w}_{ck}. \tag{4.75}$$

Although the constrained LMS algorithm can easily be implemented, the basic form given by (4.75) exhibits a suppression of the desired signal in real environments. The constraint $\boldsymbol{d}_k(\Phi_d)^{\mathrm{H}}\boldsymbol{w}_k = 1$ can hardly be met in practical situations due to microphone tolerances, microphone position errors, and most important, errors of the desired direction. If we modify the adaptive algorithm in order to achieve a large robustness against these influences, suppression of the desired signal can be avoided. By modeling the errors as uncorrelated white noise signals at the microphone inputs, we observe that the variances of these errors are amplified by $\boldsymbol{w}_k^{\mathrm{H}}\boldsymbol{w}_k$. Thus, limiting $\boldsymbol{w}_k^{\mathrm{H}}\boldsymbol{w}_k = \|\boldsymbol{w}_k\|^2$ will reduce the influence of these errors. A detailed discussion on making the Frost beamformer more robust can be found in [21].

The weight vector norm constraint can conveniently be included in the adaptive algorithm, if we split the weight vector into $\boldsymbol{w}_k(m) = \boldsymbol{v}_k(m) + \boldsymbol{w}_{ck}$ and recognize $\boldsymbol{P}_k\boldsymbol{w}_{ck} = \mathbf{0}$ (see (4.73), (4.74)). With upper bound B_k, the norm constraint can be expressed as

$$\|\boldsymbol{w}_k(m)\|^2 = \|\boldsymbol{v}_k(m)\|^2 + \|\boldsymbol{w}_{ck}\|^2 \leq B_k. \tag{4.76}$$

It follows that the norm of the variable component $\boldsymbol{v}_k(m)$ of $\boldsymbol{w}_k(m)$ must be limited by

$$\|\boldsymbol{v}_k(m)\| \leq \sqrt{B_k - \|\boldsymbol{w}_{ck}\|^2} = b_k. \tag{4.77}$$

Parameter b_k does not depend on frame index m and can be pre-computed for every frequency index k. Therefore, we get the final adaptive algorithm:

$$\text{Initialization:}\quad \boldsymbol{w}_{ck} = \boldsymbol{C}_k\left(\boldsymbol{C}_k^{\mathrm{H}}\boldsymbol{C}_k\right)^{-1}\boldsymbol{f} \tag{4.78}$$

$$\boldsymbol{P}_k = \boldsymbol{I} - \boldsymbol{C}_k\left(\boldsymbol{C}_k^{\mathrm{H}}\boldsymbol{C}_k\right)^{-1}\boldsymbol{C}_k^{\mathrm{H}} \tag{4.79}$$

$$b_k = \sqrt{B_k - \|\boldsymbol{w}_{ck}\|^2} \tag{4.80}$$

$$\boldsymbol{v}_k(0) = \mathbf{0}. \tag{4.81}$$

For each frame index m: (4.82)

$$\tilde{\boldsymbol{v}}_k(m+1) = \boldsymbol{P}_k\Big[\boldsymbol{v}_k(m) - \mu_{\mathrm{LMS}}\,\boldsymbol{x}_k(m)Y_k^*(m)\Big] \tag{4.83}$$

$$\boldsymbol{v}_k(m+1) = \begin{cases} \tilde{\boldsymbol{v}}_k(m+1) & \text{if } \|\tilde{\boldsymbol{v}}_k(m+1)\| \leq b_k \\ \dfrac{b_k\tilde{\boldsymbol{v}}_k(m+1)}{\|\tilde{\boldsymbol{v}}_k(m+1)\|} & \text{if } \|\tilde{\boldsymbol{v}}_k(m+1)\| > b_k \end{cases} \tag{4.84}$$

$$\boldsymbol{w}_k(m+1) = \boldsymbol{v}_k(m+1) + \boldsymbol{w}_{ck} \tag{4.85}$$

$$k = 0, 1, \ldots, N_f.$$

In general, this adaptive algorithm requires a substantial amount of memory due to storage of matrix $\boldsymbol{P}_k$ and vector $\boldsymbol{w}_{ck}$ for each FFT frequency index. However, for special cases like broadside arrays (azimuth $\Phi = 90°$), all of these vectors and matrices are equal. In addition, memory savings are also possible in case of symmetries regarding the location of specified nulls in the array pattern. If no specified null is present, matrix inversion in (4.73) and (4.74) reduces to scalar division because constraint matrix $\boldsymbol{C}_k$ is equal to the steering vector $\boldsymbol{d}_k(\Phi_d)$. This important case occurs at arrays for speaker tracking where fixed nulls in the beamformer pattern are not desired.

The step size μ_{LMS} of the adaptive algorithm must be selected with some care. As shown in [16], convergence of the constrained LMS algorithm is ensured if

$$0 < \mu_{\text{LMS}} < \frac{2}{3\,E\{\boldsymbol{x}_k^{\text{H}}\boldsymbol{x}_k\}}\,. \tag{4.86}$$

Therefore, a proper normalization of the step size μ_{LMS} will improve the convergence behavior of the adaptive algorithm. With such a modification, the convergence speed is independent on the signal magnitudes. In accordance to the normalized LMS algorithm, the modified weight vector update is then given by

$$\tilde{\boldsymbol{v}}_k(m+1) = \boldsymbol{P}_k \left[\boldsymbol{v}_k(m) - \frac{\mu}{\|\boldsymbol{x}_k(m)\|^2 + \varepsilon}\,\boldsymbol{x}_k(m)Y_k^*(m)\right]. \tag{4.87}$$

Typically, the new step size μ should be chosen between 0.001 and 0.02 to ensure a stable convergence of the adaptive algorithm.

Another important design parameter of the constrained LMS algorithm is the upper bound B_k. We get a sensitive superdirective array with $B_k > 10$. On the other hand, a robust delay-and-sum beamformer is obtained with small values ($B_k < 1$). In addition, B_k must be frequency dependent in order to achieve a flat beamformer frequency response not only in the exact desired direction but also at small deviations thereof. In principle, the frequency dependency of B_k can be optimized to obtain a flat frequency response. However, a tolerance analysis of perturbated arrays shows that the following set of limits works very well at a sampling frequency of $f_s = 16$ kHz [22]:

$$10\log_{10} B_k = \begin{cases} 10\text{ dB} & 0 < f \le 250\text{ Hz} \\ 8\text{ dB} & 250\text{ Hz} < f \le 450\text{ Hz} \\ 2\text{ dB} & 450\text{ Hz} < f \le 700\text{ Hz} \\ -2\text{ dB} & 700\text{ Hz} < f \le 1000\text{ Hz} \\ -4\text{ dB} & 1000\text{ Hz} < f \le 2000\text{ Hz} \\ -6\text{ dB} & 2000\text{ Hz} < f \le 4000\text{ Hz} \\ -7.5\text{ dB} & 4000\text{ Hz} < f \le 8000\text{ Hz}. \end{cases} \tag{4.88}$$

Note that the frequency index k of the N_f-point FFT is given by $k = \text{round}\left(N_f \frac{f}{f_s}\right)$.

We can combine the adaptive filterbank beamformer with a source localization subsystem as shown in Fig. 4.14. This augmented system is capable to

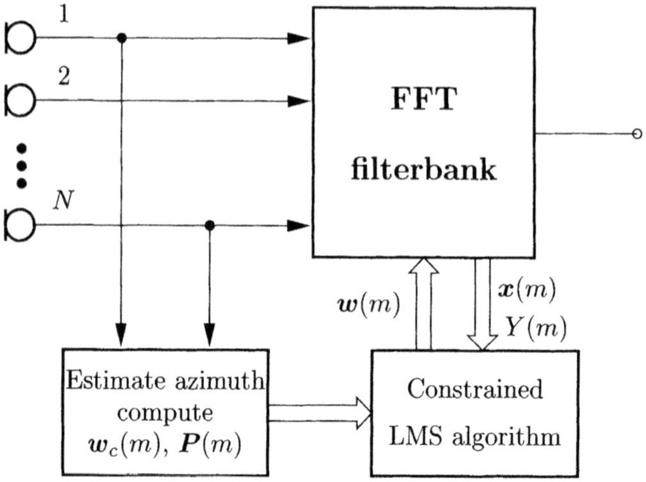

Fig. 4.14. Adaptive FFT filterbank beamformer combined with source localization to be used for automatic speaker tracking (frame index m).

focus the main lobe of the beam pattern to a moving speaker by re-computing the parameters $\boldsymbol{P}_k$, $\boldsymbol{w}_{ck}$, b_k at multiples of the frame index. With a typical frame length of 512, frames are processed every $512/4 = 128$ samples, i. e. every 8 ms at 16 kHz sampling frequency. This is the minimum time period to re-compute $\boldsymbol{P}_k$, $\boldsymbol{w}_{ck}$, b_k based on azimuth estimation. It can barely be used because the adaptive filter typically needs several 100 ms to converge. The starting solution $\boldsymbol{w}_k(0) = \boldsymbol{w}_{ck}$ corresponds to a delay-and-sum beamformer and offers an adequate beam pattern during fast movements of the speaker. Adaptation will begin after the speaker position has been settled. It should be noted, however, that there is no need to reset the adaptive filter weight vectors $\boldsymbol{w}_k$ at new azimuth estimates.

For azimuth estimation, all of the previously presented source localization algorithms can efficiently be implemented in the frequency domain. Therefore, we can directly use the already available FFTs of the microphone signals (and not the signals themselves, as shown in Fig. 4.14). We have implemented the adaptive beamformer using an array of 8 microphones, a sampling frequency of 16 kHz, and an FFT length of 512 with Hann windowing of input frames. With a frame hop size of $512/4 = 128$ samples, we obtain a filterbank oversampling by a factor of 4. This oversampling factor guarantees that distortions due to multirate filterbank processing are not audible.

Both uniform and non-uniform array geometries have been investigated. As an example, the layout of a non-uniform microphone array is sketched in Fig. 4.15. This configuration requires fewer sensors than a comparable uniform array and offers a good tradeoff between main lobe width and side lobe amplitudes over the whole frequency range. Due to the use of a linear array,

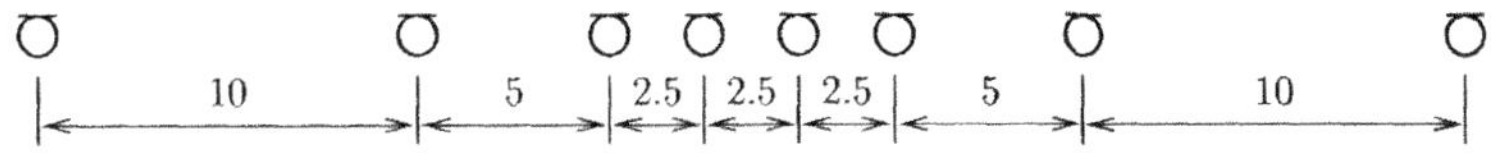

Fig. 4.15. Microphone array geometry in cm (total size 37.5 cm).

the azimuth range is confined to a 180° field-of-view. If a 360° field-of-view is required, a circular array geometry should be preferred [23].

To avoid spatial aliasing, the input signal must be bandlimited to 6400 Hz. There is no need for additional low pass filters in the microphone channels, if we set the respective frequency bins of the FFTs to zero. This will also reduce the size of vectors and matrices needed by the adaptive algorithm.

The PHAT-GCC algorithm is used for automatic speaker tracking. To provide sufficient time for convergence of the adaptive algorithm, parameters $\boldsymbol{P}_k$, $\boldsymbol{w}_{ck}$ are held constant during speech pauses and during speaker movements with changes in azimuth less than 2°.

In order to visualize the functioning of the adaptive beamformer with speaker tracking, we show a representative array pattern in Fig. 4.16 and Fig. 4.17 at a frequency of 1 kHz. We use the same speaker movement as in the source localization experiments. The speaker's position starts at broadside

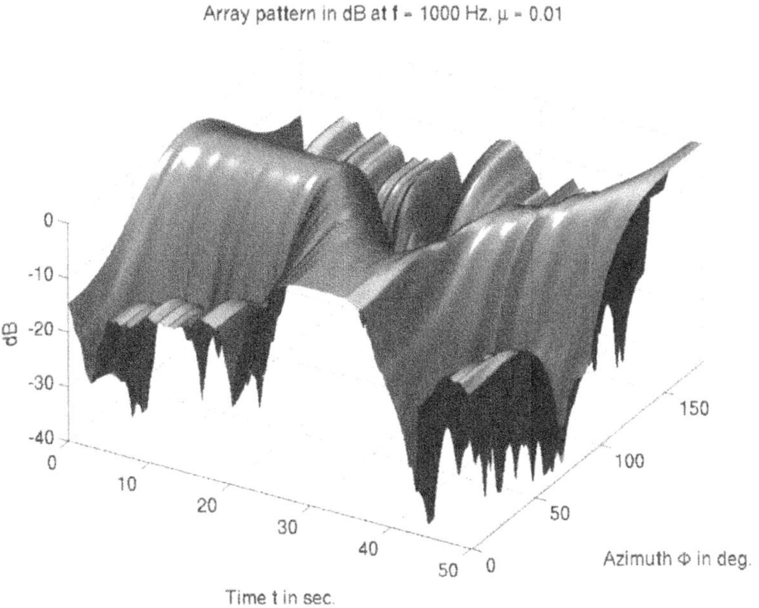

Fig. 4.16. Log-scale array pattern at $f = 1$ kHz of the adaptive beamformer automatically steered to a moving speaker.

($\Phi = 90°$), moves on towards $\Phi = 0°$, and continues to move back to $\Phi = 90°$, and finally $\Phi = 180°$. The main lobe of the array pattern follows this movement. The estimated azimuth trace is overlayed in the image plots shown in Fig. 4.17 at a frequency of 1 kHz, and in Fig. 4.18 at 3 kHz, respectively.

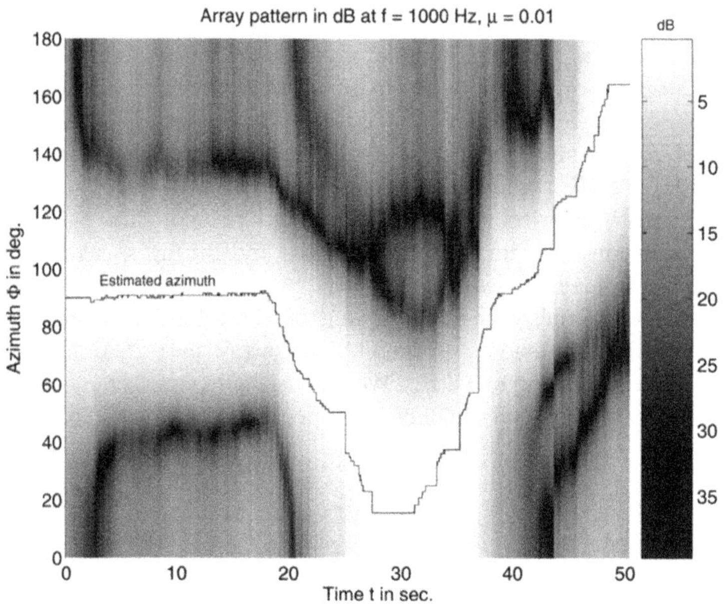

Fig. 4.17. Array pattern at $f = 1$ kHz of the automatically steered adaptive beamformer with superimposed estimated azimuth trace of a moving speaker.

The main lobe is clearly visible as a white region following the trace of the estimated azimuth. The settling period of the adaptive algorithm can be observed at the beginning where the speaker position remains constant at $\Phi = 90°$. A sharper main lobe but larger side lobe maxima are present in the array pattern at $f = 3$ kHz, as compared with the pattern at $f = 1$ kHz. This reflects the behavior of a delay-and-sum beamformer which is used as the starting solution of the adaptive algorithm. It should be noted that the beamformer shows a unity gain frequency response in desired direction. Only main lobe width and side lobe patterns change with frequency. The chopped texture of the array pattern in Fig. 4.18 is due to the step-like azimuth changes after hold operations during speech pauses.

The behavior of the adaptive beamformer depends on the input signals. The array patterns shown in Fig. 4.16, 4.17, 4.18 are computed using a single moving speaker. If we use a fixed desired direction, i.e. switch off speaker tracking, the adaptive algorithm will automatically suppress interfering sounds from other directions than the desired one. This build-in feature is due to

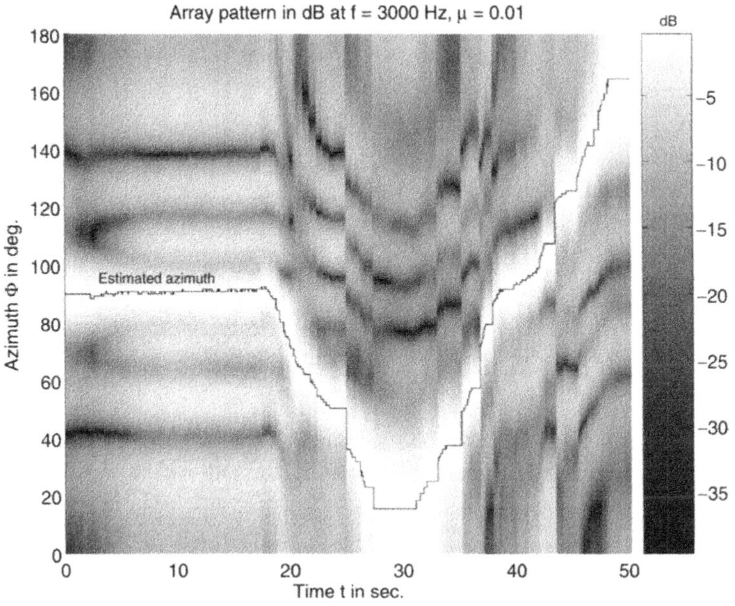

Fig. 4.18. Array pattern at $f = 3$ kHz of the automatically steered adaptive beamformer with superimposed estimated azimuth trace of a moving speaker.

the adaptive algorithm constraints by which the desired signal is emphasized. To illustrate this behavior, we show array patterns using random noise bandlimited from 300 Hz to 6400 Hz as a desired source signal. The beamformer output signal power is calculated as a function of the noise signal direction. Typical results are shown in Fig. 4.19. Four different desired directions are given. The steady-state output power is computed after the settling period of the adaptive beamformer. A sharp main lobe can be observed, especially at desired direction $\Phi = 90°$. At $\Phi = 0°$ the array is less sensitive regarding changes in the desired direction. This behavior is common to broadband adaptive beamformers based on the constrained LMS algorithm because the optimization constraint is defined for a single desired direction only. A sharp main lobe is not a disadvantage of our adaptive beamformer because the desired direction is automatically adjusted using speaker tracking.

The entire system has been simulated using a MATLAB® program which can be downloaded from the authors home page.[4] An implementation written in the C programming language runs in real-time at 16 kHz sampling frequency on any modern PC equipped with an 8 channel analog input sound system (like Terratec® EWS88MT, M-Audio® Delta 1010, or RME® Hammerfall® DSP). With CPU clock frequencies at 2 GHz, 16 microphone channels can be processed in real-time at 16 kHz sampling frequency.

[4] `www.nt.tuwien.ac.at/dspgroup/gdobling.html`

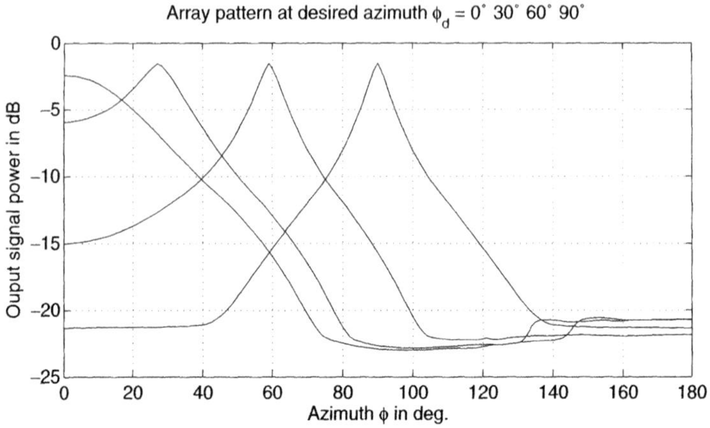

Fig. 4.19. Output signal power vs. azimuth of the adaptive array excited with random noise bandlimited from 300 Hz to 6400 Hz, and with four different desired directions.

4.7 Conclusions

We have presented an overview on different source localization techniques based on time-delay estimation using only two microphones. These algorithms are well suited for direction (azimuth) estimation and speaker tracking in real environments with moderate reverberation. The main purpose of source localization covered in this chapter is the application to speaker tracking with automatically steered microphone arrays. An efficient adaptive beamformer has been described in detail combining a multi-input overlap-add FFT filterbank, a constrained LMS algorithm, and a GCC-PHAT based source localization algorithm.

Acknowledgements

An industrial cooperation with AKG Acoustics Austria triggered my interest in acoustical beamforming and source localization. It also gave me the opportunity to consider both theoretical aspects and real-world problems in the context of microphone arrays. I particularly thank J. Granser, G. Stöbich, J. Schreiner, and G. Zach for their excellent contributions and valuable discussions.

References

[1] Brian C. J. Moore: *An Introduction to the Psychology of Hearing*, London, Great Britain: Academic Press, 2001.
[2] M. Brandstein, D. Ward (eds.): *Microphone Arrays – Signal Processing Techniques and Applications*, Berlin, Germany: Springer, 2001.
[3] S. L. Gay, J. Benesty (eds.): *Acoustic Signal Processing for Telecommunications*, Boston, MA: Kluwer, 2001.
[4] Y. Huang, J. Benesty (eds.): *Audio Signal Processing for Next-generation Multimedia Communication Systems*, Boston, MA: Kluwer, 2004.
[5] C. H. Knapp, G. C. Carter: The generalized correlation method for estimation of time delay, *IEEE Trans. Acoust. Speech Signal Process.*, **ASSP-24**(4), 320–327, 1976.
[6] Chen Liu, et al.: Localization of multiple sound sources with two microphones, *J. Acoust. Soc. Am.*, **108**(4), 1888–1905, 2000.
[7] Jacob Benesty: Adaptive eigenvalue decomposition algorithm for passive acoustic source localization, *J. Acoust. Soc. Am.*, **107**(1), 384–391, 2000.
[8] D. Li, S. E. Levinson: A linear phase unwrapping method for binaural sound source localization on a robot, in *Proc. 2002 IEEE Conf. on Robotics and Automation*, 19–23, Washington, DC, USA, 2002.
[9] I. Potamitis, H. Chen, G. Tremoulis: Tracking of multiple moving speakers with multiple microphone arrays, *IEEE Trans. Speech Audio Signal Process.*, **T-SA-12**(5), 520–529, 2004.
[10] T. Gustafsson, B. D. Rao, M. Trivedi: Source localization in reverberant environments: modeling and statistical analysis, *IEEE Trans. Speech Audio Signal Process.*, **T-SA-11**(6), 791–803, 2003.
[11] M. Omologo, P. Svaizer: Use of the crosspower-spectrum phase in acoustic event location, *IEEE Trans. Speech Audio Process.*, **T-SA-5**(3), 288–292, 1997.
[12] Jens Blauert: *Spatial hearing - revised edition, the psychophysics of human sound localization*, Cambridge MA: The MIT Press, 1996.
[13] F. A. Reed, P. L. Feintuch, N. J. Bershad: Time delay estimation using the LMS adaptive filter – static behavior, *IEEE. Trans. Acoust. Speech Signal Process.*, **ASSP-29**(3), 561–571, 1981.
[14] E. A. Ferrara: Fast implementation of LMS adaptive filters, *IEEE Trans. Acoust. Speech Signal Process.*, **ASSP-28**(4), 474–475, 1980.
[15] D. Mansour, A. H. Gray, Jr.: Unconstrained frequency-domain adaptive filter, *IEEE Trans. Acoust. Speech Signal Process.*, **ASSP-30**(5), 726–734, 1982.
[16] O. L. Frost, III: An algorithm for linearly constrained adaptive array processing, *Proc. IEEE*, **60**(8), 926–935, 1972.
[17] C. Marro, Y. Mahieux, K. U. Simmer: Analysis of noise reduction and dereverberation techniques based on microphone arrays with postfiltering, *IEEE Trans. Speech Audio Process.*, **T-SA-6**(3), 240–259, 1998.

[18] K. U. Simmer, J. Bitzer, C. Marro: Post-filtering techniques, in M. Brandstein, D. Ward (eds.), *Microphone Arrays – Signal Processing Techniques and Applications*, 39–60, Berlin, Germany: Springer, 2001.
[19] I. A. McCowan, H. Bourlard: Microphone array post-filter based on noise field coherence, *IEEE Trans. Speech Audio Process.*, **T-SA-11**(6), 709–716, 2003.
[20] R. E. Crochiere, L. R. Rabiner: *Multirate Digital Signal Processing,* Englewood Cliffs, NJ: Prentice Hall, 1983.
[21] H. Cox, R. M. Zeskind, M. M. Owen: Robust adaptive beamforming, *IEEE Trans. Acoust. Speech Signal Process.*, **ASSP-35**(10), 1365–1376, 1987.
[22] G. Stöbich: *Entwurf und Simulation eines adaptiven, zweidimensionalen Mikrofonarrays,* Vienna, Austria: Diploma Thesis, Vienna University of Technology, 2001 (in German).
[23] H. Teutsch, W. Kellermann: EB-ESPRIT: 2D localization of multiple wideband acoustic sources using eigen-beams, *Proc. ICASSP '05*, **3**, 89–92, Philadelphia, PA, USA, 2005.

Part III

Echo Cancellation

5

Adaptive Algorithms for the Identification of Sparse Impulse Responses

Jacob Benesty[1], Yiteng (Arden) Huang[2], Jingdong Chen[2], and Patrick A. Naylor[3]

[1] Université du Québec, INRS-EMT, Montreal, Canada
[2] Bell Labs, Lucent Technologies, Murray Hill, USA
[3] Imperial College, London, UK

Sparse impulse responses are encountered in many applications: network and acoustic echo cancellation, feedback cancellation in hearing aids, blind identification of acoustic impulse responses for time delay estimation, source localization, and dereverberation, etc. Recently, several adaptive algorithms have been proposed, in different contexts, which take this important information (i.e. sparseness) into account. As a result, these new adaptive filters perform (in terms of initial convergence and tracking) much better than the classical stochastic gradient, or LMS, algorithm. In this book chapter, we give an overview of the most important adaptive algorithms developed for sparse impulse responses. We show how they can be derived. We also show how they are linked to each other. Finally, we give new directions and explore how far we can go in improving performances.

5.1 Introduction

An impulse response that is sparse has a small percentage of its components with a significant magnitude while the rest are zero or small. Another definition could be the following: an impulse response is sparse if a large fraction of its energy is concentrated in a small fraction of its duration. We find sparse impulse responses in many important applications such as network and acoustic echo cancellation, feedback cancellation in hearing aids, blind identification of acoustic impulse responses for time delay estimation, source localization, and dereverberation, etc.

Classical and most used algorithms such as the normalized least-mean-square (NLMS) [1] or recursive least-squares (RLS) [2] do not take into account whether the impulse responses they try to identify are sparse or not. Intuitively however, it seems possible to improve the performance of the NLMS algorithm, for example, if the target is sparse.

Perhaps one of the first persons who exploited this intuition algorithmically was Duttweiler in the context of network echo cancellation involving a hybrid transformer in conjunction with variable network delay and where impulse responses are clearly sparse. The so-called proportionate NLMS (PNLMS) algorithm was then introduced [3]. This new algorithm converges and tracks much faster than the NLMS algorithm when the impulse response that we need to identify is sparse. PNLMS and other more sophisticated versions such as improved PNLMS (IPNLMS) [4] are success stories since they are now used in many products.

Recently, another variant of the LMS algorithm, called the exponentiated gradient algorithm with positive and negative weights (EG± algorithm), was proposed by Kivinen and Warmuth in the context of computational learning theory [5]. This new algorithm also converges much faster than the LMS algorithm when the target is sparse. The EG± algorithm has the nice feature that its update rule takes advantage of the sparseness of the impulse response to speed up its initial convergence and to improve its tracking abilities compared to LMS. In [6], a general expression of the mean squared error (MSE) is derived for the EG± algorithm showing that for sparse impulse responses, the EG± algorithm, like PNLMS, converges more quickly than the LMS for a given asymptotic MSE. Even though the EG± and PNLMS algorithms may look very different, clearly they must be linked somehow. It is quite remarkable that two equivalent algorithms, as it will be shown later, were proposed in two completely different contexts.

There are two fundamental ways to update the coefficients of an adaptive filter $\boldsymbol{h}(n)$. The linear update:

$$\boldsymbol{h}(n) = \boldsymbol{M}_1(n)\boldsymbol{h}(n-1) + \boldsymbol{m}_2(n), \tag{5.1}$$

where $\boldsymbol{M}_1(n)$ and $\boldsymbol{m}_2(n)$ are respectively a matrix and a vector independent of $\boldsymbol{h}(n-1)$, and the nonlinear update:

$$\boldsymbol{h}(n) = \boldsymbol{M}_1\big[\boldsymbol{h}(n-1)\big]\boldsymbol{h}(n-1) + \boldsymbol{m}_2(n), \tag{5.2}$$

where this time, as indicated, $\boldsymbol{M}_1[\boldsymbol{h}(n-1)]$ depends on $\boldsymbol{h}(n-1)$. All classical algorithms such as NLMS and RLS can be deduced from (5.1) and new ones can be derived from (5.2). This view gives already an answer to the important question: how can this *a priori* information (sparseness) be taken into account to improve convergence and tracking of adaptive algorithms? The study of this question is the main objective of this chapter.

5.2 Notation and Definitions

In derivations and descriptions, the following notation is used:

$$
\begin{aligned}
x(n) &= \text{ input signal},\\
y(n) &= \boldsymbol{h}_\mathrm{t}^T \boldsymbol{x}(n) + w(n), \text{ output signal plus noise},\\
\boldsymbol{x}(n) &= \left[x(n)\ x(n-1)\ \cdots\ x(n-L+1)\right]^T, \text{ excitation vector},\\
\boldsymbol{h}_\mathrm{t} &= \left[h_{\mathrm{t},0}\ h_{\mathrm{t},1}\ \cdots\ h_{\mathrm{t},L-1}\right]^T, \text{ true impulse response},\\
\boldsymbol{h}(n) &= \left[h_0(n)\ h_1(n)\ \cdots\ h_{L-1}(n)\right]^T, \text{ estimated impulse response.}
\end{aligned}
$$

Here L is the length of the adaptive filter, n is the time index, and superscript $(\cdot)^T$ denotes transpose of a vector or a matrix.

We now give some important definitions that will be used in the rest of this chapter:

$$
\begin{aligned}
e(n) &= y(n) - \widehat{y}(n)\\
&= y(n) - \boldsymbol{x}^T(n)\boldsymbol{h}(n-1), \ \textit{a priori} \text{ error signal}, \qquad (5.3)\\
\epsilon(n) &= y(n) - \boldsymbol{x}^T(n)\boldsymbol{h}(n), \ \textit{a posteriori} \text{ error signal}, \qquad (5.4)\\
e_\mathrm{n}(n) &= \left[\boldsymbol{x}^T(n)\boldsymbol{G}(n-1)\boldsymbol{x}(n)\right]^{-1/2} e(n), \qquad (5.5)\\
&\quad \text{normalized } \textit{a priori} \text{ error signal},\\
\epsilon_\mathrm{n}(n) &= \left[\boldsymbol{x}^T(n)\boldsymbol{G}(n-1)\boldsymbol{x}(n)\right]^{-1/2} \epsilon(n), \qquad (5.6)\\
&\quad \text{normalized } \textit{a posteriori} \text{ error signal},\\
\boldsymbol{e}(n) &= \boldsymbol{y}(n) - \boldsymbol{X}^T(n)\boldsymbol{h}(n-1), \ \textit{a priori} \text{ error signal vector}, \qquad (5.7)\\
\boldsymbol{\epsilon}(n) &= \boldsymbol{y}(n) - \boldsymbol{X}^T(n)\boldsymbol{h}(n), \ \textit{a posteriori} \text{ error signal vector}, \qquad (5.8)\\
\boldsymbol{e}_\mathrm{n}(n) &= \left[\boldsymbol{X}^T(n)\boldsymbol{G}(n-1)\boldsymbol{X}(n)\right]^{-1/2} \boldsymbol{e}(n), \qquad (5.9)\\
&\quad \text{normalized } \textit{a priori} \text{ error signal vector},\\
\boldsymbol{\epsilon}_\mathrm{n}(n) &= \left[\boldsymbol{X}^T(n)\boldsymbol{G}(n-1)\boldsymbol{X}(n)\right]^{-1/2} \boldsymbol{\epsilon}(n), \qquad (5.10)\\
&\quad \text{normalized } \textit{a posteriori} \text{ error signal vector},
\end{aligned}
$$

where

$$
\boldsymbol{y}(n) = \left[y(n)\ y(n-1)\ \cdots\ y(n-P+1)\right]^T
$$

is a vector containing the P more recent samples of the output signal $y(n)$,

$$
\boldsymbol{X}(n) = \left[\boldsymbol{x}(n)\ \boldsymbol{x}(n-1)\ \cdots\ \boldsymbol{x}(n-P+1)\right]
$$

is an $L \times P$ matrix of the input signal samples $x(n)$, and

$$
\boldsymbol{G}(n-1) = \mathrm{diag}\left\{ g_0(n-1)\ g_1(n-1)\ \cdots\ g_{L-1}(n-1) \right\} \qquad (5.11)
$$

is an $L \times L$ diagonal matrix, where $g_l(n-1) > 0$, $\forall n, l$. This matrix is context dependent but is usually a function of $\boldsymbol{h}(n-1)$.

5.3 Sparseness Measure

Before presenting different algorithms that have the potential to work well when impulse responses are sparse, we need first to agree somehow on what we mean by sparse. But is it possible to quantify sparseness with a number? The answer to this question is not obvious since the definition of sparseness is not obvious itself. We will argue, though, that the following measure:

$$\xi(\boldsymbol{h}) = \frac{L}{L-\sqrt{L}}\left(1 - \frac{\|\boldsymbol{h}\|_1}{\sqrt{L}\|\boldsymbol{h}\|_2}\right) \tag{5.12}$$

is a reasonable one for evaluating the sparseness of a filter $\boldsymbol{h}$ of length $L > 1$, where $\|\cdot\|_1$ and $\|\cdot\|_2$ are the 1- and 2-norm vectors, respectively. The same definition was proposed in [7].

Consider the Dirac filter,

$$\boldsymbol{h}_{\mathrm{d}} = \begin{bmatrix} 1\, 0 \cdots 0 \end{bmatrix}^T, \tag{5.13}$$

the uniform filter,

$$\boldsymbol{h}_{\mathrm{u}} = \begin{bmatrix} 1\, 1 \cdots 1 \end{bmatrix}^T, \tag{5.14}$$

and the exponentially decaying filter,

$$\boldsymbol{h}_{\mathrm{e}} = \left[1 \ \exp\left(-\tfrac{1}{\beta}\right) \ \cdots \ \exp\left(-\tfrac{L-1}{\beta}\right) \right]^T, \tag{5.15}$$

where β is a positive decay constant. The Dirac and uniform filters are particular cases of $\boldsymbol{h}_{\mathrm{e}}$. Indeed:

$$\lim_{\beta \to 0} \boldsymbol{h}_{\mathrm{e}} = \boldsymbol{h}_{\mathrm{d}}, \tag{5.16}$$

$$\lim_{\beta \to \infty} \boldsymbol{h}_{\mathrm{e}} = \boldsymbol{h}_{\mathrm{u}}. \tag{5.17}$$

While the Dirac filter is the sparsest of all possible impulse responses, the uniform filter is the most dispersive one. The filter $\boldsymbol{h}_{\mathrm{e}}$ is a good model of acoustic impulse responses where β depends on the reverberation time. For a long reverberation time (large β), $\boldsymbol{h}_{\mathrm{e}}$ will decay slowly while for a short reverberation time (small β), $\boldsymbol{h}_{\mathrm{e}}$ will decay rapidly. Having this in mind, we now give some important properties.

Properties:

$$\text{(a)} \quad 0 \leq \xi(\boldsymbol{h}) \leq 1, \tag{5.18}$$

$$\text{(b)} \quad \forall\, a \neq 0,\ \xi(a\boldsymbol{h}) = \xi(\boldsymbol{h}), \tag{5.19}$$

$$\text{(c)} \quad \xi(\boldsymbol{h}_{\mathrm{d}}) = 1, \tag{5.20}$$

$$\text{(d)} \quad \xi(\boldsymbol{h}_{\mathrm{u}}) = 0. \tag{5.21}$$

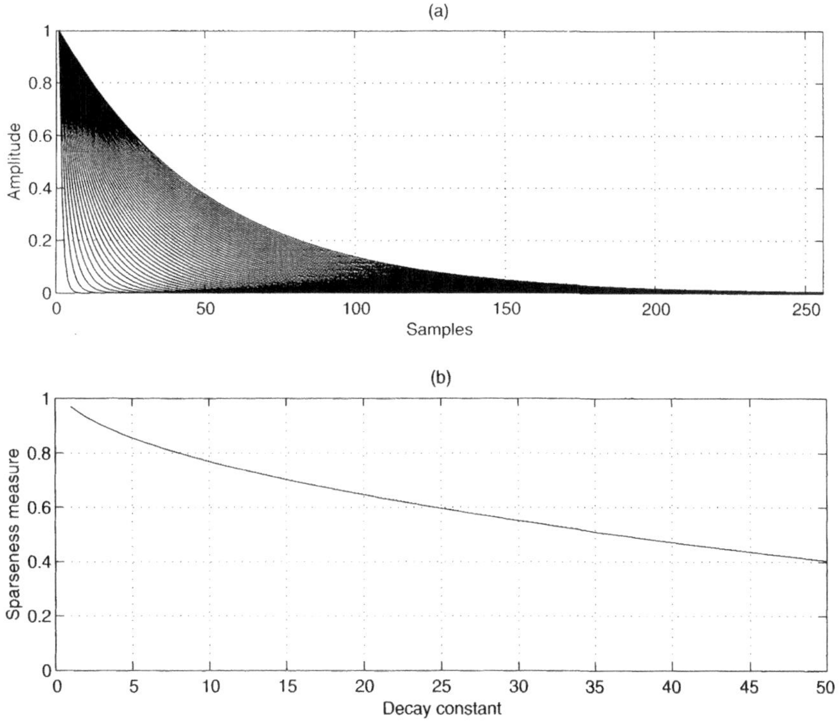

Fig. 5.1. (a) Impulse responses $\boldsymbol{h}_\mathrm{e}$ of length $L = 256$ for different values of the decay constant β (from 1 to 50). (b) Sparseness measure for $\boldsymbol{h}_\mathrm{e}$ as a function of the decay constant, β.

Proofs: We only show property (a) since (b), (c), and (d) are obvious. It is easy to check that $\|\boldsymbol{h}\|_2 \leq \|\boldsymbol{h}\|_1$, which implies that $\xi(\boldsymbol{h}) \leq 1$. It can be shown (see [21], for example) that:

$$\|\boldsymbol{h}\|_1 \leq \sqrt{L}\|\boldsymbol{h}\|_2. \tag{5.22}$$

As a result, $\xi(\boldsymbol{h}) \geq 0$.

We see from these properties that the measure is bounded and is not affected by a scaling factor. Furthermore, the closer the measure is to 1 (resp. 0), the sparser (resp. more dispersive) is the impulse response.

To further confirm that (5.12) is a good measure of sparseness, Fig. 5.1 illustrates what happens for a class of exponentially decaying filters $\boldsymbol{h}_\mathrm{e}$ of length $L = 256$. Figure 5.1(a) shows the amplitude of all those filters from $\beta = 1$ to $\beta = 50$ and Fig. 5.1(b) gives the corresponding values of $\xi(\boldsymbol{h}_\mathrm{e})$. For $\beta = 1$ (very sparse filter), $\xi(\boldsymbol{h}_\mathrm{e}) \approx 0.97$ and for $\beta = 50$ (quite dispersive filter), $\xi(\boldsymbol{h}_\mathrm{e}) \approx 0.4$. From values of the decay constant between 1 and 50, the sparseness measure decreases smoothly and follows well this decaying.

5.4 The NLMS, PNLMS, and IPNLMS Algorithms

In this section, we briefly explain the normalized least-mean-square (NLMS), proportionate NLMS (PNLMS), and improved PNLMS (IPNLMS) algorithms. Even though NLMS and IPNLMS may seem coming from the same family of adaptive filters, this is not really the case and the similarity between the two is quite deceiving.

The role of an adaptive filter, $\boldsymbol{h}(n)$, is to estimate the true impulse response, $\boldsymbol{h}_{\mathrm{t}}$, at each iteration time, n, when new samples, $x(n)$ and $y(n)$, are available. Depending on the algorithm used for this task, convergence, tracking, complexity, robustness to noise, etc, can be very different. One of the most popular adaptive filters in signal processing applications is the NLMS [1], [2], due to its simplicity and robustness. But its convergence and tracking are slow in general, especially for long impulse responses. In many situations where an adaptive algorithm is required, convergence and tracking are critical for a good performance of the entire system. While in the NLMS, the adaptation step is the same for all components of the filter, in the PNLMS [3], an adaptive individual step size is assigned to each filter coefficient. The step sizes are calculated from the last estimate of the filter coefficients in such a way that a larger coefficient receives a larger increment, thus increasing the convergence rate of that coefficient. This has the effect that active coefficients are adjusted faster than non-active coefficients (i.e. small or zero coefficients). Hence, PNLMS converges much faster than NLMS for sparse impulse responses. Unfortunately, PNLMS behaves much worse than NLMS when the impulse response is not sparse. This problem is due to the fact that the proportionate update is not very well refined. In [4], an IPNLMS was proposed where the adaptive individual step size has a better balance between the fixed step size of NLMS and the large amount of proportionality in PNLMS. As a result, IPNLMS always converges and tracks better than NLMS and PNLMS, however sparse the impulse response.

The error signal and the coefficient update equation of the three previously discussed algorithms can be written as:

$$e(n) = y(n) - \boldsymbol{h}^{\mathrm{T}}(n-1)\boldsymbol{x}(n), \tag{5.23}$$

$$\boldsymbol{h}(n) = \boldsymbol{h}(n-1) + \frac{\mu \boldsymbol{G}(n-1)\boldsymbol{x}(n)e(n)}{\boldsymbol{x}^{\mathrm{T}}(n)\boldsymbol{G}(n-1)\boldsymbol{x}(n) + \delta}, \tag{5.24}$$

where

$$\boldsymbol{G}(n-1) = \mathrm{diag}\left\{\, g_0(n-1)\ g_1(n-1)\ \cdots\ g_{L-1}(n-1)\,\right\} \tag{5.25}$$

is a diagonal matrix that adjusts the step sizes of the individual taps of the filter, μ ($0 < \mu < 1$) is the overall step-size factor, and δ is the regularization parameter.

The NLMS algorithm is obtained by taking:

$$\boldsymbol{G}(n) = \boldsymbol{I}, \tag{5.26}$$

$$\delta = \delta_{\text{NLMS}} = \text{cst} \cdot \sigma_x^2, \tag{5.27}$$

where $\boldsymbol{I}$, σ_x^2, and cst are the identity matrix, the power of the signal $x(n)$, and a small positive constant, respectively.

In the PNLMS, the diagonal elements of $\boldsymbol{G}(n) = \boldsymbol{G}_{\text{p}}(n)$ are calculated as follows [3]:

$$\gamma_{\text{p},l}(n) = \max\left\{\rho \max\left[\delta_{\text{p}},\ |h_0(n)|,\ \cdots,\ |h_{L-1}(n)|\right],\ |h_l(n)|\right\}, \tag{5.28}$$

$$g_{\text{p},l}(n) = \frac{\gamma_{\text{p},l}(n)}{\|\boldsymbol{\gamma}_{\text{p}}(n)\|_1},\ 0 \le l \le L-1, \tag{5.29}$$

where

$$\boldsymbol{\gamma}_{\text{p}}(n) = \left[\gamma_{\text{p},0}(n)\ \gamma_{\text{p},1}(n)\ \cdots\ \gamma_{\text{p},L-1}(n)\right]^{\text{T}}.$$

Parameters δ_{p} and ρ are positive numbers with typical values $\delta_{\text{p}} = 0.01$, $\rho = 0.01$. The first term in (5.28), ρ, prevents $h_l(n)$ from stalling when its magnitude is much smaller than the magnitude of the largest coefficient and δ_{p} regularizes the updating when all coefficients are zero at initialization. For the regularization parameter, we usually choose:

$$\delta_{\text{PNLMS}} = \delta_{\text{NLMS}}/L. \tag{5.30}$$

For the IPNLMS algorithm, the diagonal matrix, $\boldsymbol{G}(n) = \boldsymbol{G}_{\text{ip}}(n)$, is computed in a more elegant way [4]:

$$\gamma_{\text{ip},l}(n) = (1-\alpha)\frac{\|\boldsymbol{h}(n)\|_1}{L} + (1+\alpha)|h_l(n)|, \tag{5.31}$$

$$\begin{aligned} g_{\text{ip},l}(n) &= \frac{\gamma_{\text{ip},l}(n)}{\|\boldsymbol{\gamma}_{\text{ip}}(n)\|_1} \\ &= \frac{1-\alpha}{2L} + (1+\alpha)\frac{|h_l(n)|}{2\|\boldsymbol{h}(n)\|_1},\ 0 \le l \le L-1, \end{aligned} \tag{5.32}$$

where α $(-1 \le \alpha < 1)$ is a parameter that controls the amount of proportionality in the IPNLMS. For $\alpha = -1$, it can easily be checked that the IPNLMS and NLMS algorithms are identical. For α close to 1, IPNLMS behaves like PNLMS. In practice, a good choice for α is -0.5 or 0. With this choice and in simulations, IPNLMS always performs better than NLMS and PNLMS. As for the regularization parameter, it should be taken as:

$$\delta_{\text{IPNLMS}} = \frac{1-\alpha}{2L}\delta_{\text{NLMS}}. \tag{5.33}$$

The IPNLMS algorithm is summarized in Table 5.1.

Before finishing this section, it is worth mentioning another variant of PNLMS, called PNLMS++ [8]. In this algorithm, the adaptation of the filter

Table 5.1. The IPNLMS algorithm.

Initialization:

$$h_l(0) = 0, \; l = 0, 1, \cdots, L-1$$

Parameters:

$$-1 \leq \alpha < 1$$

$$0 < \mu < 1, \quad \delta_{\text{IPNLMS}} = \text{cst} \cdot \sigma_x^2 \frac{1-\alpha}{2L}$$

$\varepsilon > 0$ (very small number to avoid division by zero)

Error:

$$e(n) = y(n) - \boldsymbol{h}^T(n-1)\boldsymbol{x}(n)$$

Update:

$$g_{\text{ip},l}(n-1) = \frac{1-\alpha}{2L} + (1+\alpha)\frac{|h_l(n-1)|}{2\|\boldsymbol{h}(n-1)\|_1 + \varepsilon}$$

$$\mu(n) = \frac{\mu}{\sum_{j=0}^{L-1} x^2(n-j) g_{\text{ip},j}(n-1) + \delta_{\text{IPNLMS}}}$$

$$h_l(n) = h_l(n-1) + \mu(n) g_{\text{ip},l}(n-1) x(n-l) e(n)$$

$$l = 0, 1, \cdots, L-1$$

coefficients alternates between NLMS and PNLMS; as a result, PNLMS++ seems a little bit less sensitive to the assumption of a sparse impulse response than PNLMS. For a nice overview on this class of adaptive filters, see [9].

Now, a natural question arises: is it possible to find an optimization criterion that includes the sparseness information?

5.5 Universal Criterion

In this section, we show how to derive different classes of adaptive filters. As explained in [5], a reasonable adaptive algorithm must find a good balance between its needs to be conservative (retain the information it has acquired in preceding iterations) and corrective (make sure that with new information, the accuracy of the solution is increased). For that, we give a universal criterion that is the sum of two terms: one of them is a distance between the old and new weight vectors (and depending on how we define this distance, we obtain different update rules) and the other one depends on the *a posteriori* error signal. Therefore, according to this principle, one easy way to find adaptive filters that adjust the new weight vector, $\boldsymbol{h}(n)$, from the old one, $\boldsymbol{h}(n-1)$, is to minimize the following function:

$$J(n) = d\left[\boldsymbol{h}(n), \boldsymbol{h}(n-1)\right] + \boldsymbol{\epsilon}_{\mathrm{n}}^{T}(n)\boldsymbol{\epsilon}_{\mathrm{n}}(n), \tag{5.34}$$

where $d\left[\boldsymbol{h}(n), \boldsymbol{h}(n-1)\right]$ is some measure of distance from the old to the new weight vectors. Differentiating $J(n)$ with respect to $\boldsymbol{h}(n)$ and setting the resulting vector to zero, we can see that any adaptive algorithm has the form:

$$\begin{aligned} 2\boldsymbol{P}_x(n)\left[\boldsymbol{h}(n) - \boldsymbol{h}(n-1)\right] + \frac{\partial d\left[\boldsymbol{h}(n), \boldsymbol{h}(n-1)\right]}{\partial \boldsymbol{h}(n)} & \\ = 2\boldsymbol{X}(n)\left[\boldsymbol{X}^T(n)\boldsymbol{G}(n-1)\boldsymbol{X}(n)\right]^{-1} \boldsymbol{e}(n), & \end{aligned} \tag{5.35}$$

where

$$\boldsymbol{P}_x(n) = \boldsymbol{X}(n)\left[\boldsymbol{X}^T(n)\boldsymbol{G}(n-1)\boldsymbol{X}(n)\right]^{-1} \boldsymbol{X}^T(n) \tag{5.36}$$

is a projection matrix for $\boldsymbol{G}(n-1) = \boldsymbol{I}$. It has the two properties:

$$\boldsymbol{P}_x(n)\boldsymbol{G}(n-1)\boldsymbol{X}(n) = \boldsymbol{X}(n), \tag{5.37}$$

$$\boldsymbol{P}_x(n)\boldsymbol{G}(n-1)\boldsymbol{P}_x(n) = \boldsymbol{P}_x(n). \tag{5.38}$$

Clearly, the choice of the distance $d\left[\boldsymbol{h}(n), \boldsymbol{h}(n-1)\right]$ is significant. Depending on how we choose it, we may have a linear or nonlinear update equation with respect to the weight vector.

5.5.1 Linear Update

In this important category of adaptive filters, we take $\boldsymbol{G}(n-1) = \boldsymbol{I}$ and we choose for the distance:

$$d\big[\boldsymbol{h}(n), \boldsymbol{h}(n-1)\big] = \big[\boldsymbol{h}(n) - \boldsymbol{h}(n-1)\big]^T \boldsymbol{Q}_x(n)\big[\boldsymbol{h}(n) - \boldsymbol{h}(n-1)\big], \tag{5.39}$$

where the symmetric matrix $\boldsymbol{Q}_x(n)$ is positive definite and depends on the input signal $x(n)$ only. Using (5.39) in (5.35), we obtain the update equation:

$$\boldsymbol{h}(n) = \boldsymbol{h}(n-1) + \left[\boldsymbol{P}_x(n) + \boldsymbol{Q}_x(n)\right]^{-1} \boldsymbol{X}(n)\left[\boldsymbol{X}^T(n)\boldsymbol{X}(n)\right]^{-1} \boldsymbol{e}(n). \tag{5.40}$$

There are two important things to pay attention to. First, if we replace $\boldsymbol{e}(n)$ by its value in (5.40), we can see that this equation is updated linearly with respect to the estimation filter $\boldsymbol{h}(n-1)$. Second, the choice of the matrix $\boldsymbol{Q}_x(n)$ will lead to well-known algorithms and even to new ones.

Let's take:

$$\boldsymbol{Q}_x(n) = \mu^{-1}\boldsymbol{I} - \boldsymbol{P}_x(n). \tag{5.41}$$

It can easily be checked that $\boldsymbol{Q}_x(n)$ is positive definite if $0 < \mu < 1$. Replacing (5.41) in (5.40), we get the affine projection algorithm (APA) [10]:

$$\boldsymbol{h}(n) = \boldsymbol{h}(n-1) + \mu\boldsymbol{X}(n)\left[\boldsymbol{X}^T(n)\boldsymbol{X}(n)\right]^{-1} \boldsymbol{e}(n). \tag{5.42}$$

For the particular case $P = 1$, we obviously have the (non-regularized) NLMS algorithm [2].

Consider a recursive estimation of the input signal correlation matrix:

$$\boldsymbol{R}(n) = \lambda \boldsymbol{R}(n-1) + \boldsymbol{x}(n)\boldsymbol{x}^{\mathrm{T}}(n), \tag{5.43}$$

where λ $(0 < \lambda < 1)$ is an exponential forgetting factor. With $P = 1$ and plugging

$$\boldsymbol{Q}_x(n) = \frac{\boldsymbol{R}(n)}{\boldsymbol{x}^{\mathrm{T}}(n)\boldsymbol{x}(n)} - \boldsymbol{P}_x(n) \tag{5.44}$$

in (5.40), we obtain the recursive least-squares (RLS) algorithm [2]:

$$\boldsymbol{h}(n) = \boldsymbol{h}(n-1) + \boldsymbol{R}^{-1}(n)\boldsymbol{x}(n)e(n). \tag{5.45}$$

Thus, so far, we have seen how to deduce the three most classical adaptive filters that can be found in the literature: NLMS, APA, and RLS.

5.5.2 nonlinear Update

In this second category of adaptive filters, $\boldsymbol{G}(n-1)$ is now a function of the filter $\boldsymbol{h}(n-1)$ and the distance is changed accordingly:

$$d\left[\boldsymbol{h}(n), \boldsymbol{h}(n-1)\right] = \left[\boldsymbol{h}(n) - \boldsymbol{h}(n-1)\right]^{\mathrm{T}} \boldsymbol{Q}_x\left[\boldsymbol{G}(n-1)\right]\left[\boldsymbol{h}(n) - \boldsymbol{h}(n-1)\right], \tag{5.46}$$

where now the symmetric positive-definite matrix $\boldsymbol{Q}_x[\boldsymbol{G}(n-1)]$ is not only a function of the input signal $x(n)$ but also of $\boldsymbol{G}(n-1)$ [and indirectly of $\boldsymbol{h}(n-1)$] as well. Minimizing (5.34) with (5.46), we obtain a general form of the adaptive algorithm:

$$\begin{aligned}\boldsymbol{h}(n) = \boldsymbol{h}(n-1) &+ \left\{\boldsymbol{P}_x(n) + \boldsymbol{Q}_x\left[\boldsymbol{G}(n-1)\right]\right\}^{-1} \boldsymbol{X}(n) \\ &\cdot \left[\boldsymbol{X}^{\mathrm{T}}(n)\boldsymbol{G}(n-1)\boldsymbol{X}(n)\right]^{-1} e(n),\end{aligned} \tag{5.47}$$

where $\boldsymbol{P}_x(n)$ is defined in (5.36). The main difference between expressions (5.40) and (5.47) is that the former one is linearly updated with respect to the estimation filter $\boldsymbol{h}(n-1)$ while the latter one is not since $\boldsymbol{G}(n-1)$ is a function of $\boldsymbol{h}(n-1)$.

With the distance defined in (5.46), the parameter space is a curved manifold (non Euclidean). Such a space is a Riemannian space. The $L \times L$ positive-definite matrix $\boldsymbol{Q}_x[\boldsymbol{G}(n-1)]$ is called the *Riemannian metric tensor* and it depends in general on $\boldsymbol{h}(n-1)$. The Riemannian metric tensor characterizes the intrinsic curvature of a particular manifold in L-dimensional space.

Taking $P = 1$ and

$$\boldsymbol{Q}_x(n) = \mu^{-1}\boldsymbol{G}^{-1}(n-1) - \boldsymbol{P}_x(n), \tag{5.48}$$

we get the natural gradient (NG) algorithm proposed by Amari [11]:

$$\boldsymbol{h}(n) = \boldsymbol{h}(n-1) + \frac{\mu \boldsymbol{G}(n-1)\boldsymbol{x}(n)e(n)}{\boldsymbol{x}^{\mathrm{T}}(n)\boldsymbol{G}(n-1)\boldsymbol{x}(n)}. \tag{5.49}$$

Depending on the choice of $\boldsymbol{G}(n-1)$, we may obtain PNLMS [3], IPNLMS [4], or other proportionate versions of NLMS [12], [13], [14].

Now with $P > 1$ and with the same definition of $\boldsymbol{Q}_x(n)$ as in (5.48), we have the natural APA (NAPA):

$$\boldsymbol{h}(n) = \boldsymbol{h}(n-1) + \mu \boldsymbol{G}(n-1)\boldsymbol{X}(n)\left[\boldsymbol{X}^{\mathrm{T}}(n)\boldsymbol{G}(n-1)\boldsymbol{X}(n)\right]^{-1}\boldsymbol{e}(n). \tag{5.50}$$

Again, the choice of $\boldsymbol{G}(n-1)$ leads to different interesting algorithms such as proportionate APA (PAPA) [15] or improved PAPA (IPAPA) [16].

Following the same philosophy, we may derive the natural RLS (NRLS) algorithm:

$$\boldsymbol{h}(n) = \boldsymbol{h}(n-1) + \boldsymbol{G}^{1/2}(n-1)\boldsymbol{R}_h^{-1}(n)\boldsymbol{G}^{1/2}(n-1)\boldsymbol{x}(n)e(n), \tag{5.51}$$

where

$$\boldsymbol{R}_h(n) = \lambda \boldsymbol{R}_h(n-1) + \boldsymbol{x}(n)\boldsymbol{G}(n-1)\boldsymbol{x}^{\mathrm{T}}(n) \tag{5.52}$$

is an estimate of the input signal correlation matrix.

To summarize this subsection, we can say that this relatively new nonlinear framework for the update of the coefficients of the filter is very promising since it seems to fit very well the identification of sparse impulse responses: by taking this information into account, the adjustment of the coefficients of the estimated filter is done in a non-uniform manner (e.g., components with large magnitude have a larger step size than components with small magnitude) and the performance of the adaptive filter can be greatly improved. In the next section, we present another class of algorithms having the same feature.

5.6 Exponentiated Gradient Algorithms

The exponentiated gradient (EG) algorithms were first proposed by Kivinen and Warmuth in the context of computational learning theory [5]. These algorithms are highly nonlinear and can be easily derived from the criterion explained in Section 5.5, by simply using for the distance $d_{\mathrm{re}}\left[\boldsymbol{h}(n), \boldsymbol{h}(n-1)\right]$, the *relative entropy* also known as *Kullback-Leibler divergence*. Since this divergence is not really a distance, it has to be handled with care.

5.6.1 The EG Algorithm for Positive Weights

In this subsection, we assume that the components of the impulse response that we try to identify are all positive, in order that the relative entropy is meaningful.

Taking $P = 1$ and $\boldsymbol{G}(n-1) = \boldsymbol{I}$, the criterion (5.34) simplifies to:

$$\begin{aligned} J(n) &= d_{\mathrm{re}}\left[\boldsymbol{h}(n), \boldsymbol{h}(n-1)\right] + \epsilon_{\mathrm{n}}^2(n) \\ &= d_{\mathrm{re}}\left[\boldsymbol{h}(n), \boldsymbol{h}(n-1)\right] + \left[\boldsymbol{x}^T(n)\boldsymbol{x}(n)\right]^{-1} \epsilon^2(n), \end{aligned} \tag{5.53}$$

where now

$$d_{\mathrm{re}}\left[\boldsymbol{h}(n), \boldsymbol{h}(n-1)\right] = \eta^{-1} \sum_{l=0}^{L-1} h_l(n) \ln \frac{h_l(n)}{h_l(n-1)}, \tag{5.54}$$

with $\eta > 0$. With this formalism, $\boldsymbol{h}(n)$ and $\boldsymbol{h}(n-1)$ are *probability vectors*, which means that their components are nonnegative and $\|\boldsymbol{h}(n)\|_1 = \|\boldsymbol{h}(n-1)\|_1 = u > 0$, where u is a scaling factor. Therefore, we minimize $J(n)$ with the constraint that $\sum_l h_l(n) = 1$ (i.e. we take here $u = 1$). This optimization leads to:

$$\eta^{-1}\left[\ln \frac{h_l(n)}{h_l(n-1)} + 1\right] - 2x(n-l)\left[\boldsymbol{x}^T(n)\boldsymbol{x}(n)\right]^{-1}\epsilon(n) + \kappa = 0, \tag{5.55}$$
$$l = 0, 1, \cdots, L-1,$$

where κ is a Lagrange multiplier. Equation (5.55) is highly nonlinear so that solving it is very difficult if not impossible. However, if the new weight vector $\boldsymbol{h}(n)$ is close to the old weight vector $\boldsymbol{h}(n-1)$, replacing the *a posteriori* error signal, $\epsilon(n)$, in (5.55) with the *a priori* error signal, $e(n)$, is a reasonable approximation and the equation

$$\eta^{-1}\left[\ln \frac{h_l(n)}{h_l(n-1)} + 1\right] - 2x(n-l)\left[\boldsymbol{x}^T(n)\boldsymbol{x}(n)\right]^{-1}e(n) + \kappa = 0, \tag{5.56}$$
$$l = 0, 1, \cdots, L-1,$$

is much easier to solve. We then deduce the EG algorithm [5]:

$$h_l(n) = \frac{h_l(n-1)r_l(n)}{\sum_{j=0}^{L-1} h_j(n-1)r_j(n)}, \quad l = 0, 1, \cdots, L-1, \tag{5.57}$$

where

$$r_l(n) = \exp\left[\eta(n)x(n-l)e(n)\right], \tag{5.58}$$

with $\eta(n) = 2\eta\left[\boldsymbol{x}^T(n)\boldsymbol{x}(n)\right]^{-1}$. The algorithm is initialized with: $h_l(0) = c > 0, \ \forall l$.

5.6.2 The EG± Algorithm for Positive and Negative Weights

The EG algorithm is designed to work for positive weights only, due to the nature of the relative entropy definition. However, there is a simple way to generalize the idea to both positive and negative weights. Indeed, we can always find two vectors $\boldsymbol{h}^{+}(n)$ and $\boldsymbol{h}^{-}(n)$ with positive coefficients, in such a way that the vector

$$\boldsymbol{h}(n) = \boldsymbol{h}^{+}(n) - \boldsymbol{h}^{-}(n) \tag{5.59}$$

can have positive and negative components. In this case, the *a priori* and *a posteriori* error signals can be written as:

$$e(n) = y(n) - [\boldsymbol{h}^{+}(n-1) - \boldsymbol{h}^{-}(n-1)]^{\mathrm{T}}\boldsymbol{x}(n), \tag{5.60}$$

$$\epsilon(n) = y(n) - [\boldsymbol{h}^{+}(n) - \boldsymbol{h}^{-}(n)]^{\mathrm{T}}\boldsymbol{x}(n), \tag{5.61}$$

and the criterion (5.53) will change to:

$$\begin{aligned} J^{\pm}(n) = {} & d_{\mathrm{re}}[\boldsymbol{h}^{+}(n), \boldsymbol{h}^{+}(n-1)] + d_{\mathrm{re}}[\boldsymbol{h}^{-}(n), \boldsymbol{h}^{-}(n-1)] \\ & + \frac{1}{u}\left[\boldsymbol{x}^{\mathrm{T}}(n)\boldsymbol{x}(n)\right]^{-1}\epsilon^{2}(n), \end{aligned} \tag{5.62}$$

where u is a positive scaling constant. Using the Kullback-Leibler divergence plus the constraint $\sum_l [h_l^{+}(n) + h_l^{-}(n)] = u$ and the same approximation as for the EG, the minimization of (5.62) gives:

$$\eta^{-1}\left[\ln\frac{h_l^{+}(n)}{h_l^{+}(n-1)} + 1\right] - \frac{2}{u}x(n-l)\left[\boldsymbol{x}^{\mathrm{T}}(n)\boldsymbol{x}(n)\right]^{-1}e(n) + \kappa = 0, \tag{5.63}$$

$$\eta^{-1}\left[\ln\frac{h_l^{-}(n)}{h_l^{-}(n-1)} + 1\right] + \frac{2}{u}x(n-l)\left[\boldsymbol{x}^{\mathrm{T}}(n)\boldsymbol{x}(n)\right]^{-1}e(n) + \kappa = 0, \tag{5.64}$$

$$l = 0, 1, \cdots, L-1,$$

where κ is a Lagrange multiplier. From the two previous equations, we easily find the EG± algorithm [5]:

$$h_l^{+}(n) = u\frac{h_l^{+}(n-1)r_l^{+}(n)}{\sum_{j=0}^{L-1}[h_j^{+}(n-1)r_j^{+}(n) + h_j^{-}(n-1)r_j^{-}(n)]}, \tag{5.65}$$

$$h_l^{-}(n) = u\frac{h_l^{-}(n-1)r_l^{-}(n)}{\sum_{j=0}^{L-1}[h_j^{+}(n-1)r_j^{+}(n) + h_j^{-}(n-1)r_j^{-}(n)]}, \tag{5.66}$$

where

$$r_l^{+}(n) = \exp\left[\frac{\eta(n)}{u}x(n-l)e(n)\right], \tag{5.67}$$

$$\begin{aligned} r_l^{-}(n) &= \exp\left[-\frac{\eta(n)}{u}x(n-l)e(n)\right] \\ &= \frac{1}{r_l^{+}(n)}, \end{aligned} \tag{5.68}$$

Table 5.2. The EG± algorithm.

Initialization:

$$h_l^+(0) = h_l^-(0) = c > 0, \; l = 0, 1, \cdots, L-1$$

Parameters:

$$u \geq \|\boldsymbol{h}_{\mathrm{t}}\|_1$$

$$0 < \mu < 1, \quad \delta_{\mathrm{EG}} = \mathrm{cst} \cdot \sigma_x^2$$

Error:

$$e(n) = y(n) - \left[\boldsymbol{h}^+(n-1) - \boldsymbol{h}^-(n-1)\right]^T \boldsymbol{x}(n)$$

Update:

$$\mu(n) = \frac{\mu}{\boldsymbol{x}^T(n)\boldsymbol{x}(n) + \delta_{\mathrm{EG}}}$$

$$r_l^+(n) = \exp\left[L\frac{\mu(n)}{u}x(n-l)e(n)\right]$$

$$r_l^-(n) = \frac{1}{r_l^+(n)}$$

$$h_l^+(n) = u\frac{h_l^+(n-1)r_l^+(n)}{\sum_{j=0}^{L-1}\left[h_j^+(n-1)r_j^+(n) + h_j^-(n-1)r_j^-(n)\right]}$$

$$h_l^-(n) = u\frac{h_l^-(n-1)r_l^-(n)}{\sum_{j=0}^{L-1}\left[h_j^+(n-1)r_j^+(n) + h_j^-(n-1)r_j^-(n)\right]}$$

$$l = 0, 1, \cdots, L-1$$

---

with $\eta(n) = 2\eta\left[\boldsymbol{x}^T(n)\boldsymbol{x}(n)\right]^{-1}$. We can check that we always have $\|\boldsymbol{h}^+(n)\|_1 + \|\boldsymbol{h}^-(n)\|_1 = u$. This algorithm is summarized in Table 5.2.

The fact that,

$$u = \|\boldsymbol{h}^+(n)\|_1 + \|\boldsymbol{h}^-(n)\|_1 \geq \|\boldsymbol{h}^+(n) - \boldsymbol{h}^-(n)\|_1 = \|\boldsymbol{h}(n)\|_1, \quad (5.69)$$

suggests that the constant u has to be chosen such that $u \geq \|\boldsymbol{h}_{\mathrm{t}}\|_1$ in order that $\boldsymbol{h}(n)$ converges to $\boldsymbol{h}_{\mathrm{t}}$. If we take $u < \|\boldsymbol{h}_{\mathrm{t}}\|_1$, the algorithm will introduce a bias in the coefficients of the filter.

The motivation for the EG± (and EG) algorithm can be developed by taking the logarithmic of (5.65) and (5.66). This shows that the logarithmic weights use almost the same update as the NLMS algorithm. Alternatively, this can be interpreted as exponentiating the update, hence the name EG±. This has the effect of assigning larger relative updates to larger weights, thereby deemphasizing the effect of smaller weights. This is qualitatively similar to the PNLMS algorithm which makes the update *proportional* to the size of the weight. This type of behavior is desirable for sparse impulse responses where small weights do not contribute significantly to the *mean* solution but introduce an undesirable noise-like *variance*.

5.6.3 The Exponentiated RLS (ERLS) Algorithm

The RLS algorithm is optimal from a convergence point of view since its convergence does not depend on the condition number of the input signal covariance matrix. It is well known that with ill-conditioned signals (like speech) this condition number can be very large and algorithms like LMS suffer from slow convergence [2]. Thus, it is interesting to compare the RLS algorithm to the other algorithms when the impulse response to identify is sparse. The update equation (5.45) of the RLS algorithm can be rewritten as:

$$h_l(n) = h_l(n-1) + k_l(n)e(n), \ 0 \leq l \leq L-1, \tag{5.70}$$

where

$$\begin{aligned} \boldsymbol{k}(n) &= \left[\, k_0(n) \ k_1(n) \ \cdots \ k_{L-1}(n) \,\right]^{\mathrm{T}} \\ &= \boldsymbol{R}^{-1}(n)\boldsymbol{x}(n) \end{aligned} \tag{5.71}$$

is the Kalman gain. A fast RLS (FRLS) can be derived by using the *a priori* Kalman gain $\boldsymbol{k}'(n) = \boldsymbol{R}^{-1}(n-1)\boldsymbol{x}(n)$ and the forward and backward predictors. This *a priori* Kalman gain can be computed recursively with only $5L$ multiplications [2].

Following the same approach as for the EG± algorithm, we deduce the exponentiated RLS (ERLS) algorithm [17]:

$$e(n) = y(n) - [\boldsymbol{h}^{+}(n-1) - \boldsymbol{h}^{-}(n-1)]^{\mathrm{T}}\boldsymbol{x}(n), \tag{5.72}$$

$$h_l^{+}(n) = u \frac{h_l^{+}(n-1)r_l^{+}(n)}{\sum_{j=0}^{L-1} \left[h_j^{+}(n-1)r_j^{+}(n) + h_j^{-}(n-1)r_j^{-}(n)\right]}, \tag{5.73}$$

$$h_l^{-}(n) = u \frac{h_l^{-}(n-1)r_l^{-}(n)}{\sum_{j=0}^{L-1} \left[h_j^{+}(n-1)r_j^{+}(n) + h_j^{-}(n-1)r_j^{-}(n)\right]}, \tag{5.74}$$

where now:

$$\begin{aligned} r_l^{+}(n) &= \exp\left[\frac{k_l(n)}{u} e(n)\right] \\ &= \frac{1}{r_l^{-}(n)}. \end{aligned} \tag{5.75}$$

Obviously, a fast ERLS (FERLS) can easily be derived since the Kalmain gain in (5.75) is the same as the one used in the FRLS. Simulations presented later show that there is not much difference between the FRLS and FERLS for initial convergence, but for tracking, FERLS can be much better than FRLS. Hence, the FERLS algorithm may be of some interest.

5.7 The Lambert W Function Based Gradient Algorithm

As was shown in the previous section, the EG algorithm is derived from the relative entropy which is not a symmetric distance, e.g. $d_{\mathrm{re}}\left[\boldsymbol{h}(n), \boldsymbol{h}(n-1)\right] \neq d_{\mathrm{re}}\left[\boldsymbol{h}(n-1), \boldsymbol{h}(n)\right]$. Moreover, the constraint $\sum_l h_l(n-1) = \sum_l h_l(n) = 1$ needs to be added in the minimization process to ensure that $d_{\mathrm{re}}\left[\boldsymbol{h}(n), \boldsymbol{h}(n-1)\right] \geq 0$.

Consider the following symmetric distance:

$$\begin{aligned} d_{\mathrm{lw}}\Big[\boldsymbol{h}(n), \boldsymbol{h}(n-1)\Big] &= \eta'^{-1} \sum_{l=0}^{L-1} \Big[h_l(n) - h_l(n-1)\Big]\Big[\ln h_l(n) - \ln h_l(n-1)\Big] \\ &= \eta'^{-1} \sum_{l=0}^{L-1} h_l(n) \ln \frac{h_l(n)}{h_l(n-1)} \\ &\quad + \eta'^{-1} \sum_{l=0}^{L-1} h_l(n-1) \ln \frac{h_l(n-1)}{h_l(n)} \\ &= d_{\mathrm{re}}\Big[\boldsymbol{h}(n), \boldsymbol{h}(n-1)\Big] + d_{\mathrm{re}}\Big[\boldsymbol{h}(n-1), \boldsymbol{h}(n)\Big]. \end{aligned} \tag{5.76}$$

It is easy to see that $d_{\mathrm{lw}}\left[\boldsymbol{h}(n), \boldsymbol{h}(n-1)\right] \geq 0$ as long as the components $h_l(n)$ and $h_l(n-1)$ are nonnegative. This means that a criterion using $d_{\mathrm{lw}}\left[\boldsymbol{h}(n), \boldsymbol{h}(n-1)\right]$ does not need to include any constraint, which is not the case if $d_{\mathrm{re}}\left[\boldsymbol{h}(n), \boldsymbol{h}(n-1)\right]$ is used.

If we now take the general case (positive and negative components), we seek to minimize:

$$\begin{aligned} J_{\mathrm{lw}}(n) &= d_{\mathrm{lw}}\Big[\boldsymbol{h}^+(n), \boldsymbol{h}^+(n-1)\Big] + d_{\mathrm{lw}}\Big[\boldsymbol{h}^-(n), \boldsymbol{h}^-(n-1)\Big] \\ &\quad + \left[\boldsymbol{x}^T(n)\boldsymbol{x}(n)\right]^{-1} \epsilon^2(n). \end{aligned} \tag{5.77}$$

This minimization with respect to $\boldsymbol{h}^+(n)$ and $\boldsymbol{h}^-(n)$ (then approximating the *a posteriori* error with the *a priori* error) leads to the two equations:

$$1 - \eta'(n)x(n-l)e(n) = \ln \frac{h_l^+(n-1)}{h_l^+(n)} + \frac{h_l^+(n-1)}{h_l^+(n)}, \tag{5.78}$$

$$1 + \eta'(n)x(n-l)e(n) = \ln \frac{h_l^-(n-1)}{h_l^-(n)} + \frac{h_l^-(n-1)}{h_l^-(n)}, \tag{5.79}$$

$$l = 0, 1, \cdots, L-1,$$

where $\eta'(n) = 2\eta'\left[\boldsymbol{x}^T(n)\boldsymbol{x}(n)\right]^{-1}$. Exponentiating the two previous equations, we find what we call the Lambert W function based gradient (LWG) algorithm:

$$\exp\Big[1 - \eta'(n)x(n-l)e(n)\Big] = w_l^+(n) \exp w_l^+(n), \tag{5.80}$$

$$\exp\Big[1 + \eta'(n)x(n-l)e(n)\Big] = w_l^-(n) \exp w_l^-(n), \tag{5.81}$$

where

$$w_l^+(n) = \frac{h_l^+(n-1)}{h_l^+(n)}, \tag{5.82}$$

$$w_l^-(n) = \frac{h_l^-(n-1)}{h_l^-(n)}. \tag{5.83}$$

The Lambert W function is defined to be the multivalued inverse of the function $w \exp w$ [20]. Since $\exp\left[1 \pm \eta'(n)x(n-l)e(n)\right] > 0$, there is a unique value for $w_l^+(n)$ and $w_l^-(n)$.

Obviously, the complexity of the LWG algorithm is quite high since it requires, at each iteration and for each component of the filters, to find the solution of the nonlinear equation: $z = w \exp w$. Iterative algorithms exist for that and MATLAB has a function called "lambertw" to find w.

5.8 Some Important Links Among Algorithms

nonlinear algorithms like EG are not easy to analyze and even when it is possible, very often the information we can get from a tedious analysis is not that much helpful in understanding their behavior. It is sometimes more useful to link a new adaptive filter to a well-studied one such as NLMS, in order to be able to deduce its limitations and potentials.

5.8.1 Link Between NLMS and EG± Algorithms

If we initialize $h_l(0) = 0,\ l = 0, 1, \cdots, L-1$, in the NLMS algorithm, we can easily see that:

$$\begin{aligned}\boldsymbol{h}(n) &= \sum_{i=0}^{n-1} \mu(i+1)\boldsymbol{x}(i+1)e(i+1) \\ &= \mu \sum_{i=0}^{n-1} \frac{\boldsymbol{x}(i+1)e(i+1)}{\boldsymbol{x}^{\mathrm{T}}(i+1)\boldsymbol{x}(i+1)},\end{aligned} \tag{5.84}$$

where $\mu(i+1) = \mu\left[\boldsymbol{x}^{\mathrm{T}}(i+1)\boldsymbol{x}(i+1)\right]^{-1}$.

If we start the adaptation of the EG± algorithm with $h_l^+(0) = h_l^-(0) = c > 0,\ l = 0, 1, \cdots, L-1$, we can show that (5.65) and (5.66) are equivalent to [18]:

$$h_l^+(n) = u \frac{s_l^+(n)}{\sum_{j=0}^{L-1}\left[s_j^+(n) + s_j^-(n)\right]}, \tag{5.85}$$

$$h_l^-(n) = u \frac{s_l^-(n)}{\sum_{j=0}^{L-1}\left[s_j^+(n) + s_j^-(n)\right]}, \tag{5.86}$$

where

$$s_l^+(n) = \exp\left[\frac{1}{u}\sum_{i=0}^{n-1}\eta(i+1)x(i+1-l)e(i+1)\right], \tag{5.87}$$

$$\begin{aligned} s_l^-(n) &= \exp\left[-\frac{1}{u}\sum_{i=0}^{n-1}\eta(i+1)x(i+1-l)e(i+1)\right] \\ &= \frac{1}{s_l^+(n)}, \end{aligned} \tag{5.88}$$

and $\eta(i+1) = 2\eta\left[\boldsymbol{x}^{\mathrm{T}}(i+1)\boldsymbol{x}(i+1)\right]^{-1}$. Clearly, the convergence of the algorithm does not depend on the initialization parameter c (as long it is positive and nonzero). Now

$$\begin{aligned} h_l(n) &= h_l^+(n) - h_l^-(n) \\ &= u\frac{s_l^+(n) - s_l^-(n)}{\sum_{j=0}^{L-1}\left[s_j^+(n) + s_j^-(n)\right]} \\ &= u\frac{\sinh\left[\frac{1}{u}\sum_{i=0}^{n-1}\eta(i+1)x(i+1-l)e(i+1)\right]}{\sum_{j=0}^{L-1}\cosh\left[\frac{1}{u}\sum_{i=0}^{n-1}\eta(i+1)x(i+1-j)e(i+1)\right]}. \end{aligned} \tag{5.89}$$

Note that the sinh function has the effect of exponentiating the update, as previously commented.

For u large enough and using the approximations $\sinh(a) \approx a$ and $\cosh(a) \approx 1$ when $|a| \ll 1$, (5.89) becomes:

$$h_l(n) = \frac{2\eta}{L}\sum_{i=0}^{n-1}\frac{x(i+1-l)e(i+1)}{\boldsymbol{x}^{\mathrm{T}}(i+1)\boldsymbol{x}(i+1)}, \quad 0 \le l \le L-1. \tag{5.90}$$

Comparing (5.84) and (5.90), we understand that, by taking $\eta = L\mu/2$ and for u large enough, the NLMS and EG± algorithms have the same performance. Obviously, the choice of u is critical in practice: if we take $u < \|\boldsymbol{h}_{\mathrm{t}}\|_1$, the EG± will introduce a bias in the coefficients of the filter, and if $u \gg \|\boldsymbol{h}_{\mathrm{t}}\|_1$, the EG± will behave like NLMS.

5.8.2 Link Between IPNLMS and EG± Algorithms

PNLMS and IPNLMS algorithms were developed for use in network echo cancelers [19]. In comparison to the NLMS algorithm, they have very fast initial convergence and tracking when the echo path is sparse. As previously mentioned, the idea behind these "proportionate" algorithms is to update each coefficient of the filter independently of the others by adjusting the adaptation step size in proportion to the estimated filter coefficient.

How are the IPNLMS and EG± algorithms specifically related? In the rest of this subsection, we show that the IPNLMS is in fact an approximation of the EG±.

If we suppose that $\boldsymbol{h}^+(n)$ [resp. $\boldsymbol{h}^-(n)$] is close to $\boldsymbol{h}^+(n-1)$ [resp. $\boldsymbol{h}^-(n-1)$], which is usually the case in all adaptive algorithms (especially for a small step size), the two distances $d_{\mathrm{re}}[\boldsymbol{h}^+(n), \boldsymbol{h}^+(n-1)]$ and $d_{\mathrm{re}}[\boldsymbol{h}^-(n), \boldsymbol{h}^-(n-1)]$ in criterion (5.62) can be approximated as follows:

$$\begin{aligned} d_{\mathrm{re}}[\boldsymbol{h}^+(n), \boldsymbol{h}^+(n-1)] &= \eta^{-1} \sum_{l=0}^{L-1} h_l^+(n) \ln \frac{h_l^+(n)}{h_l^+(n-1)} \\ &\approx \eta^{-1} \sum_{l=0}^{L-1} h_l^+(n) \left[\frac{h_l^+(n)}{h_l^+(n-1)} - 1 \right], \end{aligned} \tag{5.91}$$

$$\begin{aligned} d_{\mathrm{re}}[\boldsymbol{h}^-(n), \boldsymbol{h}^-(n-1)] &= \eta^{-1} \sum_{l=0}^{L-1} h_l^-(n) \ln \frac{h_l^-(n)}{h_l^-(n-1)} \\ &\approx \eta^{-1} \sum_{l=0}^{L-1} h_l^-(n) \left[\frac{h_l^-(n)}{h_l^-(n-1)} - 1 \right]. \end{aligned} \tag{5.92}$$

Using (5.91) and (5.92) plus the constraint $\sum_l [h_l^+(n) + h_l^-(n)] = u$ and the same approximation as for the EG±, the minimization of (5.62) gives the approximated EG± algorithm:

$$h_l^+(n) = h_l^+(n-1) \left[1 + \frac{\eta(n)}{2u} x(n-l) e(n) - \frac{\eta(n)}{2u^2} \widehat{y}(n) e(n) \right], \tag{5.93}$$

$$h_l^-(n) = h_l^-(n-1) \left[1 - \frac{\eta(n)}{2u} x(n-l) e(n) - \frac{\eta(n)}{2u^2} \widehat{y}(n) e(n) \right], \tag{5.94}$$

so that:

$$\begin{aligned} h_l(n) &= h_l^+(n) - h_l^-(n) \\ &= h_l(n-1) + \frac{\eta(n)[h_l^+(n-1) + h_l^-(n-1)]}{2u} x(n-l) e(n) \\ &\quad - \frac{\eta(n)}{2u^2} h_l(n-1) \widehat{y}(n) e(n). \end{aligned} \tag{5.95}$$

Neglecting the last term of the right-hand side of (5.95), we get:

$$h_l(n) = h_l(n-1) + \frac{\eta(n)}{2} \frac{h_l^+(n-1) + h_l^-(n-1)}{\|\boldsymbol{h}^+(n-1)\|_1 + \|\boldsymbol{h}^-(n-1)\|_1} x(n-l) e(n). \tag{5.96}$$

If the true impulse response $\boldsymbol{h}_{\mathrm{t}}$ is sparse, it can be shown that if we choose $u = \|\boldsymbol{h}_{\mathrm{t}}\|_1$, the (positive) vector $\boldsymbol{h}^+(n-1) + \boldsymbol{h}^-(n-1)$ is also sparse after convergence. This means that the elements

$$\frac{h_l^+(n-1)+h_l^-(n-1)}{\|\boldsymbol{h}^+(n-1)\|_1+\|\boldsymbol{h}^-(n-1)\|_1}$$

in (5.96) play exactly the same role as the elements $g_{\mathrm{ip},l}(n)$ in the IPNLMS algorithm in the particular case where $\alpha = 1$ (PNLMS algorithm). As a result, we can expect the two algorithms (IPNLMS and EG±) to have similar performance. On the other hand, if $u \gg \|\boldsymbol{h}_\mathrm{t}\|_1$, it can be shown that $h_l^+(n-1)+h_l^-(n-1) \approx u/L$, $\forall l$. In this case, the EG± algorithm will behave like IPNLMS with $\alpha = -1$ (NLMS algorithm). Thus, the parameter α in IPNLMS operates like the parameter u in EG±. However, the advantage of IPNLMS is that no *a priori* information of the system impulse response is required in order to have a better convergence rate than the NLMS algorithm. Another clear advantage of IPNLMS is that it is much less complex to implement than EG±. We conclude that IPNLMS is a good approximation of EG± and is more useful in practice. Note also that the approximated EG± algorithm (5.96) belongs to the family of natural gradient algorithms [12], [13].

5.8.3 Link Between LWG and EG± Algorithms

As we have already mentioned, the complexity of the LWG is high, so this algorithm is not normally suitable for practical applications. However, it is important to understand how it is related to other algorithms.

If we make the usual assumption that $\boldsymbol{h}^+(n)$ [resp. $\boldsymbol{h}^-(n)$] is close to $\boldsymbol{h}^+(n-1)$ [resp. $\boldsymbol{h}^-(n-1)$], the LWG algorithm can be approximated as follows:

$$h_l^+(n) = \frac{h_l^+(n-1)}{1-\dfrac{\eta'(n)}{2}x(n-l)e(n)}, \tag{5.97}$$

$$h_l^-(n) = \frac{h_l^-(n-1)}{1+\dfrac{\eta'(n)}{2}x(n-l)e(n)}, \tag{5.98}$$

$$l = 0, 1, \cdots, L-1.$$

For $|a| \ll 1$, we have:

$$\frac{1}{1-a} \approx 1+a, \tag{5.99}$$

$$\frac{1}{1+a} \approx 1-a. \tag{5.100}$$

Using these approximations in (5.97) and (5.98), we obtain the approximated LWG algorithm:

$$h_l^+(n) = h_l^+(n-1)\left[1 + \frac{\eta'(n)}{2}x(n-l)e(n)\right], \tag{5.101}$$

$$h_l^-(n) = h_l^-(n-1)\left[1 - \frac{\eta'(n)}{2}x(n-l)e(n)\right], \tag{5.102}$$

$$l = 0, 1, \cdots, L-1,$$

which is equivalent to the approximated EG± algorithm. Therefore, we can expect that in practice, the EG± and LWG algorithms will perform in a very similar way.

5.9 Simulations

The objective of this section is to show, by way of simulations, how some of the algorithms presented in this chapter work in typical conditions of room acoustic impulse response identification. Comparison among the different algorithms is another important aspect we emphasize here. The aim is to give a representatives set of simulation scenarios that are relevant in this context.

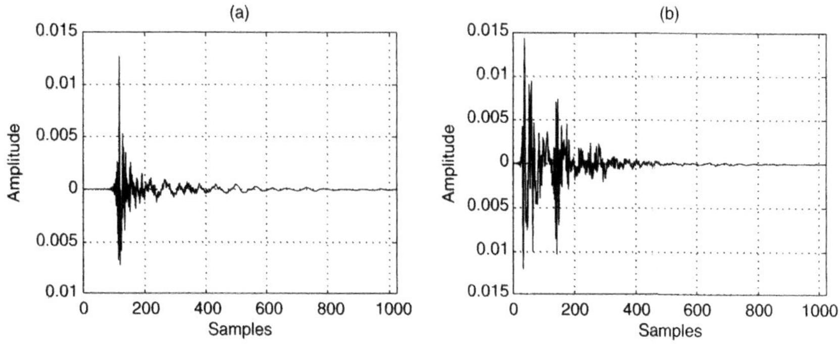

Fig. 5.2. Acoustic impulse responses used in simulations.

The two room acoustic impulse responses $\boldsymbol{h}_{\mathrm{t}}$ to be identified are shown in Fig. 5.2. The impulse response of Fig. 5.2(a) is more sparse $[\xi(\boldsymbol{h}_{\mathrm{t}}) \approx 0.69]$ than the one of Fig. 5.2(b) $[\xi(\boldsymbol{h}_{\mathrm{t}}) \approx 0.65]$. They are both of length $L = 1024$ and the same length is used for all the adaptive filters $\boldsymbol{h}(n)$. The sampling rate is 8 kHz and a white noise signal with 30 dB SNR (signal-to-noise ratio) is added to the output $y(n)$. The input signal $x(n)$ is either a white Gaussian signal or a speech signal. The parameter settings chosen (unless stated otherwise) for all the simulations are:

- $h_l(0) = 0, \;\; h_l^+(0) = h_l^-(0) = 1, \;\; l = 0, 1, \cdots, L-1,$
- $\mu = 0.3, \;\; \delta = 10\sigma_x^2,$
- $\alpha = -0.5, \;\; \varepsilon = 0.001,$

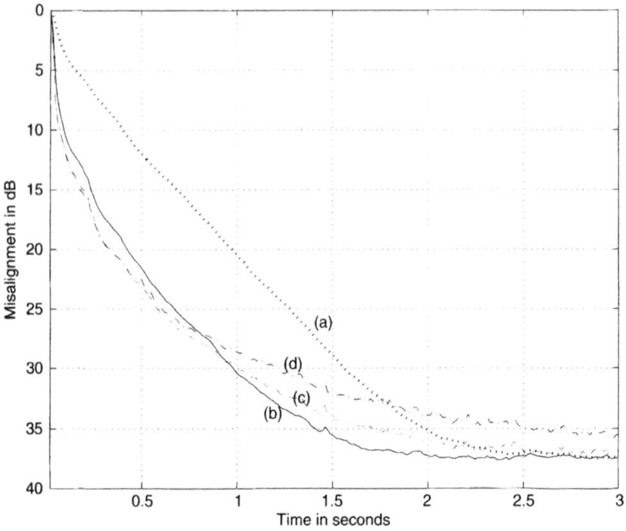

Fig. 5.3. Misalignment of the IPNLMS algorithm for different values of α with a white Gaussian noise as input signal, impulse response of Fig. 5.2(a), and using the true coefficients in $\boldsymbol{G}(n)$. (a) $\alpha = -1$ (equivalent to NLMS), (b) $\alpha = -0.5$, (c) $\alpha = 0$, and (d) $\alpha = 0.9$.

- $\rho = 0.01, \ \delta_{\mathrm{p}} = 0.01$,
- $\lambda = 1 - 1/(3L)$,
- $\delta_{\mathrm{NLMS}} = \delta_{\mathrm{EG}} = \delta, \ \delta_{\mathrm{PNLMS}} = \delta/L, \ \delta_{\mathrm{IPNLMS}} = (1-\alpha)\delta/(2L)$.

Figures 5.3–5.10 show the convergence of the normalized misalignment (in dB),

$$10 \log_{10} \frac{\|\boldsymbol{h}_{\mathrm{t}} - \boldsymbol{h}(n)\|_2}{\|\boldsymbol{h}_{\mathrm{t}}\|_2}, \tag{5.103}$$

for all the algorithms. The only simulation that was done with a speech source as excitation signal is shown in Fig. 5.10; all the others were done with a white Gaussian signal. Impulse response of Fig. 5.2(a) was used everywhere except for Fig. 5.7, where impulse response of Fig. 5.2(b) was used.

Figures 5.3 shows how IPNLMS behaves with different values of α. In this unrealistic simulation, we used for the diagonal matrix $\boldsymbol{G}(n)$, the true values of the coefficients $\boldsymbol{h}_{\mathrm{t}}$ instead of the estimated ones. This may seem like what we can do best with the "proportionate" idea. First, we see that when α approaches 1, the algorithm degrades and a good value seems to be $\alpha = -0.5$. Second, comparing Fig. 5.3 with Fig. 5.4 where this time the real IPNLMS is evaluated [with the estimated coefficients in $\boldsymbol{G}(n)$], we see that the difference is not that significant, although better when $\boldsymbol{G}(n)$ is known a priori. This observation is very important because it shows, in a very simple manner, the limits of natural gradient algorithms in general.

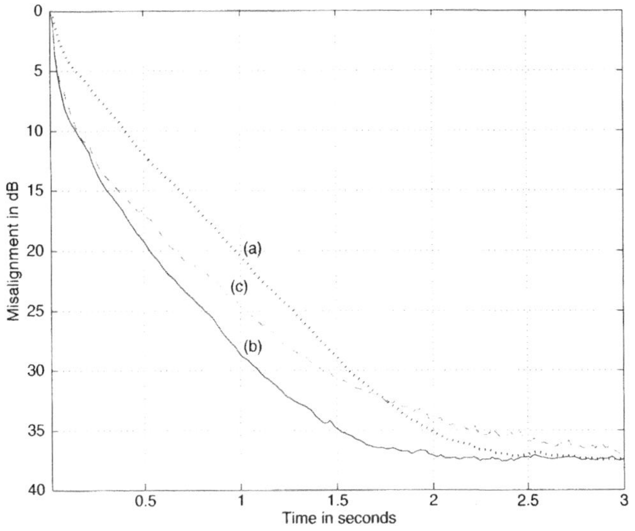

Fig. 5.4. Misalignment of the IPNLMS algorithm for different values of α with a white Gaussian noise as input signal and impulse response of Fig. 5.2(a). (a) $\alpha = -1$ (equivalent to NLMS), (b) $\alpha = -0.5$, and (c) $\alpha = 0.5$ (close to PNLMS).

Figure 5.5 presents the misalignment of the EG± algorithm for different values of u. As expected, for a large u, this algorithm coincides with NLMS and for $u = 0.5\|\boldsymbol{h}_{\rm t}\|_1$, it introduces a bias in the coefficients of the filter and, as a result, the EG± is much worse than NLMS. Clearly, the EG± algorithm is not very interesting from a practical point of view since it requires some *a priori* knowledge that we can not have.

Figure 5.6 compares the initial convergence of four algorithms (NLMS, PNLMS, IPNLMS, and EG±) with the impulse response of Fig. 5.2(a). We see on this figure that the PNLMS, IPNLMS, and EG± (with $u = 2\|\boldsymbol{h}_{\rm t}\|_1$) algorithms converge much faster than NLMS. We also see that IPNLMS and EG± are very close to each other, confirming that these two algorithms are related.

In Fig. 5.7, we compare again the initial convergence of the same four algorithms but with the impulse response of Fig. 5.2(b). While IPNLMS and EG± still perform much better than NLMS, PNLMS starts degrading very significantly after 1.2 seconds. This confirms that PNLMS is not very well optimized when the impulse response is not strongly sparse.

Tracking is another very important issue in adaptive algorithms. In applications like room acoustics, it is essential that an adaptive filter tracks fast since impulse responses are not very stationary. Figures 5.8 and 5.9 compare the algorithms in a tracking situation when after 3 seconds the sparse impulse response of Fig. 5.2(a) is shifted to the right by 12 samples. The other conditions of Fig. 5.8 are the same as that in Fig. 5.6. According to this sim-

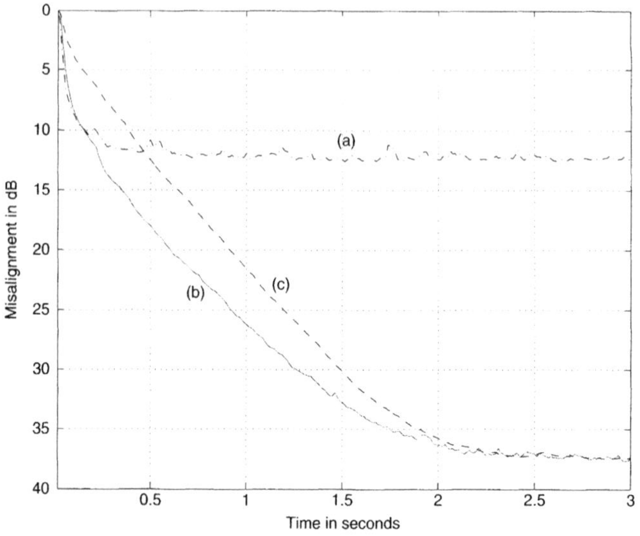

Fig. 5.5. Misalignment of the EG± algorithm for different values of u with a white Gaussian noise as input signal and impulse response of Fig. 5.2(a). (a) $u = 0.5\|\boldsymbol{h}_t\|_1$, (b) $u = 2\|\boldsymbol{h}_t\|_1$, and (c) $u = 50\|\boldsymbol{h}_t\|_1$ (equivalent to NLMS).

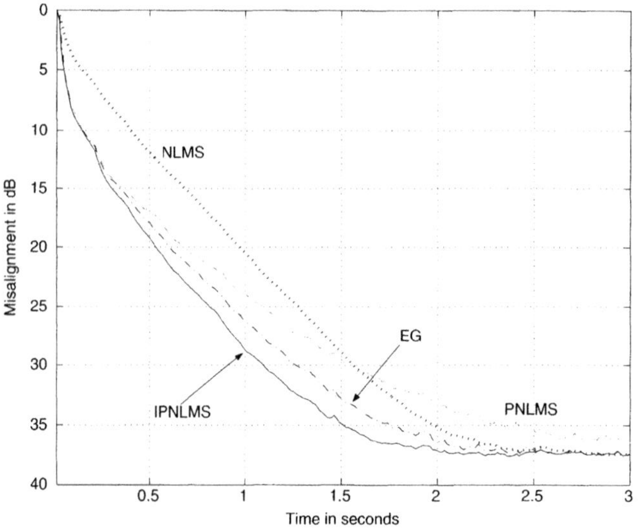

Fig. 5.6. Misalignment of the NLMS (dotted line), PNLMS (dashed line), IPNLMS (solid line), and EG± (dash-dot line) algorithms with white Gaussian noise as input signal and impulse response of Fig. 5.2(a).

ulation, the PNLMS, IPNLMS, and EG± algorithms track much better than the NLMS algorithm. In Fig. 5.9, the FRLS algorithm is compared to the

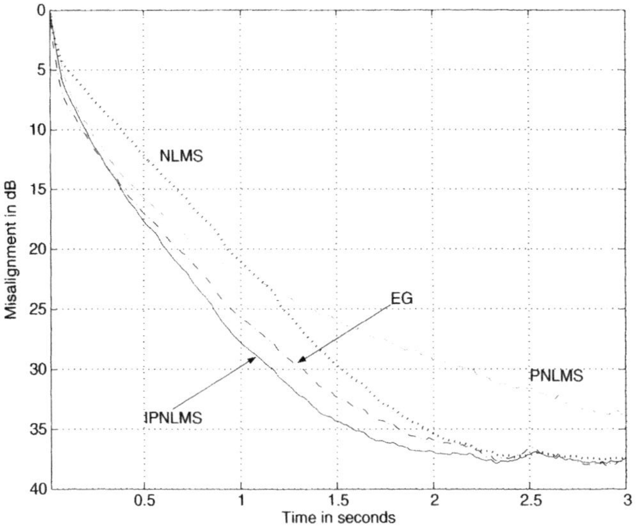

Fig. 5.7. Misalignment of the NLMS (dotted line), PNLMS (dashed line), IPNLMS (solid line), and EG± (dash-dot line) algorithms with white Gaussian noise as input signal and impulse response of Fig. 5.2(b).

FERLS algorithm with $u = 6\|\boldsymbol{h}_t\|_1$: while the initial convergence of the two algorithms is almost the same, the FERLS tracks faster than the FRLS. It is also worth noticing that IPNLMS tracks better than the FRLS and FERLS algorithms.

In Fig. 5.10, the initial convergence and tracking of the NLMS, PNLMS, IPNLMS, and EG± algorithms are compared with a speech source as input signal and impulse response of Fig. 5.2(a). Here, we changed the adaption step to $\mu = 0.5$. We notice the same trend as with a white Gaussian noise as input signal. Virtually, IPNLMS and EG± give almost the same results and they are slightly better than PNLMS; all three of them are better than NLMS in terms of initial convergence and tracking.

5.10 Conclusions

Throughout this chapter, we have shown how to use *a priori* information on sparseness in the design of adaptive algorithms in order to make them perform better (in terms of initial convergence and tracking) than classical adaptive algorithms. We have first proposed a universal criterion from which any adaptive filter can be derived. It was clearly shown that a nonlinear update with respect to the filter weights is advantageous. We have studied and compared in particular the IPNLMS and EG±, which are the two most important algorithms with nonlinear update. The IPNLMS algorithm was introduced in the

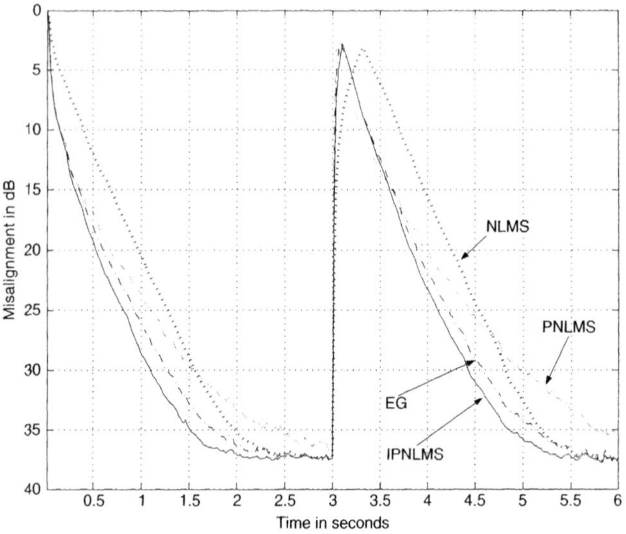

Fig. 5.8. Misalignment during impulse response change. The impulse response changes at time 3 seconds. Other conditions same as in Fig. 5.6.

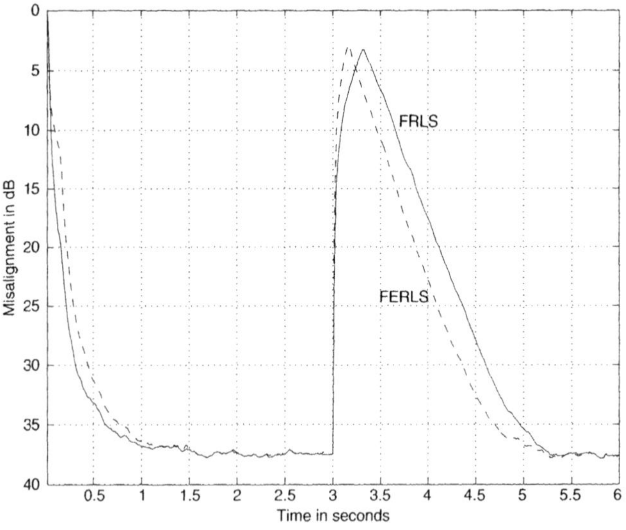

Fig. 5.9. Misalignment, during impulse response change, of the FRLS (solid line) and FERLS (dotted line) algorithms with white Gaussian noise as input signal, impulse response of Fig. 5.2(a), and $u = 6\|\boldsymbol{h}_\mathrm{t}\|_1$ for the FERLS algorithm.

context of network echo cancellation where there is a strong need to improve convergence rate and tracking. It was known for a long time that unknown echo paths in the network are most of the time sparse and there are many dif-

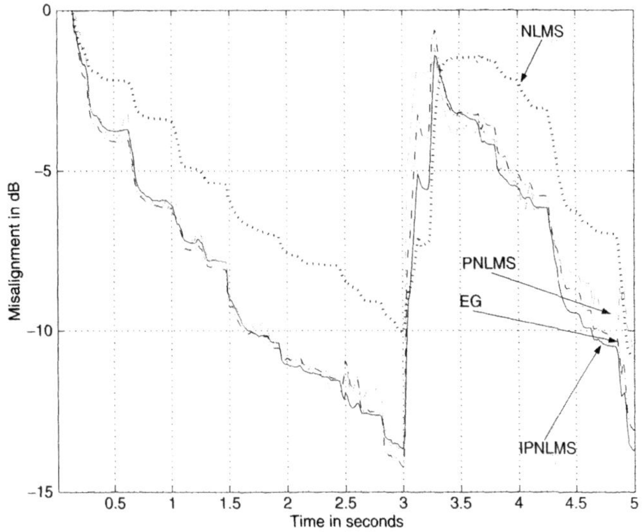

Fig. 5.10. Misalignment of the NLMS (dotted line), PNLMS (dashed line), IPNLMS (solid line), and EG± (dash-dot line) algorithms with a speech source as input signal and impulse response of Fig. 5.2(a). The impulse response changes at time 3 seconds.

ferent intuitions on how one should take advantage of that. Kivinen and Warmuth [5] derived the EG± algorithm in the context of computational learning theory. We have shown here that a good approximation of the EG± leads to the IPNLMS. As a result, the two algorithms have very similar performance in all the simulations we have investigated. We have also shown some links between the EG± and NLMS algorithms, so that with appropriate choice of some parameters, the two algorithms can be identical. We have also proposed a new algorithm called LWG by simply doubling the Kullback-Leibler divergence to have a symmetric distance. In fact, the LWG and EG± are almost equivalent.

Finally, all the ideas presented here can be generalized to blind identification of multichannel systems with sparse channels. Some possibilities are presented in [17] and [22].

References

[1] B. Widrow, S. D. Stearns: *Adaptive Signal Processing*, Englewood Cliffs, NJ, USA: Prentice Hall, 1985.

[2] S. Haykin: *Adaptive Filter Theory*, 4th ed., Englewood Cliffs, NJ, USA: Prentice Hall, 2002.

[3] D. L. Duttweiler: Proportionate normalized least mean square adaptation in echo cancelers, *IEEE Trans. Speech Audio Processing*, **8**, 508–518, September 2000.

[4] J. Benesty, S. L. Gay: An improved PNLMS algorithm, *Proc. ICASSP '02*, 1881–1884, Orlando, FL, USA, 2002.
[5] J. Kivinen, M. K. Warmuth: Exponentiated gradient versus gradient descent for linear predictors, *Inform. Comput.*, **132**, 1–64, January 1997.
[6] S. I. Hill, R. C. Williamson: Convergence of exponentiated gradient algorithms, *IEEE Trans. Signal Process.*, **49**, 1208–1215, June 2001.
[7] P. O. Hoyer: Non-negative matrix factorization with sparseness constraints, *J. of Machine Learning Res.*, **5**, 1457–1469, November 2004.
[8] S. L. Gay: An efficient, fast converging adaptive filter for network echo cancellation, *Proc. Asilomar '98*, **1**, 394–398, Asilomar, CA, USA, 1998.
[9] P. A. Naylor, J. Cui, M. Brookes: Adaptive algorithms for sparse echo cancellation, *Signal Process.*, submitted.
[10] K. Ozeki, T. Umeda: An adaptive filtering algorithm using an orthogonal projection to an affine subspace and its properties, *Electron Commun. Japan*, **67**(A), 19–27, 1984.
[11] S. Amari: Natural gradient works efficiently in learning, *Neural Computation*, **10**, 251–276, February 1998.
[12] S. L. Gay, S. C. Douglas: Normalized natural gradient adaptive filtering for sparse and nonsparse systems, *Proc. ICASSP '02*, Orlando, FL, USA, 1405–1408, 2002.
[13] R. K. Martin, W. A. Sethares, R. C. Williamson, C. R. Johnson, Jr.: Exploiting sparsity in adaptive filters, *IEEE Trans. Signal Process.*, **50**, 1883–1894, August 2002.
[14] R. E. Mahony, R. C. Williamson: Prior knowledge and preferential structures in gradient descent learning algorithms, *J. of Machine Learning Res.*, **1**, 311–355, Sept. 2001.
[15] T. Gaensler, J. Benesty, S. L. Gay, M. M. Sondhi: A robust proportionate affine projection algorithm for network echo cancellation, *Proc. ICASSP '00*, **2**, 793-796, Istanbul, Turkey, 2000.
[16] O. Hoshuyama, R. Goubran, A. Sugiyama: A generalized proportionate variable step-size algorithm for fast changing acoustic environments, *Proc. ICASSP '04*, 161–164, Montreal, Canada, 2004.
[17] J. Benesty, Y. Huang, D. R. Morgan: On a class of exponentiated adaptive algorithms for the identification of sparse impulse responses, in J. Benesty, Y. Huang (eds.): *Adaptive Signal Processing: Applications to Real-World Problems*, Berlin, Germany: Springer, 2003, Chapter 1, 1–22.
[18] J. Benesty, Y. Huang: The LMS, PNLMS, and exponentiated gradient algorithms, *Proc. EUSIPCO '04*, 721–724, Vienna, Austria, 2004.
[19] J. Benesty, T. Gänsler, D. R. Morgan, M. M. Sondhi, S. L. Gay: *Advances in Network and Acoustic Echo Cancellation*, Berlin, Germany: Springer, 2001.
[20] R. M. Corless, G. H. Gonnet, D. E. G. Hare, D. J. Jeffrey, D. E. Knuth: On the Lambert W Function, *Advances in Computational Mathematics*, **5**, 329–359, 1996.

[21] G. H. Golub, C. F. Van Loan: *Matrix Computations*, Baltimore, MD: The Johns Hopkins University Press, 1996.

[22] J. Benesty, Y. Huang, J. Chen: An exponentiated gradient adaptive algorithm for blind identification of sparse SIMO systems, *Proc. ICASSP '04*, **2**, 829-832, Montral, Canada, 2004.

6

Selective-Tap Adaptive Algorithms for Echo Cancellation

Patrick A. Naylor and Andy W.H. Khong

Imperial College London

An unknown echo path impulse response can be estimated using system identification techniques based on adaptive filters. For acoustic echo cancellation, in which adaptive filters with up to several thousand taps are required, there is a strong motivation to seek out methods for reducing the computational complexity of adaptation. One such method is to employ selective-tap adaptive algorithms in which only a subset of taps are updated at each iteration. In this chapter, it will be shown how the use of selective-tap algorithms is equivalent to imposing an approximation of sparseness on the input signal and that such an approximation only causes a graceful degradation in echo cancellation performance. The Normalized LMS algorithm will initially be employed to illustrate the main concepts underpinning selective-tap algorithms. A convergence analysis will be presented for time-varying systems. The concept of tap selection will then be applied to both Affine Projection and Recursive Least Squares adaptive filters. Simulation results are given to illustrative the performance of these algorithms for acoustic echo cancellation. Selective-tap algorithms typically suffer from a computational overhead in determining which taps should be adapted at each iteration, often in the form of a sorting operation. Fast sorting procedures will be described which substantially alleviate this computational overhead. Recently, a new use of selective-tap adaptive filters has been proposed for multichannel algorithms which exploits the sparseness approximation not to reduce complexity but instead to decorrelate the input signals in, for example, stereophonic acoustic echo cancellers. Such decorrelation will be shown to address the well known misalignment problem in stereo systems and significant performance enhancements will be demonstrated. The chapter will end with concluding remarks on selective-tap algorithms applied to echo cancellation.

6.1 Introduction

Adaptive system identification of echo responses has been, and continues to be, a topic of significant interest both in the telecommunications industry and the research communities that support it [30]. Whether the echo path arises due to electrical or acoustic coupling in a telecommunications terminal strongly affects the nature of the echo cancellation problem. The well-known technical challenges that must be faced in the design of acoustic echo cancellers include (i) the long duration of the unknown echo path response, which typical requires several thousand adaptive coefficients to model accurately, (ii) the highly nonstationary behaviour of the echo response, particularly in its later coefficients and (iii) the need to train the adaptive echo canceller using the speech signal itself, which is nonideal in terms of its spectrum, persistence and sample amplitude distribution.

The motivation for the introduction of selective-tap adaptive algorithms can be explained by considering the high computational load of adaptive algorithms with several thousand coefficients. The Normalized Least Mean Squares (NLMS) algorithm [31] for an adaptive filter of length L requires approximately $2L$ multiply-accumulate (MAC) operations per sampling period of the signal. In the past, this rate of operation was considered high for typical telecommunications end-user equipment and researchers were therefore motivated to seek techniques that could reduce the computational complexity of adaptation without significantly degrading effectiveness in terms of its convergence rate or final misadjustment. More recently, the computational capability of low-cost processing hardware has increased very rapidly so that a typical NLMS implementation would not be seen as a heavy computational demand. However, new pressures on product design have emerged - the increase of user mobility imposes a requirement of low power consumption for portable battery powered equipment; the growth of telecommunications usage imposes a requirement of high density implementation for infrastructure equipment so that the number of simultaneous echo cancellers of given tap length that can be run within a specified MIP-budget (millions of instructions per second) is maximized. Both these requirements renew the motivation for low computational complexity, even with today's high speed processors.

The basic method employed for achieving low complexity in selective-tap adaptive algorithms is to compute the coefficient update calculations at a rate lower than the sampling rate. Algorithms differ in the criteria used for selecting which coefficients to update at each iteration. The Sequential-LMS and Periodic-LMS algorithms [19] employ tap selection schemes that are independent of the input data. In contrast, data dependent tap selection criteria are employed in later algorithms including Max-LMS [18] and MMax-NLMS [2,3]. Block-based and transform domain algorithms have also been proposed, for example [15, 16].

The evaluation of selective-tap adaptive algorithms normally involves quantification of the trade-off between the computational complexity of the algorithm and its performance. It is usually not possible to make general conclusions concerning computational complexity since it depends heavily on the implementation details and architecture of the hardware employed. Instead, it is common practice to quantify complexity in terms of the number of MACs and, if appropriate, comparisons. Algorithm performance is normally measured in terms of the time evolution and asymptotic behaviour of measures such as the output Mean Square Error (MSE), the Echo Return Loss Enhancement (ERLE) and the Weight Error Vector Norm (WEVN) defined elsewhere in this book. It is normal to expect that as the number of coefficients updated per iteration is reduced, the computation complexity is also reduced but at the expense of some loss of performance. Hence the goal of the designers of selective-tap algorithms is to find ways to reduce the number of coefficients updated per iteration in a manner which degrades algorithm performance as little as possible. It will be seen that tap selection criteria have been proposed that enable as few as 50% of coefficients to be updated per iteration without significant loss of performance.

Whereas selective-tap adaptive filters were originally proposed with the aim of reducing computational complexity in single channel applications, multichannel selective-tap algorithms have recently been proposed that make use of other properties introduced by tap selection. In multichannel acoustic echo cancellation [5], standard adaptive filters converge poorly because of high levels of interchannel coherence between the input signals. It has been shown [34] that tap selection can be employed to reduce significantly the interchannel coherence, thereby improving convergence. Although not their main aim, such approaches also yield some reduction in complexity.

In this Chapter, we will initially review adaptive algorithms employing data independent tap selection. Selective-tap algorithms will then been described in which the tap selection is made dependent on the input data and it will be shown that such algorithms employ an assumption of sparseness of the input signal. A third class of tap selection will then be introduced that extends the tap selection to be dependent not only on the input data but also dependent on the unknown system estimate at each iteration. The final part of this Chapter will consider the multichannel case and will focus specifically on the application of selective-tap adaptive algorithms to the important example of stereophonic acoustic echo cancellation (SAEC).

6.2 Sequential and Periodic Tap Selection

The Periodic-LMS and Sequential-LMS algorithms were proposed in [19] and perform tap selection in a data independent manner. In the Periodic-LMS algorithm, reduction in computation is achieved at each time iteration n by updating filter coefficients periodically using the $N\lfloor n/N \rfloor^{\text{th}}$ instantaneous gradient estimate where $\lfloor \cdot \rfloor$ is defined as the truncation operator and $N \in \{1, 2, \cdots, L\}$. In addition, defining $l = 0, 1, \ldots, L-1$ as the tap indices, only taps satisfying the condition $(n+l) \bmod N = 0$ are updated. Combining these two features and defining an $L \times L$ tap selection matrix

$$\begin{aligned}\boldsymbol{Q}(n) &= \mathrm{diag}\{q_0(n), q_1(n), \ldots, q_{L-1}(n)\} \\ &= \begin{bmatrix} q_0(n) & 0 & \cdots & 0 \\ 0 & q_1(n) & \ddots & \vdots \\ \vdots & \ddots & \ddots & 0 \\ 0 & \cdots & 0 & q_{L-1}(n) \end{bmatrix}_{L \times L}\end{aligned} \tag{6.1}$$

the Periodic-LMS update can be expressed as

$$\widehat{\boldsymbol{h}}(n+1) = \widehat{\boldsymbol{h}}(n) + \mu \boldsymbol{Q}(n)\boldsymbol{x}(j)e(j) , \tag{6.2}$$

where $j = N\lfloor n/N \rfloor$. The tap selection elements for $l = 0, 1, \ldots, L-1$ are given as

$$q_l(n) = \begin{cases} 1, & \text{if } (n+l) \bmod N = 0, \\ 0, & \text{otherwise,} \end{cases} \tag{6.3}$$

while the error signal $e(n)$ is expressed as

$$e(n) = d(n) - \boldsymbol{x}^{\mathrm{T}}(n)\widehat{\boldsymbol{h}}(n). \tag{6.4}$$

It can be seen that at each time iteration, L/N filter coefficients are updated such that after N iterations all the filter coefficients are updated once. For $N = 1$, the Periodic-LMS algorithm reduces to the LMS algorithm.

In contrast to Periodic-LMS, Sequential-LMS employs the instantaneous gradient estimate at each time iteration for updating the coefficients. The filter coefficients satisfying the condition $(n - l + 1) \bmod N = 0$ only are updated. The Sequential-LMS tap-update is expressed by

$$\widehat{\boldsymbol{h}}(n+1) = \widehat{\boldsymbol{h}}(n) + \mu \boldsymbol{Q}(n)\boldsymbol{x}(n)e(n) , \tag{6.5}$$

where the tap selection elements are now given as

$$q_l(n) = \begin{cases} 1, & \text{if } (n-l+1) \bmod N = 0, \\ 0, & \text{otherwise.} \end{cases} \tag{6.6}$$

Similarly to Periodic-LMS, the Sequential-LMS algorithm is equivalent to the LMS algorithm when $N = 1$. The computational complexity in terms of number of multiplications per iteration for the Periodic-LMS and Sequential-LMS is given as $(2L+1)/N + 1/N$ and $1 + (1 + 1/N)L$ respectively. Normalization of these algorithms follows exactly the approach used in NLMS.

A brief performance evaluation of Periodic-NLMS and Sequential-NLMS is given in Fig. 6.4.

6.3 MMax Tap Selection

It has been seen how the Sequential and Periodic tap selection algorithms employ tap selection criteria which are independent of the input data. In contrast, MMax tap selection employs a data-dependent criterion and gives better performance. The MMax tap selection criterion chooses $M \leq L$ coefficients to update at each iteration. Coefficients are selected for updating if they correspond to one of the M largest amplitude elements of the tap-input vector. The remaining $L - M$ coefficients corresponding to the small amplitude elements of the tap-input vector are not updated.

This approach can be justified in the context of the NLMS coefficient update equation [31]

$$\widehat{\boldsymbol{h}}(n+1) = \widehat{\boldsymbol{h}}(n) + \frac{\mu}{\|\boldsymbol{x}(n)\|_2^2 + \delta} \boldsymbol{x}(n)e(n) \tag{6.7}$$

where $\|\cdot\|_2^2$ is defined as the squared l_2-norm operator and δ is the regularization constant. The second term on the right hand side modifies the elements of $\widehat{\boldsymbol{h}}$ by an amount proportional in magnitude to the corresponding elements of the tap-input vector $\boldsymbol{x}(n)$. Therefore, it can be seen that a signal with many samples of small amplitude will give rise to correspondingly many small updates of coefficients. In MMax tap selection, these small updates are approximated as zero, which is equivalent to approximating the $M - L$ smallest elements in the tap-input vector $\boldsymbol{x}(n)$ to be zero. We refer to this as imposing a sparse approximation on the tap-input vector. The validity of this assumption is clearly data dependent but generally accepted to be a reasonable assumption for the majority of speech signals. In order to justify the use of this assumption, let us first define a signal to be sparse if a large fraction of its energy is concentrated in a small fraction of its duration [43]. As can be seen in Fig. 6.1 for a typical sentence of male speech analyzed using a frame duration of 128 ms, 50% of the speech energy in this example is contained within 16% of the frame duration.

Based on the work in [40], one of the earliest partial-update algorithms is introduced in [17] where a family of NLMS algorithms are derived by minimizing the change in adaptive coefficients using different l-norms. Defining $\|\cdot\|_1$ as the l_1-norm, it was found that by minimizing

$$\|\widehat{\boldsymbol{h}}(n+1) - \widehat{\boldsymbol{h}}(n)\|_1^2 , \tag{6.8}$$

subject to the constraint of

$$\widehat{\boldsymbol{h}}^{\mathrm{T}}(n+1)\boldsymbol{x}(n) = d(n) , \tag{6.9}$$

the adaptive algorithm degenerates to Max-NLMS [18] in which, at each time iteration, only the filter coefficient associated with the tap-input sample that has the largest magnitude in $\boldsymbol{x}(n)$ is updated.

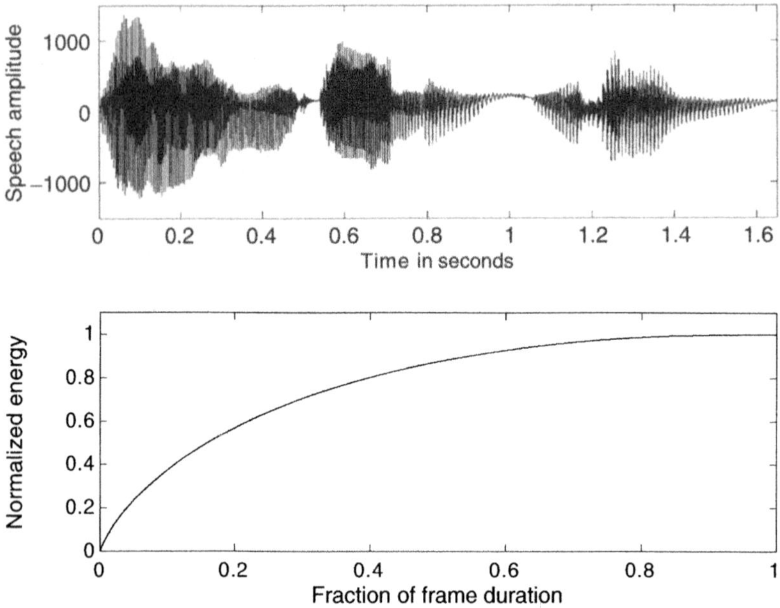

Fig. 6.1. Speech signal (upper) and sparseness (lower).

6.3.1 The MMax-NLMS Algorithm

The single channel MMax-NLMS algorithm is now briefly reviewed and the concept of MMax tap selection is then extended to the single channel affine projection (AP) and recursive least squares (RLS) algorithms in Sec. 6.3.3 and Sec. 6.3.4, respectively. The main benefit reported to motivate the introduction of AP and RLS selective-tap schemes is that they form the basis of selective-tap algorithms which are able to improve the conditioning of multi-input-multi-output (MIMO) system identification problems with correlated inputs such as occur in stereophonic acoustic echo cancellation (SAEC) [34] which we will discuss in detail in Sec. 6.8.

The single channel MMax-NLMS algorithm [2] is a direct extension of the Max-NLMS algorithm as described in [17] and [18] such that, for an adaptive filter of length L, a number of coefficients $1 \leq M \leq L$ corresponding to the M largest magnitude tap-inputs are selected for updating at each iteration. Let the subselected tap-input vector be defined

$$\widetilde{\boldsymbol{x}}(n) = \boldsymbol{Q}(n)\boldsymbol{x}(n) , \tag{6.10}$$

such that

$$\boldsymbol{Q}(n) = \mathrm{diag}\{q_0(n), q_1(n), \ldots, q_{L-1}(n)\}$$
$$= \begin{bmatrix} q_0(n) & 0 & \cdots & 0 \\ 0 & q_1(n) & \ddots & \vdots \\ \vdots & \ddots & \ddots & 0 \\ 0 & \cdots & 0 & q_{L-1}(n) \end{bmatrix}_{L\times L} \tag{6.11}$$

where for $l = 0, 1, \ldots, L-1$,

$$q_l(n) = \begin{cases} 1, & |x_l(n)| \in \{M \text{ maxima of } |\boldsymbol{x}(n)|\} \\ 0, & \text{otherwise.} \end{cases} \tag{6.12}$$

The MMax-NLMS tap-update equation is then given by

$$\widehat{\boldsymbol{h}}(n+1) = \widehat{\boldsymbol{h}}(n) + \boldsymbol{Q}(n)\frac{\mu\boldsymbol{x}(n)e(n)}{\|\boldsymbol{x}(n)\|_2^2 + \delta}, \tag{6.13}$$

where δ and μ are the regularization parameter and step size respectively. This algorithm is summarized in Table 6.2.

6.3.2 Dependence of Convergence Rate on MMax Tap Selection

For an adaptive filter of length L, the dependence of convergence rate on the number, M, of coefficients selected for updating at each iteration can be examined using the measure

$$\mathcal{M}(n) = \frac{\|\boldsymbol{Q}(n)\boldsymbol{x}(n)\|_2^2}{\|\boldsymbol{x}(n)\|_2^2}. \tag{6.14}$$

This measure quantifies the ratio of the energy of the M selected tap-inputs to the energy of the full tap-input vector so that $\mathcal{M} = 1$ corresponds to updating of all the coefficients. The study of $\mathcal{M}(n)$ provides some useful insight into the robustness of adaptive algorithms to MMax tap selection. In addition, $\mathcal{M}(n)$ will be seen to be a key feature used in the formulation of multichannel tap selection criteria as will be discussed later in Sec. 6.8.

Fig. 6.2 shows how $\mathcal{M}$ varies with the number of selected taps M for zero mean, unit variance white Gaussian noise (WGN) at a particular time iteration n. We note that $\mathcal{M}$ exhibits only a modest reduction from unity in the range $0.5L \leq M < L$. Fig. 6.3 shows the number of iterations for MMax-NLMS to achieve -20 dB normalized misalignment for various $\mathcal{M}$ and hence verifies our expectation that, over the range $0.5L \leq M < L$, a graceful reduction in convergence rate is obtained as compared to full update adaptation ($M = L$) [33].

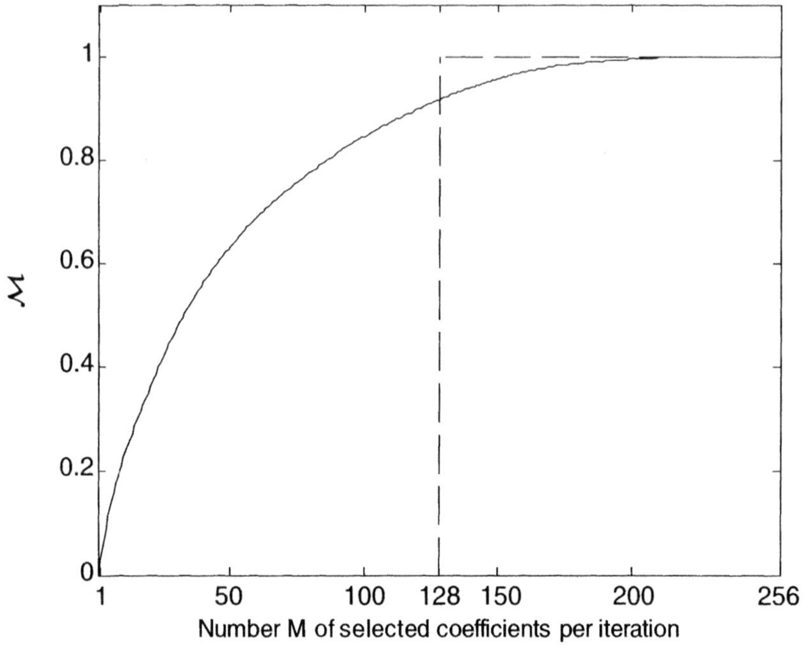

Fig. 6.2. Variation of $\mathcal{M}$ (see (6.14)) with number M of selected coefficients per iteration showing modest reduction of $\mathcal{M}$ within the region $0.5L \leq M < L$ for WGN sequence with $L = 256$.

6.3.3 The MMax Affine Projection Algorithm

The affine projection (AP) algorithm [31] incorporates multiple projections by concatenating past tap-input vectors from time iteration n to time iteration $n - K + 1$ where K is defined as the projection order. In a similar manner, our approach for MMax-AP will be to concatenate the subselected tap-input vectors such that they propagate consistently from each iteration to the next. To formulate the MMax-AP algorithm [42], we first define the subselected and full tap-input matrices of dimensions $K \times L$ respectively as

$$\widetilde{\boldsymbol{X}}(n) = \Big[\widetilde{\boldsymbol{x}}(n), \widetilde{\boldsymbol{x}}(n-1), \ldots, \widetilde{\boldsymbol{x}}(n-K+1)\Big]^{\mathrm{T}}, \tag{6.15}$$

$$\boldsymbol{X}(n) = \Big[\boldsymbol{x}(n), \boldsymbol{x}(n-1), \ldots, \boldsymbol{x}(n-K+1)\Big]^{\mathrm{T}}. \tag{6.16}$$

The tap-update for the MMax-AP algorithm is then given by

$$\widehat{\boldsymbol{h}}(n+1) = \widehat{\boldsymbol{h}}(n) + \mu \widetilde{\boldsymbol{X}}^{\mathrm{T}}(n)\Big[\boldsymbol{X}(n)\boldsymbol{X}^{\mathrm{T}}(n) + \delta \boldsymbol{I}\Big]^{-1} \mathbf{e}(n), \tag{6.17}$$

where $\boldsymbol{I}$ is a $K \times K$ identity matrix, $\mathbf{e}(n) = [e(n), e(n-1), \ldots, e(n-K+1)]^{\mathrm{T}}$ and δ is the regularization parameter. Thus for projection order $K = 1$,

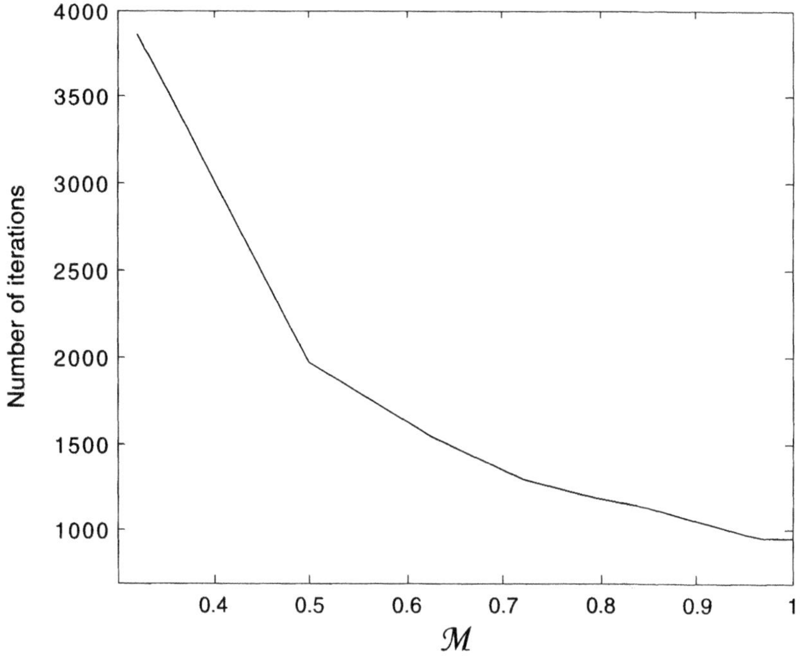

Fig. 6.3. Number of iterations to converge to -20 dB normalized misalignment as a function of $\mathcal{M}$ (see (6.14)) for $L = 256$.

MMax-AP is equivalent to MMax-NLMS. We note that MMax-AP in general cannot be classified as a partial-update algorithm since the tap-update vector $\widetilde{\boldsymbol{X}}^{\mathrm{T}}(n)\left[\boldsymbol{X}(n)\boldsymbol{X}^{\mathrm{T}}(n)+\delta\boldsymbol{I}\right]^{-1}\boldsymbol{e}(n)$ is fully populated and therefore every coefficient in $\widehat{\boldsymbol{h}}(n)$ will be updated at each iteration. Consequently, we classify MMax-AP instead as a selective-tap algorithm.

6.3.4 The MMax Recursive Least Squares Algorithm

One of the main disadvantages of the NLMS algorithm is the dependence of convergence rate on the eigenvalue spread of $\boldsymbol{R}_{\boldsymbol{xx}} = \mathrm{E}\left\{\boldsymbol{x}(n)\boldsymbol{x}^{\mathrm{T}}(n)\right\}$. Specifically, input signals having a small eigenvalue spread exhibit higher rate of convergence compared to those having larger eigenvalue spread [31]. This affects the performance of speech applications where the eigenvalue spread can be very significant (of the order of several hundred times higher than for a WGN input). We shall next derive the recursive least squares (RLS) algorithm employing MMax tap selection.

The tap-update equation of the RLS algorithm is given by [31]

$$\widehat{\boldsymbol{h}}(n+1) = \widehat{\boldsymbol{h}}(n) + \boldsymbol{k}(n)e(n)\ , \tag{6.18}$$

where $\boldsymbol{k}(n) = \boldsymbol{\Psi}^{-1}(n)\boldsymbol{x}(n)$ is defined as the Kalman gain. The time averaged autocorrelation matrix with forgetting factor λ, $0 \ll \lambda < 1$ is given by

$$\boldsymbol{\Psi}(n) = \sum_{j=1}^{n} \lambda^{n-j} \boldsymbol{x}(j)\boldsymbol{x}(j)^{\mathrm{T}} \tag{6.19}$$

and $\boldsymbol{\Psi}^{-1}(n)$ can be found recursively using [31]

$$\boldsymbol{\Psi}^{-1}(n) = \frac{\lambda^{-1}\boldsymbol{\Psi}^{-1}(n-1)}{1 + \lambda^{-1}\boldsymbol{x}^{\mathrm{T}}(n)\boldsymbol{\Psi}^{-1}(n-1)\boldsymbol{x}(n)} . \tag{6.20}$$

We note that direct extension of the MMax tap selection approach achieved by sorting the magnitude of $\boldsymbol{k}(n)$ in (6.18) will not give the desired convergence behaviour especially for statistically non-stationary signals such as speech. This is because the Kalman gain depends on previous values of the time-averaged autocorrelation matrix [34].

Our approach will be to subsample the tap-input vectors at each time iteration based on the MMax tap selection criterion such that $\boldsymbol{\Psi}(n)$ is computed from $\widetilde{\boldsymbol{x}}(n)$ giving $\widetilde{\boldsymbol{\Psi}}(n)$. This ensures that the subselected tap-input vectors propagate consistently through the memory of the RLS algorithm.

The MMax-RLS algorithm [42] solves the least-squares normal equation formed from $\widetilde{\boldsymbol{x}}(n)$ given as

$$\widehat{\boldsymbol{h}}(n) = \widetilde{\boldsymbol{\Psi}}^{-1}(n)\widetilde{\boldsymbol{\Theta}}(n) \tag{6.21}$$

where

$$\widetilde{\boldsymbol{\Psi}}(n) = \sum_{j=1}^{n} \lambda^{n-j} \widetilde{\boldsymbol{x}}(j)\widetilde{\boldsymbol{x}}^{\mathrm{T}}(j) , \tag{6.22}$$

$$\widetilde{\boldsymbol{\Theta}}(n) = \sum_{j=1}^{n} \lambda^{n-j} \widetilde{\boldsymbol{x}}(j) d(j) , \tag{6.23}$$

where $d(j)$ is the receiving room's microphone signal at the jth iteration.

We may now express (6.22) recursively as

$$\begin{aligned}\widetilde{\boldsymbol{\Psi}}(n) &= \widetilde{\boldsymbol{X}}(n)\boldsymbol{\Lambda}(n)\widetilde{\boldsymbol{X}}^{\mathrm{T}}(n) \\ &= \lambda\widetilde{\boldsymbol{\Psi}}(n-1) + \widetilde{\boldsymbol{x}}(n)\widetilde{\boldsymbol{x}}^{\mathrm{T}}(n) ,\end{aligned} \tag{6.24}$$

where $\widetilde{\boldsymbol{X}}(n) = [\widetilde{\boldsymbol{x}}(1), \widetilde{\boldsymbol{x}}(2), \ldots, \widetilde{\boldsymbol{x}}(n)]$ and $\boldsymbol{\Lambda}(n) = \mathrm{diag}\{[\lambda^n, \lambda^{n-1}, \ldots, \lambda]\}$. As before, the subselected tap-input vector is given as $\widetilde{\boldsymbol{x}}(n) = \boldsymbol{Q}(n)\boldsymbol{x}(n)$ where the diagonal elements of the MMax tap selection matrix $\boldsymbol{Q}(n)$ are defined in (6.12). In a similar manner, the cross-correlation vector in (6.23) may be expressed recursively as

$$\begin{aligned}\widetilde{\boldsymbol{\Theta}}(n) &= \widetilde{\boldsymbol{X}}(n)\boldsymbol{\Lambda}(n)\boldsymbol{d}(n) \\ &= \lambda\widetilde{\boldsymbol{\Theta}}(n-1) + \widetilde{\boldsymbol{x}}(n)d(n)\end{aligned} \tag{6.25}$$

with $\boldsymbol{d}(n) = [d(1), d(2), \ldots, d(n)]^{\mathrm{T}}$.

Like the RLS algorithm, the MMax-RLS utilizes the matrix inversion lemma [31] to compute $\widetilde{\boldsymbol{\Psi}}^{-1}(n)$ efficiently as

$$\widetilde{\boldsymbol{\Psi}}^{-1}(n) = \frac{1}{\lambda}\left[\widetilde{\boldsymbol{\Psi}}^{-1}(n-1) - \widetilde{\boldsymbol{k}}(n)\widetilde{\boldsymbol{x}}^{\mathrm{T}}(n)\widetilde{\boldsymbol{\Psi}}^{-1}(n-1)\right], \tag{6.26}$$

where the modified Kalman gain is now given by

$$\begin{aligned}\widetilde{\boldsymbol{k}}(n) &= \frac{\lambda^{-1}\widetilde{\boldsymbol{\Psi}}^{-1}(n-1)\widetilde{\boldsymbol{x}}(n)}{1+\lambda^{-1}\widetilde{\boldsymbol{x}}^{\mathrm{T}}(n)\widetilde{\boldsymbol{\Psi}}^{-1}(n-1)\widetilde{\boldsymbol{x}}(n)} \\ &= \lambda^{-1}\left[\widetilde{\boldsymbol{\Psi}}^{-1}(n-1) - \widetilde{\boldsymbol{k}}(n)\widetilde{\boldsymbol{x}}^{\mathrm{T}}(n)\widetilde{\boldsymbol{\Psi}}^{-1}(n-1)\right]\widetilde{\boldsymbol{x}}(n) \\ &= \widetilde{\boldsymbol{\Psi}}^{-1}(n)\widetilde{\boldsymbol{x}}(n)\,. \end{aligned} \tag{6.27}$$

The recursive solution to the normal equation given in (6.21) can be obtained by substituting the recursive form of $\widetilde{\boldsymbol{\Theta}}(n)$ and $\widetilde{\boldsymbol{\Psi}}(n)$ in (6.25) and (6.26) into (6.21) and using (6.27), the MMax-RLS tap-update equation is then expressed by

$$\widehat{\boldsymbol{h}}(n+1) = \widehat{\boldsymbol{h}}(n) + \widetilde{\boldsymbol{k}}(n)e(n)\,. \tag{6.28}$$

Similar to the MMax-AP algorithm as described in Sec. 6.3.3, the MMax-RLS algorithm in general updates all the taps at each iteration since the Kalman gain vector $\widetilde{\boldsymbol{k}}(n)$ is a fully populated column vector. We choose therefore to denote this a selective-tap algorithm rather than a partial-update algorithm.

6.3.5 Computational Complexity

The MMax tap selection procedure selects the M largest tap-inputs at each time iteration. This sorting operation can be achieved efficiently using, for example, the SORTLINE [44] or the Short-sort [41] routines given in Table 6.6. The Short-sort MMax procedure operates by considering a short segment of the tap-input vector $\boldsymbol{x}_{\mathrm{ss}}(n) = [x(n), x(n-1), \ldots, x(n-S+1)]$ of length $S \ll L$. Once every S iterations, an efficient insertion sort [35] is performed on $\boldsymbol{x}_{\mathrm{ss}}(n)$, as shown in [41], and A coefficients are selected corresponding to the elements of $\boldsymbol{x}_{\mathrm{ss}}(n)$ with largest magnitude. This tap selection is propagated through the filter by incrementing the indices of the selected coefficients by one at each sample period. Thus the worst-case comparison load using Short-sort is $(1+S-A)A/S$ comparisons per iteration compared to $2 + 2\log_2 L$ used in the SORTLINE procedure.

We now consider the computational complexity of the Short-sort MMax NLMS (SM-NLMS), MMax-NLMS, MMax-AP and MMax-RLS algorithms. For the purpose of this comparison, we define complexity as the total number of multiplications and comparisons per sample period. Thus MMax-NLMS employing the SORTLINE procedure requires at most $L + M + 3 + 2\log_2 L$ operations whereas $L + S + (S + 1 - A)A/S$ operations are required for SM-NLMS.

The complexity of AP using the generalized Levinson algorithm is $2LK + 7K^2$ multiplies per sample period [25]. The MMax-AP algorithm employing the SORTLINE procedure requires an additional $2 + 2\log_2 L$ sorting operations in each channel for $\widetilde{\boldsymbol{x}}(n)$. However, due to a reduction in multiplications required when computing the sparse vector $\widetilde{\boldsymbol{X}}^{\mathrm{T}}(n)\left[\boldsymbol{X}(n)\boldsymbol{X}^{\mathrm{T}}(n) + \delta\boldsymbol{I}\right]^{-1}$, the complexity for MMax-AP is $(M + L)K + 7K^2 + 2 + 2\log_2 L$ operations per sample period.

The number of multiplications required for the RLS algorithm is $4L^2 + 3L + 2$ per adaptive filter where an additional L multiplications are required for the tap-updates. Due to the subselection of input vector $\widetilde{\boldsymbol{x}}(n)$, the number of multiplications required for computing $\widetilde{\boldsymbol{\Psi}}(n)$ for the MMax-RLS is $(M + L)L + 1$ while $L^2 + M$ multiplications are required for computing the Kalman gain. Hence the number of operations required for the MMax-RLS is at most $L(L + 3M + 2) + M + 3 + 2\log_2 L$ per sample period. The computational complexity of the algorithms described is summarized in Table 6.1.

Table 6.1. Computational Complexity of MMax Algorithms

Algorithm	**Sort Procedure**	**Multiplications and Comparisons**
SM-NLMS	Short-sort	$L + S + (S + 1 - A)A/S$
MMax-NLMS	SORTLINE	$L + M + 3 + 2\log_2 L$
MMax-AP	SORTLINE	$(M + L)K + 7K^2 + 2 + 2\log_2 L$
MMax-RLS	SORTLINE	$L(L + 3M + 2) + M + 3 + 2\log_2 L$

6.4 Selective Partial Update Tap Selection

As with the other partial update algorithms so far discussed, the objective of Selective Partial Update NLMS (SPU-NLMS) [16] is to reduce computational complexity of the adaptive filter by updating only a subset of filter coefficients at each interaction. A key feature of SPU-NLMS is the partitioning of tap-input vector $\boldsymbol{x}(n) = [x_0(n), x_1(n), \ldots, x_{L-1}(n)]^{\mathrm{T}}$ and corresponding coefficient vector into $\mathcal{B}$ blocks so that

$$\boldsymbol{x}(n) = \left[\boldsymbol{x}_1^{\mathrm{T}}(n), \boldsymbol{x}_2^{\mathrm{T}}(n), \ldots, \boldsymbol{x}_{\mathcal{B}}^{\mathrm{T}}(n)\right]^{\mathrm{T}} \tag{6.29}$$

$$\widehat{\boldsymbol{h}}(n) = \left[\widehat{\boldsymbol{h}}_1^{\mathrm{T}}(n), \widehat{\boldsymbol{h}}_2^{\mathrm{T}}(n), \ldots, \widehat{\boldsymbol{h}}_{\mathcal{B}}^{\mathrm{T}}(n)\right]^{\mathrm{T}}, \tag{6.30}$$

from which the update of block i is

$$\widehat{\boldsymbol{h}}_i(n+1) = \widehat{\boldsymbol{h}}_i(n) + \frac{\mu \boldsymbol{x}_i(n) e(n)}{\|\boldsymbol{x}_i(n)\|_2^2 + \delta}, \tag{6.31}$$

and is derived as the solution to the constrained minimization problem [28]

$$\min_{1 \leq i \leq \mathcal{B}} \quad \min_{\widehat{\boldsymbol{h}}_i(n+1)} \|\widehat{\boldsymbol{h}}_i(n+1) - \widehat{\boldsymbol{h}}_i(n)\|_2^2 \tag{6.32}$$

$$\text{subject to the constraint } \widehat{\boldsymbol{h}}^{\mathrm{T}}(n+1)\boldsymbol{x}(n) = d(n). \tag{6.33}$$

A decision can then be made at each iteration n on which B out of $\mathcal{B}$ blocks to update. For $B = 1$, it is shown that the block, i, with the smallest squared Euclidean norm in (6.32) should be updated and this is found from the minimization

$$\begin{aligned}
i &= \arg\min_{1 \leq j \leq \mathcal{B}} \|\widehat{\boldsymbol{h}}_j(n+1) - \widehat{\boldsymbol{h}}_j(n)\|_2^2 \\
&= \arg\min_{1 \leq j \leq \mathcal{B}} \left\| \frac{\boldsymbol{x}_j(n) e(n)}{\|\boldsymbol{x}_j(n)\|_2^2} \right\|_2^2 \\
&= \arg\min_{1 \leq j \leq \mathcal{B}} \frac{1}{\|\boldsymbol{x}_j(n)\|_2^2} \\
&= \arg\max_{1 \leq j \leq \mathcal{B}} \|\boldsymbol{x}_j(n)\|_2^2.
\end{aligned} \tag{6.34}$$

To update more than one block, $1 < B \leq \mathcal{B}$, the set $\mathcal{I}_B = \{i_1, i_2, \ldots, i_B\}$ is defined to contain the indices of the blocks to be updated such that

$$\boldsymbol{x}_{\mathcal{I}_B}(n) = \left[\boldsymbol{x}_{i_1}^{\mathrm{T}}(n), \boldsymbol{x}_{i_2}^{\mathrm{T}}(n), \ldots, \boldsymbol{x}_{i_B}^{\mathrm{T}}(n)\right]^{\mathrm{T}}. \tag{6.35}$$

The SPU-NLMS algorithm is then given as

$$\begin{aligned}
\widehat{\boldsymbol{h}}_{\mathcal{I}_B}(n+1) &= \widehat{\boldsymbol{h}}_{\mathcal{I}_B}(n) + \frac{\mu \boldsymbol{x}_{\mathcal{I}_B}(n) e(n)}{\|\boldsymbol{x}_{\mathcal{I}_B}(n)\|_2^2 + \delta} \\
\mathcal{I}_B &= \{i \text{ for which } \|\boldsymbol{x}_i(n)\|_2^2 \text{ is one of the} \\
&\qquad B \text{ greatest of } \|\boldsymbol{x}_1(n)\|_2^2, \ldots, \boldsymbol{x}_{\mathcal{B}}(n)\|_2^2\}.
\end{aligned} \tag{6.36}$$

For $\mathcal{B} = L$, the tap selection criteria used in SPU-NLMS and MMax-NLMS are equivalent.

Extension of the selective-partial-update approach to include the affine projection adaptive algorithm is presented in [16]. Further discussion and analysis of the algorithm is presented in [49]. Bounds on the step size μ are derived for convergence in the mean squared sense and it is shown that an instantaneous estimate for μ giving the fastest convergence rate is

$$\hat{\mu} = \frac{\|\boldsymbol{x}_{\mathcal{I}_B}(n)\|_2^2}{\|\boldsymbol{x}(n)\|_2^2} \tag{6.37}$$

which implies normalization by the l_2-norm of the complete tap-input vector as in the MMax-NLMS algorithm. Such normalization has been employed in the comparative simulations in Sec. 6.5. In addition, [49] employs the concept of set-membership adaptive filters [27] jointly with the partial updating scheme to obtain a set-membership partial update NLMS algorithm.

6.5 Performance Comparison for Single-Channel Selective-Tap algorithms

We now compare the convergence performance of the fully updated NLMS algorithms to the selective-tap Sequential-NLMS, Periodic-NLMS, SPU-NLMS and MMax-NLMS algorithms. Performance evaluation of the MMax-AP and MMax-RLS algorithms is given later in Sec. 6.8 in the context of multichannel techniques since this is the manner in which they would more normally be deployed.

In this single channel example, the echo path impulse response $\boldsymbol{h}$ is generated at $f_s = 8$ kHz sampling frequency using the method of images [1] and is of length $L_{\mathrm{R}} = 1024$. The adaptive filter is chosen to be of length $L = 512$. The MMax-NLMS algorithm is tested with $M = L/2$ and $M = L/4$. For both the Sequential-NLMS and Periodic-NLMS we have used $N = 2$. For the SPU-NLMS algorithm, $\mathcal{B} = 32$ and $B = 16$ so that $L/2$ taps are updated at each iteration. The step size for each algorithm is chosen so that all algorithms achieve the same asymptotic performance in terms of final misalignment. This corresponds to $\mu_{\mathrm{NLMS}} = 0.7$ for NLMS, $\mu_{\mathrm{Periodic}} = 0.7$ for Periodic-NLMS, $\mu_{\mathrm{MMax}} = 0.7$ for MMax-NLMS, $\mu_{\mathrm{SPU}} = 0.6$ for SPU-NLMS and $\mu_{\mathrm{Sequential}} = 0.5$ for Sequential-NLMS. The SNR in this experiment is 30 dB. It can be seen from Fig. 6.4 that fully updated NLMS achieves the highest rate of convergence. For the case of MMax-NLMS with $M = 0.5L$, the convergence is close to that of NLMS.

6.6 Convergence Analysis

In this section we analyse the effect of MMax tap selection on convergence of the NLMS algorithm. Since acoustic echo path systems are time varying, we have employed a non-stationary system model which will be initially described. Subsequently, we present a steady-state misalignment analysis of the NLMS and MMax-NLMS algorithms under non-stationary system conditions.

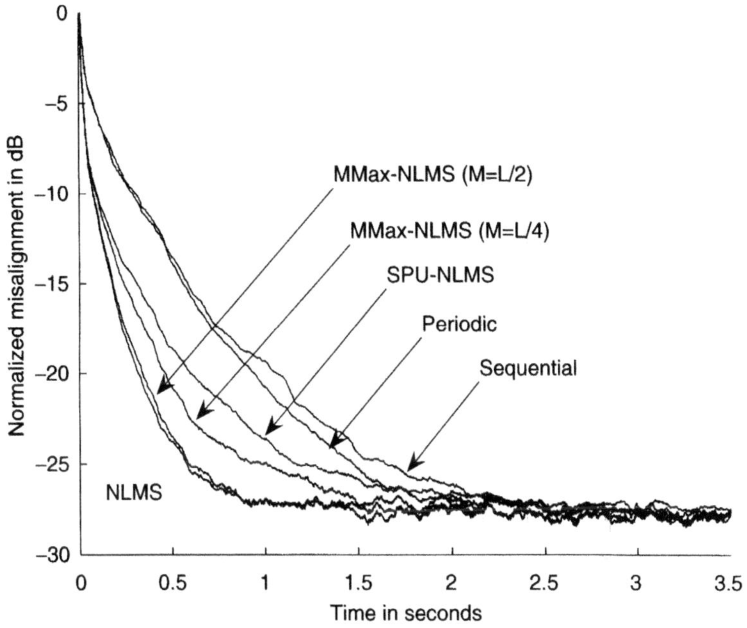

Fig. 6.4. Normalized misalignment comparison for single channel algorithms [$L = 512$, $L_\mathrm{R} = 1024$, $f_\mathrm{s} = 8$ kHz, $\mu_\mathrm{NLMS} = 0.7$, $\mu_\mathrm{Periodic} = 0.7$, $\mu_\mathrm{MMax} = 0.7$, $\mu_\mathrm{SPU} = 0.6$, $\mu_\mathrm{Sequential} = 0.5$ and SNR = 30 dB].

6.6.1 Non-stationary System Model

To introduce a time-varying unknown system model, the modified first-order Markov model [9] ,

$$\boldsymbol{h}(n+1) = \xi \boldsymbol{h}(n) + \sqrt{1-\xi^2}\, \boldsymbol{s}(n) \tag{6.38}$$

is employed, where $\boldsymbol{h}(n)$ is the impulse response of the unknown system and $\boldsymbol{s}(n)$ is a noise process drawn from the normal distribution $\mathcal{N}(0, \sigma_s^2)$. As shown in [9], this model has the key features that the single parameter $0 \ll \xi < 1$ controls the relative contributions to the instantaneous values of the coefficients of 'system memory' (the term $\xi \boldsymbol{h}(n)$) and 'innovations' (the term $\sqrt{1-\xi^2}\boldsymbol{s}(n)$). In addition, the average power of the norm of the coefficients is independent of ξ. It is subsequently shown in [9] that the system variation, measured in terms of the difficulty of tracking by an adaptive filter, is a monotonic decreasing function of ξ.

For the purpose of this analysis we assume that $\mathrm{E}\{\, \boldsymbol{h}(n) \,\} = \boldsymbol{0}$, $\mathrm{E}\{\, w(n) \,\} = 0$ and that $\boldsymbol{h}(n)$ and $w(n)$ are independent, where $w(n)$ is measurement noise of zero mean and variance σ_w^2. We also assume that the dimension of $\widehat{\boldsymbol{h}}(n)$ has been chosen to match the dimension of $\boldsymbol{h}(n)$. We define the misalignment vector

$$\boldsymbol{v}(n) = \widehat{\boldsymbol{h}}(n) - \boldsymbol{h}(n) \tag{6.39}$$

which results in the error signal given by

$$e(n) = w(n) - \boldsymbol{x}^{\mathrm{T}}(n)\boldsymbol{v}(n). \tag{6.40}$$

Consider algorithms of the form

$$\widehat{\boldsymbol{h}}(n+1) = \widehat{\boldsymbol{h}}(n) + \Gamma(n)\boldsymbol{x}(n)e(n) \tag{6.41}$$

where $\boldsymbol{\Gamma}(n)$ is the $L \times L$ diagonal tap selection control matrix, such that for NLMS and MMax-NLMS,

$$\boldsymbol{\Gamma}_{\mathrm{NLMS}}(n) = \frac{2\mu}{\|\boldsymbol{x}(n)\|_2^2 + \delta}\boldsymbol{I} \tag{6.42}$$

and

$$\boldsymbol{\Gamma}_{\text{MMax-NLMS}}(n) = \frac{2\mu}{\|\boldsymbol{x}(n)\|_2^2 + \delta}\boldsymbol{Q}(n) \tag{6.43}$$

respectively where $\boldsymbol{Q}(n)$ is given in (6.12). Using (6.13) and (6.38) - (6.40) we obtain

$$\begin{aligned}
\boldsymbol{v}(n+1) &= \widehat{\boldsymbol{h}}(n+1) - \boldsymbol{h}(n+1) \\
&= \widehat{\boldsymbol{h}}(n) - \xi\boldsymbol{h}(n) - \sqrt{1-\xi^2}\,\boldsymbol{s}(n) + \boldsymbol{\Gamma}(n)\boldsymbol{x}(n)e_(n) \\
&= \boldsymbol{v}(n) + (1-\xi)\boldsymbol{h}(n) + \boldsymbol{\Gamma}(n)\boldsymbol{x}(n)w(n) \\
&\quad -\sqrt{1-\xi^2}\,\boldsymbol{s}(n) - \boldsymbol{\Gamma}(n)\boldsymbol{x}(n)\boldsymbol{x}^{\mathrm{T}}(n)\boldsymbol{v}(n)
\end{aligned} \tag{6.44}$$

from which

$$\begin{aligned}
\boldsymbol{R}_{\boldsymbol{vv}}(n+1) &= \mathrm{E}\left\{\, \boldsymbol{v}(n+1)\boldsymbol{v}^{\mathrm{T}}(n+1) \,\right\} \\
&= \boldsymbol{R}_{\boldsymbol{vv}}(n) + 2(1-\xi)\sigma_s^2\boldsymbol{I} + \sigma_w^2\,\mathrm{E}\left\{\, \boldsymbol{\Gamma}(n)\boldsymbol{x}(n)\boldsymbol{x}^{\mathrm{T}}(n)\boldsymbol{\Gamma}^{\mathrm{T}}(n) \,\right\} \\
&\quad -\boldsymbol{R}_{\boldsymbol{vv}}(n)\,\mathrm{E}\left\{\, \boldsymbol{\Gamma}(n)\boldsymbol{x}(n)\boldsymbol{x}^{\mathrm{T}}(n) \,\right\} \\
&\quad -\boldsymbol{R}_{\boldsymbol{vv}}(n)\,\mathrm{E}\left\{\, \boldsymbol{x}(n)\boldsymbol{x}^{\mathrm{T}}(n)\boldsymbol{\Gamma}^{\mathrm{T}}(n) \,\right\} \\
&\quad +\mathrm{E}\left\{\, \boldsymbol{\Gamma}(n)\boldsymbol{x}(n)\boldsymbol{x}^{\mathrm{T}}(n)\boldsymbol{v}(n)\boldsymbol{v}^{\mathrm{T}}(n)\boldsymbol{x}(n)\boldsymbol{x}^{\mathrm{T}}(n)\boldsymbol{\Gamma}^{\mathrm{T}}(n) \,\right\}
\end{aligned} \tag{6.45}$$

where we have also made use of the following relations

$$\begin{aligned}
\mathrm{E}\left\{\, \boldsymbol{v}(n)\boldsymbol{v}^{\mathrm{T}}(n) \,\right\} &= \boldsymbol{R}_{\boldsymbol{vv}}(n) \\
\mathrm{E}\left\{\, w^2(n) \,\right\} &= \sigma_w^2 \\
\mathrm{E}\left\{\, \boldsymbol{v}(n)\boldsymbol{v}^{\mathrm{T}}(n)\boldsymbol{x}(n)\boldsymbol{x}^{\mathrm{T}}(n)\boldsymbol{\Gamma}^{\mathrm{T}}(n) \,\right\} &= \boldsymbol{R}_{\boldsymbol{vv}}(n)\mathrm{E}\left\{\, \boldsymbol{x}(n)\boldsymbol{x}^{\mathrm{T}}(n)\boldsymbol{\Gamma}^{\mathrm{T}}(n) \,\right\}
\end{aligned}$$

and, from the definition of the first-order Markov model as shown in (6.38),

$$\mathrm{E}\left\{ \boldsymbol{h}(n)\boldsymbol{h}^{\mathrm{T}}(n) \right\} = \mathrm{E}\left\{ \boldsymbol{s}(n)\boldsymbol{s}^{\mathrm{T}}(n) \right\} = \sigma_s^2 \boldsymbol{I}. \tag{6.46}$$

Following the approach adopted in [31], we assume the time variations of $\boldsymbol{h}(n)$ are sufficiently slow that the adaptive filter is able to track the unknown system to within a time lag and, after convergence, $\boldsymbol{v}(n)$ is fluctuating around its mean $\forall\, n$ and thus $\mathrm{E}\left\{ \boldsymbol{v}(n)\boldsymbol{v}^{\mathrm{T}}(n) \right\} = \boldsymbol{R}_{\boldsymbol{vv}}(n)$. We may then define $\boldsymbol{R}_{\boldsymbol{vv}}$ as the approximate time-invariant autocorrelation matrix of the mean weight error vector and write the misalignment $\eta = \mathrm{tr}\{\boldsymbol{R}_{\boldsymbol{vv}}\}$ where $\mathrm{tr}\{\cdot\}$ is the trace operator.

6.6.2 Mean Square Misalignment for NLMS with $M = L$

We first consider a fully updated algorithm such that $\boldsymbol{Q}(n) = \boldsymbol{I}$. Hence $\boldsymbol{\Gamma}(n) = \boldsymbol{\Gamma}$, $\forall\, n$, is time-invariant and statistically stationary for inputs $\boldsymbol{x}(n)$. Using the factorization property of independent Gaussian variables as shown in Appendix II [31] and denoting $\boldsymbol{R}_{\boldsymbol{xx}} = \mathrm{E}\left\{ \boldsymbol{x}(n)\boldsymbol{x}^{\mathrm{T}}(n) \right\}$ as the autocorrelation matrix of the input signal, the expectations in (6.45) can be evaluated using the terms

$$\begin{aligned}
\mathrm{E}\left\{ \boldsymbol{\Gamma}(n)\boldsymbol{x}(n)\boldsymbol{x}^{\mathrm{T}}(n)\boldsymbol{\Gamma}^{\mathrm{T}}(n) \right\} &= \boldsymbol{\Gamma}\boldsymbol{R}_{\boldsymbol{xx}}\boldsymbol{\Gamma}^{\mathrm{T}} \\
\mathrm{E}\left\{ \boldsymbol{\Gamma}(n)\boldsymbol{x}(n)\boldsymbol{x}^{\mathrm{T}}(n) \right\} &= \boldsymbol{\Gamma}\boldsymbol{R}_{\boldsymbol{xx}} \\
\mathrm{E}\left\{ \boldsymbol{x}(n)\boldsymbol{x}^{\mathrm{T}}(n)\boldsymbol{\Gamma}^{\mathrm{T}}(n) \right\} &= \boldsymbol{R}_{\boldsymbol{xx}}\boldsymbol{\Gamma}^{\mathrm{T}} \\
\mathrm{E}\left\{ \boldsymbol{\Gamma}(n)\boldsymbol{x}(n)\boldsymbol{x}^{\mathrm{T}}(n)\boldsymbol{v}(n)\boldsymbol{v}^{\mathrm{T}}(n)\boldsymbol{x}(n)\boldsymbol{x}^{\mathrm{T}}(n)\boldsymbol{\Gamma}^{\mathrm{T}}(n) \right\} &= \\
\boldsymbol{\Gamma}\left[2\boldsymbol{R}_{\boldsymbol{xx}}\boldsymbol{R}_{\boldsymbol{vv}}(n)\boldsymbol{R}_{\boldsymbol{xx}} + \boldsymbol{R}_{\boldsymbol{xx}}\mathrm{tr}\left\{ \boldsymbol{R}_{\boldsymbol{xx}}\boldsymbol{R}_{\boldsymbol{vv}}(n) \right\} \right] \boldsymbol{\Gamma}^{\mathrm{T}}&.
\end{aligned} \tag{6.47}$$

Substituting (6.47) into (6.45), we obtain

$$\begin{aligned}
\boldsymbol{R}_{\boldsymbol{vv}}(n+1) = \boldsymbol{R}_{\boldsymbol{vv}}(n) &- \boldsymbol{R}_{\boldsymbol{vv}}(n)\boldsymbol{\Gamma}\boldsymbol{R}_{\boldsymbol{xx}} - \boldsymbol{R}_{\boldsymbol{vv}}(n)\boldsymbol{R}_{\boldsymbol{xx}}\boldsymbol{\Gamma}^{\mathrm{T}} \\
&+ \boldsymbol{\Gamma}\left[2\boldsymbol{R}_{\boldsymbol{xx}}\boldsymbol{R}_{\boldsymbol{vv}}(n)\boldsymbol{R}_{\boldsymbol{xx}} + \boldsymbol{R}_{\boldsymbol{xx}}\mathrm{tr}\left\{ \boldsymbol{R}_{\boldsymbol{xx}}\boldsymbol{R}_{\boldsymbol{vv}}(n) \right\} \right] \boldsymbol{\Gamma}^{\mathrm{T}} \\
&+ \boldsymbol{\Gamma}\boldsymbol{R}_{\boldsymbol{xx}}\boldsymbol{\Gamma}^{\mathrm{T}}\sigma_w^2 + 2(1-\xi)\sigma_s^2\boldsymbol{I}.
\end{aligned} \tag{6.48}$$

We proceed by considering $\boldsymbol{\Gamma} = c\boldsymbol{I}$ where $c = 2\mu/(L\sigma_x^2)$ for NLMS and Gaussian input with variance σ_x^2 giving $\boldsymbol{R}_{\boldsymbol{xx}} = \sigma_x^2\boldsymbol{I}$. Hence we can simplify (6.48) and write the steady-state misalignment η given by

$$\eta = \mathrm{tr}\left\{ \boldsymbol{R}_{\boldsymbol{vv}} \right\} = \frac{c\sigma_w^2 L}{2\phi} + \frac{(1-\xi)L\sigma_s^2}{c\sigma_x^2\phi} \tag{6.49}$$

where

$$\phi = 1 - c\sigma_x^2(1 + \frac{L}{2}) \, . \tag{6.50}$$

The first term in (6.49) corresponds to the *estimation variance* [36] and is dependent on measurement noise $w(n)$. The second term in (6.49) corresponds to the *lag variance* [36] and is due to system time variation ξ. We note from (6.49) that these two terms are uncoupled.

For the LMS case, $c = 2\mu$ and hence

$$\eta_{\mathrm{LMS}} = \frac{\mu\sigma_w^2 L}{\phi} + \frac{(1-\xi)L\sigma_s^2}{2\mu\sigma_x^2\phi}. \tag{6.51}$$

The estimation variance term of this result is, as expected, linear in μ and consistent with that presented in [31] for which it is assumed $\phi \approx 1$. However, the analysis presented here needs no such assumption. The lag variance term is inversely proportional to μ and linearly dependent on the system variation parameter ξ.

For NLMS, $c = 2\mu/(L\sigma_x^2)$ and as a result,

$$\eta_{\mathrm{NLMS}} = \frac{\mu\sigma_w^2}{\sigma_x^2\phi} + \frac{(1-\xi)L^2\sigma_s^2}{2\mu\phi}. \tag{6.52}$$

We may proceed to evaluate the step size, μ_{mis}, which achieves the lowest misalignment under time-varying conditions by letting

$$\gamma = 2(1 + L/2)/L \tag{6.53}$$

and differentiating (6.52) with respect to μ to obtain

$$\begin{aligned}\frac{d\,\eta_{\mathrm{NLMS}}}{d\,\mu} &= \frac{\sigma_w^2}{\sigma_x^2}\left[\frac{\gamma\mu}{(1-\gamma\mu)^2} + \frac{1}{1-\gamma\mu}\right] \\ &\quad + \frac{(1-\xi)L^2\sigma_s^2}{2}\left[\frac{2\gamma\mu - 1}{\mu^2(1-\gamma\mu)^2}\right].\end{aligned}$$

Setting $d\,\eta_{\mathrm{NLMS}}/d\,\mu = 0$, we obtain a quadratic equation in terms of μ_{mis}. Under the condition that $0 < \mu_{\mathrm{mis}} \leq 1$, we may solve for μ_{mis} giving

$$\begin{aligned}\mu_{\mathrm{mis}} = 0.5\frac{\sigma_x^2}{\sigma_w^2}\Bigg[&-(1-\xi)L^2\sigma_s^2\gamma \\ &+ \sqrt{\left[(1-\xi)L^2\sigma_s^2\gamma\right]^2 + 2\left(\frac{\sigma_w^2}{\sigma_x^2}\right)(1-\xi)L^2\sigma_s^2}\,\Bigg].\end{aligned} \tag{6.54}$$

We may now see the well known result that as $\xi \to 1$, $\mu_{\mathrm{mis}} \to 0$ and hence a smaller step size achieves a lower final misalignment, though at the expense of convergence rate. Hence we note that if $\mu_{\mathrm{mis}} < \mu \leq 1$ under the condition $\xi < 1$, convergence rate increases with μ but at the expense of poorer final misalignment.

6.6.3 Mean Square Misalignment for MMax-NLMS with $M \neq L$

For convergence in the mean square, we start by considering (6.45) and the evaluation of $\mathrm{E}\left\{ \boldsymbol{\Gamma}(n)\boldsymbol{x}(n)\boldsymbol{x}^{\mathrm{T}}(n) \right\}$. We note that the tap selection elements $q_l(n)$, $l = 0, \ldots, L-1$ are not independent of $x_l(n) = x(n-l)$ as they ensure that only the M largest $|x_l(n)|$ are selected. The M selected samples of $\widetilde{\boldsymbol{x}}(n)$ are assumed to have zero mean and exploiting the mean ergodic theorem [31], the variance of $\widetilde{\boldsymbol{x}}(n)$ is defined as

$$\widetilde{\sigma}_x^2 = \frac{1}{L}\sum_{l=0}^{L-1} \widetilde{x}_l^2(n). \tag{6.55}$$

Assuming that $\boldsymbol{x}(n)\boldsymbol{x}^{\mathrm{T}}(n)$ is diagonal and defining $\boldsymbol{\Gamma}(n) = \mu(n)\boldsymbol{Q}(n)$ such that $\mu(n) = 2\mu/(L\sigma_x^2)$ and $\mathrm{E}\{\mu(n)\} = c$, a scalar constant, we can evaluate

$$\begin{aligned}
\mathrm{E}\left\{ \boldsymbol{\Gamma}(n)\boldsymbol{x}(n)\boldsymbol{x}^{\mathrm{T}}(n) \right\} &= \mathrm{E}\left\{ \boldsymbol{x}(n)\boldsymbol{x}^{\mathrm{T}}(n)\boldsymbol{\Gamma}^{\mathrm{T}}(n) \right\} \\
&= \mathrm{E}\{\mu(n)\}\,\mathrm{E}\left\{ \boldsymbol{Q}(n)\boldsymbol{x}(n)\boldsymbol{x}^{\mathrm{T}}(n) \right\} \\
&= \frac{M}{L}c\widetilde{\sigma}_x^2\boldsymbol{I}.
\end{aligned} \tag{6.56}$$

The condition $\mathrm{E}\{\boldsymbol{\Gamma}(n)\} = \boldsymbol{\Gamma}$ implicit in (6.47) is not valid in this case. However, we can proceed to evaluate $\mathrm{tr}\{\boldsymbol{R}_{\boldsymbol{vv}}(n)\}$ using

$$\begin{aligned}
&\mathrm{tr}\left\{ \mathrm{E}\left\{ \boldsymbol{\Gamma}(n)\boldsymbol{x}(n)\boldsymbol{x}^{\mathrm{T}}(n)\boldsymbol{v}(n)\boldsymbol{v}^{\mathrm{T}}(n)\boldsymbol{x}(n)\boldsymbol{x}(n)^{\mathrm{T}}\boldsymbol{\Gamma}^{\mathrm{T}}(n) \right\} \right\} \\
&\qquad = \mathrm{tr}\left\{ c^2\mathrm{E}\left\{ \boldsymbol{Q}(n)\boldsymbol{x}(n)\boldsymbol{x}^{\mathrm{T}}(n)\boldsymbol{v}(n)\boldsymbol{v}^{\mathrm{T}}(n)\boldsymbol{x}(n)\boldsymbol{x}^{\mathrm{T}}(n) \right\} \right\} \\
&\qquad = c^2\mathrm{tr}\left\{ \boldsymbol{R}_{\boldsymbol{vv}}(n)(L+2)\frac{M}{L}\widetilde{\sigma}_x^2\sigma_x^2\boldsymbol{I} \right\} \\
&\qquad = c^2\mathrm{tr}\{\boldsymbol{R}_{\boldsymbol{vv}}(n)\}(L+2)\frac{M}{L}\widetilde{\sigma}_x^2\sigma_x^2\,, \\
&\mathrm{tr}\left\{ \mathrm{E}\left\{ \boldsymbol{\Gamma}(n)\boldsymbol{x}(n)\boldsymbol{x}^{\mathrm{T}}(n)\boldsymbol{\Gamma}^{\mathrm{T}}(n) \right\} \right\} \\
&\qquad = \mathrm{tr}\left\{ \frac{M}{L}c^2\widetilde{\sigma}_x^2\boldsymbol{I} \right\} \\
&\qquad = \frac{M}{L}c^2\widetilde{\sigma}_x^2 L.
\end{aligned} \tag{6.57}$$

Substituting (6.56), (6.57) and (6.46) into (6.45),

$$\begin{aligned}\mathrm{tr}\Big\{\boldsymbol{R}_{\boldsymbol{vv}}(n+1)\Big\} &= \mathrm{tr}\Big\{\boldsymbol{R}_{\boldsymbol{vv}}(n)\Big\} - 2\mathrm{tr}\Big\{\boldsymbol{R}_{\boldsymbol{vv}}(n)\Big\}\frac{M}{L}c\widetilde{\sigma}_x^2 \\ &\quad + c^2\mathrm{tr}\Big\{\boldsymbol{R}_{\boldsymbol{vv}}(n)\Big\}(L+2)\frac{M}{L}\widetilde{\sigma}_x^2\sigma_x^2 \\ &\quad + \frac{M}{L}c^2\widetilde{\sigma}_x^2\sigma_w^2 L + 2(1-\xi)\sigma_s^2 L \\ &= \mathrm{tr}\Big\{\boldsymbol{R}_{\boldsymbol{vv}}(n)\Big\}\left[1 - 2\frac{M}{L}c\widetilde{\sigma}_x^2 + (L+2)\frac{M}{L}c^2\widetilde{\sigma}_x^2\sigma_x^2\right] \\ &\quad + Mc^2\widetilde{\sigma}_x^2\sigma_w^2 + 2(1-\xi)L\sigma_s^2. \end{aligned} \tag{6.58}$$

The misalignment for MMax-NLMS can be found from (6.58) and using the approach of (6.49) as

$$\mathrm{tr}\Big\{\boldsymbol{R}_{\boldsymbol{vv}}\Big\}\left[2\frac{M}{L}c\widetilde{\sigma}_x^2 - (L+2)\frac{M}{L}c^2\widetilde{\sigma}_x^2\sigma_x^2\right] = Mc^2\widetilde{\sigma}_x^2\sigma_w^2 + 2(1-\xi)L\sigma_s^2 \tag{6.59}$$

resulting in

$$\mathrm{tr}\Big\{\boldsymbol{R}_{\boldsymbol{vv}}\Big\} = \frac{c\sigma_w^2 L}{2-(L+2)c\sigma_x^2} + \frac{2(1-\xi)L^2\sigma_s^2/M}{2c\widetilde{\sigma}_x^2 - (L+2)c^2\widetilde{\sigma}_x^2\sigma_x^2}. \tag{6.60}$$

For MMax-NLMS where $c = 2\mu/(L\sigma_x^2)$, the steady-state misalignment is then expressed as

$$\eta_{\mathrm{MMax-NLMS}} = \frac{\mu\sigma_w^2}{\sigma_x^2\phi} + \frac{L\sigma_x^2}{\widetilde{\sigma}_x^2 M}\frac{(1-\xi)L^2\sigma_s^2}{2\mu\phi} \tag{6.61}$$

where the term ϕ is defined in (6.50).

Comparing (6.61) with (6.52) we can see the additional factor of $L\sigma_x^2/(\widetilde{\sigma}_x^2 M)$ in MMax-NLMS compared to NLMS. Hence we note that if $L\sigma_x^2/(\widetilde{\sigma}_x^2 M) > 1$, we can expect the *lag variance* of the misalignment in MMax-NLMS to be greater than in NLMS by an amount inversely proportional to μ, the fraction of taps updated and the variance of the tap vector for the updated taps. However, the *estimation variance* term is identical to that of NLMS such that for a time-invariant system with $\xi = 1$, $\eta_{\mathrm{MMax-NLMS}} = \eta_{\mathrm{NLMS}}$. We also note that for $M = L$, $\widetilde{\sigma}_x^2 = \sigma_x^2$ and as a consequence, $\eta_{\mathrm{MMax-NLMS}} = \eta_{\mathrm{NLMS}}$ for each case of ξ.

We may further proceed to evaluate, for each ξ, the step size of MMax-NLMS which achieves the lowest misalignment μ_{mis} by writing for clarity

$$\gamma = 2(1+L/2)/L \tag{6.62}$$

and

$$\psi = L\sigma_x^2/(\widetilde{\sigma}_x^2 M). \tag{6.63}$$

Differentiating (6.61) with respect to μ and solving solving the quadratic equation for μ_{mis} we obtain

$$\mu_{\text{mis}} = 0.5\frac{\sigma_x^2}{\sigma_w^2}\Bigg[-\psi(1-\xi)L^2\sigma_s^2\gamma$$
$$+\sqrt{\left[\psi(1-\xi)L^2\sigma_s^2\gamma\right]^2 + 2\psi\left(\frac{\sigma_w^2}{\sigma_x^2}\right)(1-\xi)L^2\sigma_s^2}\,\Bigg]. \tag{6.64}$$

Similar to the NLMS algorithm, we note that for the case of MMax-NLMS, if $\mu_{\text{mis}} < \mu \leq 1$ under the condition $\xi < 1$, the convergence rate increases with μ but at the expense of poorer steady-state misalignment.

6.6.4 Simulation Results for single channel NLMS and MMax-NLMS

We first present single channel NLMS and MMax-NLMS simulations to support the theoretical normalized misalignment analysis for time-varying system identification. We employ the normalized misalignment η' defined as

$$\eta'(n) = \frac{\|\widehat{\boldsymbol{h}}(n) - \boldsymbol{h}(n)\|_2^2}{\|\boldsymbol{h}(n)\|_2^2}. \tag{6.65}$$

6.6.4.1 Effect of Non-stationarity

Fig. 6.5 shows NLMS normalized misalignment results for a time-invariant system, obtained using $\xi = 1$, and three time-varying systems, obtained using $\xi = \{0.999999, 0.99999, 0.9999\}$, where smaller values of ξ indicate higher degrees of time-variation. In this simulation, the adaptive filter is of length $L = 64$ while the adaptive step size $\mu = 0.1$ is used. The values have been chosen arbitrarily for the purposes of these illustrations. The ξ values used in these tests deviate by only a small amount because $\mathbf{s}(n)$ has been chosen to be large in amplitude such that $\sigma_s^2 = \sigma_x^2 = 1$. This allow the NLMS algorithm to track the unknown system. The learning curves are averaged over 8 independent trials and the theoretical values of η'_{NLMS} given by (6.65) and (6.52) are superimposed as straight horizontal lines.

Fig. 6.6 shows the results of an equivalent experiment for MMax-NLMS with $L = 64$ and $M = 8$. The theoretical values of $\eta'_{\text{MMax-NLMS}}$ given by (6.65) and (6.61) are superimposed as straight horizontal lines. For comparison purposes, the corresponding theoretical values of η'_{NLMS} from the previous experiment are also included in Fig. 6.6 as dashed lines. For both experiments, white Gaussian measurement noise $w(n)$ is added such that the SNR is 35 dB.

The results show that both NLMS and MMax-NLMS are sensitive to time-variation of the unknown system in that the misalignment performance degrades with increasing deviation of ξ from unity. The MMax-NLMS algorithm example can be seen to perform around 3 to 4 dB worse, in terms of steady-state normalized misalignment, than NLMS under these time-varying conditions. For a time-invariant system, $\xi = 1$, both MMax-NLMS and NLMS achieve the same steady-state misalignment as can be seen by (6.52) and (6.61). The MMax-NLMS algorithm however has a lower rate of convergence compared to that of NLMS as expected.

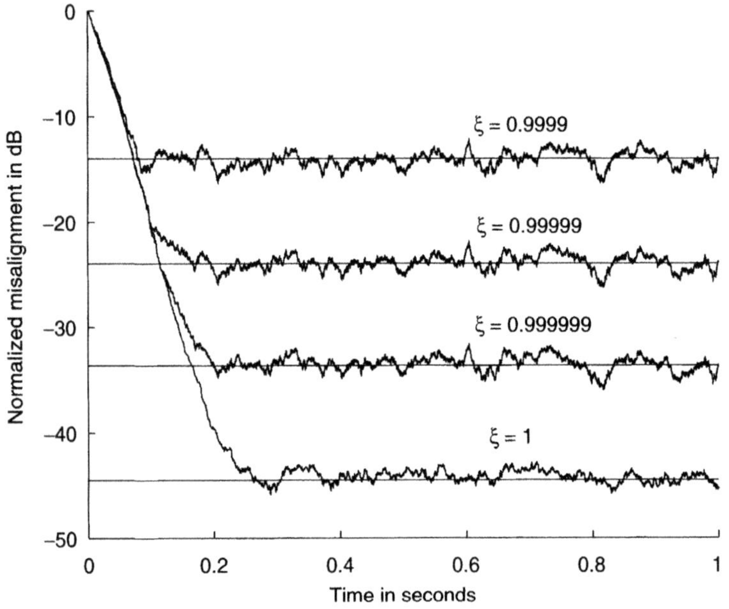

Fig. 6.5. NLMS normalized misalignment for varying ξ with $\mu = 0.1$, $\sigma_x^2 = \sigma_s^2 = 1$, $L = 64$, SNR= 35 dB.

6.6.4.2 Effect of Tap Selection on Normalized Misalignment

We now compare the effect of tap selection on the normalized misalignment under time-varying conditions of the unknown system for the MMax-NLMS algorithm. Fig. 6.7 shows the variation of average normalized misalignment with M for MMax-NLMS. The length of the adaptive filter is $L = 32$ while $8 \leq M \leq 24$ and $\xi = 0.9999$ with $\mu = 0.1$. The normalized misalignment is averaged over 5 independent trials and for each trial an SNR $= 40$ dB is used. We see that the normalized misalignment reduces with increasing M under the condition $\xi = 0.9999$ such that there is an improvement of approximately 1.5 dB as M is increased from 8 to 24. The mean error between theoretical and experimental results in this simulation is 0.0045 dB hence verifying our analysis.

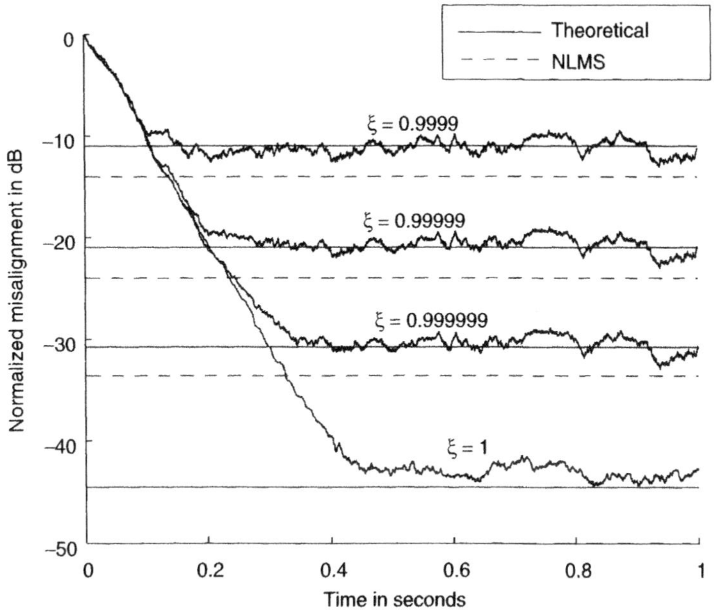

Fig. 6.6. MMax-NLMS normalized misalignment for varying ξ with $\mu = 0.1$, $\sigma_x^2 = \sigma_s^2 = 1$, $L = 64$, $M = 8$, SNR= 35 dB.

6.6.4.3 Effect of SNR on Normalized Misalignment

We next investigate the effect of SNR on the normalized misalignment for MMax-NLMS under the non-stationary unknown system condition of $\xi = 0.99999$. The experimental parameters for this simulation setup were $L = 128$, $M = 64$, $\mu = 0.1$. The normalized misalignment for each algorithm is averaged over 5 independent trials.

Fig. 6.8 shows the variation of MMax-NLMS normalized misalignment with SNR. We note that the normalized misalignment improves with increasing SNR as expected. When SNR is increased from 10 to 40 dB, the final misalignment performance is improved by approximately 4 dB. The mean error between our theoretical and experimental results is 0.061 dB.

6.6.4.4 Effect of Step-Size on Normalized Misalignment

Fig. 6.9 shows the effect of variation of μ on the final misalignment for NLMS under stationary ($\xi = 1$) and time-varying ($\xi = 0.99999$) cases. In this experiment, the filter length is $L = 128$ and $w(n)$ is added such that an SNR of 40 dB is achieved. The average final normalized misalignment is obtained from 5 independent trials.

We observe that for the stationary case $\xi = 1$, the final normalized misalignment is approximately linear in μ. As μ increases, the final normalized

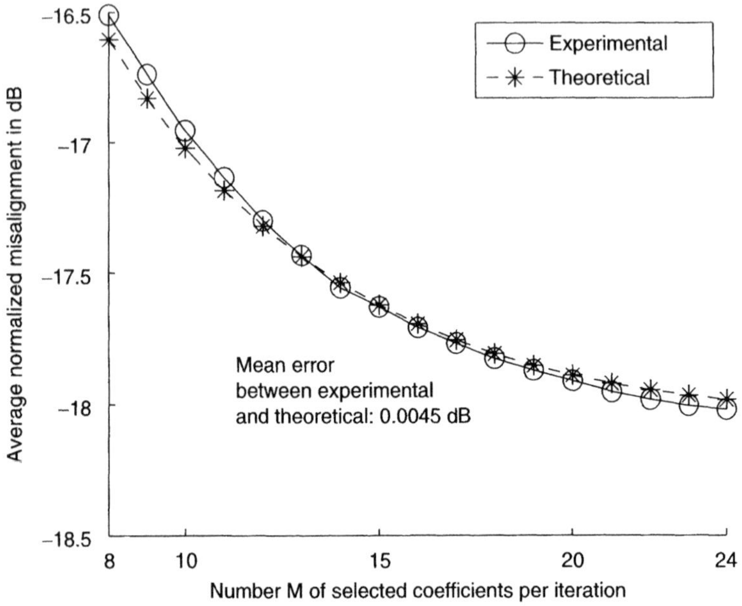

Fig. 6.7. Variation of MMax-NLMS average normalized misalignment with number M of selected coefficients per iteration for $L = 32$, $\mu = 0.1$, $\xi = 0.9999$, $\sigma_x^2 = \sigma_s^2 = 1$, SNR= 40 dB.

misalignment increases as expected. In this simulation example, the mean difference between the experimental and theoretical final normalized misalignment is 0.164 dB. For the case of $\xi = 0.99999$, we note that there exists a μ_{mis} such that the lowest misalignment can be achieved. The theoretical value of $\mu_{\text{mis}} = 0.475$, computed using (6.54), is shown by the vertical dotted line. The mean difference between the experimental and the theoretical normalized misalignment is 0.407 dB.

Fig. 6.10 shows the effect of step size on MMax-NLMS under the condition $\xi = 1$ and $\xi = 0.99999$ with $L = 128$ and $M = 64$. We have simulated this experiment using 30 dB SNR. Similar to the case of NLMS, we observe that for $\xi = 1$, the final normalized misalignment is approximately linear in μ. For the case of $\xi = 0.99999$, there exists a $\mu_{\text{mis}} = 0.384$ governed by (6.64) which is plotted as a vertical line. The mean difference between the experimental and theoretical final misalignment for the case of $\xi = 1$ and $\xi = 0.99999$ is 0.13 and 0.034 dB respectively. Note that the final misalignment for NLMS in Fig. 6.9 is generally lower than that of MMax-NLMS in Fig. 6.10 since a higher SNR is used in the former experiment.

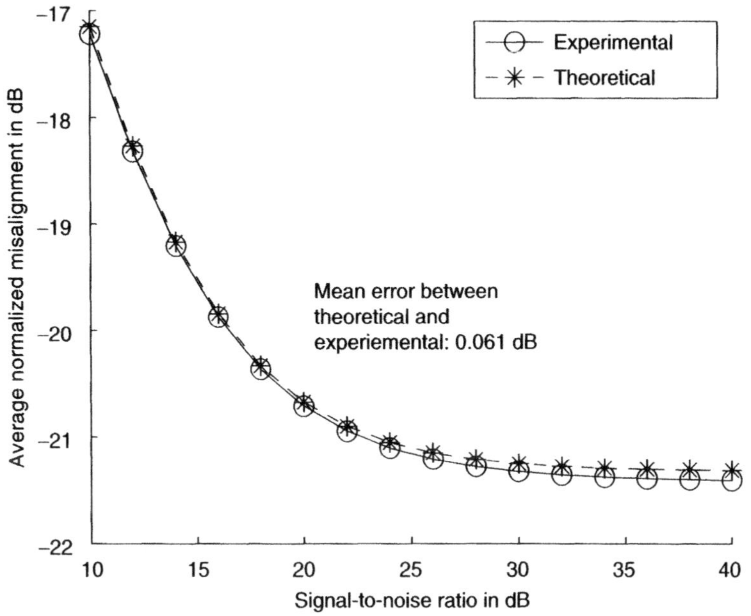

Fig. 6.8. Variation of MMax-NLMS average normalized misalignment with signal-to-noise ratio (SNR) for $L = 128$, $M = 64$, $\mu = 0.1$, $\xi = 0.99999$, $\sigma_x^2 = \sigma_s^2 = 1$.

6.7 Sparse Partial Update NLMS

An echo path impulse response may be said to exhibit sparseness, using the same definition as in Sec. 6.3, if a large fraction of its energy is concentrated in a small fraction of its duration [43]. A degree of sparseness can arise in acoustic echo cancellation for handsfree systems if, for example, the direct path acoustic propagation time from the loudspeaker to the microphone is such as to give a significant number of leading zeros in the impulse response. Alternatively, if the talker moves unexpectedly close to the microphone of the handsfree system, then the impulse response will likely be overmodelled by the adaptive echo canceller, with the effect that a significant number of trailing zeros may occur in the impulse response. Sparseness is also a very important characteristic affecting the design of network echo cancellers, particular for packet-switched networks, and is discussed further in, for example, [11] and the references contained therein.

It has been shown [43, 47] that standard adaptive algorithms perform poorly in such cases in terms of convergence. Several improvements to standard adaptive algorithms have been proposed, many of which are based on the concept of proportionate updating. For example, the Proportionate NLMS (PNLMS) algorithm, and improved versions including [8, 11, 22], adjust the adaptive step size, on a tap-by-tap basis, to make it proportional to the mag-

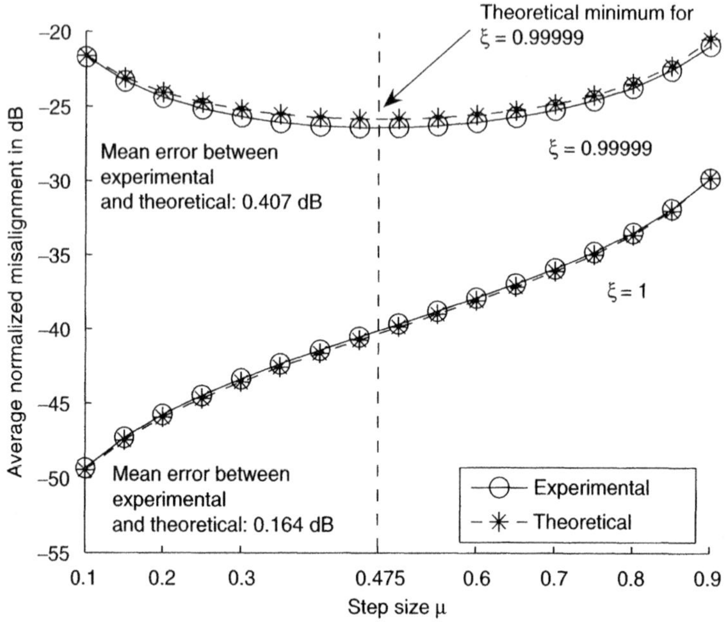

Fig. 6.9. Variation of NLMS average normalized misalignment with step size μ for $\sigma_x^2 = \sigma_s^2 = 1$, $L = 128$, SNR$=$ 40 dB.

nitude of the corresponding coefficient. In this manner, large magnitude coefficients are adapted with large steps whereas small coefficients take correspondingly small steps. Further details of these techniques are presented in Chapter 5 of this book. Proportionate updating schemes are similar in concept to the estimation of a sparse approximation of the true impulse response in which the coefficients with small magnitude are approximated as zero. Consequently, these algorithms can be thought of as exploiting sparseness in the impulse response. This is in contrast to the MMax-based selective-tap adaptive algorithms that have been discussed above which exploit (approximate) sparseness in the tap-input vector.

The Sparse Partial Update NLMS algorithm (SPNLMS) is developed in [12–14]. This algorithm is able to exploit both sparseness in the tap-input vector and also sparseness in the impulse response by employing a tap selection criterion that considers the product of the tap-input sample and the corresponding coefficient. In this case the tap selection criterion is

$$q_l(n) = \begin{cases} 1, & \text{if } \left|x_l(n)\widehat{h}_l(n)\right| \in \left\{ M \text{ maxima of } \left|\boldsymbol{x}(n) \odot \widehat{\boldsymbol{h}}(n)\right| \right\}, \\ 0, & \text{otherwise}, \end{cases} \tag{6.66}$$

for $l = 0,\ 1,\ \ldots,\ L-1$ where $\odot$ represents the element-by-element vector product.

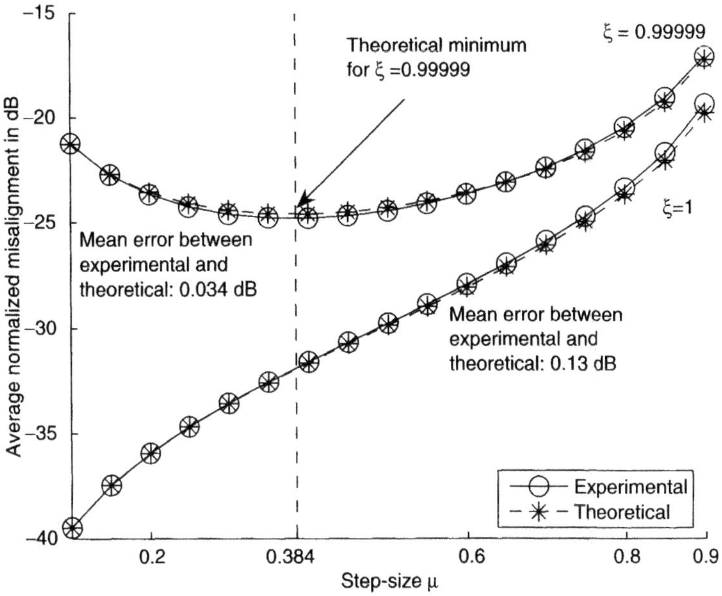

Fig. 6.10. Variation of MMax-NLMS average normalized misalignment with step size μ for $\sigma_x^2 = \sigma_s^2 = 1$, $L = 128$, $M = 64$, SNR= 30 dB.

This data dependent tap selection criterion results in convergence performance that is dependent on sparse properties of the echo path impulse response. However, comparative evaluations [43] show good performance when the echo response is suitably sparse.

6.8 multichannel Selective-Tap Algorithms for Stereophonic Acoustic Echo Cancellation

6.8.1 Overview and Rationale

Applications such as desktop conferencing and hands-free telephony benefit from multichannel audio. For example, users can localize multiple talkers in teleconference meetings using stereophonic perception. The stereophonic acoustic echo canceller (SAEC) as shown in Fig. 6.11 suppresses the echo returned to the transmission room so as to enable undisturbed communication between the rooms.

In general, the solutions for the adaptive filters in SAEC are non-unique and depend both on the transmission and receiving rooms' impulse responses [5]. In the practical case in which $L < L_{\mathrm{T}}$, where L_{T} is the length of the transmission room impulse response, the problem of non-uniqueness is ameliorated to some degree by the 'tail' effect [5]. However, the system identification problem remains ill-conditioned due to the high interchannel coherence between the two channels' tap-input vectors [5,20], resulting in very slow convergence. Several techniques have been developed to decorrelate the two input signals. One of the most effective methods of achieving interchannel decorrelation uses a nonlinear (NL) preprocessor [5] with the level of nonlinearity controlled by the nonlinearity factor $0 < \alpha \leq 0.5$. Other approaches include the use of spectrally shaped random noise [24, 46], comb filtering [6], leaky extended LMS [32] and alternating fixed-point [21] algorithms. The common aim of these algorithms is to achieve decorrelation of input signals $x_1(n)$ and $x_2(n)$ without affecting the quality or stereophonic image of the speech.

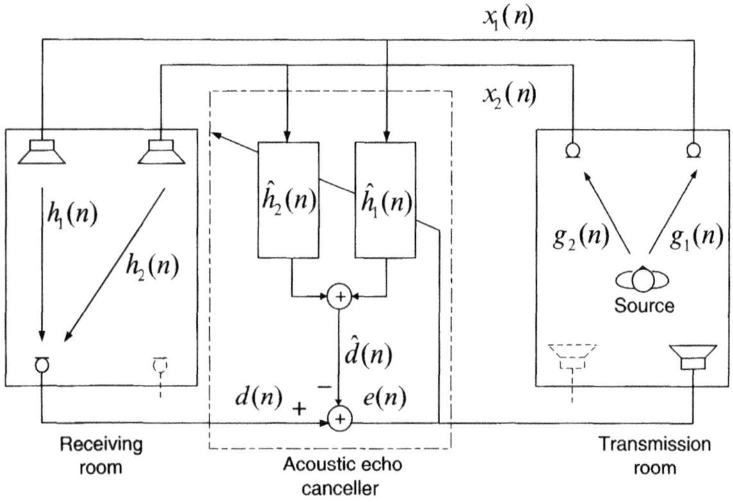

Fig. 6.11. Schematic diagram of stereophonic acoustic echo cancellation (after [5]). Only one channel of the return path is shown for simplicity.

It has been seen earlier in this chapter that selective-tap schemes were introduced with the aim of reducing the complexity of adaptive filters. In this section we consider an alternative motivation for the use of selective-tap schemes: the reduction of interchannel coherence in multichannel adaptive filtering algorithms, and we shall discuss this in terms of the SAEC problem.

6.8.2 Reducing Interchannel Coherence using Tap Selection

In order to examine the effect of tap selection on interchannel coherence in SAEC, we first employ the squared coherence function

$$C_{\boldsymbol{x}_1\boldsymbol{x}_2}(\Omega) = \frac{\left|P_{\boldsymbol{x}_1\boldsymbol{x}_2}(\Omega)\right|^2}{P_{\boldsymbol{x}_1\boldsymbol{x}_1}(\Omega)\, P_{\boldsymbol{x}_2\boldsymbol{x}_2}(\Omega)}\,, \tag{6.67}$$

where $P_{\boldsymbol{x}_1\boldsymbol{x}_2}(\Omega)$ is the cross power spectrum between the two channels and Ω is the normalized frequency.

As an illustrative example, we consider the system of Fig. 6.11 for the case when the source signal is zero mean unit variance WGN and $\boldsymbol{g}_1$ and $\boldsymbol{g}_2$ are highly correlated impulse responses each of length 1024. This results in highly correlated tap-input vectors $\boldsymbol{x}_1(n)$ and $\boldsymbol{x}_2(n)$ with $\boldsymbol{x}_j(n) = [x_j(n), x_j(n-1), \ldots, x_j(n-L+1)]^{\mathrm{T}}$ and we choose for this illustration $L = 512$. In this example, $\boldsymbol{g}_1$ is generated using the method of images [1] while $\boldsymbol{g}_2$ is formed using the following relation

$$\boldsymbol{g}_2 = \gamma \boldsymbol{g}_1 + (1-\gamma)\boldsymbol{b} , \tag{6.68}$$

where $\boldsymbol{b}$ is an independent WGN sequence also with zero mean and $0 \leq \gamma \leq 1$ controls the amount of independent WGN added to $\boldsymbol{g}_1$. To reflect the high interchannel correlation found in practice, we have used $\gamma = 0.9$, giving a correlation coefficient of 0.904.

The highly correlated tap-input vectors give rise to a squared coherence close to one across most of the frequency band as shown in Fig. 6.12(a). In the case shown in Fig. 6.12(b), taps are selected according to the MMax selection criterion with $M = 0.5L$. It can be seen clearly that MMax tap selection does not provide any significant decorrelation. This is because the MMax criterion selects nearly identical tap-indices in both filters for updating, due to the high coherence between the two channel tap-input vectors. This does not achieve our desired effect of decorrelating the signals.

In contrast, Fig. 6.12(c) shows the result obtained from an exclusive tap selection criterion such that selection of the same tap-index in both channels is not permitted. A simple, but not useful, example of such an exclusive case with $M = 0.5L$ is to select the taps corresponding to the M largest magnitude tap-inputs in the first channel and the exclusive set of taps in the second channel. The mean interchannel coherence is seen to be significantly reduced from 0.88 to 0.52, providing the motivation for further study of tap selection for multichannel adaptive algorithms.

Exclusive tap selection can be seen as a method for improving the conditioning of the input autocorrelation matrix by considering the case where $\boldsymbol{x}_1(n)$ and $\boldsymbol{x}_2(n)$ are highly correlated Gaussian inputs. Defining for the two channel case, $\boldsymbol{x}(n) = [\boldsymbol{x}_1^{\mathrm{T}}(n), \boldsymbol{x}_2^{\mathrm{T}}(n)]^{\mathrm{T}}$, the autocorrelation matrix can be expressed as

$$\begin{aligned} \boldsymbol{R}_{\boldsymbol{x}\boldsymbol{x}} &= \mathrm{E}\left\{ \boldsymbol{x}(n)\boldsymbol{x}^{\mathrm{T}}(n) \right\} \\ &= \begin{bmatrix} \boldsymbol{R}_{11} & \boldsymbol{R}_{12} \\ \boldsymbol{R}_{21} & \boldsymbol{R}_{22} \end{bmatrix} . \end{aligned} \tag{6.69}$$

After exclusive tap selection, the resulting sparse vectors $\widetilde{\boldsymbol{x}}_1(n) = \boldsymbol{Q}_1(n)\boldsymbol{x}_1(n)$ and $\widetilde{\boldsymbol{x}}_2(n) = \boldsymbol{Q}_2(n)\boldsymbol{x}_2(n)$ give rise to $\boldsymbol{R}_{\widetilde{\boldsymbol{x}}\widetilde{\boldsymbol{x}}}$. The diagonals and some off-diagonal elements of $\boldsymbol{R}_{12}$ and $\boldsymbol{R}_{21}$ are zero. This improves on the conditioning of $\boldsymbol{R}_{\boldsymbol{x}\boldsymbol{x}}$ and in the limit where $\widetilde{\boldsymbol{x}}_1$ and $\widetilde{\boldsymbol{x}}_2$ are perfectly uncorre-

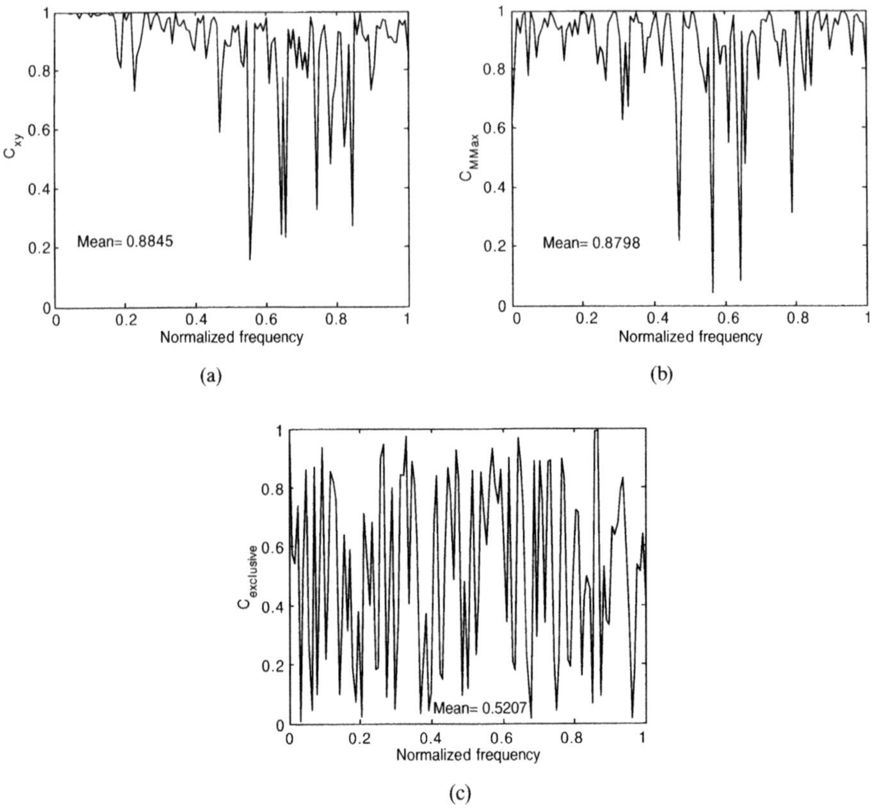

Fig. 6.12. Squared coherence for (a) $M = L = 512$ (b) $M = 0.5L$ with MMax tap selection (c) $M = 0.5L$ with exclusive tap selection.

lated and white, the autocorrelation matrix is a diagonal matrix $\boldsymbol{R}_{\widetilde{\boldsymbol{x}\boldsymbol{x}}} = \text{diag}\{[\tilde{\sigma}_1^2, \ldots, \tilde{\sigma}_1^2\ ,\ \tilde{\sigma}_2^2, \ldots, \tilde{\sigma}_2^2]\}$ with a 2-norm condition number of

$$\|\boldsymbol{R}_{\widetilde{\boldsymbol{x}\boldsymbol{x}}}\|_2 \|\boldsymbol{R}_{\widetilde{\boldsymbol{x}\boldsymbol{x}}}{}^{-1}\|_2 = \frac{\max(\tilde{\sigma}_1^2, \tilde{\sigma}_2^2)}{\min(\tilde{\sigma}_1^2, \tilde{\sigma}_2^2)}, \tag{6.70}$$

where $\tilde{\sigma}_j^2$ is the jth channel subselected tap-input variance.

Fig. 6.13 shows the variation of mean condition number of the autocorrelation matrices $\boldsymbol{R}_{\boldsymbol{x}\boldsymbol{x}}$ and $\boldsymbol{R}_{\widetilde{\boldsymbol{x}\boldsymbol{x}}}$ as a function of γ. Both the autocorrelation matrices are formed from $\boldsymbol{x}_1$ and $\boldsymbol{x}_2$ generated by convolving a WGN sequence with $\boldsymbol{g}_1$ and $\boldsymbol{g}_2$ governed by (6.68) with the additional exclusive tap selection criterion imposed when generating $\boldsymbol{R}_{\widetilde{\boldsymbol{x}\boldsymbol{x}}}$. For each case of γ, the average 2-norm condition number for 50 trials is computed and plotted as shown in Fig. 6.13(a) and (b) for $\boldsymbol{R}_{\boldsymbol{x}\boldsymbol{x}}$ and $\boldsymbol{R}_{\widetilde{\boldsymbol{x}\boldsymbol{x}}}$ respectively. We see that as γ is reduced, $\boldsymbol{x}_1(n)$ and $\boldsymbol{x}_2(n)$ become less correlated and hence a reduction of mean condition number for both $\boldsymbol{R}_{\boldsymbol{x}\boldsymbol{x}}$ and $\boldsymbol{R}_{\widetilde{\boldsymbol{x}\boldsymbol{x}}}$ is exhibited. In addition, for each case of γ, $\boldsymbol{R}_{\widetilde{\boldsymbol{x}\boldsymbol{x}}}$ has a lower mean condition number than $\boldsymbol{R}_{\boldsymbol{x}\boldsymbol{x}}$ and hence exclusive tap selection gives rise to a better conditioned autocorrelation matrix which in turn reduces the misalignment problem in SAEC.

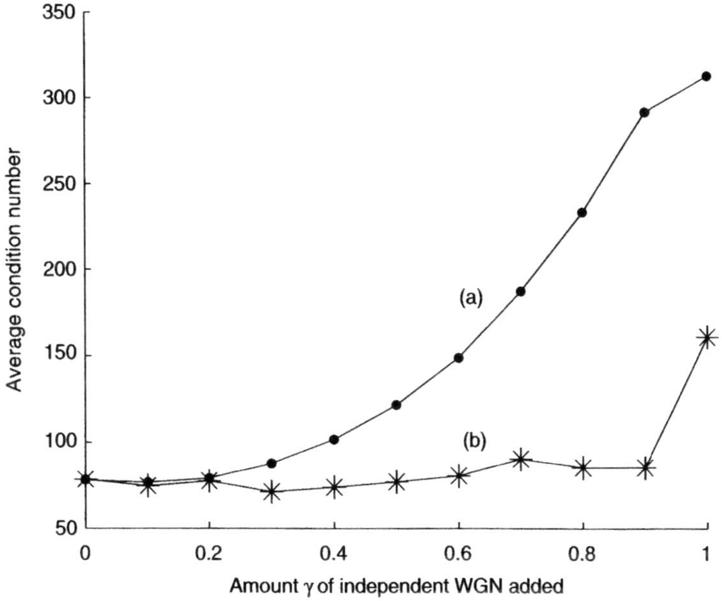

Fig. 6.13. Effect of exclusive tap selection on average condition number for amount γ of independent WGN added (see (6.68)) (a) without tap selection (b) with exclusive tap selection.

6.9 Exclusive Maximum Tap Selection

6.9.1 Formulation and Realization using Exhaustive Search

It has been shown in Sec. 6.8.2 that exclusive tap selection can reduce the interchannel coherence and hence improve the conditioning of the adaptive filtering in SAEC. We wish to develop a selective-tap adaptive filtering scheme which makes use of this concept without degrading convergence due to partial adaptation. As discussed in Sec. 6.3.2, since convergence rate can be seen to increase monotonically with $\mathcal{M}$, we propose that any degradation in convergence performance due to subselection of taps can be minimized by selecting taps so as to maximize $\mathcal{M}$. We now therefore formulate the joint optimization problem of maximizing the MMax criterion, determined by $\mathcal{M}$, and minimizing the interchannel coherence under the control of tap selection. This is done using two variables: magnitude weighting, w_{m}, to describe the 'closeness' of the tap selection to that of the MMax scheme, and coherence weighting, $w_\mathrm{c} = 1 - w_\mathrm{m}$, to describe interchannel coherence between the subsampled tap-input vectors. A magnitude weighting of $w_\mathrm{m} = 1$ corresponds to selecting coefficients based on the MMax tap selection criterion only.

We begin by considering ${}^L C_M$ combinations of selecting $M = 0.5L$ taps from each channel's adaptive filter of length L. Let the combinations be indexed $k, r = 1, 2, \ldots, {}^L C_M$ giving tap selection sets $\{\beta_k\}$ and $\{\beta_r\}$ for channel 1 and 2 respectively and define $\{\beta_{kr}(n)\}$ as the combined two channel tap selection set. Let $\widetilde{\boldsymbol{x}}_k(n)$ be defined as the subselected input vector using tap selection set $\{\beta_k(n)\}$. We next define, at each time iteration n, $\boldsymbol{A}(n)$ and $\boldsymbol{C}(n)$ as square matrices with elements

$$a_{kr}(n) = \left\| \left|\widetilde{\boldsymbol{x}}_k(n)\right| + \left|\widetilde{\boldsymbol{x}}_r(n)\right| \right\|_1 , \tag{6.71}$$

$$c_{kr}(n) = \left\langle \frac{\left|P_{\widetilde{\boldsymbol{x}}_k \widetilde{\boldsymbol{x}}_r}(\Omega)\right|^2}{P_{\widetilde{\boldsymbol{x}}_k \widetilde{\boldsymbol{x}}_k}(\Omega)\, P_{\widetilde{\boldsymbol{x}}_r \widetilde{\boldsymbol{x}}_r}(\Omega)} \right\rangle \tag{6.72}$$

respectively such that $a_{kr}(n)$ denotes the absolute sum of the selected tap-inputs in a particular tap selection set $\beta_{kr}(n)$ and $c_{kr}(n)$ is the squared coherence, with $< \cdot >$ indicating averaging over frequency, of the two tap-input vectors with $L - M$ unselected inputs in each channel set to zero.

Since the elements of matrix $\boldsymbol{A}(n)$ are the magnitude sums, of which we require the maximum, an integer cost is first associated with each of the elements $a_{kr}(n)$ such that the *least cost* is allocated to the element having the *largest magnitude* in $\boldsymbol{A}(n)$. We now denote this new magnitude cost matrix as $\underline{\boldsymbol{A}}(n)$. In a similar manner, each element in $\boldsymbol{C}(n)$ will be allocated an integer cost such that element corresponding to the *minimum coherence* is allocated the *least cost*. We denote this new coherence cost matrix as $\underline{\boldsymbol{C}}(n)$. Hence matrices $\underline{\boldsymbol{A}}(n)$ and $\underline{\boldsymbol{C}}(n)$ now contain integer cost values depending on the magnitude sum and interchannel coherence. A total cost matrix $\boldsymbol{V}(n)$ is then given by

$$\boldsymbol{V}(n) = w_\mathrm{m} \underline{\boldsymbol{A}}(n) + w_\mathrm{c} \underline{\boldsymbol{C}}(n) . \tag{6.73}$$

We define $\{\beta_\mathrm{min}\} = \{\beta_{k_\mathrm{min}, r_\mathrm{min}}\}$, for each time iteration n, as the tap selection set having minimum cost in matrix $\boldsymbol{V}(n)$ and search for $\{\beta_\mathrm{min}\}$ such that

$$k_\mathrm{min}, r_\mathrm{min} = \arg\min_{k,r} \left[\boldsymbol{V}(n)\right] \quad k, r = 1, 2, \ldots, {}^L C_M . \tag{6.74}$$

For small L and letting $\widehat{\boldsymbol{h}}(n) = \left[\widehat{\boldsymbol{h}}_1^\mathrm{T}(n), \widehat{\boldsymbol{h}}_2^\mathrm{T}(n)\right]^\mathrm{T}$ and $\boldsymbol{x}(n) = \left[\boldsymbol{x}_1^\mathrm{T}(n), \boldsymbol{x}_2^\mathrm{T}(n)\right]^\mathrm{T}$, $\boldsymbol{V}(n)$ can be searched exhaustively such that, at each time iteration, the tap selection set β_min can then be incorporated into NLMS adaptation as

$$\widehat{\boldsymbol{h}}(n+1) = \widehat{\boldsymbol{h}}(n) + \boldsymbol{Q}(n) \frac{\mu \boldsymbol{x}(n) e(n)}{\|\boldsymbol{x}(n)\|_2^2 + \delta} , \tag{6.75}$$

with $\boldsymbol{Q}(n) = \mathrm{diag}\{[\boldsymbol{q}_1^\mathrm{T}(n), \boldsymbol{q}_2^\mathrm{T}(n)]\}$ being the two channel selection matrix such that, at each time iteration n, element u of $\boldsymbol{q}_1(n)$ and element v of $\boldsymbol{q}_2(n)$ are defined for $u, v = 1, 2, \ldots, L$ as

$$\{q_{1,u}, q_{2,v}\} = \begin{cases} 1, & \text{if } u,v \in \{\beta_{\min}\}, \\ 0, & \text{otherwise.} \end{cases}$$

Fig. 6.14 shows simulation results for the normalized misalignment with different values of magnitude weighting ($w_{\rm m} = 0.1,\ 0.7,\ 0.9,\ 1.0$). In this example, the input is a zero mean WGN sequence with adaptive filters having 6 taps per channel and for every iteration, 3 taps are updated ($L = 6,\ M = 3$). The relationship between impulse responses $\boldsymbol{g}_1$ and $\boldsymbol{g}_2$ with lengths $L_{\rm T} = 12$ is again determined by (6.68) with $\gamma = 0.9$. The impulse responses $\boldsymbol{h}_1$ and $\boldsymbol{h}_2$ are taken from a WGN sequence and are of lengths $L_{\rm R} = 6$. This choice of $L_{\rm T}$ and $L_{\rm R}$ allows us to study the adaptive filters which uniquely determine the unknown system whilst minimizing the misalignment caused by undermodelling. The normalized misalignment for only one of the two channels is plotted for each case of $w_{\rm m}$ for reasons of clarity. Uncorrelated measurement noise is added to $d(n)$ such that an SNR of 40 dB is achieved.

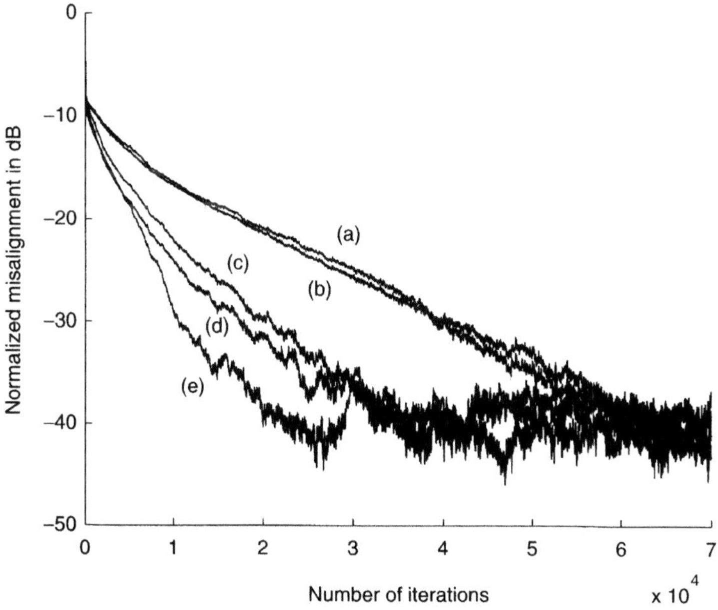

Fig. 6.14. Normalized misalignment for (a) $w_{\rm m} = 1$, (b) NLMS, (c) $w_{\rm m} = 0.9$, (d) 0.7, (e) 0.1. L=6, M=3, $\mu = 0.6$, $\gamma = 0.9$, SNR= 40 dB.

The simulation result shows that $w_{\rm m} = 1$ coincides with MMax-NLMS where performance is close to that of the fully updated NLMS as expected. The highest convergence rate can be seen when $w_{\rm m} = 0.1$ ($w_{\rm c} = 0.9$) where there is a high weighting given to the minimization of interchannel coherence. Upon further investigation, it was found that for $w_{\rm m} = 0.1$, all the tap selection sets satisfy the exclusive criterion across all time iterations, such that combinations k and r contain no tap-indices in common i.e

$$\beta_k(n) \cap \beta_r(n) = \{\phi\} \ , \ \forall \ n \ , \tag{6.76}$$

where $\{\phi\}$ is a null set. Therefore we can redefine our optimization problem in the simpler form of a search such that (6.76) is satisfied whilst maximizing $\mathcal{M}(n)$ at each iteration.

6.9.2 Efficient Realization

The exhaustive search of $\boldsymbol{V}(n)$ for the optimum exclusive maximum tap selection is computationally expensive for adaptive filters of higher orders. We now therefore propose an efficient alternative to the exhaustive search. In the following, we shall temporarily omit the dependence of variables on n for brevity such that $\boldsymbol{x}_j = [x_j(0), x_j(1), \ldots, x_j(L-1)]^{\mathrm{T}}$.

Let us define, at each time iteration, the interchannel tap-input magnitude difference vector

$$\boldsymbol{p} = |\boldsymbol{x}_1| - |\boldsymbol{x}_2| \ , \tag{6.77}$$

and

$$\breve{\boldsymbol{p}} = \left[\breve{p}(1), \ldots, \breve{p}(L)\right]^{\mathrm{T}}, \ \breve{p}(1) > \breve{p}(2) > \ldots > \breve{p}(L) \tag{6.78}$$

as $\boldsymbol{p}$ sorted in descending order. Let $\breve{x}_1(k)$ and $\breve{x}_2(k)$ denote the kth tap-input samples of channel 1 and 2, ordered according to the sorting of $\breve{\boldsymbol{p}}$ such that $\breve{p}(k) = |\breve{x}_1(k)| - |\breve{x}_2(k)|$, $k = 1, 2, \ldots, L$. In this two channel case $\mathcal{M}$ is defined as

$$\mathcal{M} = \frac{\|\boldsymbol{Q}\boldsymbol{x}\|_2^2}{\|\boldsymbol{x}\|_2^2} \tag{6.79}$$

with $\boldsymbol{Q} = \mathrm{diag}\{[\boldsymbol{q}_1^{\mathrm{T}}, \boldsymbol{q}_2^{\mathrm{T}}]\}$ and $\boldsymbol{x} = [\boldsymbol{x}_1^{\mathrm{T}}, \boldsymbol{x}_2^{\mathrm{T}}]^{\mathrm{T}}$. Utilizing the robustness of the NLMS algorithm to MMax tap selection for $0.5L \leq M \leq L$, we consider $M = 0.5L$.

As will be shown in the following paragraph, the tap selection set that maximizes $\mathcal{M}$ jointly for both channels contains the M largest elements of $\boldsymbol{p}$ from channel 1 and the M smallest elements of $\boldsymbol{p}$ from channel 2, i.e.

$$\left\{\breve{x}_1(1), \ldots, \breve{x}_1(M), \breve{x}_2(M+1), \ldots, \breve{x}_2(L)\right\}. \tag{6.80}$$

Hence at each iteration, element u of $\boldsymbol{q}_1$ and element v of $\boldsymbol{q}_2$ are defined for $u, v = 1, \ 2 \ , \ldots, L$ where

$$q_1(u) = \begin{cases} 1, & p(u) \in \{M \text{ maxima of } \boldsymbol{p}\}, \\ 0, & \text{otherwise}, \end{cases}$$

$$q_2(v) = \begin{cases} 1, & p(v) \in \{M \text{ minima of } \boldsymbol{p}\}, \\ 0, & \text{otherwise}. \end{cases} \tag{6.81}$$

To verify that the exclusive tap selection set given by (6.81) maximizes $\mathcal{M}$ jointly for both channels at each time iteration, we consider whether the absolute sum given by $\sum_{i=1}^{M}|\breve{x}_1(i)|+\sum_{i=M+1}^{L}|\breve{x}_2(i)|$ is greater than the absolute sum obtained from all ${}^{L}C_M - 1$ other exclusive combinations of tap-inputs. We start by testing whether

$$\sum_{i=1}^{M}|\breve{x}_1(i)| + \sum_{i=M+1}^{L}|\breve{x}_2(i)| > \sum_{i=M+1}^{L}|\breve{x}_1(i)| + \sum_{i=1}^{M}|\breve{x}_2(i)| \tag{6.82}$$

holds. Simplifying (6.82), we obtain

$$\sum_{i=1}^{M}\Big[|\breve{x}_1(i)| - |\breve{x}_2(i)|\Big] > \sum_{i=M+1}^{L}\Big[|\breve{x}_1(i)| - |\breve{x}_2(i)|\Big]$$

$$\sum_{i=1}^{M}\breve{p}(i) > \sum_{i=M+1}^{L}\breve{p}(i) \tag{6.83}$$

which is valid from the definition of $\breve{\mathbf{p}}$. We now consider the ${}^{L}C_M - 2$ other possible cases. Suppose for example, we select tap-indices in the set $\{\breve{x}_1(2i), \breve{x}_2(2i-1)\}$ where $i = 1, 2, \ldots, M$ for which we must test whether

$$\sum_{i=1}^{M}|\breve{x}_1(i)| + \sum_{i=M+1}^{L}|\breve{x}_2(i)| > \sum_{i=1}^{M}|\breve{x}_1(2i)| + \sum_{i=1}^{M}|\breve{x}_2(2i-1)| \tag{6.84}$$

holds. Rewriting (6.84) we obtain

$$\sum_{i=1}^{M}|\breve{x}_1(i)| - \sum_{i=1}^{M}|\breve{x}_2(2i-1)| > \sum_{i=1}^{M}|\breve{x}_1(2i)| - \sum_{i=M+1}^{L}|\breve{x}_2(i)|$$

and hence we can show that

$$\sum_{i=1}^{M/2}|\breve{x}_1(2i)| + \sum_{i=\varphi}^{L}|\breve{x}_2(2i-1)| + \sum_{i=1}^{M/2}\breve{p}(2i-1) >$$

$$\sum_{i=1}^{M/2}\breve{p}(2i+M) + \sum_{i=1}^{M/2}|\breve{x}_1(2i)| + \sum_{i=\varphi}^{L}|\breve{x}_2(2i-1)|$$

$$\sum_{i=1}^{M/2}\breve{p}(2i-1) > \sum_{i=1}^{M/2}\breve{p}(2i+M) \tag{6.85}$$

where $\varphi = 1 + M/2$. Since $M \geq 0$, (6.85) is valid from the definition of $\breve{\mathbf{p}}$. Similar analysis can then be used to verify the remaining cases.

As an illustration, consider an SAEC system with channels $j = 1, 2$, adaptive filters each of length $L = 4$ and tap-input vectors

$$\boldsymbol{x}_j = \begin{bmatrix} x_j(1) \; x_j(2) \; x_j(3) \; x_j(4) \end{bmatrix}^{\mathrm{T}}.$$

The vector $\boldsymbol{p}$ may then be expressed as

$$\begin{bmatrix} p(1) \\ p(2) \\ p(3) \\ p(4) \end{bmatrix} = \begin{bmatrix} |x_1(1)| \\ |x_1(2)| \\ |x_1(3)| \\ |x_1(4)| \end{bmatrix} - \begin{bmatrix} |x_2(1)| \\ |x_2(2)| \\ |x_2(3)| \\ |x_2(4)| \end{bmatrix}. \tag{6.86}$$

Consider the example case $p(3) > p(2) > p(1) > p(4)$, for a particular time instant. Since $p(3) + p(2) > \ldots > p(1) + p(4)$, it can be shown that

$$|x_1(3)| + |x_1(2)| + |x_2(1)| + |x_2(4)| > ... > |x_1(1)| + |x_1(4)| + |x_2(2)| + |x_2(3)|, \tag{6.87}$$

where ... refers to all other pair-wise combinations of elements p. Thus the tap selection corresponding to inputs $x_1(3), x_1(2), x_2(1)$ and $x_2(4)$ maximizes $\mathcal{M}$ with the minimum coherence constraint satisfied by the exclusivity of the tap selection at each time iteration.

In this way, the exclusive maximum (XM) tap selection criterion efficiently selects the best exclusive sets of taps where best here is defined as nearest to MMax jointly for both channels. This is achieved by maximizing the $\mathcal{M}(n)$ measure computed using the taps from both channels. Because of the exclusivity constraint, neither channel in general attains a tap selection as good as MMax and some degradation in convergence performance is therefore to be expected. Nevertheless, our results indicate that such degradation is small compared to the improvement in convergence due to the decorrelating property of XM tap selection.

As a final comment, we note that it is irrelevant to consider other tap selection sets since they have smaller magnitude sum. This approach allows us to eliminate ${}^{L}C_M \times^{L} C_M - 1$ possible combinations thus allowing efficient implementation of the exclusive maximum tap selection which we denote XM.

6.10 Exclusive Maximum Adaptive Filters

As has been shown in Sec. 6.8.2, XM tap selection can improve the conditioning of $\boldsymbol{R}_{\boldsymbol{x}}$ and hence improved convergence is expected. The effect of tap selection for the AP and RLS cases on the autocorrelation matrix will be seen to be similar to that which occurs in the NLMS case. The XM approach relies on the existence of a unique solution for the adaptive filter coefficients which is the case for $L < L_{\mathrm{T}}$. As will be shown through simulations, XM tap selection in combination with the nonlinear (NL) preprocessor leads to better conditioning than the use of the NL-preprocessor alone. This combination of XM and NL approaches, which we refer to as XMNL, is highly effective for the cases we have studied and therefore we focus on this combined structure for our later experiments. Fig. 6.15 shows the schematic diagram of the XMNL-based SAEC structure.

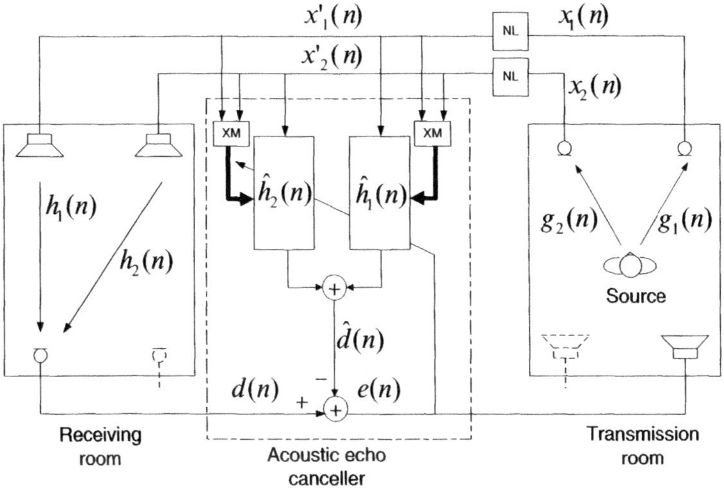

Fig. 6.15. Schematic diagram of XMNL preprocessor in stereophonic echo canceller. Bold arrows indicate tap selection control.

6.10.1 XM-NLMS Algorithm

The XM tap selection technique may be incorporated into the NLMS by selecting taps corresponding to the $M = 0.5L$ largest elements of the input magnitude difference vector $\boldsymbol{p}(n)$ in the first channel and the M smallest elements of $\boldsymbol{p}(n)$ in the second channel as shown in (6.81). Tap-indices are then updated using (6.13).

6.10.2 XMNL-NLMS Algorithm

The nonlinear (NL) preprocessor [5] is one of the most effective methods of achieving signal decorrelation without significantly affecting stereophonic perception and is written, using α as the nonlinearity constant

$$\boldsymbol{x}_1'(n) = \boldsymbol{x}_1(n) + 0.5\alpha\Big[\boldsymbol{x}_1(n) + \big|\boldsymbol{x}_1(n)\big|\Big] , \qquad (6.88)$$

$$\boldsymbol{x}_2'(n) = \boldsymbol{x}_2(n) + 0.5\alpha\Big[\boldsymbol{x}_2(n) - \big|\boldsymbol{x}_2(n)\big|\Big] . \qquad (6.89)$$

We refer to the use of the NL preprocessor with NLMS adaptation as NL-NLMS. Several workers [7, 39, 45] have proposed algorithms in combination with the NL preprocessor so as to achieve low misalignment. In the same manner, a combined algorithm has been proposed [34] employing XM tap selection to improve the conditioning of the autocorrelation matrix and hence improve the convergence rate compared to the use of the NL preprocessor alone. Hence we denote this XM tap selection for the NL processed signals as XMNL. The XMNL-NLMS algorithm is summarized in Table 6.3.

6.10.3 XMNL-AP Algorithm

We may extend the single channel MMax-AP algorithm derived in Sec. 6.3.3 for the SAEC application by using the tap selection matrix

$$\boldsymbol{Q}(n) = \text{diag}\Big\{\big[\boldsymbol{q}_1^{\mathrm{T}}(n),\, \boldsymbol{q}_2^{\mathrm{T}}(n)\big]\Big\} \tag{6.90}$$

such that elements of $\boldsymbol{q}_1(n)$ and $\boldsymbol{q}_2(n)$ are given by (6.81). The filter update equation is then given in (6.17). The resulting XMNL-AP algorithm is given in Table 6.4.

6.10.4 XMNL-RLS Algorithm

Similar to the XMNL-AP, we can extend the single channel MMax-RLS algorithm as derived in Sec. 6.3.4. Using (6.81), the XMNL-RLS algorithm is summarized in Table 6.5.

6.11 SAEC Simulation Results

6.11.1 Experimental Setup

In all our simulations, impulse responses $\boldsymbol{g}_1$, $\boldsymbol{g}_2$, $\boldsymbol{h}_1$ and $\boldsymbol{h}_2$ are generated using the method of images [1] . Two microphones are placed 1 m apart in the centre of both the transmission and receiving rooms each of dimension 3×4×5 m. The source is then positioned 1 m away from each microphone in the transmission room. Tap-input vectors $\boldsymbol{x}_1'(n)$ and $\boldsymbol{x}_2'(n)$ are obtained by convolving the source with two impulse responses $\boldsymbol{g}_1$ and $\boldsymbol{g}_2$ and then applying the nonlinear preprocessor defined in (6.88) and (6.89). The desired response $d(n)$ in the receiving room is obtained by summing $\boldsymbol{h}_1^{\mathrm{T}}\boldsymbol{x}_1'(n)$ and $\boldsymbol{h}_2^{\mathrm{T}}\boldsymbol{x}_2'(n)$. For clarity, the normalized misalignment of only one channel is plotted in each experiment.

6.11.2 NLMS Simulations

We examine the performance of XM tap selection and the NL preprocessor in combination with NLMS adaptation. In this experiment, the lengths of the adaptive filters are $L = 256$ while the lengths of the transmission and receiving rooms' impulse responses are $L_{\mathrm{T}} = 1600$ and $L_{\mathrm{R}} = 256$ respectively. Fig. 6.16 shows the normalized misalignment plot for (a) NLMS, (b) NL-NLMS, (c) XM-NLMS and (d) XMNL-NLMS. A WGN input signal with a sampling frequency of $f_{\mathrm{s}} = 8$ kHz is used with $M = 128$ and a step size of $\mu = 0.4$ is chosen for each algorithm. A nonlinear distortion factor of $\alpha = 0.5$ is used [5] and WGN sequence is added to $d(n)$ such that an SNR of 30 dB is achieved. We see that NLMS has the slowest convergence. The convergence rate of XM-NLMS and NL-NLMS increases significantly due to the XM and NL pre-processors respectively. The XMNL-NLMS algorithm shows even further improvement compared to NL-NLMS due to the additional improvement in conditioning caused by XM tap selection. Alternatively, XMNL-NLMS could achieve the same rate of convergence as NL-NLMS but with a lower value of α [34], hence reducing the nonlinear distortion.

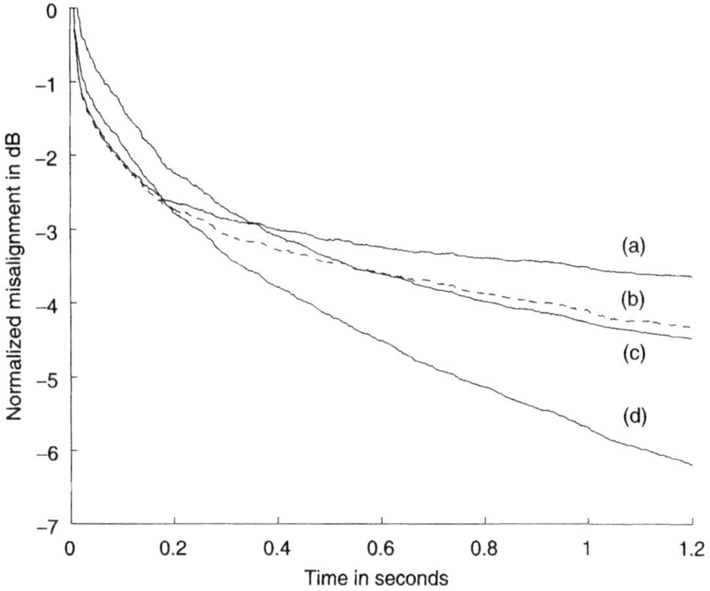

Fig. 6.16. Normalized misalignment for WGN sequence (a) NLMS, (b) NL-NLMS (c) XM-NLMS and (d) XMNL-NLMS [$L = 256$, $L_T = 1200$, $L_R = 256$, $M = 128$, $f_s = 8$ kHz, $\mu = 0.1$, $\alpha = 0.5$ and SNR = 30 dB].

6.11.3 AP Simulations

The performance of the XMNL-AP algorithm is compared with that of the AP algorithm in combination with the NL preprocessor (NL-AP) for a speech signal. The impulse responses are chosen to be of length $L_T = L_R = 1200$, adaptive filters of length $L = 512$ and $M = 256$ are used. We have used a sampling frequency of $f_s = 8$ kHz and an additive WGN is added to the desired signal such that an SNR of 30 dB is achieved. The adaptive step size for each algorithm is chosen such that they achieve approximately the same final normalized misalignment. A nonlinearity constant of $\alpha = 0.5$ and affine projection order $K = 3$ are used.

We see from Fig. 6.17 that the rate of convergence of XMNL-AP is significantly higher than that of the NL-AP. This is again due to the additional improvement in conditioning caused by XM tap selection. For the arbitrary

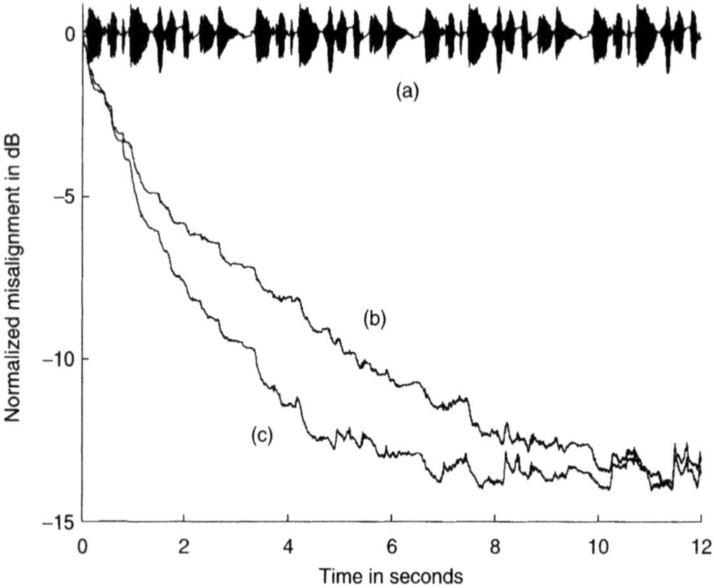

Fig. 6.17. (a) Speech signal and normalized misalignment for (b) NL-AP and (c) XMNL-AP [$L = 512$, $L_{\mathrm{T}} = L_{\mathrm{R}} = 1200$, $M = 256$, $f_{\mathrm{s}} = 8$ kHz, $\mu_{\mathrm{NL-AP}} = 0.5$, $\mu_{\mathrm{XMNL-AP}} = 0.4$, $\alpha = 0.5$, $K = 3$ and SNR $= 30$ dB].

choice of $\mu_{\mathrm{NL-AP}} = 0.5$, it was found that $\mu_{\mathrm{XMNL-AP}} = 0.4$ gives approximately the same final normalized misalignment.

6.11.4 RLS Simulations

In Fig. 6.18, we compare the performance of XMNL-RLS with that of the RLS incorporating the NL preprocessor (NL-RLS) [5]. We have used $L_{\mathrm{T}} = L_{\mathrm{R}} = 800$, $L = 256$, $M = 128$ and a speech input sequence with sampling frequency of $f_{\mathrm{s}} = 8$ kHz. As before, the nonlinearity constant is $\alpha = 0.5$ and a WGN sequence is added to the desired signal such that an SNR of 30 dB is achieved. A forgetting factor of $\lambda_{\mathrm{XMNL-RLS}} = 1 - [1/(10L)] = 0.99961$ [10] is used for XMNL-RLS while for NL-RLS, $\lambda_{\mathrm{NL-RLS}} = 0.99957$ is used such that both algorithms achieve approximately the same final normalized misalignment.

In Fig. 6.18, the XMNL-RLS algorithm shows a significant improvement in convergence rate over NL-RLS.

6.12 Discussion and Conclusion

Selective-tap schemes enable the computational complexity of updating the coefficients of an adaptive filter to be reduced without necessarily reducing

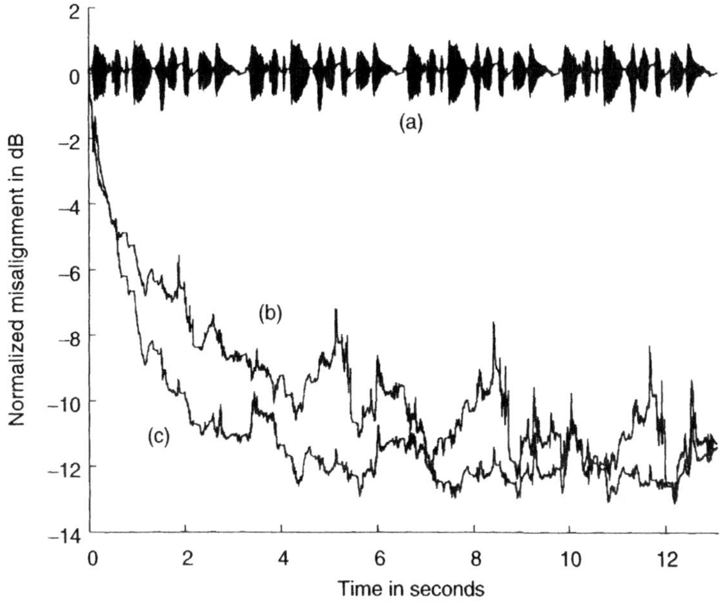

Fig. 6.18. (a) Speech signal and normalized misalignment for (b) NL-RLS and (c) XMNL-RLS [$L = 256$, $L_{\mathrm{T}} = L_{\mathrm{R}} = 800$, $M = 128$, $f_{\mathrm{s}} = 8$ kHz, $\lambda_{\mathrm{XMNL-RLS}} = 0.99961$, $\lambda_{\mathrm{NL-RLS}} = 0.99957$, $\alpha = 0.5$ and SNR = 30 dB].

the order of the filter. This is particularly useful in applications in which the number of coefficients is large such as, for example, acoustic echo cancellation for which several thousand coefficients may be required in order to model the echo path with sufficient accuracy. Several alternative techniques for selecting the set of coefficients to update at each iteration have been discussed. The main objectives in the design of tap selection criteria is to enable the number of tap-updates performed at each iteration to be reduced with minimal degradation in performance and with minimal computational overhead in performing the tap selection itself.

It has been seen that tap selection criteria that are dependent on properties of the tap-input vector, such as are employed in the MMax-NLMS and SPU-NLMS algorithms, are generally more effective than criteria that are data-independent, such as are employed in Periodic-NLMS and Sequential-NLMS. An analysis of the convergence and tracking properties of the MMax-NLMS algorithm for a time-varying system model has been presented. The SORT-LINE and Short-sort algorithms, or block-based techniques, can be used to reduce the computation overhead in performing the tap selection that would otherwise be incurred by the data-dependent schemes.

It has been seen through simulation results that updating $L/2$ taps introduces an insignificant degradation in performance when using, for example,

MMax tap selection. This has been explained intuitively by considering sparseness properties of the input signal. For speech signals, which have been seen to exhibit sparse characteristics, a significant number of the elements of the tap-input vector are small and give rise to correspondingly small tap-updates, the omission of which has negligible effect on convergence. In addition to considering sparseness of the input signal, it has been seen that sparseness of the impulse response can also be taken into account. Both types of sparseness are used simultaneously in the Sparse Partial Update algorithm.

Although selective-tap algorithms were originally introduced for the purpose of computational complexity reduction, it has been shown that they can also be used to improve the performance of multichannel adaptive filters, as used for SAEC, in which the input signals are highly correlated. The exclusive tap selection criterion has been shown to reduce the interchannel coherence of the tap-input vectors and hence improve the conditioning of the autocorrelation matrix, leading to an improvement in convergence rate. The efficient XM tap selection technique has been developed as an optimization of the MMax criterion subject to an exclusivity constraint between the tap selection sets of the two channels. This XM tap selection has been applied in combination with a nonlinear preprocessor to the NLMS, AP and RLS algorithms. Simulation results have shown a significant improvement in convergence compared with algorithms that use the NL-preprocessor alone.

A Appendices

A.1 Algorithm Summary Tables

Table 6.2. Tap selection schemes.

Single Channel MMax Tap Selection

$$M \in \{1, 2, \ldots, L\}$$

$$\boldsymbol{Q}(n) = \text{diag}\Big\{q_0(n), q_1(n), \ldots, q_{L-1}(n)\Big\}$$

$$q_l(n) = \begin{cases} 1, & \big|x_l(n)\big| \in \Big\{M \text{ maxima of } \big|\boldsymbol{x}(n)\big|\Big\} \\ 0, & \text{otherwise} \end{cases}$$

Stereophonic XMNL Tap Selection

$$M = \frac{L}{2}$$

$$\widehat{\boldsymbol{h}}(n) = \Big[\widehat{\boldsymbol{h}}_1^{\mathrm{T}}(n)\,,\,\widehat{\boldsymbol{h}}_2^{\mathrm{T}}(n)\Big]^{\mathrm{T}}$$

$$\widehat{\boldsymbol{k}}(n) = \Big[\widehat{\boldsymbol{k}}_1^{\mathrm{T}}(n)\,,\,\widehat{\boldsymbol{k}}_2^{\mathrm{T}}(n)\Big]^{\mathrm{T}}$$

$$\boldsymbol{x}_1'(n) = \boldsymbol{x}_1(n) + 0.5\,\alpha\Big[\boldsymbol{x}_1(n) + \big|\boldsymbol{x}_1(n)\big|\Big]$$

$$\boldsymbol{x}_2'(n) = \boldsymbol{x}_2(n) + 0.5\,\alpha\Big[\boldsymbol{x}_2(n) - \big|\boldsymbol{x}_2(n)\big|\Big]$$

$$\boldsymbol{x}(n) = \Big[\boldsymbol{x}'^{\mathrm{T}}_1(n)\;\;,\,\boldsymbol{x}'^{\mathrm{T}}_2(n)\Big]^{\mathrm{T}}$$

$$\boldsymbol{q}_j(n) = \Big[q_{j,0}(n), q_{j,1}(n), \ldots, q_{j,L-1}(n)\Big]^{\mathrm{T}}, \; j = 1, 2$$

$$\boldsymbol{p}(n) = \big|\boldsymbol{x}_1'(n)\big| - \big|\boldsymbol{x}_2'(n)\big|$$

$$\boldsymbol{Q}(n) = \text{diag}\Big\{\Big[\boldsymbol{q}_1^{\mathrm{T}}(n)\,,\,\boldsymbol{q}_2^{\mathrm{T}}(n)\Big]\Big\}$$

$$q_{1,u}(n) = \begin{cases} 1, & p_u(n) \in \Big\{M \text{ maxima of } \boldsymbol{p}(n)\Big\} \\ 0, & \text{otherwise} \end{cases}$$

$$q_{2,v}(n) = \begin{cases} 1, & p_v(n) \in \Big\{M \text{ minima of } \boldsymbol{p}(n)\Big\} \\ 0, & \text{otherwise} \end{cases}$$

Table 6.3. MMax-NLMS and XMNL-NLMS Algorithms

$$e(n) = d(n) - \widehat{\boldsymbol{h}}^{\mathrm{T}}(n)\boldsymbol{x}(n)$$

$$\widehat{\boldsymbol{h}}(n+1) = \widehat{\boldsymbol{h}}(n) + \mu \frac{\boldsymbol{Q}(n)\boldsymbol{x}(n)e(n)}{\|\boldsymbol{x}(n)\|_2^2 + \delta}$$

Table 6.4. MMax-AP and XMNL-AP Algorithms

$$\boldsymbol{X}(n) = \left[\boldsymbol{x}(n), \boldsymbol{x}(n-1), \ldots, \boldsymbol{x}(n-K+1)\right]^{\mathrm{T}}$$

$$\widetilde{\boldsymbol{x}}(n) = \boldsymbol{Q}(n)\boldsymbol{x}(n)$$

$$\widetilde{\boldsymbol{X}}(n) = \left[\widetilde{\boldsymbol{x}}(n), \widetilde{\boldsymbol{x}}(n-1), \ldots, \widetilde{\boldsymbol{x}}(n-K+1)\right]^{\mathrm{T}}$$

$$\boldsymbol{d}(n) = \left[d(n), d(n-1), \ldots, d(n-K+1)\right]^{\mathrm{T}}$$

$$\boldsymbol{e}(n) = \boldsymbol{d}(n) - \boldsymbol{X}(n)\widehat{\boldsymbol{h}}(n)$$

$$\widehat{\boldsymbol{h}}(n+1) = \widehat{\boldsymbol{h}}(n) + \mu \widetilde{\boldsymbol{X}}^{\mathrm{T}}(n)\left[\boldsymbol{X}(n)\boldsymbol{X}^{\mathrm{T}}(n) + \delta \boldsymbol{I}\right]^{-1}\boldsymbol{e}(n)$$

Table 6.5. MMax-RLS and XMNL-RLS Algorithms

Initialize:

$$\widetilde{\boldsymbol{\Psi}}_0^{-1} = \delta^{-1}\boldsymbol{I}$$

Algorithm:

$$\widetilde{\boldsymbol{x}}(n) = \boldsymbol{Q}(n)\boldsymbol{x}(n)$$

$$\widetilde{\boldsymbol{k}}(n) = \frac{\widetilde{\boldsymbol{\Psi}}^{-1}(n-1)\,\widetilde{\boldsymbol{x}}(n)}{\lambda + \widetilde{\boldsymbol{x}}^{\mathrm{T}}(n)\,\widetilde{\boldsymbol{\Psi}}^{-1}(n-1)\,\widetilde{\boldsymbol{x}}(n)}$$

$$e(n) = d(n) - \widehat{\boldsymbol{h}}^{\mathrm{T}}(n)\boldsymbol{x}(n)$$

$$\widehat{\boldsymbol{h}}(n+1) = \widehat{\boldsymbol{h}}(n) + \widetilde{\boldsymbol{k}}(n)e(n)$$

$$\widetilde{\boldsymbol{\Psi}}^{-1}(n) = \frac{1}{\lambda}\left[\widetilde{\boldsymbol{\Psi}}^{-1}(n-1) - \widetilde{\boldsymbol{k}}(n)\widetilde{\boldsymbol{x}}^{\mathrm{T}}(n)\widetilde{\boldsymbol{\Psi}}^{-1}(n-1)\right]$$

Table 6.6. Short-sort Algorithm

Parameters:

S : sort window length ($< N$)
A : # samples to select ($< S$)
c : counter
$\boldsymbol{q} = \left[q_0, q_1, ..., q_{A-1}\right]$: storage
m : smallest q_i, $i = 0, 1, ..., A-1$
m_i : index in $\boldsymbol{q}$ of smallest sample

Algorithm:

```
for n = 0, 1, 2, ...
c = n mod S
if (c = 0) then
  m = ∞
endif
if (c < A) then
  q_c = c
  if m > |x(n)| then
    m = |x(n)|
    m_i = q_c
  endif
else
  if m < |x(n)|
    q_{m_i} = c
    m = min value in [x(n − q_i)], i = 0, 1, ..., A − 1
    m_i = value of i for [x(n − q_i)] = m, i = 0, 1, ..., A − 1
  endif
endif
```

A.2 Fourth-order Factorization for Zero Mean Gaussian Variables

For an i.i.d. Gaussian distributed signal $x(n)$, the matrix $\Psi = \mathrm{E}\{\, \boldsymbol{x}_n \boldsymbol{x}_n^{\mathrm{T}} \boldsymbol{x}_n \boldsymbol{x}_n^{\mathrm{T}} \,\}$ has elements

$$\Psi_{k,l} = \mathrm{E}\left\{ x(n-k) \sum_{i=1}^{L} x^2(n-i) x(n-l) \right\}$$

where $\boldsymbol{x}_n = [x(n), x(n-1), \ldots, x(n-L+1)]^{\mathrm{T}}$. The factorization property of real zero-mean Gaussian variables is that

$$\begin{aligned} \mathrm{E}\{\, x(i)x(j)x(k)x(l) \,\} = {} & \mathrm{E}\{\, x(i)x(j) \,\} \mathrm{E}\{\, x(k)x(l) \,\} \\ & + \mathrm{E}\{\, x(i)x(k) \,\} \mathrm{E}\{\, x(j)x(l) \,\} \\ & + \mathrm{E}\{\, x(i)x(l) \,\} \mathrm{E}\{\, x(j)x(k) \,\} \end{aligned}$$

from which

$$\begin{aligned} \mathrm{E}\{\, \boldsymbol{x}_n \boldsymbol{x}_n^{\mathrm{T}} \boldsymbol{x}_n \boldsymbol{x}_n^{\mathrm{T}} \,\}_{kl} = {} & 2 \sum_{i=1}^{L} \mathrm{E}\{\, x(n-k)x(n-i) \,\} \\ & \times \mathrm{E}\{\, x(n-l)x(n-i) \,\} \\ & + \mathrm{E}\{\, x(n-k)x(n-l) \,\} \sum_{i=1}^{L} \mathrm{E}\{\, x^2(n-i) \,\} . \end{aligned}$$

From the above it can be seen that, for the complete matrix, $\Psi = 2\boldsymbol{R}^2 + \boldsymbol{R}\,\mathrm{tr}\{\boldsymbol{R}\}$. Now for $x(n)$ i.i.d Gaussian variables

$$\mathrm{E}\{\, x(n-i)x(n-j) \,\} = \begin{cases} 0. & i \neq j, \\ \sigma_x^2, & i = j, \end{cases}$$

so that $\Psi = (L+2)\sigma_x^4 \boldsymbol{I}$.

References

[1] J. B. Allen, D. A. Berkley: Image method for efficiently simulating small-room acoustics, *J. Acoust. Soc. Amer.*, **65**(4), 943–950, 1979.
[2] T. Aboulnasr, K. Mayyas: Selective coefficient update of gradient-based adaptive algorithms, *Proc. ICASSP '97,* **3**, 1929–1932, Munich, Germany, 1997.
[3] T. Aboulnasr, K Mayyas: MSE analysis of the M-Max NLMS adaptive algorithm, *Proc. ICASSP '98,* **3**, 1669–1672, Washington, DC, USA, 1998.
[4] T. Aboulnasr, K. Mayyas: Complexity reduction of the NLMS algorithm via selective coefficient update, *IEEE Trans. Signal Process.*, **47**(5), 1421–1424, 1999.

[5] J. Benesty, D. R. Morgan, M. M. Sondhi: A Better Understanding and an Improved Solution to the Specific Problems of Stereophonic Acoustic echo Cancellation, *IEEE Trans. Speech Audio Process.*, **T-SA-6**(2), 156–165, 1998.

[6] J. Benesty, D. R. Morgan, J. L. Hall, M. M. Sondhi: Stereophonic acoustic echo cancellation using nonlinear transformations and comb filtering, *Proc. ICASSP '98*, **6**, 3673–3676, Seattle, Washington, USA, 1998.

[7] J. Benesty, T. Gänsler, D. R. Morgan, M. M. Sondhi, S. L. Gay: *Advances in Network and Acoustic Echo Cancellation*, Berlin, Germany: Springer, 2001.

[8] J. Benesty, S. L. Gay: An improved PNLMS algorithm, *Proc. ICASSP '02*, 1881–1884, Orlando, FL, USA, 2002.

[9] N. J. Bershad, S. McLaughlin, C. F. N. Cowan: Performance comparison of RLS and LMS algorithms for tracking a first order Markov communications channel, *Proc. IEEE Int. Symposium on Circuits and Systems*, **1**, 266–270, New Orleans, LA, UAS, 1990.

[10] C. Breining, P. Dreiseitel, E. Hänsler, A. Mader, B. Nitsch, H. Puder, T. Schertler, G. Schmidt, J. Tilp: Acoustic echo control: An application of very-high-order adaptive filter, *IEEE Signal Process. Mag.*, **16**(4), 42–69, 1999.

[11] J. Cui, P. A. Naylor, D. T. Brown: An improved IPNLMS algorithm for echo cancellation in packet-switched networks, *Proc. ICASSP '04*, **4** iv-141–iv-144, Montreal, Canada, 2004.

[12] H. Deng, M. Doroslovacki: Modified PNLMS Adaptive Algorithm for Sparse Echo Path Estimation, *Proc. Infor., Sciences, Systems*, 1072–1077, Mar, 2004.

[13] H. Deng, M. Doroslovacki: New Sparse Adaptive algorithms using partial update, *Proc. ICASSP '04*, 845–848, Montreal, Canada, 2004.

[14] H. Deng, M. Doroslovacki: Improving convergence of the PNLMS algorithm for sparse impulse response identification, *IEEE Signal Process. Letters*, **12**(3), 181–184, 2005.

[15] K. Dogancay, O. Tanrikulu: Selective-partial-update NLMS and affine projection algorithms for acoustic echo cancellation, *Proc. ICASSP '00*, **1**, 448–451, Istanbul, Turkey, 2000.

[16] K. Dogancay, O. Tanrikulu: Adaptive filtering algorithms with selective partial updates, *IEEE Trans. Circuits Systems II*, **48**(8), 762–769, 2001.

[17] S. C. Douglas: A family of normalized LMS algorithms, *IEEE Signal Process. Letters*, **1**(3), 49–51, 1994.

[18] S. C. Douglas: Analysis and implementation of the max-NLMS adaptive filter, *Conference Record of the Twenty-Ninth Asilomar Conference on Signals, Systems and Computers*, **1**(30), 659–663, 1995.

[19] S. C. Douglas: Adaptive filters employing partial updates, *IEEE Trans. Circuits Syst.*, **44**(3), 209–216, 1997.

[20] P. Eneroth, S. L. Gay, T. Gänsler, J. Benesty: A real-time implementation of a stereophonic acoustic echo canceller, *IEEE Trans. Speech Audio Process.,* **9**(5), 513–523, 2001.
[21] N. T. Forsyth, J. A. Chambers, P. A. Naylor: An alternating fixed-point algorithm for stereophonic acoustic echo cancellation, *Proc. Vision, Image and Signal Process. '02,* **149**(1), 1–9, 2002.
[22] S. L. Gay: An efficient, fast converging adaptive filter for network echo cancellation, *Conference Record of the Thirty-Second Asilomar Conference on Signals, Systems and Computers,* 394–398, 1998.
[23] T. Gänsler, J. Benesty: An adaptive nonlinearity solution to the uniqueness problem of stereophonic echo cancellation, *Proc. ICASSP '02,* **2**, 1885–1888, Orlando, FL, USA, 2002.
[24] A. Gilloire, V. Turbin: Using auditory properties to improve the behaviour of stereophonic acoustic echo cancellers, *Proc. ICASSP '98,* **6**, 3681–3684, Seattle, WA, USA, 1998.
[25] S. L. Gay, J. Benesty: *Acoustic Signal Process. for Telecommunication,* Boston, MA, USA: Kluwer Academic Publishers, 2001.
[26] A. Gilloire, M. Vetterli: Adaptive filtering in subbands with critical sampling: Analysis, experiements, and application to acoustic echo cacnellation, *IEEE Trans. Signal Process.,* **40**(8), 1862–1875, 1992.
[27] S. Gollamudi, S. Nagaraj, S. Kapoor, Y.-F. Huang: Set-membership filtering and a set-membership normalized LMS algorithm with an adaptive step-size, *IEEE Signal Process. Letters,* **5**(5), 111–114, 1998.
[28] G. C. Goodwin, K. S. Sin: *Adaptive Filtering, Prediction and Control,* Englewood Cliffs, NJ, USA: Prentice-Hall, 1984.
[29] E. Hänsler: Hands-free telephones- joint control of echo cancellation and postfiltering, *Signal Process.,* **80**(11), 2295–2305, 2000.
[30] E. Hänsler, G. Schmidt: *Acoustic Echo and Noise Control: A Practical Approach,* Hoboken, NJ, USA: Wiley, 2004.
[31] S. Haykin: *Adaptive Filter Theory,* 4th ed., Englewood Cliffs, NJ, USA: Prentice Hall, 2002.
[32] T. Hoya, Y. Loke, J. A. Chambers, P. A. Naylor: Application of the leaky extended LMS algorithm in stereophonic acoustic echo cancellation, *Signal Process.,* **64**, 87–91, 1998.
[33] A. W. H. Khong, P. A. Naylor: Reducing inter-channel coherence in stereophonic acoustic echo cancellation using partial update adaptive filters, *Proc. EUSIPCO '04,*, 405–408, Vienna, Austria, 2004.
[34] A. W. H. Khong, P. A. Naylor: Selective-tap adaptive algorithms in the solution of the nonuniqueness problem for stereophonic acoustic echo cancellation, *IEEE Signal Process. Letters,* **12**(4), 269–272, 2005.
[35] D. E. Knuth: *The Art of Computer Programming, Vol. 3, Sorting and Searching,* Boston, MA, USA: Addison Wesley, 1973.
[36] O. Macchi: Optimization of adaptive identification for time-varying filters, *IEEE Trans. Automat. Contr.,* **31**(3), 283–287, 1986.

[37] R. Martin, J. Altenhöner: Coupled adaptive filters for acoustic echo control and noise reduction, *Proc. ICASSP '95*, **5**, 3043–3046, Detriot, USA, 1995.

[38] K. Mayyas, T. Aboulnasr, T. Eldos: A study of the robustness of the MMax NLMS adaptive algorithm, *Proc. IEEE Int. Symposium on Circuits and Systems '99*, **3**, 154–157, Orlando, FL, USA, 1999.

[39] K. Mayyas: Stereophonic acoustic echo cancellation using lattice orthogonalization, *IEEE Trans. Speech Audio Process.*, **10**(7), 517–525, 2002.

[40] J. Nagumo, A. Noda: A learning method for system identification, *IEEE Trans. Automat. Contr.*, **12**(3), 282–287, 1967.

[41] P. A. Naylor, W. Sherliker: A short-sort M-Max NLMS partial update adaptive filter with applications to echo cancellation, *Proc. ICASSP '03*, **5**, 373–376, Hong Kong, 2003.

[42] P. A. Naylor, A. W. H. Khong: Affine projection and recursive least squares adaptive filters employing partial updates, *Conference Record of the Thirty-Eighth Asilomar Conference on Signals, Systems and Computers*, **1**, 950–954, 2004.

[43] P. A. Naylor, Jingjing Cui, Mike Brookes: Adaptive algorithms for sparse echo cancellation, *Signal Process.*, to appear, 2005.

[44] I. Pitas: Fast algorithms for running ordering and max/min calculation, *IEEE Trans. Circuits Syst.*, **36**(6), 795–804, 1989.

[45] S. Shimauchi, Y. Haneda, S. Makino, Y. Kaneda: New configuration for a stereo echo canceller with nonlinear preprocessing, *Proc. ICASSP '98*, **6**, 3685–3688, Seattle, Washington, USA, 1998.

[46] T. Tangsangiumvisai, J. A. Chambers, A. G. Constantinides: Time-varying allpass filters using spectral-shaped noise for signal decorrelation in stereophonic acoustic echo cancellation, *Proc. 14th Int. Conf. Digital Signal Process.*, 1273–1276, Santorini, Greece, 2002.

[47] O. Tanrikulu, K. Dogancay: Selective-partial-update proportionate normalized least-mean-squares algorithm for network echo cancellation, *Proc. ICASSP '02*, **2**, 1889–1892, Orlando, FL, USA, 2002.

[48] V. Turbin, A. Gilloire, P. Scalart: Comparison of three post-filtering algorithms for residual acoustic echo reduction, *Proc. ICASSP '97*, **1**, 307–310, Munich, Germany, 1997.

[49] S. Werner, M. L. R. de Campos, P. S. R. Diniz: Partial Update NLMS algorithms with data-selective updating, *IEEE Trans. Signal Process.*, **52**(4), 938–949, 2004.

7

Nonlinear Acoustic Echo Cancellation

Fabian Küch and Walter Kellermann

Telecommunications Laboratory, University Erlangen-Nuremberg, Germany

While standard approaches for acoustic echo cancellation in telecommunication systems assume that the echo path to be identified can be modeled by a linear system, in practice, many loudspeaker systems involve non-negligible nonlinearities, e.g., caused by overloaded amplifiers due to low battery voltage of mobile communication receivers, or nonlinearities in the electroacoustic transduction as common with low-cost loudspeakers driven at high volume. Above a certain degree of nonlinear distortion, purely linear approaches are not able to provide a sufficient echo attenuation and nonlinear echo cancellers become desirable. Based on a nonlinear discrete-time model for the acoustic echo path we discuss different nonlinear adaptive structures for nonlinear acoustic echo cancellation and verify their effectiveness by measurements in real-world environments. While the frequency-dependent nonlinear behaviour of common electrodynamic loudspeakers can be modeled by Volterra filters, power filters are well suited to compensate memoryless saturation-type nonlinearities as they occur with overloaded amplifiers and miniaturized loudspeakers, e.g., in mobile phones.

7.1 Introduction

Linear adaptive filtering plays an important role in statistical signal processing and respective theoretical and practical results are well established [14]. In practice, however, nonlinear adaptive filtering often becomes desirable if the considered systems exhibit nonlinear behaviour. Acoustic echo cancellation represents an important example for such situations.

Standard approaches for the cancellation of acoustic echoes rely on the assumption that the echo path can be modeled by a linear system [5]. Accordingly, the acoustic echo canceller (AEC) is implemented as a linear filter. Since the echo path is unknown and, moreover, can change during operation of the echo canceller, the linear filter has to be realized adaptively. Unfortunately, the simple assumption of a linear echo path does not always hold in practice, as it does not include the behaviour of nonlinear audio hardware. The nonlinearly distorted components of the echo signal can not be captured by a linear AEC and, thus, are transmitted back to the far-end speaker who perceives an annoying copy of his own voice. Consequently, any non-negligible nonlinear distortion of the echo signal leads to a reduction of the echo attenuation achievable by purely linear approaches and, thus, impairs the quality of speech communication systems.

Possible sources for nonlinear distortion in the echo path are, e.g., small loudspeakers driven at high volume or overloaded amplifiers [32, 42, 43]. The problem of nonlinearly distorted echoes is especially common in mobile communication devices where high sound levels are desired with only low battery voltage available. For instance, in case of mobile phones operated in their hands-free mode, consumers usually prefer a nonlinear distortion of the loudspeaker signal over reduced output levels. Nonlinear echoes also occur in hands-free teleconferencing systems that include small-sized loudspeakers. If the consumer sets the loudspeaker system to its maximum volume, linear behaviour of small and/or cheap loudspeakers can not be expected anymore. The listening tests presented in [41] show that the accepted level of nonlinear distortion of speech is sufficiently high to cause annoying nonlinear echoes which can not be compensated by linear AECs.

To surpass echo cancellation performance of purely linear approaches, nonlinear methods have to be taken into consideration, where basically two approaches can be applied:

- nonlinear preprocessing of the loudspeaker signal,
- nonlinear adaptive filtering in the AEC.

The first approach aims at a linearization of the audio hardware components via nonlinear preprocessing of the received far-end signal. Then, the overall echo path to be modeled by the AEC consists of the acoustic echo path which is extended by the nonlinear preprocessing stage. In case of an ideal preprocessing of the loudspeaker signal, this overall echo path is linear and, thus, the AEC can also be realized as a linear filter. This approach can include methods known from the linearization of loudspeakers [8] and/or techniques that are used to compensate for the nonlinear distortion introduced by overloaded power amplifiers in digital communication systems [24]. Another method is to intentionally limit the excitation signal of the loudspeaker in order to avoid nonlinear behaviour of the loudspeaker and its amplifier. Note that in this case, the linear AEC has to be adapted with respect to the preprocessed signal. A major drawback imposed by these approaches is the required exact *a priori* knowledge of the nonlinearities of the loudspeaker system. This, however, implies that the nonlinear preprocessor of the echo cancellation unit can be designed only if the audio components of the loudspeaker system are accessible.

Here, we consider the more general approach of nonlinear adaptive filtering in the AEC in order to be more independent of the actual hardware in the loudspeaker system. It turns out that then only the type of nonlinearity in the acoustic echo path has to be known but not its exact properties, i.e., its parameter values. Here, we distinguish between two different types of nonlinear behaviour:

- nonlinearities with memory, as in case of a small loudspeaker driven at high volume,
- memoryless nonlinearities such as the saturation characteristic of overloaded amplifiers.

In the following, we apply certain polynomial filters [28] such as Volterra filters, truncated Taylor series expansions, and linear filters in order to model the behaviour of each of these nonlinearities. Based on these models, corresponding nonlinear filters can be developed which sufficiently model the overall nonlinear acoustic echo path. The goal is then to derive suitable adaptive algorithms for these nonlinear filters in order to provide a satisfying echo cancellation performance for the case that nonlinear audio hardware is included in telecommunication systems.

This chapter is organized as follows: In Sec. 7.2, we consider the properties of acoustic echo paths. After a discussion of nonlinear audio components, we introduce a discrete-time model of the acoustic echo path based on a nonlinear cascaded structure. The discussion of suitable adaptive approaches is divided into the following two sections. On the one hand, adaptive Volterra filters are considered in Sec. 7.3 and address nonlinearities with memory [28]. On the other hand, Sec. 7.4 focusses on the situations where the nonlinearity in the echo path can be considered as memoryless. It turns out that so-called power filters are more suitable than Volterra filters in this case [21]. The effectiveness of the discussed approaches in nonlinear acoustic echo cancellation is confirmed by experiments using real hardware.

7.2 Nonlinear Acoustic Echo Paths

For the design of nonlinear acoustic echo cancellers, it is essential to have sufficient knowledge about the properties of the underlying physical echo path. Therefore, we initially investigate the main components of typical acoustic echo paths. These results can then be used to obtain suitable nonlinear models for the identification of real echo paths.

The general structure of an acoustic echo path is illustrated in Fig. 7.1 and is common for hands-free telephone sets or mobile phones. The respective signal path is a cascade of digital-to-analog (D/A) converter, amplifier, loudspeaker, microphone, microphone preamplifier, and analog-to-digital (A/D) converter. Additionally, it comprises the acoustic propagation path of the speech signal between loudspeaker and microphone.

In general, the propagation path between loudspeaker and microphone can be considered as a linear system. It is commonly modeled by a linear FIR filter representing the room impulse response [5].

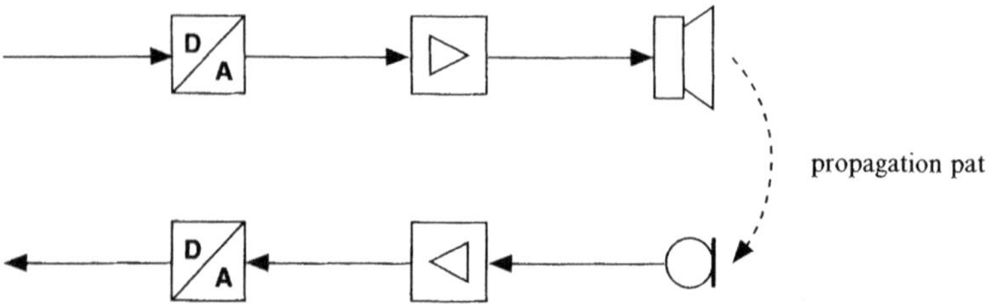

Fig. 7.1. Block diagram of the general structure of an acoustic echo path.

The microphone signals that are common with hands-free and mobile telephony have only moderate excitation levels. Thus, it is reasonable to assume a linear behaviour for the microphone and its preamplifier which is in accordance with the observations reported in [42].

From a system theoretical point of view, an ideal D/A converter can be described by an impulse response of a linear filter [34]. In practice, non-ideal hardware components can lead to a nonlinear mapping of the digital input signal to the analog output of the D/A converter [1, 34]. The same applies to A/D converters which, in addition, imply quantization of analog signals due to finite word lengths used for representation of digital signals. Early publications [1, 6] address the problem of nonlinear network echo cancellation resulting from nonlinear D/A and A/D converters, respectively. With the modern, high-resolution converters used in todays telecommunication systems, it is mostly acceptable to neglect both, quantization errors, and any other nonlinear mapping characteristic caused by non-ideal signal conversion.

In this chapter we consider two sources for nonlinear distortion: the loudspeaker and its amplifier. The properties of these nonlinear system components are discussed next.

Amplifier nonlinearities are especially present in mobile communication devices. There, the dilemma arises to provide high signal levels while having only low battery voltage available. The consumers usually prefer an overloading of the amplifier over a reduction of the sound volume. The nonlinear behaviour of amplifiers can therefore be described as saturation characteristic with a soft clipping of large amplitude values [42]. Due to the limited bandwidth of telephone signals, amplifiers applied to audio applications can in general be considered as memoryless.

Many research efforts aimed at the characterization of the nonlinearities of electrodynamic loudspeakers [16, 37]. Summarizingly, one can distinguish between three different parts of the loudspeaker which may introduce nonlinear distortion: the acoustical part, the electromagnetic part, and the mechanical part. Nonlinearities in the acoustical part, such as nonlinear wave propagation play an important role in the modeling and linearization of horn loudspeakers [17]. However, as this type of loudspeaker is generally used in public announcement systems only, any nonlinearities caused by sound radiation are not considered here.

The nonlinearities in the electromagnetic part (also referred to as motor part) are mainly caused by the asymmetries of the magnetic flux, and its decay outside the air gap of the motor. Thus, the driving force on the voice coil is a nonlinear function of its position. Additionally, the self-inductance of the voice coil depends on its displacement, too.

In the mechanical part, the nonlinear dependency of the stiffness of the spider and the outer rim on the position of the voice coil has to be taken into account. It is worth mentioning that the characteristic of this nonlinearity is slowly time-varying, as the mechanical properties of the spider and the rim are changing in time due to changes in temperature and aging effects of the used materials.

A common approach to incorporate the above-mentioned nonlinearities into a loudspeaker model is to approximate the various nonlinear characteristics by a truncated Taylor series expansion for the decisive parameters. Then, the approximated parameters are introduced into the differential equations that describe the behaviour of the loudspeaker [16, 37]. However, such a representation of the loudspeaker is out of scope of this chapter, as this is not suited for adaptive realizations as required for the echo cancellation application. Consequently, we exploit the main result of [16, 37], i.e., the nonlinear behaviour of loudspeakers will be modeled by an appropriate Volterra filter. More precisely, we consider the loudspeaker as a black box, the input/output relation of which can sufficiently well be approximated by a second-order Volterra filter. On the other hand, the results presented in [23] indicate that for acoustic echo cancellation in mobile phones, a saturation-type behaviour of the miniaturized loudspeakers can be expected. In this case, specialized third-order polynomial filters with less memory support represent a more suitable choice.

Other sources for nonlinear distortion in loudspeaker systems are given by rattling and vibration effects caused by a strong physical coupling between loudspeaker, microphone, and their enclosure, as, e.g., common in mobile phones. However, this distortion can hardly be modeled or predicted, as it is of chaotic nature [3]. It should rather be considered as uncorrelated noise (analogously to any background noise) and, thus, be processed accordingly. The problem of vibrating system components is, however, not further considered here.

Furthermore, mechanical clipping can be observed in the loudspeaker for very high excitation levels if the available displacement range for the voice coil is not sufficiently large [37, 40]. This problem, however, is also not further considered here.

Discrete-Time Model for the Echo Path

As a result of the above discussion, we are now able to derive a discrete-time model for the acoustic echo path that is common in acoustic echo cancellation. Such a model can consist of the cascade of different linear and nonlinear components as illustrated by the block diagram Fig. 7.2. Obviously, the cascaded structure is a direct consequence of the cascade of the different system components shown in Fig. 7.1.

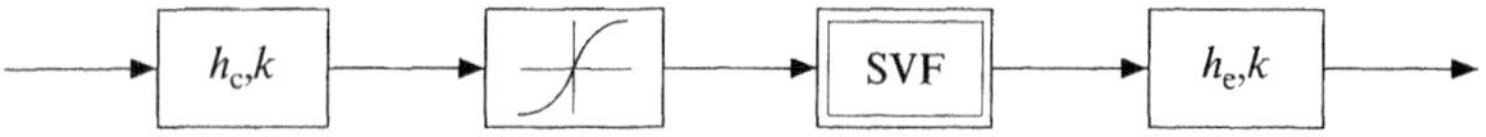

Fig. 7.2. Block diagram of the nonlinear model of the acoustic echo path.

The first block in Fig. 7.2 is the linear FIR filter $h_{c,k}$ representing the combination of all linear filtering steps involved in the D/A-conversion. Following [42], we assume that the loudspeaker amplifier can be regarded as memoryless soft clipping and, thus, be approximated by a truncated Taylor series expansion. Although a seventh-order Taylor series has been applied in [42] to model the behaviour of the amplifier, simulation results have shown that a third-order polynomial can sufficiently reproduce the influence of the amplifier nonlinearity. The corresponding block in Fig. 7.2 illustrates the soft clipping by showing a corresponding mapping characteristic. Note that for mobile phones, the level of nonlinear distortion introduced by the amplifier may depend on the charge level of the battery that provides the power supply. Thus, the parameters of the polynomial representing the amplifier have to be at least slowly time-variant.

The third block in Fig. 7.2 represents a second-order Volterra filter (SVF) that is used to simulate the loudspeaker nonlinearities. As already mentioned, the mechanical contribution to the nonlinear distortion is not constant over time due to fatigue of material. Consequently, the coefficients of the Volterra filter should also be at least slowly time-variant. In case of mobile phones, the nonlinear behaviour of miniaturized loudspeakers exhibit a memoryless saturation characteristic [23]. Then, the Volterra filter can be replaced by a Taylor series expansion. The cascade of the loudspeaker and its amplifier can thus be modeled by using only a single truncated Taylor series expansion, i.e., by discarding the third block in Fig. 7.2.

The last block comprises three cascaded linear models, representing the sound propagation path between loudspeaker and microphone, the microphone characteristic (including its preamplifier), and the A/D converter, respectively. As the linear FIR filter $h_{e,k}$ includes the room impulse response, it may be rapidly time-variant.

It can be shown that every parallel/cascaded combination of linear filters, truncated Taylor series expansions, and Volterra filters can be replaced by a corresponding Volterra filter exhibiting the same input/output relation. It is straightforward to verify that for the above model of the nonlinear echo path, this results in a fifth-order Volterra filter if the amplifier is represented by a third-order polynomial and a second-order Volterra filter is used as loudspeaker model. This approach, however, is not practicable due to the enormous number of required coefficients for higher-order Volterra filters [28].

Simplifications of the general model of the echo path according to Fig. 7.2 can be achieved if any *a priori* knowledge about the properties of the system to be identified can be exploited. In the following we assume that at least one of the nonlinear components in Fig. 7.2 can be neglected: In Sec. 7.3 we look at the case where solely the second-order Volterra filter is included in the actual model of the acoustic echo path, whereas in Sec. 7.4 the Taylor series expansion is considered as the only nonlinear component. Furthermore, the cascaded nature of the echo path can be taken into account for deriving more efficient overall models of the nonlinear echo path.

7.3 Volterra Filters

For the case that the small-sized loudspeaker of a hands-free telecommunication device represents the main source for nonlinear distortion in the echo path, it has to be modeled by a second-order Volterra filter. In this section we therefore discuss Volterra filters and corresponding adaptive realizations in both, time domain and frequency domain. The basic concepts of adaptive Volterra filtering are presented for the general case of P-th order Volterra filters. In more specific parts such as the control of the adaptation or the evaluation of the presented algorithms, we explicitly refer to the acoustic echo cancellation application and limit ourselves to second-order Volterra filters.

The output $d(n)$ of a P-th order Volterra filter is composed of the sum of the outputs of all kernels up to order P:

$$d(n) = \sum_{p=1}^{P} d_p(n). \tag{7.1}$$

Most commonly, the input/output relation of the p-th order kernel is expressed by [28]

$$d_p(n) = \sum_{k_{p,1}=0}^{N_p-1} \sum_{k_{p,2}=k_{p,1}}^{N_p-1} \cdots \sum_{k_{p,p}=k_{p,p-1}}^{N_p-1} h_{\boldsymbol{k}_p} \prod_{i=1}^{p} x(n - k_{p,i}), \tag{7.2}$$

where the memory lengths N_p of the Volterra kernels can in general be different for each order p. In Eq. 7.9, the index vector

$$\boldsymbol{k}_p = [k_{p,1},\, k_{p,2},\, \ldots,\, k_{p,p}] \tag{7.3}$$

can be interpreted as reference to a certain coefficient $h_{\boldsymbol{k}_p}$ of the p-th order Volterra kernel in a p-dimensional Cartesian coordinate system. Thus, Eq. 7.2 is referred to as Cartesian coordinate representation (CCR) of Volterra filters [35]. As can be noticed from Eq. 7.2, there is a strong relation between *multidimensional linear filtering* and Volterra filters in CCR [28].

In the following we consider an alternative representation of Volterra filters which turns out to be more useful in the nonlinear echo cancellation context featuring nonlinear cascaded structures. Regarding [35], we apply the following change of coordinates:

$$k_{p,1} = k, \quad 0 \leq k \leq N_p - 1, \tag{7.4}$$

$$k_{p,i} = r_{p,i-1} + k, \quad 2 \leq i \leq p. \tag{7.5}$$

For interpreting the above coordinate transform, we recall that the set of indices $k_{p,1}, k_{p,2}, \ldots, k_{p,p}$ can be considered as Cartesian coordinates which corresponds to the p-dimensional sampled hypercube representing the p-th order Volterra kernel. The combination of the indices $r_{p,1}, r_{p,2}, \ldots, r_{p,p-1}$, and k can then be understood as reference to the kernel coefficients lying on a straight line which is parallel to the main diagonal of the Cartesian coordinate system. Following [35], we refer to these straight lines as diagonals, where the main diagonal is defined by setting $k_{p,1} = k_{p,2} = \ldots = k_{p,p}$ which implies $r_{p,1} = r_{p,2} = \ldots = r_{p,p-1} = 0$. Based on these interpretations, we consider the new set of indices $r_{p,1}, r_{p,2}, \ldots, r_{p,p-1}$, and k as coordinates of the so-called diagonal coordinate system. The relation between the CCR and the diagonal coordinate representation (DCR) is illustrated in Fig. 7.3 for the quadratic Volterra kernel. As an example, the diagonal corresponding to

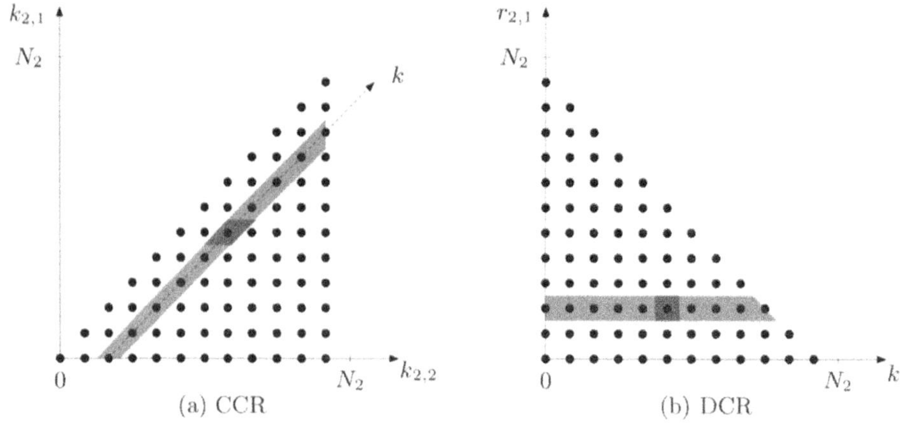

Fig. 7.3. Illustration of the relation between the CCR and DCR for a quadratic Volterra kernel ($p = 2$). Each • corresponds to a kernel coefficient.

$r_{2,1} = 2$ is highlighted in both figures. Additionally, the dark quadrangles mark the indices $(k_{2,1}, k_{2,2}) = (5, 7)$ and $(r_{2,1}, k) = (2, 5)$, respectively, which reference the same kernel coefficient.

Analogously to the coefficient index vector $\boldsymbol{k}_p$ in Eq. 7.3, we introduce the two coefficient vectors

$$\boldsymbol{r}_p = [r_{p,1}, r_{p,2}, \ldots, r_{p,p-1}], \tag{7.6}$$

$$\boldsymbol{r}_{p,k} = [k, r_{p,1} + k, \ldots, r_{p,p-1} + k], \tag{7.7}$$

where for the linear kernel $\boldsymbol{r}_1 = [\,]$ and $\boldsymbol{r}_{1,k} = k$. The diagonal index vector $\boldsymbol{r}_p$ references a certain diagonal, whereas the corresponding coefficient index vector $\boldsymbol{r}_{p,k}$ references a certain kernel coefficient on that diagonal. Note that $\boldsymbol{r}_p$ has the length $p-1$, while $\boldsymbol{r}_{p,k}$ consists of p elements. The diagonal index vector associated with the main diagonal is obviously given by $\boldsymbol{r}_p = \mathbf{0}$. Thus, the index vector elements $r_{p,i}$ of $\boldsymbol{r}_p$ determine the distance of that diagonal from the main diagonal of the p-th order Cartesian coordinate system. It should be pointed out that with the definition of the coefficient index vector $\boldsymbol{r}_{p,k}$ in Eq. 7.7, the notation of the kernel coefficients have been kept unchanged, i.e.,

$$h_{\boldsymbol{r}_{p,k}} = h_{\boldsymbol{k}_p}, \quad \text{if } \boldsymbol{r}_{p,k} = \boldsymbol{k}_p. \tag{7.8}$$

The desired form of the input/output relation of the p-th order kernel is obtained by introducing the new index vectors $\boldsymbol{r}_p$ and $\boldsymbol{r}_{p,k}$ into Eq. 7.2 and by additionally changing the order of summation:

$$d_p(n) = \sum_{r_{p,1}=0}^{N_p-1} \cdots \sum_{r_{p,p-1}=r_{p,p-2}}^{N_p-1} \sum_{k=0}^{L_{\boldsymbol{r}_p}-1} h_{\boldsymbol{r}_{p,k}}\, x(n-k) \prod_{i=1}^{p-1} x(n - r_{p,i} - k), \tag{7.9}$$

where

$$L_{\boldsymbol{r}_p} = N_p - r_{p,p-1}, \tag{7.10}$$

obviously depends on both, the kernel order p, and the actual value of $r_{p,p-1}$. For the linear kernel ($p = 1$) we have $L_{\boldsymbol{r}_1} = N_1$. As proposed in [35], we refer to Volterra filters featuring the above form for computing the output of the p-th order Volterra kernel as Volterra filters in diagonal coordinate representation (DCR).

In the following we examine the relation between Volterra filters in DCR and *multichannel linear filtering*. For a deeper insight into the internal multichannel structure of the DCR, we introduce the input signal of the diagonal $\boldsymbol{r}_p$ according to

$$x_{\boldsymbol{r}_p}(n) = x(n) \prod_{i=1}^{p-1} x(n - r_{p,i}), \tag{7.11}$$

where $x_{\boldsymbol{r}_1}(n) = x(n)$. The corresponding output $d_{\boldsymbol{r}_p}(n)$ of the diagonal with index $\boldsymbol{r}_p$ is then given by

$$d_{\boldsymbol{r}_p}(n) = \sum_{k=0}^{L_{\boldsymbol{r}_p}-1} h_{\boldsymbol{r}_{p,k}}\, x_{\boldsymbol{r}_p}(n-k). \tag{7.12}$$

Obviously, $d_{\boldsymbol{r}_p}(n)$ can be considered as the output of the linear FIR filter $h_{\boldsymbol{r}_{p,k}}$ of length $L_{\boldsymbol{r}_p}$ with input $x_{\boldsymbol{r}_p}(n)$. In other words, $d_{\boldsymbol{r}_p}(n)$ results from the convolution of $x_{\boldsymbol{r}_p}(n)$ with the linear filter $h_{\boldsymbol{r}_{p,k}}$ and can therefore be expressed by

$$d_{\boldsymbol{r}_p}(n) = h_{\boldsymbol{r}_{p,n}} * x_{\boldsymbol{r}_p}(n), \tag{7.13}$$

where $*$ denotes convolution. The above definitions are used to rewrite the output $d_p(n)$ of the p-th order kernel according to Eq. 7.9:

$$d_p(n) = \sum_{r_{p,1}=0}^{N_p-1} \cdots \sum_{r_{p,p-1}=r_{p,p-2}}^{N_p-1} d_{\boldsymbol{r}_p}(n). \tag{7.14}$$

From the specific form of Eq. 7.14 we notice that $d_p(n)$ can be interpreted as the output of a linear multiple input/single output (MISO) system, where each diagonal with index vector $\boldsymbol{r}_p$ corresponds to one linear channel with input $x_{\boldsymbol{r}_p}(n)$. Extending this interpretation to the computation of the output of the Volterra filter according to Eq. 7.1, $d(n)$ can be considered as the output of a special MISO system featuring a combination of P multichannel structures, where each channel corresponds to one particular diagonal of the DCR.

Aiming at a compact vector notation for the computation of $d_p(n)$, we introduce the input signal vectors $\boldsymbol{x}_p(n)$ associated to the p-th order kernel vector $\boldsymbol{h}_p$ according to

$$\boldsymbol{x}_p(n) = \left[\ldots, x(n) \prod_{i=1}^{p-1} x(n - r_{p,i}), \ldots\right]^{\mathrm{T}}, \tag{7.15}$$

$$\boldsymbol{h}_p = \left[\ldots, h_{\boldsymbol{r}_{p,k}}, \ldots\right]^{\mathrm{T}}. \tag{7.16}$$

The $\binom{N_p+p-1}{p}$ elements of $\boldsymbol{x}_p(n)$ and $\boldsymbol{h}_p$ can in principle be arranged arbitrarily according to any given preferences. Of course, the elements in $\boldsymbol{x}_p(n)$ and $\boldsymbol{h}_p$ have to be arranged consistently such that

$$d_p(n) = \boldsymbol{h}_p^{\mathrm{T}} \boldsymbol{x}_p(n). \tag{7.17}$$

With the definitions of the vectors

$$\boldsymbol{x}_{\mathrm{VF}}(n) = \left[\boldsymbol{x}_1^{\mathrm{T}}(n), \boldsymbol{x}_2^{\mathrm{T}}(n), \ldots, \boldsymbol{x}_P^{\mathrm{T}}(n)\right]^{\mathrm{T}}, \tag{7.18}$$

$$\boldsymbol{h}_{\mathrm{VF}} = \left[\boldsymbol{h}_1^{\mathrm{T}}, \boldsymbol{h}_2^{\mathrm{T}}, \ldots, \boldsymbol{h}_P^{\mathrm{T}}\right]^{\mathrm{T}}, \tag{7.19}$$

we can finally extend the vector notation also to the computation of the overall Volterra filter output $d(n)$:

$$d(n) = \boldsymbol{h}_{\mathrm{VF}}^{\mathrm{T}} \boldsymbol{x}_{\mathrm{VF}}(n). \tag{7.20}$$

We notice that Eq. 7.20 reflects the linearity of the output $d(n)$ with respect to the Volterra filter coefficients which are summarized in $\boldsymbol{h}_{\mathrm{VF}}$. This formal analogy to linear filtering can be exploited in order to straightforwardly extend adaptive algorithms known from linear adaptive filtering to Volterra filters.

7.3.1 Application to Cascaded Structures

In the following, we look at the configuration according to Fig. 7.4 in more detail which consists of the cascade of a Volterra filter $h_{\boldsymbol{r}_{p,k}}$ followed by a linear filter c_k.

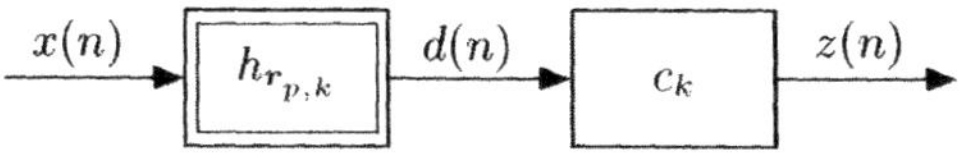

Fig. 7.4. Cascaded structure consisting of a second-order Volterra filter $h_{\boldsymbol{r}_{p,k}}$ followed by a linear FIR filter c_k.

First, we recall the assumption that the model of the echo path Fig. 7.2 can be simplified to the cascade of a linear filter, a second-order Volterra filter, and another linear filter. The cascade of a linear filter followed by a Volterra filter can be represented by a corresponding Volterra filter of the same order but with increased memory length. Thus, the assumed simplified model of the acoustic echo path represents a special case of the configuration shown in Fig. 7.4.

From an efficiency point of view, one might tend to directly use an adaptive implementation of the two cascaded units for realizing the nonlinear acoustic echo canceller. This approach has already been proposed in [11] for acoustic echo cancellation including nonlinearly distorting loudspeakers. However, it is challenging to assure convergence to the optimum solution or even assure a stable adaptation behaviour for cascaded structures. This problem has also been observed by the authors of [11]. As a remedy, they suggest to adapt the Volterra filter only after the linear postfilter has 'sufficiently' converged. In order to circumvent any sophisticated adaptation control as required for the adaptation of cascades, we consider an equivalent Volterra model as a parallelized realization of the cascaded structure instead. It turns out that the DCR provides an elegant representation of such equivalent Volterra models.

As the convolution is a linear operation, the computation of the output of thecascaded structure directly follows from Eqs. 7.1, 7.13, 7.14:

$$z(n) = \sum_{p=1}^{P} z_p(n), \tag{7.21}$$

$$z_p(n) = \sum_{r_{p,1}=0}^{N_p-1} \sum_{r_{p,2}=r_{p,1}}^{N_p-1} \cdots \sum_{r_{p,p-1}=r_{p,p-2}}^{N_p-1} z_{\boldsymbol{r}_p}(n), \tag{7.22}$$

where the outputs $z_{\boldsymbol{r}_p}(n)$ of the respective DCR-channel read

$$\begin{aligned} z_{\boldsymbol{r}_p}(n) &= c_n * h_{\boldsymbol{r}_{p,n}} * x_{\boldsymbol{r}_p}(n) \\ &= g_{\boldsymbol{r}_{p,n}} * x_{\boldsymbol{r}_p}(n). \end{aligned} \tag{7.23}$$

We note from Eqs. 7.22, 7.23 that $z_p(n)$ can be considered as the output of a special p-th order Volterra kernel $g_{\boldsymbol{r}_{p},k}$ with input $x(n)$, where the number and the position of the diagonals are not changed compared to the Volterra kernel $h_{\boldsymbol{r}_{p},k}$. However, the length of the filter in each DCR-channel with index vector $\boldsymbol{r}_p$ is increased according to

$$\widetilde{L}_{\boldsymbol{r}_p} = L_{\boldsymbol{r}_p} + N_c - 1, \tag{7.24}$$

where N_c denotes the length of the linear filter c_k. Obviously, the corresponding CCR of the kernel $g_{\boldsymbol{r}_{p},k}$ has the overall memory length $N_p + N_c - 1$, where only a corridor with respect to the main diagonal of width N_p has non-zero coefficients. The resulting region of support of the Volterra kernels, i.e., the non-zero coefficients, is illustrated in Fig. 7.5 for the quadratic kernel and the special case $N_2 = 4$ and $N_c = 16$. Comparing Fig. 7.5 with Fig. 7.3(b), the

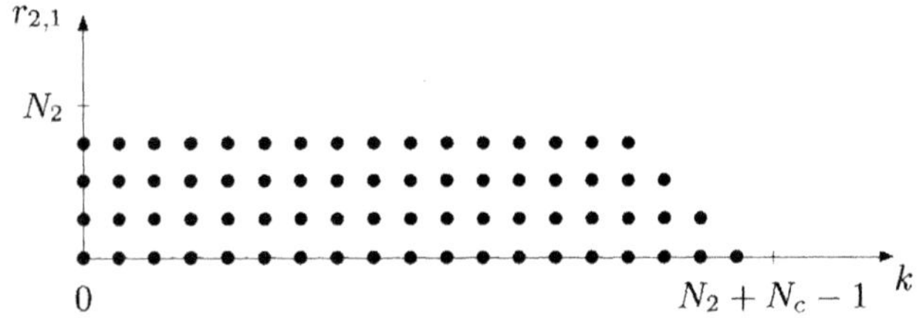

Fig. 7.5. Illustration of the quadratic Volterra kernel corresponding to the cascaded structure according to Fig. 7.4 for the special case $N_2 = 4$ and $N_c = 16$. Each • corresponds to a non-zero kernel coefficient.

specific shape of the region of support becomes clear.

The reduced region of support as required for Volterra models of cascaded structures according to Fig. 7.4, can easily be taken into account by appropriately modifying Eq. 7.14:

$$d_p(n) = \sum_{r_{p,1}=0}^{R_p-1} \sum_{r_{p,2}=r_{p,1}}^{R_p-1} \cdots \sum_{r_{p,p-1}=r_{p,p-2}}^{R_p-1} d_{\boldsymbol{r}_p}(n). \tag{7.25}$$

The parameter R_p is used here to specify the maximum distance of a diagonal with respect to the main diagonal. Introducing Eq. 7.12 into Eq. 7.25 yields

$$d_p(n) = \sum_{r_{p,1}=0}^{R_p-1} \sum_{r_{p,2}=r_{p,1}}^{R_p-1} \cdots \sum_{r_{p,p-1}=r_{p,p-2}}^{R_p-1} \sum_{k=0}^{L_{\boldsymbol{r}_p}-1} h_{\boldsymbol{r}_{p},k}\, x_{\boldsymbol{r}_p}(n-k). \tag{7.26}$$

The output $d_{\boldsymbol{r}_p}(n)$ of the diagonal with index vector $\boldsymbol{r}_p$ is still computed according to Eq. 7.12, implying that the linear filter of the corresponding channel has the memory length $L_{\boldsymbol{r}_p} = N_p - r_{p,p-1}$. Choosing $R_p < N_p$ yields the desired reduced region of support compared to the case $R_p = N_p$ (as imposed by Eq. 7.14). The possibility to reduce the width R_2 of the quadratic kernel without impairing the echo cancellation performance of an adaptive second-order Volterra filter is exemplified in Sec. 7.3.4 for a real acoustic echo path.

Throughout the rest of this chapter, we always refer to the extended definition Eq. 7.26 when considering Volterra filters in DCR. Moreover, we restrict ourselves to the case $R_p \leq N_p$. Then, N_p still represents the *memory length*[1] of the p-th order kernel, whereas R_p is referred to as its *width*. It should be emphasized that the distinction between R_p and N_p does not imply an additional degree of freedom for the design of Volterra filters: We only explicitly exclude certain coefficients of the corresponding CCR which are *a priori* assumed to be zero.

The number of diagonals $N_{\text{diag},p}$ included in the p-th order kernel with width R_p is given by

$$N_{\text{diag},p} = \binom{R_p + p - 2}{p - 1}. \tag{7.27}$$

Note that only for $p \leq 2$, i.e., for the linear and the quadratic kernel, the width R_p equals to the number of diagonals. For the linear kernel R_1 obviously always equals one. The number of coefficients $N_{\text{coeff},p}$ of the p-th order kernel with memory length N_p and width R_p is obtained as

$$N_{\text{coeff},p} = \binom{R_p + p - 1}{p} + (N_p - R_p) N_{\text{diag},p} \tag{7.28}$$

with $N_{\text{diag},p}$ according to Eq. 7.27.

7.3.2 Time-domain Adaptive Volterra Filters

The fundamental problem of adaptive Volterra filtering in acoustic echo cancellation is illustrated in Fig. 7.6. From Fig. 7.6 we notice that the require-

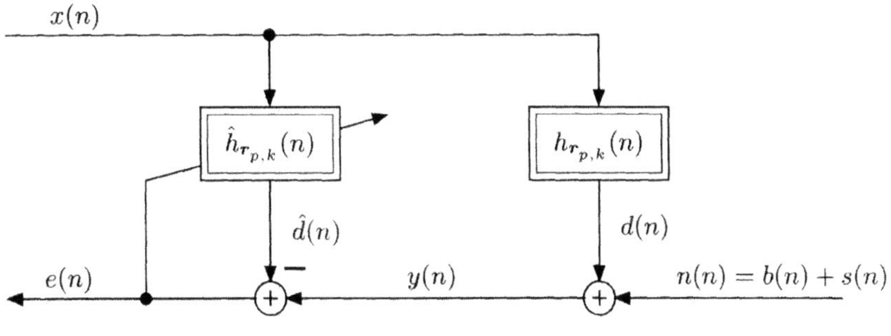

Fig. 7.6. General configuration for adaptive Volterra filtering in acoustic echo cancellation.

[1] Strictly speaking, the memory length is $N_p - 1$.

ments for adaptive Volterra filtering are basically equivalent to the task of a linear adaptive filtering in the echo cancellation context: The coefficients $\hat{h}_{\boldsymbol{r}_{p,k}}(n)$ of the adaptive Volterra filter have to be determined such that $\hat{d}(n)$ matches the output of the unknown system $d(n)$. As already indicated in Fig. 7.6, we assume in the following that the unknown system (i.e., the acoustic echo path) can can be characterized by the Volterra filter coefficients $h_{\boldsymbol{r}_{p,k}}(n)$.

Using the notation of Fig. 7.6, the error $e(n)$ is defined as

$$e(n) = y(n) - \hat{d}(n), \tag{7.29}$$

where the output of the P-th order adaptive Volterra filter reads

$$\hat{d}(n) = \hat{\boldsymbol{h}}_{\mathrm{VF}}^{\mathrm{T}}(n)\,\boldsymbol{x}_{\mathrm{VF}}(n). \tag{7.30}$$

The coefficient vector $\hat{\boldsymbol{h}}_{\mathrm{VF}}(n)$ is defined analogously to Eqs. 7.16, 7.19, but contains the kernel coefficients $\hat{h}_{\boldsymbol{r}_{p,k}}(n)$ of the adaptive Volterra filter instead of $h_{\boldsymbol{r}_{p,k}}(n)$. The output $d(n)$ of the unknown system is given by Eqs. 7.1, 7.26, i.e., we assume that it can be completely modeled by a P-th order Volterra filter. In the following we additionally assume that the order P and the memory lengths N_p are equal for both, the adaptive Volterra filter, and the unknown system.

The observed microphone signal $y(n)$ is given by

$$y(n) = d(n) + b(n) + s(n), \tag{7.31}$$

where $d(n)$ represents the actual echo signal. The external distortions $b(n)$ and $s(n)$ represent background noise and local speech, respectively, and are summarized to

$$n(n) = b(n) + s(n). \tag{7.32}$$

The residual echo $\varepsilon(n)$ is given by

$$\varepsilon(n) = d(n) - \hat{d}(n). \tag{7.33}$$

Similarly to linear adaptive filtering, the LMS algorithm represents the most commonly used adaptation algorithm for Volterra filters [27,28] mainly because of its simplicity and robustness. This is especially important since Volterra filters imply a huge number of coefficients to be adapted, as can be noticed from Eq. 7.28. The update equation for the coefficient vector $\hat{\boldsymbol{h}}_{\mathrm{VF}}(n)$, applying the LMS algorithm, is given by [27]

$$\hat{\boldsymbol{h}}_{\mathrm{VF}}(n+1) = \hat{\boldsymbol{h}}_{\mathrm{VF}}(n) + \mu_{\mathrm{LMS}}(n)\,e(n)\boldsymbol{x}_{\mathrm{VF}}(n). \tag{7.34}$$

The step size control parameter $\mu_{\mathrm{LMS}}(n)$ can be chosen to vary for different coefficients, as discussed later in this section.

The standard NLMS algorithm for Volterra filters is obtained by normalizing the step size parameter $\mu_{\mathrm{LMS}}(n)$ according to

$$\mu_{\mathrm{LMS}}(n) = \frac{\alpha_{\mathrm{NLMS}}(n)}{\boldsymbol{x}_{\mathrm{VF}}^{\mathrm{T}}(n)\boldsymbol{x}_{\mathrm{VF}}(n)}, \quad 0 < \alpha_{\mathrm{NLMS}}(n) < 2. \tag{7.35}$$

The given range for the step size $\alpha_{\mathrm{NLMS}}(n)$ indicates the range where stable convergence can be expected [27].

Although the above normalization formally yields the NLMS algorithm for Volterra filters, it might not always be useful in practice. From the definition of $\boldsymbol{x}_{\mathrm{VF}}(n)$ in Eq. 7.18 it follows that the denominator of Eq. 7.35 is composed of the sum over different moments of $x(n)$, up to order $2P$. In general, the orders of magnitude of these moments significantly differ and, thus, the joint normalization of all Volterra kernels according to Eq. 7.35 is not suitable for higher-order Volterra filters. Considering

$$\boldsymbol{x}_p^{\mathrm{T}}(n)\boldsymbol{x}_p(n) \ll \max_i \left\{\boldsymbol{x}_i^{\mathrm{T}}(n)\boldsymbol{x}_i(n)\right\}, \tag{7.36}$$

we realize that the adaptation of the coefficients of the p-th order kernel almost freezes for a joint normalization of the step size $\alpha_{\mathrm{NLMS}}(n)$.

There exist also more sophisticated algorithms for the adaptation of Volterra filters which can be used to circumvent the inherent slow convergence of the LMS algorithm. Prominent examples are the Affine Projection Algorithm (APA) [7] or the RLS algorithm [27] which are also well known in linear adaptive filtering [9]. Note that due to the huge number of coefficients that are associated with higher-order Volterra filters, the APA and the RLS algorithm are usually not realizable in practice.

There is another major difference between the matrix representation of linear filters and the matrix representation of Volterra filters: The input vector $\boldsymbol{x}_{\mathrm{VF}}(n)$ does not exhibit the tapped delay line structure of the input vector $\boldsymbol{x}(n) = \boldsymbol{x}_1(n)$ as used in linear filtering. Assuming stationary input $x(n)$, the autocorrelation matrix

$$\boldsymbol{R}_{\boldsymbol{xx}} = \mathrm{E}\left\{\, \boldsymbol{x}(n)\boldsymbol{x}^{\mathrm{T}}(n) \,\right\}, \tag{7.37}$$

associated with the input vector $\boldsymbol{x}(n)$ for linear filters, features a Toeplitz form, whereas this is not true for the autocorrelation matrix

$$\boldsymbol{R}_{\boldsymbol{x}_{\mathrm{VF}}\boldsymbol{x}_{\mathrm{VF}}} = \mathrm{E}\left\{\, \boldsymbol{x}_{\mathrm{VF}}(n)\boldsymbol{x}_{\mathrm{VF}}^{\mathrm{T}}(n) \,\right\}, \tag{7.38}$$

corresponding to the input vector $\boldsymbol{x}_{\mathrm{VF}}(n)$ of Volterra filters. Unfortunately, computationally efficient versions of the RLS and the APA for linear adaptive filters explicitly exploit the Toeplitz structure of $\boldsymbol{R}_{\boldsymbol{xx}}$ [9]. Therefore, these methods cannot be applied to adaptive Volterra filters in a straightforward manner. Nevertheless, there are still structural features in the input vector $\boldsymbol{x}_{\mathrm{VF}}(n)$ on which fast versions of the RLS algorithm for Volterra filters can be based on, as has been shown in [25] for the second-order case.

Adaptation Control for Second-order Volterra Filters

The performance of an adaptive algorithm strongly depends on the control of the adaptation. This is especially true for the LMS algorithm, as it implies rather slow convergence for correlated input signals. In the following, we therefore present a coefficient-dependent adaptation control for the LMS algorithm for second-order Volterra filters that corresponds to the approach proposed in [20].

Since we are aiming at a coefficient-dependent step size, we consider the LMS update equation for a single coefficient $\hat{h}_{\boldsymbol{r}_{p,k}}(n)$. From Eq. 7.34 we obtain

$$\hat{h}_{\boldsymbol{r}_{p,k}}(n+1) = \hat{h}_{\boldsymbol{r}_{p,k}}(n) + \mu_{\boldsymbol{r}_{p,k}}(n)\, e(n)\, x_{\boldsymbol{r}_p}(n-k). \tag{7.39}$$

For the following discussion, we assume that the input $x(n)$ is an independent, identically distributed (IID) random process, where the probability density function (PDF) of the amplitudes of $x(n)$ is an even function. Additionally, we assume that the coefficients of the adaptive Volterra filter are independent of the input signal. The assumed properties of $x(n)$ imply that the input of the linear kernel and the output of the quadratic kernel of both, the adaptive Volterra filter, and the unknown Volterra filter are orthogonal, i.e., $\mathrm{E}\{\, y_2(k)x(n-k)\,\} = \mathrm{E}\{\, \hat{y}_2(k)x(n-k)\,\} = 0$. It can be shown that then, the optimum filter coefficients of the linear adaptive kernel $\hat{h}_{\boldsymbol{r}_{1,k}}(n)$ are equal to the corresponding filter coefficients $h_{\boldsymbol{r}_{1,k}}(n)$. Correspondingly, it can be shown that the optimum coefficients for the quadratic kernel $\hat{h}_{\boldsymbol{r}_{2,k}}(n)$ are given by $h_{\boldsymbol{r}_{2,k}}(n)$. Thus, we introduce the coefficient errors of the linear and the quadratic kernels according to

$$m_{\boldsymbol{r}_{p,k}}(n) = h_{\boldsymbol{r}_{p,k}}(n) - \hat{h}_{\boldsymbol{r}_{p,k}}(n), \quad p \in \{1,2\}. \tag{7.40}$$

Following [20], the optimality criterion for determining the optimum coefficient-dependent step size $\mu_{\mathrm{opt},\boldsymbol{r}_{p,k}}(n)$ is chosen as the mean squared error between the actual coefficient error and the corresponding LMS update term:

$$J_{\mu_{\boldsymbol{r}_{p,k}}}(n) = \mathrm{E}\left\{ \left[m_{\boldsymbol{r}_{p,k}}(n) - \mu_{\boldsymbol{r}_{p,k}}(n)\, e(n)\, x_{\boldsymbol{r}_p}(n-k)\right]^2 \right\}, \quad p \in \{1,2\}. \tag{7.41}$$

As shown in [20], the optimum step size, which minimizes the cost function Eq 7.41, is obtained as

$$\mu_{\mathrm{opt},\boldsymbol{r}_{p,k}}(n) = \frac{\mathrm{E}\left\{ m^2_{\boldsymbol{r}_{p,k}}(n) \right\}}{\mathrm{E}\left\{ \varepsilon^2(n) + b^2(n) + s^2(n) \right\}}, \quad p \in \{1,2\}, \tag{7.42}$$

where it has been assumed that the input $x(n)$, the background noise $b(n)$, and the speech signal of the near-end talker $s(n)$ are mutually statistically independent processes. Interestingly, the form of the optimum step size is identical for both kernels, which results from the linearity of the output with respect to the coefficients of any kernel. Considering that the error signal can be expressed as

$$e(n) = \varepsilon(n) + b(n) + s(n), \tag{7.43}$$

we notice that the denominator in Eq. 7.42 can be identified as the second-order moment of the error signal $e(n)$. Since $e(n)$ is observable, its second-order moment can in general be estimated reliably. However, the mean squared coefficient errors $\mathrm{E}\left\{ m_{\boldsymbol{r}_{p,k}}^2(n) \right\}$ are not observable in practice, making a straightforward realization of the optimum step size impossible. Therefore, we apply the approach proposed in [20] in order to obtain models for the estimation of the unknown statistical terms.

For the following discussion it will be useful to distinguish between the residual echoes associated to each single Volterra kernel, and define

$$\varepsilon_p(n) = d_p(n) - \hat{d}_p(n). \tag{7.44}$$

The overall residual echo for second-order Volterra filters can then be expressed by

$$\varepsilon(n) = \varepsilon_1(n) + \varepsilon_2(n). \tag{7.45}$$

Note that for the assumed input signal $x(n)$, the residual echoes of the different Volterra kernels are orthogonal, and, thus,

$$\mathrm{E}\left\{ \varepsilon^2(n) \right\} = \mathrm{E}\left\{ \varepsilon_1^2(n) \right\} + \mathrm{E}\left\{ \varepsilon_2^2(n) \right\}. \tag{7.46}$$

As the residual echoes $\varepsilon_p(n)$ result from the misadjustment of the corresponding kernel coefficients, we rewrite Eq. 7.44 for the linear and the quadratic kernel in terms of the coefficient errors:

$$\varepsilon_1(n) = \sum_{k=0}^{N_1-1} m_{\boldsymbol{r}_{1,k}}(n) x(n-k), \tag{7.47}$$

$$\varepsilon_2(n) = \sum_{r_{2,1}=0}^{R_2-1} \sum_{k=0}^{L_{\boldsymbol{r}_2}-1} m_{\boldsymbol{r}_{2,k}}(n)\, x_{\boldsymbol{r}_2}(n-k). \tag{7.48}$$

For the considered zero-mean IID input $x(n)$, the mean squared residual echo of the linear kernel can then be expressed by

$$\mathrm{E}\left\{ \varepsilon_1^2(n) \right\} = \sum_{k=0}^{N_1-1} \mathrm{E}\left\{ m_{\boldsymbol{r}_{1,k}}^2(n) \right\} \mathrm{E}\left\{ x^2(n-k) \right\}. \tag{7.49}$$

As discussed in [20], in the echo cancellation context it is reasonable to apply a corresponding approximation of the mean squared residual echo of the quadratic kernel:

$$\mathrm{E}\left\{ \varepsilon_2^2(n) \right\} \approx \sum_{r_{2,1}=0}^{R_2-1} \sum_{k=0}^{L_{\boldsymbol{r}_2}-1} \mathrm{E}\left\{ m_{\boldsymbol{r}_{2,k}}^2(n) \right\} \mathrm{E}\left\{ x_{\boldsymbol{r}_2}^2(n-k) \right\}. \tag{7.50}$$

For a better understanding of the optimum step size, we follow [20] and introduce the kernel-independent auxiliary step-size factors

$$\alpha_{\mathrm{dt}}(n) = \frac{\mathrm{E}\{\varepsilon^2(n) + b^2(n)\}}{\mathrm{E}\{\varepsilon^2(n) + n^2(n) + s^2(n)\}}, \tag{7.51}$$

$$\alpha_{\mathrm{bn}}(n) = \frac{\mathrm{E}\{\varepsilon^2(n)\}}{\mathrm{E}\{\varepsilon^2(n) + b^2(n)\}}. \tag{7.52}$$

Note that for the definition of $\alpha_{\mathrm{dt}}(n)$ and $\alpha_{\mathrm{bn}}(n)$, the mutual statistical independence of the input signal, the background noise, and the near-end speech has been used. Additionally, we introduce the kernel-dependent auxiliary step-size factors

$$\alpha_{\varepsilon_p}(n) = \frac{\mathrm{E}\{\varepsilon_p^2(n)\}}{\mathrm{E}\{\varepsilon_1^2(n)\} + \mathrm{E}\{\varepsilon_2^2(n)\}}, \quad p \in \{1, 2\}, \tag{7.53}$$

where the orthogonality property according to Eq. 7.46 has been exploited. Furthermore, we define coefficient-dependent step-size factors

$$\alpha_{\boldsymbol{r}_{1,k}}(n) = \frac{\mathrm{E}\left\{m_{\boldsymbol{r}_{1,k}}^2(n)\right\}}{\sum_{l=0}^{N_1-1} \mathrm{E}\left\{m_{\boldsymbol{r}_{1,l}}^2(n)\right\}\mathrm{E}\left\{x^2(n-l)\right\}} \tag{7.54}$$

for the adaptation of the linear kernel. The corresponding step sizes for the coefficients of the quadratic kernel are given by

$$\alpha_{\boldsymbol{r}_{2,k}}(n) = \frac{\mathrm{E}\left\{m_{\boldsymbol{r}_{2,k}}^2(n)\right\}}{\sum_{r_{2,1}=0}^{R_2-1} \sum_{l=0}^{L_{\boldsymbol{r}_2}-1} \mathrm{E}\left\{m_{\boldsymbol{r}_{2,l}}^2(n)\right\} \mathrm{E}\left\{x_{\boldsymbol{r}_2}^2(n-l)\right\}}. \tag{7.55}$$

Note that the auxiliary step-size factors $\alpha_{\boldsymbol{r}_{1,k}}(n)$ and $\alpha_{\boldsymbol{r}_{2,k}}(n)$ are based on Eq. 7.49 and the approximation in Eq. 7.50, respectively.

With the above auxiliary step sizes, the optimum step size according to Eq. 7.42 can be approximated by a factorized version [20]:

$$\mu_{\mathrm{opt},\boldsymbol{r}_{p,k}}(n) \approx \alpha_{\mathrm{dt}}(n)\,\alpha_{\mathrm{bn}}(n)\,\alpha_{\varepsilon_p}(n)\,\alpha_{\boldsymbol{r}_{p,k}}(n), \quad p \in \{1, 2\}. \tag{7.56}$$

The influence of the different step-size parameters on the control of the adaptation is discussed next.

According to its definition $\alpha_{\mathrm{dt}}(n)$ accounts for double-talk (dt) situations, where $s(n) \neq 0$. In the echo cancellation context it is reasonable to implement $\alpha_{\mathrm{dt}}(n)$ as an on/off switch in combination with a double-talk detector [2], i.e., $\alpha_{\mathrm{dt}}(n) = 0$ if a near-end talker is active, in order to avoid divergence of the adaptive filter coefficients, and $\alpha_{\mathrm{dt}}(n) = 1$, otherwise.

The step-size factor $\alpha_{\mathrm{bn}}(n)$ controls the adaptation of the acoustic echo canceller with respect to the distortion introduced by background noise (bn) $b(n)$. Methods for the estimation of the product $\alpha_{\mathrm{dt}}(n)\alpha_{\mathrm{bn}}(n)$ have been thoroughly discussed in [26] and are not further considered here.

For an interpretation of $\alpha_{\varepsilon_p}(n)$ we note that the error introduced by a misadjusted linear kernel acts as an interference for the adaptation of the quadratic kernel, and vice versa. Hence, the step-size factors $\alpha_{\varepsilon_p}(n)$ can be regarded as an adaptation control with respect to interferences caused by the misadjusted Volterra kernels. As follows from Eq. 7.53, the computation of $\alpha_{\varepsilon_p}(n)$ requires knowledge of at least the ratio of the second-order moments of $\varepsilon_p(n)$ and $\varepsilon(n)$ which is in general not accessible. Therefore, a model for estimating the respective second-order moments is required. More precisely, we assume that the second-order moment of the residual echo of the linear kernel, i.e., $\varepsilon_1(n)$, is proportionate to the output of the adaptive linear kernel, i.e., $\hat{d}_1(n)$. Analogously, the second-order moment of $\varepsilon_2(n)$ is assumed to be proportionate to $\hat{d}_2(n)$:

$$\mathrm{E}\left\{\,\varepsilon_p^2(n)\,\right\} \approx \gamma_\varepsilon(n)\left[\delta_{\varepsilon_p} + \overline{\left|\hat{d}_p(n)\right|}\right], \quad p \in \{1, 2\}, \tag{7.57}$$

where $\overline{\left|\hat{d}_p(n)\right|}$ denotes a smoothed version of the magnitude of $\hat{d}_p(n)$. This estimation model can be regarded as the first term of a Taylor series expansion of the mean squared residual echoes with respect to the magnitude of the output of the corresponding kernel. The smoothing of the output is used to avoid significant variations of the estimates due to strongly varying amplitudes of the output signal. The offset term δ_{ε_p} can be used in Eq. 7.57 to manipulate the dependency of the kernel-dependent step size $\alpha_{\varepsilon_p}(n)$ on the corresponding kernel output $\hat{d}_p(n)$. Note that δ_{ε_p} is required especially in the beginning of the adaptation, where $\hat{d}_p(n) = 0$ if the Volterra coefficients were initialized with zero. The proportionality factor $\gamma_\varepsilon(n)$ represents the general convergence properties of the Volterra kernels, i.e., $\gamma_\varepsilon(n)$ decreases for a stable adaptation. However, the actual values $\gamma_\varepsilon(n)$ are not required explicitly, as the fraction appearing in the definition of $\alpha_{\varepsilon_p}(n)$ can be reduced correspondingly.

The coefficient-dependent step size $\alpha_{\boldsymbol{r}_{p,k}}(n)$ can be used to speed-up the adaptation of coefficients that cause large coefficient errors. However, the coefficient errors are not known and, therefore, we have to use models for estimating the respective second-order moments. A common assumption is that large coefficient magnitudes also cause large error magnitudes [36]. Consequently, we assume that the second-order moment of a certain coefficient error is proportionate to the magnitude of the corresponding adaptive coefficient:

$$\mathrm{E}\left\{\,m_{\boldsymbol{r}_{p,k}}^2(n)\,\right\} \approx \gamma_{m,p}(n)\left[\beta_{m,p}(n) + \left|\hat{h}_{\boldsymbol{r}_{p,k}}(n)\right|\right], \quad p \in \{1, 2\}. \tag{7.58}$$

This estimation model can be considered as the first term of a Taylor series expansion of the mean squared coefficient error with respect to the magnitude of the corresponding coefficient of the adaptive Volterra filter. The time-variant proportionality factor $\gamma_{m,p}(n)$ reflects the reduction of the magnitude of the coefficient errors during the convergence of the adaptive filter. The actual value of $\gamma_{m,p}(n)$ does not have to be specified explicitly, as the fractions in Eq. 7.54 and Eq. 7.55 can be reduced respectively. The parameter $\beta_{m,p}(n)$ can be used to adjust the influence which the coefficients of the adaptive Volterra filter have on their associated LMS update term. Note that $\beta_{m,p}(n)$ should not equal zero in the beginning of the adaptation if all coefficients were initialized with zeroes. Otherwise, the coefficients remain at their initial values. For the computation of Eq. 7.54 and Eq. 7.55, we additionally replace the expectations with respect to the input by the corresponding instantaneous values.

It should be mentioned that there is a strong link between the coefficient-dependent step size presented above and the proportionate NLMS (PNLMS) for second-order Volterra filters proposed in [18]: If the parameters $\beta_{m,p}(n)$ are chosen according to

$$\beta_{m,1}(n) = \frac{\beta_{c,1}}{N_1} \sum_{k=0}^{N_1-1} \left| \hat{h}_{\mathbf{r}_{1,k}}(n) \right|, \tag{7.59}$$

$$\beta_{m,2}(n) = \frac{\beta_{c,2}}{N_{\text{coeff},p}} \sum_{r_{2,1}=0}^{R_2-1} \sum_{k=0}^{L_{\mathbf{r}_2}-1} \left| \hat{h}_{\mathbf{r}_{2,k}}(n) \right|, \tag{7.60}$$

the PNLMS according to [18] results.

Simulations

To evaluate the performance of the step size control algorithm presented above, we present simulation results obtained for a second-order adaptive Volterra filter. In the experiment, the input has been wide-sense stationary coloured noise with a power spectral density (PSD) corresponding to the long-term PSD of speech. The nonlinear echo path has been modeled by a second-order Volterra filter in DCR, where the memory length of the linear kernel has been $N_1 = 320$. To account for the cascaded nature of nonlinear acoustic echo paths, the memory length of the quadratic kernel has been $N_2 = 64$, while its width is only $R_2 = 20$. The same region of support has also been chosen for the second-order adaptive Volterra filter. As double-talk detection algorithms are outside the scope of this chapter, we set $s(n) = 0$ in the following which implies $\alpha_{\text{dt}} = 1$. An SNR of 30 dB has been preset with respect to $b(n)$ and $d(n)$. Since we are mainly interested in the improvements resulting from the kernel-dependent and the coefficient-dependent step size parameters, a fixed value $\alpha_{\text{bn}} = 0.3$ has been used. The performance is measured using the *Echo Return Loss Enhancement* (ERLE) which is defined by

$$\text{ERLE} = 10\log_{10}\frac{\text{E}\left\{d^2(n)\right\}}{\text{E}\left\{\varepsilon^2(n)\right\}}\ [\text{dB}]. \tag{7.61}$$

The level of nonlinear distortion has been preset such that the maximum achievable ERLE of a purely linear approach is limited to approximately 20 dB.

In Fig. 7.7, the ERLE graphs obtained for a second-order adaptive Volterra Filter (VF) applying two different realizations of its step size are compared to a purely linear approach. Here, we look at the case where the step size para-

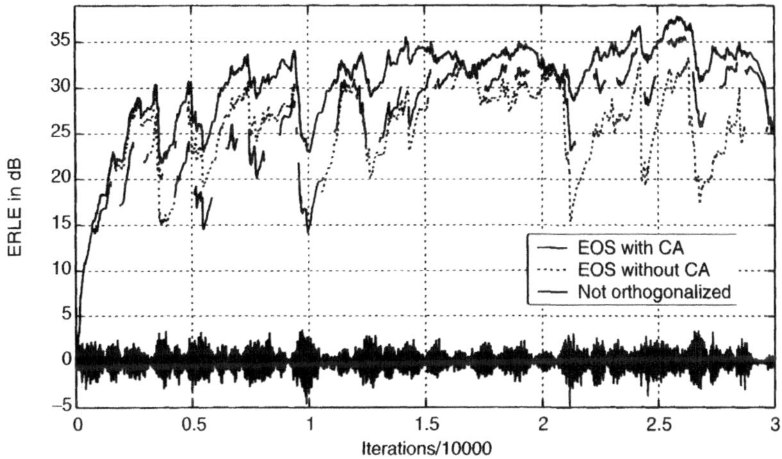

Fig. 7.7. ERLE obtained for a second-order adaptive Volterra filter (VF) applying the LMS algorithm with the proposed coefficient-dependent step size, and the kernel-independent NLMS, together with a linear approach for wide-sense stationary coloured noise input.

meters $\alpha_{\varepsilon_p}(n)$ and $\alpha_{r_{p,k}}(n)$ are estimated by means of the models Eq. 7.57 and Eq. 7.58. The constant parameter δ_{ε_p} of the model for the residual errors have been chosen to $\delta_{\varepsilon_p} = 0.001$ for both kernels. For the model of the coefficient errors, Eqs. 7.59 and 7.60 has been used, where $\beta_{c,1} = \beta_{c,2} = 1$. Note that this choice implies the practically realizable PNLMS algorithm for second-order Volterra filters [18]. Furthermore, Fig. 7.7 shows the ERLE-graph obtained for the kernel-independent NLMS with a fixed step size $\alpha_{\text{NLMS}}(n) = 0.3$. Fig. 7.7 additionally depicts the ERLE of a linear adaptive filter which has been implemented analogously to the linear kernel of the adaptive Volterra filter with coefficient-dependent step-size control.

As can be seen, the second-order Volterra filter with coefficient-dependent step size clearly outperforms the corresponding Volterra filter applying a kernel-independent NLMS algorithm. We also notice from Fig. 7.7 that the achievable echo attenuation of the purely linear approach is limited due to the nonlinear distortion in the echo path.

7.3.3 Multidelay Adaptive Volterra Filters

DFT-domain approaches are very popular in linear adaptive filtering, as they allow for an increased convergence speed, while at the same time reducing the computational complexity [30, 38]. Corresponding DFT-domain methods which exploit fast convolution techniques via block processing are therefore desirable in adaptive Volterra filtering, too. For the derivation of such algorithms, we can basically distinguish between two different approaches for exploiting fast convolution methods. They are based on

- linear multidimensional filtering,
- linear multichannel filtering.

The first approach exploits the relation between linear multidimensional systems and Volterra filters in CCR [15, 22]. These methods are most efficient if the entire region of support of the Volterra kernels (i.e., $R_p = N_p$) has to be included. The second approach bases on the interpretation of Volterra filters in DCR as special linear MISO systems [19, 29], as already discussed above. The linear multidelay filter [30] applies partitioned block techniques which can be exploited to allow for different memory lengths for each kernel in the corresponding generalization to Volterra filters. This is especially attractive with acoustic echo cancellation for nonlinear loudspeaker systems, where it has been observed that the required memory length for the linear kernel is larger than that of the quadratic kernel [43]. Therefore, the restriction to a uniform DFT length for all Volterra kernels, as imposed in the DFT-domain approaches according to [15, 29], leads to inefficient system configurations, making the approaches [19, 22] more attractive for the acoustic echo cancellation application.

The DFT-domain algorithm presented in this section corresponds to [19] and represents an extension of the linear adaptive multidelay filter according to [30] to Volterra filters in DCR. An advantage of the resulting multidelay Volterra filter is that it preserves the flexibility with respect to choosing a desired region of support, as featured by the DCR.

Following the linear approach [30], a block-partitioned version of Eq. 7.13 is obtained by partitioning the linear filter $h_{\boldsymbol{r}_{p,k}}$ into $B_{\boldsymbol{r}_p}$ blocks of length N. In the following we assume that the memory lengths N_p of all Volterra kernels are integer multiples of the partition length N. From the definition of $L_{\boldsymbol{r}_p}$ in Eq. 7.10 we recall that the memory lengths of the filters in different channels are in general not uniform but can take on any value in the range $N_p - R_p < L_{\boldsymbol{r}_p} < N_p$. Consequently, the number of partitions $B_{\boldsymbol{r}_p}$ has to be chosen depending on the memory length $L_{\boldsymbol{r}_p}$ of the corresponding diagonal such that

$$\left(B_{\boldsymbol{r}_p} - 1\right) N < L_{\boldsymbol{r}_p} \leq B_{\boldsymbol{r}_p} N, \quad B_{\boldsymbol{r}_p}, N \in \mathbb{N}. \tag{7.62}$$

Aiming at a partitioned block version of the overlap/save method, we introduce zero-padded partitions of memory length $M = 2N$ according to

$$h_{\boldsymbol{r}_{p,b,l}} = \begin{cases} h_{\boldsymbol{r}_{p,bN+l}}, & 0 \le l < N \wedge 0 \le b < B_{\boldsymbol{r}_p}, \\ 0, & N \le l < M \wedge 0 \le b < B_{\boldsymbol{r}_p}. \end{cases} \tag{7.63}$$

It is important to note that in addition to the *explicit* zero-padding for $N \le l < M$, the definition Eq. 7.63 also includes an *implicit* zero-padding in case of $L_{\boldsymbol{r}_p} < B_{\boldsymbol{r}_p} N$. For the last partition with index $b = B_{\boldsymbol{r}_p} - 1$, we additionally have to regard that

$$h_{\boldsymbol{r}_{p,bN+l}}\Big|_{b=B_{\boldsymbol{r}_p}-1} = 0, \quad \text{for } l \ge L_{\boldsymbol{r}_p} - (B_{\boldsymbol{r}_p} - 1)N. \tag{7.64}$$

The partitioning and the zero-padding of the channel filters $h_{\boldsymbol{r}_{p,k}}$ according to Eq. 7.63 is illustrated in Fig. 7.8 for $B_{\boldsymbol{r}_p} = 6$, where $h_{\boldsymbol{r}_{p,b,l}}$ is shown for $b = 1$ and $b = 5$. Note that only the shaded areas contain nonzero coefficients. To

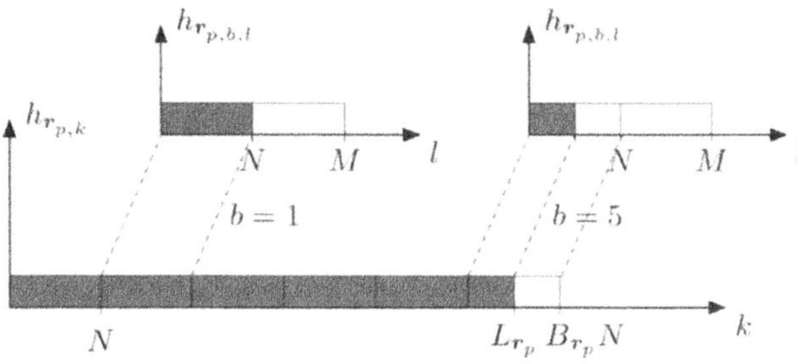

Fig. 7.8. Illustration of the partitioning according to Eq. 7.63 for $B_{\boldsymbol{r}_p} = 6$ and $L_{\boldsymbol{r}_p} \le B_{\boldsymbol{r}_p} N$. The implicit zero-padding according to Eq. 7.64 occurs for $b = 5$.

exemplify the implicit zero-padding for $b = B_{\boldsymbol{r}_p} - 1$ according to Eq. 7.64, the memory length $L_{\boldsymbol{r}_p}$ has been chosen smaller than $B_{\boldsymbol{r}_p} N$.

The input signal of each partition $h_{\boldsymbol{r}_{p,b,l}}$ is defined as

$$x_{\boldsymbol{r}_p,b}(n) = x_{\boldsymbol{r}_p}(n - bN). \tag{7.65}$$

Introducing the definitions Eq. 7.63, Eq. 7.65 into Eq. 7.13 yields a block-partitioned version for computing the output of the channel $\boldsymbol{r}_p$:

$$d_{\boldsymbol{r}_p}(n) = \sum_{b=0}^{B_{\boldsymbol{r}_p}-1} h_{\boldsymbol{r}_{p,b,n}} * x_{\boldsymbol{r}_p,b}(n). \tag{7.66}$$

Furthermore, we introduce a block-time index ν and signal blocks of length $M = 2N$ for the input signals of each partition according to

$$x_{\boldsymbol{r}_p,b}(\nu, \kappa) = x_{\boldsymbol{r}_p,b}(\nu L + \kappa - N), \quad \text{for } 0 \le \kappa < M, \tag{7.67}$$

in order to account for the block processing of the overlap/save method. Following [30], the block time shift L, representing the number of new samples of successive signal blocks, is defined using an overlap factor ρ, so that

$$L = \frac{N}{\rho}, \quad L \in \mathbb{N}. \tag{7.68}$$

Introducing the block time index ν into Eq. 7.66 yields

$$d_{\boldsymbol{r}_p}(\nu, \kappa) = \sum_{b=0}^{B_{\boldsymbol{r}_p}-1} h_{\boldsymbol{r}_p,b,\kappa} * x_{\boldsymbol{r}_p,b}(\nu, \kappa), \tag{7.69}$$

where the convolution is performed with respect to κ. The DFT-domain correspondence of Eq. 7.69 is given by

$$D_{\boldsymbol{r}_p}(\nu, m) = \sum_{b=0}^{B_{\boldsymbol{r}_p}-1} H_{\boldsymbol{r}_p,b,m} X_{\boldsymbol{r}_p,b}(\nu, m), \tag{7.70}$$

where $H_{\boldsymbol{r}_p,b,m}$ and $X_{\boldsymbol{r}_p,b}(\nu, m)$ denote the M-point DFT of $h_{\boldsymbol{r}_p,b,l}$ and $x_{\boldsymbol{r}_p,b}(\nu, \kappa)$, respectively. Analogously to the time domain, the computation of the DFT-domain output $D(\nu, m)$ of the Volterra filter is performed according to

$$D_p(\nu, m) = \sum_{r_{p,1}=0}^{R_p-1} \sum_{r_{p,2}=r_{p,1}}^{R_p-1} \cdots \sum_{r_{p,p-1}=r_{p,p-2}}^{R_p-1} D_{\boldsymbol{r}_p}(\nu, m), \tag{7.71}$$

$$D(\nu, m) = \sum_{p=1}^{P} D_p(\nu, m). \tag{7.72}$$

Let $\breve{d}(\nu, \kappa)$ represent the M-point inverse DFT of $D_p(\nu, m)$, i.e.,

$$\breve{d}(\nu, \kappa) = \mathcal{F}_M^{-1}\Big\{D(\nu, m)\Big\}, \tag{7.73}$$

where $\mathcal{F}_M^{-1}\{\cdot\}$ denotes the M-point inverse DFT. Taking the relation between circular and linear convolution into account [34], we notice that the first N elements of $\breve{d}(\nu, \kappa)$ are corrupted by time-domain aliasing, while the last N elements of $\breve{d}(\nu, \kappa)$ are equal to the desired output signal block of the P-th order Volterra filter at block time index ν. The output sequence $d(n)$ is finally obtained by applying the overlap/save method [34], i.e., by discarding the first N elements of $\breve{d}(\nu, \kappa)$ and setting

$$d(n) = \breve{d}(\nu, n - \nu L + N), \quad \nu L \leq n < \nu L + N\,, \tag{7.74}$$

for the last N elements of $\breve{d}(\nu, \kappa)$. Note that for an overlapping factor $\rho > 1$, only the last L elements represent new values of $d(n)$, whereas the remaining $N - L$ elements have already been computed in previous block time steps. However, choosing $\rho > 1$ is beneficial for the adaptive implementation of the Volterra filter, as then, the adaptation of the kernel coefficients is performed ρ times more frequently, resulting in an increased convergence speed of the adaptive algorithm.

It should be mentioned that for the special case that all kernels have the same memory length (i.e., $N_i = N_j$) and no partitioning is applied (i.e., $B_{\boldsymbol{r}_p} = 1$), the above algorithm reduces to the approach presented in [29].

The following discussion of the adaptation of the multidelay Volterra filter is based on the notation according to Fig. 7.6. The DFT-domain output of the adaptive Volterra filter $\hat{h}_{\boldsymbol{r}_{p,k}}(n)$ is computed analogously to Eqs. 7.70–7.72:

$$\widehat{D}_{\boldsymbol{r}_p}(\nu, m) = \sum_{b=0}^{B_{\boldsymbol{r}_p}-1} \widehat{H}_{\boldsymbol{r}_{p,b,m}}(\nu)\, X_{\boldsymbol{r}_p,b}(\nu, m), \tag{7.75}$$

$$\widehat{D}_p(\nu, m) = \sum_{r_{p,1}=0}^{R_p-1} \sum_{r_{p,2}=r_{p,1}}^{R_p-1} \cdots \sum_{r_{p,p-1}=r_{p,p-2}}^{R_p-1} \widehat{D}_{\boldsymbol{r}_p}(\nu, m), \tag{7.76}$$

$$\widehat{D}(\nu, m) = \sum_{p=1}^{P} \widehat{D}_p(\nu, m), \tag{7.77}$$

where $\widehat{H}_{\boldsymbol{r}_{p,b,m}}(\nu)$ denotes the DFT-domain correspondence of the adaptive Volterra filter coefficients $\hat{h}_{\boldsymbol{r}_{p,k}}(n)$. Furthermore, we define the DFT-domain error signal

$$\breve{E}(\nu, m) = D(\nu, m) - \widehat{D}(\nu, m). \tag{7.78}$$

As indicated by the symbol $\breve{\ }$, $\breve{E}(\nu, m)$ results from using the output of the Volterra filter based on the circular convolution according to Eq. 7.73, instead of the output based on linear convolution. Therefore, we introduce the windowed time-domain error signal according to [30]

$$e(\nu, \kappa) = \begin{cases} 0, & 0 \le \kappa < N, \\ \breve{e}(\nu, \kappa), & N \le \kappa < M, \end{cases} \tag{7.79}$$

where $\breve{e}(\nu, \kappa)$ denotes the M-point inverse DFT of $\breve{E}(\nu, m)$. The adaptation of the DFT-domain coefficients $\widehat{H}_{\boldsymbol{r}_{p,b,m}}(\nu)$ is then based on the DFT-domain correspondence of $e(\nu, \kappa)$, i.e.,

$$E(\nu, m) = \mathcal{F}_M\Big\{e(\nu, \kappa)\Big\}. \tag{7.80}$$

Regarding [19], the LMS-type update equation for the DFT-domain adaptive coefficients $\widehat{H}_{\boldsymbol{r}_{p,b,m}}(\nu)$ is given by

$$\widehat{H}_{\boldsymbol{r}_{p,b,m}}(\nu{+}1) = \widehat{H}_{\boldsymbol{r}_{p,b,m}}(\nu){+}\mu_p(\nu, m)\, \mathcal{F}_M\Big\{w_{\boldsymbol{r}_{p,b,l}}\; \mathcal{F}_M^{-1}\Big\{E(\nu, m) X^*_{\boldsymbol{r}_p,b}(\nu, m)\Big\}\Big\}. \tag{7.81}$$

The time-domain window function $w_{\boldsymbol{r}_{p,b,l}}$ is used to explicitly enforce the zero-padding of the time-domain partitions according to Eq. 7.63. For the definition of $w_{\boldsymbol{r}_{p,b,l}}$ we have to distinguish between different values of the partition index b. In case of $b < B_{\boldsymbol{r}_p} - 1$, the window function $w_{\boldsymbol{r}_{p,b,l}}$ is defined by

$$w_{\boldsymbol{r}_p,b,l} = \begin{cases} 1, & 0 \le l < N \;\wedge\; b < B_{\boldsymbol{r}_p} - 1, \\ 0, & N \le l < M \;\wedge\; b < B_{\boldsymbol{r}_p} - 1, \end{cases} \tag{7.82}$$

whereas in case of $b = B_{\boldsymbol{r}_p} - 1$ we use the definition according to

$$w_{\boldsymbol{r}_p,b,l} = \begin{cases} 1, & 0 \le l < L_{\boldsymbol{r}_p} \;\wedge\; b = B_{\boldsymbol{r}_p} - 1, \\ 0, & L_{\boldsymbol{r}_p} \le l < M \;\wedge\; b = B_{\boldsymbol{r}_p} - 1. \end{cases} \tag{7.83}$$

The differences in the definition of $w_{\boldsymbol{r}_p,b,l}$ in dependence on the partition index b is due to the implicit zero-padding of $\hat{h}_{\boldsymbol{r}_p,b,l}(\nu)$ resulting from Eq. 7.64. Due to the these time-domain constraints included in the update equation, Eq. 7.81 is commonly referred to as *constrained* adaptation algorithm.

For an implementation of the adaptive multidelay Volterra filter, the required distinction with respect to the definition of the time-domain window function $w_{\boldsymbol{r}_p,b,l}$ is rather inconvenient. Therefore, it is beneficial to quantize the filter lengths $L_{\boldsymbol{r}_p}$ to integer multiples of the partition length N. In this case, $w_{\boldsymbol{r}_p,b,l}$ can be replaced by the single window function

$$w_l = \begin{cases} 1, & 0 \le l < N, \\ 0, & N \le l < M, \end{cases} \tag{7.84}$$

which is then used for the adaptation of all partitions of each channel.

Step-size Normalization and Control for Second-Order Volterra Filters

A major advantage of linear DFT-domain adaptive filtering is the possibility to apply a frequency-dependent normalization of the step size [30, 38]. This approach is motivated by the approximate orthogonality property of the DFT, implying

$$\mathrm{E}\{\, X^*_{\boldsymbol{r}_1,b}(\nu, i) X_{\boldsymbol{r}_1,b}(\nu, j) \,\} \approx 0, \quad \text{for } i \neq j, \tag{7.85}$$

if the DFT length M is sufficiently large [10]. In the following we assume that the corresponding orthogonality property also holds for the input of both, linear and nonlinear channels:

$$\mathrm{E}\Big\{\, X^*_{\boldsymbol{r}_p,a}(\nu, i) X_{\boldsymbol{s}_q,b}(\nu, j) \,\Big\} \approx 0, \quad \text{for } i \neq j, \quad \forall\, \boldsymbol{r}_p,\, \boldsymbol{s}_q,\, a,\, b, \tag{7.86}$$

where $\boldsymbol{r}_p$ and $\boldsymbol{s}_q$ denote index vectors of certain diagonals of the Volterra filter and a and b denote a certain partition of these diagonals. It then directly follows from Eq. 7.86 that the normalization of the step size can be performed DFT bin-wise in case of adaptive multidelay Volterra filters, too. In the echo cancellation context it is reasonable to assume that the input signal of the linear kernel and the input signals of the diagonals of the quadratic kernel are orthogonal, implying

$$\mathrm{E}\{\, X^*_{\boldsymbol{r}_1,a}(\nu, m) X_{\boldsymbol{r}_2,b}(\nu, m) \,\} = 0, \quad \forall\, a, b. \tag{7.87}$$

It can be shown that this orthogonality property is always fulfilled if the input signal $x(n)$ is a so-called spherically invariant random process (SIRP). Since SIRPs represent a realistic model for bandlimited speech [4], Eq. 7.87 in general holds in the echo cancellation application. Thus, the normalization of the step size $\mu_p(\nu, m)$ in Eq. 7.81 can be performed frequency- *and* kernel-dependently:

$$\mu_p(\nu, m) = \frac{\alpha_p(\nu, m)}{S_{X,p}(\nu, m)}, \quad p \in \{1, 2\} \tag{7.88}$$

Following the linear approach [39], the normalization factor $S_{X,1}(\nu, m)$ of the linear kernel partitions is computed recursively according

$$S_{X,1}(\nu, m) = \lambda S_{X,1}(\nu, m) + (1 - \lambda) \sum_{b=0}^{B_{r_1}-1} \left|X_{r_1,b}(\nu, m)\right|^2, \tag{7.89}$$

with the forgetting factor λ in the range $0 \leq \lambda < 1$. The normalization factor $S_{X,2}(\nu, m)$ used for the adaptation of the quadratic kernel is given by

$$S_{X,2}(\nu, m) = \lambda S_{X,2}(\nu, m) + (1 - \lambda) \sum_{r_{2,1}=0}^{R_2-1} \sum_{b=0}^{B_{r_2}-1} \left|X_{r_2,b}(\nu, m)\right|^2. \tag{7.90}$$

The step-size normalization according to Eq. 7.88 implies that $\alpha_p(\nu, m)$ has to control the adaptation with respect to the local distortions such as double-talk and background noise. Additionally, it has to take the interaction between different Volterra kernels into account, as a misadjusted linear kernel affects the adaptation of the quadratic kernel and vice versa. As in the time domain, we apply the factorization method for implementing $\alpha_p(\nu, m)$ and introduce

$$\alpha_p(\nu, m) = \alpha_{\mathrm{dt}}(\nu L)\, \alpha_{\mathrm{bn}}(\nu, m)\, \alpha_{\mathcal{E}_p}(\nu, m). \tag{7.91}$$

As in the time domain, step-size factor $\alpha_{\mathrm{dt}}(n)$ accounts for double-talk situations and has already been defined in Eq. 7.51. It should be mentioned that the double-talk detection used to implement $\alpha_{\mathrm{dt}}(\nu L)$ as an on/off switch, can also be performed in the DFT domain [2].

If $b(n)$ represents coloured noise, the level of distortion introduced by the background noise is in general different for each DFT bin. In this case it is advantageous to implement $\alpha_{\mathrm{bn}}(\nu, m)$ individually for each DFT bin. Regarding the derivation for linear DFT-domain adaptive filters in [31], the DFT-domain correspondence of Eq. 7.51 is given by

$$\alpha_{\mathrm{bn}}(\nu, m) = \frac{\mathrm{E}\left\{ \left|\mathcal{E}(\nu, m)\right|^2 \right\}}{\mathrm{E}\left\{ \left|\mathcal{E}(\nu, m)\right|^2 \right\} + \mathrm{E}\left\{ \left|B(\nu, m)\right|^2 \right\}}, \tag{7.92}$$

where $B(\nu, m)$ denotes the DFT-domain representation of the background noise $b(n)$ at block time index ν, and $\mathcal{E}(\nu, m)$ is the DFT-domain correspondence of the residual echo $\varepsilon(n)$. Methods for estimating the statistical terms required for computing $\alpha_{\mathrm{bn}}(\nu, m)$ are also presented in [31] and, thus, are not further considered here.

Analogously to its time-domain counterpart Eq. 7.53, the kernel-dependent auxiliary step size $\alpha_{\mathcal{E}_p}(\nu, m)$ is given by

$$\alpha_{\mathcal{E}_p}(\nu, m) = \frac{\mathrm{E}\left\{\left|\mathcal{E}_p(\nu, m)\right|^2\right\}}{\mathrm{E}\left\{\left|\mathcal{E}_1(\nu, m)\right|^2\right\} + \mathrm{E}\left\{\left|\mathcal{E}_2(\nu, m)\right|^2\right\}}, \quad p \in \{1, 2\}. \tag{7.93}$$

For estimating the mean squared magnitude of the DFT-domain residual echos $\mathcal{E}_p(\nu, m)$, it is possible to use the proportionality model according to Eq. 7.57 in its DFT-domain version:

$$\mathrm{E}\left\{\left|\mathcal{E}_p(\nu, m)\right|^2\right\} \approx \gamma_{\mathcal{E}}(\nu, m)\left[\delta_{\mathcal{E}_p} + \overline{\left|\widehat{D}_p(\nu, m)\right|}\right], \quad p \in \{1, 2\}, \tag{7.94}$$

where $\overline{|\widehat{D}_p(\nu, m)|}$ represents a smoothed version of the magnitude of $\widehat{D}_p(\nu, m)$. The meaning of the remaining parameters in Eq. 7.94 is equivalent to the corresponding parameters of the time-domain model Eq. 7.57. Simulation results indicate that in general the even simpler assumption of uniform mean squared residual echos, i.e,

$$\mathrm{E}\left\{\left|\mathcal{E}_1(\nu, m)\right|^2\right\} \approx \mathrm{E}\left\{\left|\mathcal{E}_2(\nu, m)\right|^2\right\} \tag{7.95}$$

can be used without loss in performance. Then, Eq. 7.94 reduces to a kernel-independent factor

$$\alpha_{\mathcal{E}_p}(\nu, m) \approx \frac{1}{2}, \quad p \in \{1, 2\}, \tag{7.96}$$

implying a kernel-independent step size $\alpha_p(\nu, m)$.

It should finally be mentioned that applying a coefficient-dependent DFT-domain step-size control does not further improve the performance of the adaptive multidelay Volterra filter. This results from the fact that the DFT already yields a sufficient decoupling of the adaptation of the kernel coefficients for different DFT bins.

Simulations

In the following we present simulation results in order to evaluate the performance of adaptive multidelay Volterra filters in the acoustic echo cancellation context. The nonlinear echo path has been modeled by a second-order Volterra filter in DCR with a memory length of $N_1 = 320$ taps for the linear kernel and a memory length of $N_2 = 64$ taps for the quadratic kernel. The width of the quadratic kernel has been set to $R_2 = 20$. The input signal has been wide-sense stationary coloured Gaussian noise. There has been no active near-end talker, i.e., $s(n) = 0$, and the variance of the additive white noise signal $b(n)$ has been chosen such that an SNR of 30 dB is obtained with respect to the variance of the echo signal $d(n)$.

The adaptive multidelay Volterra filter (MDVF) has been implemented with a partition length of $N = 64$ and an overlap factor $\rho = 4$. The number of partitions of the linear kernel has been $B_{\boldsymbol{r}_1} = 5$, implying $N_1 = 320$. In accordance with the echo path model, the memory length and the width of the quadratic kernel have been chosen to $N_2 = 64$ and $R_2 = 20$, respectively. Since the block length N matches the memory length of the quadratic kernel, no partitioning is applied to any diagonal of the quadratic kernel (implying $B_{\boldsymbol{r}_2} = 1$). The adaptation of the multidelay Volterra filter has been performed using the kernel-dependent normalization of the step size according to Eq. 7.88, where the fixed, kernel-independent value $\alpha_p(\nu, m) = 0.3$ has been used for both kernels.

Additionally, we consider a time-domain adaptive second-order Volterra filter that has the same region of support as the multidelay Volterra filter described above. The adaptation has been controlled applying the coefficient-dependent step size as already used for the simulations shown in Fig. 7.7.

The echo cancellation performance of the different approaches is illustrated in Fig. 7.9, where the ERLE obtained for a linear adaptive multidelay filter (MDF) that corresponds to the linear kernel of the adaptive multidelay Volterra filter is shown, too[2]. As can be clearly noticed, the convergence speed of the DFT-domain Volterra filter is significantly faster compared to the corresponding time-domain approach. This result shows the capability of DFT-domain methods to improve the convergence behaviour of adaptive Volterra filters. It can also be seen from Fig. 7.9 that the performance of the linear approach is clearly limited due to the nonlinear distortion in the echo path.

7.3.4 Application to Real Systems

Second-order Volterra filters have been introduced as a model for the nonlinear behaviour of loudspeakers. In the following we examine the suitability of this Volterra filter model when applied to real acoustic echo paths.

The experimental approach is divided into two parts, i.e., signal acquisition followed by simulations with recorded data: First, the echo signal is recorded in a room with low reverberation that also exhibits a low level of background noise. The actual experiments with respect to acoustic echo cancellation are then performed using the stored audio files of the input signal and the corresponding recording of the microphone signal. To provide an increased level

[2] Note that due to the definition of the ERLE in Eq. 7.61, the ERLE-values can be larger than the SNR with respect to the echo signal and the noise signal.

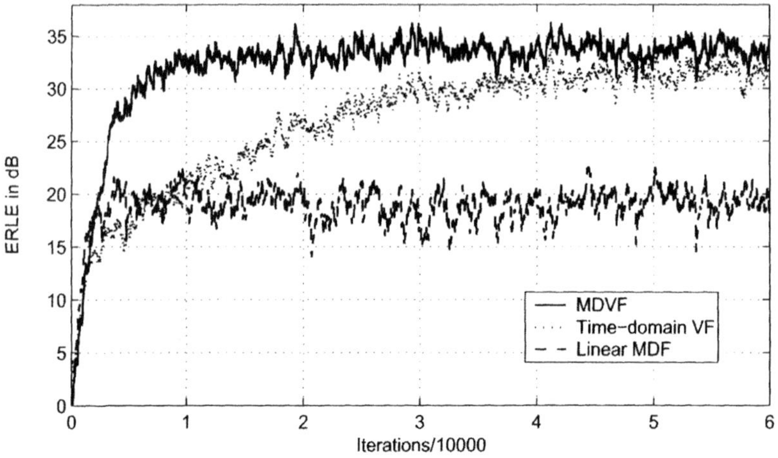

Fig. 7.9. ERLE obtained for a second-order adaptive multidelay Volterra filter (MDVF), a time-domain second-order adaptive Volterra filter, and a linear adaptive multidelay filter (MDF) for wide-sense stationary coloured noise input.

of background noise, an artificial noise signal is added to the recording of the microphone signal.

The loudspeaker used in the following experiments is a small electrodynamic loudspeaker with a diameter of 3.5 centimeters that is mounted in a closed box with a volume of about one liter. During the measurements it has been assured that the amplifier of the loudspeaker does not introduce significant nonlinear distortion. This allows for the desired isolated analysis of the nonlinear behaviour of the loudspeaker.

The coefficients of the linear and the quadratic kernels of the adaptive Volterra filters obtained from measurements of the considered loudspeaker with white Gaussian noise input are shown in Fig. 7.10. For illustrative reasons, we have used the Cartesian coordinate representation of the quadratic kernel, here. The zero-coefficients of the quadratic kernel corresponding to the indices $k_{2,2} < k_{2,1}$ result from the triangular representation of the Volterra filter according to Eq. 7.2. Note that the zero-valued coefficients for small values of $k_{2,1}$ correspond to the initial delay of the linear kernel. This initial delay results from the propagation of the echo signal on the direct path from the loudspeaker to the microphone. We further notice that the magnitudes of the quadratic kernel coefficients decay rapidly for increasing values of the coefficient indices $k_{2,1}$ and $k_{2,2}$. Nevertheless, the coefficients of the quadratic kernel have nonnegligible amplitudes within a large region in the $(k_{2,1}, k_{2,2})$-plane which confirms that the loudspeaker nonlinearities can not be considered as memoryless.

For the experimental results shown in Fig. 7.11, a speech signal sampled at 8 kHz has been used as input. The nonlinear echo canceller has been implemented as a second-order adaptive MDVF with memory lengths $N_1 = 320$ for

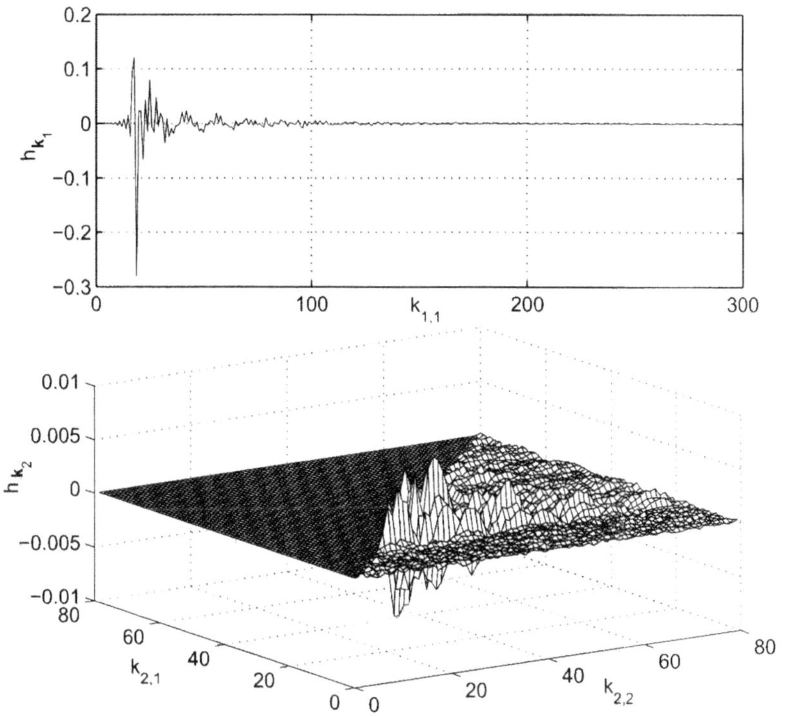

Fig. 7.10. Coefficients of the linear and quadratic Volterra kernels in Cartesian coordinate representation obtained from measurements in a room with low reverberation.

the linear kernel and $N_2 = 64$ for the quadratic kernel. The DFT length has been chosen to $M = 128$, implying that no partitioning has been applied to the quadratic kernel, whereas the linear kernel has been divided into $B_1 = 5$ partitions. To achieve faster convergence, an overlap factor of $\rho = 4$ has been used. The adaptation is performed according to Eq. 7.81, where the kernel-dependent normalization is applied. For both kernels, the normalized step size has been fixed to $\alpha_p(\nu, m) = 0.3$. For the simulation, a white Gaussian noise signal has been added to the recording of the microphone signal. The variance of the noise has been adjusted to provide an SNR of 30 dB relative to the microphone signal. The ERLE values obtained for the different kernel widths $R_2 = N_2 = 64$ and $R_2 = 15$ are shown in Fig. 7.11. For comparison, the result of a linear DFT-domain adaptive filter that corresponds to the linear kernel of the Volterra filters is given, too. As can be seen from Fig. 7.11, the performance of the purely linear approach is severely affected by the nonlinear distortion caused by the small-sized loudspeaker. We further notice that a remarkable increase of the echo attenuation can be achieved by both im-

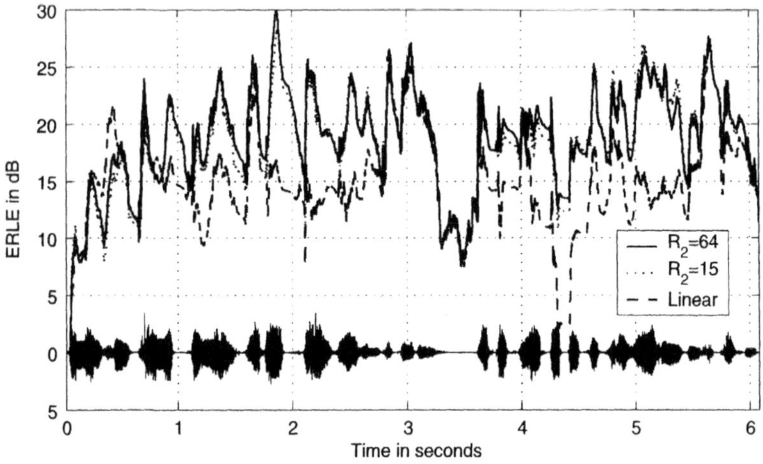

Fig. 7.11. ERLE obtained for second-order adaptive MDVFs with different width of the quadratic kernel together with the speech input. For comparison, the ERLE of a corresponding linear approach is shown, too.

plementations of the second-order Volterra filter. Especially during periods of high excitation levels, the adaptive MDVFs are able to improve the ERLE by 5 to 10 dB.

The DCR of Volterra filters has been motivated by its suitability to efficiently represent the cascaded structure that has been used as a simplified model for the acoustic echo path in case of nonlinearly distorting loudspeakers. The possibility to reduce the region of support of the adaptive Volterra filter without impairing the performance of the echo canceller is also shown in Fig. 7.11: The width of the quadratic kernel can be reduced to $R_2 = 15$ without any significant loss in achievable echo attenuation. The reduction of R_2 implies that the number of coefficients of the quadratic kernel is considerably decreased from $N_{\mathrm{coeff},2} = 2080$ to $N_{\mathrm{coeff},2} = 855$ in case of using only $R_2 = 15$ diagonals instead of $R_2 = 64$. Accordingly, the MDVF with reduced region of support increases the computational complexity compared to the linear approach only by a factor of four, whereas in case of $R_2 = 64$ a factor of thirteen results. Although not shown here, a further decrease of the width of the quadratic kernel yields a significant reduction of the achievable echo attenuation.

The experimental results according to Fig. 7.11 confirm the capability of second-order adaptive Volterra filters to cope with nonlinearly distorting loudspeakers. Furthermore, they illustrate that the advantageous structural features of the DCR allow for an efficient representation of the corresponding nonlinear acoustic echo path and according computational savings.

7.4 Power Filters

The nonlinear filters considered in this section are called power filters. They differ from general Volterra filters as they do not include nonlinear combinations of input samples taken at different time instances, while still representing a nonlinear system with memory. As shown later in this section, power filters represent a parallelized approximation of the echo path model according to Fig. 7.2 if the Volterra filter (SVF) is discarded. In other words, power filters represent a suitable approximation of the acoustic echo path if the nonlinear audio components can be considered as memoryless.

The block diagram shown in Fig. 7.12 illustrates the multichannel structure of a P-th order power filter. The input signal $x(n)$ is passed into P different

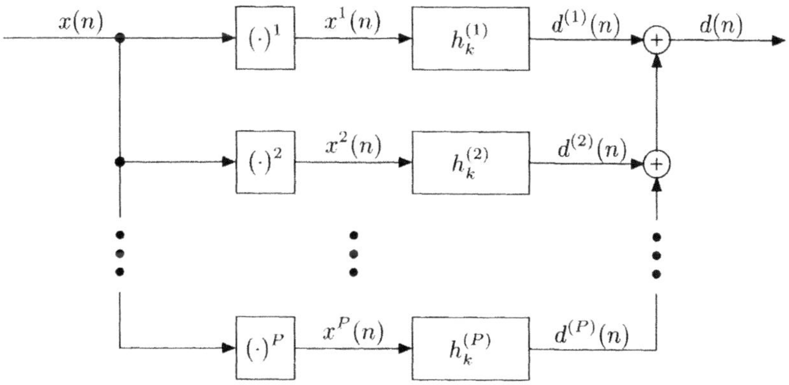

Fig. 7.12. Block diagram of a P-th order power filter.

channels. In the p-th channel, the input sample $x(n)$ is taken to the p-th power, and then passed through a linear filter $h_k^{(p)}$. The overall output $d(n)$ of the power filter is obtained by the summation over all channel outputs $d^{(p)}(n)$:

$$d(n) = \sum_{p=1}^{P} d^{(p)}(n). \tag{7.97}$$

The output of the p-th channel results from the linear convolution of $x^p(n)$ with the filter coefficients $h_k^{(p)}$, i.e.,

$$d^{(p)}(n) = \sum_{k=0}^{N_p-1} h_k^{(p)}\, x^p(n-k). \tag{7.98}$$

Obviously, power filters can be interpreted as linear MISO systems, where the input of the p-th channel is given by the p-th power of $x(n)$. Comparing

Eq. 7.98 with Eq. 7.26, we notice that there is a strong relation between power filters and Volterra filters in diagonal coordinate representation: The p-th channel of a power filter corresponds to the main diagonal of a p-th order Volterra kernel in DCR. Thus, power filters can be considered as a special type of Volterra filters in DCR, where all kernels have width $R_p = 1$. Setting

$$h_k^{(p)} = h_{\boldsymbol{r}_{p,k}} \Big|_{\boldsymbol{r}_{p,k} = [k,k,\ldots,k]^{\mathrm{T}}} \tag{7.99}$$

shows the equivalency of Eq. 7.98 and Eq. 7.26 in case of $R_p = 1$.

For compactness, we rewrite Eq. 7.98 in vector notation:

$$d^{(p)}(n) = \boldsymbol{h}^{(p)^{\mathrm{T}}} \boldsymbol{x}^{(p)}(n), \tag{7.100}$$

where the input vector $\mathbf{x}^{(p)}(n)$ and the coefficient vector $\mathbf{h}^{(p)}$ are defined by

$$\boldsymbol{x}^{(p)}(n) = \left[x^p(n),\, x^p(n-1),\, \ldots,\, x^p(n-N_p+1)\right]^{\mathrm{T}}, \tag{7.101}$$

$$\boldsymbol{h}^{(p)} = \left[h_0^{(p)},\, h_1^{(p)},\, \ldots,\, h_{N_p-1}^{(p)}\right]^{\mathrm{T}}. \tag{7.102}$$

The DFT-domain implementation of power filters can in principle be obtained from Section 7.3.3 for the special case $R_p = 1$. For presentational convenience in upcoming sections, we assume in the following that no partitioning is applied to the channel filters $h_k^{(p)}$. The block length N is then chosen according to the maximum memory length of all channels, i.e.,

$$N = \max_p N_p. \tag{7.103}$$

The length of the DFT is $M = 2N$. The DFT-domain input vector

$$\boldsymbol{X}^{(p)}(\nu) = \left[X^{(p)}(\nu,0),\, X^{(p)}(\nu,1),\, \ldots,\, X^{(p)}(\nu,M-1)\right]^{\mathrm{T}} \tag{7.104}$$

corresponding to the time-domain input vector $\boldsymbol{x}_p(n)$ of the p-th channel is obtained from

$$\boldsymbol{X}^{(p)}(\nu) = \boldsymbol{F}_{M\times M} \left[x^p(\nu L - N),\, x^p(\nu L - N + 1),\, \ldots,\, x^p(\nu L + N - 1)\right]^{\mathrm{T}}, \tag{7.105}$$

where ν represents the block time index $n = \nu L$. In Eq. 7.105, $\boldsymbol{F}_{M\times M}$ is defined as the $M \times M$ DFT matrix which has elements of the form $e^{-j2\pi\kappa m/M}$. The block time shift $L = N/\rho$ has been introduced in Eq. 7.68. The DFT-domain coefficient vector corresponding to the p-th channel is given by

$$\boldsymbol{H}^{(p)} = \boldsymbol{F}_{M\times M} \left[\boldsymbol{h}^{(p)\mathrm{T}}\, \boldsymbol{0}^{\mathrm{T}}_{(M-N_p)\times 1}\right]^{\mathrm{T}}. \tag{7.106}$$

The DFT-domain representation for the output $\boldsymbol{D}^{(p)}(\nu)$ of the p-th channel is given by

$$\boldsymbol{D}^{(p)}(\nu) = \operatorname{diag}\left\{\boldsymbol{H}^{(p)}\right\} \boldsymbol{X}^{(p)}(\nu). \tag{7.107}$$

As in the time domain, the overall DFT-domain output vector $\boldsymbol{D}(\nu)$ is finally obtained by the summation over all channel outputs:

$$\boldsymbol{D}(\nu) = \sum_{p=1}^{P} \boldsymbol{D}^{(p)}(\nu). \tag{7.108}$$

The relation between the DFT-domain output vector $\boldsymbol{D}(\nu)$ (of length M) and the corresponding time-domain output block

$$\boldsymbol{d}(\nu) = \left[d(\nu L),\, d(\nu L+1),\, \ldots,\, d(\nu L+N-1)\right]^{\mathrm{T}} \tag{7.109}$$

of length N results from the overlap/save method and reads

$$\boldsymbol{d}(\nu) = \left[\boldsymbol{0}_{N\times N} \quad \boldsymbol{I}_{N\times N}\right] \boldsymbol{F}_{M\times M}^{-1}\, \boldsymbol{D}(\nu). \tag{7.110}$$

Here, $\boldsymbol{0}_{N\times N}$ represents the $N\times N$ zero matrix and $\boldsymbol{I}_{N\times N}$ denotes the $N\times N$ identity matrix. Note that the matrix notation Eq. 7.110 corresponds to the element-wise notation of the overlap/save method according to Eqs. 7.73 and 7.74.

The discussion of power filters in the sequel is organized as follows: The application of adaptive power filters to nonlinear acoustic echo cancellation is motivated in Section 7.4.1 by showing that for certain applications the nonlinear echo path can be approximated by power filters. Orthogonalized versions of power filters in both, time domain and frequency domain are introduced in Section 7.4.2 in order to provide better performance of corresponding adaptive implementations. In Section 7.4.3, we apply adaptive orthogonalized power filters to real audio systems, including a nonlinear amplifier and the nonlinear loudspeaker of a mobile phone.

7.4.1 Application to Cascaded Structures

In the following we consider the cascaded structure shown in Fig. 7.13. It

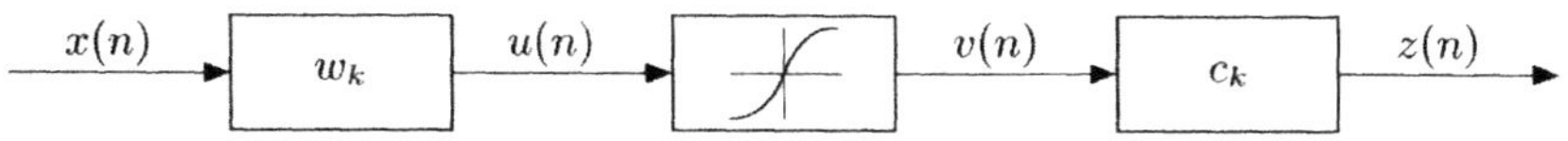

Fig. 7.13. Block diagram of the considered nonlinear cascaded structure.

consists of the cascade of a linear filter w_k, a memoryless nonlinearity, and a second linear filter c_k. Comparing Fig. 7.13 with the model of the acoustic echo path according to Fig. 7.2, we notice that these two cascaded structures are equivalent if the Volterra filter (SVF) representing the nonlinear behaviour of the loudspeaker is discarded in Fig. 7.2. In practice, there are two cases, where Fig. 7.13 in fact models the nonlinear acoustic echo path well:

- Loudspeakers can be regarded as almost linear if the required output sound level is well below the maximum output level. Then, the only source of nonlinear distortion is given by the amplifier, and the model of the acoustic echo path reduces to Fig. 7.13.
- The nonlinear behaviour of miniaturized loudspeakers operating close to the maximum level can be modeled sufficiently well by a memoryless saturation characteristic [23]. If both, the nonlinearity of the amplifier and the nonlinear characteristic of the loudspeaker are approximated by a truncated Taylor series expansion, their cascade can be modeled by a single Taylor series expansion, too. Again, the simplified model of the echo path according Fig. 7.13 results.

Note that the model according to Fig. 7.13 basically coincides with the model for the nonlinear echo path proposed in [32]. However, in [32] the authors use a continuously differentiable saturation characteristic based on a parametric function as an alternative model for the Taylor series expansion used here.

Using the notation given in Fig. 7.13, the output $v(n)$ of the memoryless nonlinearity yields

$$v(n) = \sum_{p=1}^{P} a_p \, u^p(n), \tag{7.111}$$

where a_p denote the coefficients of the truncated Taylor series expansion of the nonlinearity. The overall output $z(n)$ of the nonlinear cascade is then given by

$$z(n) = \sum_{p=1}^{P} \sum_{k=0}^{N_c-1} a_p c_k \, u^p(n-k), \tag{7.112}$$

where N_c denotes the filter length of c_k. Comparing Eq. 7.112 with Eq. 7.97 and Eq. 7.98 shows that $z(n)$ can be considered as the output of a P-th order power filter, having $u(n)$ as input. The coefficients of the linear filter associated to the p-th channel are obviously given by $c_k^{(p)} = a_p c_k$. Thus, we can rewrite Eq. 7.112 using the power filter model:

$$z(n) = \sum_{p=1}^{P} z^{(p)}(n), \tag{7.113}$$

$$z^{(p)}(n) = \sum_{k=0}^{N_c-1} c_k^{(p)} \, u^p(n-k). \tag{7.114}$$

Note that this interpretation of the computation of $z(n)$ corresponds to [23], where power filters are considered as parallelized implementation of the cascade of a memoryless nonlinearity and a linear filter.

Let us now consider the computation of the terms $u^p(n)$ which are required for computing $z^{(p)}(n)$. As $u(n)$ is the output of the linear filter w_k with memory length N_w, we obtain

$$u^p(n) = \sum_{k_1=0}^{N_w-1} \sum_{k_2=0}^{N_w-1} \cdots \sum_{k_p=0}^{N_w-1} \prod_{i=1}^{p} w_{k_i}\, x(n-k_i). \tag{7.115}$$

Due to the commutativity of the product terms in $\prod_{i=1}^{p} w_{k_i}\, x(n-k_i)$, we can rewrite Eq. 7.115 by changing the lower limits on its right hand side:

$$u^p(n) = \sum_{k_1=0}^{N_w-1} \sum_{k_2=k_1}^{N_w-1} \cdots \sum_{k_p=k_{p-1}}^{N_w-1} \Gamma(k_1, k_2, \ldots, k_p) \prod_{i=1}^{p} w_{k_i}\, x(n-k_i), \tag{7.116}$$

where $\Gamma(k_1, k_2, \ldots, k_p)$ denotes the number of possible distinct permutations of the indices $k_1, k_2, \ldots, k_p$. Comparing Eq. 7.116 with Eq. 7.2 we notice that $u^p(n)$ can be considered as the output of a specific p-th order Volterra kernel. The coefficients $w_{\mathbf{k}_p}$ of the corresponding Volterra kernel are obtained by equating $w_{\mathbf{k}_p} = \Gamma(k_1, k_2, \ldots, k_p) \prod_{i=1}^{p} w_{n_i}$. From the results in Section 7.3.1 it follows that the configuration according to Fig. 7.13 can exactly be represented by an appropriately chosen P-th order Volterra filter in DCR having memory length of $N_w + N_c - 1$ and width N_w. Note that the illustration in Fig. 7.5 can also serve as an example for the region of support of the corresponding quadratic kernel if we set $N_w = 4$ and $N_c = 16$. In general, such a Volterra model for acoustic echo paths is not practicable due to the enormously large required region of support of higher order kernels as can be noticed from Eq. 7.28. Thus, we look for an approximation of the equivalent Volterra filter by a corresponding power filter. For illustrative reasons, we decompose $u^p(n)$ into two parts:

$$u^p(n) = u^{(p)}(n) + u_{\text{res},p}(n), \tag{7.117}$$

where, the first term on the right hand side of Eq. 7.117 is defined by

$$u^{(p)}(n) = \sum_{k=0}^{N_w-1} w_k^p\, x^p(n-k), \tag{7.118}$$

i.e., it results from linear filtering of $x^p(n)$ with the coefficients w_k^p. Note that $u^{(p)}(n)$ represents the output of the main diagonal of the p-th order Volterra kernel corresponding to Eq. 7.116. Discarding the residual term $u_{\text{res},p}(n)$ in Eq. 7.117 yields an approximation for the computation of $z^{(p)}(n)$ according to

$$z^{(p)}(n) \approx \sum_{k=0}^{N_c-1} c_k^{(p)}\, u^{(p)}(n-k). \tag{7.119}$$

The approximation underlying Eq. 7.119 is illustrated in Fig. 7.14. As can be seen, the cascade of the linear filter w_k and a p-th order potentiator is replaced by the cascade of the potentiator followed by a linear filter with coefficients w_k^p. Using Fig. 7.5, this approximation can be illustrated for $p = 2$: All coefficients of the quadratic Volterra kernel are discarded expect for those lying on the main diagonal $r_{2,1} = 0$. Note that the approximation according to Fig. 7.14

Fig. 7.14. Illustration of the approximation applied in Eq. 7.119.

represents an equality if the prefilter w_k is only a single delay. In this case, the model of the echo path according to Fig. 7.13 can be simplified to a corresponding cascade of a memoryless nonlinearity followed by a linear filter as proposed in [42]. Obviously, Eq. 7.119 *exactly* holds for the linear channel.

In case of the approximation Eq. 7.119, $z^{(p)}(n)$ can be interpreted as the output of the cascade of the linear filters $c_k^{(p)}$ and w_k^p, having $x^p(n)$ as input. If we finally introduce Eq. 7.118 in Eq. 7.119, we obtain the desired approximation of the nonlinear cascaded structure by a corresponding power filter:

$$z^{(p)}(n) \approx \sum_{k=0}^{N_g-1} g_k^{(p)}\, x^p(n-k), \tag{7.120}$$

where the coefficients $g_k^{(p)}$ of the power filter are given by

$$g_k^{(p)} = \sum_{l=0}^{N_c-1} c_l^{(p)}\, w_{k-l}^p. \tag{7.121}$$

The memory length of $g_k^{(p)}$ is $N_g = N_c + N_w - 1$.

The approximation of the nonlinear echo path model according to Fig. 7.13 using power filters can be regarded as a compromise between model accuracy and convergence behaviour of a corresponding adaptive implementation: The authors of [32] propose to realize the echo canceller by applying the same cascaded structure as used for the echo path model. However, it is challenging to assure convergence to the optimum solution or even assure stable adaptation behaviour for cascaded structures. This is especially true for the case that multiple linear filters are involved. The improvements with respect to convergence properties which should result from the inherent parallel nature of power filters are only achieved if a mutual orthogonalization of the channel inputs $x^p(n)$ is applied.

The approximation of Fig. 7.13 by power filters also represents a compromise between an exact model of the echo path and the approximation proposed in [42]. While [42] completely discards the prefilter, the power filter model includes part of the influence of w_k on the echo signal, as implied by Eq. 7.121. Experimental results indicate, however, that this increase in model accuracy does not improve the performance of corresponding adaptive implementations with respect to achievable echo attenuation.

Note that power filters realize the linear component of the echo path with only one single linear filter, whereas the approach [32] implicitly uses the cascade of two. This is an important property, since acoustic echo paths are usually only weakly nonlinear, i.e., the major contribution to the echo signal results from linear filtering of the input. Adaptive power filters therefore circumvent convergence problems that can not be excluded with adaptive structures that consist of the cascade of multiple linear filters.

7.4.2 Adaptive Orthogonalized Power Filters

The actual goal of the considerations presented in this section is the derivation of efficient adaptive implementations of power filters for their application to acoustic echo cancellation. The main obstacle to this is here that the input signals of the different channels of power filters, i.e., $x(n)$, $x^2(n)$, ..., $x^P(n)$ are not mutually orthogonal. Therefore, a direct adaptive implementation of the power filter structure according to Fig. 7.12 suffers from slow convergence as the adaptation of different channels interacts. To improve the performance of adaptive power filters, we discuss corresponding orthogonalized versions in the following.

Orthogonalization of the Input Signals

Following [23], we introduce a new set of mutually orthogonal input signals $x_{\mathrm{o}}^{(p)}(n)$ according to

$$x_{\mathrm{o}}^{(1)}(n) = x(n), \tag{7.122}$$

$$x_{\mathrm{o}}^{(p)}(n) = x^p(n) + \sum_{i=1}^{p-1} q_{p,i}\, x^i(n), \quad 1 < p \leq P. \tag{7.123}$$

The orthogonalization coefficients $q_{p,i}$ are chosen such that

$$\mathrm{E}\left\{ x_{\mathrm{o}}^{(i)}(n)\, x_{\mathrm{o}}^{(j)}(n) \right\} = 0, \quad \text{for } i \neq j. \tag{7.124}$$

A well-known approach for determining the orthogonalization coefficients $q_{p,i}$ is given by the Gram-Schmidt orthogonalization method [33]. The $p-1$ coefficients $q_{p,i}$ which are required for orthogonalizing the input of the p-th order channel can be obtained by solving

$$\begin{bmatrix} m_x^{(2)} & m_x^{(3)} & \dots & m_x^{(p)} \\ m_x^{(3)} & m_x^{(4)} & \dots & m_x^{(p+1)} \\ \vdots & \vdots & \ddots & \vdots \\ m_x^{(p)} & m_x^{(p+1)} & \dots & m_x^{(2p-2)} \end{bmatrix} \begin{bmatrix} q_{p,1} \\ q_{p,2} \\ \vdots \\ q_{p,p-1} \end{bmatrix} = - \begin{bmatrix} m_x^{(p+1)} \\ m_x^{(p+2)} \\ \vdots \\ m_x^{(2p-1)} \end{bmatrix}, \tag{7.125}$$

where $m_x^{(k)}$ denotes the k-th order moment of $x(n)$, i.e.,

$$m_x^{(k)} = \mathrm{E}\left\{ x^k(n) \right\}. \tag{7.126}$$

If $x(n)$ is a stationary process and its statistics are *a priori* known, the orthogonalization coefficients $q_{p,i}$ are constant in time and can be calculated in advance. In practice, however, $m_x^{(k)}$ has to be replaced by corresponding time-variant estimates $\widehat{m}_x^{(k)}(n)$, especially if $x(n)$ is non-stationary. The estimation of $m_x^{(k)}$ can be performed, e.g., by applying the first order recursion

$$\widehat{m}_x^{(k)}(n) = \lambda \widehat{m}_x^{(k)}(n-1) + (1-\lambda)x^k(n). \tag{7.127}$$

The forgetting factor λ is in the range $0 \leq \lambda < 1$ and can be adjusted in order to adapt the estimation to the statistics of the input signal $x(n)$. Obviously, the orthogonalization coefficients $q_{p,i}(n)$ always depend on time if they are determined from Eq. 7.125 but based on time-variant estimates $\widehat{m}_x^{(k)}(n)$ of $m_x^{(k)}$.

For presentational convenience we assume in the following that all channels of the power filter have uniform memory length, i.e., $N_p = N$. The matrix representation for a block of N input samples corresponding to Eq. 7.122 and Eq. 7.123 reads

$$\boldsymbol{x}_{\mathrm{o}}^{(1)}(n) = \boldsymbol{x}^{(1)}(n), \tag{7.128}$$

$$\boldsymbol{x}_{\mathrm{o}}^{(p)}(n) = \boldsymbol{x}^{(p)}(n) + \sum_{i=1}^{p-1} \boldsymbol{Q}_{p,i}(n)\, \boldsymbol{x}^{(i)}(n), \quad 1 < p \leq P. \tag{7.129}$$

The orthogonalized signal vectors $\boldsymbol{x}_{\mathrm{o}}^{(p)}(n)$ are defined analogously to Eq. 7.101, i.e.,

$$\boldsymbol{x}_{\mathrm{o}}^{(p)}(n) = \left[x_{\mathrm{o}}^{(p)}(n),\, x_{\mathrm{o}}^{(p)}(n-1),\, \ldots,\, x_{\mathrm{o}}^{(p)}(n-N+1)\right]^{\mathrm{T}}. \tag{7.130}$$

In Eq. 7.129, $\boldsymbol{Q}_{p,i}(n)$ represents the diagonal orthogonalization matrix

$$\boldsymbol{Q}_{p,i}(n) = \mathrm{diag}\left\{\left[q_{p,i}(n),\, q_{p,i}(n-1),\, \ldots,\, q_{p,i}(n-N+1)\right]\right\}. \tag{7.131}$$

The definition of $\boldsymbol{Q}_{p,i}(n)$ already includes the possible time-variance of its elements $q_{p,i}(n)$. Note that the orthogonalization Eq. 7.123 is performed in a sample-based manner: The coefficients $q_{p,i}(n)$ are determined such that the instantaneous orthogonality property Eq. 7.124 holds. The vectors $\boldsymbol{x}_{\mathrm{o}}^{(p)}(n)$, however, are in general mutually orthogonal, i.e.,

$$\mathrm{E}\left\{\boldsymbol{x}_{\mathrm{o}}^{(i)}(n)\boldsymbol{x}_{\mathrm{o}}^{(j)\mathrm{T}}(n)\right\} = \boldsymbol{0}, \quad \text{for } i \neq j, \tag{7.132}$$

only if $x(n)$ is an IID random process. In case of correlated input, Eq. 7.132 is generally not satisfied. Nevertheless, we assume for the following that in practice Eq. 7.132 is met sufficiently well. For illustration we consider a zero-mean, first-order stationary Laplacian Markov process $x(n)$ with an autocorrelation function $\mathrm{E}\{x(n)x(n-k)\} = 0.9^{|k|}$. The corresponding normalized crosscorrelation function

$$c_{1,3}(k) = \frac{\mathrm{E}\{x(n)x^3(n-k)\}}{\sqrt{\mathrm{E}\{x^2(n)\}\,\mathrm{E}\{x^6(n)\}}} \tag{7.133}$$

between $x(n)$ and $x^3(n)$ is shown in Fig. 7.15 together with the normalized crosscorrelation function

$$c_{\mathrm{o},1,3}(k) = \frac{\mathrm{E}\left\{x_{\mathrm{o}}^{(1)}(n)x_{\mathrm{o}}^{(3)}(n-k)\right\}}{\sqrt{\mathrm{E}\left\{\left(x_{\mathrm{o}}^{(1)}(n)\right)^2\right\}\mathrm{E}\left\{\left(x_{\mathrm{o}}^{(3)}(n)\right)^2\right\}}} \tag{7.134}$$

between the orthogonalized signals $x_{\mathrm{o}}^{(1)}(n)$ and $x_{\mathrm{o}}^{(3)}(n)$. As indicated by

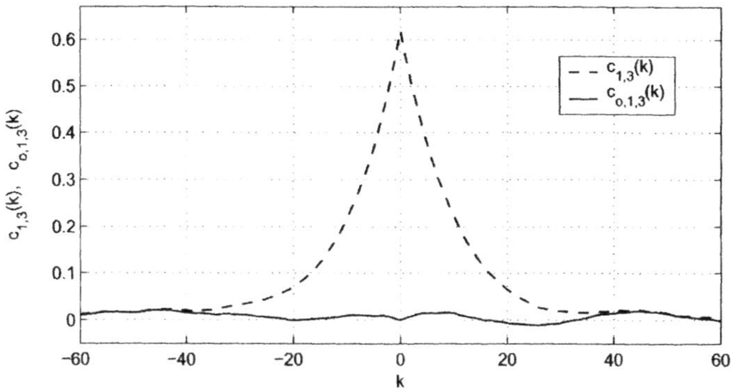

Fig. 7.15. Normalized crosscorrelation functions $c_{1,3}(k)$ and $c_{\mathrm{o},1,3}(k)$ between $x(n)$, $x^3(n)$ and $x_{\mathrm{o}}^{(1)}(n)$, $x_{\mathrm{o}}^{(3)}(n)$, respectively.

Fig. 7.15, the orthogonality property Eq. 7.132 is valid for the considered example. Thus, it is also reasonable to assume that Eq. 7.132 is sufficiently satisfied for speech input, since long-term properties of speech are commonly modeled by a Laplacian process [4].

Note that, for correlated input, the orthogonalization according to Eq. 7.129 does not orthogonalize ('whiten') the samples within each input vector $\mathbf{x}_{\mathrm{o}}^{(p)}(n)$: Although the input vector of different channels are mutually orthogonal, in general

$$\mathrm{E}\left\{x_{\mathrm{o}}^{(p)}(n)x_{\mathrm{o}}^{(p)}(n-k)\right\} \neq 0, \quad 0 \leq k < N_p, \tag{7.135}$$

holds as an immediate consequence of the (auto-)correlation of $x(n)$. A *quasi-complete* orthogonalization can be achieved by considering the asymptotic orthogonalization property of the DFT for large transform lengths [10].

The DFT-domain correspondence of Eq. 7.128 and Eq. 7.129 yields

$$\boldsymbol{X}_{\mathrm{o}}^{(1)}(\nu) = \boldsymbol{X}^{(1)}(\nu), \tag{7.136}$$

$$\boldsymbol{X}_{\mathrm{o}}^{(p)}(\nu) = \boldsymbol{X}^{(p)}(\nu) + \sum_{i=1}^{p-1} \boldsymbol{\Phi}_{p,i}(\nu)\, \boldsymbol{X}^{(i)}(\nu), \quad 1 < p \leq P. \tag{7.137}$$

Regarding the definition of the DFT-domain input vectors $\boldsymbol{X}^{(p)}(\nu)$ in Eq. 7.105, the DFT-domain orthogonalization matrices $\boldsymbol{\Phi}_{p,i}(\nu)$ are given by

$$\boldsymbol{\Phi}_{p,i}(\nu) = \boldsymbol{F}_{M\times M}\, \mathrm{diag}\Big\{ \Big[q_{p,i}(\nu L - N),\, \ldots,\, q_{p,i}(\nu L + N - 1) \Big] \Big\}\, \boldsymbol{F}_{M\times M}^{-1}. \tag{7.138}$$

It is important to note that in contrast to the time-domain orthogonalization matrices $\boldsymbol{Q}_{p,i}(n)$, their DFT-domain counterparts $\boldsymbol{\Phi}_{p,i}(\nu)$ are in general not diagonal. With the DFT-domain vectors $\boldsymbol{X}_{\mathrm{o}}^{(p)}(\nu)$ we achieve a quasi-complete orthogonalization of power filters: On the one hand, the above discussion of Eq. 7.132 with respect to correlated input implies

$$\mathrm{E}\Big\{ \boldsymbol{X}_{\mathrm{o}}^{(i)}(\nu)\boldsymbol{X}_{\mathrm{o}}^{(j)\mathrm{H}}(\nu) \Big\} = \mathbf{0}, \quad \text{for } i \neq j. \tag{7.139}$$

On the other hand, the asymptotic orthogonalization property of the DFT additionally implies orthogonality of the DFT-domain input vector elements within each channel:

$$\mathrm{E}\Big\{ X_{\mathrm{o}}^{(p)}(\nu,k)X_{\mathrm{o}}^{(p)^*}(\nu,m) \Big\} \approx 0, \quad \text{for } k \neq m, \tag{7.140}$$

if the DFT length M is sufficiently large.

Equivalent Orthogonalized Structure

When using the orthogonalized channel inputs $x_{\mathrm{o}}^{(p)}(n)$ for computing the output of power filters, the coefficients of the corresponding orthogonalized versions have to be adjusted accordingly. In the following we show the relation between the coefficients of the original power filter and their orthogonalized counterparts. Furthermore, we discuss how a time-variant orthogonalization of the input affects the coefficients of the equivalent orthogonalized structure of power filters.

The output $d(n)$ of a P-th order power filter can be computed by using the orthogonalized input vectors $\boldsymbol{x}_{\mathrm{o}}^{(p)}(n)$, i.e.,

$$d(n) = \sum_{p=1}^{P} d_{\mathrm{o}}^{(p)}(n), \tag{7.141}$$

$$d_{\mathrm{o}}^{(p)}(n) = \boldsymbol{h}_{\mathrm{o}}^{(p)\mathrm{T}} \boldsymbol{x}_{\mathrm{o}}^{(p)}(n). \tag{7.142}$$

Obviously, $d(n)$ can equivalently be expressed by either using combinations of $\boldsymbol{h}^{(p)}$ and $\boldsymbol{x}^{(p)}(n)$, or using the corresponding pairs of vectors $\boldsymbol{h}_{\mathrm{o}}^{(p)}(n)$ and

$\boldsymbol{x}_{\mathrm{o}}^{(p)}(n)$. Following [23], we refer to the combination of the orthogonalized input vectors $\boldsymbol{x}_{\mathrm{o}}^{(p)}(n)$ and the corresponding filter coefficient vectors $\boldsymbol{h}_{\mathrm{o}}^{(p)}(n)$ as *equivalent orthogonalized structure (EOS)* of power filters. For determining the coefficients of the EOS, we notice that the right hand sides of Eq. 7.97 and Eq. 7.141 have to be equal, implying

$$\sum_{p=1}^{P} \boldsymbol{h}^{(p)\mathrm{T}} \boldsymbol{x}^{(p)}(n) = \sum_{p=1}^{P} \boldsymbol{h}_{\mathrm{o}}^{(p)\mathrm{T}} \boldsymbol{x}_{\mathrm{o}}^{(p)}(n). \tag{7.143}$$

Introducing the definition of the orthogonalized input vectors Eq. 7.128 and Eq. 7.129 into Eq. 7.143 and solving for $\boldsymbol{h}_{\mathrm{o}}^{(p)}(n)$ for each p (starting with $p = P$) leads to the relation between the original filter coefficients $\boldsymbol{h}^{(p)}$ and the coefficients of the corresponding EOS:

$$\boldsymbol{h}_{\mathrm{o}}^{(P)} = \boldsymbol{h}^{(P)}, \tag{7.144}$$

$$\boldsymbol{h}_{\mathrm{o}}^{(p)}(n) = \boldsymbol{h}^{(p)} - \sum_{i=p+1}^{P} \boldsymbol{Q}_{i,p}(n) \boldsymbol{h}_{\mathrm{o}}^{(i)}(n), \quad 1 \leq p < P. \tag{7.145}$$

We notice that due to the orthogonalization of the input vectors, all channels of order $i > p$ contribute to the p-th channel of the corresponding EOS. Note that Eq. 7.145 implies that for time-varying orthogonalization matrices $\boldsymbol{Q}_{i,p}(n)$ the coefficients of the EOS $\boldsymbol{h}_{\mathrm{o}}^{(p)}(n)$ will generally be time-variant, although the coefficients $\boldsymbol{h}^{(p)}$ may be constant in time.

For the discussion of the DFT-domain EOS of power filters we introduce the diagonal matrix

$$\boldsymbol{H}_{\mathrm{diag}}^{(p)} = \mathrm{diag}\left\{\boldsymbol{H}^{(p)}\right\}. \tag{7.146}$$

Then, Eq. 7.107 can be rewritten according to

$$\boldsymbol{D}^{(p)}(\nu) = \boldsymbol{H}_{\mathrm{diag}}^{(p)} \boldsymbol{X}^{(p)}(\nu). \tag{7.147}$$

Analogously to Eq. 7.108 and Eq. 7.147, the computation of the DFT-domain output vector $\boldsymbol{D}(\nu)$ can alternatively be expressed by using the orthogonalized input vectors $\boldsymbol{X}_{\mathrm{o}}^{(p)}(\nu)$:

$$\boldsymbol{D}(\nu) = \sum_{p=1}^{P} \boldsymbol{D}_{\mathrm{o}}^{(p)}(\nu), \tag{7.148}$$

$$\boldsymbol{D}_{\mathrm{o}}^{(p)}(\nu) = \boldsymbol{H}_{\mathrm{diag,o}}^{(p)} \boldsymbol{X}_{\mathrm{o}}^{(p)}(\nu). \tag{7.149}$$

The matrices $\boldsymbol{H}_{\mathrm{diag,o}}^{(p)}$ represent the DFT-domain EOS of the corresponding power filter. The relation between the coefficient matrices $\boldsymbol{H}_{\mathrm{diag,o}}^{(p)}$ of the EOS and the original coefficient matrices $\boldsymbol{H}_{\mathrm{diag}}^{(p)}$ is given by

$$\boldsymbol{H}^{(P)}_{\text{diag,o}} = \boldsymbol{H}^{(P)}_{\text{diag}}, \tag{7.150}$$

$$\boldsymbol{H}^{(p)}_{\text{diag,o}}(\nu) = \boldsymbol{H}^{(p)}_{\text{diag}} - \sum_{i=p+1}^{P} \boldsymbol{\Phi}_{i,p}(\nu)\boldsymbol{H}^{(i)}_{\text{diag,o}}(\nu), \quad 1 \leq p < P. \tag{7.151}$$

In accordance to the time-domain EOS, the DFT-domain EOS has to be time-variant due to time-variant orthogonalization matrices $\boldsymbol{\Phi}_{i,p}(\nu)$.

Since the orthogonalization matrices $\boldsymbol{\Phi}_{i,p}(\nu)$ are in general not diagonal, it follows from Eq. 7.151 that this is also true for the DFT-domain coefficient matrices $\boldsymbol{H}^{(p)}_{\text{diag,o}}(\nu)$. Thus, the DFT-domain EOS of a power filter requires a set of $M \times M$ coefficient matrices, although it can be completely described by the original $M \times 1$ DFT-domain coefficient vectors $\boldsymbol{H}^{(p)}$. Obviously, this DFT-domain EOS of power filters constitutes a very inefficient way to represent power filters. This problem can be circumvented by performing the orthogonalization of the DFT-domain input vectors in a 'block time'-based manner. Thereby, the orthogonalization coefficients $q_{i,p}(n)$ are updated only once per block time index ν. Note that this implies the assumption of a short time stationary input $x(n)$. The desired diagonal orthogonalization matrices are then obtained by modifying their definition Eq. 7.138 according to

$$\begin{aligned} \boldsymbol{\Phi}_{p,i}(\nu) &= \boldsymbol{F}_{M\times M}\,\text{diag}\Big\{\Big[q_{p,i}(\nu L),\ \ldots,\ q_{p,i}(\nu L)\Big]\Big\}\,\boldsymbol{F}^{-1}_{M\times M} \\ &= q_{p,i}(\nu L)\,\boldsymbol{I}_{M\times M}. \end{aligned} \tag{7.152}$$

With these diagonal orthogonalization matrices, we can simplify Eq. 7.150 and Eq. 7.151 to a vector-based representation. The coefficient vectors of the DFT-domain EOS corresponding to the original vectors $\boldsymbol{H}^{(p)}$ are finally obtained as

$$\boldsymbol{H}^{(P)}_{\text{o}} = \boldsymbol{H}^{(P)}, \tag{7.153}$$

$$\boldsymbol{H}^{(p)}_{\text{o}}(\nu) = \boldsymbol{H}^{(p)} - \sum_{i=p+1}^{P} q_{i,p}(\nu L)\,\boldsymbol{H}^{(i)}_{\text{o}}(\nu), \quad 1 \leq p < P. \tag{7.154}$$

Obviously, the introduction of the block time index for determining the orthogonalization matrices is not only suggested by the inherent block processing of DFT-domain approaches, but it also leads to a more efficient implementation. Therefore, we restrict ourselves to diagonal orthogonalization matrices $\boldsymbol{\Phi}_{p,i}(\nu)$ according to Eq. 7.152 throughout the rest of this chapter.

Accounting for the block processing of DFT-domain power filters, the estimation of the k-th order moments $m_x^{(k)}$ can be performed via block averaging, i.e.,

$$\widehat{m}_x^{(k)}(\nu L + l) = \frac{1}{M}\sum_{i=0}^{M-1} x^k(\nu L - N + i), \quad \text{for } 0 \leq l < L. \tag{7.155}$$

The estimates $\widehat{m}_x^{(k)}(\nu L + l)$ are then introduced into Eq. 7.125 for computing the orthogonalization coefficients $q_{i,p}(\nu L)$.

Let us now look at the inevitable adjustment of the filter coefficients of the EOS arising from the time-varying input orthogonalization. For the derivation of the required adjustment, we solve Eq. 7.145 for the time instant $n-1$ with respect to the original coefficient vector $\boldsymbol{h}^{(p)}$:

$$\boldsymbol{h}^{(p)} = \boldsymbol{h}_{\mathrm{o}}^{(p)}(n-1) + \sum_{i=p+1}^{P} \boldsymbol{Q}_{i,p}(n-1)\boldsymbol{h}_{\mathrm{o}}^{(i)}(n-1). \tag{7.156}$$

Let us now consider the changes in Eq. 7.156 that occur for the next time instant. Due to the time-variance of the EOS for time-variant orthogonalization matrices, Eq. 7.156 becomes

$$\boldsymbol{h}^{(p)} = \boldsymbol{h}_{\mathrm{o}}^{(p)}(n) + \sum_{i=p+1}^{P} \boldsymbol{Q}_{i,p}(n)\boldsymbol{h}_{\mathrm{o}}^{(i)}(n). \tag{7.157}$$

Assuming that the original coefficients of the power filter are constant in time, we can replace $\boldsymbol{h}^{(p)}$ in Eq. 7.156 by the right hand side of Eq. 7.157. Solving for $\boldsymbol{h}_{\mathrm{o}}^{(p)}(n)$ finally leads to the required coefficient adjustment: After each change of the orthogonalization matrices $\boldsymbol{Q}_{i,p}(n)$, the coefficients vectors $\boldsymbol{h}_{\mathrm{o}}^{(p)}(n)$ are recursively recomputed according to

$$\boldsymbol{h}_{\mathrm{o}}^{(p)}(n) = \boldsymbol{h}_{\mathrm{o}}^{(p)}(n-1) + \sum_{i=p+1}^{P} \left[\boldsymbol{Q}_{i,p}(n-1)\boldsymbol{h}_{\mathrm{o}}^{(i)}(n-1) - \boldsymbol{Q}_{i,p}(n)\boldsymbol{h}_{\mathrm{o}}^{(i)}(n)\right], \tag{7.158}$$

starting with $p = P-1$. From Eq. 7.144 we notice that no adjustment is required for the P-th order channel, i.e., $\boldsymbol{h}_{\mathrm{o}}^{(P)}(n) = \boldsymbol{h}_{\mathrm{o}}^{(P)}(n-1)$.

The necessity of this coefficient adjustment becomes obvious when regarding that each set of orthogonalization matrices $\boldsymbol{Q}_{i,p}(n)$ yields a corresponding EOS. This implies that after each change of the orthogonalization matrices both, a new set of input vectors and a new set of associated coefficient vectors have to be determined.

The above time-domain result can directly be used to obtain a corresponding adjustment for the DFT-domain EOS. With the definition of the simplified DFT-domain EOS according to Eq. 7.153 and Eq. 7.154, the DFT-domain counterpart to Eq. 7.158 is given by

$$\boldsymbol{H}_{\mathrm{o}}^{(p)}(\nu) = \boldsymbol{H}_{\mathrm{o}}^{(p)}(\nu-1) + \sum_{i=p+1}^{P} \left[q_{i,p}(\nu L - L)\boldsymbol{H}_{\mathrm{o}}^{(i)}(\nu-1) - q_{i,p}(\nu L)\boldsymbol{H}_{\mathrm{o}}^{(i)}(\nu)\right], \tag{7.159}$$

where we start with $p = P-1$. As in the time domain, no adjustment is required for the channel with the highest order, i.e., $\boldsymbol{H}_{\mathrm{o}}^{(P)}(\nu) = \boldsymbol{H}_{\mathrm{o}}^{(P)}(\nu-1)$.

Adaptation of Orthogonalized Power Filters

Since we consider the adaptation of *orthogonalized* power filters we express the output $\hat{d}(n)$ of the adaptive power filter according to Eq. 7.141 and Eq. 7.142:

$$\hat{d}(n) = \sum_{p=1}^{P} \hat{d}_{\mathrm{o}}^{(p)}(n), \tag{7.160}$$

$$\hat{d}_{\mathrm{o}}^{(p)}(n) = \hat{\boldsymbol{h}}_{\mathrm{o}}^{(p)\mathrm{T}}(n)\boldsymbol{x}_{\mathrm{o}}^{(p)}(n). \tag{7.161}$$

In the following we use the same notation for the signals as introduced in Fig. 7.6, i.e., the observed signal $y(n)$ is composed of the echo signal $d(n)$, background noise $b(n)$, and local speech $s(n)$. The error signal $e(n) = y(n) - \hat{d}(n)$ is the difference between the observed signal and the output of the adaptive power filter. The LMS update equation for the coefficients of the adaptive EOS is then given by

$$\hat{h}_{\mathrm{o},k}^{(p)}(n+1) = \hat{h}_{\mathrm{o},k}^{(p)}(n) + \mu_{\mathrm{o},k}^{(p)}(n)\, e(n) x_{\mathrm{o}}^{(p)}(n-k). \tag{7.162}$$

The control of the adaptation by appropriately choosing the step size $\mu_{\mathrm{o},k}^{(p)}(n)$ is discussed later in this section.

Note that the coefficient adjustment according to Eq. 7.158 is carried out first, and then the coefficients of the EOS are adapted subsequently by applying Eq. 7.162.

For deriving a DFT-domain adaptation of power filters we recall that they can be considered as linear multichannel system. This interpretation obviously also applies for the EOS of power filters. Thus, we can directly use the results of Section 7.3.3 that have been obtained for DFT-domain Volterra filters in DCR. Assuming diagonal orthogonalization matrices, the DFT-domain output $\widehat{D}(\nu, m)$ of the adaptive power filter is given by

$$\widehat{D}(\nu, m) = \sum_{p=1}^{P} \widehat{D}_{\mathrm{o}}^{(p)}(\nu, m), \tag{7.163}$$

$$\widehat{D}_{\mathrm{o}}^{(p)}(\nu, m) = \widehat{H}_{\mathrm{o}}^{(p)}(\nu, m) X_{\mathrm{o}}^{(p)}(\nu, m). \tag{7.164}$$

The adaptation algorithm for power filters immediately follows from the corresponding update equation Eq. 7.81 for multidelay Volterra filters:

$$\widehat{H}_{\mathrm{o}}^{(p)}(\nu+1, m) = \widehat{H}_{\mathrm{o}}^{(p)}(\nu, m) + \mu_{\mathrm{o}}^{(p)}(\nu, m)\, \mathcal{F}_M \left\{ w_l\, \mathcal{F}_M^{-1} \left\{ E(\nu, m) X_{\mathrm{o}}^{(p)^*}(\nu, m) \right\} \right\}. \tag{7.165}$$

The time-domain window function w_l is introduced to assure the zero-padding of the time-domain coefficient vectors according to Eq. 7.106 and has been defined in Eq. 7.84. Since we have assumed uniform memory length $N_p = N$ for all channels, the same window function can be applied for each order p.

Adaptation Control

First, we look at the control of the step size parameter $\mu_{o,n}^{(p)}(n)$ of the time-domain LMS algorithm according to Eq. 7.162 which corresponds to the coefficient-dependent step size for P-th order Volterra filters presented in Section 7.3.2. Due to the mutual orthogonality of all channel inputs of the EOS, the reasoning applied in Section 7.3.2 for second-order Volterra filters can correspondingly be applied for the derivation of an optimum coefficient-dependent step size for the adaptive EOS of P-th order power filters.

The coefficient error $m_{o,n}^{(p)}(k)$ with respect to the time-domain EOS of power filters is defined by

$$m_{o,n}^{(p)}(n) = h_{o,k}^{(p)}(n) - \hat{h}_{o,k}^{(p)}(n). \tag{7.166}$$

Analogously to Section 7.3.2, we use the mean squared error between the actual coefficient error and the corresponding LMS update term as optimality criterion for determining the optimum value of the step size $\mu_{o,k}^{(p)}(n)$, i.e.,

$$J_{\mu_{o,k}^{(p)}}(n) = \mathrm{E}\left\{ \left[m_{o,k}^{(p)}(n) - \mu_{o,k}^{(p)}(n) e(n) x_o^{(p)}(n-k) \right]^2 \right\}. \tag{7.167}$$

As in Section 7.3.2, we assume that the input $x(n)$ is an IID random process with an even PDF, i.e., the orthogonality property Eq. 7.132 holds. We further assume that the adaptive coefficients $\hat{h}_{o,k}^{(p)}(n)$ are statistically independent of the input. Applying the same reasoning as in [20] for Volterra filters, it is straightforward to show that Eq. 7.42 correspondingly holds for orthogonalized power filters:

$$\mu_{\mathrm{opt},o,k}^{(p)}(n) = \frac{\mathrm{E}\left\{ \left[m_{o,k}^{(p)}(n) \right]^2 \right\}}{\mathrm{E}\left\{ \varepsilon^2(n) + b^2(n) + s^2(n) \right\}}. \tag{7.168}$$

Aiming at a factorized version of Eq. 7.168, we introduce the residual echo $\varepsilon_o^{(p)}(n)$ of the p-th channel of the EOS according to

$$\varepsilon_o^{(p)}(n) = d_o^{(p)}(n) - \hat{d}_o^{(p)}(n). \tag{7.169}$$

The overall residual echo $\varepsilon(n) = d(n) - \hat{d}(n)$ can then be written as

$$\varepsilon(n) = \sum_{p=1}^{P} \varepsilon_o^{(p)}(n). \tag{7.170}$$

Since the orthogonality property Eq. 7.132 holds for the assumed input, we can express the mean square of the residual echo by

$$\mathrm{E}\left\{ \varepsilon^2(n) \right\} = \sum_{p=1}^{P} \mathrm{E}\left\{ \left[\varepsilon_o^{(p)}(n) \right]^2 \right\}. \tag{7.171}$$

The expression Eq. 7.171 for computing the mean squared residual echo can be exploited to derive a factorized version of Eq. 7.168. The desired factorized version of Eq. 7.168 is obtained as

$$\mu_{\mathrm{opt,o},k}^{(p)}(n) = \alpha_{\mathrm{dt}}(n)\,\alpha_{\mathrm{bn}}(n)\,\alpha_{\varepsilon_{\mathrm{o}}}^{(p)}(n)\,\alpha_{\mathrm{o},k}^{(p)}(n). \tag{7.172}$$

The auxiliary step sizes $\alpha_{\mathrm{dt}}(n)$ and $\alpha_{\mathrm{bn}}(n)$ have been defined in Eq. 7.51 and Eq. 7.52, respectively, and account for double-talk and background noise. The channel-dependent step size parameter $\alpha_{\varepsilon_{\mathrm{o}}}^{(p)}(n)$ is used to control the adaptation with respect to mutual interferences caused by misadjusted channel filters. Analogously to Eq. 7.53, it is defined by

$$\alpha_{\varepsilon_{\mathrm{o}}}^{(p)}(n) = \frac{\mathrm{E}\left\{\left[\varepsilon_{\mathrm{o}}^{(p)}(n)\right]^2\right\}}{\sum\limits_{i=1}^{P}\mathrm{E}\left\{\left[\varepsilon_{\mathrm{o}}^{(i)}(n)\right]^2\right\}}. \tag{7.173}$$

The coefficient-dependent step size parameter $\alpha_{\mathrm{o},k}^{(p)}(n)$ is finally given by

$$\alpha_{\mathrm{o},k}^{(p)}(n) = \frac{\mathrm{E}\left\{\left[m_{\mathrm{o},k}^{(p)}(n)\right]^2\right\}}{\sum\limits_{l=0}^{N_p-1}\mathrm{E}\left\{\left[m_{\mathrm{o},l}^{(p)}(n)\right]^2\right\}\mathrm{E}\left\{\left[x_{\mathrm{o}}^{(p)}(n-l)\right]^2\right\}}, \tag{7.174}$$

which obviously corresponds to the coefficient-dependent step size for second-order Volterra filters according to Eqs. 7.54 and 7.55, respectively.

Analogously to Eq. 7.57, the second-order moments of the residual echoes $\varepsilon_{\mathrm{o}}^{(p)}(n)$ can be estimated using the model

$$\mathrm{E}\left\{\left[\varepsilon_{\mathrm{o}}^{(p)}(n)\right]^2\right\} \approx \gamma_{\varepsilon}(n)\left[\delta_{\varepsilon_p} + \left|\hat{d}_{\mathrm{o}}^{(p)}(n)\right|\right]. \tag{7.175}$$

For realizing the coefficient-dependent step size $\alpha_{\mathrm{o},k}^{(p)}(n)$, we apply the proportionality model Eq. 7.58 for estimating the mean square of the coefficient errors, i.e.,

$$\mathrm{E}\left\{\left[m_{\mathrm{o},k}^{(p)}(n)\right]^2\right\} \approx \gamma_{m,p}(n)\left[\beta_{m,p}(n) + \left|\hat{h}_{\mathrm{o},k}^{(p)}(n)\right|\right]. \tag{7.176}$$

The meaning of the parameters appearing in the estimation models Eq. 7.175 and Eq. 7.176 have already been discussed in Section 7.3.2.

The derivation of a step-size control for the DFT-domain EOS of power filters follows the kernel-dependent approach for DFT-domain Volterra filters according to Section 7.3.3. Thus, we first introduce a channel-dependent normalization of $\mu_{\mathrm{o}}^{(p)}(\nu, m)$ according to

$$\mu_{\mathrm{o}}^{(p)}(\nu,m)=\frac{\alpha_{\mathrm{o}}^{(p)}(\nu,m)}{\widehat{S}_{\mathrm{o},X}^{(p)}(\nu,m)}. \tag{7.177}$$

The normalization factor $\widehat{S}_{\mathrm{o},X}^{(p)}(\nu,m)$ represents an estimate of

$$S_{\mathrm{o},X}^{(p)}(\nu,m)=\mathrm{E}\left\{\left|X_{\mathrm{o}}^{(p)}(\nu,m)\right|^{2}\right\}, \tag{7.178}$$

i.e., of the PSD of the input signal of the p-th channel of the EOS. $\hat{S}_{\mathrm{o},X}^{(p)}(\nu,m)$ can be obtained, e.g., analogously to Eq. 7.89.

The normalized step size $\alpha_{\mathrm{o}}^{(p)}(\nu,m)$ is implemented by using a corresponding factorized version according to

$$\alpha_{\mathrm{o}}^{(p)}(\nu,m)=\alpha_{\mathrm{dt}}(\nu L)\,\alpha_{\mathrm{bn}}(\nu,m)\,\alpha_{\mathcal{E}_{\mathrm{o}}}^{(p)}(\nu,m). \tag{7.179}$$

The auxiliary step-size parameters $\alpha_{\mathrm{dt}}(\nu L)$ and $\alpha_{\mathrm{bn}}(\nu,m)$ have already been discussed in Section 7.3.3 and are not further considered here.

The channel-dependent step size $\alpha_{\mathcal{E}_{\mathrm{o}}}^{(p)}(\nu,m)$ represents the DFT-domain correspondence of $\alpha_{\varepsilon_{\mathrm{o}}}^{(p)}(n)$. Regarding Eq. 7.139, it is obvious that the orthogonality property Eq. 7.171 also holds in the DFT-domain. The mean squared magnitude of the DFT-domain residual echo can then be written as

$$\mathrm{E}\left\{|\mathcal{E}(\nu,m)|^{2}\right\}=\sum_{i=1}^{P}\mathrm{E}\left\{\left|\mathcal{E}_{\mathrm{o}}^{(i)}(\nu,m)\right|^{2}\right\}. \tag{7.180}$$

Here, $\mathcal{E}_{\mathrm{o}}^{(p)}(\nu,m)$ represents the DFT-domain correspondence of the time-domain residual echo $\varepsilon_{\mathrm{o}}^{(p)}(n)$. Consequently, the channel-dependent step size $\alpha_{\mathcal{E}_{\mathrm{o}}}^{(p)}(\nu,m)$ can be defined correspondingly to Eq. 7.173:

$$\alpha_{\mathcal{E}_{\mathrm{o}}}^{(p)}(\nu,m)=\frac{\mathrm{E}\left\{\left|\mathcal{E}_{\mathrm{o}}^{(p)}(\nu,m)\right|^{2}\right\}}{\sum_{i=1}^{P}\mathrm{E}\left\{\left|\mathcal{E}_{\mathrm{o}}^{(i)}(\nu,m)\right|^{2}\right\}}. \tag{7.181}$$

As in case of DFT-domain Volterra filters, the assumption of uniform mean squared magnitudes of the residual echoes for all p yields a good performance for DFT-domain adaptive power filter, too. Then, Eq. 7.181 simplifies to

$$\alpha_{\mathcal{E}_{\mathrm{o}}}^{(p)}(\nu,m)\approx\frac{1}{P}, \tag{7.182}$$

i.e., to its channel-independent form.

Simulations

The following simulations illustrate the effect of time-variant orthogonalization of adaptive power filters on their performance and show the necessity of an appropriate adjustment of the coefficients of the EOS.

For the simulations, the echo path has been modeled by the cascade of a third-order memoryless polynomial and a linear filter of length $N_c = 200$, i.e., it can exactly be represented by a third-order power filter. The input signal has been a zero-mean, uncorrelated, non-stationary Laplacian process [4]. A white noise signal $b(n)$ has been added to the echo signal $d(n)$, where the noise variance yields an SNR of 30 dB with respect to the variance of $d(n)$.

The echo canceller has been realized as a third-order time-domain power filter, where the memory length of each channel has also been chosen to $N_p = 200$. Fig. 7.16 shows the ERLE graphs obtained for the EOS of the power filter with coefficient adjustment (CA) according to Eq. 7.158, the EOS without CA, and the corresponding non-orthogonalized power filter. The or-

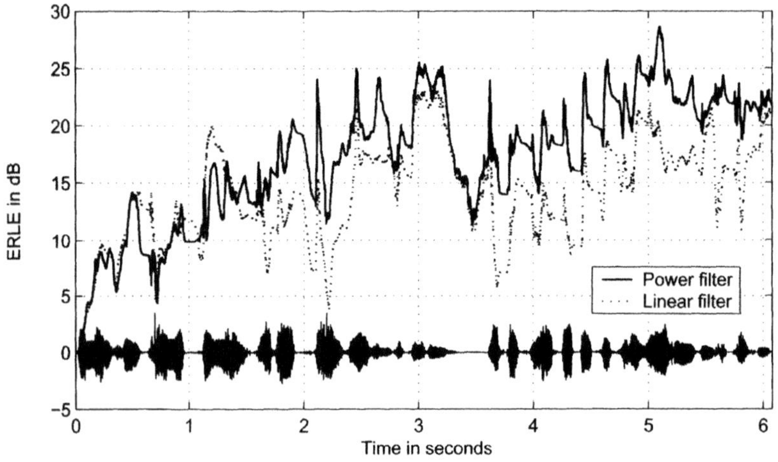

Fig. 7.16. ERLE obtained for different implementations of third-order adaptive power filters together with the uncorrelated, non-stationary input signal.

thogonalization of the input has been performed signal-adaptively, using the recursive estimation of the moments according to Eq. 7.127 with a forgetting factor $\lambda = 0.97$. The step-size control has been realized according to Eq. 7.172 with a fixed value of $\alpha_{\mathrm{dt}}(n)\alpha_{\mathrm{bn}}(n) = 0.3$. The models Eqs. 7.175, 7.176 have been applied to approximate the channel-dependent step size $\alpha_{\varepsilon_o}^{(p)}(n)$ and the coefficient-dependent step size $\alpha_{o,k}^{(p)}(n)$, respectively. The model parameter δ_{ε_p} for estimating the mean squared residual echoes has been chosen to $\delta_{\varepsilon_p} = 0.001$ for all channels. The model for the coefficient-dependent auxiliary step sizes has been realized analogously to Eq. 7.59 and Eq. 7.60, where

$\beta_{c,1} = \beta_{c,2} = 1$. To allow for a fair comparison of all approaches, the same adaptation control has been applied to all algorithms. It should, however be kept in mind that in case of non-orthogonalized power filters, this choice does not assure stable convergence, although it is advantageous with respect to convergence speed.

The limitation of achievable echo attenuation of the adaptive EOS without coefficient adjustment can clearly be seen in Fig. 7.16. This result confirms the importance of the coefficient adjustment required for time-variant orthogonalization matrices. We further notice from Fig. 7.16 that the EOS with coefficient adjustment outperforms the non-orthogonalized version in both, convergence speed and achievable echo attenuation.

7.4.3 Application to Real Systems

Power filters have been introduced as an approximation of the nonlinear cascaded model of the acoustic echo path according to Fig. 7.13. In this model, a Taylor series expansion has been used to approximate the nonlinear behaviour of the amplifier and the loudspeaker of a mobile phone, respectively. In this section we examine the suitability of these approximations when modeling real acoustic echo paths. Thereby, we look at the case of nonlinear distortion introduced by the amplifier and also consider the influence of the nonlinear behaviour of the loudspeaker of a mobile phone.

Nonlinear Amplifier

The experimental setup used for the following experiments consists of a commercial one-chip amplifier connected to an electro-dynamic loudspeaker with a diameter of six centimeters which has been placed in a room with low reverberation and a low background noise level. The power supply of the amplifier has been adjusted such that for high input levels the amplifier causes nonlinear distortion. Throughout the experiments, the nonlinear distortion introduced by the loudspeaker is negligible at the considered excitation levels.

The nonlinear echo canceller has been implemented as the adaptive EOS of a third-order power filter. The memory length of the linear kernel has been $N_1 = 256$ (implying a DFT length $M = 512$), whereas $N_2 = N_3 = 100$ has been chosen for the quadratic and the cubic channel. The orthogonalization of the channel inputs is performed block time-based, where the required moments of the input are estimated via block averaging according to Eq. 7.155. The DFT-domain EOS has been adapted applying using the fixed normalized step size $\alpha_{o}^{(p)}(\nu, m) = 0.1$ for all channels. The input has been a speech signal sampled at 8 kHz. To simulate a higher level of background noise, a white Gaussian noise signal has been added to the recording of the real echo signal. The noise variance has been adjusted to give an SNR of 30 dB with respect to the measured microphone signal.

In Fig. 7.17, the ERLE obtained for the third-order adaptive EOS is compared to a linear approach which corresponds to the linear channel of the power filter. Except for the initial convergence phase, the adaptive power filter

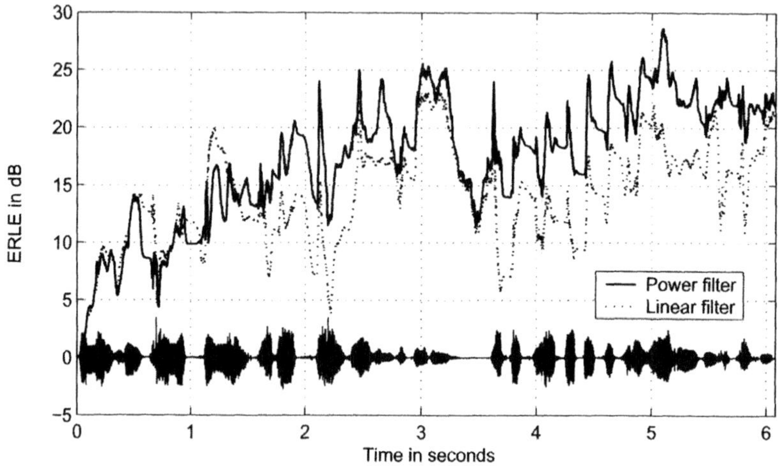

Fig. 7.17. ERLE obtained for the adaptive EOS of a third-order power filter and a linear approach together with the speech input.

continuously provides an improvement of the echo cancellation performance, especially during periods of high excitation levels. Thereby, a gain of about 6 dB compared to the linear approach is well possible. Note that the applied third-order power filter increases the required number of multiplications only by a factor of 2.5 compared to the linear approach.

Nonlinear Loudspeaker of a Mobile Phone

The nonlinear behaviour of a moderately-sized electro-dynamic loudspeaker has successfully been modeled by second-order Volterra filters. In this section we examine the very small electro-dynamic loudspeaker of a mobile phone. Due to its limited dimensions, the nonlinear behaviour of this type of loudspeakers is different from that used in Section 7.3.4 [23].

For the recordings, the loudspeaker has been mounted in the handset, while the microphone has been separated from it to avoid undesired vibration effects due to physical coupling of the loudspeaker and the microphone. During the measurements it has been assured that there is no nonlinear distortion introduced by overloading of the amplifier, i.e., the nonlinearity in the acoustic echo path is solely caused by the loudspeaker. The echo signal used for the simulations has been recorded in a room with low reverberation. The input has been a speech segment sampled at 8 kHz. A white Gaussian noise signal has been added to the recording of the microphone signal in order to simulate a background noise level corresponding to an SNR of 30 dB with respect to the acoustic echo. Since an algorithmic delay is not desirable in mobile phones, we consider the time-domain implementation of the EOS, where the memory length $N_1 = 250$ for the linear channel. Here, the memory lengths $N_2 = N_3 = 100$ for both, the quadratic and cubic channel are already sufficient. The orthogonalization of the channel inputs has been performed signal-adaptively, where the moments are estimated recursively according to Eq. 7.127 with a forgetting factor $\lambda = 0.97$.

In Fig. 7.18, the echo cancellation performance of the adaptive EOS of the third-order power filter is compared to a linear approach which corresponds to the linear channel of the power filter. As can be noticed, the performance

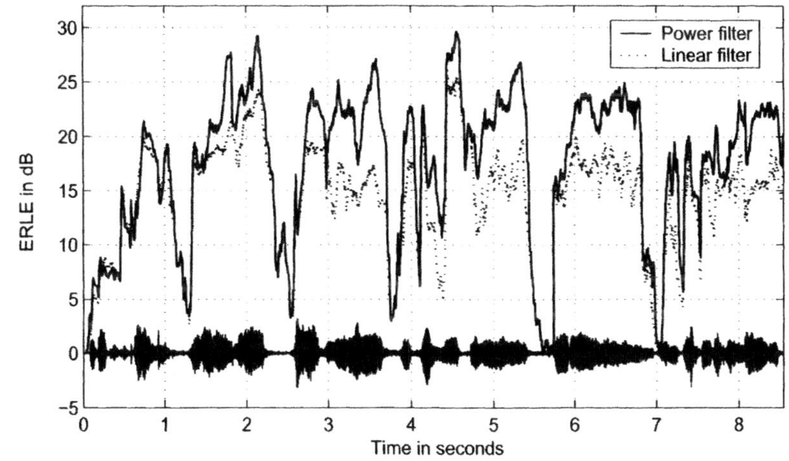

Fig. 7.18. ERLE obtained for the adaptive EOS of a third-order power filter and a corresponding linear approach for speech input.

of the linear adaptive filter is clearly inferior due to the nonlinear distortion introduced by the loudspeaker. The third-order power filter succeeds in improving the level of echo attenuation during almost the whole simulation period. Especially for speech segments that exhibit high excitation levels the increase of the ERLE is significant. Note that due to the short filters in the nonlinear channels, the computational complexity of the considered orthogonalized power filter is only two times higher than that of the linear filter.

7.5 Conclusions

In todays telecommunication devices often cheap audio hardware is included which introduces non-negligible nonlinear distortion into the loudspeaker signal. In case of hands-free telephone systems or mobile communication devices, these nonlinear audio components cause nonlinearly distorted acoustic echoes that can not be sufficiently attenuated by purely linear AECs. In this chapter, we have focused on special types of adaptive nonlinear filters which require only little *a priori* knowledge about the audio hardware actually included in the telecommunication device, i.e., Volterra filters and power filters.

If moderately-sized loudspeakers represent the only source of nonlinear distortion, second-order Volterra filters have been used to model the nonlinear acoustic echo path. Due to the assumption that only the loudspeaker introduces nonlinear distortion, the model of the acoustic echo path simplifies to a cascade of a second-order Volterra filter followed by a linear filter. It has been shown in Section 7.3 that the overall model of this cascade can be represented by a corresponding second-order Volterra filter that has a reduced region of support for the quadratic kernel. The DCR of Volterra filters allows for an elegant way to exploit this *a priori* knowledge about the acoustic echo path: The width of the quadratic kernel is simply chosen smaller than its memory length. By doing so, coefficients that are known (or assumed) to be zero can be explicitly excluded and inefficient system configurations can be avoided.

The DCR has also led to the interpretation of Volterra filters as a special type of linear multichannel systems. Based on that, efficient DFT-domain methods known from linear adaptive filtering could straightforwardly be extended to adaptive Volterra filters, too. The proposed MDVF does not affect the multichannel structure of the DCR and, therefore, preserves its advantageous features.

Experimental results obtained for a real loudspeaker system have been presented in order to verify the suitability of adaptive Volterra filters. In a realistic acoustic echo cancellation scenario, the echo attenuation has been improved by about 5 up to 10 dB compared to a linear approach. Due to the reduced width of the quadratic kernel, the computational complexity has only been increased by a factor of approximately four compared to the linear approach. Thus, second-order Volterra filters can be considered as a well suited approach to cope with nonlinear loudspeakers in hands-free telecommunication systems.

In Section 7.4 we have considered the case that only the amplifier or the miniaturized loudspeaker of a mobile phone cause the nonlinear distortion in the echo path. For this scenario, the model of the acoustic echo path simplifies to the cascade of a linear filter, a memoryless nonlinearity (modeled by a Taylor series expansion), and a second linear filter. It has been shown that power filters represent an efficient parallelized approximation of this overall model of the echo path. Since the saturation characteristics imply power filters of orders higher than two, the input signals of different channels are not mutually orthogonal anymore. In order to improve the performance of corresponding adaptive implementations, a method to signal-adaptively orthogonalize the inputs of the different channels of the power filter has been discussed.

Experimental results based on measurements with an overloaded amplifier have shown that the considered adaptive third-order power filter has been able to improve the echo attenuation of a purely linear AEC by about 6 dB. For the case that the loudspeaker of a mobile phone causes the nonlinear distortion in the echo path, third-order power filters are able to increase the achievable echo attenuation of a linear approach by approximately 5 dB.

Although the proposed nonlinear approaches provide significant improvements over purely linear adaptive filters, the achieved level of echo attenuation might not be sufficient in some applications. A common method in linear echo cancellation is to further suppress the residual echo that remains after the echo cancellation step. Usually, such methods apply postfiltering of the residual echo based on Wiener filtering techniques [12, 13] which, however, have to be appropriately extended to account for nonlinear acoustic echo paths.

References

[1] O. Agazzi, D. G. Messerschmitt, D. A. Hodges: Nonlinear echo cancellation of data signals, *IEEE Trans. on Communications,* **30**(11), 2421–2433, Nov. 1982.

[2] J. Benesty, D. R. Morgan, J. H. Cho: A new class of doubletalk detectors based on cross-correlation, *IEEE Trans. on Acoustics, Speech, and Signal Processing,* **8**(2), 168–172, March 2000.

[3] A. N. Birkett, R. A. Goubran: Limitations of handsfree acoustic echo cancellers due to nonlinear loudspeaker distortion and enclosure vibration effects, *Proc. WASPAA '95,* 13–16, New Paltz, NY, USA, Oct. 1995.

[4] H. Brehm, W. Stammler: Description and generation of spherically invariant speech-model signals, *Signal Processing,* **12**(2), 119–141, March 1987.

[5] C. Breining, P. Dreiseitel, E. Hänsler, A. Mader, B. Nitsch, H. Puder, T. Schertler, G. Schmidt, J. Tilp: Acoustic echo control, *IEEE Signal Processing Magazine,* **16**(4), 42–69, July 1999.

[6] J. Chen, J. Vandevalle: Study of adaptive nonlinear echo canceller with Volterra expansion, *Proc. Processing ICASSP '89,* 1376–1379, Glasgow, UK, May 1989.

[7] A. Fermo, A. Carini, G. L. Sicuranza: Simplified Volterra filters for acoustic echo cancellation in GSM receivers, *Proc. EUSIPCO '00,* 2413–2416, Tampere, Finland, Sept. 2000.

[8] W. A. Frank: An efficient approximation to the quadratic Volterra filter and its application in realtime loudspeaker linearization, *Signal Processing,* **45**(1), 97–113, July 1995.

[9] G. O. Glentis, K. Berberidis, S. Theodoridis: Efficient least squares adaptive algorithms for FIR transversal filtering: a unified view, *IEEE Signal Processing Magazine,* **16**(4), 13–41, July 1999.

[10] R. M. Gray: On the asymptotic eigenvalue distribution of Toeplitz matrices, *IEEE Trans. on Information Theory,* **18**(6), 725–730, Nov. 1972.

[11] A. Guérin, G. Faucon, R. Le Bouquin-Jeannès: Nonlinear acoustic echo cancellation based on Volterra filters, *IEEE Trans. on Acoustics, Speech, and Signal Processing,* **11**(6), 672–683, Nov. 2003.

[12] S. Gustafsson, R. Martin, P. Vary: Combined acoustic echo control and noise reduction for hands-free telephony, *Signal Processing,* **64**(1), 21–32, 1998.

[13] E. Hänsler, G. Schmidt: *Acoustic Echo and Noise Control: A Practical Approach,* Hoboken, NJ, USA: Wiley, 2004.

[14] S. Haykin: *Adaptive Filter Theory,* Englewood Cliffs, NJ, USA: Prentice Hall, 1996.

[15] S. Im, E. J. Powers: A fast method of discrete third-order Volterra filtering, *IEEE Trans. on Signal Processing,* **44**(9), 2195–2208, Sept. 1996.

[16] W. Klippel: Dynamic measurement and interpretation of the nonlinear parameters of electrodynamic loudspeakers, *J. Audio Eng. Soc.,* **38**(12), 944–955, Dec. 1990.

[17] W. Klippel: Filter structures to compensate for nonlinear distortion of horn loudspeakers, *J. Audio Eng. Soc.,* Preprint 4102, 1995.

[18] F. Kuech, W. Kellermann: Proportionate NLMS algorithm for second-order Volterra filters and its application to nonlinear echo cancellation, *Proc. IWAENC '03,* 75–78, Kyoto, Japan, Sept. 2003.

[19] F. Kuech, W. Kellermann: A novel multidelay adaptive algorithm for Volterra filters in diagonal coordinate representation, *Proc. ICASSP '04,* **2**, 869-872, Montreal, Canada, May 2004.

[20] F. Kuech, W. Kellermann: Coefficient-dependent step-size for adaptive second-order Volterra filters, *Proc. EUSIPCO '04,* 1805–1808, Vienna, Austria, Sept. 2004.

[21] F. Kuech, W. Kellermann: Orthogonalized power filters for nonlinear acoustic echo cancellation, *Signal Processing,* submitted Jan. 2005.

[22] F. Kuech, W. Kellermann: Partitioned block frequency-domain adaptive second-order Volterra filter, *IEEE Trans. on Signal Processing,* **53**(2), 564–575, Feb. 2005.

[23] F. Kuech, A. Mitnacht, W. Kellermann: Nonlinear acoustic echo cancellation using adaptive orthogonalized power filters, *Proc. ICASSP '05,* **3**, 105–108, Philadelphia, PA, USA, March 2005.

[24] G. Lazzarin, S. Pupolin, A. Sarti: Nonlinearity compensation in digital radio systems, *IEEE Trans. on Communications,* **42**(2), 988–999, Feb. 1994.

[25] J. Lee, V. J. Mathews: A fast recursive least squares adaptive second order Volterra filter and its performance analysis, *IEEE Trans. on Signal Processing,* **41**(3), 1087–1102, March 1993.

[26] A. Mader, H. Puder, G. U. Schmidt: Step-size control for acoustic echo cancellation filters - an overview, *Signal Processing,* **80**(9), 1697–1719, Sept. 2000.

[27] V. J. Mathews: Adaptive polynomial filters, *IEEE Signal Processing Magazine,* **8**(3), 10–26, July 1991.

[28] V. J. Mathews, G. L. Sicuranza: *Polynomial Signal Processing,* New York, USA: John Wiley and Sons, 2000.
[29] M. Morháč: A fast algorithm of nonlinear Volterra filtering, *IEEE Trans. on Signal Processing,* **39**(10), 2353–2356, Oct. 1991.
[30] E. Moulines, O. A. Amrane, Y. Grenier: The generalized multidelay adaptive filter: Structure and convergence analysis,*IEEE Trans. on Signal Processing,* **43**(1), 14–28, Jan. 1995.
[31] B. H. Nitsch: A frequency-selective stepfactor control for an adaptive filter algorithm working in the frequency domain, *Signal Processing,* **80**(9), 1733–1745, Sept. 2000.
[32] B. S. Nollett, D. L. Jones: Nonlinear echo cancellation for hands-free speakerphones, *Proc. NSIP '97,* Mackinac Island, MI, USA, Sept. 1997.
[33] A. Papoulis, S. U. Pillai: *Probability, Random Variables and Stochastic Processes,* New York, USA: McGraw-Hill, 2002.
[34] J. G. Proakis, D. G. Manolakis: *Digital Signal Processing: Principles, Algorithms and Applications,* Englewood Cliffs, NJ, USA: Prentice Hall, 1996.
[35] G. M. Raz, B. D. Van Veen: Baseband Volterra filters for implementing carrier based nonlinearities, *IEEE Trans. on Signal Processing,* **46**(1), 103–114, Jan. 1998.
[36] M. Rupp, J. Cezanne: Robustness conditions of the LMS algorithm with time-variant matrix step-size, *Signal Processing,* **80**(9), 1787–1794, Sept. 2000.
[37] H. Schurer: *Linearization of Electroacoustic Transducers,* Enschede, Netherlands: Print Partners Ipskamp, 1997.
[38] J. J. Shynk: Frequency domain and multirate adaptive filtering, *IEEE Signal Processing Magazine,* **9**(1), 14–37, Jan. 1992.
[39] J.-S. Soo, K. K. Pang: Multidelay block frequency domain adaptive filter, *IEEE Trans. on Acoustics, Speech, and Signal Processing,* **38**(2), 373–376, Feb. 1990.
[40] M. Soria-Rodriguez, M. Gabbouj, N. Zacharov, M. S. Hämäläinen, K. Koivuniemi: Modeling and real-time auralization of electrodynamic loudspeaker non-linearities, *Proc. ICASSP '04,* **4**, 17–21, Montreal, Canada, May 2004.
[41] A. Stenger:*Kompensation akustischer Echos unter Einfluss von nichtlinearen Audiokomponenten,* Aachen, Germany: Shaker, 2001 (in German).
[42] A. Stenger, W. Kellermann: Adaptation of a memoryless preprocessor for nonlinear acoustic echo cancelling, *Signal Processing,* **80**(9), 1741–1760, Sept. 2000.
[43] A. Stenger, R. Rabenstein: Adaptive Volterra filters for acoustic echo cancellation, *Proc. NSIP '99,* **2**, 679–683, Antalya, Turkey, June 1999.

8

Intelligent Control Strategies for Hands-Free Telephones

Christina Breining[1] and Andreas Mader[2]

[1] Siemens AG, Ulm, Germany
[2] Smiths-Heimann, Wiesbaden, Germany

8.1 Introduction

In a system that is intended to improve the acoustic characteristics of a hands-free telephone, there are two core issues from a scientific point of view: the algorithms for noise reduction and those for echo control or echo cancellation. A large number of both kinds of algorithms have been discussed together in [17]. However, using these systems in industrial applications very often requires a number of these algorithms to be applied in parallel. An acoustic echo canceller cannot guarantee for the echo attenuation required in the ITU recommendations [24] when the filter has not yet been adapted or has diverged. It might therefore be implemented together with an automatic loss control, which can add the missing attenuation in those situations. In a car, noise reduction will also most certainly be part of the system, maybe an indoor communication unit is also added.

In themselves, these methods have critical states, such as double-talk in echo cancellation, or local speech for noise estimation. A large number of scientific publications deal with the problem of detecting these critical states, usually in order to improve one specific algorithm. However, if all the components of a system are taken into account when it is designed, this knowledge can be used to optimize the whole system with respect to both performance and computational efficiency. In the case of a noise reduction algorithm, very detailed information on the background noise characteristics is inherently available and could be applied to control the step size in the echo canceller. Similarly, the speech detection methods required for the echo canceller and loss control could be re-used in the noise estimator. The aim of this chapter is to investigate possible methods to exploit these synergies.

8.1.1 State Representation of a Hands-Free Telephone

In general, a number of characteristics are of interest for the classification of each speech sample:

- Is there background noise?
- Is the local speaker talking?
- Is the far-end speaker talking?
- How much residual echo must be expected, i.e. has the echo canceller reached sufficient convergence?

All these four characteristics are independent of each other. They can occur in all possible combinations, e.g. there is background noise (car), the far-end speaker is active, but the local speaker is mute, and the adaptation quality is still bad (convergence not sufficient). The adaptive algorithm of an echo canceller would be working at low speed to take care of the noise and still achieve convergence. A noise reduction unit would deduce that there is some noise to be compensated, but that it will be mixed with contributions from the far-end speech, so that the noise estimation update has to be done very carefully. A loss control unit would use the information about the speech sources to open the far-end speaker's channel and insert the missing attenuation for the local noise and echo. A speech recognition unit would be stopped in order to reduce the probability of misinterpreting the far-end speech for commands, and so on for additional speech processing algorithms that might be included.

The possible combination of states for a hands-free telephone is shown in Fig. 8.1. In order to visualize the relationships between the states, we reduced the four-dimensional representation to a three-dimensional cube. All the states with inactive far-end speaker are displayed as an inner cube. This is the typical reduction used in echo cancellation. The resulting step size for each of the states is also qualitatively displayed in the form of the shading of the states.

We can identify $2^4 = 16$ possible states for the hands-free telephone. These states, or sub-sets of them, have to be distinguished in order to control the overall system in a satisfying manner. It seems to be straightforward to use one centralized algorithm to derive the states and afterwards evaluate the state information separately for each speech quality enhancement algorithm. Most algorithms are more sensitive to some wrong state decisions than to others, and these weak spots are different for each algorithm. An echo canceller must immediately slow down the adaptation when a local speaker starts talking (double-talk) because the high-power highly correlated interference of the local speaker can spoil the attained adaptation quality in very short time. However, if in a single-talk situation (no local speech) the state is not detected as fast, the adaptation quality usually suffers less because the achieved attenuation is maintained, only its increase is slowed down. However, in the case of an abrupt change of the room impulse response (system change) in a rather short time, a misclassification of the residual echo as local speech (erroneous double-talk detection) will stop the adaptation. This can lead to freezing of the adaptation or very slow recovery, and can keep the achieved attenuation too small for a long time. For a noise estimation algorithm, priorities will be different: usually the noise estimation shall only be adapted when there is no local speech signal

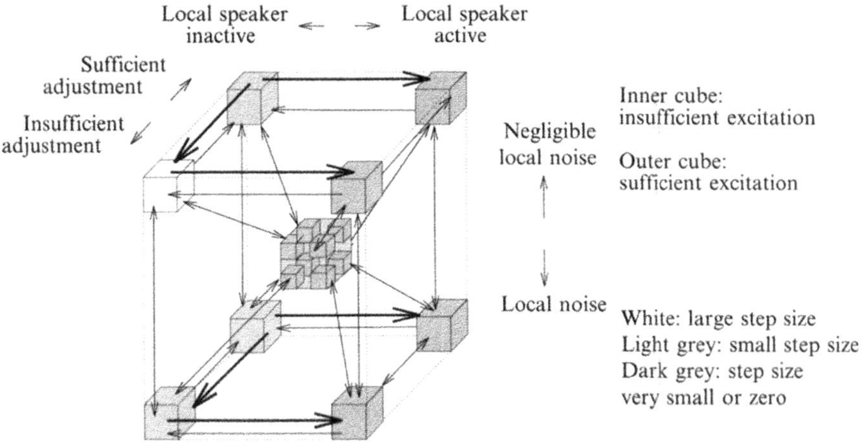

Fig. 8.1. The states of a hands-free telecommunications unit displayed in the state space. States with "no far-end speech" are located on the inner cube. The arrows are only inserted between the "far-end speaker active" subset of states. This reduction is useful for an application in echo cancellation. The bold arrows indicate transitions that require fast actions of the control circuits.

and little residual echo. In case of a system change, the noise estimation should be stopped. If this happens a little too late, the noise reduction will also reduce parts of the speech signal, and the speech quality decrease. At least, the noise reduction unit will not run into a deadlock situation like the echo canceller.

Therefore, a central control unit may collect possible state information from various detection methods, but the distinction between the states will have to be tailored for each algorithm separately, e.g. by using different thresholds for the different algorithms. This means that several input data streams serve to produce several outputs, and is called a multiple-input multiple-output-(MIMO-)System.

A number of methods for state detections in the context of acoustic echo cancellation and noise reduction have been presented in [17]. They are mostly designed to detect only one transition, e.g. "noise" ↔ "no noise" or "single-talk" ↔ "double-talk". Since the components of local and far-end signals (local /far-end speech, local/ far-end noise, echo) are not known, the detectors assume underlying statistic processes and try to estimate their characteristics in order to separate the components. These methods are obviously based on statistical estimations and therefore prone to errors. Just as the original speech quality enhancement algorithms, most of the estimation methods tend to fail in certain states: a double-talk detection might be fooled by unstationary background noise, or by a change of the room impulse response that increases the residual microphone signal. Both will lead to the unwanted stop of the adaptation ("freezing") that had been mentioned before. For this reason, a number of methods must be combined so that they can cover each other's weak spots. Even knowing the state of the last sample can be used to improve the parameters of the state estimation methods. Knowledge of the state will make the estimation algorithm itself more reliable, e.g. if from the noise estimation it can be deduced that the acoustic system is situated in a running car, the double-talk detection can adapt its decision threshold to that background noise.

8.1.2 Combination of Control Algorithms

The detection and estimation algorithms which are used for those control tasks are often rather complicated, like the cepstral distance [17] or nonlinear correlation methods. In most detectors, continuous estimation results are compared to a threshold to distinguish between the states. For example, a noise estimator calculates the background noise level which can have any value, but after comparison with a threshold, this will be assigned to the states "noise" $\leftrightarrow$ "no noise". Normally, the algorithms and their parameters and thresholds are manually optimized on a certain number of situations, either by simulation or by analyzing a real-time implementation in a realistic (but still exemplary) environment. This is possible for one algorithm, but it becomes tiresome when several methods with a number of parameters each are combined. Moreover, it can become almost impossible to achieve a sufficient solution in reasonable time.

Most of the control methods have not been as deeply investigated in theory as the main speech quality enhancement algorithms. One reason is that while the performance of echo cancellation and noise reduction algorithms can be calculated on the basis of ideal signal properties for benchmarking, the control algorithms often exploit the special characteristics of the speech and noise signals, especially their spectral density and their non-stationarity. Moreover, a large number of these methods employ nonlinearities and thresholds. This makes theoretical analyzes extremely complicated. Therefore, the optimization of the control algorithms has mostly been heuristic. This is obviously not very satisfying, especially taking into account the amount of research that went into the speech quality enhancement algorithms, which may lose a great part of their performance due to badly tuned control units. This imbalance becomes more significant as the original speech enhancement algorithms are improved: the faster the adaptation algorithms get, the more crucial their control becomes, and the faster its decision must be [14].

In today's highly complex systems, a large part of the effort must be spent on the control methods and their adaptation to the requirements defined for that system. Dependent on the circumstances, a suitable set of control and speech enhancement algorithms must be selected. This choice must be guided by performance requirements on one hand, and by limitations on processing time, computational cost, accuracy of the digital representation, or memory space, on the other hand. Due to the number of possible requirements, this very crucial and complex choice will be left to the system designer. In this chapter, we will focus on concepts about how such a given set of control methods can be combined and evaluated. The following section discusses the concept of fuzzy systems for that purpose.

8.2 Fuzzy Systems

Usually, knowledge about the behavior of control methods is available from experience in qualitative linguistic rules, like "The better the adaptation quality, the lower the step size should be".

When this kind of expert knowledge is supposed to enter a technical system, these rules have to be formulated explicitly as mathematical functions. Normally, the experts for the application try to find a mathematical representation that approximates their experience and is at the same time easy to use and implement. For complex systems, though, these relations may not be very well known or too complicated to fulfill these requirements, which will result in a loss of performance.

In order to come to a representative model without having to find the mathematical description by hand, the theory of fuzzy logic and fuzzy systems is a helpful tool.

Fuzzy logic has been invented especially for the purpose of representing human reasoning [47,48]. In classic knowledge bases, the condition in the rules can either be true or false. But fuzzy logic allows for in-between states, so that a rule is true and false to a certain degree. This makes it suitable for the "the higher-the lower" kind of rule, i.e. for continuous inputs and outputs of a fuzzy system in contrast of binary inputs and outputs in a classic knowledge based system. The system designer defines what inputs and outputs he wants to be involved. He defines how many and which distinctions he wants to make about the inputs and outputs, like "LOW", "MEDIUM", "HIGH", the so-called fuzzy sets. Fuzzy sets are defined by their membership functions, the parameters of which are also set by the expert. Each input value has a corresponding membership degree for each set. It can e.g. belong to "LOW" with membership 0.1 and to "MEDIUM" with 0.9, meaning that it is "a bit smaller than medium". Determining these membership degrees is part of the "fuzzification". The designer also specifies a number of rules ("rule base") that characterize the relations between inputs and outputs. A certain situation, represented by an input vector, will fulfill each of the given rules to some degree. The calculation of these degrees is called "inference", and includes algorithms that represent the meaning of "AND", "OR" and "NOT" in order to combine the influence of the inputs. The degree by which the rule is fulfilled determines how much the output of this rules influences the total output of the fuzzy system. This is calculated in the so-called "defuzzification".

Note that all possible input values must belong to at least one class and be covered by at least one rule. The form of the membership function can be defined by the user according to his needs, but it must be convex on the interval belonging to the class. It is common but not required to limit membership degrees in [0, 1], and it helps for the interpretation of the rule base to make the membership degrees of one input value sum up to 1. Triangles and trapezoidal functions are widely in use for the membership functions, because they make it easy to calculate the rule inference and results. However, in some applications it might be more appropriate to use Gaussian functions or other differentiable forms, which are also useful as the basis for automatic optimization methods.

The idea has been very well elaborated [27, 47, 48, 50] and is often used for automation and control applications. This chapter only shows one small part of the complete concept. It can incorporate the experience of the experts without requiring clear knowledge of the mathematical functionality that lies behind the general concepts, and can thus speed up the modelling process. Moreover, it makes algorithms robust both to noisy input data and to parameterization errors by eliminating fixed thresholds. And even a very complex mathematical model will stay transparent to the user through its description by linguistic rules.

8.2.1 Classic Versus Fuzzy Detector – Example: The Correlation Coefficient

In the following section, the concept of fuzzy logic and fuzzy systems will be explained on the example of the correlation coefficient as a double-talk detection method. This coefficient is calculated from correlation between the excitation signal $x(n)$ and the microphone signal $y(n)$ (see Fig. 8.2):

$$u(n) = \max_{l} \frac{\left[\sum\limits_{i=0}^{M} x(n-i-l)y(n-i)\right]^2}{\sum\limits_{i=0}^{M} \left[x(n-i-l)y(n-i)\right]^2}, \tag{8.1}$$

where M is the length of the estimation window.

For cost-efficient implementation, the squares can be replaced by absolute values [21]. The correlation coefficient can as well be calculated for excitation and error signal. During the initial adaptation, its performance is worse, but it becomes more reliable when good adaptation has been reached.

For a classic detector, a threshold is defined by which single-talk and double-talk are distinguished. Depending on that decision, the adaptation step size will be set to either a high or a low value. In that threshold system, high correlation of the microphone and the loudspeaker signal would be stated for values above 0.9, low correlation would then be from 0.9 downwards, in order to avoid misclassifications of double-talk as single-talk.

As opposed to the threshold approach, a fuzzy system design might make high correlation start from about 0.8, membership increasing for higher values. However, 0.8 would already represent the concept "LOW" with a high

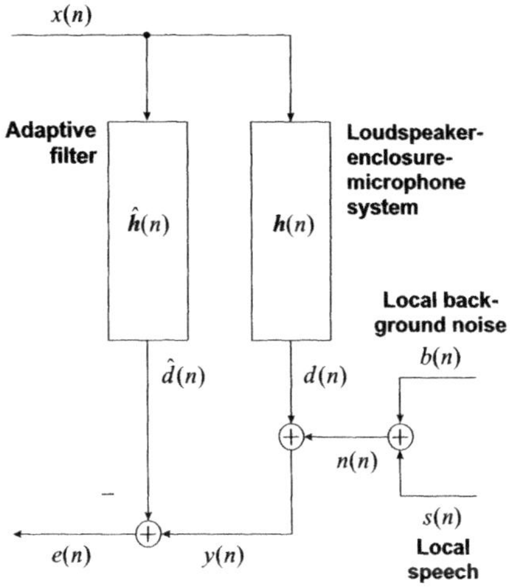

Fig. 8.2. Diagram of an echo cancelling system.

membership degree. "LOW" would then start at e.g. 0.9, with its membership values increasing as the correlation decreases. Such a fuzzification is shown in Fig. 8.3.

Our rule base consists of just one rule and its complementary rule in order to cover all possible inputs:

IF correlation $u(n)$ is LOW, THEN step size $\mu(n)$ is SMALL.

IF correlation $u(n)$ is HIGH, THEN step size $\mu(n)$ is LARGE.

In our simple case, the degree to which a rule is fulfilled is equal to the membership of the correlation value to the fuzzy sets "LOW" and "HIGH". Fig. 8.3 also visualizes two ways to apply this degree in the inference: the max-min and the max-prod interference (for details see [50]). Both limit the membership to the output sets. Defuzzification combines the resulting fuzzy sets so that the space they occupy in these graphs represents the influence they have on the output: in the center-of-gravity method, the center of gravity of the weighted fuzzy sets is calculated, either taking the union of both fuzzy sets, or averaging the centers of gravity of each set. Note that for these methods, the fuzzy sets of the output need to be bounded. The upper bound needs to be higher than the highest desired output, since that highest output is equal to the center of gravity of the fuzzy set representing the highest values, and vice versa for the lower bound. A special case is the assignment of just one value to the output instead of a fuzzy set ("singleton"). This reduces the degrees of freedom for the designer, but also facilitates parameterization and defuzzification. This simplification will be used in Sec. 8.2.2.

In this fuzzy system, the correlation value of 1 still corresponds to a high step size. In contrast to the classic rule base that discards the uncertainty information of the double-talk detection, we can even define the fuzzy sets so that it corresponds to 1. For a correlation value between 0.8 and 0.9, however, we will have a membership smaller than 1 to the first rule, and also a membership greater than 0 to the second rule. The actual value of that step size will be somewhere in [0, 1], depending on the design of the fuzzy sets.

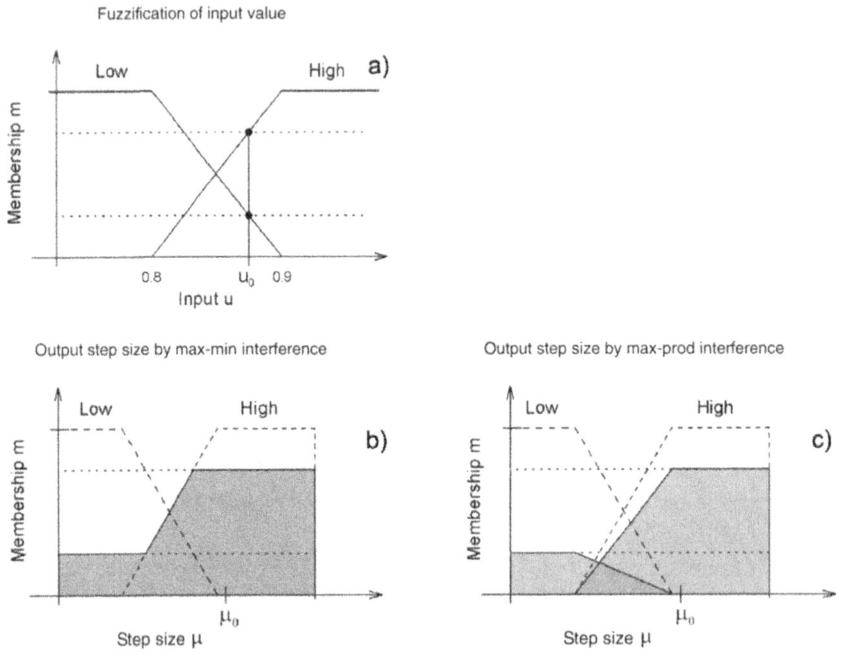

Fig. 8.3. Fuzzification of the input value u_0 (a). Inference by max-min (b) and max-prod method (c). Output μ_0 calculated with center-of-gravity method.

Obviously, the fuzzy system eliminates the threshold, so that the resulting step size mirrors the uncertainty of the state decision and makes the adaptation more robust. This effect becomes even more important when the system is more complex, dealing with several complicated rules and multiple inputs. If a hard decision, e.g. a state detection, is required as a result, it still helps that the threshold can be introduced at the end of the algorithm instead of in the beginning, so that the uncertainty information is available as long as possible.

8.2.2 Application of Fuzzy Systems in a step-size Control for an Echo Canceller

As an example from the field of acoustic echo cancellation, we choose a very cheap and simple set of just two control methods to construct a robust step-size control.

Theoretically, the optimum adaptation step size is dependent on both the interference level and the adaptation quality. The formula is quite simple under some constraints, as derived in Sec. 8.4.1. It reduces to:

$$\mu_{\text{opt}}(n) = \frac{\mathrm{E}\Big\{\varepsilon^2(n)\Big\}}{\mathrm{E}\Big\{e^2(n)\Big\}} \tag{8.2}$$

where $\varepsilon(n) = d(n) - \widehat{d}(n)$ is the adaptation error due to the mismatch of $\boldsymbol{h}(n)$ and $\widehat{\boldsymbol{h}}(n)$ (see Fig. 8.2) at sample n, and $e(n) = y(n) - \hat{d}(n)$ is the error plus interference term $n(n)$ that is actually accessible. In an environment with static background noise levels and without local speech, the step size should decrease as the adaptation quality increases.

This step size can be estimated by means of the precursor coefficient method as presented first in [46]. However, we need a couple of preconditions to be met again:

1. The mismatches on all individual filter coefficients $\widehat{h}_i(n)$ at a given time sample n have the same statistical characteristics.
2. The contribution of each filter coefficient to the variance of the residual error is about equal.
3. The vector of the excitation signal

   $$\boldsymbol{x}(n) = \Big[x(n)\,,\, x(n-1)\,,\,\ldots\,,\, x(n-N+1)\Big]^{\mathrm{T}}, \tag{8.3}$$

   where N is the number of coefficients of the echo cancelling filter, and the remaining coefficient vector

   $$\boldsymbol{h}_{\boldsymbol{\Delta}}(n) = \boldsymbol{h}(n) - \widehat{\boldsymbol{h}}(n), \tag{8.4}$$

   with

   $$\boldsymbol{h}(n) = \Big[h_0(n)\,,\, h_1(n)\,,\,\ldots\,,\, h_{n-N+1}(n)\Big]^{\mathrm{T}}, \tag{8.5}$$

   and $\widehat{\boldsymbol{h}}(n)$ respectively, of a given time sample n are uncorrelated.

If these assumptions are true, the excitation signal power and the mismatch at any part of the coefficient vector indicates the residual error power:

$$\mathrm{E}\Big\{\varepsilon^2(n)\Big\} \approx \frac{1}{N_{\text{prec}}}\,\mathrm{E}\Big\{\big\|\boldsymbol{h}_{\boldsymbol{\Delta},\text{prec}}(n)\big\|^2\,\big\|\boldsymbol{x}(n)\big\|^2\Big\} \tag{8.6}$$

(for the meaning of "prec" see below). Since the excitation signal power is easily accessible, the only missing information is the mean coefficient mismatch. For this purpose, we can use the fact that the audio signal needs some time to be transmitted over the room between the loudspeaker and the microphone. The coefficients of the LEM system corresponding to this span of time – here marked by the index "prec" – are supposed to be always equal to zero. Assuming a minimum distance between the loudspeaker and the microphone, we can initialize the corresponding filter coefficients unequal to zero. The artefacts will be masked by the bad overall adaptation quality in the initial phase. After a short time, the amplitudes of these coefficients, which are at the same time their mismatch, will mirror the general adaptation state.

With no special knowledge about the system, the minimum delay of the signal has to be assumed as very short, so that only few coefficients can contribute to the estimation of the mismatch which makes it unreliable. There are three possibilities to enlarge the number of precursor coefficients and thus the quality of the estimation:

- Delay the acoustic signal on its way through the filter. This is the straightforward solution, but it is sometimes impossible due to the delay limits set by the ITU [24].
- Store the outgoing error signal and delay it only internally in a parallel internal filter in order to add some delay and thus enlarge the number of precursor coefficients. This optimum solution avoids additional delay, but at the cost of high computational effort [10, 40].
- Use the latest ("tail") coefficients as benchmark for mismatch estimation, because the variance of the coefficients shrinks with their delay with respect to the direct path. This workaround is easy and computationally efficient, but since these coefficients are not supposed to converge to zero, an offset compensation needs to be inserted, which introduces some quality loss due to an additional estimation process. The amount of quality loss depends on how well the environment can be predicted, and is bigger if the offset compensation must fit different kinds of enclosures [7].

It is up to the system designer to chose the appropriate method according to the requirements for his system. In general, precursor/tail coefficients have proven to be one of the most reliable step-size estimators. However, two effects can be observed:

- In the beginning of the adaptation, the adaptation error estimation tends to be exaggerated. For this reason, the conventional method foresees dividing the resulting step size by 2 and limiting it in $[0, 1]$.
- In the case of an abrupt change in the room impulse response, this estimator is likely to freeze the adaptation because the adaptation error is then estimated much too low (see Fig. 8.4).

The performance of this step-size estimator could therefore be improved considerably if we could scale the estimation according to the actual state.

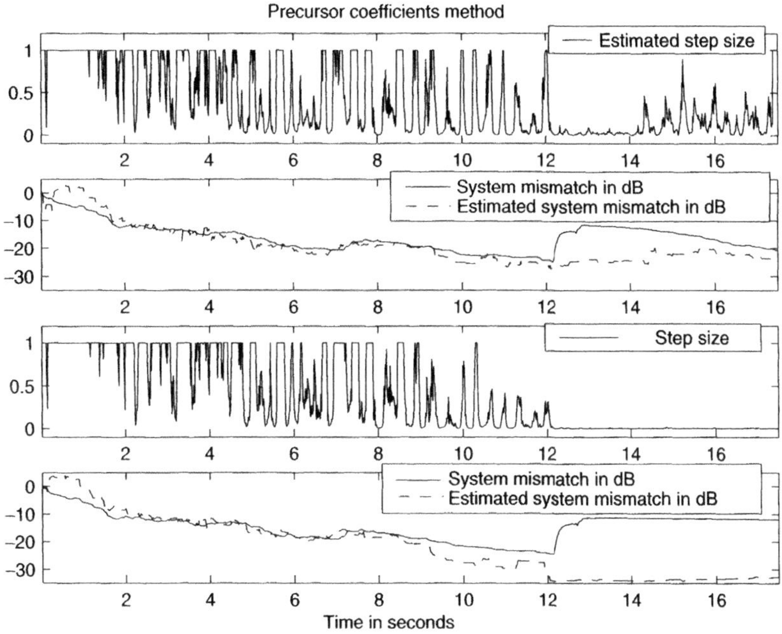

Fig. 8.4. Step size estimated by the precursor coefficients method and system mismatch $\|\boldsymbol{h}_\Delta(n)\|^2$ in dB for adaptation at a sudden change of the room impulse response. Above: System recovers. Below: Adaptation freezes.

This requires a detector for sudden changes of the room impulse response, especially when the system is well adapted. In that situation, the precursor coefficients stay at very small amplitudes, so that the adaptation freezes. One detector for system changes looks for a change in the frequency characteristics of the error signal over periodic intervals [34]. A change of the LEM system leads to more error power in the higher frequency band, but not so much in the lower band, while local speech enhances the lower band of the error signal. Therefore, we introduce the quotient

$$Q_{\mathrm{LP}}(n) = \frac{\widehat{\sigma^2_{e_{\mathrm{LP}}}}(n)}{\widehat{\sigma^2_{y_{\mathrm{LP}}}}(n)} \tag{8.7}$$

in order to distinguish between these two situations. This quotient is obtained by lowpass filtering of the error $e(n)$,

$$e_{\mathrm{LP}}(n) = e(n) \, * \, h_{\mathrm{LP}}(n), \tag{8.8}$$

and the microphone signal $y(n)$,

$$y_{\mathrm{LP}}(n) = y(n) * h_{\mathrm{LP}}(n), \tag{8.9}$$

(where "$*$" means convolution) and comparing the recursively smoothed power of the resulting error signal

$$\widehat{\sigma^2_{e_{LP}}}(n) = \lambda \widehat{\sigma^2_{e_{LP}}}(n-1) + (1-\lambda)e^2_{LP}(n) \tag{8.10}$$

to that of the lower-band microphone signal

$$\widehat{\sigma^2_{y_{LP}}}(n) = \lambda \widehat{\sigma^2_{y_{LP}}}(n-1) + (1-\lambda)y^2_{LP}(n) \tag{8.11}$$

with $0.9 < \lambda < 1$.

For the complete detector as proposed in [34], a similar quotient for the higher band would be required, and the filtering should be done by using elliptic filters for a pass band of 500 Hz. Its impulse response is denoted here as $h_{LP}(n)$. The full detector will be used later on in Sec. 8.3, but for this low-cost step-size control, we use only the lower-band part.

We can now try and put these experimental results in a fuzzy system with one rule and its complementary:

IF $Q_{LP}(n)$ SMALL AND $\mu_{opt}(n)$ SMALL, THEN $a(n)$ LARGE.
IF $Q_{LP}(n)$ LARGE OR $\mu_{opt}(n)$ LARGE, THEN $a(n)$ SMALL.

The resulting step size is then calculated as:

$$\mu_{res}(n) = a(n) \cdot \mu_{opt}(n) \tag{8.12}$$

$Q_{LP}(n)$ and $\mu_{opt}(n)$ are assigned the membership functions as depicted in Fig. 8.5, based on experience, and optimized according to experimental results with training data. The resulting step-size control is applied to a different set of speech data and room impulse response. The averaged system distance obtained in four runs is shown in Fig. 8.6.

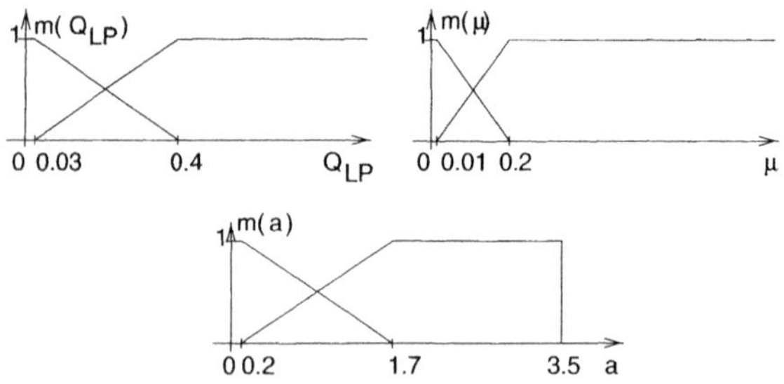

Fig. 8.5. Fuzzy sets for input and output of the fuzzy system for adaptation step-size control of the echo canceller.

This very easy system helped improve the performance of an echo canceller considerably, at only slightly increased processing power. Using a fuzzy system instead of a fixed threshold helped a lot since the additional detector $Q_{LP}(n)$ has been shown to be rather unreliable and its threshold depends on the background noise [5].

However, the fuzzy system that is obtained is as good as the expert who designed it. A mathematical optimization is not incorporated in the concept.

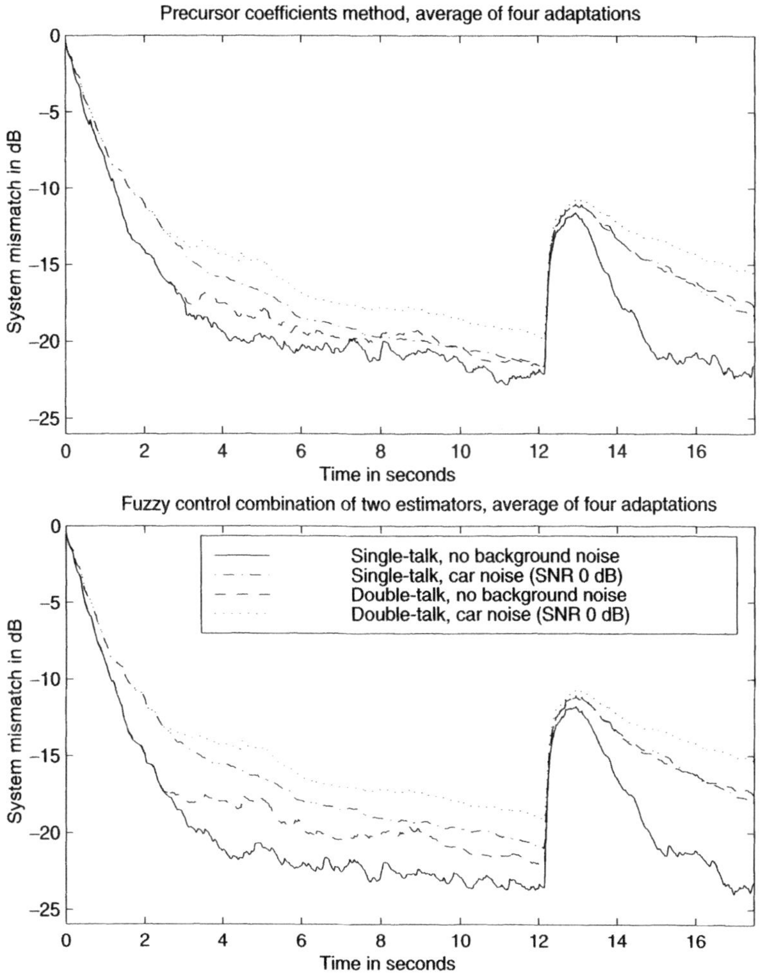

Fig. 8.6. Resulting system mismatch $||\boldsymbol{h}_{\Delta}(n)||^2$ with the precursor coefficients method (above) and the fuzzy control system (below), averaged over four adaptations, with a different room and set of speakers.

Several approaches have been presented to put optimization on top of that method [18]. In the following sections, we will discuss two of them that seem appropriate for our original control problem: One is learning vector quantization which can deal with a large number of input values and which can be enhanced by using fuzzy rules for the learning algorithm. The second is a neural network which can be initialized based on expert knowledge as described e.g. by a fuzzy rule base.

8.3 Learning Vector Quantization

We already stated in the introduction to this chapter that the control of several speech quality enhancement algorithms in one system can be regarded as a classification problem. Classification problems are very common e.g. in speech or speaker recognition or in image processing. The difference to normal detection problems is that there are usually multiple inputs that provide information. In consequence, there are more resulting classes to be selected from.

One widely used classification method is vector quantization [15]: the input values are interpreted as an m-dimensional vector that describes a point in the m-dimensional space spanned by the characteristics of the input sample. All classes are described by a "typical" reference vector, or prototype. For the very popular type of so-called nearest neighbor or Voronoi quantization, an input vector is detected as belonging to the class to whose reference vector it has the smallest distance. This reference vector is called the winner prototype. This method divides the m-dimensional space into sub-spaces around the reference vectors, also called Voronoi regions.

In the case of two input characteristics, the input vectors span a plain, and the borders of the classes are defined by half the distance to the nearest reference vector of another class, as can be seen in the left part of Fig. 8.7. However, this is only correct if the representation of the characteristics are chosen so that the border between the classes is located in the middle between the two prototypes. If the samples belonging to one class are very well concentrated around their center, while those of the other class are not, the border should be driven closer to the concentrated class in order to produce the least cost for erroneous decisions. This is illustrated in the right part of Fig. 8.7.

The correct placing of the reference vectors is a crucial point for the classification quality. The classic way to find these vectors is to analyze reference data, where the class of each sample is known. This usually requires manual labelling of the reference samples. The vectors belonging to the same class are averaged in order to obtain the reference vector of that class. For this method, a representative selection of these reference data is very important. Averaging can be done at once over the complete set of reference data, or in the form of a recursive first-order filtering, or exponential window method [45].

If such a set of representative reference data is not available, it is also possible to start with an educated guess for the reference vectors. These are adapted to the input vectors $\boldsymbol{u}(n)$ of the application during operation of the system: When the vector $\boldsymbol{u}(n)$ has been assigned to a class w, the reference vector $\boldsymbol{p}_w(n)$ of that class is drawn a little bit – according to the value of μ_{LVQ} – into the direction of that vector in order to incorporate it into its averaging procedure – very similar to an exponential window in time:

$$\boldsymbol{p}_w(n+1) = \boldsymbol{p}_w(n) + \mu_{\mathrm{LVQ}} \cdot \Big(\boldsymbol{u}(n) - \boldsymbol{p}_w(n)\Big). \tag{8.13}$$

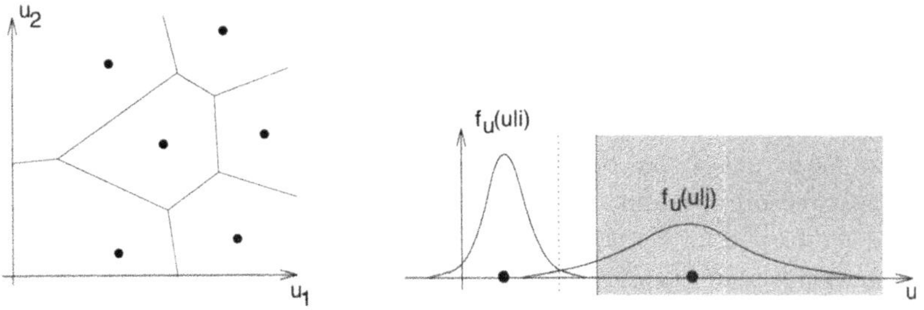

Fig. 8.7. Left: Input space spanned by u_1 and u_2 and its separation in so-called Voronoi regions (classes). Right: These regions can be suboptimal if the vectors of the classes have different distributions around their centers. The figure shows the conditional probability density functions for two classes i and j. Broken line: optimal border between the classes according to Bayes if all kinds of errors are weighted equally and all classes have equal probability of occurrence. The Voronoi region belonging to the right class is marked in grey.

The general concept of these algorithms is called learning vector quantization (LVQ), if no information is at hand before the adaptation. The algorithms have been described in several publications, e.g. [15, 29, 45, 49].

If the input data are uncorrelated and represent all possible input vectors in realistic proportions, the resulting set of reference vectors is automatically adapted to the situation under consideration. This condition is very important, since otherwise all vectors are drawn to where most of the data are located, even if these data do not properly represent the classes. This behavior is similar to the signal adaptation problem in adaptive filtering, where for real system adaptation a persistent excitation signal is required [20].

If there exists no idea whatsoever about the location of the classes in the input space, it is possible to start from a sufficiently large set of arbitrarily distributed vectors as initial reference vectors. This technique is known as self-organizing map and belongs to the field of artificial intelligence and neural networks. It was presented by Kohonen in [29] as a representation of the human brain. The reference vectors will move to where most of the data are clustered. In this approach, a class can possess more than one reference vector. The concept also includes negative feedback learning, i.e. all non-winning prototypes are pushed away slightly from the incoming vector. This prevents the reference vectors from clustering as in LVQ.

The less we know about the system at the start, the more the adaptation of the reference vectors becomes important to the quality of the classification. In addition to the greater need for adaptation, errors with badly adapted reference vectors, i.e. misclassifications, influence future classifications. As classification in itself is highly nonlinear at the borders between the classes, small deviations can lead to important errors.

In order to limit these effects, a fuzzy approach could be used which softens the borders and thus the consequences of a misclassification. In that case, membership to a class decreases with the distance to that class: A vector found at the same or very similar coordinates as a reference vector very probably represents a sample belonging to the same class. Therefore, its membership is high, whereas at the borders of that class or beyond, the membership is low. In consequence, the output for an input vector on the border of a class will also take into account the result for the neighboring classes, and thereby reduce the impact of the erroneous classification.

If a membership grade is defined, then it is straightforward to adapt the step size for the learning algorithm to that membership grade, as it describes the certainty of the classification. The membership function shall be convex, i.e. it shall decrease with increasing distance between the input vector and the nearest reference vector, or winner prototype:

$$\boldsymbol{p}_w(n+1) = \boldsymbol{p}_w(n) + h\Big(\boldsymbol{u}(n), \boldsymbol{p}_w(n)\Big) \cdot \Big(\boldsymbol{u}(n) - \boldsymbol{p}_w(n)\Big) \tag{8.14}$$

where $\boldsymbol{u}(n)$ is the input vector, and $\boldsymbol{p}_w(n)$ the winner prototype at time n. $h(\boldsymbol{i}(n), \boldsymbol{p}_w(n))$ can be chosen arbitrarily, but should decline monotonically with increasing distance $\|\boldsymbol{u}(n) - \boldsymbol{p}_w(n)\|$, and for easier interpretation be bounded by 1 as its maximum value [26].

In our hands-free telephone application, where the classes represent the states as introduced in Sec. 8.1, states like single-talk stay the same for a large number of samples during normal operation. In order to prevent all the prototypes from being drawn into the single-talk cluster, which is bound to happen with the presented algorithm, we modify the algorithms similarly to Kohonen's proposal [28] to move the prototypes of the other "non-winner" classes away from the input vector:

$$\boldsymbol{p}_w(n+1) = \boldsymbol{p}_w(n) + \mu_{\mathrm{LVQ,w}}\, \boldsymbol{u}(n), \tag{8.15}$$

$$\boldsymbol{p}_i(n+1) = \boldsymbol{p}_i(n) - \mu_{\mathrm{LVQ,l}} \frac{\Big\|\boldsymbol{p}_w(n) - \boldsymbol{u}(n)\Big\|^2}{\Big\|\boldsymbol{p}_i(n) - \boldsymbol{u}(n)\Big\|^2}\, \boldsymbol{u}(n) \qquad \forall \;\; i \neq w\,, \tag{8.16}$$

$$\text{with} \quad \mu_{\mathrm{LVQ,l}} \ll \mu_{\mathrm{LVQ,w}}. \tag{8.17}$$

The corresponding step size is chosen much smaller than that for the winner prototypes, as not winning is much more often for each prototype than winning. The resulting behavior is illustrated in Fig. 8.8.

8.3.1 Example: Fuzzy LVQ for State Detection in a Hands-free Telephone

We applied the presented method of modified fuzzy LVQ to the problem of state detection for a hands-free telephone set, especially for the purpose of echo cancellation. In that case, the state diagram of Fig. 8.1 applies, for which we want to distinguish between eight states of activation and one of initial adaptation required to start the adaptation with sufficient step size.

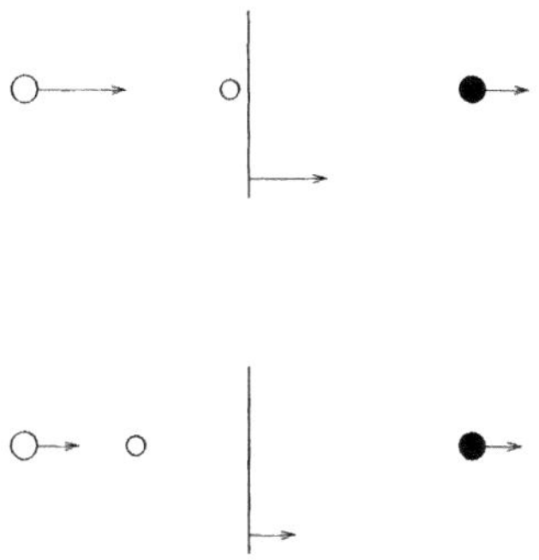

Fig. 8.8. Step size of the prototypes for different distances to the input vector with modified LVQ. The input vector (small white circle) belongs to the class of the left prototype (reference vector marked with large white circle). The vertical line marks the border, arrows visualize the objects' direction and step size of movement.

States as shown in the following figures are denoted in discrete numbers as shown in table 8.1. The state 0 without far-end speech is listed for completeness, but was cut out of our simulations in order to save time. Assuming that the noise is sufficiently stationary, this is relatively easy to do even in a real-time application.

Table 8.1. Notation for the states

State no.	Description of the state		
	Local speaker	Local noise	Filter adjustement
0	*no far-end speech*		
1	inactive (single-talk)	low	sufficient
2	active (double-talk)	low	sufficient
3	inactive (single-talk)	high	sufficient
4	active (double-talk)	high	sufficient
5	inactive (single-talk)	low	insufficient
6	active (double-talk)	low	insufficient
7	inactive (single-talk)	high	insufficient
8	active (double-talk)	high	insufficient
9	*beginning of the adaptation (initial condition)*		

The training of the states was carried out in several states. In a first step, the states were defined based on an ideal adaptation with the optimum step size. The samples were labelled so that supervised training could be performed.

In a second step, the states were assigned corresponding step size, which were then optimized according to the results shown by the cost function. It turned out that the step size had to be changed very much because neighboring states interfered considerably. In a third step, the resulting step size was used in an unsupervised training session where the step size from the LVQ controlled the adaptation. This third step was introduced in order to fine-tune the states, and had to be carried out very carefully with low step sizes. The performance of the state recognition is shown in Fig. 8.9. Single-talk can reliably be distinguished from distorted situations. These, however, are not always clearly separated, and the source of the interference is not reliably detected. This is partly due to car noise being concentrated in low frequencies, similar to speech. The double-talk detectors used for state detection have not designed with focus on that distinction. Still, the adaptation is performed reasonably well, and freezing was avoided in our experiments. The resulting adaptation quality is displayed in Fig. 8.10.

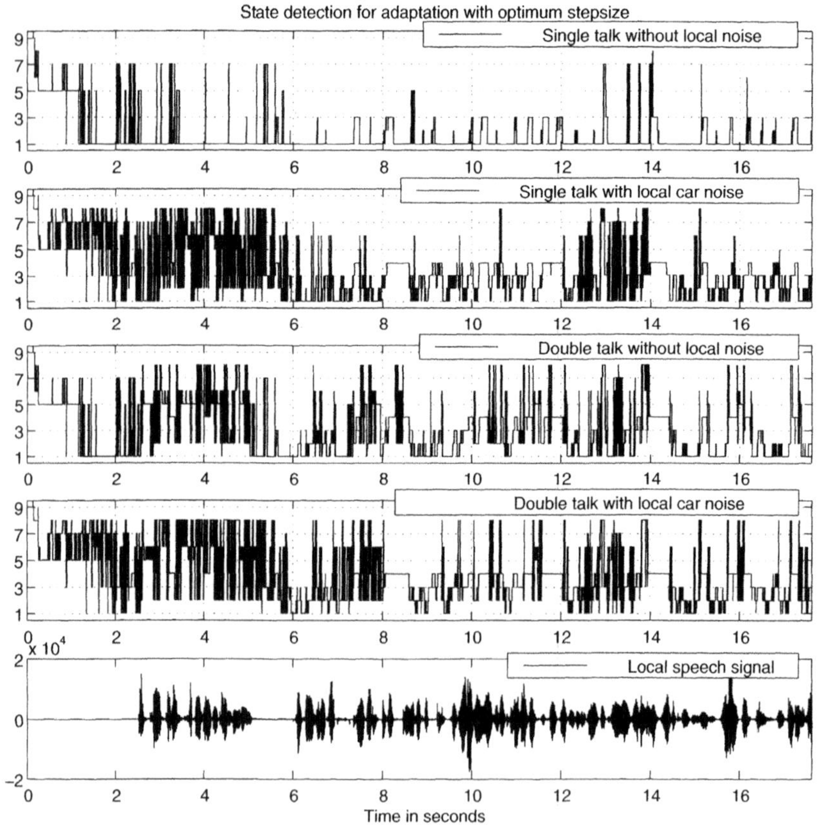

Fig. 8.9. State detection during adaptation with the optimum step size. Numbering according to table 8.1

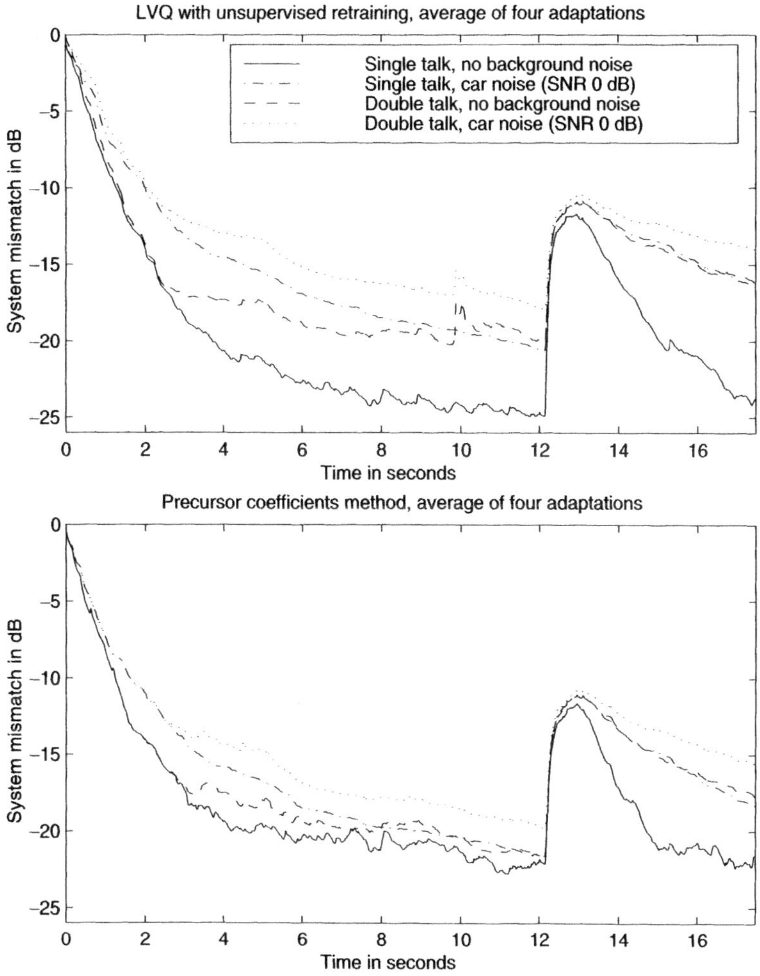

Fig. 8.10. System mismatch $\|\boldsymbol{h}_{\Delta}(n)\|^2$ for the precursor coefficients method (below) and the modified LVQ system, after supervised training followed by unsupervised training in adaptations using its own step size, averaged over four runs.

If the membership functions are continuous and differentiable mathematical functions, they can automatically be optimized by means of adaptive algorithms. However, optimization needs input about the residual error and its significance for the system. The following section will deal with deriving this information with focus on echo cancellation control.

8.4 Prerequisites for Automatic Optimization of Control Algorithms: Optimum Step Size and Cost Function

Automatic optimization is normally performed by some kind of neural network. Generally speaking, the most commonly used neural networks can be regarded as adaptation algorithms capable of approximating nonlinear relations between the input and the output signals. Similarly to the linear adaptive algorithms, like e.g. LMS or RLS, the training procedure is derived from a cost function based on the difference between the result of the network and the optimal output.

In case of acoustic echo cancellation, which will serve as our example here, the output of the network is supposed to be the step size of the adaptive filter. We will train this network to calculate the best possible step size for each time sample based on preprocessed speech, distortion, and error signals, i.e. results from estimators and double-talk detectors, as in Sec. 8.3.1. In order to train the network, we need to define the goal, in this case the reference or optimum step size for that sample. The difference between the network-generated step size and the optimum will be fed back into the network in order to control further adaptation steps, just like in linear adaptation algorithms.

8.4.1 Optimum Step size for Network Training

We can derive the expected value of an optimal step size from maximizing the improvement in expected convergence during the new adaptation step, based on what has been achieved in the step before:

$$\begin{aligned} \mathrm{E}\Big\{\Delta(n)\Big\} &= \mathrm{E}\Big\{\big\|\boldsymbol{h}-\widehat{\boldsymbol{h}}(n)\big\|^2\Big\} - \mathrm{E}\Big\{\big\|\boldsymbol{h}-\widehat{\boldsymbol{h}}(n+1)\big\|^2\Big\} \\ &= \mathrm{E}\Big\{\big\|\boldsymbol{h}_\Delta(n)\big\|^2\Big\} - \mathrm{E}\Big\{\big\|\boldsymbol{h}_\Delta(n+1)\big\|^2\Big\} \rightarrow \max . \end{aligned} \tag{8.18}$$

Replacing the newest coefficient vector by the adaptation equation of the NLMS algorithm for step $n+1$,

$$\boldsymbol{h}_\Delta(n+1) = \boldsymbol{h}_\Delta(n) + 2\mu(n)\frac{e(n)\boldsymbol{x}(n)}{\big\|\boldsymbol{x}(n)\big\|^2}\,, \tag{8.19}$$

and using the abbreviation $\varepsilon(n) = \boldsymbol{h}_\Delta(n)^T\boldsymbol{x}(n)$ it follows:

$$\begin{aligned}\left.\frac{\partial \mathrm{E}\{\Delta(n)\}}{\partial \mu(n)}\right|_{\mu(n)=\mu_{\mathrm{opt},1}(n)} &= \frac{\partial}{\partial \mu(n)}\,\mathrm{E}\Bigg\{-2\mu(n)\frac{e(n)\varepsilon(n)}{\|\boldsymbol{x}(n)\|^2} \\ &\qquad \left. +\mu^2(n)\,\frac{e^2(n)}{\|\boldsymbol{x}(n)\|^2}\Bigg\}\right|_{\mu(n)=\mu_{\mathrm{opt},1}(n)} \\ &= \left.\mathrm{E}\left\{-2\frac{e(n)\varepsilon(n)}{\|\boldsymbol{x}(n)\|^2}+2\,\mu(n)\frac{e^2(n)}{\|\boldsymbol{x}(n)\|^2}\right\}\right|_{\mu=\mu_{\mathrm{opt},1}(n)} \\ &= 0.\end{aligned} \tag{8.20}$$

For this calculation, we can solve for the optimal step size as:

$$\mu_{\mathrm{opt},1}(n) = \frac{\mathrm{E}\left\{\dfrac{e(n)\varepsilon(n)}{\|\boldsymbol{x}(n)\|^2}\right\}}{\mathrm{E}\left\{\dfrac{e^2(n)}{\|\boldsymbol{x}(n)\|^2}\right\}}. \tag{8.21}$$

If the filter length can be assumed as sufficiently long, and if the input signal is an ergodic process, we can suppose that the norm of the input vector is nearly a constant, and exclude it from the calculation of the expected value:

$$\mu_{\mathrm{opt},2}(n) = \frac{e(n)\varepsilon(n)}{e^2(n)} \tag{8.22}$$

Further assuming that the noise and the adaptation error, emitted by different sources, are uncorrelated, we obtain the simplified optimum step size:

$$\mu_{\mathrm{opt},3}(n) = \frac{\varepsilon^2(n)}{e^2(n)} \tag{8.23}$$

Although the conditions required to derive this step size are not very realistic, this step size has the big advantage of being easy to interpret:

- The higher the noise and distortion level, the lower the step size.
- The smaller the remaining undistorted error $\varepsilon(n)$ in relation to the noise and distortion term, the lower the step size.

This knowledge is the basis for most common step size control methods, as presented in Sec. 8.2.

In real applications, however, the expected values of the excitation, the undistorted error, and the distortion signal are not known: first, the undistorted error is not accessible, since due to the nature of our problem we do not know the real LEM system. It is only available in offline simulations. Second, even if the signals are known, we do not know their expected values at a given time n, since we do not know the process that lies beneath the signals. Maybe it would be possible to describe some kinds of machine noise, like PC coolers, but already the motor noise of a car poses big problems for deriving the statistics of the noise at a given time, due to the varying speed which is usually unknown to the echo canceller. But the most difficult part are all the involved speech signals: Their signals change a lot over time. The parts of the signal that might be regarded as pseudo-stationary are usually very short, and estimation on the statistical properties of these short signal parts is then quite unreliable. Moreover, there are a lot of transient states, i.e. periods of time in between two pseudo-stationary periods, like plosives between two vowels, for which an estimation is almost impossible. In Fig. 8.11, it is shown that the estimation method and its artefacts in nonstationary environment can spoil the convergence, thus proving the impossibility of estimating the optimal step size. For this demonstration, we used the step size $\mu_{\text{opt},1}(n)$ which is theoretically the best step size in a statistical sense without any restrictions about signal properties.

In order to observe the impact of using averages instead of expected values in a non-stationary environment, the expected values are estimated by first-order recursive filtering of their arguments:

$$\hat{\mathrm{E}}\left\{\frac{e(n)\varepsilon(n)}{\|\boldsymbol{x}(n)\|^2}\right\} = \lambda\hat{\mathrm{E}}\left\{\frac{e(n-1)\varepsilon(n-1)}{\|\boldsymbol{x}(n-1)\|^2}\right\} + (1-\lambda)\frac{e(n-1)\varepsilon(n-1)}{\|\boldsymbol{x}(n-1)\|^2} \tag{8.24}$$

and

$$\hat{\mathrm{E}}\left\{\frac{e^2(n)}{\|\boldsymbol{x}(n)\|^2}\right\} = \lambda\hat{\mathrm{E}}\left\{\frac{e^2(n-1)}{\|\boldsymbol{x}(n-1)\|^2}\right\} + (1-\lambda)\frac{e^2(n)}{\|\boldsymbol{x}(n)\|^2}\,. \tag{8.25}$$

Since this kind of filtering is computationally very efficient and yields similar results as a linear average over a certain time interval in the case of stationary processes, it is very common and is also used in Eq. 8.11.

In the case of the NLMS algorithm for the echo canceller, we have to guess a step size for the coming time sample $n+1$ from the past samples of the signals. Therefore, we assume that the underlying processes are stationary, and we derive a step size based on the characteristics of these processes. Therefore, the result is based on expected values of the signals.

However, knowing that these expected values are difficult to obtain in the real application, but that we can easily access all signal values all the time in an offline simulation, we can as well concentrate on the real signals instead of on their statistical properties. We can derive an optimal step size for the NLMS algorithm as

$$\mu_{\text{opt},4}(n) = \frac{\varepsilon(n)}{e(n)}. \tag{8.26}$$

and use this value at time $n+1$.

This optimal step size for time $n+1$ is now only dependent on the samples of undistorted and distorted error at time n. This facilitates the definition of it for a training algorithm, as estimation of the statistical properties of the

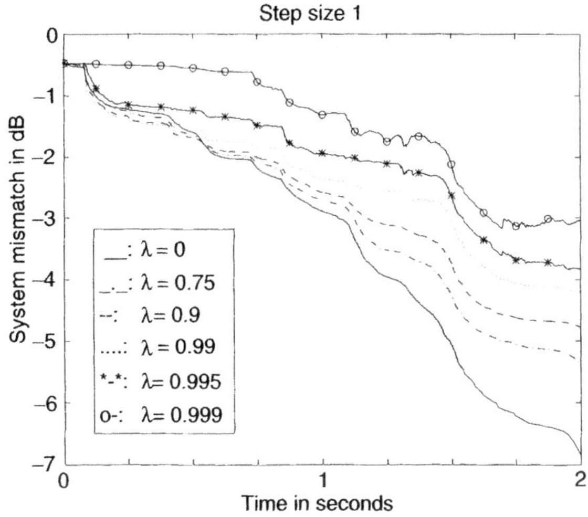

Fig. 8.11. Comparison of the convergence in double-talk with speech signals (averaged over 10 adaptations). The step size $\mu_{\mathrm{opt},1}(n)$ used for the adaptation was calculated using recursive first-order filters as in 8.11 with forgetting factor λ for the estimation of the expected values.

signals is not needed any longer. Moreover, we already show in Fig. 8.11 that the convergence of an adaptation algorithm without filtering, i.e. with $\lambda = 0$, is much faster than with the step size composed by the expected values – $\mu_{\mathrm{opt},4}(n)$ is the special case of $\mu_{\mathrm{opt},1}(n)$.

8.4.2 Cost Function

It seems obvious to feed the difference between output and reference back to the network in order to improve the results. This is proposed in the back-propagation algorithm of [38] for the so-called multi-layer perceptrons. In the linear world, it is very similar to the LMS algorithm.

However, we know that for speech processing, the NLMS algorithm has proven much more effective. The reason is that for unstationary signals, the LMS moves in bigger steps for signal periods with higher power levels. Unfortunately, higher power levels in speech signals normally indicate vowels with large power concentration on the lower frequencies. The higher frequencies are only present in fricatives and plosives, which have lower power levels. Therefore, the echo attenuation for higher frequencies is worse with an LMS algorithm. In contrast to this, the NLMS algorithm normalizes the step size with respect to the input signal power level. This gives more importance to the higher frequencies and speeds up the convergence of the filter.

We generated two input signals: $x_1(n)$ is white Gaussian noise, $x_2(n)$ consists of periods of low-frequency signals with high power ("vowels") and high-frequency signals with low power ("voiceless speech"), as depicted in Fig. 8.12. These signals were used for the adaptation of a filter with Gaussian distributed coefficients and approximately constant transfer function. The results are shown in Fig. 8.13 and support the explanation given above.

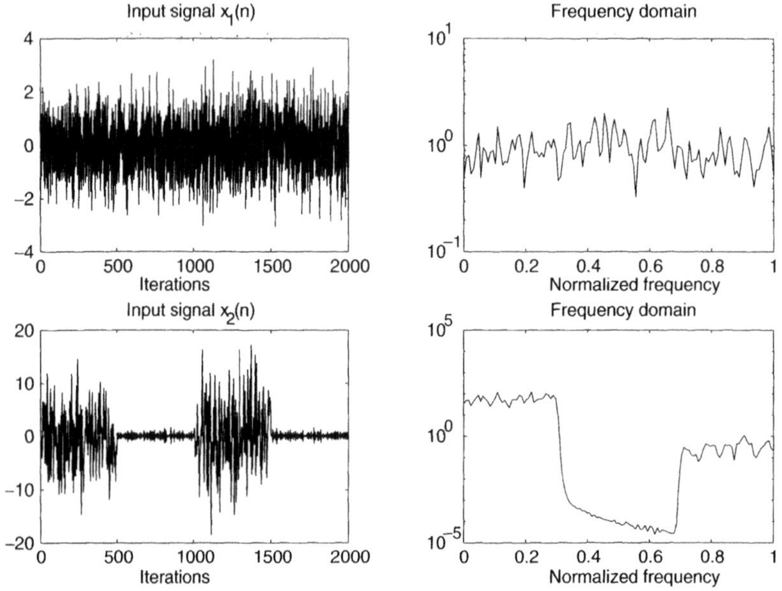

Fig. 8.12. Input signals $x_1(n)$ and $x_2(n)$ over time (left) and in the frequency domain (right).

The main differences between the LMS and the NLMS algorithms lie in the constructions of the cost functions from which they are derived. For the LMS algorithm, it is the simple function:

$$J = \mathrm{E}\Big\{e^2(n)\Big\} \rightarrow \min. \tag{8.27}$$

For the NLMS algorithm, however, the cost function is:

$$J = \mathrm{E}\left\{\frac{e^2(n)}{\left\|\boldsymbol{x}(n)\right\|^2}\right\} \rightarrow \min. \tag{8.28}$$

A cost function thus influences the way in which the convergence is reached. In a real application, the convergence will never be perfect, and convergence time will be of great importance. The cost function accounts for the characteristics of the remaining error and for the convergence speed, and should, therefore, be selected carefully.

These insights have to be taken into account for the design of the neural network adaptation algorithm. In the echo cancellation application, we do not

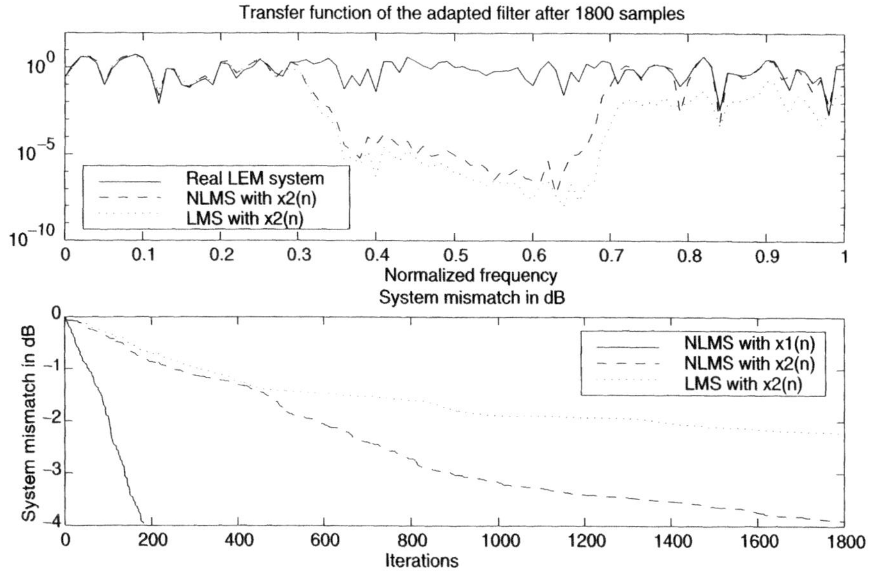

Fig. 8.13. Above: Transfer functions of the filters after 1800 samples. Below: system mismatch $\|\boldsymbol{h}_\Delta(n)\|^2$ in dB

care about the absolute or squared error of the adaptation step size, but about what it does to the convergence of the NLMS algorithm in the given situation. Therefore, we have a close look at the improvement of convergence that can be achieved based on the latest state:

$$\begin{aligned} \mathrm{E}\Big\{\Delta(n)\Big\} &= \mathrm{E}\Big\{\|\boldsymbol{h}_\Delta(n)\|^2\Big\} - \mathrm{E}\Big\{\|\boldsymbol{h}_\Delta(n+1)\|^2\Big\} \\ &= \mathrm{E}\left\{2\,\mu(n)\frac{\varepsilon(n)e(n)}{\|\boldsymbol{x}(n)\|^2} - \mu^2(n)\frac{e^2(n)}{\|\boldsymbol{x}(n)\|^2}\right\}. \end{aligned} \tag{8.29}$$

Replacing the step size with

$$\mu(n) = \mu_{\mathrm{opt}}(n) + \delta_\mu(n) \tag{8.30}$$

yields

$$\begin{aligned} E\Big\{\Delta(n)\Big\} &= E\Big\{\Delta_{\mathrm{opt}}(n)\Big\} - \\ & E\left\{2\,\delta_\mu(n)\frac{\varepsilon(n)e(n) - \mu_{\mathrm{opt}}(n)e^2(n)}{\|\boldsymbol{x}(n)\|^2} - \delta_\mu^2(n)\frac{e^2(n)}{\|\boldsymbol{x}(n)\|^2}\right\}. \end{aligned} \tag{8.31}$$

The term in the second expectation marks the deviation at sampling instant n due to a non optimal step size. Its sum over M sampling instants can serve as cost function:

$$J = \sum_{n=1}^{M} J_n$$
$$= \sum_{n=1}^{M} \delta_\mu(n) \frac{e^2(n)}{\|\boldsymbol{x}(n)\|^2} \left[\delta_\mu(n) + 2\,\mu_{\text{opt}}(n) - 2\,\frac{\varepsilon(n)}{e(n)} \right] . \tag{8.32}$$

Inserting the optimal step size $\mu_{\text{opt}}(n) = \mu_{\text{opt},4}(n)$ (Eq. 8.26) to simplify interpretation of the cost function finally leads to

$$J = \sum_{n=1}^{M} J_n = \sum_{n=1}^{M} \delta_\mu^2(n) \frac{e^2(n)}{\|\boldsymbol{x}(n)\|^2} . \tag{8.33}$$

Beside its importance for the adaptation of a neural network that models the step size, this cost function can also be used to analyze the weak points of a step size control method. This is illustrated in Fig. 8.14. Whenever the cost function is high, the step size is not well chosen for that situation.

The results of the cost function can now be used for the training of an arbitrary kind of neural network, e.g. a multi-layer perceptron [5, 6]. In this chapter, another approach will be investigated which combines the state detection, fuzzy logic, and the automatic training in the form of a radial basis function network.

8.5 Radial Basis Function Network for Step-Size Control

This section deals with the application of a radial basis function network for automatic step-size control of an acoustic echo canceller as part of a hands-free telephone set. The aim is to achieve a robust and generalized control mechanism while automatically tuning the network parameters. That is, for the step-size control no more optimization by hand is necessary. Further on, the extraction of expert knowledge, like information on the current state, should be supported. The RBF network combines the main advantages of the learning vector quantization (LVQ) and the multilayer perceptron (MLP): The state information of the LVQ approach is preserved while an automatic tuning of parameters like the MLP approach can be achieved [30].

8.5.1 Radial Basis Function Network – A Short Overview

RBF networks are three layer neural networks [19], whereas all neurons of one layer are directly connected to the neurons of the next layer (see Fig. 8.15). The only nonlinear functions are implemented in the RBF kernels in the hidden layer, typically Gaussian functions are used [37].

The input-output mapping performed by a RBF network can be written as

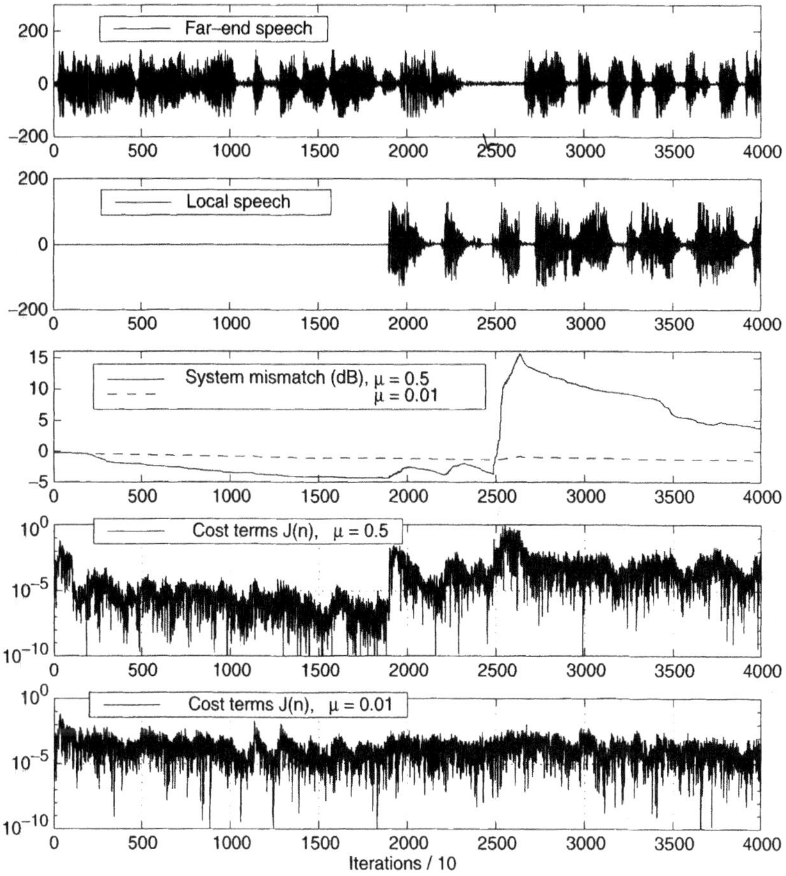

Fig. 8.14. This figure shows the cost function $J(n)$ during adaptation with a speech signal for two different values for the step size μ.

$$\widetilde{f}_v(\boldsymbol{u}) = \sum_{i \in \mathcal{U}_{\mathrm{H}}} w_{v,i}\, h_i(\boldsymbol{u},\, \boldsymbol{c}_i) \qquad v \in \mathcal{U}_{\mathrm{O}}, \tag{8.34}$$

where here we restrict $h_i(\boldsymbol{u},\, \boldsymbol{c}_i)$ to

$$h_i\Big(\big\|\boldsymbol{u} - \boldsymbol{c}_i\big\|_M\Big) = \exp\left(-\frac{1}{2}(\boldsymbol{u} - \boldsymbol{c}_i)^{\mathrm{T}}\, \boldsymbol{\Sigma}_i^{-1}\, (\boldsymbol{u} - \boldsymbol{c}_i)\right),$$

with $\mathcal{U}_{\mathrm{H}}, \mathcal{U}_{\mathrm{O}}$ the set of neurons in the hidden and the output layer. $h_i(\boldsymbol{u},\, \boldsymbol{c}_i)$ is the nonlinear transfer function of RBF kernel i and calculates a "distance" between the input vector $\boldsymbol{u}$ and center $\boldsymbol{c}_i$. $\|\boldsymbol{u} - \boldsymbol{c}_i\|_M$ is the Mahalanobis distance [22], which is used for this application. In general, other RBF kernels or distance measures can be applied. The matrix $\boldsymbol{\Sigma}_i$ influences the shape, size and orientation of RBF kernel i. Finally, the output of the hidden layer is linearly transformed to the output space by the weighting factors $w_{v,i}$ of output neuron v and RBF kernel i.

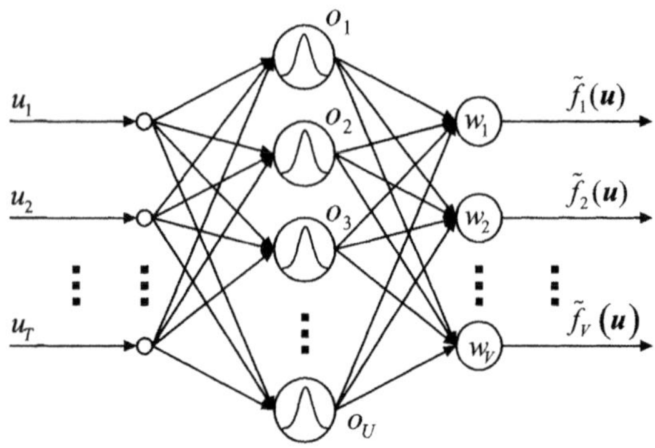

Fig. 8.15. General structure of a radial basis function (RBF) network. The network consists of three layers, whereas the input layer includes T neurons, the one and only hidden layer U neurons and the output layer V neurons. The hidden layer contains the nonlinear radial basis function.

The RBF network is known to do a local classification due to the spatially limited support of the RBF kernels. To improve the abilities of generalization and extrapolation a normalized RBF network can be applied [9]. With Eq. 8.34 the normalized RBF network can be expressed as

$$f_v(\boldsymbol{u}) = \frac{\sum\limits_{i\,\in\,\mathcal{U}_\mathrm{H}} w_{v,i}\, h_i(\boldsymbol{u},\, \boldsymbol{c}_i)}{\sum\limits_{i\,\in\,\mathcal{U}_\mathrm{H}} h_i(\boldsymbol{u},\, \boldsymbol{c}_i)} \qquad v \in \mathcal{U}_\mathrm{O}. \tag{8.35}$$

Beside the basic RBF network we have researched two kinds of dynamic RBF networks by the inclusion of memory – that are a feedback network and a network using delay-units for the input signals. For this application, slight advantages for the feedback network structure can be achieved [32]; hence we confine to discuss only this nonlinear infinite impulse response (NIIR) network structure here.

8.5.2 Applied Network Structure

As described in Sec. 8.1, the problem of step-size control can be regarded as classification problem – that is the detection of the states of the system. Here, a RBF network is applied for classification. Inputs of the network are the signals of various detection and speech enhancement algorithms; in the following these classification features are denoted as *detectors*. For this application of step-size control and state classification considered here the following detectors are used [17, 31]:

- correlation analysis between excitation $x(n)$ and microphone signal $y(n)$ [3, 21],

- estimation of filter mismatch of the acoustic echo canceller using "delay" coefficients [46],
- estimation of optimal step size using "delay" coefficients [46],
- cepstral analysis of microphone signal $y(n)$ and output signal of the adaptive filter $\hat{d}(n)$ [13],
- lowpass power ratio of error $e(n)$ and microphone signal $y(n)$ [34],
- difference between logarithmic high- and lowpass power ratio of error and microphone signal [34],
- slightly smoothed lowpass power ratio of error and microphone signal [8],
- background noise estimation based on minimum statistics [33],
- shadow filter in parallel to the existing echo cancelling filter for detecting enclosure dislocations [39].

The outputs of the detectors are normalized to a range of 0 to 1 to simplify the training of the network parameters and to allow extraction of meaningful information.

The states which are supposed to characterize the system adequately are listed in Table 8.1. Due to prior knowledge about the separation of the input data into certain classes (states of the system), the RBF network can be setup accordingly (Fig. 8.16): Each state $(1-9)$ of the system is assigned to a single

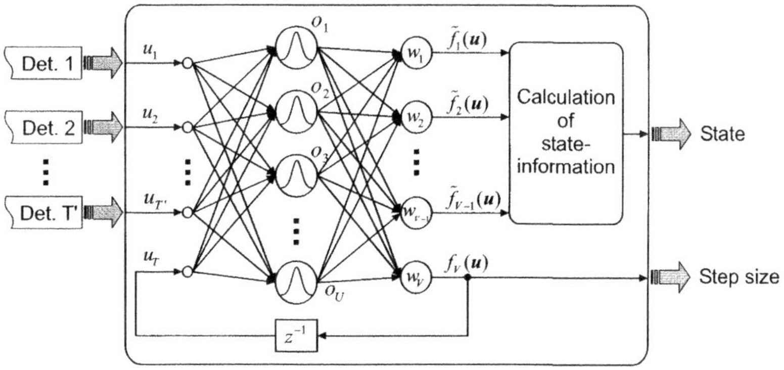

Fig. 8.16. Applied structure of the radial basis function (RBF) network. Input of the network are the different detectors. Output is a signal of the current state of the communication system and the step size for the acoustic echo cancelling filter. By adding a feedback-unit with the step size as NIIR-part, the network is extended to include memory.

RBF neuron in the hidden layer. The far-end speech activity (corresponding to state number 0) can be detected reliably using a separate speech activity detector [17]. Therefore, only nine RBF neurons representing the nine states of the system are necessary. The neurons of the input layer are assigned to the output of the detectors – here, nine neurons plus one for the feedback are necessary. The outputs of the network are used to calculate both a state information and a step size for the adaptive echo cancelling filter; hence, nine neurons are applied for state detection and one neuron delivers the step size.

For simplification, statistical independent input signals from the detectors are assumed. Thereby, the orientation of the radial-basis functions coincides with the direction of the axis of the input space and the covariance matrix $\boldsymbol{\Sigma}_i$ simplifies to a diagonal matrix. The RBF network can be rewritten as in Eq. 8.35 where

$$\begin{aligned} h_i(\boldsymbol{u}, \boldsymbol{c}_i) &= \prod_{t \in \mathcal{U}_{\mathrm{I}}} \exp\left(-\frac{1}{2} \frac{(u_t - c_{i,t})^2}{\sigma_{i,t}^2}\right) \\ &= \exp\left(-\frac{1}{2} \sum_{t \in \mathcal{U}_{\mathrm{I}}} \frac{(u_t - c_{i,t})^2}{\sigma_{i,t}^2}\right), \end{aligned} \tag{8.36}$$

with $\mathcal{U}_{\mathrm{I}}$ the set of neurons in the input layer. The remaining parameters of the RBF network – that are the centers and widths for each RBF kernel and the output weights – have to be adapted in a training process.

As already shown in Sec. 8.4.2, a cost function can be calculated that takes into account the mismatch of the adaptive echo cancelling filter and the difference between the actual step size $\mu(n)$ and its optimal value $\mu_{\mathrm{opt}}(n)$. It can be expanded to

$$J_n = \delta_\mu(n) \frac{e^2(n)}{\|\boldsymbol{x}(n)\|^2} \left[\delta_\mu(n) + 2\,\mu_{\mathrm{opt}}(n) - 2\,\frac{\varepsilon(n)}{e(n)}\right], \tag{8.37}$$

with

$$\delta_\mu(n) = \mu(n) - \mu_{\mathrm{opt}}(n) \tag{8.38}$$

the deviance of the actual and an optimal step size. In order to utilize this robust cost function, a supervised training process in applied for the RBF network.

As depicted in Fig. 8.17, the whole system has a feedback loop – the step size generated by the network influences the detectors and therefore itself. Due to a poor initialized network in the beginning, a two-stage training is required: In a first stage the adaptation of the acoustic echo canceller is driven by an external step size – namely a pseudo-optimal step size (Sec. 8.4.1, [17]). This leads to a reasonable initialization of the RBF network. In the second stage the RBF network is trained with a closed feedback loop. Hence, the network is adapted with representative input values. Due to the feedback loop, an online training can be implemented only.

The following subsections mainly deal with the application of the RBF network for step-size control – the emphasis is placed on the last output neuron. In Sec. 8.6 the task of state classification will be discussed in more detail.

8.5.3 Training of Radial Basis Function Parameters

For the applied RBF network, three parameter sets have to be adapted: The centers and widths of the Gaussian transfer functions (RBF kernels) and the weight factors of the last output neuron (number V) utilized as step size (the training process of the other output neurons utilized for state detection are discussed in Sec. 8.6). The adaptation process of these network parameters discussed in the following also contains network step sizes. These step sizes have – quite similar to the acoustic echo canceller step size – a large impact on the whole adaptation quality. Thus, we give a short outlook on the control of the network step sizes at the end of this chapter.

In Sec. 8.5.2, the initialization of the RBF network with one RBF neuron per class (that represents one state of the system) was shown. However, in Sec. 8.5.4 a growing network approach will be presented, leading to more than one RBF neuron per class. For this reason the training processes described in this section are in general form taking several neurons per class into account, in which the network with one neuron per class is a special case.

8.5.3.1 Centers of Radial Basis Function Neurons

Due to positive experiences for this application, the learning vector quantization (LVQ) and self-organizing maps (SOM) are applied for the adaptation process of the centers (see also Sec. 8.3 and [5]).

First of all, the centers $c_{i,t}(n)$ for each RBF neuron i are initialized to the center of the expected input space. Such an initialization seems reasonable

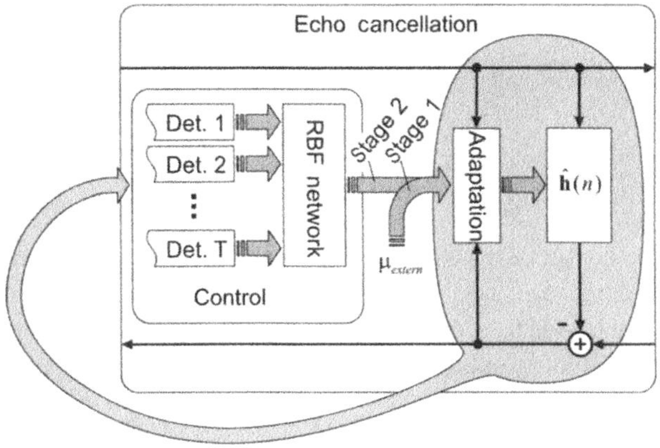

Fig. 8.17. Principle of the two-stage training. Depicted are the adaptive echo cancelling filter, its adaptation process and the control structure given by the detectors and the RBF network. The hatched area shows the feedback of the adaptive filter to the detectors and therefore to the RBF network, too. In the first stage, the feedback loop is open and the network is controlled by an external step size.

since no prior knowledge of the position of the radial basis function exists. Recalling that the detectors are normalized to a range of 0 to 1, we get

$$c_{i,t}(0) = 0.5, \quad i \in \mathcal{U}_{\mathrm{H}},\ t \in \mathcal{U}_{\mathrm{I}}. \tag{8.39}$$

With $\mathcal{U}_{\mathrm{x}} \subset \mathcal{U}_{\mathrm{H}}$ the subset of neurons belonging to the current class (recall: supervised training) we have to determine the neuron $w \in \mathcal{U}_{\mathrm{x}}$ with the smallest distance to the current input vector $\boldsymbol{u}(n)$ (*winner* neuron):

$$w = \arg\min_{i} \left\{ \left\| \boldsymbol{u}(n) - \boldsymbol{c}_i(n) \right\|^2 \right\}, \quad i \in \mathcal{U}_{\mathrm{x}}. \tag{8.40}$$

Then, the centers of all RBF neurons are adapted with a modified LVQ-SOM-approach [29]:

$$\begin{aligned} \boldsymbol{c}_i(n+1) &= \boldsymbol{c}_i(n) + \Delta\boldsymbol{c}_i(n) \\ &= \boldsymbol{c}_i(n) + \begin{cases} \boldsymbol{\eta}_i^{(\mathrm{cw})}(n)\, \nu_{wi}(n) \left[\boldsymbol{u}(n) - \boldsymbol{c}_i(n)\right], & i \in \mathcal{U}_{\mathrm{x}} \\ -\boldsymbol{\eta}_i^{(\mathrm{cl})}(n)\, \nu_{wi}(n) \left[\boldsymbol{u}(n) - \boldsymbol{c}_i(n)\right], & i \in (\mathcal{U}_{\mathrm{H}} \setminus \mathcal{U}_{\mathrm{x}}) \end{cases} \\ &\quad \text{with } \boldsymbol{\eta}_i^{(\mathrm{cl})}(n) \ll \boldsymbol{\eta}_i^{(\mathrm{cw})}(n)\,. \end{aligned} \tag{8.41}$$

The amplitude of the adaptation steps are influenced by the network step sizes

$$\boldsymbol{\eta}_i^{(\mathrm{cw})}(n) = \left[\eta_{i,1}^{(\mathrm{cw})}(n),\ \eta_{i,2}^{(\mathrm{cw})}(n),\ \cdots,\ \eta_{i,T}^{(\mathrm{cw})}(n)\right]^{\mathrm{T}} \quad \text{and} \tag{8.42}$$

$$\boldsymbol{\eta}_i^{(\mathrm{cl})}(n) = \left[\eta_{i,1}^{(\mathrm{cl})}(n),\ \eta_{i,2}^{(\mathrm{cl})}(n),\ \cdots,\ \eta_{i,T}^{(\mathrm{cl})}(n)\right]^{\mathrm{T}}, \tag{8.43}$$

as well as the neighborhood function

$$\nu_{wi}(n) = \frac{\left\|\boldsymbol{c}_w(n) - \boldsymbol{u}(n)\right\|^2}{\left\|\boldsymbol{c}_i(n) - \boldsymbol{u}(n)\right\|^2}. \tag{8.44}$$

Obviously the centers for the winner neurons are moved towards the current input vector $\boldsymbol{u}(n)$, and the centers of all other (*looser*) neurons are slightly moved into the opposite direction. The movement into the opposite direction shout prohibit an accumulation of all neurons at the current class due to the dispersion of the detectors [5].

8.5.3.2 Widths of Radial Basis Function Neurons

The widths of all RBF neurons are initialized to

$$\sigma_{i,t}^2(0) = 0.04,\ i \in \mathcal{U}_{\mathrm{H}},\ t \in \mathcal{U}_{\mathrm{I}}. \tag{8.45}$$

The Gaussian activation functions have an amplitude of 60% in a distance of 0.16 around the centers with this value. Due to this, the radial basis functions do not overlap too much, considering the normalization of input values to a range of 0 to 1.

Similar to the adaptation of the centers, the winner neuron $w \in \mathcal{U}_{\mathrm{x}}$ with the smallest distance to the current input vector $\boldsymbol{u}(n)$ is determined. Then, a recursive smoothing of the width of this neuron towards the distance between the current input vector and the center of this neuron is calculated:

$$\begin{aligned}\sigma_{w,t}^2(n+1) &= \sigma_{w,t}^2(n) + \Delta\sigma_{w,t}^2(n), \qquad t \in \mathcal{U}_{\mathrm{I}} \\ &= \sigma_{w,t}^2(n) + \eta_{w,t}^{(\sigma)}(n)\left[\Big(u_t(n) - c_{w,t}(n)\Big)^2 - \sigma_{w,t}^2(n)\right] \qquad (8.46) \\ &= \left(1 - \eta_{w,t}^{(\sigma)}(n)\right)\sigma_{w,t}^2(n) + \eta_{w,t}^{(\sigma)}(n)\Big(u_t(n) - c_{w,t}(n)\Big)^2 .\end{aligned}$$

Once again, the amplitude of the adaptation step is influenced by the time-variant network step size $\eta_{w,t}^{(\sigma)}(n)$. An appropriate control mechanism will be discussed in Sec. 8.5.3.4. The widths of all other (*looser*) neurons are not adapted.

8.5.3.3 Weights of Output Neuron Number V

The adaptation process for the weights of the output neurons (1 to $V-1$) for state classification is discussed in 8.6. The last output neuron V (utilized as step size for the acoustic echo cancelling filter) is initialized by

$$\boldsymbol{w}_V(0) = \Big[0.3,\ 0,\ 0.2,\ 0.02,\ 0.8,\ 0.05,\ 0.7,\ 0.1,\ 1.0\Big]^{\mathrm{T}}. \qquad (8.47)$$

These values are chosen according to prior knowledge of appropriate step sizes for the several states of the system (that are represented by the RBF neurons). E.g. for state number 9 ("initialization"), a large step size around 1.0 is supposed to be suitable. When generalizing the network to more than one RBF neuron per class (state), all weights belonging to the RBF neurons of the same state are initialized by the same values.

For the adaptation of weight $\boldsymbol{w}_V(n)$ a gradient approach is used, in which the cost function J_n (see Eq. 8.37) is minimized. The negative gradient of J_n with respect to the weights $w_{V,i}(n), i \in \mathcal{U}_{\mathrm{H}}$ can be expressed as

$$\begin{aligned}w_{V,i}(n+1) &= w_{V,i}(n) + \Delta w_{V,i}(n), \quad i \in \mathcal{U}_{\mathrm{H}} \\ &= w_{V,i}(n) - \eta_{V,i}^{(\mathrm{w})}(n)\,\frac{\partial J_n}{\partial w_{V,i}(n)}, \qquad (8.48)\end{aligned}$$

with the network step size $\eta_{V,i}^{(\mathrm{w})}(n)$, that effects the amplitudes of the adaptation steps. Using Eq. 8.37, the negative gradient reads as follows:

$$
\begin{aligned}
-\frac{\partial J_n}{\partial w_{V,i}(n)} &= -\frac{\partial}{\partial w_{V,i}(n)} \left[\delta_\mu(n) \frac{e^2(n)}{\|\boldsymbol{x}(n)\|^2} \left(\delta_\mu(n) + 2\,\mu_{\text{opt}}(n) - 2\,\frac{\varepsilon(n)}{e(n)}\right)\right] \\
&= -2\,\frac{e^2(n)}{\|\boldsymbol{x}(n)\|^2} \left[\delta_\mu(n) + \mu_{\text{opt}}(n) - \frac{\varepsilon(n)}{e(n)}\right] \frac{\partial\,\delta_\mu(n)}{\partial w_{V,i}(n)} \qquad (8.49) \\
&= -2\,\frac{e^2(n)}{\|\boldsymbol{x}(n)\|^2} \left[\mu(n) - \frac{\varepsilon(n)}{e(n)}\right] \frac{\partial\,\mu(n)}{\partial w_{V,i}(n)}.
\end{aligned}
$$

The step size $\mu(n)$ is given by the output $f_V(\boldsymbol{u})$ of the network. Therefore, the partial derivative of the step size can be dissolved to:

$$
\frac{\partial\,\mu(n)}{\partial w_{V,i}(n)} = \frac{\partial}{\partial w_{V,i}(n)} \left[\frac{\sum\limits_{i' \in \mathcal{U}_H} w_{V,i'}(n)\, h_{i'}(\boldsymbol{u},\, \boldsymbol{c}_{i'})}{\sum\limits_{i' \in \mathcal{U}_H} h_{i'}(\boldsymbol{u},\, \boldsymbol{c}_{i'})}\right], \quad i \in \mathcal{U}_O. \qquad (8.50)
$$

Considering that $h_i(\boldsymbol{u},\, \boldsymbol{c}_i)$ does not depend on $w_{V,i}(n)$, the term can be simplified to:

$$
\frac{\partial\,\mu(n)}{\partial w_{V,i}(n)} = \frac{h_i(\boldsymbol{u},\, \boldsymbol{c}_i)}{\sum\limits_{i' \in \mathcal{U}_H} h_{i'}(\boldsymbol{u},\, \boldsymbol{c}_{i'})}. \qquad (8.51)
$$

Taking Eq. 8.48 into account, the complete adaptation rule for the weights $\boldsymbol{w}_V(n)$ results in:

$$
\begin{aligned}
\Delta w_{V,i}(n) &= -\eta_{V,i}^{(\text{w})}(n)\, \frac{\partial J_n}{\partial w_{V,i}(n)} \qquad (8.52) \\
&= -2\,\eta_{V,i}^{(\text{w})}(n)\, \underbrace{\frac{e^2(n)}{\|\boldsymbol{x}(n)\|^2}}_{1.}\, \underbrace{\left[\mu(n) - \frac{\varepsilon(n)}{e(n)}\right]}_{2.}\, \underbrace{\frac{h_i(\boldsymbol{u},\, \boldsymbol{c}_i)}{\sum\limits_{i' \in \mathcal{U}_H} h_{i'}(\boldsymbol{u},\, \boldsymbol{c}_{i'})}}_{3.}.
\end{aligned}
$$

The rule mainly consists of three terms, which can be interpreted as follows:

1. The first term considers the impact of a step size deviance on the adaptation quality of the acoustic echo canceller.
2. The second term qualifies the step size error itself.
3. The third term takes the normalized activation of the current RBF neuron u into account. If the input vector $\boldsymbol{u}(n)$ is outside the sphere of this neuron, the adaptation step of the corresponding weight is quite small.

8.5.3.4 Network Step size

The network step sizes $\eta_{i,t}^{(\text{cw})}(n)$, $\eta_{i,t}^{(\text{cl})}(n)$, $\eta_{i,t}^{(\sigma)}(n)$, and $\eta_{V,i}^{(\text{w})}(n)$ are locally adapted, which means that each parameter is adapted for itself. Here, an approach based on the Super SAB (Super self-adjusting back-propagation algorithm) [42] extended by an momentum term is applied; an likewise approach in [11] is called Jacobs heuristic. For some general examinations the parameter $\chi(n)$, which is representative for the network parameters $c_{i,t}(n)$, $\sigma^2_{u,i}(n)$ and $w_{V,i}(n)$, is introduced. The adaptation rule is formulated:

- Every network parameter $\chi(n)$ has its own network step size $\eta^{(\chi)}(n)$.
- The non-weighted adaptation step

$$\Delta'\chi(n) = \frac{\Delta\chi(n)}{\eta^{(\chi)}(n)} \tag{8.53}$$

  is analyzed for each parameter.
- The network step size is enlarged if the non-weighted adaptation step $\Delta'\chi(n)$ has the same sign for several consecutive iterations.
- The network step size is decreased if the sign of the non-weighted adaptation step is alternating for several consecutive iterations. That is due to the assumption, that the minima of the cost function is close by.

Taking these considerations into account, the adaptation rule can be expressed as

$$\eta^{(\chi)}(n+1) = \begin{cases} \eta_+^{(\chi)}\,\eta^{(\chi)}(n), & \overline{\Delta'}\chi(n)\,\Delta'\chi(n) > 0, \\ \eta_-^{(\chi)}\,\eta^{(\chi)}(n), & \overline{\Delta'}\chi(n)\,\Delta'\chi(n) < 0, \\ \eta^{(\chi)}(n), & \text{else}, \end{cases} \tag{8.54}$$

with $\overline{\Delta'}\chi(n)$ the estimation of the mean adaptation step, calculated as first-order recursive filter:

$$\overline{\Delta'}\chi(n) = \big(1 - \vartheta^{(\chi)}\big)\overline{\Delta'}\chi(n-1) + \vartheta^{(\chi)}\,\Delta'\chi(n). \tag{8.55}$$

The smoothing factor $\vartheta^{(\chi)}$ is proposed to range $0.3 < \vartheta^{(\chi)} < 0.9$ in [11]. Nevertheless, for this application with very slow alternating classes (duration of one state about 2000 – 8000 samples) a bigger smoothing factor is appropriate.

The network step sizes for adaptation of the *centers* of the *winner neurons* $\eta_{i,t}^{(\mathrm{cw})}(n)$ are adapted according to Eq. 8.54, with $\chi(n) = \eta_{i,t}^{(\mathrm{cw})}(n)$. The parameters are chosen to

$$\eta_+^{(\mathrm{cw})} = 1.000001, \tag{8.56}$$

$$\eta_-^{(\mathrm{cw})} = 0.99996, \tag{8.57}$$

$$\vartheta^{(\mathrm{cw})} = 0.00005, \tag{8.58}$$

and the step sizes are initialized by

$$\eta_{i,t}^{(\mathrm{cw})}(0) = 0.0005, \qquad i \in \mathcal{U}_{\mathrm{H}},\ t \in \mathcal{U}_{\mathrm{I}}. \tag{8.59}$$

The network step sizes for the adaptation of the *centers* of all other *(looser) neurons* $\eta_{i,t}^{(\mathrm{cl})}(n)$ are chosen proportional to the "winner" step size:

$$\eta_{i,t}^{(\mathrm{cl})}(n) = \frac{\eta_{i,t}^{(\mathrm{cw})}(n)}{\kappa_\eta}, \tag{8.60}$$

with an proportional factor of $\kappa_\eta = 2000$ found adequately.

For the adaptation of the *width step sizes* $\eta_{i,t}^{(\sigma)}(n)$ not only the sign of successive adaptation steps $\Delta'\sigma_{i,t}^2(n)$, but also the amplitudes of these steps are considered:

$$\eta_{i,t}^{(\sigma)}(n+1) = \begin{cases} \left(1 + (\eta_+^{(\sigma)} - 1)\left[\dfrac{\Delta'\sigma_{i,t}^2(n)}{\overline{\Delta'\sigma}_{i,t}^2(n)}\right]_{-10}^{10}\right)\eta_{i,t}^{(\sigma)}(n), & \overline{\Delta'\sigma}_{i,t}^2(n)\,\Delta'\sigma_{i,t}^2(n) > 0, \\ \left(1 - (\eta_-^{(\sigma)} - 1)\left[\dfrac{\Delta'\sigma_{i,t}^2(n)}{\overline{\Delta'\sigma}_{i,t}^2(n)}\right]_{-10}^{10}\right)\eta_{i,t}^{(\sigma)}(n), & \overline{\Delta'\sigma}_{i,t}^2(n)\,\Delta'\sigma_{i,t}^2(n) < 0, \\ 0, & \text{else}, \end{cases} \tag{8.61}$$

$$i \in \mathcal{U}_\mathrm{H},\ t \in \mathcal{U}_\mathrm{I},$$

with the estimation for the mean adaptation step

$$\overline{\Delta'\sigma}_{i,t}^2(n) = \left(1 - \vartheta^{(\sigma)}\right)\overline{\Delta'\sigma}_{i,t}^2(n-1) + \vartheta^{(\sigma)}\,\Delta'\sigma_{i,t}^2(n). \tag{8.62}$$

The notation $[\,\cdot\,]_a^b$ denotes the limitation

$$[x]_a^b = \min\Big\{\max\{x, a\}, b\Big\}. \tag{8.63}$$

For the free parameters, adequate values for this application yield to

$$\eta_+^{(\sigma)} = 1.000005, \tag{8.64}$$

$$\eta_-^{(\sigma)} = 0.99998, \tag{8.65}$$

$$\vartheta^{(\sigma)} = 0.00005. \tag{8.66}$$

The step size is initialized by

$$\eta_{i,t}^{(\sigma)}(0) = 0.0001, \qquad i \in \mathcal{U}_\mathrm{H},\ t \in \mathcal{U}_\mathrm{I}. \tag{8.67}$$

The *output weight step sizes* $\eta_{V,i}^{(\mathrm{w})}(n)$ are adapted similar to the width step sizes $\eta_{i,t}^{(\sigma)}(n)$, despite the fact of different smoothing parameters

$$\eta_{+}^{(\mathrm{w})} = 1.000003, \tag{8.68}$$

$$\eta_{-}^{(\mathrm{w})} = 0.999995, \tag{8.69}$$

$$\vartheta^{(\mathrm{w})} = 0.00002, \tag{8.70}$$

and a different initialization:

$$\eta_{V,i}^{(\mathrm{w})}(0) = 0.001, \qquad i \in \mathcal{U}_{\mathrm{H}}. \tag{8.71}$$

8.5.4 Growing Network Structure

Generally, the classification performance of RBF networks can be improved by using more than one radial basis function per class. However, the number of neurons per class mostly are unpredictable when initializing the network. For this reason a growing network structure, automatically adding neurons [36], is applied in this approach. Since the number of input and output ports are constant, only RBF neurons in the hidden layer have to be added.

The procedure for adding RBF neurons can be split up into two steps: First, the decision for expanding the network has to be made; secondly the position and other parameters have to be assigned to the new neuron.

8.5.4.1 Decision for New Neurons

The decision for adding a neuron is based upon the quality measurement function J_n in Eq. 8.37. The aim is to add new neurons at these positions that are producing high cost.

For each hidden neuron z a decision (threshold) function $\zeta_z^{(\mathrm{th})}(n)$ and a position (center) function $\boldsymbol{\zeta}_z^{(\mathrm{c})}(n)$ is introduced. Illustrately, an RBF neuron is added at position $\boldsymbol{\zeta}_z^{(\mathrm{c})}(n)$ when the decision function $\zeta_z^{(\mathrm{th})}(n)$ exceeds a certain threshold $\widetilde{\zeta}^{(\mathrm{th})}$. The new neuron then is assigned to class z which is producing the highest cost. Both functions are updated in parallel to the adaptation of the parameters of the hidden neurons representing the class $\mathcal{K}_u$ of the current input vector $\boldsymbol{u}$:

$$\zeta_z^{(\mathrm{th})}(n+1) = \Big(1 - \gamma_\zeta(n)\Big)\, \zeta_z^{(\mathrm{th})}(n) + \gamma_\zeta(n)\, \widehat{\zeta}^{(\mathrm{th})}, \qquad z \in \mathcal{K}_u \tag{8.72}$$

$$\boldsymbol{\zeta}_z^{(\mathrm{c})}(n+1) = \Big(1 - \gamma_\zeta(n)\Big)\, \boldsymbol{\zeta}_z^{(\mathrm{c})}(n) + \gamma_\zeta(n)\, \boldsymbol{u}(n), \tag{8.73}$$

whereas the functions are not adapted for all other neurons $z \notin \mathcal{K}_u$. For $\widehat{\zeta}^{(\mathrm{th})}$ and $\widetilde{\zeta}^{(\mathrm{th})}$ see Eqs. 8.78 and 8.79.

Both functions are initialized by zero:

$$\zeta_z^{(\mathrm{th})}(0) = 0, \tag{8.74}$$

$$\boldsymbol{\zeta}_z^{(\mathrm{c})}(0) = 0. \tag{8.75}$$

The performance of the whole algorithm fundamentally depends on the control of the time-variant smoothing factor $\gamma_\zeta(n)$. For a large time factor the position function is moving fast towards the input vector $\boldsymbol{u}$ of the current class, meanwhile the decision function increases rapidly. Such a behavior is desired for classes that cannot be adequately represented by the network. Specifically these are the situations producing large cost measured by J_n. For this reason the smoothing factor is chosen proportional to the cost function:

$$\gamma_\zeta(n) = \kappa_1 \underbrace{\Big[J_n\Big]_0^{2\cdot 10^{-4}}}_{1.} \underbrace{\Big[\kappa_2^{\kappa_3\,\Delta(n)}\Big]_{0.01}^{100}}_{2.} \underbrace{\Big(1 - \nu_z(n)\Big)}_{3.}, \qquad (8.76)$$

$$\text{with } \kappa_2 > 1,$$

in which $\Delta(n)$ denotes the change of the system mismatch vector (of the acoustic echo cancelling filter)

$$\Delta(n) = \left\|\boldsymbol{h}_\Delta(n+1)\right\|^2 - \left\|\boldsymbol{h}_\Delta(n)\right\|^2, \qquad (8.77)$$

and $\nu_z(n)$ is the membership function for class z. This function is generated by the first $V-1$ output neurons and discussed in Sec. 8.6. $[\,\cdot\,]_a^b$ qualifies a limitation (see Eq. 8.63). The adaptation rule for the smoothing factor consists of three terms:

1. The cost function increases the smoothing factors when measuring large costs. However, the influence on the smoothing factor is limited to a value of $2 \cdot 10^{-4}$ which is typical for a bad initialized acoustic echo cancelling filter.
2. The second term accelerates the adding of new neurons for situations while the acoustic echo cancelling filter diverges. Experiments yield to favorable parameters of about $\kappa_3 \approx 5000$ and $\kappa_2 \approx 1.1$.
3. If the current class is already well represented by the network, a (large) membership function close to one decreases the smoothing factor and thereby slows down the process for adding neurons. By contrast a small membership function suggests a poor representation of the current state of the system; hence, a neuron should be added quite fast.

The parameter κ_1 influences the overall dimension of the smoothing factor – experiments yield to a reasonable value of $\kappa_1 \approx 10$. The threshold and the final value of the decision function are chosen to

$$\widehat{\zeta}^{(\text{th})} = 1.0, \qquad (8.78)$$

$$\widetilde{\zeta}^{(\text{th})} = 0.9. \qquad (8.79)$$

That means, a neuron is added if the decision function has reached 90 % of its final value.

In order to react only on the current situation of the acoustic echo cancelling system, the adaptation of the decision function in Eq. 8.72 is extended by a forgetting factor $\gamma_\zeta^{(\text{down})}$

$$\zeta_z^{(\mathrm{th})}(n+1) = \begin{cases} \gamma_\zeta^{(\mathrm{down})} \left[\left(1-\gamma_\zeta(n)\right) \zeta_z^{(\mathrm{th})}(n) + \gamma_\zeta(n)\, \widehat{\zeta}^{(\mathrm{th})} \right], & z \in \mathcal{K}_u, \\ \gamma_\zeta^{(\mathrm{down})}\, \zeta_z^{(\mathrm{th})}(n), & z \in (\mathcal{K} \setminus \mathcal{K}_u). \end{cases} \tag{8.80}$$

Previous situations, that are not sufficient enough for adding a new neuron, are no longer taken into account. The forgetting factor is chosen according to a time constant of about 10000 samples, which is approximately equivalent to the duration of one situation:

$$\gamma_\zeta^{(\mathrm{down})} = 0.9999\,. \tag{8.81}$$

The general functionality of the decision function is explained by a simulation with an insufficiently adapted acoustic echo canceller in Fig. 8.18.

In Fig. 8.19, the algorithm for adding new neurons is illustrated by a flow diagram.

8.5.4.2 Parameters of the New Neuron

The parameters of the new neuron should be chosen cleverly to keep the impact on the other neurons slightly. Following, the neuron to be added is denoted *new*.

- The center is initialized by the position function:

$$\boldsymbol{c}_{\mathrm{new}}(n) = \boldsymbol{\zeta}_z^{(\mathrm{c})}(n)\,. \tag{8.82}$$

  Due to the same smoothing factor for the decision- and the position function the new neuron is added in a region where the input vector are generating large cost.
- The widths are chosen to minimize the overlapping with other radial basis functions while improving the network output for the current situation. Thus, we have found an adequate initialization

$$\sigma_{\mathrm{new},t}(n) = \sqrt{T}\, \frac{\left| c_{\mathrm{cl},t}(n) - c_{\mathrm{new},t}(n) \right|}{4}, \qquad t \in \mathcal{U}_{\mathrm{I}}, \tag{8.83}$$

  with the RBF neuron "cl" that has the smallest distance to neuron "new" and T the dimension of the input space $\mathcal{U}_{\mathrm{T}}$.
- The dimension of the weights of the output neurons $\boldsymbol{w}_V(n)$ are enlarged by one (for the new hidden neuron). The weight corresponding to the new neuron is initialized by averaging all weights representing the class $\mathcal{K}_u$ of the current input vector u:

$$w_{V,\mathrm{new}}(n) = \frac{\sum\limits_{i \in \mathcal{K}_u} w_{V,i}(n)}{\sum\limits_{i \in \mathcal{K}_u} 1}\,. \tag{8.84}$$

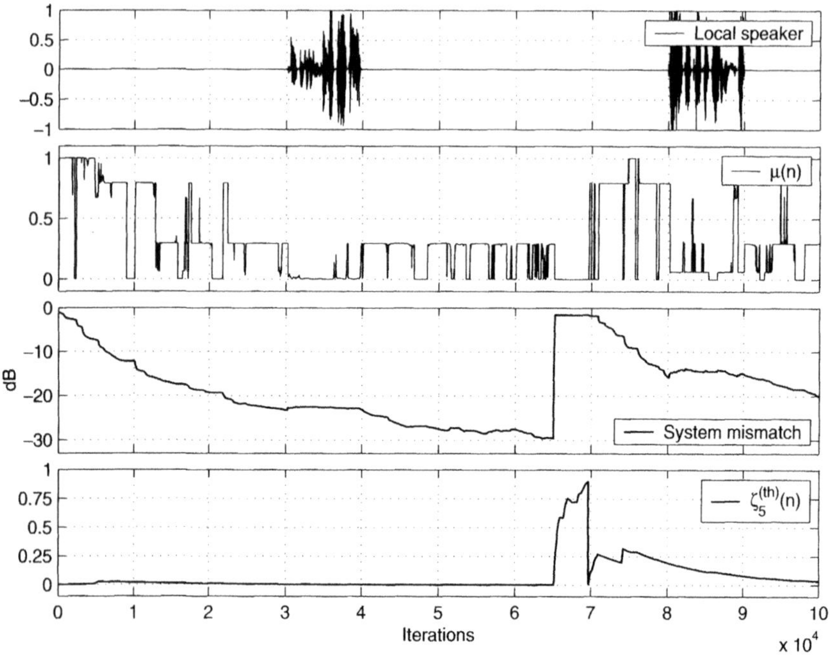

Fig. 8.18. Simulation to explain the operation of the decision function $\zeta_z^{(\text{th})}(n)$ for adding new neurons. Depicted are (top down): speech signal of local speaker, adaptation step size $\mu(n)$ of acoustic echo canceller, system mismatch of echo cancelling filter, decision function $\zeta_5^{(\text{th})}(n)$ for state number 5 ("filter adjustment insufficient, ..."). An enclosure dislocation occurs at sample 65000. The RBF network (for controlling the step size) is not well adapted, hence the adaptive filter stalls. Following, the decision function increases correctly at this point and a neuron is added at sample 70000. As a result of the new neuron the step size increases and the readaptation of the acoustic echo canceller is induced.

- Likewise, the new network step sizes are initialized by averaging all existing step sizes for the current class $\mathcal{K}_u$

$$\eta_{\text{new},t}^{(\text{cw})}(n) = \frac{\sum\limits_{i\in\mathcal{K}_u} \eta_{i,t}^{(\text{cw})}(n)}{\sum\limits_{i\in\mathcal{K}_u} 1}, \qquad t \in \mathcal{U}_\text{I} \tag{8.85}$$

$$\eta_{\text{new},t}^{(\sigma)}(n) = \frac{\sum\limits_{i\in\mathcal{K}_u} \eta_{i,t}^{(\sigma)}(n)}{\sum\limits_{i\in\mathcal{K}_u} 1}, \qquad t \in \mathcal{U}_\text{I} \tag{8.86}$$

$$\eta_{V,\text{new}}^{(\text{w})}(n) = \frac{\sum\limits_{i\in\mathcal{K}_u} \eta_{V,i}^{(\text{w})}(n)}{\sum\limits_{i\in\mathcal{K}_u} 1}. \tag{8.87}$$

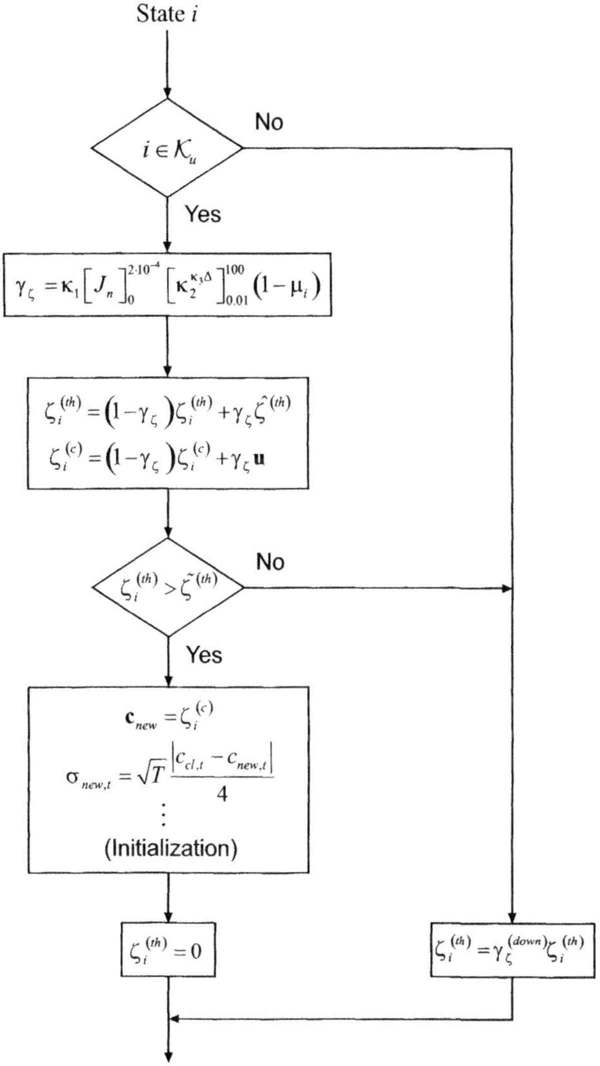

Fig. 8.19. Flow diagram for adding neurons in the growing network. The algorithm is shown for class i, nevertheless it is processed for all classes (that represent the states of system) in parallel. The main path is passed for class $\mathcal{K}_u$ of the current input vector u. When the decision function $\zeta_z^{(\mathrm{th})}(n)$ exceeds a certain threshold, a new neuron for class $\mathcal{K}_u$ is added at position $\zeta_z^{(\mathrm{c})}(n)$. The time argument is omitted for sake of clarity.

8.5.5 Results

The system is trained and verified in a simulation environment; however, real-world signals and systems are utilized. That are the speech signals recorded by different speakers in different languages, the background noise signals from different environments (office, car, etc.) and the room impulse response measured in different enclosures. Here, a sampling frequency of $f_s = 8$ kHz is used. The room impulse response is simulated with the length of 2000 samples. For the adaptive echo cancelling filter 1024 plus 40 samples are applied (40 samples for the delay coefficients [17]). In order to achieve meaningful results for the presented algorithm, the whole simulation data is split up into 70 % training data and 30% verification data. Hereby, an over-fitting of the RBF network on the training data can be detected. Furthermore we achieve an independent validation of the adaptation quality. Following, some selected results of the verification simulation for the presented method are shown.

The main task of the acoustic echo cancelling filter is to match the impulse response of the loudspeaker-enclosure-microphone system [17]. Hence, we evaluate the quality of the whole system by the system mismatch parameter. In Fig. 8.20 the simulation result for a verification simulation with low noise level is depicted (including the system mismatch). It should be mentioned that all RBF network parameters are fixed in the verification simulations – the parameters are adapted only during the training process. A verification simulation

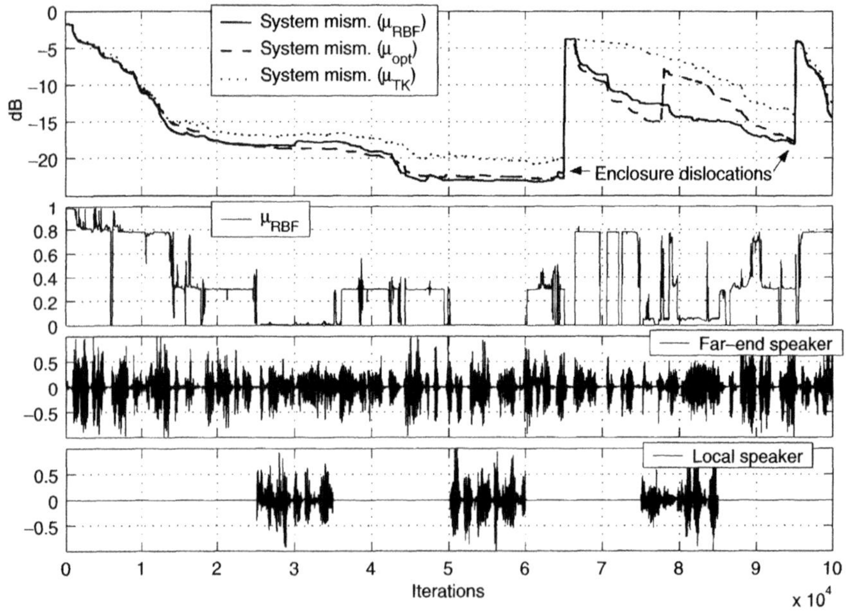

Fig. 8.20. Simulation example for 40dB SNR (low background noise). Depicted are (top down): The system mismatch for a step-size control applying the RBF network ($\mu_{\text{RBF}}(n)$), a pseudo-optimal step-size control ($\mu_{\text{opt}}(n)$) and a classical approach utilizing the delay-coefficient method ($\mu_{\text{TK}}(n)$). Beneath, the step size generated by the RBF network is depicted. The excitation signal of the far-end speaker and the disturbance signal of the local speaker are shown below. Two enclosure dislocations occur at samples 65000 and 95000.

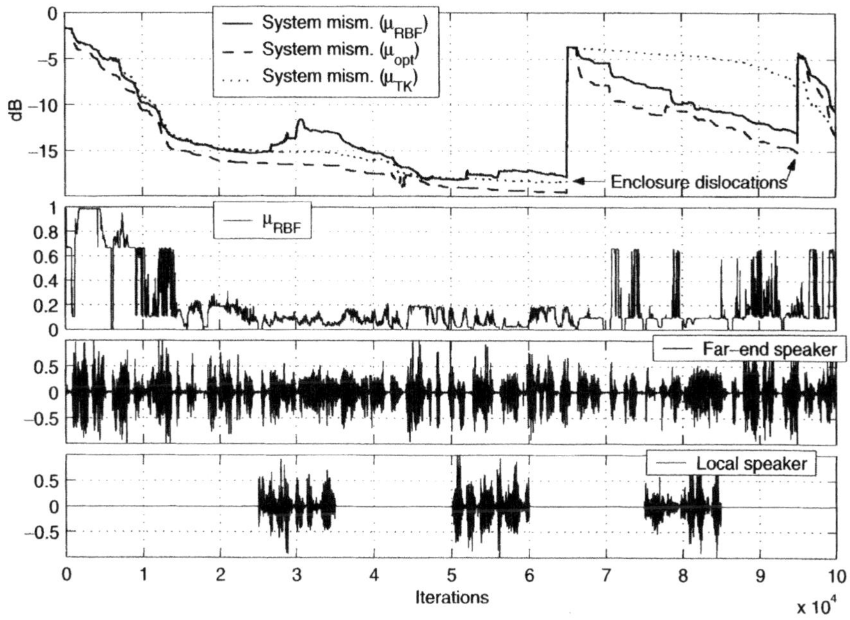

Fig. 8.21. Simulation example for 0 dB SNR (background noise level close to level of local speech signal). Depicted are (top down): The system mismatch for a step-size control applying the RBF network ($\mu_{\text{RBF}}(n)$), a pseudo-optimal step-size control ($\mu_{\text{opt}}(n)$) and a classical approach utilizing the delay-coefficient method ($\mu_{\text{TK}}(n)$). Beneath, the step size generated by the RBF network is depicted. The excitation signal of the far-end speaker and the disturbance signal of the local speaker are shown below (note: the local background noise is not included in the local signal shown here). Two enclosure dislocations occur at samples 65000 and 95000.

of the same system (particularly the RBF parameters are identical) with a high local background noise level is shown in Fig. 8.21. It can be realized that the RBF network satisfactorily handles with both situations and provides a well adjusted step size for the acoustic echo canceller in almost all situations. Specifically the enclosure dislocations are detected rapidly in contrast to the "classical" approach utilizing the delay-coefficient method only. Also, the step size is correctly reduced for situations with large background noise. Finally, the system copes very well with double-talk situations (when both the local and the far-end speaker are active), too: The step size is rapidly decreased for theses situations in order to prohibit the adaptive filter from diverging.

8.6 Radial Basis Function Network for State Detection

In Sec. 8.5 the idea of a RBF network for step-size control of an acoustic echo canceller is presented. The state-space representation of the whole hands-free system and its application as prior-knowledge for the RBF network is introduced. That is, one Gaussian function represents a single class or state of the system. In case of the growing network approach (Sec. 8.5.4) further RBF neurons are added for each state leading to an improved state classification. The state information is not only interesting for control of the echo canceller but can be useful for all other speech enhancement algorithms of a hands-free system. For example, a noise reduction can process the state information on local background noise and adjust its parameters.

Here, the application of the RBF network for state detection is considered in more detail. Further on, the decision for a certain state of the system is analyzed using means of detection theory [44]. Theses results are used to evaluate the reliability of the applied detectors.

8.6.1 State Classification

Classification problems can be regarded as follows: a feature space, whose axes coincide with the rows of measurable feature vectors, can be partitioned into several regions. These regions are denoted as classes. Measuring a new feature or pattern vector $\boldsymbol{u}$, the task is to make a decision to which class $z \in \mathcal{K}$ the vector belongs. For this application, the classes are given by the states of the system (as listed in Tab. 8.1). For classification, the conditional probability $P(z|\boldsymbol{u})$ of class z for given feature vector $\boldsymbol{u}$ is analyzed [2]. According to Bayes theorem the association probability can be derived from the conditional probability density function (PDF) $f_{\boldsymbol{u}}(\boldsymbol{u}|z)$ of the feature vector $\boldsymbol{u}$ conditioned on class z and from the a priori probability $P(z)$ of class z through

$$P(z|\boldsymbol{u}) = \frac{P(z)\, f_{\boldsymbol{u}}(\boldsymbol{u}|z)}{\sum\limits_{z'=1}^{N_Z} P(z')\, f_{\boldsymbol{u}}(\boldsymbol{u}|z')}, \tag{8.88}$$

with N_Z the number of classes (states). The denominator is the a priori PDF $f_{\boldsymbol{u}}(\boldsymbol{u})$ of feature vector $\boldsymbol{u}$. The following section deals with the estimation of the conditional PDF $f_{\boldsymbol{u}}(\boldsymbol{u}|z)$ and the decision for a certain state based on Bayes theorem.

8.6.1.1 Estimation of Probability of States

For estimation of probability density functions parametric or non-parametric methods can be utilized [4, 43]. For example, multidimensional histograms measuring the frequency distribution are non-parametric approaches. However, the complexity increases by the square of the input dimension and is not practicable for dimensions larger than two for this reason. For parametric approaches mixture-models, calculated by the superposition of several functions, can be applied. The estimation of the PDF is performed by estimating the parameters of the underlying function – hence the method is called parametric. Assuming the functions are probability density functions for itself, the mixture-model for estimation of $\widehat{f}_{\boldsymbol{u}}(\boldsymbol{u}|z)$ is given by

$$\widehat{f}_{\boldsymbol{u}}(\boldsymbol{u}|z) = \sum_{i=1}^{N_z} \pi_{iz}\, \widehat{f}_{\boldsymbol{u}}(\boldsymbol{u}|i,z), \qquad z \in \mathcal{K}, \tag{8.89}$$

with the mixing coefficient π_{iz} stating the a priori probability that $\boldsymbol{u}$ is represented by the conditional PDF $\widehat{f}_{\boldsymbol{u}}(\boldsymbol{u}|i,z)$. N_z is the number of superposed functions. The mixing coefficient has to fulfill the condition

$$\sum_{i=1}^{N_z} \pi_{iz} = 1, \qquad z \in \mathcal{K}. \tag{8.90}$$

Utilizing Gaussian functions for $\widehat{f}_{\boldsymbol{u}}(\boldsymbol{u}|i,z)$, Eq. 8.89 is called Gaussian mixture model (GMM). With an adequately number of Gaussian functions the GMM can be used to model arbitrary densities [1]. Here, we are using the GMM to estimate the conditional PDF $f_{\boldsymbol{u}}(\boldsymbol{u}|z)$.

Regarding Eq. 8.89 a large similarity of the GMM with the RBF network (Eq. 8.34) can be realized. Actually, a GMM can be implemented by an RBF network [2,4,35,41]. Following, we analyze the applied growing RBF network with respect to an estimation of the conditional PDF $f_{\boldsymbol{u}}(\boldsymbol{u}|z)$.

Inserting Gaussian functions into Eq. 8.89 and assuming uncorrelated elements of the feature vector $\boldsymbol{u}$, the GMM yields

$$\widehat{f}_{\boldsymbol{u}}(\boldsymbol{u}|z) = \sum_{i=1}^{N_z} \pi_{iz} \prod_{t=1}^{T} \left[\frac{1}{\sqrt{2\pi}\,\sigma_{(z,i),t}} \exp\left(-\frac{1}{2} \frac{\left(u_t - c_{(z,i),t}\right)^2}{\sigma^2_{(z,i),t}} \right) \right], \tag{8.91}$$

which can be further expanded to

$$\widehat{f}_{\boldsymbol{u}}(\boldsymbol{u}|z) = \sum_{i=1}^{N_z} \pi_{iz} \prod_{t=1}^{T} \left[\frac{1}{\sqrt{2\pi}\,\sigma_{(z,i),t}} \right] \prod_{t=1}^{T} \left[\exp\left(-\frac{1}{2} \frac{(u_t - c_{(z,i),t})^2}{\sigma^2_{(z,i),t}} \right) \right]. \tag{8.92}$$

The second product term can be interpreted as output $h_j(\boldsymbol{u})$ of RBF neuron j, whereas the neuron j models the Gaussian function (z,i). With $\mathcal{U}_z$ the set of all RBF neurons used for modelling the density $\widehat{f}_{\boldsymbol{u}}(\boldsymbol{u}|z)$ we obtain

$$\widehat{f}_{\boldsymbol{u}}(\boldsymbol{u}|z) = \sum_{j\in\mathcal{U}_z} \pi_{jz} \prod_{t=1}^{T} \frac{1}{\sqrt{2\pi}\,\sigma_{j,t}}\, h_j(\boldsymbol{u}). \tag{8.93}$$

Choosing the output weight $w_{v,i}$ of the RBF network according to the first term in Eq. 8.92

$$w_{v,j} = \begin{cases} \pi_{jz} \displaystyle\prod_{t=1}^{T} \frac{1}{\sqrt{2\pi}\,\sigma_{j,t}}, & v = z,\ j \in \mathcal{U}_z, \\ 0, & \text{else}, \end{cases} \qquad v \in (\mathcal{U}_{\mathrm{O}} \setminus V),\ j \in \mathcal{U}_{\mathrm{H}}, \tag{8.94}$$

the output neuron v can be utilized for estimation of the conditional PDF $\widehat{f}_{\boldsymbol{u}}(\boldsymbol{u}|z)$. That is, only these weights $w_{v,j}$ corresponding to state z (represented by neuron v) are chosen different to zero.

Hence, the structure of the RBF network can be used to estimate the PDF $\widehat{f}_{\boldsymbol{u}}(\boldsymbol{u}|z)$ by modelling an GMM. However, it has to be ensured that the given RBF network is able to model the statistical properties of the feature vector adequately. Furthermore, the parameters of the RBF network have to be adjusted. Here, we apply a recursive smoothing for the adaptation of the centers and widths parameters (see Sec. 8.5.3.1 and Sec. 8.5.3.2). Assuming the feature data to be obtained from a stationary random process, the adaptation rule for the center in Eq. 8.41 complies with the estimation of the conditional mean value given feature data belonging to neuron j:

$$c_{j,t}(n) \overset{n\to\infty}{\approx} m_{u_t|j} = \mathrm{E}\Big\{u_t(n)\big|j\Big\}. \tag{8.95}$$

Likewise, the adaptation rule for the widths (Eq. 8.46) is an estimation for the conditional variance of the feature data:

$$\sigma_{j,t}^2(n) \overset{n\to\infty}{\approx} \mathrm{E}\left\{\Big(u_t(n) - m_{u_t|j}\Big)^2\bigg|j\right\}. \tag{8.96}$$

As shown above, neuron j models the Gaussian function (z, i), which means the center and width parameters can be used for an estimation of the Gaussian mean and variance parameters.

The mixing parameter π_{jz} in Eq. 8.94 controls the influence of the superposed Gaussian function to the estimation of the conditional PDF $\widehat{f}_{\boldsymbol{u}}(\boldsymbol{u}|z)$. Here, we estimate these mixing coefficients by measuring the accumulation of the feature data close to the winner neurons. Neurons with high frequency of occurrence get larger mixing parameters.

Due to the growing network structure, the given RBF network cannot only fit Gaussian distributions but also models any distribution of the feature data. That is, new superimposed Gaussian functions are added at the regions where the feature data accumulates but the conditional PDF (calculated by the RBF network) has no amplitude yet. Hence, further neurons are added until the PDF can be modelled arbitrarily close.

With the presumption of reliable estimations of the conditional PDFs $\widehat{f}_{\boldsymbol{u}}(\boldsymbol{u}|z)$, an estimation of the association probability for state z given feature vector $\boldsymbol{u}$ can be formulated:

$$\widehat{P}(z|\boldsymbol{u}) = \frac{\widehat{P}(z)\,\widehat{f}_{\boldsymbol{u}}(\boldsymbol{u}|z)}{\sum\limits_{z'=1}^{N_Z} \widehat{P}(z')\,\widehat{f}_{\boldsymbol{u}}(\boldsymbol{u}|z')}. \tag{8.97}$$

The a priori probability $\widehat{P}(z)$ can be estimated by the frequency of occurrence of state z in a training set [41]

$$\widehat{P}(z) = \frac{L_z}{L}, \qquad z \in \mathcal{K}, \tag{8.98}$$

with L the total number of training data and L_z the number of training data belonging to state z. Here, L is the number of iteration steps during training, and L_z the number of iterations for adapting parameters of state z. However, one has to ensure that the distribution of the training data with respect to the states complies to the real-world distribution.

Following, the decision rule for a certain state z is deduced from detection theory [44]. A hypothesis $H_z \in \{H_1, H_2, \ldots, H_9\}$ is established for each state z, whereas each hypothesis means that the related state appears. A decision is made in accepting a single hypothesis and consequently rejecting all other hypotheses. Several approaches are know for optimization of the decision [44]. A decision with respect to the minimization of the mean cost is the aim of a Bayes decision. The cost for all decision possibilities are defined for this purpose, whereas cost C_{ij} corresponds to a decision for hypothesis H_j while hypothesis H_i is true. Presuming the a priori probability $P(H_z) = P(z)$ of each hypothesis, the total cost of a decision is given by

$$C = \sum_{i=1}^{N_Z} P(i) \sum_{j=1}^{N_Z} C_{ij} \int\limits_{\boldsymbol{U}_j} f_{\boldsymbol{u}}(\boldsymbol{u}|H_i)\, d\boldsymbol{u}, \tag{8.99}$$

with $\boldsymbol{U}_j$ being the region of the input space in which a decision for hypothesis H_j is made. For representation of the decision rules $N_z - 1$ likelihood ratios are defined:

$$\begin{aligned} \Lambda_2 &= \frac{f_{\boldsymbol{u}}(\boldsymbol{u}|H_2)}{f_{\boldsymbol{u}}(\boldsymbol{u}|H_1)}, \\ \Lambda_3 &= \frac{f_{\boldsymbol{u}}(\boldsymbol{u}|H_3)}{f_{\boldsymbol{u}}(\boldsymbol{u}|H_1)}, \\ &\ \vdots \qquad \vdots \\ \Lambda_9 &= \frac{f_{\boldsymbol{u}}(\boldsymbol{u}|H_9)}{f_{\boldsymbol{u}}(\boldsymbol{u}|H_1)}, \end{aligned} \tag{8.100}$$

with $f_{\boldsymbol{u}}(\boldsymbol{u}|H_i) = f_{\boldsymbol{u}}(\boldsymbol{u}|z_i)$. A $(N_z - 1)$ dimensional decision space is spanned up by the likelihood ratios Λ_i, segmented into several regions by the decision rules. For simplification the same cost $C = 1$ for all false decisions and the cost $C = 0$ for all correct decisions are assumed:

$$C_{ij} = \begin{cases} 0, & i = j, \\ 1, & i \neq j. \end{cases} \tag{8.101}$$

Hence, the decision rules can be formulated by $(N_z - 1)N_z/2 = 36$ (for $N_z = 9$) different comparisons, each discarding one hypothesis [44]:

$$\Lambda_i \underset{\overline{H}_i}{\overset{\overline{H}_1}{\gtrless}} \frac{P(1)}{P(i)} \rightarrow P(i)\, f_{\boldsymbol{u}}(\boldsymbol{u}|H_i) \underset{\overline{H}_i}{\overset{\overline{H}_1}{\gtrless}} P(1)\, f_{\boldsymbol{u}}(\boldsymbol{u}|H_1), \quad i = 2, ..., 9 \quad (8.102)$$

$$\frac{\Lambda_i}{\Lambda_2} \underset{\overline{H}_i}{\overset{\overline{H}_2}{\gtrless}} \frac{P(2)}{P(i)} \rightarrow P(i)\, f_{\boldsymbol{u}}(\boldsymbol{u}|H_i) \underset{\overline{H}_i}{\overset{\overline{H}_2}{\gtrless}} P(2)\, f_{\boldsymbol{u}}(\boldsymbol{u}|H_2), \quad i = 3, ..., 9$$

$$\frac{\Lambda_i}{\Lambda_3} \underset{\overline{H}_i}{\overset{\overline{H}_3}{\gtrless}} \frac{P(3)}{P(i)} \rightarrow P(i)\, f_{\boldsymbol{u}}(\boldsymbol{u}|H_i) \underset{\overline{H}_i}{\overset{\overline{H}_3}{\gtrless}} P(3)\, f_{\boldsymbol{u}}(\boldsymbol{u}|H_3), \quad i = 4, ..., 9$$

$$\vdots \qquad\qquad\qquad \vdots$$

$$\frac{\Lambda_9}{\Lambda_8} \underset{\overline{H}_9}{\overset{\overline{H}_8}{\gtrless}} \frac{P(8)}{P(9)} \rightarrow P(9)\, f_{\boldsymbol{u}}(\boldsymbol{u}|H_9) \underset{\overline{H}_9}{\overset{\overline{H}_8}{\gtrless}} P(8)\, f_{\boldsymbol{u}}(\boldsymbol{u}|H_8)\,,$$

where $\overline{H}_i$ means rejection of H_i. Regarding the right hand side in Eq. 8.102, the search of the maximum $P(i)\, f_{\boldsymbol{u}}(\boldsymbol{u}|H_i)$, $i = 1, ..., 9$ yields to an equivalent decision for hypothesis H_z:

$$z = \arg\max_{z^*} P(z^*)\, f_{\boldsymbol{u}}(\boldsymbol{u}|H_{z^*}). \tag{8.103}$$

If the statistical parameters are not available the decision has to be based on estimated values:

$$\hat{z} = \arg\max_{z^*} \widehat{P}(z^*)\, \widehat{f}_{\boldsymbol{u}}(\boldsymbol{u}|H_{z^*}). \tag{8.104}$$

The right hand side can be multiplied by $1/\sum_{z'=1}^{N_Z} \widehat{P}(z')\, \widehat{f}_{\boldsymbol{u}}(\boldsymbol{u}|z')$ without restrictions of any kind. So finally we get the Bayes decision rule for hypothesis H_z (and state z) with

$$\begin{aligned} z \approx \hat{z} &= \arg\max_{z^*} \frac{\widehat{P}(z^*)\, \widehat{f}_{\boldsymbol{u}}(\boldsymbol{u}|z^*)}{\sum\limits_{z'=1}^{N_Z} \widehat{P}(z')\, \widehat{f}_{\boldsymbol{u}}(\boldsymbol{u}|z')} \\ &= \arg\max_{z^*} \widehat{P}(z^*|\boldsymbol{u}), \end{aligned} \tag{8.105}$$

which is the maximum association probability for state z given feature vector $\boldsymbol{u}$ (see Eq. 8.97).

The performance of the given state classification is shown by means of an simulation example in Fig. 8.22. The constraints of the simulation are similar to the ones specified in Sec. 8.5.5. Hence, a simulation example with well adapted but fixed parameters is analyzed. When taking the definitions of states in Tab. 8.1 into account, a reliable state detection for both low and high local background noise can be noticed. The states for insufficient filter convergence are correctly detected at the beginning of the adaptation and – even more important – after the two enclosure dislocations. Likewise, the three double-talk situations during sufficient and insufficient filter convergence are successfully passed. Some small outlier only occur during the first double-talk situation for large background noise (at about sample 30000). Furthermore, a small misclassification of the initialization-state at the very beginning of the simulation can be noticed. This is no weakness of the classification rule but is due to the fact that the RBF network is still optimized for step-size control and not for state classification.

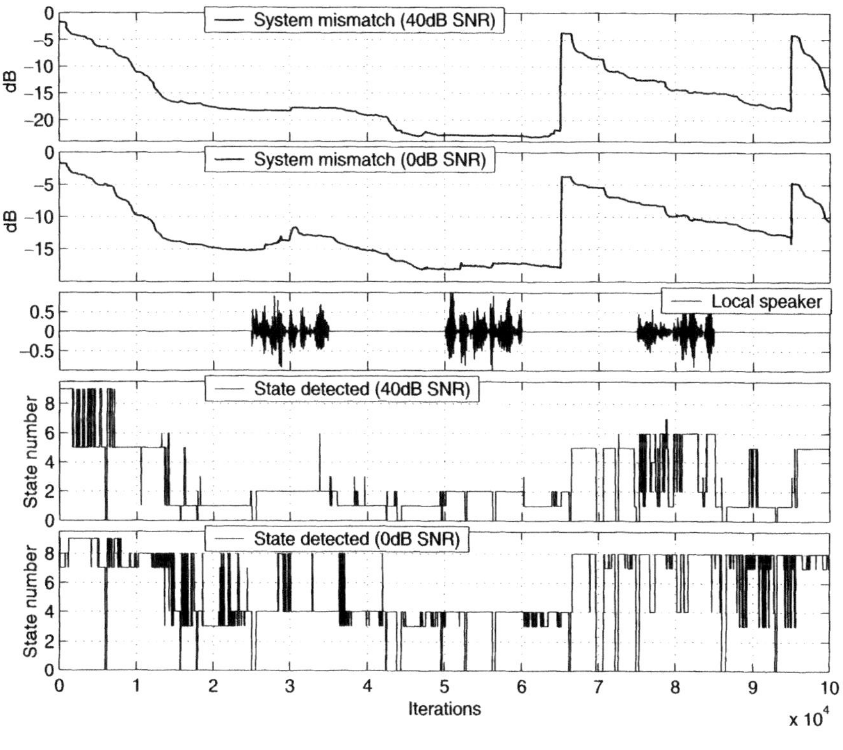

Fig. 8.22. Simulation examples for classification of states. A well optimized RBF network is used for both step-size control and state estimation. Two enclosure dislocations occur at sample 65000 and 95000. Depicted are (top down): System mismatch for simulations with low and high local background noise. The local speaker signal is shown underneath. Beneath, the states detected for both simulations with low and high background noise are depicted. The enumeration of states corresponds to Tab. 8.1.

8.6.1.2 Probability of State Characteristics

For many speech enhancement algorithms not only the discrete state of the system z but also the state characteristics $\tilde{z}$ are of interest. The state characteristics are the axes (local speech activity, local background noise, filter convergence, initialization) of the state space. In contrast to the classification problem we have to solve a binary decision problem here; e.g. local speaker active: yes/no. With the notation of a state characteristic $\tilde{z}$ we introduce distinct characteristics $\tilde{z}_1$:

$$\tilde{z}_1 \in \widetilde{\mathcal{K}} = \{ \text{"local speaker active", "large background noise",} \\ \text{"good filter convergence", "initialization"} \}. \tag{8.106}$$

The complementary state characteristics $\tilde{z}_0$ denote non-distinct characteristics:

$$\tilde{z}_0 \in \overline{\widetilde{\mathcal{K}}} = \{ \text{"local speaker inactive", "low background noise",} \\ \text{"poor filter convergence", "no initialization"} \}. \tag{8.107}$$

Assuming same cost and a priori probabilities for both characteristics $\tilde{z}_1$ and $\tilde{z}_0$, the activity of a certain characteristic can be decided according to Bayes:

$$\widehat{P}(\tilde{z}_1|\boldsymbol{u}) > \widehat{P}(\tilde{z}_0|\boldsymbol{u}). \tag{8.108}$$

Considering Bayes theorem (Eq. 8.88), the estimation of $f_{\boldsymbol{u}}(\boldsymbol{u}|\tilde{z})$ has to be determined for the calculation of $\widehat{P}(\tilde{z}|\boldsymbol{u})$. Likewise to Sec. 8.6.1.1, this can be done by use of GMM. Replacing state z by state characteristic $\tilde{z}$ in Eq. 8.89 the PDF can be modelled as superposition of $N_{\tilde{z}}$ conditional PDFs:

$$\widehat{f}_{\boldsymbol{u}}(\boldsymbol{u}|\tilde{z}) = \sum_{i=1}^{N_{\tilde{z}}} \pi_{i\tilde{z}}\, \widehat{f}_{\boldsymbol{u}}(\boldsymbol{u}|i,\tilde{z}), \qquad \tilde{z} \in \widetilde{\mathcal{K}}. \tag{8.109}$$

For the implemented RBF network follows in like manner to Eq. 8.93:

$$\widehat{f}_{\boldsymbol{u}}(\boldsymbol{u}|\tilde{z}) = \sum_{j\in\mathcal{U}_{\tilde{z}}} \pi_{j\tilde{z}} \prod_{t=1}^{T} \frac{1}{\sqrt{2\pi}\,\sigma_{j,t}}\, h_j(\boldsymbol{u}), \tag{8.110}$$

with neuron j modelling Gaussian function $(\tilde{z}, i)$. $\mathcal{U}_{\tilde{z}}$ is the number of neurons representing a given state characteristic $\tilde{z}$:

$$\mathcal{U}_{\tilde{z}} = \bigcup_{z\in\mathcal{K}_{\tilde{z}}} \mathcal{U}_z. \tag{8.111}$$

$\mathcal{K}_{\tilde{z}}$ is the set of states incorporating state characteristic $\tilde{z}$. The conditional PDFs of the distinct and non-distinct characteristics $\tilde{z}_1$ and $\tilde{z}_0$ can be formulated in the same way.

For calculation of the two characteristics $\tilde{z}_1$ and $\tilde{z}_0$ the output layer of the implemented RBF network has to be enlarged by twice the number of state characteristics. Hence, eight output neurons $v \in \mathcal{U}_{\widetilde{O}}$ with the weights

$$w_{v,i} = \begin{cases} \pi_{i\tilde{z}_1} \prod_{t=1}^{T} \frac{1}{\sqrt{2\pi}\,\sigma_{i,t}}, & v = \tilde{z}_1,\ i \in \mathcal{U}_{\tilde{z}_1}, \\ \pi_{i\tilde{z}_0} \prod_{t=1}^{T} \frac{1}{\sqrt{2\pi}\,\sigma_{i,t}}, & v = \tilde{z}_0,\ i \in \mathcal{U}_{\tilde{z}_0}, \\ 0, & \text{else}, \end{cases} \qquad v \in \mathcal{U}_{\widetilde{O}},\ i \in \mathcal{U}_{\mathrm{H}}\,, \tag{8.112}$$

with $\mathcal{U}_{\mathrm{H}}$ the set of neurons in the hidden layer, have to be added. The implemented RBF network can then be used for estimating conditional PDFs of both characteristics:

$$\widehat{f}_{\boldsymbol{u}}(\boldsymbol{u}|\tilde{z}_1) = f_{\tilde{z}_1}(\boldsymbol{u})\,, \qquad \tilde{z}_1 \in \mathcal{U}_{\widetilde{O}}\,,$$

$$\widehat{f}_{\boldsymbol{u}}(\boldsymbol{u}|\tilde{z}_0) = f_{\tilde{z}_0}(\boldsymbol{u})\,, \qquad \tilde{z}_0 \in \mathcal{U}_{\widetilde{O}}\,.$$

Recalling Bayes theorem (Eq. 8.88)

$$\widehat{P}(\tilde{z}_1|\boldsymbol{u}) = \frac{\widehat{P}(\tilde{z}_1)\,\widehat{f}_{\boldsymbol{u}}(\boldsymbol{u}|\tilde{z}_1)}{\widehat{f}_{\boldsymbol{u}}(\boldsymbol{u})}\,, \tag{8.113}$$

$$\widehat{P}(\tilde{z}_0|\boldsymbol{u}) = \frac{\widehat{P}(\tilde{z}_0)\,\widehat{f}_{\boldsymbol{u}}(\boldsymbol{u}|\tilde{z}_0)}{\widehat{f}_{\boldsymbol{u}}(\boldsymbol{u})}\,, \tag{8.114}$$

$$\text{with } \widehat{f}_{\boldsymbol{u}}(\boldsymbol{u}) = \widehat{P}(\tilde{z}_1)\,\widehat{f}_{\boldsymbol{u}}(\boldsymbol{u}|\tilde{z}_1) + \widehat{P}(\tilde{z}_0)\,\widehat{f}_{\boldsymbol{u}}(\boldsymbol{u}|\tilde{z}_0), \quad \tilde{z} \in \widetilde{\mathcal{K}}, \tag{8.115}$$

and Eq. 8.108, a Bayes decision can be made by use of these estimates:

$$\widehat{P}(\tilde{z}_1|\boldsymbol{u}) \underset{H_0^{(\tilde{z})}}{\overset{H_1^{(\tilde{z})}}{\gtrless}} \widehat{P}(\tilde{z}_0|\boldsymbol{u}), \qquad \tilde{z} \in \widetilde{\mathcal{K}}, \tag{8.116}$$

with the two hypothesis $H_1^{(\tilde{z})}$ and $H_0^{(\tilde{z})}$ stating that state characteristic $\tilde{z}$ is distinct or not:

$$\begin{aligned} \tilde{z}_1 &\rightarrow H_1^{(\tilde{z})} \text{ true}, \\ \tilde{z}_0 &\rightarrow H_0^{(\tilde{z})} \text{ true}, \end{aligned} \qquad \tilde{z} \in \widetilde{\mathcal{K}}. \tag{8.117}$$

Although a basic RBF network with one neuron per class can be used for the estimation, obviously better approximations can be achieved with the growing network structure.

For demonstration of the given Bayes estimator, a simulation of the a posteriori probabilities $\widehat{P}(\tilde{z}|\boldsymbol{u})$ considering the two state characteristics "local speaker active" and "poor filter convergence" as example is depicted in Fig. 8.23. It can be noticed that both estimated probabilities are quite reliable, although the estimation of double talk for large background noise is fairly disturbed.

The Bayes decision assuming the same cost for all decisions and the same a priori probabilities for both characteristics is symmetric to $\widehat{P}(\tilde{z}|\boldsymbol{u}) = 0.5$.

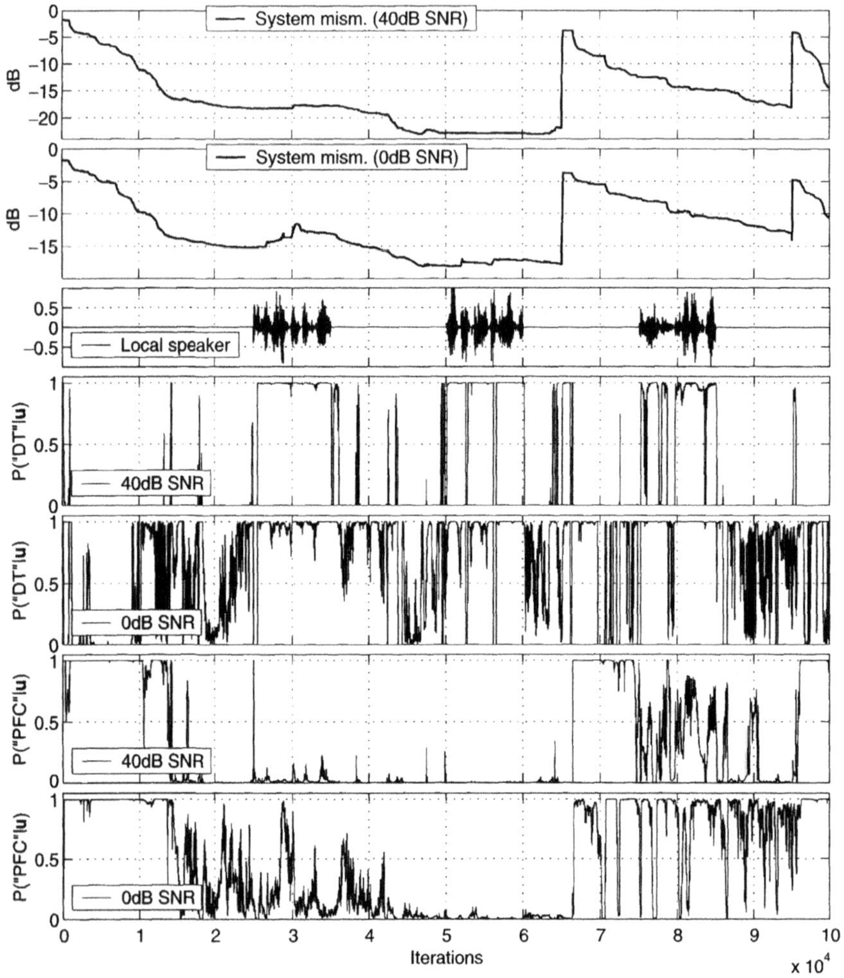

Fig. 8.23. Examples for probability of two state characteristics "local speaker active" (together with far-end speech activity: DT – double talk) and "poor filter convergence" (PFC). Enclosure dislocations occur at samples 65000 and 95000. Two simulations for low and high background noise are performed. Depicted are the estimated a posteriori probabilities $\widehat{P}(\text{"DT"}|\boldsymbol{u})$ and $\widehat{P}(\text{"PFC"}|\boldsymbol{u})$ for both background noise levels, respectively.

However, other decision rules can be applied. A Minimax test can be used for example if the a priori probabilities $\widehat{P}(\tilde{z})$ cannot be obtained. Finally, if no assumptions on the cost can be established, a Neyman-Pearson test will be applicable [44].

8.6.2 Reliability of Detectors

The previous sections deal with the application of RBF networks for nonlinear mapping of detector outputs on a (reliable) step size. The network was extended to calculated some state information, too. However, all the information processing can only use the detector signals for input. That is, the reliabilities of the step size and state estimations are mainly based upon the reliabilities of the detectors. For this reason it is quite helpful to obtain information on the reliabilities of various detectors; more precisely, determine the reliability of a certain detector for the interesting state characteristics. By this knowledge an overall control strategy can be arranged by selecting and integrating those detectors that perfectly fit together in mutually compensating their weaknesses on certain state characteristics.

Consequently, it is the task to qualitatively benchmark the detectors concerning the classification of the interesting state characteristics. The evaluation of the receiver operating characteristics (ROC) for each detector is one possibility we want to analyze here in more detail. First of all, the conditional detection probability $P_{\mathrm{D}}^{(\tilde{z})}$ (detecting $\tilde{z}_1$ correctly when $\tilde{z}_1$ is true) and the conditional false-alarm probability $P_{\mathrm{F}}^{(\tilde{z})}$ (detecting $\tilde{z}_0$ when $\tilde{z}_1$ is true) can be defined:

$$P_{\mathrm{D}}^{(\tilde{z})} = \int_{\boldsymbol{U}_1^{(\tilde{z})}} f_{\boldsymbol{u}}\left(\boldsymbol{u}\middle|H_1^{(\tilde{z})}\right) d\boldsymbol{u}, \qquad (8.118)$$

$$P_{\mathrm{F}}^{(\tilde{z})} = \int_{\boldsymbol{U}_1^{(\tilde{z})}} f_{\boldsymbol{u}}\left(\boldsymbol{u}\middle|H_0^{(\tilde{z})}\right) d\boldsymbol{u},$$

with $\boldsymbol{U}_1^{(\tilde{z})}$ the region in the feature space leading to a decision for $H_1^{(\tilde{z})}$.

The reliability of detectors can be qualified in analyzing the ratio of the two conditional probabilities. However, this ratio depends on the decision rules and cannot be obtained directly. The approach of a ROC is to determine $P_{\mathrm{D}}^{(\tilde{z})}$ in dependency of $P_{\mathrm{F}}^{(\tilde{z})}$. According to Eq. 8.118, the interrelationship between both probabilities is given by the discrimination threshold. The likelihood ratio of a Bayes optimization [44] can be defined as threshold parameter λ and leads to

$$\Lambda^{(\tilde{z})} = \frac{f_{\boldsymbol{u}}\left(\boldsymbol{u}\middle|H_1^{(\tilde{z})}\right)}{f_{\boldsymbol{u}}\left(\boldsymbol{u}\middle|H_0^{(\tilde{z})}\right)} = \lambda, \qquad \tilde{z} \in \widetilde{\mathcal{K}}. \qquad (8.119)$$

Remembering that we are not interested in the reliability of all detectors together but in the reliability of each single detector yields to

$$\frac{f_{\boldsymbol{u}}\left(u_t\middle|H_1^{(\tilde{z})}\right)}{f_{\boldsymbol{u}}\left(u_t\middle|H_0^{(\tilde{z})}\right)} = \lambda, \qquad \tilde{z} \in \widetilde{\mathcal{K}},\, t \in \mathcal{U}_{\mathrm{I}}. \qquad (8.120)$$

Calculating $P_{\mathrm{D}}^{(\tilde{z})}$ as a function of $P_{\mathrm{F}}^{(\tilde{z})}$ for different values of λ leads to the intended relation, whereas the two conditional PDFs can be estimated by the given RBF network, as shown in Eq. 8.110. The ROCs for the given RBF network and the applied nine detectors (see Sec. 8.5.2) are depicted in Fig. 8.24. Four curves are drawn for the interesting state characteristics for

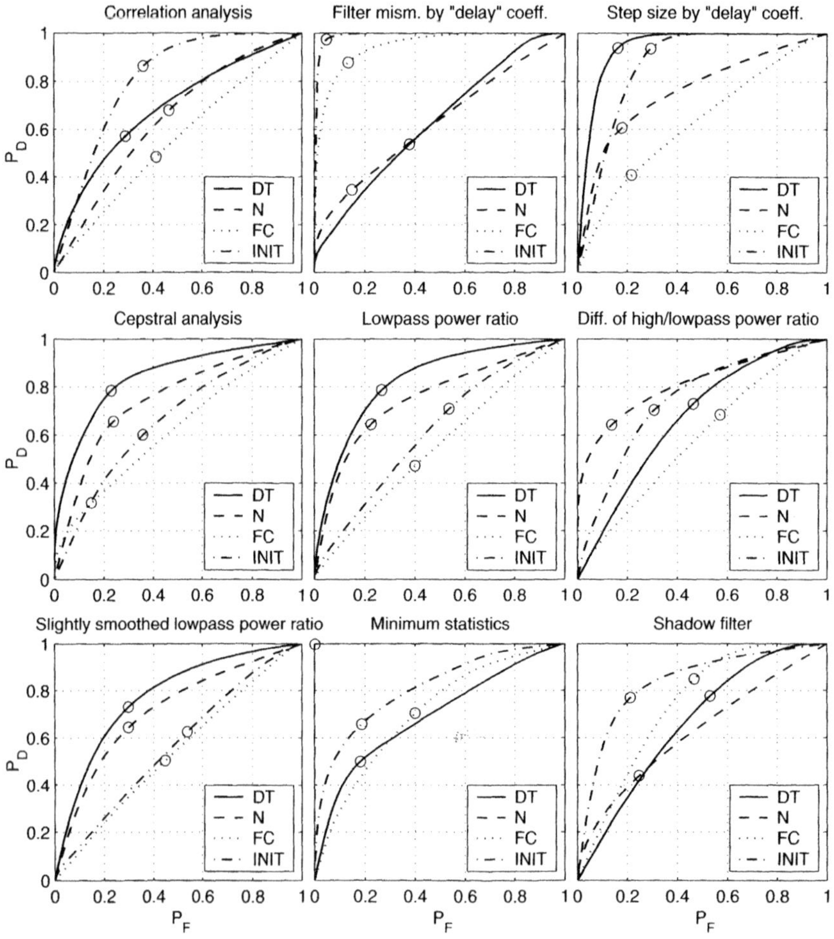

Fig. 8.24. Receiver operating characteristics (ROCs) – conditional detection probability $P_{\mathrm{D}}^{(\tilde{z})}$ as a function of the conditional false-alarm probability $P_{\mathrm{F}}^{(\tilde{z})}$ – for all nine applied detectors (see Sec. 8.5.2). Depicted are curves for the four state characteristics "double talk" (DT), "background noise" (N), "filter convergence" (FC) and "initialization" (INIT). Marked are the points for $\lambda = 1$, representing the optimal decision boundary according to Bayes for equal cost of false detections and equal a priori probabilities. Note: in case of minimum statistics the ROC for state characteristic "background noise" passes the (optimal) upper left corner.

each detector, respectively. The position of $\lambda = 1$ is marked; it complies with the Bayes decision bound for equal cost for all false detections and same a priori probabilities (see Eq. 8.116). A ROC will ideally pass the upper left corner of the diagram, that corresponds with the operating point of $P_{\mathrm{D}}^{(\tilde{z})} = 1$ with no false alarms ($P_{\mathrm{F}}^{(\tilde{z})} = 0$). By contrast, an unreliable detector features the bisecting line as characteristic – which means $P_{\mathrm{D}}^{(\tilde{z})} = P_{\mathrm{F}}^{(\tilde{z})}$ and the decision is purely random.

Analyzing the ROCs for the nine detectors in Fig. 8.24 shows large differences between the curves, as expected. Really high reliability can be achieved for classifying background noise by use of the minimum statistics (detector number 8) for example: 100% detection without false alarms. However, it is obviously that this detector is not able to classify the other state characteristics, since these curves are quite close to the bisection line. All other detectors yield a more or less strong reliability for classification of the four state characteristics, even though none of them passes the optimal operation point (which is quite seldom). For instance, the detector "filter mismatch by delay coefficients" can be effectively used for detecting the initialization state and the filter convergence. Conversely, reliable information on double-talk situations can be achieved by use of the "step size by delay coefficients" detector. Quite interesting is the fact, that the "correlation analysis" detector proves as moderately reliable only, even though this detector is often (successfully) applied for double-talk detection [21]. The reason can be found in the large dispersion of the estimated signals. For application of double-talk detectors, some additional signal processing is performed on the noisy signals. In our simulation, we only used the raw data of the detector.

8.6.3 Conclusions

In this chapter, we focussed on the control of speech quality improvement algorithms in hands-free units. Analyzing past research on the subjects reveals a huge number of approaches to that problem, which are justified by the conditions under which the hands-free unit is to operate – regarding the acoustic environment as well as customer requirements, such as low-cost in contrast to high-end quality. But we also found that control is usually implemented separately for each single algorithm, like noise-reduction, echo cancellation, and loss control.

The aim of this chapter was therefore to discuss combinations of various detectors and estimation normally in place for the various algorithms, in order to make the control decisions more reliable. A state model was presented to illustrate the synergies, and we applied fuzzy logic, which works from expert knowledge but is no means for optimization itself.

We also presented generic methods to optimize this combination of detectors and estimators, and the parameters for the control unit, for a given application, focussing on acoustic echo cancellation as the algorithm most sensible to the quality of its control unit. A number of possibly optimum step sizes was discussed, and from the result an evaluation of the quality of a step size with respect to echo cancellation was derived, the so-called cost function.

These results were applied in the design of neural network-like combination methods, comprising multilayer perceptrons as well as learning vector quantization.

The second part of the chapter showed the application of RBF networks not only for step-size calculation but also for evaluating the reliability of the applied detectors. The estimation of the receiver operating characteristics was based on the conditional probability density functions for the state characteristics. These PDFs could be reliably estimated by the RBF network, since a growing network structure, that facilitates the representation of any density, is applied. However, one has to keep in mind that the RBF network is learning by training samples only. Hence, the overall quality of the system, including its generalization ability, is mainly influenced by the compilation of training and verification data. We made a great effort to set up some generalized training data and to compile several training samples, even though we did not discuss this topic here. This cost some time during the design process, but it payed off by optimized performance at reduced cost during the operation of the hands-free unit.

This chapter should set the reader in the position of having a complete tool set at hand in order to design an efficient control unit for his hands-free unit, which is as indispensable as selecting and optimizing the audio algorithm itself.

References

[1] J. L. Alba, L. Docio, D. Docampo, O. W. Marquez: Growing Gaussian mixture network for classification applications, *Signal Processing*, **76**(1), 43–60, 1999.

[2] S. Albrecht, J. Busch, M. Kloppenburg, F. Metze, P. Tavan: Generalized radial basis function networks for classification and novelty detection: Self-organization of optimal Bayesian decision, *Neural Networks*, **13**, 1075–1093, 2000.

[3] J. Benesty, D. R. Morgan, J. H. Cho: A familiy of doubletalk detectors based on cross-correlation, *Proc. IWAENC '99,* 108–111, Pocono Manor, NJ, USA, 1999.

[4] C. Bishop: *Neural Networks for Pattern Recognition,* Oxford, UK: Clarendon Press, 1997.

[5] C. Breining: *Steuerung eines Kommunikationsterminals mit Freisprecheinrichtung,* Düsseldorf, Germany: Fortschr.-Ber., VDI-Reihe 10(570), VDI Verlag, 1998 (in German).

[6] C. Breining: Applying a neural network for step-size control in echo cancellation, *Proc. IWAENC '97,* London, UK, 1997.

[7] C. Breining, T. Schertler: Delay-free low-cost step gain estimation for adaptive fiters in acoustic echo cancellation, *Signal Processing* **80**(9), 1721–1731, 2000.
[8] C. Breining: A Robust fuzzy logic-based step-gain control for adaptive filters in acoustic echo cancellation, *IEEE Trans. Speech Audio Process.*, **T-SA-9**(2), 162–167, 2001.
[9] G. Bugmann: Normalized Gaussian radial basis function networks, *Neurocomputing*, **20**, 97–110, 1998.
[10] T. Burger: Practical application of adaptation control for NLMS-algorithms used for echo cancellation with speech signals, *Proc. IWAENC '95*, 87–90, Røros, Norway, 1995.
[11] A. Cichocki, R. Unbehauen: *Neural Networks for Optimization and Signal Processing,* New York, NY, USA: Wiley, 1993.
[12] G. Cybenko: Approximation by superpositions of a sigmoidal function, *Mathematics of Control, Signals, and Systems*, **2**, 303–314, 1989.
[13] R. Frenzel: *Freisprechen in gestörter Umgebung*, Düsseldorf, Germany: Fortschr.-Ber. VDI-Reihe 10(228), VDI Verlag, 1992 (in German).
[14] T. Gänsler, J. Benesty: The fast normalized cross-correlation double-talk detector, *Signal Processing*, accepted for publication, 2006.
[15] A. Gersho, R. M. Gray: *Vector Quantization and Signal Compression*, Boston, MA, USA: Kluwer, 1992.
[16] F. Girosi, T. Poggio: Networks and the best approximation property, *Biological Cybernetics*, **63**, 169–176, 1990.
[17] E. Hänsler, G. Schmidt: *Adaptive Echo and Noise Control – A Practical Approach*, New York, NY, USA: Wiley, 2004.
[18] S. Halgamuge: *Advanced Methods for Fusion of Fuzzy Systems and Neural Networks in Intelligent Data Processing*, Düsseldorf, Germany: Fortschr.-Ber., VDI-Reihe 10(401), VDI Verlag, 1996.
[19] S. Haykin: *Neural Networks,* 2nd ed., Englewood Cliffs, NJ, USA: Prentice-Hall, 1999.
[20] S. Haykin: *Adaptive Filter Theory,* 4th ed., Englewood Cliffs, NJ, USA: Prentice Hall, 2002.
[21] P. Heitkämper: An adaptation control for acoustic echo cancellers, *IEEE Signal Process. Letters,* **4**(6), 170–172, 1997.
[22] C. C. Holmes, B. K. Mallick: Bayesian radial basis functions of variable dimension, *Neural Computation*, **10**, 1217–1233, 1998.
[23] K. Hornik, M. Stinchcombe, H. White: Multilayer feedforward networks are universal apprixomators, *Neural Networks,* **2**, 359–366, 1989.
[24] ITU-T Recommendation G.167: *General Characteristics of International Telephone Connections and International Telephone Circuits – Acoustic echo Controllers,* Helsinki, Finland, 1993.
[25] N. B. Karayiannis, C. Bezdek, et al: Repairs to GLVQ: A new family of competitive learning schemes, *IEEE Trans. on Neural Networks*, **7** (5), 1062–1071, 1996.

[26] N. B. Karayiannis, Pin-I Pai: Fuzzy algorithms for learning vector quantization, *IEEE Trans. on Neural Networks* **7** (5), 1196–1211, 1996.

[27] G. J. Klir, B. Yuan: *Fuzzy Sets and Fuzzy Logic: Theory and Applications*, New Jersey, NY, USA: Prentice Hall, 1995.

[28] T. Kohonen: The self-organizing map, *Proc. IEEE*, **78**, 1464–1480.

[29] T. Kohonen: *Self-Organizing Maps*, 2nd ed., Berlin, Germany: Springer, 1997.

[30] A. Mader: Automatic optimization for a control of hands-free telephone sets, *Proc. IWAENC '99*, 120–123, Pocono Manor, NJ,USA, 1999.

[31] A. Mader, H. Puder, G. Schmidt: Step-size control for acoustic echo cancellation filters – an overview, *Signal Processing*, **80**(9), 1697–1719, 2000.

[32] A. Mader: *Radiale-Basisfunktionen-Netz zur automatischen Optimierung der Steuerung einer Freisprecheinrichtung*, Düsseldorf, Germany: Fortschr.-Ber. VDI-Reihe 10(711), VDI Verlag, 2002 (in German).

[33] R. Martin: An Efficient algorithm to estimate the instantaneous SNR of speech signals, *Proc. EUROSPEECH '93*, **3**, 1093–1096, Berlin, Germany, 1993.

[34] J. Marx: *Akustische Aspekte der Echokompensation in Freisprecheinrichtungen*, Düsseldorf, Germany: Fortschr.-Ber. VDI-Reihe 10(400), VDI Verlag, 1996 (in German).

[35] D. J. Miller, H. S. Uyar: Combined learning and use for a mixture model equivalent to the RBF classifier, *Neural Computation*, **10**, 281–293, 1998.

[36] J. Platt: A resource-allocating network for function interpolation, *Neural Computation*, **3**, 213–225, 1991.

[37] T. Poggio, F. Girosi: A theory of networks for approximation and learning, *A. I. Memo No. 1140, Artificial Intelligence Laboratory*, Massachusetts Institute of Technology, USA, 1989.

[38] D. E. Rumelhart, J. L. McClelland (eds.): *Parallel Distributed Processing*, **1**, Cambridge, MA, USA: MIT Press, 1986.

[39] G. Schmidt: Step-size control in subband echo cancellation Systems, *Proc. IWAENC '99*, 116–119, Pocono Manor, NJ, USA, 1999.

[40] U. Schultheiss: *Über die Adaption eines Kompensators für akustische Echos*, Düsseldorf, Germany: Fortschr.-Ber. VDI-Reihe 10(90), VDI Verlag, 1988 (in German).

[41] M. K. Titsias, A. C. Likas: Shared kernel models for class conditional density estimation, *IEEE Trans. Neural Networks*, **12**(5), 987–997, 2001.

[42] T. Tollenaere: Super SAB: Fast adaptive backpropagation with good scaling properties, *Neural Networks*, **3**, 561–573, 1990.

[43] H. G. C. Traven: A neural network approach to statistical pattern classification by "semiparametric" estimation of probability density functions, *IEEE Trans. Neural Networks*, **2**(3), 366–377, 1991.

[44] H. L. v. Trees: *Detection, Estimation, and Modulation Theory, Part I*, New York, NY, USA: Wiley, 2003.

[45] E. Yair, K. Zeger, A. Gersho: Competitive learning and soft competition for vector quantizer design, *IEEE Trans. on Signal Process.*, **40**(2), 294–309, 1992.
[46] S. Yamamoto, S. Kitayama: An adaptive echo canceller with variable step gain method, *Trans. IECE Jpn.*, **E65**(1), 1–8, 1982.
[47] L. A. Zadeh: Fuzzy sets, *Information and Control*, **8**, 338–353, 1965.
[48] L. A. Zadeh, J. Kacprzyk (eds.), *Fuzzy Logic for the Management of Uncertainty*, New York, NY, USA: Wiley, 1992.
[49] C. Zhu, J. Wang, T. Wang: Analysis of learning vector quantization algorithms for pattern classification, *Proc. ICASSP-95*, **5**, 3471–3474, Detroit, MI, USA, 1995.
[50] H. J. Zimmermann. *Fuzzy Set Theory and its Applications*, 4th ed., Boston, MA, USA: Kluwer, 2001.

Part IV

Noise Reduction

9

Noise Reduction

Ulrich Heute

Chair for Circuits & Systems, Faculty of Engineering
Christian-Albrecht University Kiel, Germany

9.1 Introduction

9.1.1 Noise and Speech

In a telephone connection, the microphones do not only capture the speakers' voices, but also some of the surrounding acoustic signals, often summarized as – implicitly: undesired – "noise". The type of such disturbances may vary considerably: A "classical" handset in a home or office might record some standard, relatively low-level background events, like other people's voices ("babble"), fan noise from a computer nearby, or wind and remote-vehicle humming, audible through an open window. Such effects are amplified, however, relative to the actual speech sound, when a hands-free telephone is used, with a much larger distance between the speaker's mouth and a microphone. Even worse and stronger effects appear when the telephone is applied in more open environments, e.g., a mobile phone in a hands-free mode inside a moving car or beside a lively street. Furthermore, also the speakers' voices themselves may appear as an interference, in case of echos and reverberation due to a corresponding acoustic environment.

The additional signals may, to a certain extent, also carry some information for a listener, namely, on the actual conditions under which the speaker is uttering her or his words. Nevertheless, the "noise" is mainly perceived as a disturbance, since it causes difficulties in a telephone connection, as the listener is, of course, interested mainly in the words of the speaker and not in the sound of a car passing by. Even more problems arise for an automatic speaker- or speech-recognition (ASR) system which is supposed to derive some reaction from the spoken utterance.

The annoyance may be increased, if the speech-plus-noise signal is transmitted through some data-compressing system, like in a mobile-radio channel or also a modern wire-bound, low-rate link: Here, codecs reduce the data rate from the well-known 64 $kbit/s$ of PCM / ISDN down to the 13 $kbit/s$ of "full-rate GSM", to 5...6 *kbit/s* in more recent ITU-T standards for "multi-media communication", or even to 2 *kbit/s* like in some MPEG-4 variants (see, e.g., [31]). Such codecs rely massively on models of the *one speech* signal to be encoded. This basis is destroyed by additive, completely different components as well as by competing speech or speech-like parts, and this may cause strongly non-linear effects due to the system's malfunction.

A removal or, at least, reduction and control of additive disturbances is therefore often highly desirable.

9.1.2 Types of Disturbances, Aim of Reduction

In this chapter, additive *noise* will be dealt with, only.

This includes more or less white noise, as produced, approximately, by the wind through an open window in a moving car; it includes also coloured noise, as generated by the car's mechanical rolling ("lowpass noise"), or even noise with some harmonic components, possibly stemming from the vehicle's engine. It may include "babble", but not directly competing speech signals of other speakers close to the same microphone: The so-called "cocktail-party effect" (see, e.g., [4]) is not covered.

Echo reduction, as important as noise reduction, is dealt with in Chapters 2, 5, 6, and 7 of this book.

The effects of noise on an ASR can be quite strong: For example, the recognition rate of a single-letter recognizer may drop, from some 87% with "clean" speech, by some 10% even at relatively good noise conditions with an average signal-to-noise ratio (SNR) of $\sim 15\ dB$, or even by some 70% at $SNR \approx 0\ dB$ (see, e.g., [50]). This problem will, however, not be in the focus of this chapter: The central point of this section will be an *enhanced speech-signal reproduction for a human listener.*

Still, this may be useful also for ASR, with the simple philosophy that a "better-sounding" signal should also be "more understandable", both for a machine and for a listener. Although this is not necessarily true, sound-optimized noise-reduction front-ends for ASR are quite commonly used.

With the same reasoning, such a pre-processing is used prior to data compression: It is assumed that a speech signal with an improved sound is also good for the codec. Therefore, in this section, we shall not look into the details of noise influencing the coding parameters and the avoidance of especially bad effects.

9.1.3 Noise-Reduction Approaches

There is a common discrimination of single-, two-, and multi-channel noise-reduction techniques. These terms simply refer to the number of microphones.

The step from "two" to "multi" seems to be quite arbitrary. It has, however, a justification: Two microphones can be placed, with respect to the speaker, either beside or behind each other, where any non-zero angle makes the difference vanish. An alternative is to use one microphone close to the speaker, the other one "far enough away", such that it catches mainly (or only) the noise. (In this case, the latter microphone could also be exchanged for some other sensor delivering information about the noise source - e.g., about the frequency and the amplitude of motor activities, allowing to estimate the resulting motor noise in the cabin.) With at least three microphones, however, a larger *multi*tude of possible positions can be imagined, opening also more variants of signal combinations in the processing scheme.

In this chapter, *single-channel methods* will be treated. In Sections 2, 3, and 4 of this book, techniques with more sensors are covered in depth.

One classical single-microphone approach is based on a minimum-mean-square error (MMSE), leading to a Wiener filter. Its generalization, the Kalman filter, is out of this section's scope; it is covered in Chapter 10. A second, also classical approach is that of spectral subtraction; it is based on the simple idea that a noise that is additive should be subtracted from the disturbed signal, and as this turns out to be impossible in the time domain, it is done in the spectrum.

9.1.4 Wiener Filter and Spectral Subtraction

In short, the above-named MMSE approach can be summarized as follows: Let a microphone signal $y(n)$ consist of a clean speech wave $s(n)$ spoilt by additive noise $b(n)$:

$$y(n) = s(n) + b(n). \tag{9.1}$$

Then design a filter with a frequency response $H(e^{j\Omega})$ and an impulse response $h(n) = \mathcal{F}^{-1}\{H(e^{j\Omega})\}$ such that its output signal $\hat{s}(n)$ approximates $s(n)$ in the MMSE sense (see Fig. 9.1).

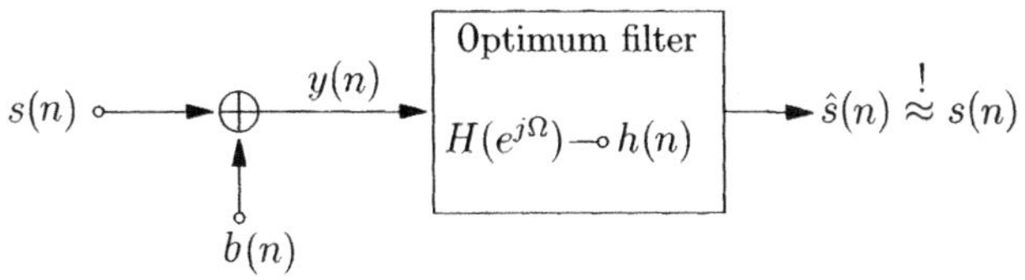

Fig. 9.1. Basic structure of a Wiener filter.

The frequency variable

$$\Omega \doteq 2\pi \cdot \frac{f}{f_s} = 2\pi \cdot f \cdot T_s \tag{9.2}$$

is the angular frequency $2\pi \cdot f$ normalized to the sampling frequency $f_\mathrm{s} = 1/T_\mathrm{s}$. The optimized filter is often termed Wiener filter, since its idea is already found in a 1949 publication by N. Wiener [78]. It may not only be described, but also realized both in the time and the frequency domains: The output signal is found by the convolution

$$\begin{aligned} \hat{s}(n) &= h(n) * y(n) = h(n) * \left[s(n) + b(n)\right] \\ &= \sum_{\kappa=-\infty}^{\infty} h(n-\kappa) \cdot \left[s(\kappa) + b(\kappa)\right] \\ &= \sum_{\kappa=-\infty}^{\infty} h(\kappa) \cdot \left[s(n-\kappa) + b(n-\kappa)\right], \end{aligned} \tag{9.3}$$

or by the inverse Fourier transformation of its spectrum

$$\begin{aligned} \hat{S}\left(e^{j\Omega}\right) &= H\left(e^{j\Omega}\right) \cdot Y\left(e^{j\Omega}\right) \\ &= H\left(e^{j\Omega}\right) \cdot \left[S\left(e^{j\Omega}\right) + B\left(e^{j\Omega}\right)\right], \end{aligned} \tag{9.4}$$

where $Y(e^{j\Omega})$, $S(e^{j\Omega})$, and $B(e^{j\Omega})$ are the Fourier transforms of the noisy signal $y(n)$, the clean signal $s(n)$, and the noise $b(n)$, respectively.

In practice, the filter is adapted to the short-time situation in blocks of M data samples, and it is kept constant during corresponding time periods of length

$$T_{adapt} = M \cdot T_s = M/f_s. \tag{9.5}$$

For example, in a typical telephone case, we might have

$$f_s = 8\ kHz, \tag{9.6}$$

$$M = 128, \tag{9.7}$$

and, according to (9.5),

$$T_{adapt} = 16\ ms. \tag{9.8}$$

So, in (9.4), $Y(e^{j\Omega})$, $S(e^{j\Omega})$, and $B(e^{j\Omega})$ are to be understood as short-time spectra, valid for M values and a duration T_{adapt}.

Starting from the (short-time) spectrum $Y(e^{j\Omega}) = S(e^{j\Omega}) + B(e^{j\Omega})$ of the noisy signal $y(n)$, a subtraction of $B(e^{j\Omega})$ is obviously equivalent to a removal of the noise $b(n)$. Since, however, only $y(n)$ is available, $B(e^{j\Omega})$ is unknown, a priori. But it can be assumed that, by some means to be discussed, at least an approximate magnitude $|\tilde{B}(e^{j\Omega})|$ of the noise spectrum can be measured or "estimated". Assuming, furthermore, that a phase distortion is not too important for speech signals [44], a modified subtraction can be applied:

$$\hat{S}\left(e^{j\Omega}\right) \doteq Y\left(e^{j\Omega}\right) - \tilde{B}\left(e^{j\Omega}\right)$$
$$\doteq \left[\left|Y\left(e^{j\Omega}\right)\right| - \left|\tilde{B}\left(e^{j\Omega}\right)\right|\right] \cdot e^{j \cdot arg\{Y(e^{j\Omega})\}}. \tag{9.9}$$

Here, only a spectral-*magnitude* subtraction is carried out, while the noisy signal's phase is left unchanged.

A slightly different argumentation starts from the assumption that $s(n)$ and $b(n)$ are uncorrelated and that, therefore, their power (-density) spectra (PDS) add:

$$S_{yy}(\Omega) = S_{ss}(\Omega) + S_{bb}(\Omega).$$

Assuming again that spectra can be replaced by short-time measurements, i.e. here: PDS terms by squared magnitudes of the above used descriptions, an estimated noise-power spectrum $|\tilde{B}(e^{j\Omega})|^2$ has then to be subtracted from $|Y(e^{j\Omega})|^2$. Variants using $|\tilde{B}(e^{j\Omega})|^\beta$ and $|Y(e^{j\Omega})|^\beta$, with a suitably chosen parameter β, are also known.

Eq. (9.9) may be re-written:

$$\hat{S}\left(e^{j\Omega}\right) = Y\left(e^{j\Omega}\right) \cdot \left[1 - \tilde{B}\left(e^{j\Omega}\right)/Y\left(e^{j\Omega}\right)\right] \doteq Y(e^{j\Omega}) \cdot H_{sps}(e^{j\Omega}) \tag{9.10}$$

As $\hat{S}(e^{j\Omega})$ maintains the phase $arg\{Y(e^{j\Omega})\}$, according to (9.9), the factor $H_{sps}(e^{j\Omega})$ denotes a real-valued, even-symmetrical, zero-phase frequency response:

$$H_{sps}\left(e^{j\Omega}\right) \doteq W_o(\Omega) = W_o(-\Omega) \in \mathbf{R}. \tag{9.11}$$

This means that the spectral subtraction can be interpreted as a spectral weighting. The close link to the Wiener-Filter description in (9.4) is evident; of course, the relation between the optimized function $H(e^{j\Omega})$ and the "abbreviation" $H_{sps}(e^{j\Omega})$ in (9.10) has to be investigated.

The above two, closely linked approaches to speech enhancement will be in this chapter's focus. Variants of the "subtraction rule" in (9.9) and the "noise estimation" or, equivalently, "filter definition" according to (9.10) and (9.4) are described. In particular, possible ways from the given signals to their short-time spectra and back to time signals, i.e., various spectral-analysis and spectral-synthesis systems are addressed.

9.2 Optimum-Filter Design in the Time Domain

9.2.1 Mean-Square Error Minimization

As stated in Sec. 9.1.4, the filter's output signal shall approximate the clean-speech signal $s(n)$ in the MMSE sense. If all signals are viewed as random sequences, the mean-square error is given by the expectation

$$\varepsilon = E\Big\{\big[\hat{s}(n) - s(n)\big]^2\Big\}. \tag{9.12}$$

With (9.3), ε depends on the infinitely many impulse-response samples $h(\kappa)$. Minimization of ε with regard to these degrees of freedom means that all corresponding partial derivatives have to be zero:

$$\partial\varepsilon/\partial h(\lambda) = 0, \quad \lambda \in \mathbf{Z}. \tag{9.13}$$

This leads to infinitely many linear equations with the unknown values $h(n)$. With the usual definition of a correlation sequence for two real-valued signals $u(n)$ and $v(n)$, belonging to stationary processes, by

$$r_{uv}(\lambda) = E\big\{u(n) \cdot v(n+\lambda)\big\}, \tag{9.14}$$

these equations have the following form:

$$\sum_{\kappa=-\infty}^{\infty} h(\kappa) \cdot r_{yy}(\lambda - \kappa) = r_{ys}(\lambda), \quad \lambda \in \mathbf{Z}. \tag{9.15}$$

Based on (9.1), Eq. (9.15) is re-written as

$$\sum_{\kappa=-\infty}^{\infty} h(\kappa) \cdot \big[r_{ss}(\lambda - \kappa) + r_{bb}(\lambda - \kappa) + r_{sb}(\lambda - \kappa) + r_{bs}(\lambda - \kappa)\big]$$
$$= r_{ss}(\lambda) + r_{bs}(\lambda), \quad \lambda \in \mathbf{Z}.$$

This can be simplified, if speech and noise are assumed to have zero mean values and to be uncorrelated. Then, $r_{sb}(\lambda) = r_{bs}(\lambda) \equiv 0 \ \forall \lambda$ holds, and we have

$$\sum_{\kappa=-\infty}^{\infty} h(\kappa) \cdot \big[r_{ss}(\lambda - \kappa) + r_{bb}(\lambda - \kappa)\big] = r_{ss}(\lambda), \quad \lambda \in \mathbf{Z}. \tag{9.16}$$

9.2.2 Approximate FIR-Filter Solution

The filter can now be restricted to have a finite-length impulse response (FIR). A simple choice is

$$h(\kappa) \text{ optimized for } \kappa \in \big\{-N_F, \ldots, -1, 0, 1, \ldots, N_F\big\},$$
$$h(\kappa) = 0 \text{ else.} \tag{9.17}$$

Using now, for example, the central $2N_F + 1$ equations of (9.16), with $\lambda \in \{-N_F, \ldots, -1, 0, 1, \ldots, N_F\}$, a finite-dimension matrix / vector formulation is found instead of (9.16):

$$\boldsymbol{R}_{yy} \cdot \boldsymbol{h} = \boldsymbol{r}_{ss}. \tag{9.18}$$

Due to the well-known symmetries of auto-correlation sequences, namely, $r_{uu}(-\lambda) = r_{uu}(\lambda)$, also $\boldsymbol{R}_{yy}$ and $\boldsymbol{r}_{ss}$ possess symmetries: The correlation vector $\boldsymbol{r}_{ss}$ is even with respect to its central point $r_{ss}(0)$ according to

$$\boldsymbol{r}_{ss} \doteq \left[r_{ss}(N_F), \ldots, r_{ss}(1), r_{ss}(0), r_{ss}(1), \ldots, r_{ss}(N_F)\right],$$

and the correlation matrix $\boldsymbol{R}_{yy}$ is a symmetrical Toeplitz matrix:

$$\boldsymbol{R}_{yy} = \begin{bmatrix} r_{yy}(0) & r_{yy}(1) & \ldots & r_{yy}(2N_F-1) & r_{yy}(2N_F) \\ r_{yy}(1) & r_{yy}(0) & \ldots & r_{yy}(2N_F-2) & r_{yy}(2N_F-1) \\ \vdots & \vdots & \vdots & \vdots & \vdots \\ r_{yy}(2N_F-1) & r_{yy}(2N_F-2) & \ldots & r_{yy}(0) & r_{yy}(-1) \\ r_{yy}(2N_F) & r_{yy}(2N_F-1) & \ldots & r_{yy}(1) & r_{yy}(0) \end{bmatrix}.$$

Therefore, also the solution vector

$$\boldsymbol{h} \doteq \left[h(-N_F), \ldots, h(0), \ldots, h(N_F)\right] = \left[\boldsymbol{R}_{yy}\right]^{-1} \cdot \boldsymbol{r}_{ss} \tag{9.19}$$

of (9.18) is real-valued and symmetrical with respect to its center at $\kappa = 0$:

$$h(-\kappa) = h(\kappa).$$

Thus, the corresponding FIR filter has an even, real-valued, zero-phase frequency response

$$H\left(e^{j\Omega}\right) = \mathcal{F}\{h(n)\} \doteq H_o^{\mathrm{TD}}(\Omega) = H_o^{\mathrm{TD}}(-\Omega) \in \mathbf{R}. \tag{9.20}$$

A realizable, causal version of this filter is found by simply shifting $h(n)$ according to

$$h_1(n) \doteq h(n - N_{\mathrm{F}}), \quad n \in \{0, 1, 2, \ldots, 2N_{\mathrm{F}} - 1, 2N_{\mathrm{F}}\}, \tag{9.21}$$

which leads to a linear-phase frequency response

$$H_1\left(e^{j\Omega}\right) = \mathcal{F}\{h_1(n)\} = H_{\mathrm{o}}^{\mathrm{TD}}(\Omega) \cdot e^{-jN_{\mathrm{F}}\Omega}.$$

The additional phase term says that the "approximately optimum" output signal is now $\hat{s}(n - N_F)$, delayed by N_F samples.

For the solution (9.19), the knowledge or estimation of the auto-correlations of both the noisy and the clean speech signals is needed – especially, the latter being an obvious problem, as only $y(n)$ is available. This question will be dealt with in Sec. 9.7.

9.3 Wiener-Filter Description in the Frequency Domain

9.3.1 Optimum Frequency Response

Eq. (9.16) describes a convolution:

$$h(\lambda) * \left[r_{ss}(\lambda) + r_{bb}(\lambda)\right] = r_{ss}(\lambda), \qquad \lambda \in \mathbf{Z}.$$

With the PDS being a correlation's Fourier transform, i.e., generally,

$$S_{uv}(\Omega) \doteq \mathcal{F}\{r_{uv}(\lambda)\},$$

an equivalent description in the frequency domain is found by means of the convolution theorem:

$$H\left(e^{j\Omega}\right) \cdot \left[S_{ss}(\Omega) + S_{bb}(\Omega)\right] = H\left(e^{j\Omega}\right) \cdot S_{yy}(\Omega) = S_{ss}(\Omega).$$

This is easily solved for the Wiener-filter frequency response:

$$H\left(e^{j\Omega}\right) = \frac{S_{ss}(\Omega)}{S_{yy}(\Omega)} = \frac{S_{ss}(\Omega)}{S_{ss}(\Omega) + S_{bb}(\Omega)} \doteq H_{\mathrm{o}}(\Omega). \tag{9.22}$$

Due to the well-known symmetries of a PDS, also this function $H_{\mathrm{o}}(\Omega)$ is real-valued and even in Ω, and it describes a zero-phase system, like $H_{\mathrm{o}}^{TD}(\Omega)$ in (9.20). It has to be noted, however, that, despite the striking similarity in (9.22) and (9.19), the filters are not identical: For (9.19), an FIR filter was assumed a priori, and an arbitrary, *limited section* of the autocorrelation was used; here, the Fourier transformation is applied to infinitely long correlation sequences. There is no justification to assume that $H(e^{j\Omega})$ of (9.22) corresponds to a finite-length inverse transform. It is not even clear whether there is any rational transfer function $H(z)$ behind (9.22) to be realized in one of the usual digital-filter structures.

9.3.2 Approximate FIR-Filter Solution

A relatively simple step, however, leads to an approximate, zero-phase FIR filter: $H(e^{j\Omega})$ may be sampled in M equi-spaced frequencies $\Omega_i = i \cdot 2\pi/M, i \in \{0, 1, \ldots, M-1\}$; then, a length-$M$ inverse Discrete Fourier Transformation (IDFT) yields a time sequence

$$\check{h}_2(n) = IDFT_M\left\{H\left(e^{j\Omega_i}\right)\right\} = \frac{1}{M} \cdot \sum_{i=0}^{M-1} H\left(e^{j\Omega_i}\right) \cdot e^{jni\cdot\frac{2\pi}{M}},$$
$$n \in \{0, 1, \ldots, M-1\}, \tag{9.23}$$

which is real-valued and even-symmetrical:

$$\check{h}_2(M-n) = \check{h}_2(n), \quad n \in \{0, 1, \ldots, M/2\}.$$

Due to the implicit periodicity of DFT and IDFT, a realizable, causal FIR filter is created, if $\check{h}_2(n)$ is shifted by N_F samples again and, then, the resulting values for $n \in \{0, 1, \ldots, 2N_\mathrm{F}\}$ are used:

$$h_2(n) \doteq \check{h}_2(n - N_\mathrm{F}), \; n \in \{0, 1, \ldots, 2N_\mathrm{F}\}. \tag{9.24}$$

Again, the delayed result $\hat{s}(n - N_\mathrm{F})$ is only "approximately optimal": The impulse response $h_2(n)$ of (9.24) is a time-domain aliased, finite-length version of the infinitely long inverse Fourier transform of the frequency response $H(e^{j\Omega})$ in (9.22). It is, therefore, not the same approximate solution found as $h_1(n)$ in (9.21) from a-priori time restrictions.

9.4 Examples and Filtering Effects

With the help of a few theoretical, constructed cases, the above filter design and the general influences of Wiener filters shall be explained, for ease of understanding.

9.4.1 "Low-Pass Signal" plus "Band-Stop Noise"

As a constructed example, a clean signal $s(n)$ is assumed to have a constant, but band-limited PDS $S_{ss}(\Omega)$. It is spoilt by additive wide-band noise $b(n)$ with a band gap in its PDS $S_{bb}(\Omega)$ (see Fig. 9.2a). The resulting "stair-case" PDS $S_{yy}(\Omega) = S_{ss}(\Omega) + S_{bb}(\Omega)$ as well as the Wiener-filter frequency response according to (9.22) are shown in Fig. 9.2b.

Following the time-domain design of Sec. 9.2, by calculating the correlation terms $r_{ss}(\lambda)$ and $r_{yy}(\lambda)$ from the above spectra and choosing, e.g., $N_\mathrm{F} = 30$, yields an impulse response $h_1(n)$ – see (9.21) – as depicted in Fig. 9.3a. The real frequency response $H_\mathrm{o}(\Omega)$ corresponding to $h_1(n)$ is depicted in Fig. 9.4a, together with the desired, theoretical Wiener filter. In Fig. 9.4b, $S_{ss}(\Omega)$ and $S_{bb}(\Omega)$ are shown once more, in comparison with the PDS $S_{\hat{s}\hat{s}}(\Omega) = S_{yy}(\Omega) \cdot [H_\mathrm{o}^\mathrm{TD}(\Omega)]^2$ of the filter-output signal $\hat{s}(n)$.

Obviously, the MMSE optimized filter yields a noise reduction indeed – the upper band of $S_{bb}(\Omega)$ is "zeroed" in Fig. 9.4b! – but also a very noticeable signal (-spectrum) distortion. We shall come back to this below.

Following the frequency-domain approach of Sec. 9.3 now, sampling the ideal Wiener-filter frequency response of (9.22) and reordering the IDFT results as in (9.24), leads to the impulse response $h_2(n)$ given in Fig. 9.3b. It is evident that $h_1(n) \neq h_2(n)$: As said in Sec. 9.3 already, the two FIR approximations do not give identical filters. Also the filter frequency responses are different as well as the results of the filtering processes, as can be seen from a comparison between Fig. 9.4 and Fig. 9.5.

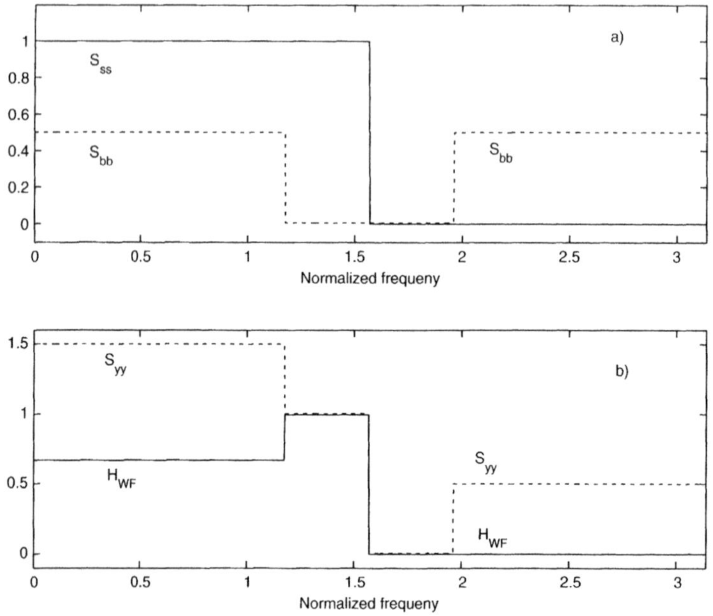

Fig. 9.2. Constructed example of a) PDS of a clean signal $s(n)$ and a noise signal $b(n)$, b) the resulting noisy-signal PDS and the Wiener-filter theoretical frequency response H_{WF}.

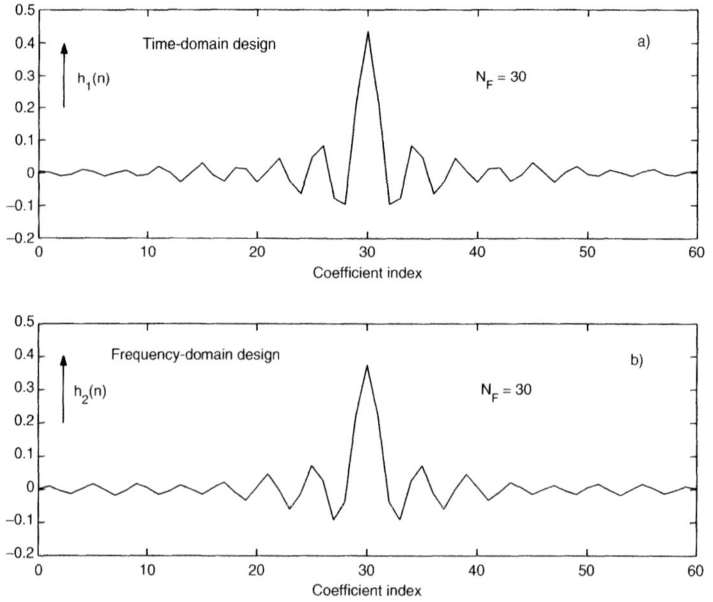

Fig. 9.3. Impulse responses of FIR Wiener-filter approximations: a) $h_1(n)$ of the time-domain design, b) $h_2(n)$ of the frequency-domain approach, both of length $2N_{\mathrm{F}} + 1 = 61$.

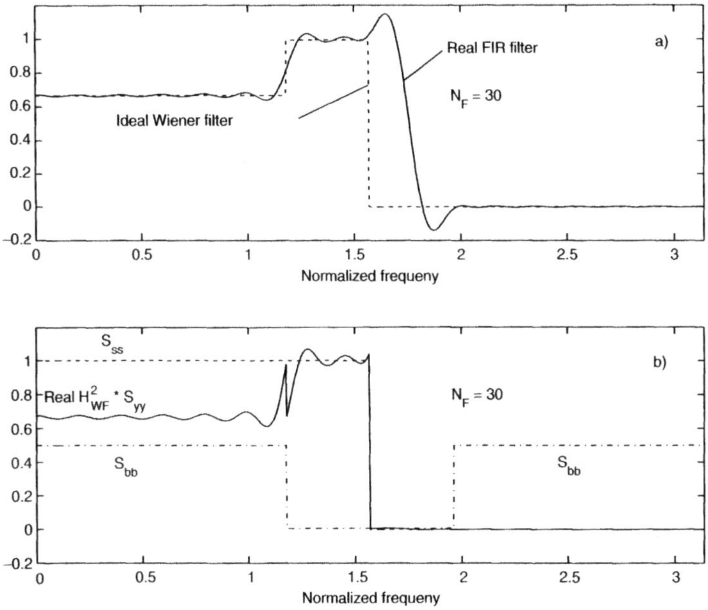

Fig. 9.4. a) Ideal and time-domain approximated Wiener filter; b) PDS of signal, noise, and resulting output signal $\hat{s}(n)$.

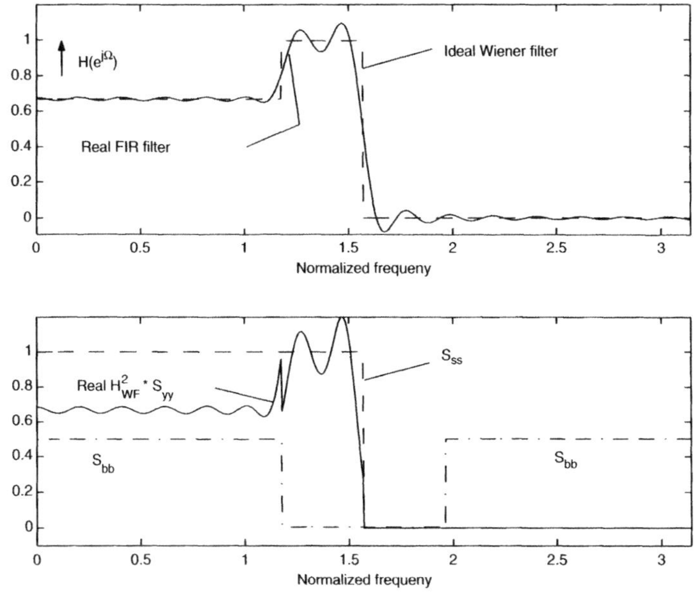

Fig. 9.5. a) Ideal and frequency-domain approximated Wiener filter; b) PDS of signal, noise, and resulting output signal $\hat{s}(n)$.

Both Figs. 9.4 and 9.5 reveal that the above-named strong deviation between $S_{ss}(\Omega)$ and $S_{\hat{s}\hat{s}}(\Omega)$ has only to some extent to do with the FIR approximations. The distortion is severe also in the low-frequency band in which both approaches follow the ideal curve quite closely (see Figs. 9.4a and 9.5a). In fact, a similar, large difference would even appear if the ideal filter was applied. The reason lies in the fact that $S_{yy}(\Omega)$ is multiplied by $H^2_{\mathrm{WF}}(\Omega)$:

$$S_{\hat{s}\hat{s}}(\Omega) = S_{yy}(\Omega) \cdot \left[\frac{S_{ss}(\Omega)}{S_{yy}(\Omega)}\right]^2 = \frac{S_{ss}(\Omega)}{S_{yy}(\Omega)} \cdot S_{ss}(\Omega)$$
$$= \frac{S_{ss}(\Omega)}{S_{ss}(\Omega) + S_{bb}(\Omega)} \cdot S_{ss}(\Omega) \qquad (9.25)$$

Eq. (9.25) says that the original spectrum will be damped at all frequencies wherever there is an additional noise component $S_{bb}(\Omega) > 0$.

9.4.2 Decaying Spectrum plus Wide-Band Noise

The spectral attenuation of (9.25) is the stronger, the higher the noise PDS is relative to the signal PDS, i.e., the lower the local SNR. This means that a Wiener filter designed (by either of the above two approaches) for some white noise and a signal with a PDS that decays towards higher frequencies (like, e.g., speech!), will show a low-pass behaviour, and, thus, the resulting, filtered signal $\hat{s}(n)$ has a more pronounced decay in its PDS than $s(n)$ (see Fig. 9.6). If the disturbances exhibit some positive slope for increasing frequencies, the spectral distortion is even stronger (see Fig. 9.7).

The observation, by the way, that, for signals like speech, an optimum filter acts as a low-pass is a justification of the simplest noise-reduction idea: If white noise is added to a low-pass signal, an appropriate low-pass filter will reduce the noise more than the signal and, thus, improve the SNR. Since it is known, however, that, for speech signals, the SNR is not a good estimate for the perceived quality (see, e.g., [57, 61]), the resulting distortion may cause problems; this has to be kept in mind.

A simple idea for a possibly "somewhat reduced" distortion at the price of a "little more" noise slightly modifies the Wiener filter: Instead of $H(e^{j\Omega})$, a power-η term $H_\eta(e^{j\Omega})$ is used:

$$H_\eta\left(e^{j\Omega}\right) \doteq \left[H\left(e^{j\Omega}\right)\right]^\eta, \quad \eta \in (0, 1]. \qquad (9.26)$$

For $\eta < 1$, strong attenuations in regions of low SNR are reduced, leaving a "little more" of the small signal PDS here.

9.5 Wiener-Filter Realizations

With either the time-domain or the frequency-domain approach, we have now an FIR-filter impulse response $h_1(n)$ or $h_2(n)$ (or a zero-phase equivalent $h(n)$)

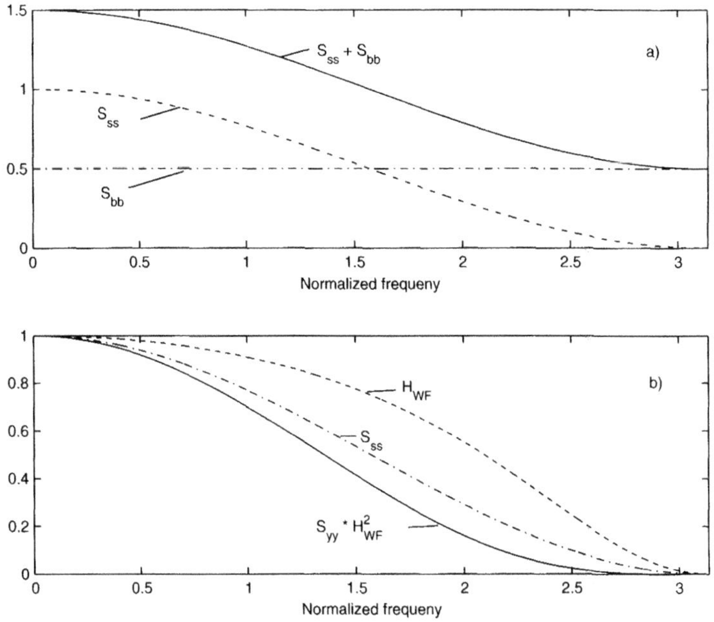

Fig. 9.6. a) Smoothly decaying PDS plus-white noise PDS; b) resulting Wiener-filter frequency response H_{WF}, ideal signal PDS, and filter-output PDS.

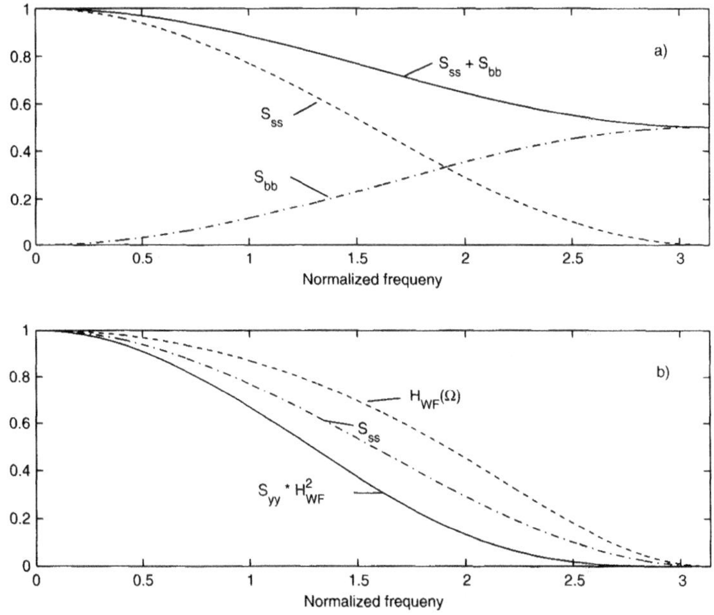

Fig. 9.7. a) Smoothly decaying PDS plus PDS of "pre-emphasis-type" noise; b) resulting Wiener-filter frequency response H_{WF}, ideal-signal PDS, and filter-output PDS.

which approximately corresponds to the optimal filter. Causal FIR systems can be simply realized in the time domain, as is well known, e.g., in the 2-nd canonical (direct) form of a non-recursive digital filter, also termed "tapped delay line" (see, e.g., [55, 65, 73]).

It is well-known also, from the same literature, that an equivalent solution takes the de-tour via the frequency domain: Both the input signal and the impulse response are transformed by means of DFT's, the spectral values are then multiplied pointwise, and finally the product sequence is subject to an IDFT creating the output signal. This de-tour can be computationally attractive, i.e., faster than the direct way, if the DFT's and the IDFT are realized efficiently via a Fast Fourier Transformation (FFT), giving rise to the notation as "fast convolution".

In order to make this really functional, some additional measures have to be taken. In our case, the procedure would be as follows:

- Define signal blocks of length L each:
  $\{y_n(\kappa) \doteq y(n-\kappa),\ \kappa \in \{0,1,\ldots,L-1\}\}$
- Create signal blocks of length $M > L$ each, by padding $y_n(\kappa)$ with $M-L$ zeros:
  $\{\tilde{y}_n(\kappa) = y_n(\kappa),\ \kappa \in \{0,1,\ldots,L-1\}, \tilde{y}_n(\kappa) = 0 \text{ for } \kappa \in \{L,\ldots,M-1\}\}$.
- Create a length-M set $\tilde{h}_{1,2}(\kappa)$ from one of the FIR sequences $h_{1,2}(\kappa)$ by zero-padding them with $M-(2N_\mathrm{F}+1)$ zeros:
  $\{\tilde{h}_{1,2}(\kappa) = h_{1,2}(\kappa), \kappa \in \{0,1,\ldots,2N_\mathrm{F}\}, \tilde{h}_{1,2}(\kappa) = 0 \text{ for } \kappa \in \{2N_\mathrm{F}+1,\ldots,M-1\}$.
- Compute the length-M DFT's $\{\tilde{Y}_n(i)\}$ from $\{\tilde{y}_n(\kappa)\}$ and $\{\tilde{H}(i)\}$ from $\{\tilde{h}_{1,2}(\kappa)\}$.
- Multiply the two DFT sets pointwise, creating $\{\tilde{X}_n(i), i \in \{0,1,\ldots,M-1\}\}$.
- Apply an IDFT of length M to $\{\tilde{X}(i)\}$, yielding time-signal blocks $\{\tilde{x}_n(\kappa),\ \kappa \in \{0,1,\ldots,M-1\}\}$.
- Create output signal blocks $\{\hat{s}_n(\kappa), \kappa \in \{0,1,\ldots,L-1\}$ from $\{x_n(\kappa)\}$ by using an overlap-add process (see, e.g., [55, 65]).
- Also, input-signal blocks may consist of overlapping parts of $y(n)$; this must, of course, be taken into account in the overlap-add step.

This procedure is depicted graphically in Fig. 9.8.

Seemingly, there is a wasteful step in this realization, if $h_2(n)$ is used, as found from the frequency-domain approach of Sec. 9.3: The inverse DFT in (9.23) of the optimum frequency response (9.22) could be saved, since now $h_2(n)$ undergoes a DFT. So, samples of $H(e^{j\Omega})$ could be directly used for the spectral multiplication in Fig. 9.8. Seemingly also, the (I)DFT length M is of less importance now, because no FIR sequence needs to be picked from a longer signal $\check{h}_2(n)$; M could simply be chosen such that it covers the signal-block length L and is suitable in terms of the signal's short-time stationarity and the acceptable algorithmic delay of the system.

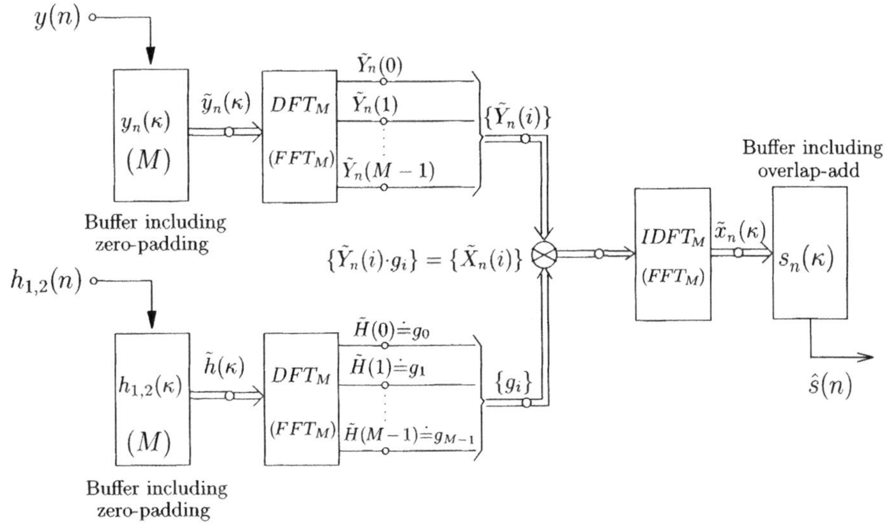

Fig. 9.8. Block diagram of a Wiener filtering realized by a fast convolution, using spectral weighting factors $g_i = \tilde{H}(i)$.

It has to be noted, however, that following these ideas is *not* the same as the described fast convolution: For a "large" value of M, the *implicitly applied* impulse response is, in general, longer than some "chosen" number $2N_F + 1$ (namely, filling all M points with possibly small, but non-zero samples). For a "small" choice of M, time-domain aliasing of $\check{h}_2(n)$ occurs, and the convolution becomes noticeably cyclic.

Still, using weights $g_i \doteq H(e^{j\Omega_i}), i \in \{0, 1, \ldots, M-1\}$, with a "suitably chosen" number M, is a common realization. In fact, some investigations concerning possible improvements by avoiding the circularity have shown only marginal effects [2, 12].

Another problem, to be seen already in (9.22), is equivalent to that one mentioned in Sec. 9.2, after finding the solution in (9.19): For (9.22), the knowledge of $S_{ss}(\Omega)$ and $S_{bb}(\Omega)$ is needed – both are not known a priori, as only $y(n)$ is available and, thereby, $S_{yy}(\Omega)$. As said before, this question will be dealt with in Section 9.7.

9.6 Spectral Subtraction: Principles and Realization

9.6.1 Definition and Variants

A basic definition was already given in (9.9), in the introduction, together with a filtering interpretation by Eq. (9.10). It was also mentioned that the magnitude subtraction of (9.9) could be replaced by a power-spectral subtraction. In a more general formulation, this reads:

$$\hat{S}\left(e^{j\Omega}\right) = \left[\left|Y\left(e^{j\Omega}\right)\right|^{\beta} - \left|\tilde{B}\left(e^{j\Omega}\right)\right|^{\beta}\right]^{1/\beta} \cdot e^{j arg\{Y(e^{j\Omega})\}}. \qquad (9.27)$$

With $\beta = 1$, Eq. (9.9) results, and with $\beta = 2$, power spectra are subtracted. In the latter case, a subtraction of auto-correlations $r_{\hat{s}\hat{s}}(\lambda) = r_{yy}(\lambda) - r_{bb}(\lambda)$ would be equivalent, as addressed in [43].

Let us assume that squared-magnitude spectra denote, in practice, estimated PDS functions, i.e.,

$$\left|Y\left(e^{j\Omega}\right)\right| \approx \sqrt{S_{yy}(\Omega)}, \qquad \left|\tilde{B}\left(e^{j\Omega}\right)\right| \approx \sqrt{S_{bb}(\Omega)},$$
$$\left|S\left(e^{j\Omega}\right)\right| \approx \sqrt{S_{ss}(\Omega)}, \qquad \left|\hat{S}\left(e^{j\Omega}\right)\right| \approx \sqrt{S_{\hat{s}\hat{s}}(\Omega)};$$

then, the actual spectral-subtraction part of (9.27) reads:

$$S_{\hat{s}\hat{s}}^{1/2}(\Omega) = \left[S_{yy}^{\beta/2}(\Omega) - S_{bb}^{\beta/2}(\Omega)\right]^{1/\beta}. \qquad (9.28)$$

Once more, with $\beta = 1$, Eq. (9.9) results, and with $\beta = 2$, power spectra are dealt with.

The corresponding zero-phase filter description, following (9.10) and (9.11), is then written as

$$W_{\mathrm{o}}(\Omega) = \left\{1 - \left[\frac{S_{bb}(\Omega)}{S_{yy}(\Omega)}\right]^{\beta/2}\right\}^{1/\beta}. \qquad (9.29)$$

A further generalization may be introduced, following the thoughts leading to (9.26): As the weighting factor $W_{\mathrm{o}}(\Omega)$ becomes smaller with decreasing local SNR, possibly resulting distortions may be diminished, at the expense of more remaining noise, by applying

$$W_{\mathrm{o}}^{\alpha}(\Omega) = \left\{1 - \left[\frac{S_{bb}(\Omega)}{S_{yy}(\Omega)}\right]^{\beta/2}\right\}^{\alpha/\beta} \doteq W_{\alpha}(\Omega) \qquad (9.30)$$

9.6.2 Relation with Wiener Filtering

The qualitative similarity between the optimum-filter approach and the idea of subtracting an additive disturbance in the spectral domain was already observed in Sec. 9.1.4. Now, this relation can be formalized.

Let $\beta \doteq 2$. From (9.29) and with (9.22), we find:

$$H_{\mathrm{sps}}(e^{j\Omega}) = W_{\mathrm{o}}(\Omega) = \sqrt{1 - \frac{S_{bb}(\Omega)}{S_{yy}(\Omega)}} = \sqrt{H_{\mathrm{o}}(\Omega)}, \qquad (9.31)$$

and with $\alpha \doteq 2\eta$, $\beta \doteq 2$, Eq. (9.30) gives

$$W_{2\eta}(\Omega) = W_{\text{o}}^{2\eta}(\Omega) = H_\eta(\Omega); \tag{9.32}$$

the choice of $\alpha = \beta = 2$ yields

$$W_\alpha(\Omega) = W_2(\Omega) = 1 - \frac{S_{bb}(\Omega)}{S_{yy}(\Omega)} = H_{\text{o}}(\Omega). \tag{9.33}$$

So, the generalized spectral-subtraction description allows to *include* Wiener filtering by $H_{\text{o}}(\Omega)$ as well as its generalization $H_\eta(\Omega)$ of Eq. (9.26).

Of special interest is the power-subtraction variant, i.e., the case with $\beta \doteq 2$, and $\alpha = \eta \doteq 1$, which leads to the "root-Wiener" filter of (9.31). If this is applied to our theoretically constructed example of Sec. 9.4.1, Figs. 9.9 and 9.10 result.

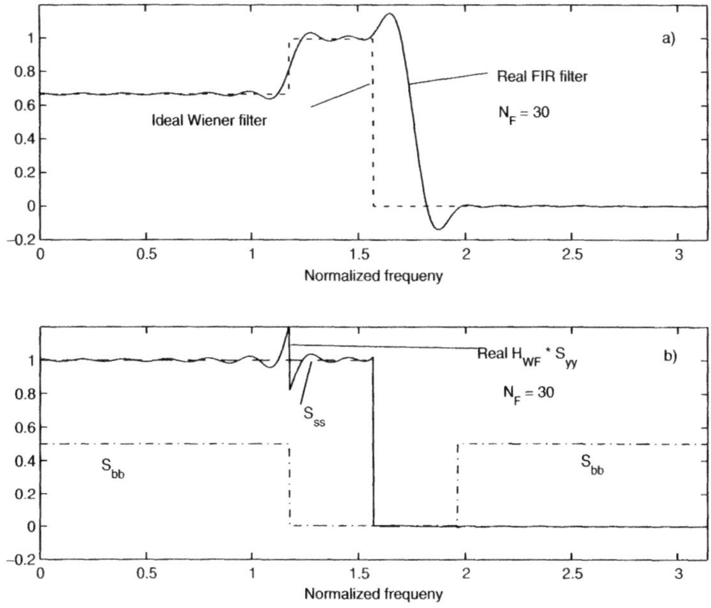

Fig. 9.9. a) Ideal and time-domain approximated Wiener filter; b) PDS of signal, noise, and resulting output signal $\hat{s}(n)$ after filtering with $\sqrt{H_{\text{WF}}}$.

In parts b) of both figures, the resulting output PDS is compared to the actual input PDS $S_{ss}(\Omega)$. A much closer match is to be seen now in comparison to that of Figs. 9.4 and 9.5.

The comparison is generalized in Fig. 9.11: Here, the relation

$$H_{\text{sq}}(\Omega) \doteq \frac{S_{\hat{s}\hat{s}}(\Omega)}{S_{ss}(\Omega)}$$

is displayed as a function of the local SNR at some arbitrary frequency Ω. While the spectral-power subtraction indeed reconstructs the clean-signal

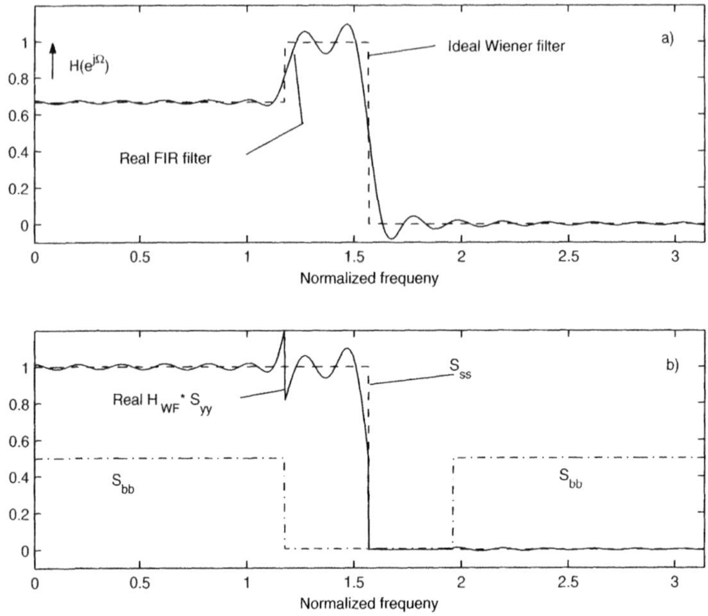

Fig. 9.10. a) Ideal and frequency-domain approximated Wiener filter; b) PDS of signal, noise, and resulting output signal $\hat{s}(n)$ after filtering with $\sqrt{H_{\mathrm{WF}}}$.

PDS ideally, under the given assumptions, the Wiener filter leaves quite strong deviations for $SNR \lesssim 10\ dB$; the spectral-*magnitude* subtraction, however, is even worse in this respect, with strong differences at $SNR \lesssim 35\ dB$.

It has to be noted, however, that the improved PDS reconstruction is a *subjective* observation – no optimality criterion is fulfilled by choosing some combination of α, β, and η, in general.

Some further discussions on the variants achievable by changing the above parameters are to be found in [73].

9.6.3 Realization

The notation of $W_\alpha(\Omega)$ as a weighting function does, in fact, already show how a spectral subtraction in any of its generalized forms can be realized: For time intervals as described in (9.5) – (9.8), short-time spectra have to be measured from $b(n)$ and $y(n)$, from which PDS values have to be derived, they are inserted into (9.30), the noisy-signal spectrum has to be multiplied by $W_\alpha(\Omega)$, and a spectral re-synthesis yields the de-noised output signal $\hat{s}(n)$, finally.

For example, a length-M DFT can be applied to transform blocks of $y(n)$. The resulting spectral samples $\widetilde{Y}_n(i)$ are multiplied by factors

$$g_i \doteq W_\alpha(i \cdot 2\pi/M), i \in 0, 1, \ldots, M-1, \tag{9.34}$$

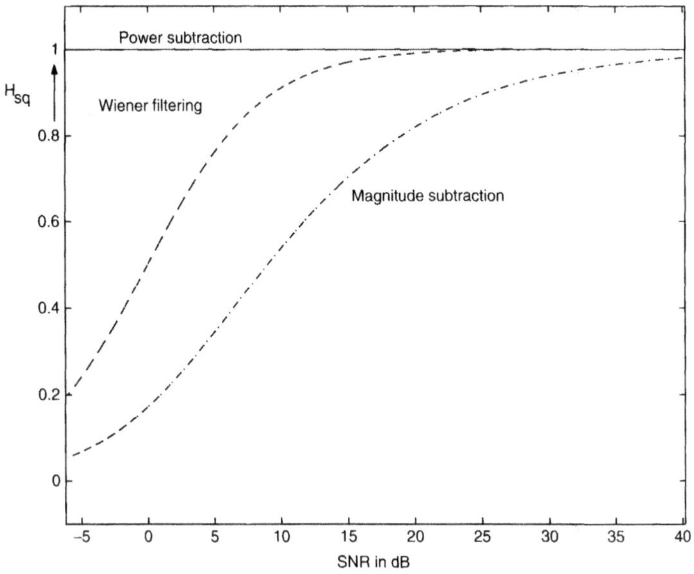

Fig. 9.11. Comparison of the PDS reconstruction capabilities of power-spectral and magnitude-spectral subtractions and the Wiener filter.

and a length-M IDFT creates blocks of $\hat{s}(n)$. This procedure is depicted graphically in Fig. 9.12. The similarity with the Wiener-filter realization of Fig. 9.8 is evident, which is not surprising after Sec. 9.6.2. Differences exist in the determination of the weighting factors g_i and some "suitably chosen" parameters α, β, and η; but the principle function of optimal filtering and spectral subtraction is the same.

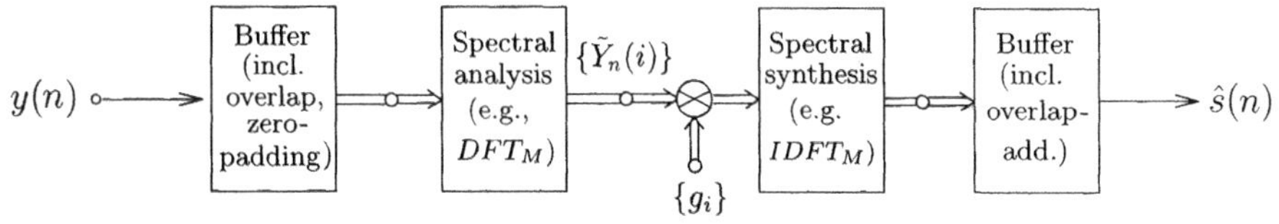

Fig. 9.12. Block diagram of a spectral-subtraction realized by spectral weighting with factors g_i.

9.7 Noise Power Density Spectrum Estimation

In all variants of noise subtraction or filtering, some spectral or, equivalently, correlation informations are needed which, as mentioned, are not available directly, as only the noisy signal $y(n)$ of Eq. (9.1) can be observed. This concerns the (short-time estimated) power densities $S_{ss}(\Omega)$ and $S_{bb}(\Omega)$ or the auto-correlations $r_{ss}(\lambda)$ and $r_{bb}(\lambda)$. Based on the above assumed uncorrelatedness of speech and disturbance, however, the problem can be focussed on one necessary information: With

$$S_{yy}(\Omega) = S_{ss}(\Omega) + S_{bb}(\Omega), \quad r_{yy}(\lambda) = r_{ss}(\lambda) + r_{bb}(\lambda),$$

the desired unavailable clean-signal characteristic can be replaced by

$$S_{ss}(\Omega) = S_{yy}(\Omega) - S_{bb}(\Omega), \quad r_{ss}(\lambda) = r_{yy}(\lambda) - r_{bb}(\lambda).$$

With available terms $S_{yy}(\Omega)$ or $r_{yy}(\lambda)$, this leaves us with the task to measure or estimate the *noise-only* description, as needed also in (9.30). A considerable part of the noise-reduction work of the past years has been devoted to this problem, closely linked to the subtraction or weighting rules yielding the factors g_i in Fig. 9.12.

9.7.1 Noise Measurement in Speech Pauses

A straightforward solution is based on the fact that speech is interrupted by pauses usually, some of which are longer than a common block length as given in (9.5) or (9.8). In such segments, we have $y(n) = b(n)$. So, if a speech pause can be identified, the noise PDS can be estimated (e.g., by the squared magnitudes $|\tilde{Y}_n(i)|^2$ of the block-DFT values, i.e., the periodogram), and this can be used in the succeeding speech-plus-noise segments. A detection of breaks, however, is not so easy; at least, in cannot be simply based on an energy thresholding, especially in low-SNR situations, where the power in noise-only blocks can be quite close to that in segments with speech and distortion. More refined voice-activity detectors (VAD) are necessary, particularly with respect to robustness against disturbing signals. Several variants were investigated, for example, in [3]; here, it was found that the GSM-standard VAD algorithm [17] is quite a good (and meanwhile wide-spread, well tested) choice. It can be outperformed, however, to a certain extent, by special methods optimized for special criteria [60].

An extremely simple idea foregoes a detection: It can be assumed that before any speech communication really begins, there is a reasonably long time without a signal, but with the noise already present.

All pause-based methods, however, have one severe drawback: Between two (detected) breaks, there might be quite long pause-less speech-signal sequences. A constant estimate of $S_{bb}(\Omega)$ is applied during this time interval, while the true noise behaviour might – at least: slowly – vary with time. This means that a noise stationarity is presumed which perhaps does not hold. Especially, an estimation before the conversation onset, kept constant thereafter, is useful only for strictly stationary conditions, given only in simple cases like that of a fan near the microphone.

9.7.2 Continuous Noise Measurements

The problems of both reliable VAD and – at least: slow – instationarity are avoided by a number of techniques estimating the noise PDS continuously, without waiting for a pause (e.g., [1, 18, 33]). In the simplest case, it is assumed that noise is present at all times and its characteristics vary slowly, while speech appears in "bursts". Thus, a time-averaged spectral magnitude at some frequency follows, more or less, the noise, and "speech-evoked peaks" are smoothed out. This turns out to be too simplistic, and, therefore, modifications have been proposed, like a limitation of the influence of a newly found value on the averaged size, an exclusion of values above some heuristic, adaptive threshold, or a decision on the "true" estimate based on a histogram of occurring values.

9.7.3 Minimum Statistics

One approach, following also the above ideas, but in an elaborate, statistically substantiated way became a standard, with later amendments, within the past ten years [51, 52]. The following reasoning includes some thoughts also used in the methods named in Sec. 9.7.2:

While *total* speech-signal pauses may appear too seldom, some narrow frequency bands may contain only noise more often, even during speech activity. This holds particularly for voiced sounds with their periodicity and, therefore, line-spectral structure. The lines are separated by a distance f_0, termed "fundamental" or often "pitch frequency" of the speaker, such that there are gaps in the spectrum with no (or, at least: low) signal components, but noise contributions, if noise is present. *If* there is noise only, at some i-th frequency point, then the spectral size will be smaller than or, at most, equal to that occurring when a speech component is also present at this point. So, vice versa, if a component *is* small compared to other time instances, it can be assumed to describe the noise only. Since f_0 varies with time, it is possible to update the noise-PDS estimation at all frequencies more often, without waiting for the next signal break, and some tracking becomes feasible.

One way of implementing this idea consists of the following steps:

- Calculate the short-time spectrum $\tilde{Y}_n(i), i \in \{0, 1, \ldots, M-1\}$, e.g., via a DFT, from (possibly overlapping and zero-padded) signal blocks $y_n(\kappa), \kappa \in \{0, 1, \ldots, M-1\}$.
- Calculate time-smoothed estimations $\overline{|\tilde{Y}_n(i)|^2}$ of the short-time spectral powers in each frequency point i.
- Assume that the minimum $\min(\overline{|\tilde{Y}_n(i)|^2})$ of these values within a predefined time window of duration T_w describes the short-time noise PDS.
- Scale this value by a constant "over-estimation" factor *omin*, correcting an observed (and theoretically derived [51]) bias.

- Apply the result, finally, to determine the weighting factor g_i for the i-th frequency according to Figs. 9.8 and 9.12.

The necessary correction factor *omin* depends on the smoothing time-constant and the regarded time-window length T_{w}. An investigation in [24] has shown that $omin \in (1.0, 1.6)$ is useful, together with $T_{\mathrm{w}} \geq 0.8\ s$, and a smoothing over some 20 signal segments. An "optimal" combination can only be derived theoretically with assumptions not truely fulfilled in practice; so, a heuristical choice has to be made, based on listening experiments.

In Fig. 9.13, a result of the investigations in [24] is displayed.

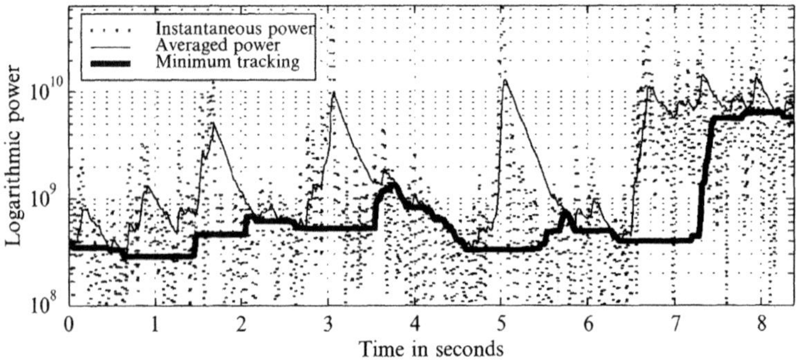

Fig. 9.13. Short-time power spectral values $|\tilde{Y}_n(i)|^2$ of noisy speech in a narrow frequency band near 1 kHz, the time-averaged power $\overline{|\tilde{Y}_n(i)|^2}$, and the PDS-value estimation for the minimum-statistics approach.

On a log scale, the instantaneous power, the averaged power, and the "minimum-tracking" result are plotted over time. There are some speech activities visible at ca. 1.6 s, 3 s, and 5 s, where the smoothed-power contour stays below the local peaks but still reaches high values, and there is a speech pause at the end, with a strong growth of the noise, also leading to a correspondingly large size of $\overline{|\tilde{Y}_n(i)|^2}$. Obviously, the minimum- tracking successfully suppresses the non-noise, speech-activity peaks and describes the noise-PDS behaviour quite well – except for a delay of some 0.8 $s \approx T_{\mathrm{w}}$ after the noise growth. This can be explained by the fact that only after this time T_{w}, the algorithm accepts large averaged spectral values as belonging to the distortion, as, until then, there are smaller values inside the window.

Extensions of the original minimum-tracking approach were proposed in [11], yielding smoother noise-PDS contours, and in [53], using adaptive over-estimation and averaging. The above delay problem, however, remains.

9.7.4 Improved Instationarity Tracking

The delayed reaction to sudden noise-power changes is caused by the smoothing and delayed minimum decision in the above technique. Both are seemingly necessary to avoid fast random fluctuations of the noise-PDS *shape*, as expressed by $\overline{\min(|\tilde{Y}_n(i)|^2)}$ over the frequency indices $i \in \{0, 1, \ldots, M-1\}$. Such fluctuations do certainly not reflect the noise characterisation in stationary situations, but they create unnatural artifacts also in cases with relatively rapid changes, like in that of an accelerating car: The spectral shape would often still vary quite gradually, while the main effect would be an increased noise loudness, i.e., a growing *size* of the whole noise PDS.

Based on such considerations, a separation of a *smooth-shape* and a *fast-gain* estimation was proposed in [23] and discussed thoroughly in [24].

For the first part, an enhanced version of the proposal [1] was chosen, with "subjectively best results"; but also a minimum-tracking variant like those in [11, 52] was found to be applicable. No over-estimation factor *omin* was used. Instead, a "correction factor" G_n was introduced as the instantaneous gain for the data block $\{\tilde{y}_n(\kappa)\}$, taken at time n. The gain is determined in the following steps:

- Assume that the present block's PDS estimation $|\tilde{Y}_n(i)|^2$ has been calculated.
- Assume that an information on the smoothed noise-PDS shape $|B'(i)|^2$ $(\overset{e.g.}{=} \overline{\min(|\tilde{Y}_n(i)|^2)})$ is also available.
- Find the $L_{\min}$ smallest values of $|\tilde{Y}_n(i)|^2$ at the frequency points i_l, $l \in \{1, 2, \ldots, L_{\min}\}$.
- Assume that these minimum values mainly are due to the noise components in this segment; calculate their (linear or nonlinear) mean $\widehat{|\tilde{Y}_n|^2}_{L_{\min}}$.
- Compute the corresponding mean $\widehat{|B'|^2}_{L_{\min}}$ of the smoothed spectral shape description in the same frequency points i_l, $l \in \{1, 2, \ldots, L_{\min}\}$.
- Apply a corrected noise-PDS estimation $G_n \cdot |B'(i)|^2$ for the spectral subtraction or Wiener filtering, with the instantaneous gain

$$G_n = \widehat{|\tilde{Y}_n|^2}_{L_{\min}} / \widehat{|B'|^2}_{L_{\min}}. \tag{9.35}$$

Because the position of the $L_{\min}$ smallest contributions depends on the type of the noise PDS, a modification of the gain determination was proposed also in [24]. In Fig. 9.14, the successful application of the "fast-gain / slow-shape" estimation approach, including the above modification, is evident: In the same scenario as posed for Fig. 9.13, the speech-evoked peaks are still avoided, though with less smoothing, while the sudden noise-power growth at about $t = 6.5$ *s* is tracked without delay now.

A recent study showed that still some mis-detections of the local-spectrum minima used for (9.35) are observed unless the type of the *signal* spectrum is

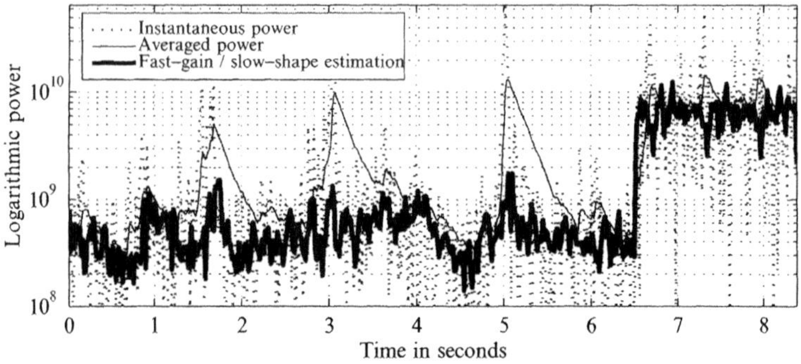

Fig. 9.14. Short-time power spectral values $|\tilde{Y}_n(i)|^2$ of noisy speech in a narrow frequency band near 1 kHz, the time-averaged power $\overline{|\tilde{Y}_n(i)|^2}$, and the PDS-value estimation for the "fast-gain / slow-shape" approach.

also taken into consideration. A modified derivation of G_n was proposed and applied successfully, using an evaluation in separate subbands [35].

9.8 Subtraction and Weighting Rules

Now, it can be assumed that the noisy-signal a well as the noise PDS estimation are available. For simplicity, we refer to them as $S_{yy}(\Omega)$ and $S_{bb}(\Omega)$, without detailing the actual type of spectral measurement or estimation; also, the time index n, defining the position of the actual signal frame, is disregarded as far as possible.

The factors g_i weighting the noisy-signal spectrum $Y(e^{j\Omega})$ in the equivalent realizations of Figs. 9.8 and 9.12 can then be determined – according to the considerations in Chap. 9.6.1, however, in different ways, to be explained now.

9.8.1 Magnitude and Power Subtraction

There are two basic, well-known articles introducing spectral subtraction. In [5], we find

$$g_i \doteq 1 - \sqrt{\frac{S_{bb}(\Omega_i)}{S_{yy}(\Omega_i)}}, \tag{9.36}$$

i.e., spectral magnitudes are subtracted. In [2], power subtraction is proposed instead, i.e.,

$$g_i \doteq \sqrt{1 - \frac{S_{bb}(\Omega_i)}{S_{yy}(\Omega_i)}}. \tag{9.37}$$

Eq. (9.37) corresponds to the "root-Wiener filter" of Eq. (9.31) or to $H_\eta(\Omega)$ in (9.26) with $\eta \doteq 1/2$.

The difference between these weighting rules was discussed in Sec. 9.6.2, as far as the reconstruction of the PDS $S_{ss}(\Omega)$ is concerned. In terms of applicability, however, both rules share a common difficulty: Since both $S_{bb}(\Omega_i)$ and $S_{yy}(\Omega_i)$ are in fact short-time *estimations* with different smoothings possibly included, $S_{bb}(\Omega_i) > S_{yy}(\Omega_i)$ may happen. This can be imagined especially well, if $S_{bb}(\Omega_i)$ stems from a measurement in a pause some time before the current signal block; but also tracking algorithms do not avoid this, in general. The naive use of (9.36) or (9.37) is then impossible, since it would correspond to a creation of negative magnitudes or powers.

Already in [5], a solution has been suggested: Instead of (9.36),

$$g_i \doteq 0, \qquad S_{bb}(\Omega_i) > S_{yy}(\Omega_i) \tag{9.38}$$

is used for these critical cases. Of course, the same "half-wave rectification" can be combined with (9.37).

9.8.2 Musical Noise

Eq. (9.38), however, remedies only a mathematical problem: Frequently, $g_i = 0$ will now set spectral values $S_{\hat{s}\hat{s}}(\Omega_i) \doteq 0$ for one block, while in the following block small, but non-zero values may often appear. Such a block-wise "on-off" switching of more or less randomly distributed small frequency components creates signal blocks with sums of random tones. While the actual input noise may be well reduced, tonal or so-called "musical noise" is artificially generated. Such artifacts can be much more annoying than the original, "natural" disturbances. So, the avoidance or, at least, limitation of this effect is a reason why so much work has been devoted for smoothly tracking, reliable noise estimation (see Sec. 9.7) and on variations of the weighting or subtraction rules.

9.8.3 Noise Floor, Over-Estimation, and Non-Linear Subtraction

The above effect can be diminished if, in cases of a very small factor $g_i \approx 0$, the down-sizing is limited: A certain, small part of the original noise is kept, termed the "spectral noise floor" [2].

As, on the other hand, in frequency points with large components $S_{yy}(\Omega_i)$ there is no danger, normally, of creating tonal artifacts, and, as it can be observed that simple as well as more refined estimations of $S_{bb}(\Omega_i)$ often give too small values (see Sec. 9.7.3), some heuristical "over-estimation" factor may allow for a stronger reduction here.

A generalization of these considerations consists of a subtraction of $[\delta \cdot S_{bb}(\Omega_i)]$, with a factor δ leaving a noise floor or over-estimating the noise contribution in dependence of the local SNR value at frequency Ω_i:

$$S_{\hat{s}\hat{s}}(\Omega_i) = S_{yy}(\Omega_i) - \delta\,(SNR_i) \cdot S_{bb}(\Omega_i).$$

Any adaptive subtraction of magnitudes or powers in the spectrum defines actually a non-linear system; still, the above "extra-non-linearity" is the reason why this has been called "non-linear spectral subtraction", in the literature [2,48].

Numerous, partly heuristic variants were proposed, including different combinations of time and frequency smoothings and adaptation rules (see, e.g., [46,47,62]).

9.8.4 Approaches Based on Statistical Models of Signal and Noise

A non-heuristic approach to a modified spectral subtraction or weighting was proposed in [6, 13, 14]. It aims at an MMSE approximation of the spectral amplitudes $|\hat{S}(e^{j\Omega_i})|$, based on the assumption that both the noise and the signal spectra consist of complex Gaussian random variables. The result is a weighting factor g_i which depends on the "a-priori SNR" ρ_i^{pri} and the "a-posteriori SNR" ρ_i^{post}. The latter term describes, for a frequency point i, the power ratio of the de-noised signal $\hat{s}(n)$ and the noise $b(n)$ by

$$\rho_i^{\text{post}} \doteq \frac{S_{yy}(\Omega_i) - S_{bb}(\Omega_i)}{S_{bb}(\Omega_i)} = \frac{S_{yy}(\Omega_i)}{S_{bb}(\Omega_i)} - 1,$$

using the PDS estimates for the current frame taken at time n. The a-priori term takes the processing in the past blocks recursively into account, and it also includes a "rectification" step in order to avoid negative weights. So, with the variable n included explicitly now, we have to compute

$$\rho_i^{\text{pri}}(n) \doteq (1-\zeta)\cdot\max\left\{\rho_i^{\text{post}}(n), 0\right\} + \zeta\cdot g_i^2(n-1)\cdot\left[\rho_i^{\text{post}}(n-1)+1\right].$$

The "forgetting factor" ζ is chosen close to one, like $\zeta = 0.98$ in [6]. From these equations, the weighting factor (for the n-th frame) is finally found:

$$g_i = \frac{\sqrt{\pi}}{2}\cdot\sqrt{\frac{1}{1+\rho_i^{\text{post}}}\cdot\frac{\rho_i^{\text{pri}}}{1+\rho_i^{\text{pri}}}}\cdot M\left(\left(1+\rho_i^{\text{post}}\right)\cdot\left(\frac{\rho_i^{\text{pri}}}{1+\rho_i^{\text{pri}}}\right)\right).$$

The dependence $M(\cdot)$ is given as

$$M(x) \doteq e^{-x/2}\cdot\left[(1+x)\cdot I_0(x/2) + x\cdot I_1(x/2)\right],$$

with $I_{0,1}$ denoting modified Bessel functions of orders 0 and 1, respectively.

This approach became a "base-line standard" for several years. It reduces the input noise considerably and avoids musical noise, if combined with a good noise estimation like minimum tracking. Therefore, it was often used for investigations of other parts of the noise-reduction task, for example, when different noise measurements or spectral-analysis / synthesis techniques had to be compared for a fixed subtraction rule (see Sec. 9.10).

A variant was proposed in [15] with an MMSE approximation of the logarithmic spectrum. Another, more recent variant, however, promises more progress towards an even better new quasi-standard [49, 79]. Here, the fact is taken into consideration that speech is not at all a Gaussian signal, and that spectral values are closer to a Laplace or Gamma distribution [54]. This is described by means of a parametric probability density function (PDF) defining a general "super-gaussian" distribution; its parameters are fitted to a measured PDF (i.e., a histogram). Due to the closed-form PDF, a theoretical derivation can be carried through; it leads to a weighting factor which maximizes the a-posteriori probability of the *complex* spectral value $\hat{S}(e^{j\Omega_i})$, given the observed components $Y(e^{j\Omega_i})$.

9.9 Spectral Analysis and Synthesis

9.9.1 DFT and IDFT

In our introduction, we addressed spectra $S(\Omega)$, $B(\Omega)$, and $Y(\Omega)$ as well as PDS functions $S_{ss}(\Omega)$, $S_{bb}(\Omega)$, and $S_{yy}(\Omega)$ as Fourier transforms of infinitely long signals or auto-correlation sequences. It was mentioned that, in practice, they should be understood as short-time estimates: They result from some short-time spectral analysis, and the output signal $\hat{s}(n)$ is created by some corresponding spectral synthesis.

The fast-convolution realization of the Wiener filter in Fig. 9.8 lead us to the idea to generally assume a block-DFT / FFT and IDFT / IFFT system in Fig. 9.12. This was, by the way, also the basis used first in [2, 5].

9.9.2 Generalizations

Of course, as also mentioned before, overlapping signal frames could be used, and a suitable time weighting or "windowing" would help to avoid block-edge effects: The spectral modification between analysis and synthesis transformations may evoke, in the time domain, largely differing signal samples close to the beginning of a block and at the end of the preceding one, audible as unnatural cracks at a distance of the frame length T_{adapt} [see Eqs. (9.5) and (9.8)]. This can be mitigated by applying, e.g., triangular, trapezoidal, or cosine-shaped windows $w(n), n \in \{0, 1, \ldots, M-1\}$, to suitably overlapping parts of the signal $y(n)$, taken at a distance of r samples, with $r \in \mathbf{N}$.

Fig. 9.15 shows some examples. Since the overlapping windows in all cases sum up to a constant value 1.0, it is clear that the summation of the inverse transforms will ideally yield $y(n)$, if no spectral manipulation or weighting has been done between analysis and synthesis. Such an arrangement is said to have a "perfect-reconstruction" (PR) property. It is also clear from inspection that the windowing may be distributed to the input-signal block before the transformation and the output-signal block after the inverse transformation, e.g., by using the square-root of $w(n)$ on both sides. Then the sum of the window products has to give a constant value 1.0, in order to have a PR system again.

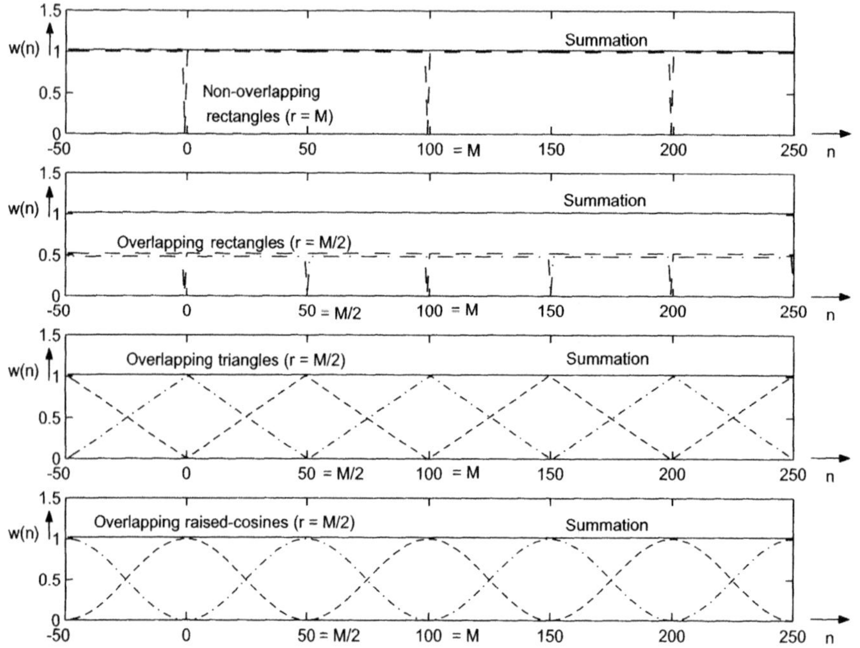

Fig. 9.15. Examples of overlapping and adjacent windows adding up to a constant value 1.0 (e.g., for $M \doteq 100$).

A choice of $r > 1$ means that the analysis is only carried out at every r-th input clock, i.e., at a smaller rate; r is termed the "rate-reduction" or "decimation" factor, in the following. For a classical block-by-block DFT without overlap, we would have $r = M$.

Furthermore, the window sequences may be chosen to have a length $L_w > M$. Then, more than M signal samples enter the calculation of M spectral components. Fig. 9.16 shows the signal-flow graph (SFG) of a correspondingly extended analysis system. With $L_w \doteq M$, it contains the "windowed DFT / FFT", and with, in addition, $w(n) = 1$ the simple DFT / FFT of the segment is also included. The rate-reduction parameter r indicates here that the DFT / FFT is started only every r clocks; obviously, also the preceding multiplications and additions can be calculated at the reduced speed.

9.9.3 Complex-Modulated Filterbank

The network in front of the DFT / FFT in Fig. 9.16 sums products of signal and window subsets. These subsets are found by sub-sampling the sequences

$y(n)$ and $w(n)$ at a distance of M samples, and they differ from each other by the index of their first element, i.e., the phase of the down-sampling. This part of the SFG is therefore called a "polyphase network" (PPN), and the whole analysis system is termed "PPN-FFT" [32,72,73].

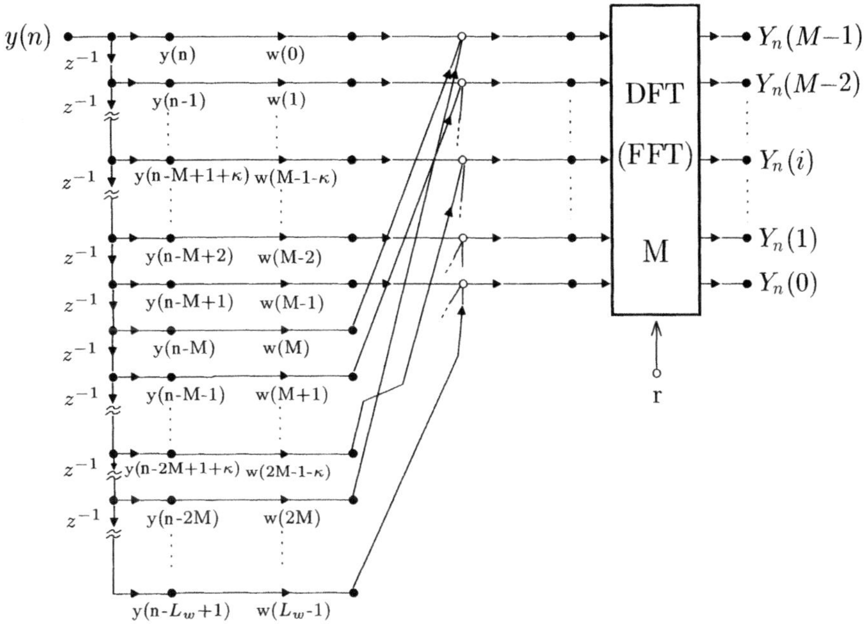

Fig. 9.16. SFG of a complex-modulated analysis filterbank consisting of a polyphase network (PPN) and an FFT.

Its output values $Y_n(i)$ are found, after a few analysis steps, as

$$Y_n(i) = e^{ji(n+1)\frac{2\pi}{M}} \cdot \sum_{\kappa=0}^{L_w-1} y^{(i)}(n-\kappa) \cdot w(\kappa).$$

Disregarding the phase term $e^{ji(n+1)\frac{2\pi}{M}}$, they result from a convolution of the window $w(n)$ with the sequences

$$y^{(i)}(n) \doteq y(n) \cdot e^{jin\frac{2\pi}{M}}.$$

This means that, in the i-th channel at a center frequency $\Omega_i = i \cdot \frac{2\pi}{M}$, the input signal is de-modulated by Ω_i and filtered thereafter. The filter's frequency response

$$H_w^{(i)}\left(e^{j\Omega}\right) \equiv H_w\left(e^{j\Omega}\right) = \mathcal{F}\{w(n)\} \quad \forall i \tag{9.39}$$

is identical for all channels. If $w(n)$ is chosen as a low-pass FIR sequence, the interpretation is that, in each channel no. i, a spectral component at frequency

Ω_i is shifted down to zero and then passed through the low-pass. Equivalently, it can be said that each channel can be described by a band-pass filter which is a copy of the "prototype" $H_w(e^{j\Omega})$ shifted up by Ω_i. Therefore, the PPN-FFT of Fig. 9.16 is called a "complex-modulated filterbank" with identically shaped band-pass filters at equi-spaced center frequencies.

Now, the advantage of the DFT extensions in Sec. 9.9.2 becomes obvious: For the pure DFT / FFT, with rectangular windowing, the well-known "poor" sinc-type function defines the analysis quality; with a length-M window $w(n)$, at least the commonly used Hamming-, Hann-, or Bartlett-functions (or some more variants) offer some choice; with $L_w > M$, however, some suitable frequency-response shape may be designed with as many degrees of freedom as desired (and affordable). Fig. 9.17 depicts a possible prototype's frequency response together with its shifted copies, i.e., the equivalent band-pass filters.

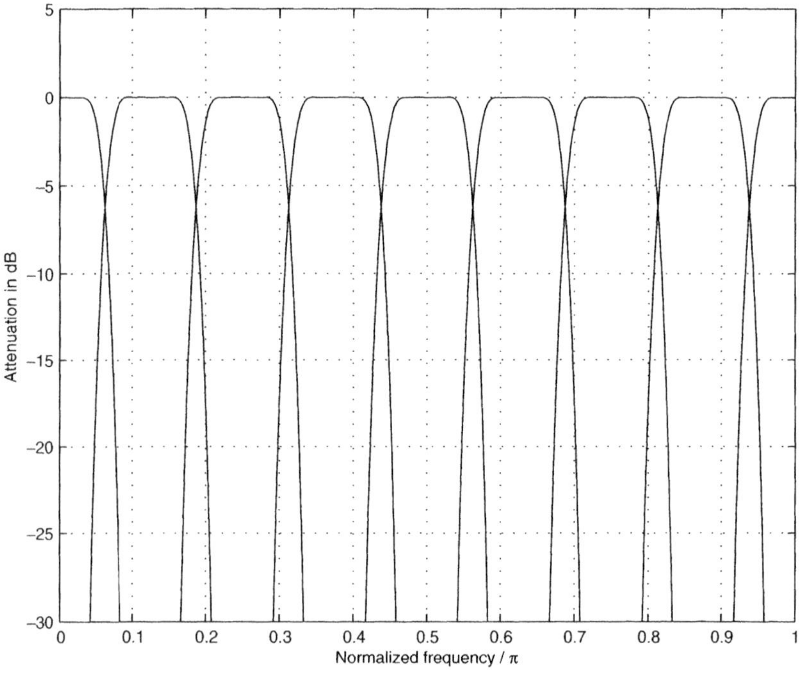

Fig. 9.17. Low-pass prototype filter attenuation $20 \cdot \log_{10} |H_w(e^{j\Omega})|$ and its first M/2 shifted, equivalent band-pass copies (for $M = 16$).

It has to be noted, however, that, in our application, an additional aspect is important for the prototype design: The analysis filterbank is followed by a spectral synthesis. It is carried out by the dual of the system in Fig. 9.16, namely, an inverse transformation, a synthesis-window weighting, and $(L_w - 1)$ delay-addition operations. If no weighting or other manipulations happen in-between, the whole analysis-synthesis system should deliver a replica of $y(n)$ at its output. So, as mentioned for the much simpler case of overlapping length-M windows in Sec. 9.9.2, the windows on both sides have to be appropriately chosen. A really *perfect* reconstruction is, however, often unnecessary: The small remaining (aliasing and linear) distortions of a "near-perfect reconstruction" (NPR) should just not dominate the residual noise, artifacts, or distortions of the noise-reduction operations.

For more details concerning PR and NPR filterbanks, their theory, and their realizations, the reader is referred to the literature (e.g., [21, 41, 55, 69, 73]).

9.9.4 Real-Valued Filterbanks

A variant of the above PPN-FFT replaces the DFT by a so-called Generalized DFT (GDFT); here, the center-frequency grid is still equi-spaced, but shifted, e.g., such that all filters become band-pass systems and there is no "low-pass channel".

Replacing the GDFT by a Generalized Discrete Cosine Transformation (GDCT) leads to cosine-modulated filterbanks. Here, the prototype frequency response $H_w(e^{j\Omega})$ appears in pairs of shifted copies at Ω_i and at $(2\pi - \Omega_i)$, due to the real-valued modulation. Also cosine-modulated filterbanks are attractive, since the DCT, like the DFT, can be realized by fast algorithms.

Other, also real-valued filterbanks may be built up by a repeated application of half-band-filter pairs in a tree structure; the rate-reduction factor $r \doteq 2^\mu$ is distributed, in this case, over the μ stages of the filterbank: A complementary pair of a low- and a high-pass filter can be realized efficiently as one filtering block. The full frequency range of the input signal, with $f \in [0, f_s/2]$ or $\Omega \in [0, \pi]$ for real-valued signals, is divided into two symmetrical halfs, with $\Omega \in [0, \pi/2]$ and $\Omega \in [\pi/2, \pi]$, respectively. Because of the bandwidth-halving, also a decimation by a factor $r = 2$ is allowed; in an appropriate (e.g., FIR) realization structure, it can already be exploited *within* the filter operation. Thereafter, on a newly normalized frequency axis

$$\Omega' = 2\pi \cdot \left[f/(f_s/2)\right] = 2 \cdot \Omega,$$

replacing Ω as defined by (9.2), each output signal covers the full bandwidth $\Omega' \in [0, \pi]$ again. Another pair of half-band filters will then separate two symmetrical bands again, while operating at a clock $f_s/4$, and so on. Fig. 9.18 depicts the principle, with $\mu = 3$ stages and $2^3 = 8$ frequency components. With or without some weighting or other modification, they can then be recombined by the dual structure of Fig. 9.18, to create an output signal with possibly PR or NPR property.

Since the high-pass transfer function is related to that of the low-pass by

$$H_{\mathrm{HP}}(z) = H_{\mathrm{LP}}(-z),$$

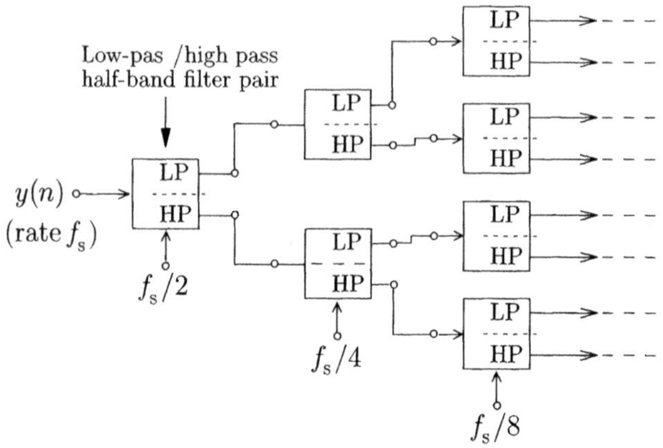

Fig. 9.18. Tree-structure filterbank built by successive band-halving low-pass / high-pass filter pairs with inherent down-sampling by $r = 2$.

the impulse responses are coupled by

$$h_{\mathrm{HP}}(n) = (-1)^n \cdot h_{\mathrm{LP}}(n).$$

This corresponds to a modulation by $(-1)^n = e^{jn\pi}$, i.e., the frequency shift by $\Omega = \pi$ which transfers the low-pass into a high-pass filter. Thus, the structure in Fig. 9.18 may well be also explained as a real-valued modulated filterbank; a closer look into the one-block realization of the filter pair shows that, beyond, it uses also the polyphase principle.

9.9.5 Non-Equispaced Frequency Bands

Up to now, spectral decompositions of $y(n)$ into M components on a uniform frequency grid were assumed, with $\Omega_i = i \cdot 2\pi/M$. This is not necessary, and there are reasons why one might prefer non-equally spaced frequency points, with, accordingly, different bandwidths of the filterbank channels:

- Voiced sounds possess line spectra, as mentioned before. The line distance is the "pitch" frequency f_0, which varies during speech production, but shows an average value $\bar{f}_0$ for a certain speaker. This mean frequency depends on gender, age, anatomic details, and some other factors; in general, however, we have $\bar{f}_0 \in (50, 400)\,Hz$.
  Between the lines, the short-time spectra of disturbed voiced sounds contain noise, which we want to remove. So, it would be good to have a narrow frequency grid able to separate lines and gaps. On the other hand, the line structure becomes less and less pronounced, if the higher-frequency range is dealt with. Here, frequency points on a narrow grid would increase the risk of "musical-noise" production. Thus, larger bandwidths would be good for the upper frequency channels.

- In the human hearing apparatus, the basilar membrane in the inner ear acts as a filterbank. On sections with a constant length of $\sim 1.5\ mm$, components within certain frequency bands are grouped and treated together. The spacing and, correspondingly, the widths of these so-called "critical bands" are, however, not uniform. A description due to [81] "numbers" the bands by the Bark frequency Θ, which is related to the physical frequency f by

$$\frac{\Theta}{Bark} = 13 \cdot arctan\left[0.76 \cdot \frac{f}{kHz}\right] + 3.5 \cdot arctan\left[\left(\frac{f}{7.5\ kHz}\right)^2\right]. \quad (9.40)$$

  In Fig. 9.20, the solid line indicates that the channel numbers follow the frequency in a log-type manner, according to (9.40).

So, using a filterbank which mimics the Bark scale would also be useful in terms of the first observation.

There are various filterbanks realizing a non-uniform frequency resolution. the simplest idea, namely, to use M separately designed and implemented band-pass systems, is discarded: There are much more efficient approaches, which is important especially if still a relatively large number of bands is to be used, like $M > 8$.

9.9.5.1 Partial Recombination

After an analysis with a large number of – say: $M = 256$ – equispaced bands, a spectral component with a higher bandwidth can be generated by a spectral synthesis from $M_i < M$ adjacent narrow components; this is exemplified in Fig. 9.19 for the case of a DFT with partial IDFTs of different lengths. If the first transformation is carried out at every rth clock, one sample appears at every DFT output after a calculation cycle; the partial IDFTs, however, deliver $M_i \geq 1$ samples per cycle. This corresponds to the increased bandwidths at their outputs, i.e., the short-time spectral values, and it means that a system with multi-rate signals has been defined. This is necessarily true for all filterbanks with non-uniform bands.

Such a system was investigated, with a full PPN-FFT instead of the pure DFT, in [26,27]. It was found to be too expensive computationally, if applied within a PPN-FFT analysis / synthesis system with general values M_i, without offering reasonable advantages on the other side. It was re-visited later, however, from two different points of view; we shall come back to this below.

9.9.5.2 Warped PPN-FFT

If in an FIR low-pass transfer function $H(z)$ the delay term z^{-1} is replaced by an allpass transfer function $A(z)$, we find that in the low-pass frequency

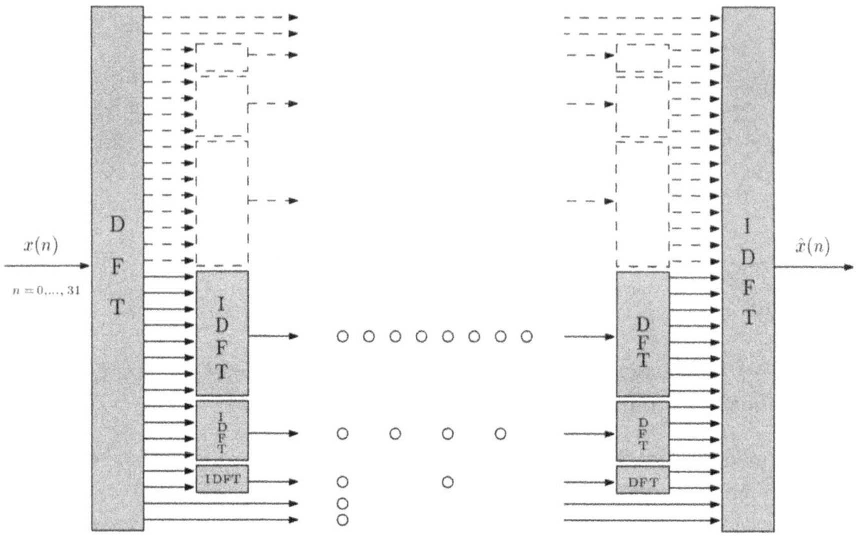

Fig. 9.19. Analysis filterbank realized by an un-windowed length-32 DFT / FFT and IDFT / IFFT, with intermediate partial syntheses of lengths $M_1 = M_2 = 1$, $M_3 = 2$, $M_4 = 4$, $M_5 = 8$, with correspondingly different numbers of time-signal samples, and with the dual arrangement on the synthesis side.

response $H(e^{j\Omega})$ the frequency variable Ω is replaced: Due to the allpass characteristic, namely,

$$A\left(e^{j\Omega}\right) = 1 \cdot e^{-j\phi(\Omega)},$$

we have

$$H\left(e^{j\Omega}\right) := H\left[A^{-1}\left(e^{j\Omega}\right)\right] = H\left[e^{j\phi(\Omega)}\right].$$

So, the frequency-response *type* is unchanged, while the frequency *axis* is "warped", depending on the allpass coefficients. For instance, a first-order allpass is described by $A(z) = (az+1)/(a+z)$; its application gives an "allpass-transformed" frequency

$$\Omega_{warp} = \phi(\Omega) = 2 \cdot arctan\left[\frac{1-a}{1+a} \cdot tan(\Omega/2)\right]. \tag{9.41}$$

A variation of the parameter a transforms, e.g., a low-pass filter into another low-pass filter with a different cutoff frequency.

This well-known filter-design tool [7,8,64] was proposed for the *implementation* of variable filters in [66]: In a hardware filter, the delay elements were to be replaced by allpass blocks with tunable coefficients. The idea was transferred to PPN-FFT realizations in [70]. Replacing the delay chain in Fig. 9.16 by a chain of allpass filters warps the filterbank-channel, i.e., the equivalent band-pass filters of the spectral analysis according to

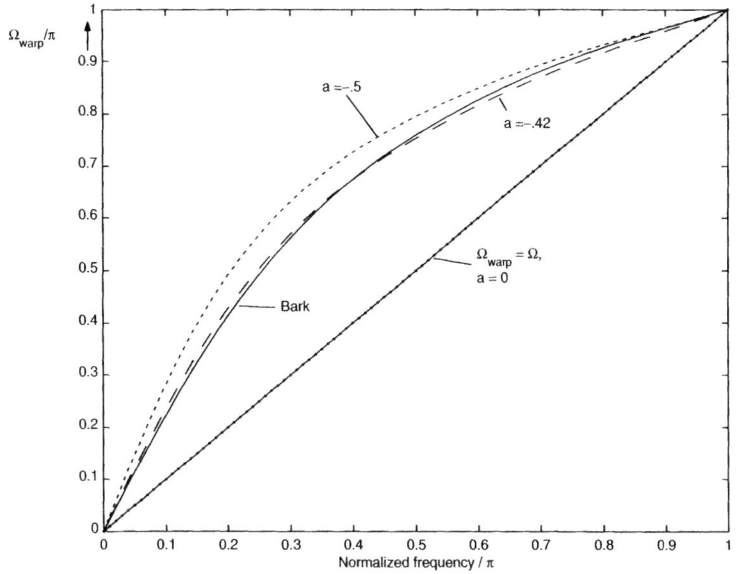

Fig. 9.20. Normalized Bark-scale variable Θ over normalized frequency Ω (solid line), compared with allpass-transformation frequency mappings Ω_{warp}.

$$H_w^{(i)}\left(e^{j\Omega}\right) = H_w\left(e^{j\left(i\cdot\frac{2\pi}{M}+\phi(\Omega)\right)}\right).$$

The band-pass filters are no more identical, shifted versions of the prototype, as was indicated in (9.39) for the standard PPN-FFT. Fig. 9.21 shows the variation for a first-order allpass with $a = -0.5$ in contrast to the uniform filterbank depicted in Fig. 9.17, with the same number of $M = 16$ channels. If $a \doteq -0.42$ is chosen, a similar picture results, with a close approximation of the "Bark-scale" warping in Eq. (9.40); this is demonstrated in Fig. 9.20.

It must be noted that, now, the modified PPN-FFT contains *recursive* structures, namely, the allpass blocks. Such structures are also necessary in the synthesis. Actually, the inverse allpass function $A^{-1}(e^{j\Omega})$ is needed, which, for a stable allpass, would become an unstable system. A synthesis with $A(z)$ itself is also found to be applicable; but this causes strong phase distortions for the output signal of the whole filterbank system, due to the product $A^{(L_w-1)}(e^{j\Omega})$ on both sides. This can be taken care of by an approximate phase-equalizing FIR filter, which, in turn, introduces a huge signal delay. Beyond, the transformed PPN-FFT is less efficiently realized: The speed-reduction factor r is limited by the largest channel bandwidth. So, some realization problems have to be kept in mind if this analysis-synthesis system is to be used.

For more details about allpass-transformed filterbanks, the reader is referred to the literature, e.g., [19, 20, 39, 40].

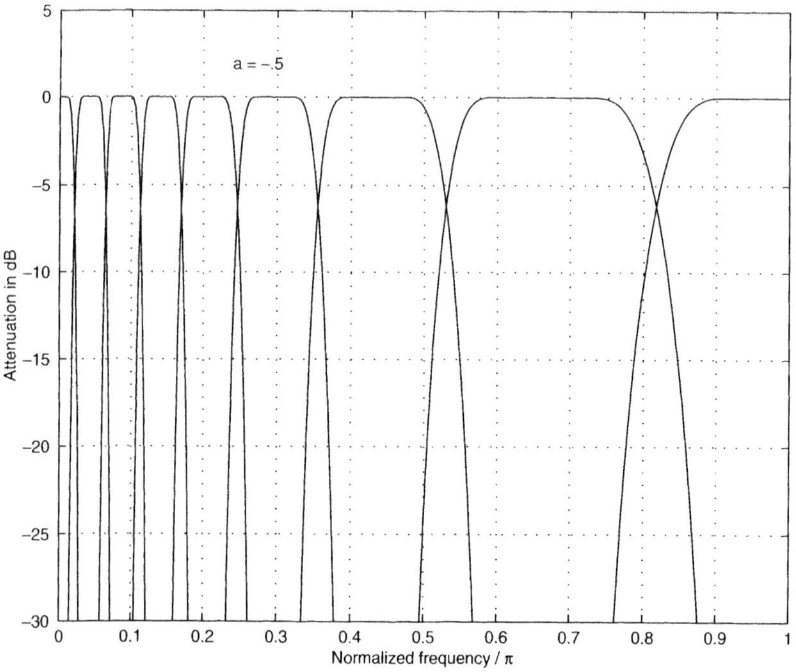

Fig. 9.21. First (M/2 + 1) equivalent band-pass frequency responses of a complex-modulated, allpass-transformed filterbank with a first-order allpass, $a = -0.5$, and $M = 16$.

9.9.5.3 Pruned Tree Structure

If, in Fig. 9.18, only the low-pass outputs of the filter pairs are processed further by a succeeding filter pair and thereby split into two half bands again, while the high-pass outputs are kept unchanged, Fig. 9.22 results. There are two analysis channels of width $(\pi/8)$ each, and one channel each covering $(\pi/4)$ and $(\pi/2)$, on both sides of the frequency axis. Further stages may be added, as also in Fig. 9.18, and with a continued "pruning", i.e., deletion of high-pass subdivisions, an "octave-band" filterbank is constructed; its bandwidths essentially double from output to output index.

If such a fast exponential growth of the covered frequency ranges is reckoned inappropriate, e.g., in comparison to the "third-band" filtering observed in the inner ear, other prunings are of course possible. Also, a sub-filtering (equally or non-equally spaced) within the octaves may be applied.

9.9.5.4 Wavelet-Related Analysis-Synthesis Systems

The continuous wavelet transformation (CWT) of a general, continuous signal $x_o(t)$ is defined by

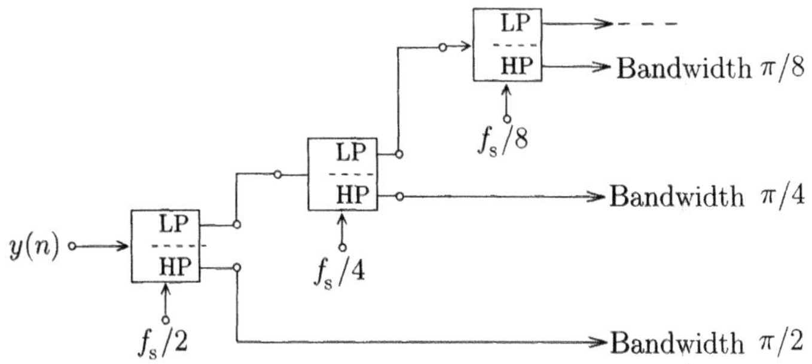

Fig. 9.22. Pruned tree structure realizing an octave-band analysis.

$$\mathcal{W}_x^{\psi}(b,a) = |a|^{-\frac{1}{2}} \int_{-\infty}^{+\infty} x_o(t)\psi^*\left(\frac{t-b}{a}\right) dt \quad , \tag{9.42}$$

where $\psi(t)$ is the chosen "prototype wavelet", which has to satisfy some theoretical constraints [9]. $\psi(t)$ can be imagined as an oscillation with some time-limited envelope; the oscillation defines the center frequency, the duration the bandwidth of the wavelet. By shifting and scaling $\psi(t)$ with the parameters a and b, all basis functions $\psi_{b,a}(t) = |a|^{-\frac{1}{2}}\psi\left(\frac{t-b}{a}\right)$ for a signal decomposition are obtained. Varying a changes the time scale of $\psi_{b,a}(t)$ and, thus, the corresponding bandwidth and center frequency.

Because the continuous wavelet transformation given by equation (9.42) is highly redundant, a discretization of a and b is necessary. Usually, a "dyadic grid" is chosen with a as a power of 2 and b depending on a such that $a = 2^m$ and $b = k2^mT_s$, $m, k \in \mathbb{Z}$. For this choice, after time discretization of the signal, the wavelet transformation described by equation (9.42) becomes

$$w_x^{\psi}(2^m k, 2^m) = 2^{-\frac{m}{2}} \sum_n x(n)\psi^*\left(2^{-m}n - k\right) . \tag{9.43}$$

Due to the factor-2 scaling with increasing m, it realizes an octave-band analysis with different sampling rates in each octave. In practical realizations, mostly the so-called Á-Trous algorithm [68] is used, because of the high computational load of the direct implementation of equation (9.43). It turns out that, for correspondingly chosen wavelets, this algorithm has the form of the pruned half-band filter tree structure of Fig. 9.22. The possible so-called "voicing" is equivalent to the further sub-filtering mentioned in Sec. 9.9.5.3.

With an application to the analysis of vibration data, another realization of the discrete wavelet transformation was proposed in [58, 59]: The convolution theorem of the Fourier transformation allows the formulation of equation (9.42) in the frequency domain as

$$W_x(b,a) = |a|^{\frac{1}{2}} \frac{1}{2\pi} \int_{-\infty}^{\infty} X_o(\omega)\Psi^*(a\omega)e^{j\omega b}\, d\omega . \tag{9.44}$$

Therefore, the wavelet transformation can be calculated by a multiplication of the spectrum $X_o(\omega)$ with $\Psi^*(a\omega)$ and an inverse Fourier transformation:

$$W_x(b,a) = |a|^{\frac{1}{2}} \mathcal{F}^{-1}\Big\{X_o(\omega)\Psi^*(a\omega)\Big\}. \quad (9.45)$$

Furthermore, the Fourier transform $\Psi(\omega)$ of the mother-wavelet $\psi(t)$ can be chosen to be constant in a limited frequency range and zero outside:

$$\Psi(\omega) = \begin{cases} 1, \text{ for} & \omega_0 - \omega_g < \omega < \omega_0 + \omega_g\,, \\ 0, \text{ otherwise.} & \end{cases} \quad (9.46)$$

The corresponding wavelet in the time-domain becomes

$$\psi(t) = \frac{\omega_g}{\pi} \frac{\sin(\omega_g t)}{\omega_g t} \cdot e^{j\omega_0 t}. \quad (9.47)$$

For the transformation of discrete signals, Eq. (9.45) becomes

$$\hat{W}_x(b,a) = |a|^{\frac{1}{2}} \mathcal{F}^{-1}\Big\{X\left(e^{j\Omega}\right)\Psi^*\left(e^{ja\Omega}\right)\Big\} \quad , \quad (9.48)$$

where $\Psi(e^{\Omega})$ is the Fourier transform of the sampled wavelet $\psi(kT_s)$.

For visualization, Fig. 9.23 shows, on the left-hand side, the frequency resolution as well as the real and imaginary parts of the prototype wavelet $\psi(n)$. On the right-hand side, the variations in frequency and time after scaling the wavelet by $a = 2$ are demonstrated.

In a practical realization of the wavelet-transformation by equation (9.48), for a finite block $x(n)$, $n \in \{0, 1, \ldots, M-1\}$, the DFT of length M is used. The spectral windowing is carried out by setting those discrete spectral values to zero which are not in the passband of the corresponding wavelet. After that, the modified spectrum will be transformed with the IDFT of the full length M. In such a way, a redundant wavelet-transformation is generated, because for every frequency band M wavelet coefficients are calculated.

For a wavelet-representation of $x(n)$ with reduced redundancy, the sampling rates can be fitted to the wavelet bandwidths. A simple way is to transform only M_i non-zero values of the modified spectrum with an IDFT of length $M_i < M$. A block diagram, realizing this wavelet transformation with a block-length $M = 32$ and one frequency band per octave, however, has already been shown in Fig. 9.19: This approach is equivalent to the "partial recombination" idea discussed in Sec. 9.9.5.1. It was mentioned that, in [26], it was found to be too expensive in terms of computational effort in the context of a PPN-FFT and general band combinations; here, with a DFT / IDFT only and a dyadic scaling, it is an attractive alternative to the above Á-Trous tree structure [25].

Of course, the values M_i can be chosen such that smaller bandwidths result than from an octave grouping. This corresponds to the above-mentioned "voicing" in the pruned-tree structure. A natural conclusion is, then, that one

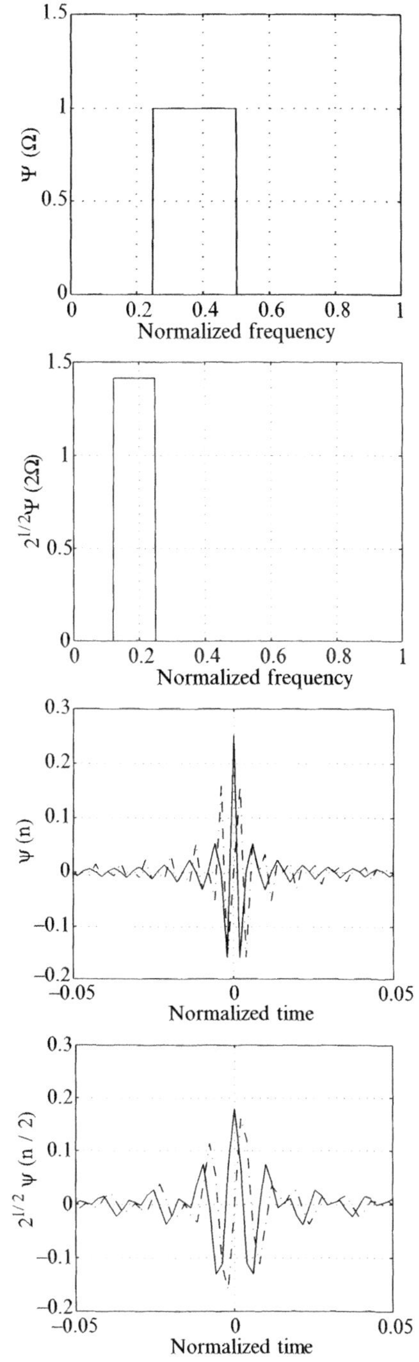

Fig. 9.23. Frequency resolution and time function of the mother-wavelet and the scaled version with $a = 2$.

might also drop the pruning of the Á-Trous scheme. This is indeed possible, and it leads us to the application of so-called "wavelet packets". With the digital wavelet-packet analysis (DWPA) [75,77], nearly arbitrary time-frequency resolutions are possible. Wavelet-packets are bases of a transformation which arise from linear combinations of wavelets. They make it possible to process the wavelet-coefficients $\hat{W}_x(k2^m, 2^m)$ of a DWT again, so that an octave can be divided into several subbands.

A realization of the discrete wavelet-packet analysis (DWPA) is shown in the upper part of Fig. 9.24, and it turns out to correspond indeed to a tree-structured filterbank.

The input signal $x(n)$ is processed with all possible cascades of the basic element, consisting of a quadrature-mirror filter pair with the highpass analysis filter $G(z)$ and the complementary lowpass analysis filter $H(z)$ followed by a factor-2 subsampling. The synthesis is done in the dual, lower half of the system.

9.9.6 Adaptive Bandwidths

9.9.6.1 Motivation

In Sec. 9.9.5, the step towards non-uniform frequency resolution was motivated by the behaviour of the human ear, on one hand, with the monotonically increasing bandwidths of the Bark-scale filters. A more refined model of the cochlear processing contains, beyond this, a signal-adaptive shift of all center frequencies. On the other hand, different frequency separations were explained as being helpful for the removal of noise in pitch-line gaps within the lower-frequency range and the avoidance of narrow-band, musical artifacts in the upper range. Since the border between "higher" and "lower" frequencies is certainly not well defined, since it varies with time as well as the line distance f_0, and since the existence and visibility of a line structure in the short-time spectra will depend on both the signal and the noise behaviour in the current time slot, it becomes evident that signal-adaptive bandwidths and center frequencies should outperform a fixed non-linear distribution.

9.9.6.2 Possibilities

All above non-equally resolving spectral-analysis / synthesis schemes have some parameter(s) which could be varied in time, controlled by some short-time signal characteristics: The partial recombination of Fig. 9.19 is easily imagined with variable numbers M_i of re-synthesized components; this will, however, make it expensive. The warped PPN-FFT of Sec. 9.9.5.2 may adapt the allpass coefficient a to the signal; this would, however, not change the monotonicity of the bandwidths, either growing or decaying; higher-order allpass transformations could mitigate this restriction [39,40], at the expense of a higher complexity. The non-pruned tree-structure in Fig. 9.18, i.e., also the

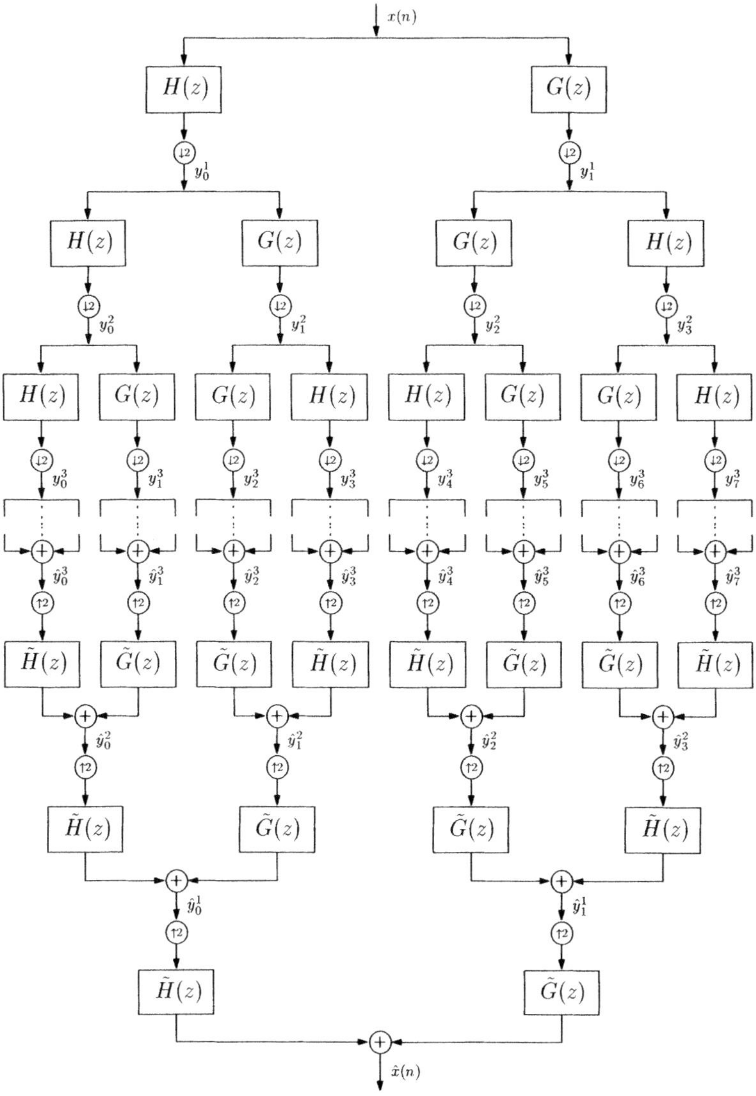

Fig. 9.24. Realization of the discrete wavelet-packet analysis and synthesis (e.g. for $M = 8$).

DWPA of Fig. 9.24, can be imagined with a succeeding signal-steered deletion of certain blocks; for a sufficiently large number μ of half-band filter stages, this is certainly a most flexible method. Also, in the pruned-tree structure, an additional "voicing" may be switched on or off at appropriate points.

9.9.6.3 Efficient Realization

Any of the above possibilities may be implemented with some minimized computational effort. All of them suffer, however, from a common drawback of all more sophisticated noise-reduction schemes: The number of parameters influencing the achievable quality of the output signal $\hat{s}(n)$ becomes very large. Particularly, in non-uniform spectral-analysis systems, multi-rate signals appear, and in the adaptive-bandwidth case, these rates vary. The parameters of the noise-estimation techniques and the subtraction rules, however, depend on the signals' time and frequency scales [24].

A simple solution was suggested in [24,25]: The actual spectral analysis and synthesis are kept constant, i.e., *non-adaptive.* Instead, the spectral weights g_i in Fig. 9.12 are adaptively grouped in frequency bands; in each band, the same factor $\bar{g}_\nu$ is applied.

A varying number of K bands with variable lower and upper frequency indices $i_{\mathrm{l},\nu}$ and $i_{\mathrm{u},\nu}$ are simulated in this way. For the weights, a simple average can be applied:

$$\bar{g}_\nu = \frac{1}{i_{\mathrm{u},\nu} - i_{\mathrm{l},\nu} + 1} \cdot \sum_{i=i_{\mathrm{l},\nu}}^{i_{\mathrm{u},\nu}} g_i, \quad \nu \in \{1, 2, \ldots, K\}. \tag{9.49}$$

The adaptive channel number K and the edge indices are derived from a tree-structured procedure, indicated in Fig. 9.25. In the first stage of the tree, the whole frequency range is divided into a relative small number $K \ll \frac{M}{2}$ of bands with equal widths $\frac{f_\mathrm{s}}{2K}$. If the signal energy within the subband ν is lower than the corresponding noise estimation, i.e., if

$$\sum_{i=i_{\mathrm{l},\nu}}^{i_{\mathrm{u},\nu}} S_{yy}(\Omega_i) < \eta \cdot \sum_{i=i_{\mathrm{l},\nu}}^{i_{\mathrm{u},\nu}} S_{bb}(\Omega_i) \quad , \quad \nu \in \{1, 2, \ldots, K\}, \tag{9.50}$$

the spectral weights are averaged for this subband, which will not be considered further in the next steps. If the signal energy is higher than the corresponding noise estimation, the resolution in this subband is increased by a factor of 2. If Eq. (9.50) is at least not satisfied in one subband, the next stage in the tree is reached, with an increased number K of channels. The corresponding limits $i_{\mathrm{l},\nu}$ and $i_{\mathrm{u},\nu}$, $\nu \in \{1, 2, \ldots, K\}$, are updated. Then the condition (9.50) is checked again for every newly generated subband, and the splitting procedure will be executed if necessary. The last stage in the tree is reached for a subband if Eq. (9.50) is satisfied (white subbands on the left-hand side of Fig. 9.25), or if the bandwidth corresponds to the original resolution of the spectral analysis. The right-hand side of Fig. 9.25 symbolizes the final resolution for the presented example: Depending on the local SNR, quite arbitrary band configurations may be found, with only one weighting factor each. In spectral regions with high SNR, the original high frequency resolution is reached, while low SNR-regions are smoothed, depending on the stage reached in the tree.

It should be mentioned that the tree-structure can be implemented efficiently by using a recursive algorithm. To avoid block effects in the frequency direction, the subbands of the first stage can be initialized by using overlap-add techniques.

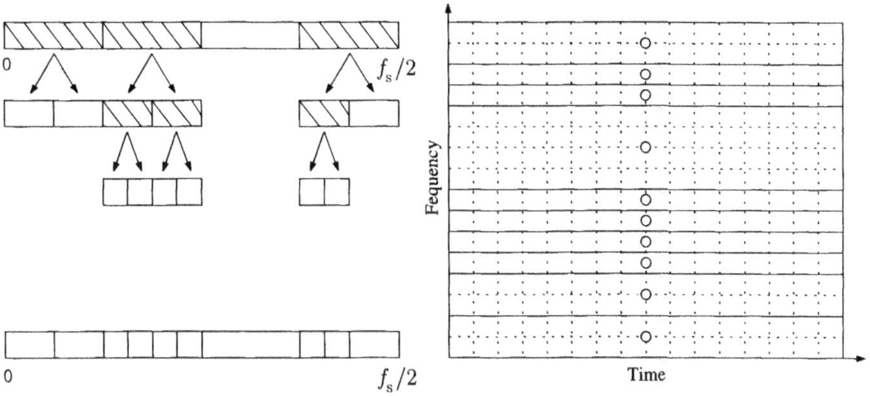

Fig. 9.25. Schematic of the adaptation of the bandwidths for a signal block and corresponding time-frequency resolution.

9.10 System Configurations, Experiments, and Comparisons

9.10.1 Status

The first proposals around 1980 [2,5], began with block-DFT / FFT analysis / synthesis systems, with magnitude or power subtraction, pause-noise estimation, and some simple non-linearities to reduce artifacts. Later, around 1985, with growing computer power, block-overlap or PPN-structures became feasible, combined with more refined non-linearities and smoothing techniques (e.g., [71]). Minimum-tracking noise estimation and signal-statistics-based subtraction rules became a certain standard in the 1990-ies. Wavelets entered the scene near the end of the century (e.g., [42, 67]), and other non-uniform band separations were tested around 2000 (e.g., [80]).

A vast variety of combinations is available from the literature. An overview of the earlier systems is found in [45], a somewhat later one in [74]. For industrial applications, with the limited resources available in a car, nowadays, multi-rate PPN-FFT-based systems with further band-division and quite complex noise estimation and subtraction algorithms are affordable and realizable in fixed-point arithmetic [63].

In the following, a few interesting configurations are to be described and analysed, in order to give a feeling of what can be achieved at which expense. Comparisons are made on the basis of spectrograms and some results of informal listening tests.

9.10.2 Examples

9.10.2.1 Uniform vs. Non-uniform Bandwidths

An earlier detailed study [56] confirmed that overlapping signal blocks in a DFT / FFT system as well as the generalization by a PPN-FFT would indeed enhance the smoothness of the synthesized output signal $\hat{s}(n)$. In a later investigation [12], therefore, many variants of uniformly and non-uniformly resolving filterbanks, including wavelet-derived ones, were examined. Figs. 9.27 to 9.30 reflect some of the results.

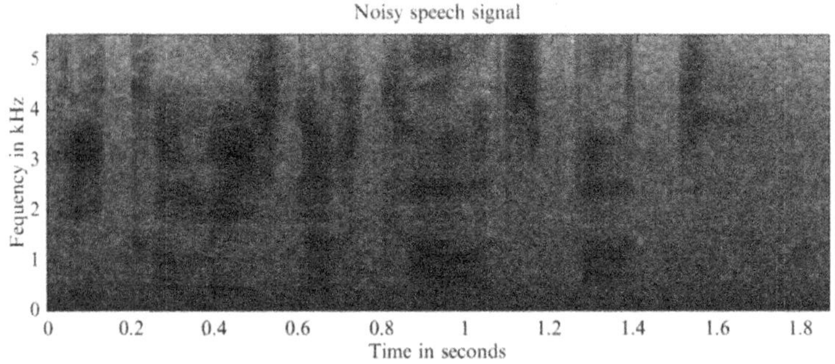

Fig. 9.26. Spectrogram of speech corrupted by noise from a moving car.

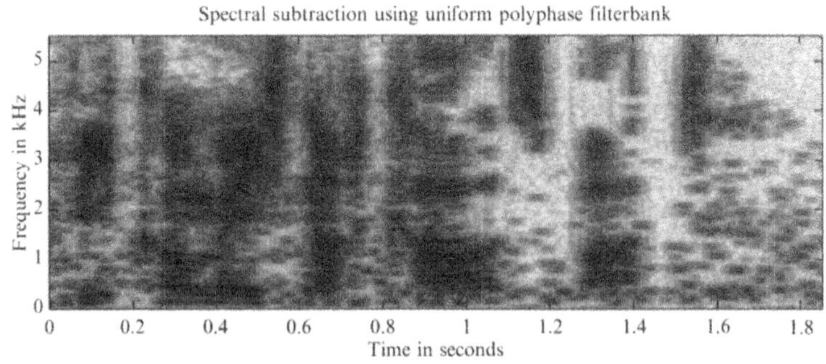

Fig. 9.27. Spectrogram after application of a simple subtraction rule with a uniform polyphase filterbank.

Fig. 9.26 shows the noisy-speech spectrogram. The relatively strong speech parts with their formants and pitch-harmonics, varying over time, are clearly visible in front of the more or less grey noise background. The disturbed signal was taken from the microphone of an hands-free telephone in a car moving at a speed of 120 km/h, and it was sampled at a rate $f_s = 11.025\ kHz$. The noise power was estimated using a minimum tracking. The de-noising was carried out with a very simple magnitude subtraction, including only a "half-wave rectification" and no other non-linearities, like over-estimation or noise floor, in order to make the effects of different spectral analyses more visible.

Fig. 9.27 proves that, indeed, noise has been removed, as visible in the zones which are "brighter" now than in Fig. 9.26. The small dark rectangles, however, distributed randomly in these regions, indicate the residual, musical artifacts. Their bandwidths and durations are more or less the same everywhere, defined by the channel number $M = 256$ and the prototype filter of length $L_w = 1024$. Fig. 9.28 clarifies the effects of an allpass transformation in the PPN: With a first-order allpass, chosen such that the Bark-scale resolution is approximated, the PPN-FFT with the same number of channels and the same prototype low-pass filter creates longer, but more narrow-band artifacts in the lower-frequency region, while the tonal effects become shorter and more broad-band (i.e.: less tonal) in the upper frequency parts. The same is visible also in Fig. 9.29. Here, an À-Trous (i.e., pruned-tree) structure with 7 octaves and 10 voices per octave, i.e., only 70 channels, was chosen. The similarity of the results indicates that in an allpass-transformed PPN-FFT also a reduced channel number $M < 256$ could have been used, reducing the extra-computational load somewhat.

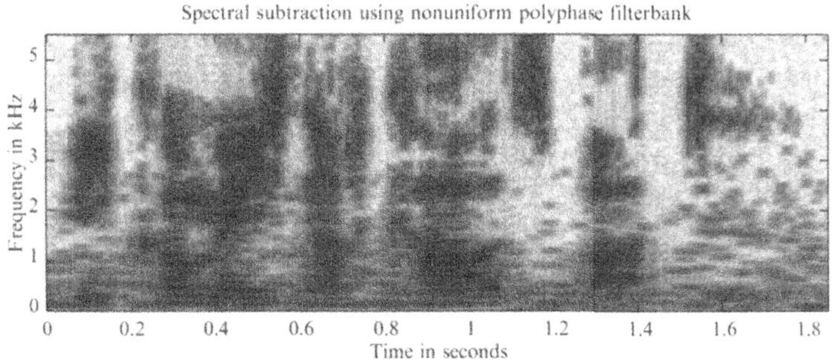

Fig. 9.28. Spectrogram after application of a simple subtraction rule with an allpass-transformed PPN-FFT.

Replacing the above simple weighting rule by the "quasi-standard" of Sec. 9.8.4 "smeares" the little rectangles in all cases; still, some differentiation between "long / narrow-band" and "short / wide-band" in lower and higher spectral ranges remains visible (see Fig. 9.30, [22]).

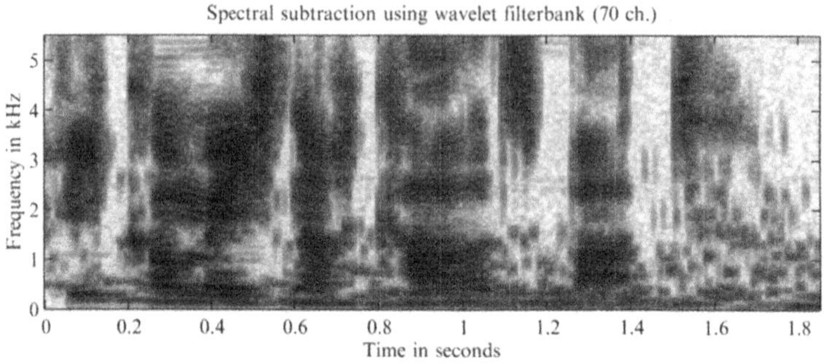

Fig. 9.29. Spectrogram after application of a simple subtraction rule with a pruned-tree filterbank.

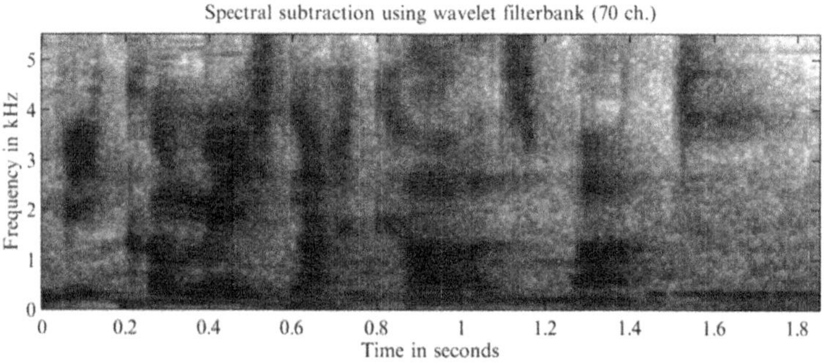

Fig. 9.30. Spectrogram after application of a refined subtraction rule with a pruned-tree filterbank.

Perceptually, both non-uniform filterbanks provide more pleasant residual disturbances plus a more natural speech sound; this holds for the simple as well as the more sophisticated subtraction rule.

Other variants of non-equispaced spectral decompositions were included into another thorough study [24, 25]. Especially, the DFT-based realization of the wavelet transformation by partial resynthesis was checked. Here, the "voicing" or non-equally spaced sub-division of octaves can be realized by grouping more, but smaller sets of original DFT lines in Fig. 9.19. For a first comparison, the above-used simple rule was applied again. The noise-PDS was now estimated in a one-step, initial-pause measurement, justified by the fact that only artificial, stationary white noise was added to the clean speech signal; the sampling rate is $f_s = 8\ kHz$ now.

Fig. 9.31 shows the corresponding corrupted-signal spectrogram, Figs. 9.32, 9.33, and 9.34 the results of denoising with a tree-structure of 7 octaves with 10 voices each, a DFT-based version with 9 octaves and 5 voices,

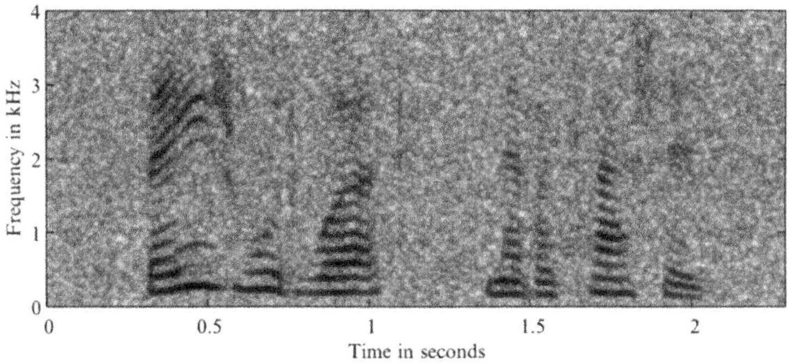

Fig. 9.31. Spectrogram of the noise-corrupted speech signal.

and a DFT-based solution with 9 octaves and 10 voices, respectively.[1] The same inverse time-frequency effects are visible in all cases, and no difference is observed perceptually. Due to the much smaller computational load, a preference of the DFT- / FFT-based realizations is concluded.

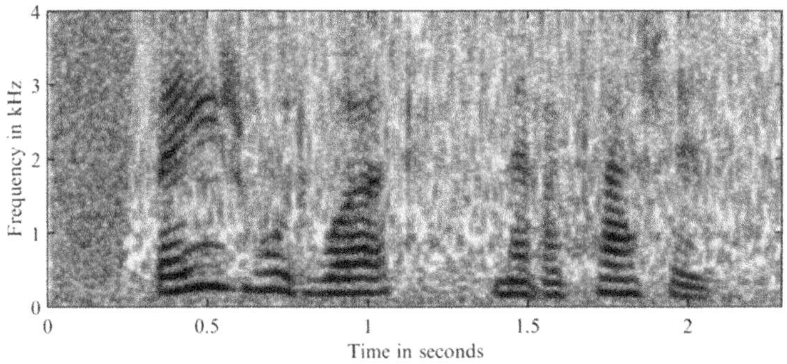

Fig. 9.32. Spectrogram after application of a simple subtraction rule with a wavelet-based tree structure.

The DWPA, exposed in Fig. 9.24, was said to offer a very high flexibility in terms of band definitions. Therefore, some heuristically appealing configurations were investigated, too, in [24, 25]. Figs. 9.35 and 9.36 give examples, where the simple and the enhanced weighting are carried out in a system

[1] The upper octave has a bandwidth of 2 kHz. If it is split into 5 voices, the lowest band-width is 0.30 kHz, the four higher ones larger by factors $(\sqrt[5]{2})^1$, ..., $(\sqrt[5]{2})^4$, such that the highest band has a width of 0.52 kHz. In the octave below, all bandwidths are halved. With 10 voices, a similar calculation is needed, with factors $(\sqrt[10]{2})^i$. In a length-M DFT-based realization, the sub-division stops at the DFT-line distance f_S/M.

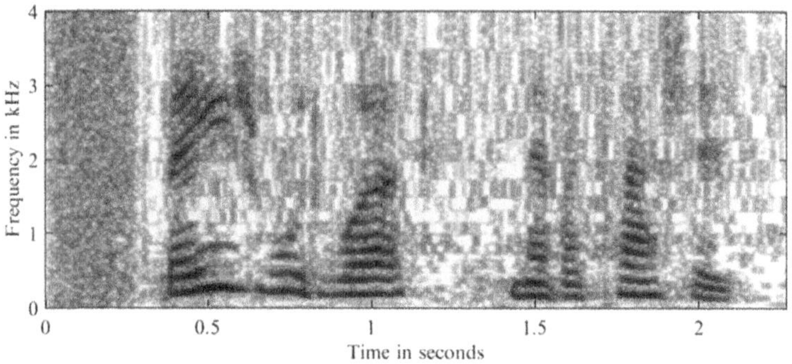

Fig. 9.33. Spectrogram after application of a simple subtraction rule with a DFT-based wavelet transformation: 9 octaves, 5 voices.

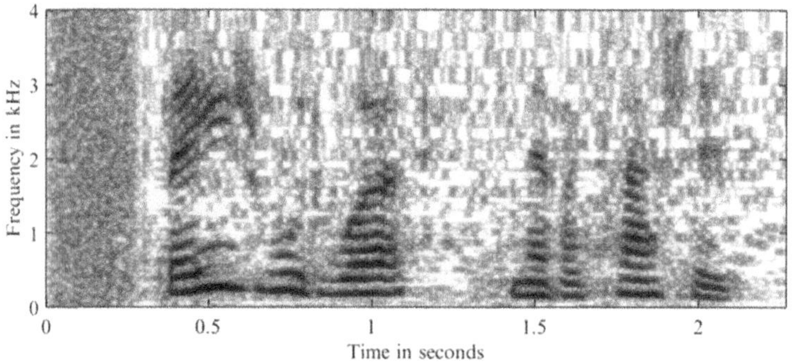

Fig. 9.34. Spectrogram after application of a simple subtraction rule with a DFT-based wavelet transformation: 9 octaves, 10 voices.

with 5 octaves, the upper 4 being sub-divided into 8 bands each, the lowest octave having 16 channels; a total of only 48 components results, though with a favourably narrow bandwidth in the lowest frequency range. As to be seen from these spectrograms and the comparison with all earlier results, noise reductions are achievable with obviously quite different residuals. No version was found in this heuristical search which would clearly outperform others, with also non-uniform, fixed bandwidths.

9.10.2.2 Fixed vs. Adaptive Bandwidths

In the investigations of [24,25], also experiments with bandwidth-adaptation were included. Additive white noise was used, and the sampling rate was $f_s = 8\ kHz$ again. For the spectral analysis, a simple DFT with $M = 256$ channels is used. So, the original spectral analysis has a resolution of ~ 31 Hz. For the initial bandwidths of the subbands in the first stage of the tree, 500 Hz ($K = 8$) are a good choice. To avoid block effects, two tree-structures with

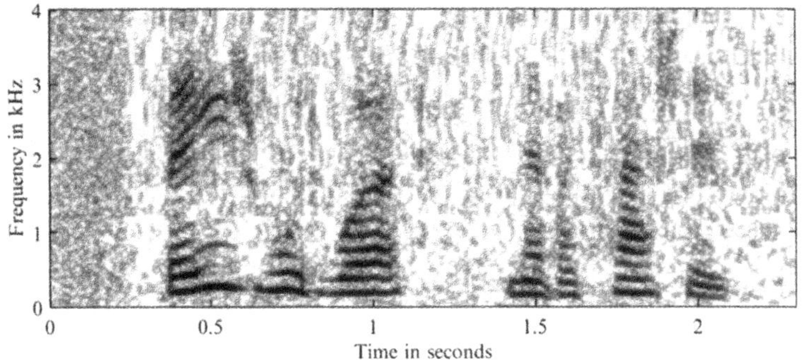

Fig. 9.35. Spectrogram after application of a simple subtraction rule with a DWPA, choosing 5 octaves.

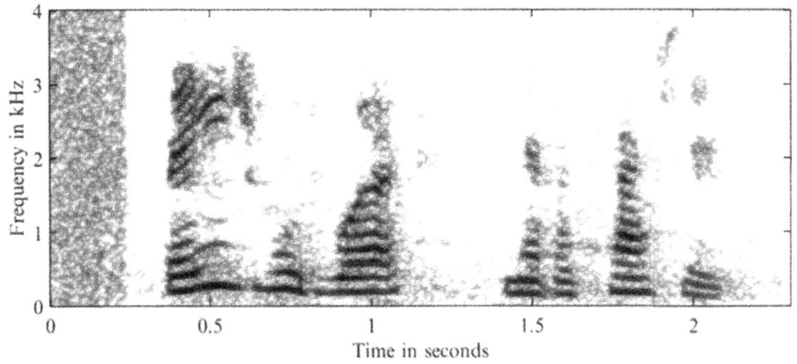

Fig. 9.36. Spectrogram after application of a refined subtraction rule with a DWPA, choosing 5 octaves.

50% overlap are used, with averaging the resulting weighting factors. First, we explain the effect of the tree-structured post-processing for a single voiced frame, as shown in Figs. 9.37, 9.38, and 9.39.

As expected, the algorithm smoothes the spectral weights in regions with a low SNR and leaves them unchanged in high-SNR regions, especially in pitch-structured regions.

The overall processing result is illustrated in Figs. 9.40, 9.41, and 9.42.

The first spectrogram contains the time-frequency representation of the enhanced signal after application of the basic spectral subtraction rule without bandwidth adaptation. Especially in speech pauses, many randomly distributed spectral peaks occur, which produce undesirable tonal noise.

In contrast, the second spectrogram shows the time-frequency representation of the enhanced signal after application of the same subtraction rule, but with tree-structured adaptivity for the subbands, as explained in Sec. 9.9.6.3. The musical tones are completely suppressed by smoothing during non-speech

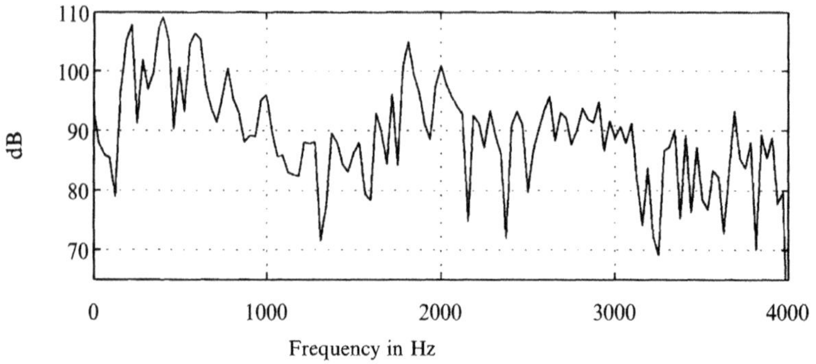

Fig. 9.37. Short-time spectrum of a vowel.

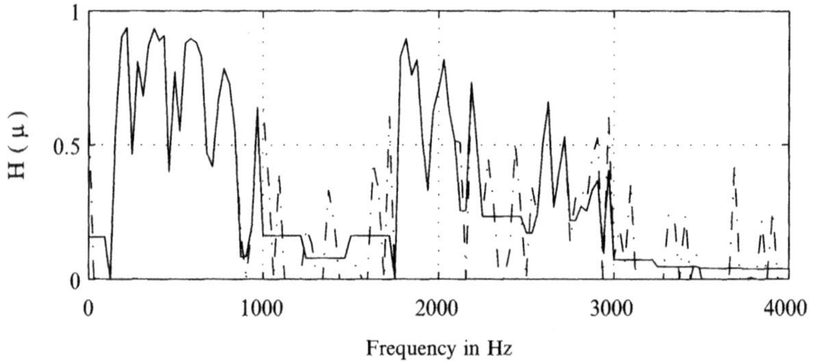

Fig. 9.38. Weighting factors g_i before (dashed line) and $\bar{g}_\nu$ after averaging (solid line).

activities, while the speech itself is not disturbed additionally. Also informal listening tests confirm this effect: The residual noise has now a nearly natural character, while the speech is not affected by the smoothing process.

Fig. 9.42, finally, visualizes the resulting different bandwidths as a coded plot: Each gray-tone symbolizes a different stage as reached in the tree-structure; the black areas represent a very high frequency resolution, namely, that of the original DFT (31 Hz), while the white areas correspond to the first stage of the tree, where the weighting factors are smoothed over bandwidths of 500 Hz. Between these two extrema, other stages in the tree occur, coded as gray-levels. It can be seen that the algorithm detects the contours of the speech activity and adapts the bandwidths.

It should be noted that no smoothing in time direction is done, so that the speech quality is not disturbed by echoes.

Experiments with the above-used, more complex subtraction rule, are found to offer no additional advantages.

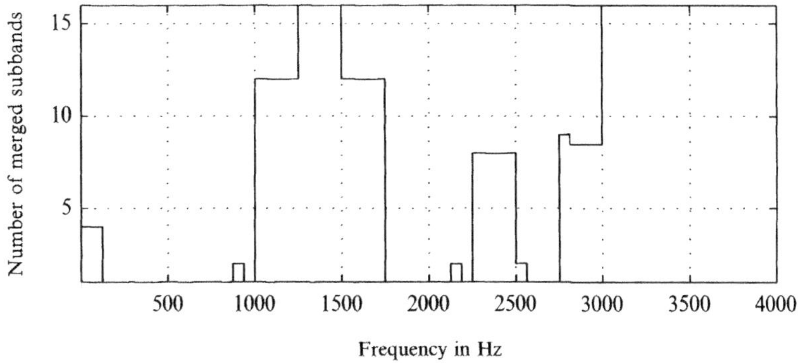

Fig. 9.39. Number of merged subbands.

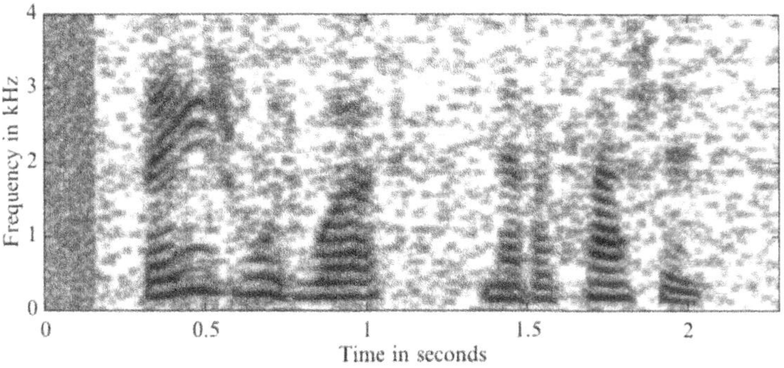

Fig. 9.40. Spectrogram after application of a simple subtraction rule with fixed-bandwidth weighting.

9.10.2.3 Noise Instationarity

Beyond the instationarities of the signal, time-variant noise characteristics are a known difficulty. Some experiments with the improved noise-tracking method, introduced in Sec. 9.7.4, show that this problem can be solved [23].

In Fig. 9.43, a speech signal is depicted with additive, cosine-modulated noise; Fig. 9.44 contains the corresponding spectrogram. As can be seen from Fig. 9.45, the standard minimum-tracking method is unable to follow the changing noise bahaviour; the gain-shape separation, however, tracks the changes well. A more detailed analysis of the estimation errors in [23] shows strong advantages oft this approach for different profiles of time-varying disturbances. Figs. 9.46 and 9.47 back the above measurement: The spectrogram (achieved with a very simple, DFT-based spectral subtraction) with the enhanced tracking shows much less unnatural signal deletions, distortions, and

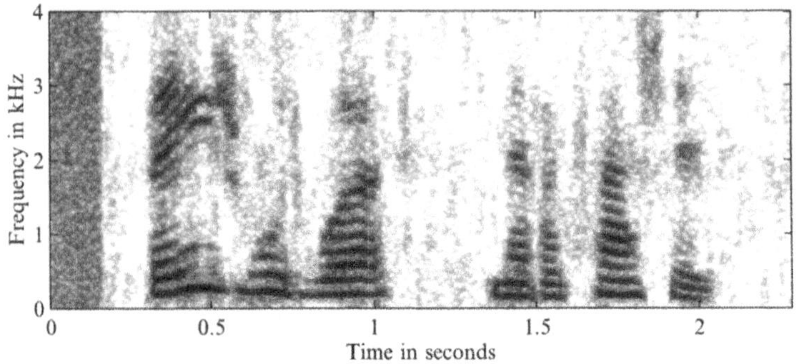

Fig. 9.41. Spectrogram after application of the simple subtraction rule with variable-bandwidth weighting.

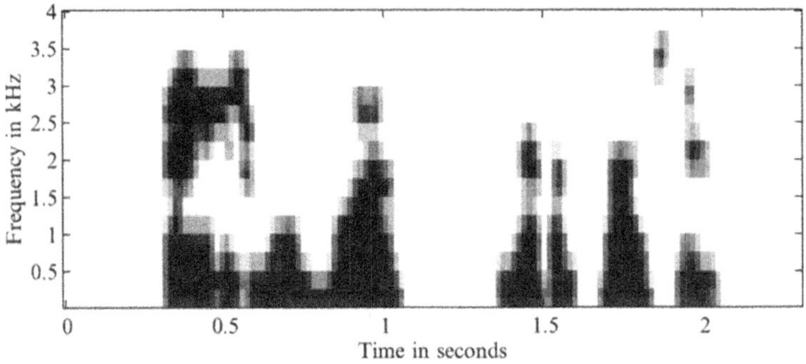

Fig. 9.42. Visualization of the time-frequency regions with different bandwidths.

remaining noise-modulation than that found after a standard minimum tracking.

9.11 Further Problems and Ideas, Concluding Remarks

According to the experiments described above, two setups are a good choice for single-channel noise reduction:

- Fixed, non-uniform-bandwidth spectral analysis / synthesis, favourably realized by a DFT-based wavelet transformation, together with a minimum-tracking noise-PDS shape estimation and a fast, separate gain tracking, combined with a statistics-based subtraction rule.
- Fixed, uniform-bandwidth spectral analysis / synthesis, e.g., by a simple DFT / FFT, together with the same noise estimation as above, with a

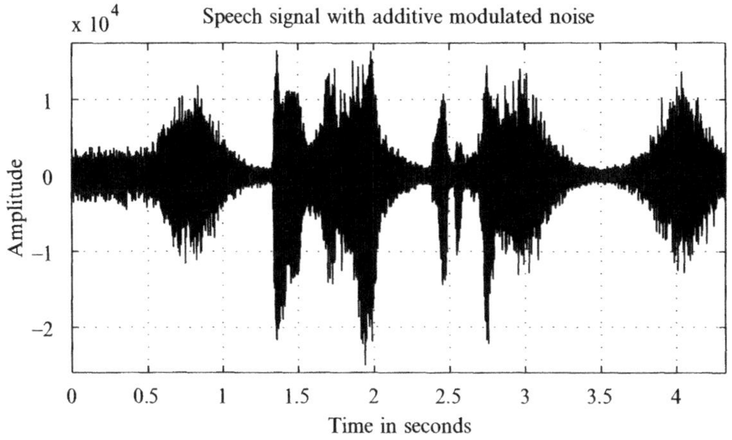

Fig. 9.43. Speech signal with additive, cosine-modulated noise.

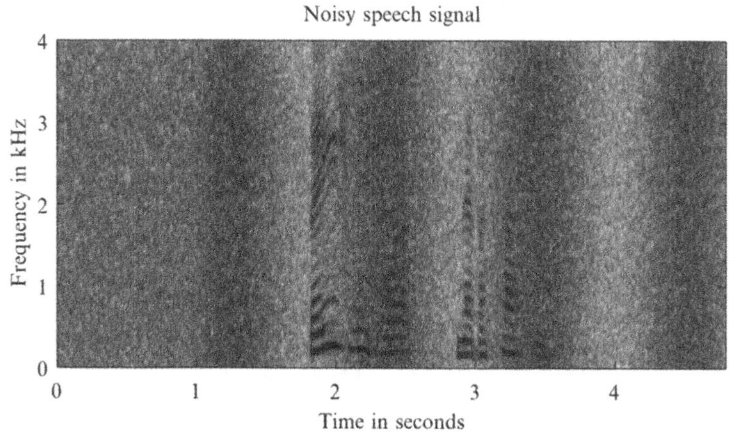

Fig. 9.44. Spectrogram of a speech signal corrupted with amplitude-modulated white noise.

much simpler weighting rule, but with adaptive-bandwidth weight smoothing.

In the former case, improvements are to be expected by applying more advanced signal- and noise-statistic models; in the latter case, the simple DFT should probably be replaced by a PPN-DFT.

A combination of non-uniform with adaptive bandwidth seems to be useless: The adaptivity would be hindered from finding the "best band separation" freely.

Above, also the combination of bandwidth adaptation with the enhanced subtraction rule was found to be unfavourable up to now. This may, however,

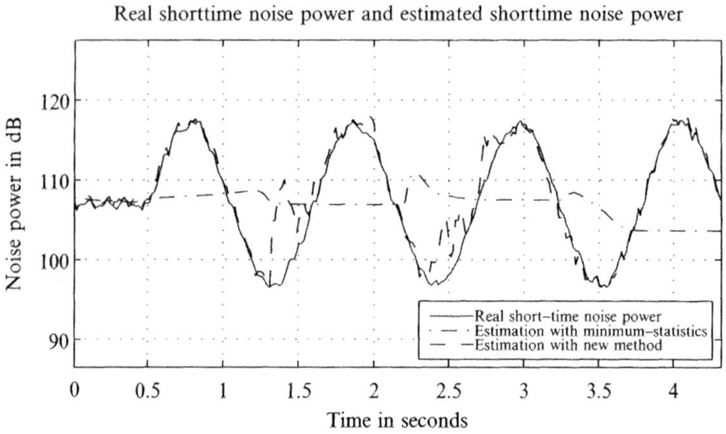

Fig. 9.45. Originally applied noise-power modulation and estimations from a minimum-tracking and the adaptive-gain method.

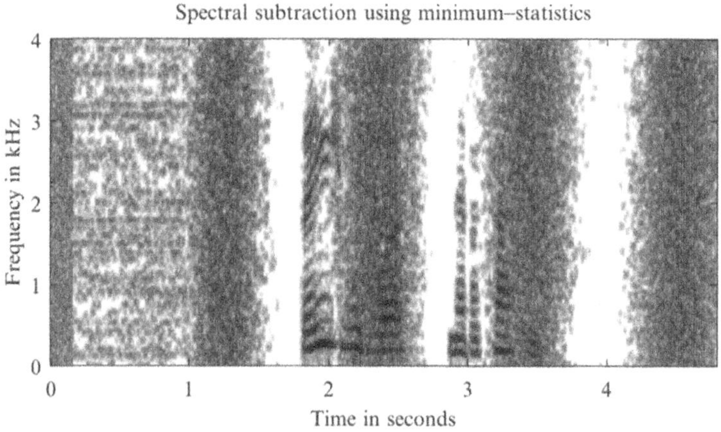

Fig. 9.46. Spectrogram after application of a simple subtraction rule with a minimum-tracking noise estimation.

be due to the difficult optimization of too many algorithmic parameters with strong interdependencies.

Some other aspects, not mentioned before, deserve some further attention. One of them is the application of *completely different* definitions of "spectra"; a signal can be decomposed into orthogonal bases which are not pre-defined as exponentials or wavelets a priori, but found, instead, from the current signal. Singular-value decomposition (SVD) is one possibility, as proposed in [10, 12, 16, 29, 38]; there are interesting filterbank interpretations of these analysis / synthesis systems, but no final break-through has been reported. A more recent idea may be the use of intrinsic-mode functions, also derived

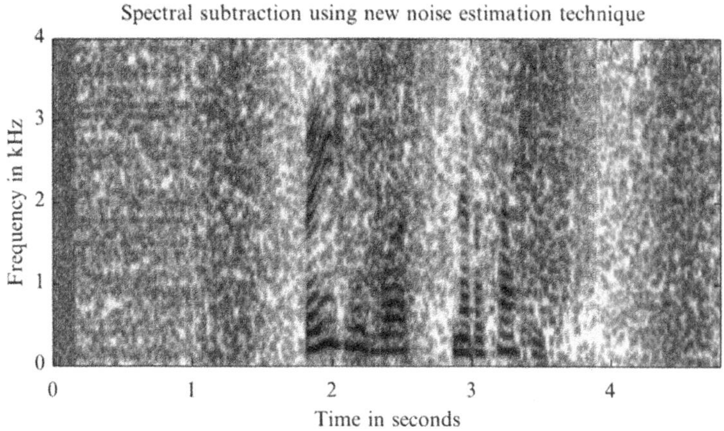

Fig. 9.47. Spectrogram after application of a simple subtraction rule with the adaptive-gain noise estimation.

adaptively from the given signal [34]. A second aspect is a practical one: With the advent of possible wide-band speech transmission, e.g., in UMTS or in voice-over-IP applications, noise reduction for speech with a bandwidth of 7 kHz will become interesting, where the perceived quality will be of even more importance than in the classical telephone-band case; it is an open question whether the present methods can be simply transferred, or whether other approaches are helpful; some experiments have shown that, e.g., in noise-PDS estimation, modifications are needed [36] and that separate treatments in a few subbands may help. As to perceived quality, finally, methods taking psychoacoustics into account have gained interest in recent years (e.g., [28, 30]); perhaps they provide a way out of the "vicious triangle" still limiting the success of single-channel noise reduction: "There is no noise suppression, there is noise *transformation*" – from noise to artifacts, reduced at the price of distortions, reduced at the expense of more noise... Hiding the residual effects for the ear should provide the solution.

A final remark is necessary: The above thoughts and discussions do, of course, rely to a large extent on the work done in the author's group and in some closely cooperating institutes. Ideas and publications of others may have been overlooked or not seen in their true importance. So, this chapter is certainly subjective – this is admitted, but probably unavoidable.

References

[1] L. Arslan, A. McCree, V. Viswanathan: New metods for adaptive noise Suppression, *Proc. ICASSP '95,* 812–815, 1995.

[2] M. Berouti, R. Schwartz, J. Makhoul: Enhancement of speech corrupted by acoustic noise, *Proc. ICASSP '79,* 208–211, 1979.
[3] M. Bierhaus: *Speech-Pause Detection in Severely Disturbed Speech Signals,* Dipl. Thesis, Ruhr-Univ. Bochum, AGDSV, Bochum, 1992 (in German).
[4] M. Bodden: Binaural Signal Processing: *Modelling the Recognition of Direction and the Cocktail-Party Effect,* Doct. Diss., Ruhr-Univ. Bochum, Aachen, Germany: Shaker, 1992 (in German).
[5] S. F. Boll: Suppression of acoustic noise in speech using spectral subtraction, *IEEE Trans. Acoust., Speech, Sig. Process.,* **T-ASSP-27**(2), 113–119, 1979.
[6] O. Cappé: Elimination of the musical noise phenomenon, *IEEE Trans. Speech Audio Process.,* **T-SA-2**(2), 345–349, 1994.
[7] A. G. Constantinides: Frequency transformations for digital filters, *IEE El. Lett.*, **3**, 487–489, 1967.
[8] A. G. Constantinides: Frequency transformations for digital filters, *IEE El. Lett.*, **4**, 115–116, 1968.
[9] I. Daubechies: *Ten lectures on Wavelets,* Philadalphia, USA: Soc. Ind. Appl. Math., 1992.
[10] M. Dendrinos, S. Bakamidis, G. Carayannis: Speech enhancement from noise: a regenerative approach, *Speech Communication,* **10**, 45–57, 1991.
[11] G. Doblinger: Computationally efficient speech enhancement by spectral minima tracking in subbands, *Proc. EUROSPEECH '95*, **2**,1513–1516, 1995.
[12] A. Engelsberg: *Transformation-Based Systems for Single-Channel Noise Reduction in Speech Signals,* Doct. Diss., Christian-Albrechts Univ. Kiel, Aachen, Germany: Shaker, 1998 (in German).
[13] Y. Ephraim, D. Malah: Speech-enhancement using optimum non-linear spectral-amplitude estimation, *Proc. ICASSP '83*, 1118–1121, 1983.
[14] Y. Ephraim: Speech-enhancement using a minimum mean-square error short-time spectral amplitude estimator, *IEEE Trans. Acoust., Speech, Sig. Process.*, **T-ASSP-32**, 1109–1121, 1984.
[15] Y. Ephraim, D. Malah: Speech-enhancement using a minimum mean-square error log-spectral amplitude estimator, *IEEE Trans. Acoust., Speech, Sig. Process.*, **T-ASSP-33**, 443–445, 1985.
[16] Y. Ephraim, H. L. van Trees: A signal-subspace approach for speech enhancement, *IEEE Trans. Speech Audio*, **T-SA-3**, 251–266, 1995.
[17] ETSI Rec. GSM 06.92: *Voice-Activity Detector,* European Telecomm. Standard. Institute, Valboune, 1989.
[18] N. Flores, S. J. Young: Continuous speech recognition in noise using spectral subtraction and HMM adaptation, *Proc. ICASSP '94*, **1**, 409–412, 1994.
[19] E. Galijasevic: *Allpass-Based Near-Perfect-Reconstruction Filterbanks,* Doct. Diss., Christian-Albrechts-Univ., Kiel, Aachen, Germany: Shaker, 2002.

[20] E. Galijasevic, J. Kliewer: Design of allpass-based non-uniform oversampled DFT filter banks, *Proc. ICASSP '02*, 1181–1184, Orlando, USA: 2002.

[21] R. Gluth: *Contributions to the Description and Realization of Digital, Non-recursive Filterbanks Based on Linear Discrete Transformations,* Doct. Diss., Ruhr-Univ. Bochum, Aachen, Germany: Shaker, 1993 (in German).

[22] T. Guelzow, A. Engelsberg, U. Heute: Comparision of a discrete wavelet transformation and a nonuniform polyphase filterbank applied to spectral–subtraction speech enhancement, *Signal Process.*, **64**(1), 5–19, 1998.

[23] T. Guelzow: Spectral-subtraction speech enhancement using a new estimation technique for non-statioary noise, *Proc. WAENC '99*, 76–79, 1999.

[24] T. Guelzow: *Enhancement of the Quality of Severely Disturbed Speech Signals - Detection of a Carrier Mismatch and Suppression of Additive Disturbances,* Doct. Diss., Christian-Albrechts-Univ. Kiel, Aachen, Germany: Shaker, 2001 (in German).

[25] T. Guelzow, T. Ludwig, U. Heute: Spectral-subtraction speech enhancement in multirate systems with and without non-uniform and adaptive bandwidths, *Signal Process.*, **83**, 1613–1631, 2003.

[26] L. C. Guendel: *Applications of Filterbanks for Speech Coding in the Frequency Domain,* Doct. Diss., Friedrich-Alexander-Univ., Erlangen, 1987 (in German).

[27] L. C. Guendel: Filterbanks with unequal-spaced channels, *Proc. EUSIPCO '90*, **1**, 581–584, 1990.

[28] S. Gustafsson, P. Jax, P. Vary: A novel psychoacoustically motivated audio-enhancement algorithm preserving background-noise characteristics, *Proc. ICASSP '98*, **1**, 397–400, 1998.

[29] P. S. K. Hansen, P. C. Hansen, S. D. Hansen, J. A. Soerensen: Experimental comparison of signal-subspace-based noise-reduction methods, *Proc. ICASSP '99*, **1**, 101–104, 1999.

[30] T. Haulick, K. Linhard, P. Schroegmeier: Residual-noise suppression using psychoacoustic criteria, *Proc. EUROSPEECH '97*, **3**, 1395–1398, 1997.

[31] U. Heute: Speech and audiocoding: aiming at high quality and low data rates, in J. Blauert (ed.), *Communication Acoustics*, Berlin, Germany: Springer, 2005, 336–393.

[32] U. Heute, P. Vary: A digital filterbank with polyphase network and FFT processor: measurements and applications, *Signal Processing*, **4**, 307–319, 1981.

[33] H. G. Hirsch, C. Ehrlicher: Noise-estimation techniques for robust speech recognition, *Proc. ICASSP '95*, **1**, 153–156, 1995.

[34] N. E. Huang, Z. Shen, S. R. Long, M. C. Wu, H. H. Shih, Q. Zheng, N.-C. Yen, C. C. Tung, H. H. Liu: The empirical mode decomposition and

the HILBERT spectrum for non-linear and non-stationary time-series analysis, *Proc. Royal Soc. London*, **454**, 903–905, London, UK, 1998.

[35] D. Janardhanan, U. Heute: Wideband speech enhancement using a robust noise estimation, *Proc. DAGA '05*, Munich, Germany: 2005.

[36] D. Janardhanan, U. Heute: Wideband speech enhancement using a modified noise estimation, *Proc. ESSV '05*, Prague, Cz, 2005.

[37] M. Jelinek, R. Salami: Noise-reduction method for wideband-speech coding, *Proc. EUSIPCO '04*, 1959–1962, Vienna, Austria, 2004.

[38] S. H. Jensen, P. C. Hansen, S. D. Hansen, J. A. Soerensen: A spectral subspace approach for noise reduction of speech signals, *Proc. EUSIPCO '94*, **2**, 1174–1177, 1994.

[39] M. Kappelan: *Characteristics of Allpass Chains and their Application for Non-equispaced Spectral Analysis and Synthesis,* Doct. Diss., RWTH Aachen, Wiss. Verlag, Mainz, Germany, 1998 (in German).

[40] M. Kappelan, B. Strauss, P. Vary: Flexible non-uniform filter banks using allpass transformations of multiple order, *Proc. EUSIPCO '96*, 1745–1748, Trieste, Italy, 1996.

[41] J. Kliewer: *Contributions to the Design of Modulated Filterbanks for Varying Sub-band Sampling Rates,* Doct. Diss., Christian-Albrechts-Univ. Kiel, Aachen, Germany: Shaker, 1999 (in German).

[42] M. Lang, H. Guo, J. E. Odegard, C. S. Burrus, R. O. Wells: Noise reduction using an undecimated Discrete Wavelet Transform, *IEEE Sig. Process. Lett.*, **3**, 10–12, 1996.

[43] J. S. Lim: Evaluation of a correlation-subtraction method for enhancing speech degraded by additive white noise, *IEEE Trans. Acoust., Speech, Sig. Process.*, **T-ASSP-26**, 471–472, 1978.

[44] D. L. Wang, J. S. Lim: The unimportance of phase in speech enhancement, *IEEE Trans. Acoust., Speech, Sig. Process.*, **T-ASSP-30**, 679–681, 1982.

[45] J. S. Lim: *Speech Enhancement,* Englewood Cliffs, USA: Prentice Hall, 1983.

[46] K. Linhard, T. Haulick: Non-linear smoothing and noise reduction for disturbed speech signals, *Proc. 9-th Aachen Colloq. Sig. Theory: Image and Speech Sig.*, RWTH, Aachen, Germany, 1997 (in German).

[47] K. Linhard, T. Haulick: Spectral noise subtraction with recursive gain curves, *Proc. ICSLP '98*, **4**, 1479–1482, Sydney, Australia, 1998.

[48] P. Lockwood, J. Boudy: Experiments with a non-linear spectral subtractor (NSS), Hidden Markov Models, and the projection, for robust speech recognition in cars, *Speech Communication*, **11**, 215–228, 1992.

[49] T. Lotter, P. Vary: Noise reduction by joint maximum a posteriori spectral amplitude and phase estimation with super-GAUSSian speech modelling, *Proc. EUSIPCO '04*, 1457–1460, Vienna, Austria, 2004.

[50] Th. Lungwitz: *Investigations of Multi-channel Adaptive Noise Reduction for Speech Recognition in Vehicles,* Doct. Diss., Christian-Albrechts Univ. Kiel, Aachen, Germany: Shaker, 1999 (in German).

[51] R. Martin: An efficient algorithm to estimate the instantaneous SNR of speech signals, *Proc. EUROSPEECH '93*, 1093–1096, 1994.

[52] R. Martin: Spectral subtraction based on minimum statistics, *Proc. EURASIP '94*, 1182–1185, Elsevier, Amsterdam, NL, 1994.

[53] R. Martin: Noise power spectral density estimation based on optimal smoothing and minimum statistics, *IEEE Trans. Speech Audio Process.*, **T-SA-9**(5), 504–512, 2001.

[54] R. Martin: Speech enhancement using MMSE short-time spectral estimation with gamma-distributed priors, *Proc. ICASSP '02*, **1**, 253–256, Orlando, FL, USA, 2002.

[55] S. K. Mitra: *Digital Signal Processing – A Computer-Based Approach*, New York, USA: McGraw-Hill, 1998.

[56] S. Moeller: *Single-Channel Noise Reduction in the Frequency Domain with Improved Spectral Analysis and Adaptivity*, Dipl. Thesis, Ruh-Univ. Bochum, 1993 (in German).

[57] S. Moeller: *Assessment and Prediction of Speech Quality in Telecommunications*, Boston, USA: Kluwer Acad. Press, 2000.

[58] D. E. Newland: Time-frequency and time-scale analysis by harmonic Wavelets, *Proc. ECSAP '87*, 53–59, Prague, Cz, 1997.

[59] D. E. Newland: Time-frequency and time-scale analysis by harmonic Wavelets, in A. Prochazka, J. Uhlir, P. j. W. Rayner, N. G. Kingsbury (eds.), *Signal Analysis and Prediction*, 3–26, Berlin, Germany: Birkhaeuser, 1998.

[60] H. Puder, O. Soffke: An approach to an optimized voice-activity detector for noisy speech signals, *Proc. EUSIPCO '02*, **1**, 243–246, Tolouse, France, 2002.

[61] S. R. Quackenbusch, T. P. Barnwell, M. A. Clements: *Objective Measures of Speech Quality*, Englewood Cliffs, USA: Prentice Hall, 1988.

[62] V. Schless, F. Class: SNR-dependent flooring and noise overestimation for joint application of spectral subtraction and model combination, *Proc. ICSLP '98*, 1495–1497, 1998.

[63] G. U. Schmidt: *Design and Realization of a Multi-Rate System for Handsfree Telephony*, Doct. Diss., Techn. Univ. of Darmstadt, Progress Rep., Duesseldorf, Germany: VDI, 2001 (in German).

[64] H. W. Schuessler: *Digital Systems for Signal Processing*, Berlin, Germany: Springer, 1973.

[65] H. W. Schuessler: *Digital Signal Processing I*, Berlin, Germany: Springer, 1994 (in German).

[66] H. W. Schuessler, W. Winkelnkemper: Variable digital filters, *Arch. El. Uebertr.*, **24**, 524–525, 1970 (in German).

[67] J. W. Seok, K. S. Bae: Speech enhancement with reduction of noise components in the wavelet domain, *Proc. ICASSP '97*, **2**, 1223–1226, 1997.

[68] M. J. Shensa: The discrete Wavelet transform: wedding the Á-Trous- and Mallat algorithms, *IEEE Trans. Sig. Process.*, **T-ASSP-40**, 2464–2482, 1992.

[69] P. P. Vaidyanathan: *Multirate Systems and Filter Banks,* Englewood Cliffs, USA: Prentice Hall, 1993.

[70] P. Vary: A Contribution to Short-time Spectral Analysis by Digital Systems, Doct. Diss., Friedrich-Alexander-Univ., Erlangen, 1978 (in German).

[71] P. Vary: Noise suppression by spectral magnitude estimation – mechanism and theoretical limits, *Signal Processing*, **8**, 387–400, 1985.

[72] P. Vary, U. Heute: A short-time spectrum analyzer with polyphase network and DFT, *Signal Processing*, **3**, 55–65, 1980.

[73] P. Vary, U. Heute, W. Hess: *Digital Speech-Signal Processing,* Stuttgart, Germany: Teubner, 1998 (in German).

[74] S. V. Vaseghi: *Advanced Signal Processing and Noise Reduction,* New York, USA / Stuttgart, Germany: Wiley / Teubner, 1997.

[75] M. Vetterli, J. Kovacevic: *Wavelets and Subband Coding,* Englewood Cliffs, USA: Prentice HAll, 1995.

[76] N. Virag: Single-channel speech enhancement based on masking properties of the human auditory system, *IEEE Trans. Speech Audio Process.*, **7**, 126–137, 1999.

[77] M. V. Wickerhauser: *Adaptive Wavelet Analysis,* Braunschweig, Germany: Vieweg, 1995.

[78] N. Wiener: *Extrapolation, Interpolation, and Smoothing of Stationary Time Series with Engineering Applications,* New York, USA: Wiley, 1949.

[79] P. J. Wolfe, S. J. Godsill: Efficient alternatives to the Ephraim-Malah suppression rule for audio-signal enhancement, *EURASIP J. Appl. Sig. Process.*, **10**, 1043–1051, 2003.

[80] X. M. Xie, S. C. Chan, T. I. Yuk: A class of perfect-reconstruction non-uniform cosine-modulated filterbanks with dynamic recombination, *Proc. EUSIPCO '02*, **2**, 549–552, 2002.

[81] E. Zwicker, H. Fastl: *Psychoacoustics – Facts and Models,* 2nd ed., Berlin, Germany: Springer, 1999.

10

Noise Reduction with Kalman-Filters for Hands-Free Car Phones Based on Parametric Spectral Speech and Noise Estimates

Henning Puder

Siemens Audiological Engineering Group, Erlangen, Germany

10.1 Introduction

For some years now, in many countries, the use of hands-free telephones in cars is only allowed with hands-free mobile phones. Instead of the telephone receiver, for these hands-free facilities, a loudspeaker and a microphone are installed in the car. Thus, the telephone user has his hands free for driving the car.

Besides the well-known and mostly solved echo problem of hands-free telephones [4, 16] a major problem of hands-free car phones is the ambient noise which superimposes to the desired speech signal and entails a less comfortable telephone conversation. This noise problem is, compared to normal receiver telephones, more severe due to the larger distance of the speaker and the microphone of the hands-free unit.

The preferred method for reducing this interference noise, in this case, is the utilization of beamformers (see Chapter 2) which perform a space selective filtering and steer a beam in the direction of the desired speaker. However, due to problems of installing several microphones which are required for beamforming or simply due to the related costs, often only one microphone can be utilized. In this case, for noise reduction, one is restricted to single channel noise reduction methods which manage with the only available noisy signal being the superposition of the desired speech and noise.

But also when beamforming methods can be applied, signal channel noise reduction methods can be utilized as additional noise reduction method applied to the single output signal of the beamformer. In Fig. 10.1 these two applications of single channel noise reduction are illustrated.

Single channel noise reduction is a research topic for many years now. Usually, Wiener filter based approaches are utilized to solve this task [3,13]. In the case of Wiener filters, the noisy signal is decomposed into many frequency components which are then weighted according to their individual signal-to-noise ratio (SNR). Besides the Wiener filter rule, alternative weighting rules

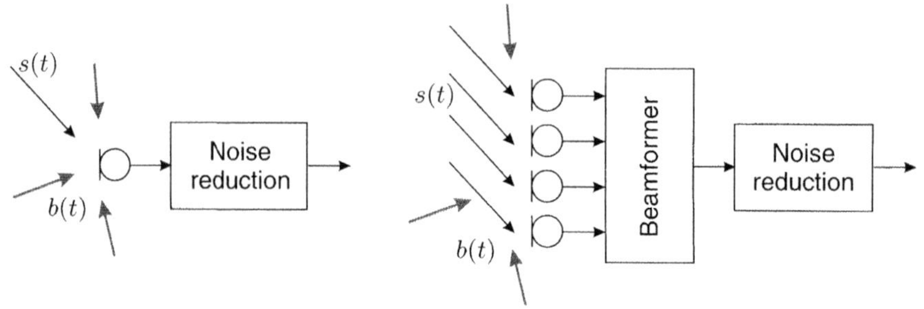

Fig. 10.1. Application of single channel noise reduction for enhancing a single noisy speech source (left) or as post-processing unit in addition to a beamformer.

have been proposed, where surely the most prominent one is the Ephraim-Malah approach [9, 10].

In this chapter we will introduce a complete alternative approach based on Kalman filters. The motivation for applying this approach is twofold:

- The Kalman filter approach allows an optimal filter design for non-stationary signals.
- The Kalman filter related parametric spectral estimation allows to incorporate a priori knowledge of speech and noise to make the estimation more reliable.

Concerning the filter design for non-stationary signals, the Kalman filter is optimum for non-stationary signals such as speech, whereas the Wiener filter is designed such as to be optimum for stationary signals, only. In order to be able to utilize this approach for noisy speech, anyhow, the Wiener filter is subsequently applied to short speech segments of approximately 20 msec which are supposed to be stationary. Nevertheless, the Wiener filter looses its optimum performance for which it is designed for.

For the spectral estimation, the Wiener and Ephraim-Malah filter based approaches estimate the power spectral density of speech and noise in dependence of frequency. In contrast, Kalman filters utilize parametric spectral models of speech and noise, for which mostly auto-recursive (AR) processes are utilized:

$$s(n) = \sum_{l=1}^{p} a_l(n)\, s(n-l) + w(n)\,. \tag{10.1}$$

The parametric spectral distribution of the signal $s(n)$ is implicitly represented by the AR parameters $a_l(n)$ and the power of the white excitation signal $w(n)$.

For estimating these AR parameters, different methods can be utilized which will be explained in the next section. These parametric estimation methods allow to incorporate a priori knowledge of the signals for which the models have to be estimated – in our case speech and noise. This especially concerns the choice of the AR model orders which are related to the required frequency resolution. It will be shown that for speech a higher model order is required than for noise. Especially, for a high-quality noise reduction, the pitch structure of the speech has to be resolved. In contrast to other publications [11,18] of Kalman filters for speech enhancement, these relations were first published in [22,23].

An additional advantage of this parametric estimation is that – compared to direct spectral estimation methods – less unknowns have to be estimated. Therefore, based on the same amount of information, these parameters can be estimated more reliably which reduces the risk of non-desired *musical tones* [2].

This chapter is organized as follows: In the following section, first an analysis of speech and car noise is performed and their main properties are extracted which are required for the design of optimum Kalman filters.

In the third section, the necessary theoretic basics of Kalman filters and the of required parametric spectral estimation are introduced.

Then, in the fourth section, a practical application of Kalman filters for noise reduction will be proposed. Here, first, a subband Kalman filter approach will be motivated based on the characteristic of speech. Then methods for the parametric estimation of speech and noise models – only based on the noisy speech signal – will be elaborated. And finally, methods will be sketched which enhance the noise reduction performance based on the a-priori knowledge of the pitch frequency.

The last section before the conclusions is dedicated to a detailed analysis of the proposed Kalman filter noise reduction method, compared with alternative approaches, i.e. methods based on direct spectral estimation such as Wiener filter and Ephraim-Malah approaches.

10.2 Speech and Car Noise Analysis

10.2.1 Car Noise Analysis

Analyzing the spectral distribution of car noise, one first can observe that car noise exhibits strong low-frequency components. After a steep decrease the power spectral density (PSD) then decreases more slowly towards higher frequency components.

These properties can, as an example, be observed in Fig. 10.2. On the left, the power spectral densities of the car noise at 50 km/h and 110 km/h are depicted.

For frequency components above approximately 300 Hz, a PSD increase of about 10 dB can be noticed due to the higher car speed. Performing a spectrogram analysis, with a resolution of 32 Hz one obtains the result depicted in Fig. 10.3.

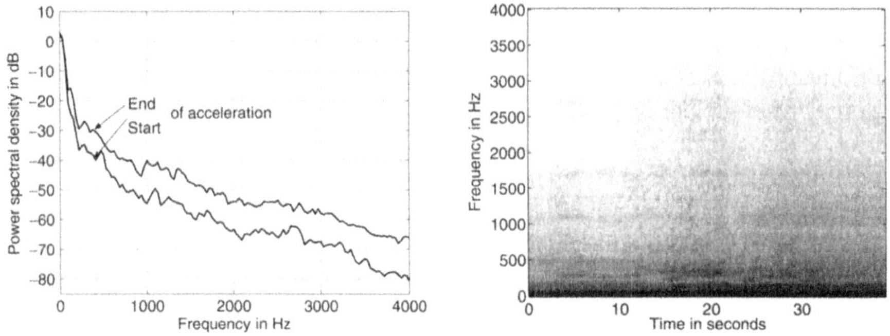

Fig. 10.2. Power spectral density at the beginning (50 km/h) and at the end (110 km/h) of an acceleration (left) with the corresponding spectrogram (right).

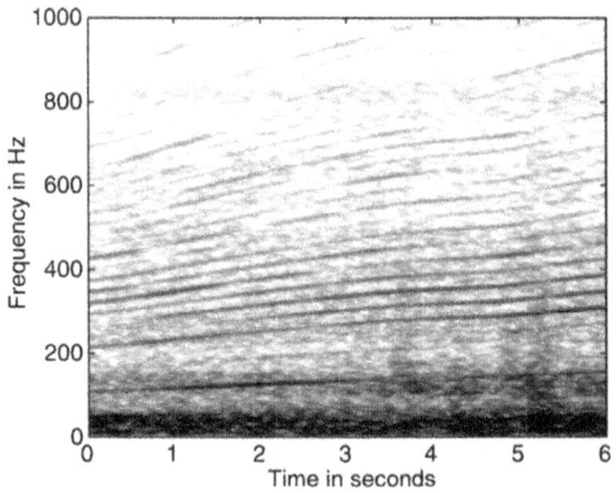

Fig. 10.3. Spectrogram of a car noise analyzed during an acceleration.

The spectral harmonics, one can observe, are harmonics of the engine noise which, in particular, occur when accelerating the car. They vary faster with the time than the other car noise components.

In total, car noise is mostly composed of several components:

- transmission,
- car body,
- engine,
- wind, and
- tyre noise,

where the last three ones are generally the dominant ones. These components will be analyzed separately in the following. A special focus will be laid on possibilities to predict the noise PSD in dependence of the car speed or the engine speed since the noise PSD is usually hard to estimate during speech activity.

10.2.1.1 Engine Noise

Analyzing engine noise separately, one observes that strong spectral components are present at multiples of half of the engine frequency. This can be observed for the engine noise example depicted in Fig. 10.4.

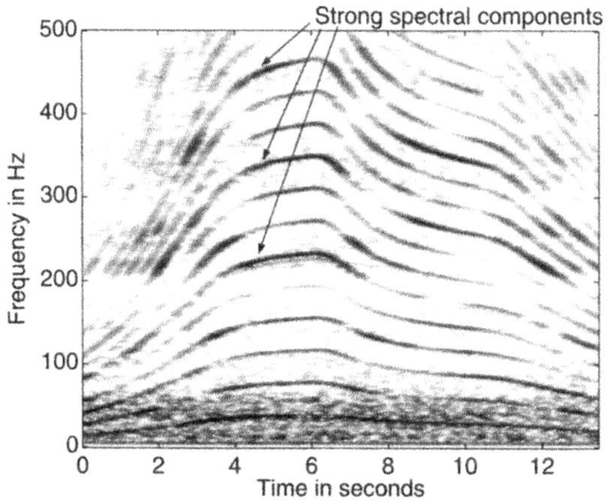

Fig. 10.4. Spectrogram of an engine noise with varying engine frequency.

This relation is valid for the usual four cycle engines independent of the number of cylinders. The only property that depends on the number of cylinders is the relative power of the spectral harmonics. For four cylinder engines every second and for six cylinders every third engine harmonic has a higher power than the others.

10.2.1.2 Wind Noise

Wind noise components mainly occur due to air turbulence at the car cabin. The usually optimized design of car cabins help to avoid whistling noise sounds.

Wind noise components depend on the car speed and change rather slowly within the time. An example of the power spectral density of wind noise is depicted in Fig. 10.5.

Since wind noise components are usually less powerful than tyre noise components, a prediction of the power spectral density in dependence of the car speed does not make sense.

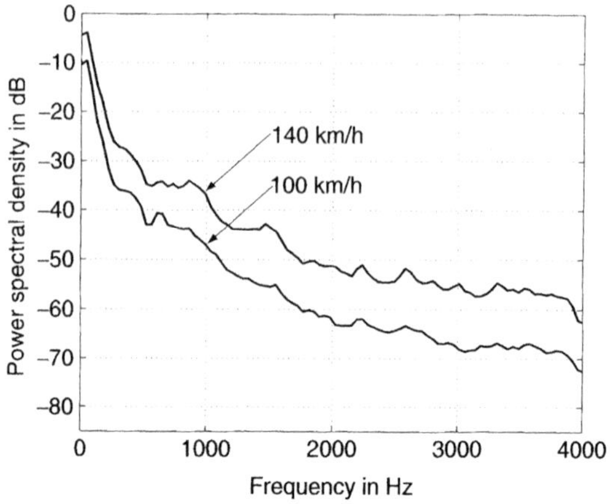

Fig. 10.5. Power spectral densities of wind noise at two different speeds measured in a wind tunnel.

10.2.1.3 Tyre Noise

Tyre noise is usually the most powerful noise component of car noise. It depends mostly on the car speed and the road surface. The power spectral density of tyre noise is depicted in Fig. 10.6 for two different road surfaces for the same car. One especially notices differences below 1 kHz.

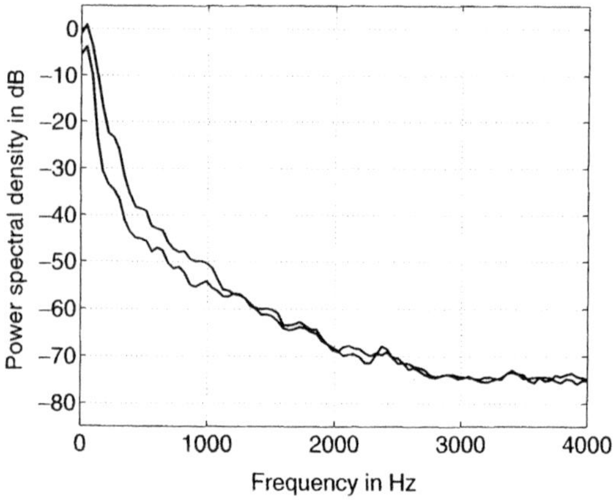

Fig. 10.6. Power spectral densities of tyre noise for two different road surfaces.

Investigating the dependence of the full-band power on the car speed, one can observe an approximately linear relation between the the full-band noise power P_N in dB and the car speed

$$P_N(v) = P_N(v_0)\, e^{K_v[v-v_0]}, \tag{10.2}$$

which is also depicted in Fig. 10.7.

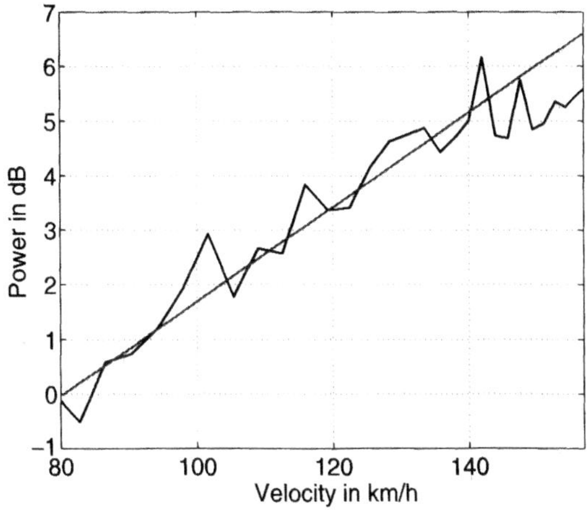

Fig. 10.7. Mean power of the tyre noise in dependence of the car speed (black). The gray graph depicts the approximately linear relation between the noise power and the car speed according to Eq. 10.2.

Performing, however, a frequency dependent analysis, one observes that

- the strength of the power increase is frequency dependent and
- the linear increase is only valid for the strongly smoothed mean, the variance, however, is very high.

Thus, a prediction of the tyre noise power spectral density in dependence of the car speed is very difficult.

Concluding, the main property of car noise is that its components show a rather smooth dependence on time and frequency. Performing a detailed analysis for the car noise components, this smoothness is especially valid for wind and tyre noise.

Engine noise components show dominant harmonic components at multiples of half of the engine frequency. This offers the potential for an special harmonic engine harmonic cancellation procedure supposing the engine frequency is available [21, 23], e.g. via the so-called *controller area network bus* (CAN bus) of a usual modern car.

10.2.2 Speech Analysis

In contrast to car noise, speech signals strongly vary with time. They are mainly composed of two different components:

- Voiced and
- unvoiced

speech frames, as well as transition frames.

Typical frames of unvoiced and voiced speech are depicted in Fig. 10.8 and Fig. 10.9, respectively.

Unvoiced speech frames occur when pronouncing phonemes such as 's', 'f', and 'sh'. These fricatives exhibit a noise like characteristic with a remarkable spectral contribution at frequency components above 2 kHz.

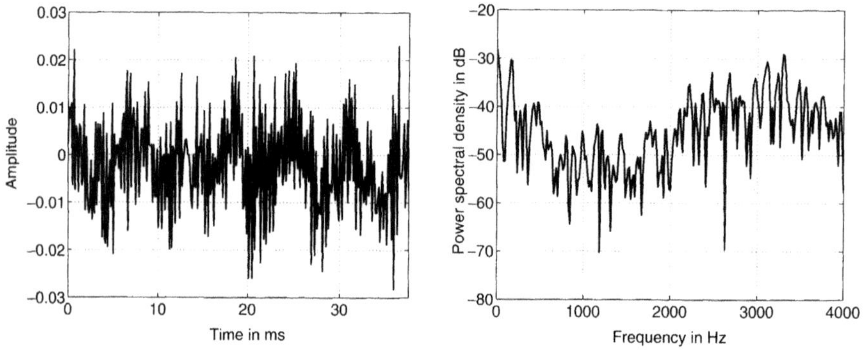

Fig. 10.8. Unvoiced frame of a speech signal, 'sh' sound (left) and corresponding spectrogram (right).

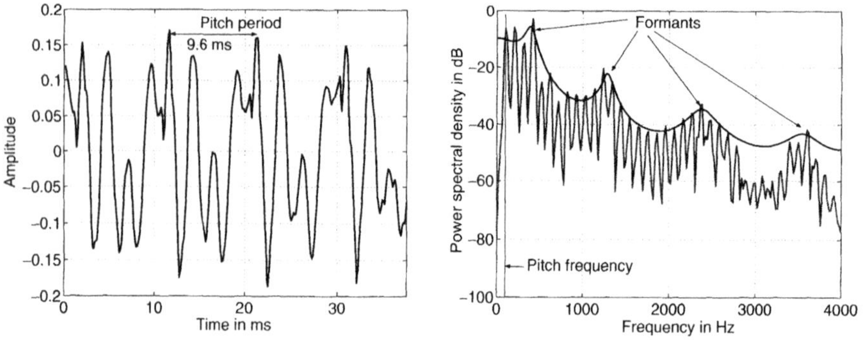

Fig. 10.9. Voiced speech frame of a 'u' sound (left) and the corresponding power spectral density (right). Pitch period, pitch frequency, and the formants are marked.

Voiced speech frames are characterized by the pitch frequency or the pitch period, respectively, as well as the formants. The pitch frequency is the frequency of the periodic excitation signal whereas the formants characterize the spectral envelope. Formants are formed by the vocal tract and are characteristic for the different vowels.

A model for the speech excitation is depicted in Fig. 10.10.

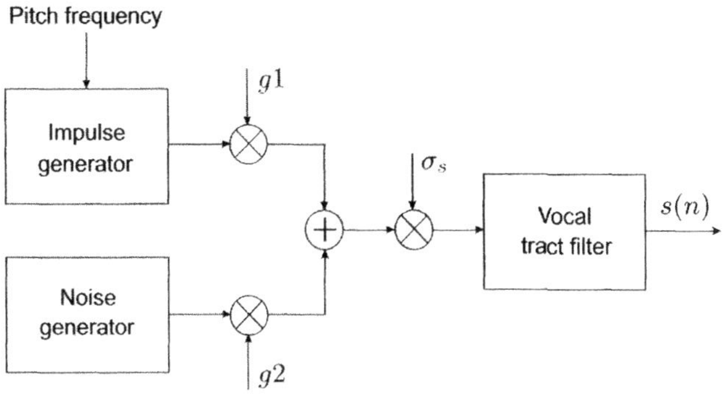

Fig. 10.10. Speech excitation model. With the factors $g1$ and $g2$ one can continuously switch between voiced and unvoiced exciation. The factor σ_s determines the speech signal power and the vocal-tract filter models the human vocal tract.

The excitation signals are different for unvoiced frames (noise generator) and voiced frames (impulse generator). With $g1$ and $g2$, both values $\in [0, 1]$ and $g1 + g2 = 1$, the ratio of the voiced and unvoiced components of the speech sound can be chosen to model transition frames.

Finally, in Fig. 10.11 the spectrogram of a male speech signal is depicted. One clearly observes unvoiced and voiced speech frames with their typical periodic pitch structure. The main property that the speech components change rapidly with respect to time and frequency is obvious.

Concluding, the analysis of car noise and speech signals as well as having in mind the noise reduction task, the following properties have to be emphasized:

- Typical car noise components exhibit dominant components at very low frequencies.
- Their spectrum is rather smooth with respect to time and frequency.
- Thus, their spectral characteristics vary slowly.
- Typical speech components show a strongly time varying spectrum.
- The power dominant voiced speech frames are characterized by a pitch-periodic spectrum.

10.3 Theoretical Basics

10.3.1 Kalman Filters for Colored Noise Signals

In this section, the Kalman filter equations are derived for the realistic case where the desired signal as well as the noise are non-white signals such as

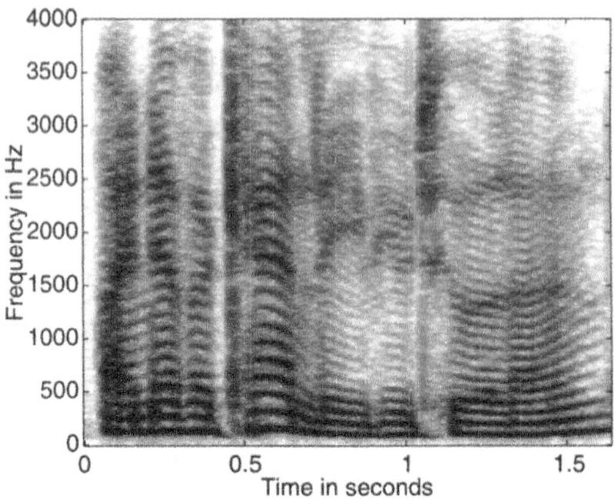

Fig. 10.11. Spectrogram of a male speech signal with a pitch frequency around 140 Hz.

for our specific application of hands-free car phones. In this case, i.e. for car noise, the usual assumption when deriving Kalman filter does not hold that the disturbing noise is white.

Describing speech $s(n)$ and noise $b(n)$ as AR processes, the following relation can be noted:

$$x(n) = s(n) + b(n), \tag{10.3}$$

$$s(n) = \sum_{k=1}^{p} a_k(n)\, s(n-k) + w(n), \tag{10.4}$$

$$b(n) = \sum_{k=1}^{q} c_k(n)\, b(n-k) + \nu(n), \tag{10.5}$$

with $\nu(n)$ and $w(n)$ being white excitation signals.

In the state-space domain this may be denoted as follows:

$$\boldsymbol{s}(n) = \boldsymbol{A}_s(n-1)\, \boldsymbol{s}(n-1) + \boldsymbol{g}_s\, w(n), \tag{10.6}$$

$$\boldsymbol{b}(n) = \boldsymbol{A}_b(n-1)\, \boldsymbol{b}(n-1) + \boldsymbol{g}_b\, \nu(n), \tag{10.7}$$

$$x(n) = \boldsymbol{h}_s^{\mathrm{T}} \boldsymbol{s}(n) + \boldsymbol{h}_b^{\mathrm{T}} \boldsymbol{b}(n) \tag{10.8}$$

with

$$\boldsymbol{s}(n) = \begin{bmatrix} s(n-p+1) \\ s(n-p+2) \\ \vdots \\ s(n) \end{bmatrix}_{p\times 1}, \quad \boldsymbol{h}_s = \boldsymbol{g}_s = \begin{bmatrix} 0 \\ 0 \\ \vdots \\ 1 \end{bmatrix}_{p\times 1},$$

$$\boldsymbol{A}_s(n) = \begin{bmatrix} 0 & 1 & \cdots & 0 \\ \vdots & \vdots & \ddots & \vdots \\ 0 & 0 & \cdots & 1 \\ a_p(n) & a_{p-1}(n) & \cdots & a_1(n) \end{bmatrix}_{p\times p}, \tag{10.9}$$

and

$$\boldsymbol{b}(n) = \begin{bmatrix} b(n-q+1) \\ b(n-q+2) \\ \vdots \\ b(n) \end{bmatrix}_{q\times 1}, \quad \boldsymbol{h}_b = \boldsymbol{g}_b = \begin{bmatrix} 0 \\ 0 \\ \vdots \\ 1 \end{bmatrix}_{q\times 1},$$

$$\boldsymbol{A}_b(n) = \begin{bmatrix} 0 & 1 & \cdots & 0 \\ \vdots & \vdots & \ddots & \vdots \\ 0 & 0 & \cdots & 1 \\ c_q(n) & c_{q-1}(n) & \cdots & c_1(n) \end{bmatrix}_{q\times q}. \tag{10.10}$$

Combining the state space equations, the following relations can be noted:

$$\boldsymbol{x}(n) = \boldsymbol{A}_x(n-1)\,\boldsymbol{x}(n-1) + \boldsymbol{G}\,\boldsymbol{v}(n), \tag{10.11}$$

$$x(n) = \boldsymbol{h}_x^{\mathrm{T}}\,\boldsymbol{x}(n), \tag{10.12}$$

with:

$$\boldsymbol{x}(n) = \begin{bmatrix} \boldsymbol{s}(n) \\ \boldsymbol{b}(n) \end{bmatrix}, \ \boldsymbol{v}(n) = \begin{bmatrix} w(n) \\ \nu(n) \end{bmatrix}, \ \boldsymbol{G} = \begin{bmatrix} \boldsymbol{g}_s & \boldsymbol{0} \\ \boldsymbol{0} & \boldsymbol{g}_b \end{bmatrix},$$

$$\boldsymbol{A}_x(n) = \begin{bmatrix} \boldsymbol{A}_s(n) & \boldsymbol{0} \\ \boldsymbol{0} & \boldsymbol{A}_b(n) \end{bmatrix}, \boldsymbol{h}_x = \begin{bmatrix} \boldsymbol{h}_s \\ \boldsymbol{h}_b \end{bmatrix}, \boldsymbol{V}(n) = \begin{bmatrix} \sigma_w^2(n) & 0 \\ 0 & \sigma_\nu^2(n) \end{bmatrix} \tag{10.13}$$

where the noisy input can be noted as

$$\begin{aligned} x(n) &= \boldsymbol{h}_x^{\mathrm{T}}\,\boldsymbol{x}(n), \\ &= \boldsymbol{h}_s^{\mathrm{T}}\,\boldsymbol{s}(n) + \boldsymbol{h}_b^{\mathrm{T}}\,\boldsymbol{b}(n). \end{aligned} \tag{10.14}$$

These notations are visualized in Fig. 10.12.

The power of the zero-mean signals $\nu(n)$ and $w(n)$ are denoted as $\sigma_w^2(n)$ and $\sigma_\nu^2(n)$.

In the following derivation of the Kalman filter, $\hat{\boldsymbol{x}}(n|n-1)$ and $\hat{\boldsymbol{x}}(n|n)$ are the estimates for the state $\hat{\boldsymbol{x}}(n)$ on the basis of the measurements $x(i)$ for $n-1$ and n signal values, respectively.

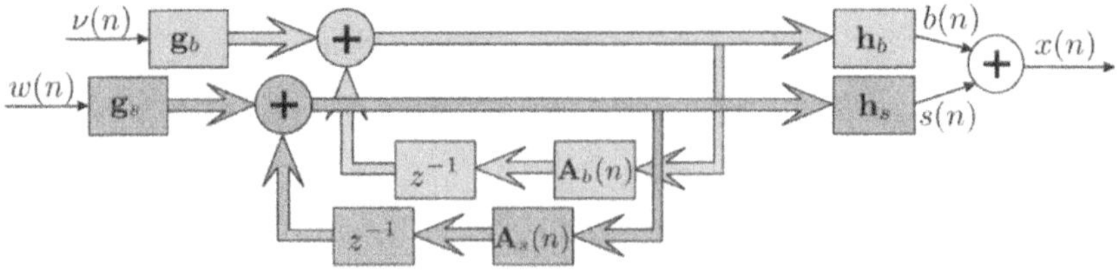

Fig. 10.12. The excitation models in state-space domain notation.

The corresponding error signals and the covariance matrices can then be noted as:

$$\begin{aligned} \boldsymbol{e}(n|n) &= \boldsymbol{x}(n) - \hat{\boldsymbol{x}}(n|n), & \boldsymbol{P}(n|n) &= \mathrm{E}\Big\{\boldsymbol{e}(n|n)\,\boldsymbol{e}^{\mathrm{H}}(n|n)\Big\}, \\ \boldsymbol{e}(n|n-1) &= \boldsymbol{x}(n) - \hat{\boldsymbol{x}}(n|n-1), & \boldsymbol{P}(n|n-1) &= \mathrm{E}\Big\{\boldsymbol{e}(n|n-1)\,\boldsymbol{e}^{\mathrm{H}}(n|n-1)\Big\} \end{aligned}$$

The derivation of the Kalman filter may now be performed in two steps: First, an estimate of the state $\hat{\boldsymbol{x}}(n|n-1)$ is determined based on the old measurement values up to $n-1$ and then the current input sample $x(n)$ is considered for the estimation of $\hat{\boldsymbol{x}}(n|n)$.

10.3.1.1 Predicted Estimate

One possibility for predicting the current state estimate is a linear modification based on the preceding estimate:

$$\hat{\boldsymbol{x}}(n|n-1) = \boldsymbol{A}_x(n-1)\,\hat{\boldsymbol{x}}(n-1|n-1). \tag{10.15}$$

In the following, it is shown that this is an appropriate approach since the mean of the error is zero. Denoting the error as

$$\begin{aligned} \boldsymbol{e}(n|n-1) &= \boldsymbol{x}(n) - \hat{\boldsymbol{x}}(n|n-1) \\ &= \boldsymbol{A}_x(n-1)\,\boldsymbol{e}(n-1|n-1) + \boldsymbol{G}\,\boldsymbol{v}(n), \end{aligned} \tag{10.16}$$

it is obvious that the expectation value of $\boldsymbol{e}(n|n-1)$ is zero since the preceding estimate has been determined such as to be unbiased, i.e. the mean of $\boldsymbol{e}(n-1|n-1)$ is zero and also the mean of $\boldsymbol{v}(n)$ is zero.

Therefore, the error covariance matrix can be determined as

$$\boldsymbol{P}(n|n-1) = \boldsymbol{A}_x(n-1)\boldsymbol{P}(n-1|n-1)\boldsymbol{A}_x^{\mathrm{H}}(n-1) + \boldsymbol{G}\,\boldsymbol{V}(n)\,\boldsymbol{G}^{\mathrm{T}}. \tag{10.17}$$

10.3.1.2 Current Estimate

For estimating the current model state $\hat{\boldsymbol{x}}(n|n)$, the approach is to linearly combine the predicted estimate and current measurement value:

$$\hat{\boldsymbol{x}}(n|n) = \tilde{\boldsymbol{K}}(n)\,\hat{\boldsymbol{x}}(n|n-1) + \boldsymbol{k}(n)\,x(n). \tag{10.18}$$

The corresponding model error $\boldsymbol{e}(n|n)$ can then be determined as:

$$\begin{aligned}
\boldsymbol{e}(n|n) &= \boldsymbol{x}(n) - \tilde{\boldsymbol{K}}(n)\,\hat{\boldsymbol{x}}(n|n-1) - \boldsymbol{k}(n)\,x(n), \\
&= \boldsymbol{x}(n) - \tilde{\boldsymbol{K}}(n)\Big[\boldsymbol{x}(n) - \boldsymbol{e}(n|n-1)\Big] - \boldsymbol{k}(n)\Big[\boldsymbol{h}_x^{\mathrm{T}}\boldsymbol{x}(n)\Big], \\
&= \Big[\boldsymbol{I} - \tilde{\boldsymbol{K}}(n) - \boldsymbol{k}(n)\boldsymbol{h}_x^{\mathrm{T}}\Big]\,\boldsymbol{x}(n) + \tilde{\boldsymbol{K}}(n)\,\boldsymbol{e}(n|n-1),
\end{aligned} \tag{10.19}$$

where the value $\tilde{\boldsymbol{K}}(n)$ is comprised of a matrix and the value $\boldsymbol{k}(n)$ denotes a vector which will be called Kalman gain later on.

The estimate $\hat{\boldsymbol{x}}(n|n)$ exhibits the desired zero-mean error if the following equation is fulfilled:

$$\tilde{\boldsymbol{K}}(n) = \boldsymbol{I} - \boldsymbol{k}(n)\,\boldsymbol{h}_x^{\mathrm{T}}. \tag{10.20}$$

Thus, the model state estimate can be written as

$$\hat{\boldsymbol{x}}(n|n) = \hat{\boldsymbol{x}}(n|n-1) + \boldsymbol{k}(n)\Big[x(n) - \boldsymbol{h}_x^{\mathrm{T}}\,\hat{\boldsymbol{x}}(n|n-1)\Big]. \tag{10.21}$$

The remaining unknown is the Kalman gain $\boldsymbol{k}(n)$.

For determining this Kalman gain, the minimization of the mean square error can be utilized as optimization criterion:

$$\begin{aligned}
\mathrm{E}\Big\{\boldsymbol{e}^{\mathrm{H}}(n|n)\,\boldsymbol{e}(n|n)\Big\} &= \mathrm{tr}\Big\{\mathrm{E}\{\boldsymbol{e}(n|n)\,\boldsymbol{e}^{\mathrm{H}}(n|n)\}\Big\}, \\
&= \mathrm{tr}\Big\{\boldsymbol{P}(n|n)\Big\} \overset{!}{=} \min,
\end{aligned} \tag{10.22}$$

with:

$$\boldsymbol{e}(n|n) = \boldsymbol{e}(n|n-1) + \boldsymbol{k}(n)\Big[x(n) - \boldsymbol{h}_x^{\mathrm{T}}\,\hat{\boldsymbol{x}}(n|n-1)\Big]. \tag{10.23}$$

Thus, the minimization of the mean square error is equivalent to the minimization of the trace of the covariance matrix $\boldsymbol{P}(n|n)$.

Writing this covariance matrix in dependence of the estimates based on the $n-1$ measurement signal values and the currently observed signal value $x(n)$, one obtains the following:

$$\begin{aligned}
\boldsymbol{P}(n|n) = {} & \boldsymbol{P}(n|n-1) \\
& + \boldsymbol{k}(n)\,\mathrm{E}\Big\{\Big[x(n) - \boldsymbol{h}_x^{\mathrm{T}}\,\hat{\boldsymbol{x}}(n|n-1)\Big]\Big[x^*(n) - \boldsymbol{h}_x^{\mathrm{T}}\,\hat{\boldsymbol{x}}^*(n|n-1)\Big]\Big\}\boldsymbol{k}^{\mathrm{H}}(n) \\
& - \mathrm{E}\Big\{\boldsymbol{e}(n|n-1)\Big[x^*(n) - \boldsymbol{h}_x^{\mathrm{T}}\,\hat{\boldsymbol{x}}^*(n|n-1)\Big]\Big\}\boldsymbol{k}^{\mathrm{H}}(n) \\
& - \boldsymbol{k}(n)\mathrm{E}\Big\{\Big[x(n) - \boldsymbol{h}_x^{\mathrm{T}}\,\hat{\boldsymbol{x}}(n|n-1)\Big]\boldsymbol{e}^{\mathrm{H}}(n|n-1)\Big\}.
\end{aligned} \tag{10.24}$$

Deriving the trace of this matrix with respect to the single elements of the Kalman gain vector, one obtains the following condition [12]:

$$\boldsymbol{k}(n)\,\mathrm{E}\Big\{\overbrace{\Big[x(n)-\boldsymbol{h}_x^{\mathrm{T}}\,\hat{\boldsymbol{x}}(n|n-1)\Big]}^{\boldsymbol{h}_x^{\mathrm{T}}[\boldsymbol{x}(n)-\hat{\boldsymbol{x}}(n|n-1)]}\Big[x^*(n)-\boldsymbol{h}_x^{\mathrm{T}}\,\hat{\boldsymbol{x}}^*(n|n-1)\Big]\Big\}$$
$$-\mathrm{E}\Big\{\underbrace{\boldsymbol{e}(n|n-1)}_{\boldsymbol{x}(n)-\hat{\boldsymbol{x}}(n|n-1)}\Big[x^*(n)-\boldsymbol{h}_x^{\mathrm{T}}\,\hat{\boldsymbol{x}}^*(n|n-1)\Big]\Big\}\overset{!}{=}\mathbf{0}, \qquad (10.25)$$

which can be written as

$$\Big[\boldsymbol{I}-\boldsymbol{k}(n)\,\boldsymbol{h}_x^{\mathrm{T}}\Big]$$
$$\cdot\mathrm{E}\Big\{\Big[\boldsymbol{x}(n)-\hat{\boldsymbol{x}}(n|n-1)\Big]\boldsymbol{h}_x^{\mathrm{T}}\Big[\boldsymbol{x}^*(n)-\hat{\boldsymbol{x}}^*(n|n-1)\Big]\Big\}\overset{!}{=}\mathbf{0}. \quad (10.26)$$

This equation may be reordered and written as

$$\Big[\boldsymbol{I}-\boldsymbol{k}(n)\,\boldsymbol{h}_x^{\mathrm{T}}\Big]\,\underbrace{\mathrm{E}\Big\{\boldsymbol{e}(n|n-1)\boldsymbol{e}^{\mathrm{H}}(n|n-1)\Big\}}_{\boldsymbol{P}(n|n-1)}\,\boldsymbol{h}_x\overset{!}{=}\mathbf{0}. \qquad (10.27)$$

Thus, the optimum Kalman gain may be denoted as

$$\boldsymbol{k}(n)=\boldsymbol{P}(n|n-1)\,\boldsymbol{h}_x\Big[\boldsymbol{h}_x^{\mathrm{T}}\boldsymbol{P}(n|n-1)\boldsymbol{h}_x\Big]^{-1}. \qquad (10.28)$$

Replacing this result in Eq. 10.24, after some reordering of the equations, one obtains the following result for the covariance matrix:

$$\boldsymbol{P}(n|n)=\Big[\boldsymbol{I}-\boldsymbol{k}(n)\,\boldsymbol{h}_x^{\mathrm{T}}\Big]\,\boldsymbol{P}(n|n-1). \qquad (10.29)$$

Concluding, all equations necessary for the Kalman filtering may be denoted as

$$\hat{\boldsymbol{x}}(n|n-1)=\boldsymbol{A}_x(n-1)\,\hat{\boldsymbol{x}}(n-1|n-1), \qquad (10.30)$$
$$\boldsymbol{P}(n|n-1)=\boldsymbol{A}_x(n-1)\boldsymbol{P}(n-1|n-1)\boldsymbol{A}_x^{\mathrm{H}}(n-1)+\boldsymbol{G}\boldsymbol{V}(n)\boldsymbol{G}^{\mathrm{T}}, \quad (10.31)$$
$$\hat{\boldsymbol{x}}(n|n)=\hat{\boldsymbol{x}}(n|n-1)+\boldsymbol{k}(n)\Big[x(n)-\boldsymbol{h}_x^{\mathrm{T}}\,\hat{\boldsymbol{x}}(n|n-1)\Big], \qquad (10.32)$$
$$\boldsymbol{k}(n)=\boldsymbol{P}(n|n-1)\,\boldsymbol{h}_x\Big[\boldsymbol{h}_x^{\mathrm{T}}\boldsymbol{P}(n|n-1)\boldsymbol{h}_x\Big]^{-1}, \qquad (10.33)$$
$$\boldsymbol{P}(n|n)=\Big[\boldsymbol{I}-\boldsymbol{k}(n)\,\boldsymbol{h}_x^{\mathrm{T}}\Big]\,\boldsymbol{P}(n|n-1). \qquad (10.34)$$

And the estimation for the clean signal $\hat{s}(n)$ can be written as follows:

$$\hat{s}(n)=\Big[\boldsymbol{h}_s^{\mathrm{T}}\mathbf{0}_{1\times q}\Big]\,\hat{\boldsymbol{x}}(n|n). \qquad (10.35)$$

Fig. 10.13 visualizes this result. Here, one clearly observes the close relation of the state space model according to Fig. 10.12 and the Kalman equations.

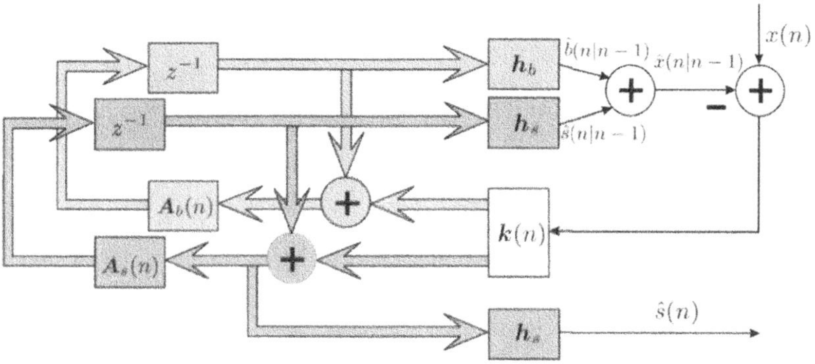

Fig. 10.13. Visualization of the Kalman filter equations when modeling speech and noise with AR models.

10.3.2 Parametric Spectral Estimation

As mentioned in Sec. 10.1, the application of Kalman filters and the denoted advantages are directly related to the description of speech and noise with parametric models.

In this section, in a compact way, the two different methods will be introduced which will be utilized for the parametric speech and noise model estimation required for the application of the Kalman filter based noise reduction in Sec. 10.4.

These two estimation methods are

- the autocorrelation method and
- the Burg method.

The methods belong to two different groups of methods for the parametric spectral estimation:

- Direct methods [12, 15] where the AR parameters of a signal's model are determined based on the estimation of the signal's autocorrelation matrix by solving the Yule-Walker equations.
- Recursive methods [12, 20] which determine the reflection coefficients of a prediction error filter in Lattice structure. These reflection coefficients can be used to easily determine the required AR parameters of the signal models.

The signal models have to be determined for signals such as speech and noise which are not stationary and can only be modeled as short-term stationary signals. Thus, the model coefficients can only be estimated for signal frames of a limited length L_{AR} for which the signal properties do not change significantly. A time-dependence of the models is the consequence.

10.3.2.1 Direct Parametric Spectral Estimation Methods

For direct methods, the model coefficients for each signal frame are determined based on one minimization step. The approach is based on the minimization of the mean square output signal of the prediction error signal $w(n)$ which is depicted in Fig. 10.14.

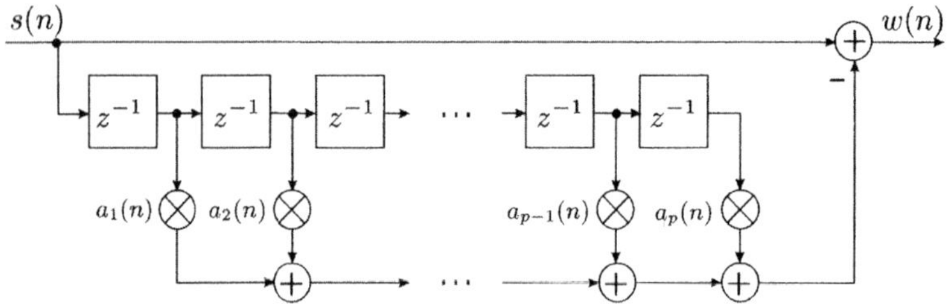

Fig. 10.14. Structure of a predictor error filter.

The different direct estimation methods such as the autocorrelation, the covariance and the modified covariance methods differ by the utilized minimization criterion. The autocorrelation method, which will be presented in the following, shows – in contrast to the other before mentioned methods – the advantage to determine AR models which are guaranteed to be stable.

The Autocorrelation Method

The autocorrelation method determines the prediction coefficients by minimizing the sum of the squared output signal samples $w(n)$. First, one assumes that the output $w(n)$ is determined for $n \to \infty$ samples. Since, however, the models should only be determined in dependence of the windowed input signal of length $L_{\rm AR}$, assumptions about the input signal values which are outside of the window are required. The assumption of the autocorrelation method is that these values are zero. Thus, the model coefficients are determined based in the windowed signal, denoted as:

$$s_n(\nu) = \begin{cases} s(\nu) : \nu \in \left[n - L_{\rm AR}/2 + 1, \ldots, n + L_{\rm AR}/2\right] \wedge [\nu > 0], \\ 0 \quad : \text{else.} \end{cases} \tag{10.36}$$

For this signal, one assumes, without loss of generality, that the signal $s_n(\nu)$ is zero for $\nu \leq 0$. The signal which has to be minimized can therefore be noted as:

$$\epsilon_p^{\rm AC} = \sum_{\nu=0}^{\infty} \left|w(\nu)\right|^2 = \sum_{\nu=0}^{\infty} \left|s_n(\nu) - \sum_{i=1}^{p} a_i(n)\, s_n(\nu - i)\right|^2. \tag{10.37}$$

Deriving this signal with respect to each coefficient, one obtains the following minimizing condition:

$$\frac{\partial \epsilon_p^{\mathrm{AC}}}{\partial a_j^*(n)} = \sum_{\nu=0}^{\infty} \frac{\partial\big[w(\nu)\, w^*(\nu)\big]}{\partial a_j^*(n)} = \sum_{\nu=0}^{\infty} w(\nu)\, \frac{\partial w^*(\nu)}{\partial a_j^*(n)}$$

$$= -\sum_{\nu=0}^{\infty} w(\nu)\, s_n^*(\nu - j) \stackrel{!}{=} 0, \tag{10.38}$$

which can be reordered as:

$$\sum_{i=1}^{p} a_i(n) \underbrace{\sum_{\nu=0}^{\infty} s_n(\nu - i)\, s_n^*(\nu - j)}_{\hat{s}_{ss,n}^{\mathrm{AC}}(j,i)} \stackrel{!}{=} \underbrace{\sum_{\nu=0}^{\infty} s_n(\nu)\, s_n^*(\nu - j)}_{\hat{s}_{ss,n}^{\mathrm{AC}}(j,0)}. \tag{10.39}$$

Noting

$$\hat{s}_{ss,n}^{\mathrm{AC}}(j+1, i+1) = \sum_{\nu=0}^{\infty} s_n\big(\nu - [i+1]\big)\, s_n^*\big(\nu - [j+1]\big),$$

$$= \sum_{\nu=-1}^{\infty} s_n(\nu - i)\, s_n^*(\nu - j),$$

$$= \hat{s}_{ss,n}^{\mathrm{AC}}(j,i) + \underbrace{s_n\big(-[i+1]\big)\, s_n^*\big(-[j+1]\big)}_{=0 \quad \text{for } i,j>0} \tag{10.40}$$

one obtains the following property for the autocorrelation coefficients:

$$\hat{s}_{ss,n}^{\mathrm{AC}}(j+1, i+1) = \hat{s}_{ss,n}^{\mathrm{AC}}(j,i). \tag{10.41}$$

Thus, it is possible to note the condition of Eq. 10.39 with only with one argument of the autocorrelation function:

$$\sum_{i=1}^{p} a_i(n)\, \hat{s}_{ss,n}^{\mathrm{AC}}(j-i) = \hat{s}_{ss,n}^{\mathrm{AC}}(j), \tag{10.42}$$

with

$$\hat{s}_{ss,n}^{\mathrm{AC}}(i) = \sum_{\nu=0}^{\infty} s_n(\nu)\, s_n^*(\nu - i) \tag{10.43}$$

In matrix-vector notation this can be denoted as follows:

$$\begin{bmatrix} \hat{s}^{\mathrm{AC}}_{ss,n}(0) & s^{\mathrm{AC},*}_{ss,n}(1) & s^{\mathrm{AC},*}_{ss,n}(2) & \cdots & s^{\mathrm{AC},*}_{ss,n}(p-1) \\ \hat{s}^{\mathrm{AC}}_{ss,n}(1) & \hat{s}^{\mathrm{AC}}_{ss,n}(0) & s^{\mathrm{AC},*}_{ss,n}(1) & \cdots & s^{\mathrm{AC},*}_{ss,n}(p-2) \\ \hat{s}^{\mathrm{AC}}_{ss,n}(2) & \hat{s}^{\mathrm{AC}}_{ss,n}(1) & \hat{s}^{\mathrm{AC}}_{ss,n}(0) & \cdots & s^{\mathrm{AC},*}_{ss,n}(p-3) \\ \vdots & \vdots & \vdots & \ddots & \vdots \\ \hat{s}^{\mathrm{AC}}_{ss,n}(p-1) & \hat{s}^{\mathrm{AC}}_{ss,n}(p-2) & \hat{s}^{\mathrm{AC}}_{ss,n}(p-3) & \cdots & \hat{s}^{\mathrm{AC}}_{ss,n}(0) \end{bmatrix} \cdot \begin{bmatrix} a_1(n) \\ a_2(n) \\ a_3(n) \\ \vdots \\ a_p(n) \end{bmatrix} = \begin{bmatrix} \hat{s}^{\mathrm{AC}}_{ss,n}(1) \\ \hat{s}^{\mathrm{AC}}_{ss,n}(2) \\ \hat{s}^{\mathrm{AC}}_{ss,n}(3) \\ \vdots \\ \hat{s}^{\mathrm{AC}}_{ss,n}(p) \end{bmatrix}. \tag{10.44}$$

Using Eq. 10.36, the estimated autocorrelation values can be denoted based on the input signal $s(\nu)$:

$$\hat{s}^{\mathrm{AC}}_{ss,n}(i) = \sum_{\nu=n-L_{\mathrm{AR}}/2+1+i}^{n+L_{\mathrm{AR}}/2} s(\nu)\, s^*(\nu - i) \quad \text{for } i = 0, \ldots, p. \tag{10.45}$$

Based on Eq. 10.45 one observes that, dependent on the index i a different number of elements is summed for estimating the autocorrelation values. On the one hand, this causes an estimation bias which is equivalent to a reduced estimation precision. On the other hand, based on these biased estimates, the matrix of Eq. 10.44 exhibits a Toeplitz structure. With the help of the Levinson-Durbin recursion [12] such a matrix can be inverted computationally efficiently. Additionally, the matrix is positive definit. Thus, the stability of the AR models determined by solving Eq. 10.44 is guaranteed.

10.3.2.2 Recursive Spectral Estimation Methods

The difference between recursive and direct spectral estimation methods is that the model coefficients of the recursive method are not determined in one step, but recursively in several minimization steps by increasing the model order. Depicting the prediction error filter in the corresponding *lattice* structure (see Fig. 10.15) the procedure becomes obvious.

Recursive methods successively minimize the sum of the squared error signals $w_j^{+/-}(n)$ of the different stages of the prediction error filter in *lattice* structure in order to determine the reflection coefficients $\Gamma_j(n)$ of the corresponding stage j. In each iteration step, only this reflection coefficient is determined, the before determined reflection coefficients do not change.

Based on the reflection coefficients and with the help of the step-up recursion, which is identical to the second part of the Levinson-Durbin recursion, the AR coefficients $a_j(n)$ can be determined.

The step-up recursion is a recursion over the order. When increasing the order from $j-1$ to j, the corresponding reflection coefficient $\Gamma_j(n)$ is used for

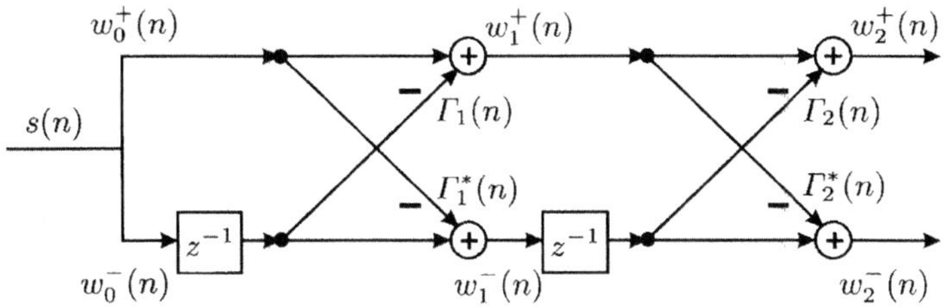

Fig. 10.15. Prediction error filter in *lattice* structure.

updating the AR parameters up to the chosen model order p for each value $i = 1, \ldots, p$.

The step-up recursion consists of the following two steps for each iteration step j:

(1) Update of the preceding AR parameters:

$$a_i(n) = a_i(n) + \Gamma_j(n) a_{j-i}^*(n) \quad \text{for } i = 1, \ldots, j-1. \tag{10.46}$$

(2) Setting the currently last AR coefficient as the reflection coefficient

$$a_j(n) = \Gamma_j(n). \tag{10.47}$$

The recursive methods differ by the utilized minimization criteria: utilizing the forward or the backward covariance method, or the sum of the squared forward $w_j^+(n)$ or the sum of the squared backward $w_j^-(n)$ prediction error signals are minimized, respectively. The Burg method, however, utilizes the sum of both squared error signals as minimization criterion for determining the reflection coefficients.

As already mentioned, the Burg method exhibits – in contrast to the forward and backward prediction method – the advantage that the AR corresponding estimated models are stable.

The Burg Method

For determining the reflection coefficients with the Burg algorithm, the sum of the squared forward and backward prediction error signal is utilized as minimization criterion:

$$\epsilon_j^{\mathrm{B}} = \epsilon_j^+ + \epsilon_j^- = \sum_{n=j}^{L_{\mathrm{AR}}-1} \left| w_j^+(n) \right|^2 + \sum_{n=j}^{L_{\mathrm{AR}}-1} \left| w_j^-(n) \right|^2. \tag{10.48}$$

Utilizing the relations of the signals according to Fig. 10.15 and deriving ϵ_j^{B} with respect to $[\Gamma_j^+(n)]^*$ one obtains the following condition:

$$\sum_{n=j}^{L_{\text{AR}}-1} \left[w_{j-1}^{+}(n) - \Gamma_j(n)\, w_{j-1}^{-}(n-1)\right] \left[w_{j-1}^{-}(n-1)\right]^{*}$$
$$+ \sum_{n=j}^{L_{\text{AR}}-1} w_{j-1}^{+}(n) \left[(w_{j-1}^{-}(n-1))^{*} - \Gamma_j(n) \left[w_{j-1}^{+}(n)\right]^{*}\right] \stackrel{!}{=} 0. \quad (10.49)$$

Resolving the above noted equation with respect to $\Gamma_j(n)$ one obtains the formula for determining the reflection coefficients optimized with the Burg method:

$$\Gamma_j^{\text{B}}(n) = \frac{2 \sum\limits_{n=j}^{L_{\text{AR}}-1} w_{j-1}^{+}(n) \left[w_{j-1}^{-}(n-1)\right]^{*}}{\sum\limits_{n=j}^{L_{\text{AR}}-1} \left\{ \left|w_{j-1}^{+}(n)\right|^2 + \left|w_{j-1}^{-}(n-1)\right|^2 \right\}}. \quad (10.50)$$

Noting the $L_{\text{AR}} - j$ elements of $w_j^{+}(n)$ and $w_j^{-}(n)$ as vectors $\boldsymbol{w}_j^{+}(n)$ and $\boldsymbol{w}_j^{-}(n)$, respectively, $\Gamma_j^{\text{B}}(n)$ can be written as

$$\Gamma_j^{\text{B}}(n) = \frac{2|\boldsymbol{w}_j^{+}(n)^T\, \boldsymbol{w}_j^{-}(n)|}{\left\|\boldsymbol{w}_j^{+}(n)\right\|^2 + \left\|\boldsymbol{w}_j^{-}(n)\right\|^2} < 1. \quad (10.51)$$

Since this result is, as indicated according to the Schwarz equation, always smaller than one the corresponding AR model is guaranteed to be stable.

10.4 Application of Kalman Filters for Noise Reduction

The goal of this section is to develop a practical realization procedure for Kalman filters based on the theoretical relations described in the preceding section. Here, the main difficulty is the estimation of the parametric spectral models for speech and noise based on the only available noisy speech samples.

This section is organized as follows: First a subband processing for the Kalman filtering is proposed. Then, methods for the AR modeling of the speech and noise subband signals are developed and their performances are analyzed, respectively.

Based on these different methods, a combined estimation procedure is developed which combines the desired properties of the analyzed AR estimation procedures. This combined procedure is able to avoid musical tones and models the spectral components of speech with high accuracy.

In a third step, it is shown that the parametric spectral estimation offers the possibility to enhance the model estimation of speech with the help of the a priori known pitch frequency. Of cause a reliable estimate for this pitch frequency has to be available in this case. Methods for the estimation of the pitch frequency were investigated in [24, 25] with the special focus on noisy speech signals. This is the reason why these pitch frequency estimation procedures are not further investigated in this chapter.

10.4.1 Subband Processing

Investigating the required order of AR models for speech and noise, one observes that especially for speech the order has to be very large. The reason is the pitch structure of the spectrum of voiced speech frames. For being able to attenuate noise components in between these pitch components, the pitch structure has to be modeled with an sufficient accuracy. This can only by achieved when modeling each pitch component with one pole of the AR model polynomial. The required order of the model is then approximately f_s/f_{pitch}, i.e. the quotient of the sampling and the pitch frequency. For male speech with a pitch frequency below 100 Hz this results in model orders larger than 80 for the usual sampling frequency of 8 kHz. This is shown in Fig. 10.16 where the power spectral noise density (PSD) of a voiced speech frame with a pitch frequency of 140 Hz is depicted. Only with the model order 60 or larger one is able to resolve the pitch components appropriately.

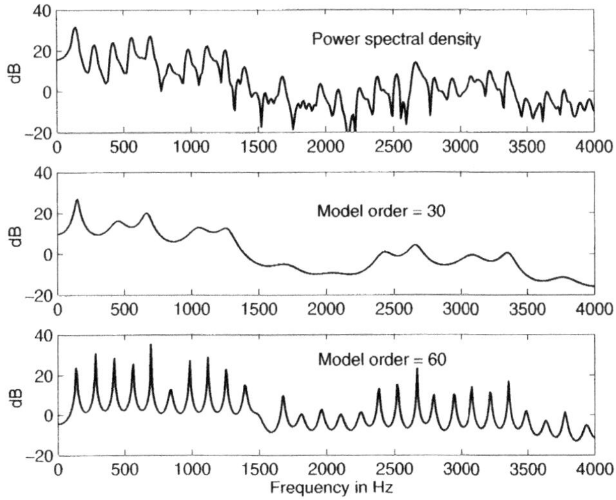

Fig. 10.16. Power spectral density of a voice speech frame (top) and the corresponding AR models of order 30 (mid) and 60 (buttom).

AR models of such an order are difficult to be estimated precisely, especially when only the disturbed speech signal is available. Satisfactory results can hardly be achieved. To reduce the model order, the noisy input signal may first be decomposed into several subbands. Applying then the Kalman filtering in these subbands (see Fig. 10.17), a reduction of the model order by the sub-sampling rate is possible for each subband - compared to the required order for the full-band signal.

For the subband decomposition, here, we propose a 16 channel filterbank with a sub-sampling rate of 12 and a 64-samples-length prototype low-pass

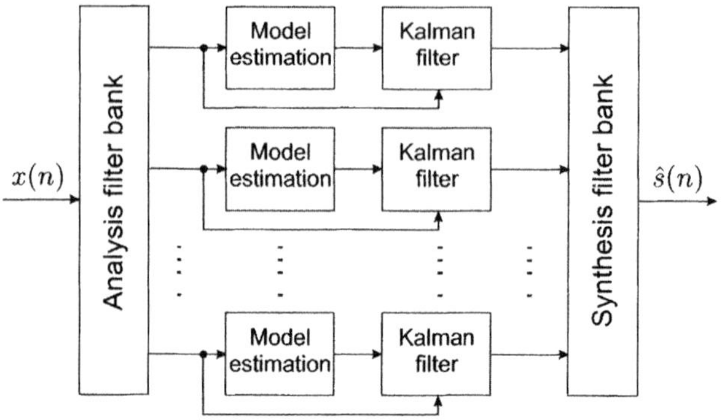

Fig. 10.17. Overview of the subband noise reduction procedure with Kalman filters.

filter which cause a delay of 8 msec. This allows to chose a reduced subband AR model order for speech of 4 to 6 which will be shown in the following.

10.4.2 AR Model Estimation for Speech and Noise

The methods proposed in Sec. 10.3.2 for AR model estimation require a signal frame of finite length. The length of this frame should not be chosen too large in order

- for being able to model speech appropriately which is only short-term stationary and in order
- to limit the processing delay of the algorithm.

On the other hand, the signal frames should be chosen large enough in order

- to reduce the variance of the estimation.

Comparative simulations for an AR modeling of clean speech showed that signal frames of length $N_{\text{model}} = 32$ samples of the sub-sampled signal (equivalent to a frame length of 48 msec for a sub-sampling factor of 12) are appropriate for a reliable estimation while respecting that speech is only short-term stationary.

For the update rate, we have chosen an update every 5th signal sample. Thus, each model is valid for 5 samples of the Kalman filter. The best results are obtained when these samples are located in the middle of the signal frame which is utilized for the AR model estimation. This is shown in Fig. 10.18. Since this estimation procedure requires for half of the signal frame "future" samples with respect to the five samples for which the model is used, this is equivalent to a half block length signal delay of 24 msec. Summing this with the delay of the sub-sampling procedure of 8 msec, in total, the proposed Kalman filter procedure causes a delay of 32 msec.

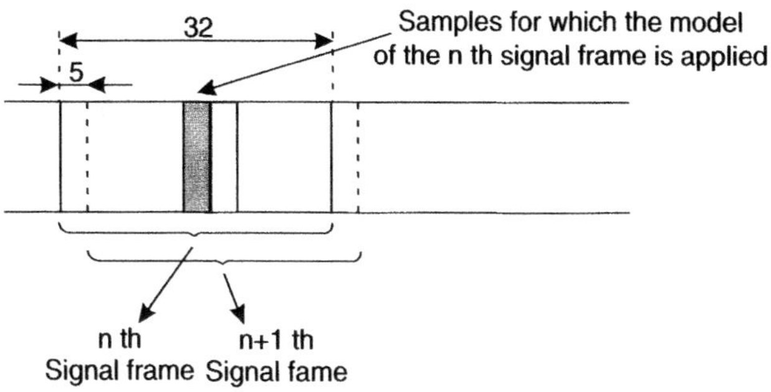

Fig. 10.18. Graph of two consecutive signal blocks of 32 samples which are utilized for the AR model estimation. The models are utilized for the Kalman filtering of the 5 signal samples located in the middle of the signal frame.

In Sec. 10.3.2 different methods for the AR modeling were analyzed, in particular the autocorrelation and the Burg method since both guarantee stable models. The Burg method exhibits the additional advantage to yield unbiased estimates. Estimation examples of a voiced speech signal frame for the second subband shown in Fig. 10.19 confirm this slight advantage. One can observe that the Burg method allows a slight better resolution of the pitch components.

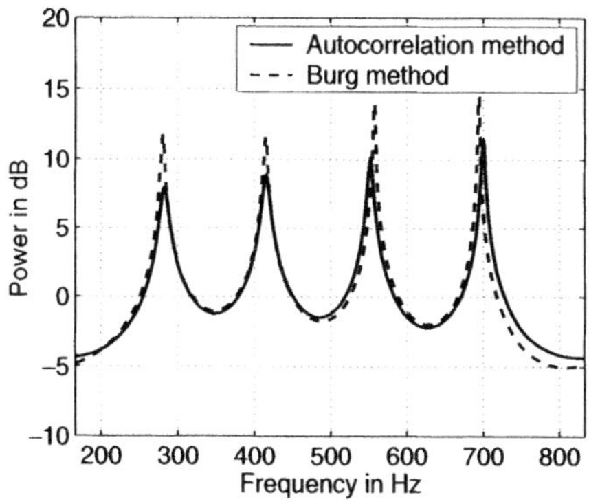

Fig. 10.19. Estimated AR models for the second subband of a voiced speech frame. The solid graph shows the model estimated with the autocorrelation method whereas the dashed graph is the result of the Burg method estimation which models the pitch components slightly better. One has to consider that due to the non-critical subsampling rate of 12 for 16 subbands the subband spectra overlap. Thus, the second subband covers a frequency range between 166 and 833 Hz.

In the following, it is shown how both methods can be successfully combined for a speech and noise model estimation.

The model estimation is performed for each subband separately and independently. In the following, for the sake of simplicity, the different subband signals are not marked explicitly.

10.4.2.1 Estimation of Speech Models Based on Noisy Speech Signals

Before developing model estimation procedures, the model orders for each subband have to be fixed. For being able to resolve the pitch components, which in particular show a dominant contribution for low subbands, it is necessary to choose a high model order. For the higher frequency bands, the model order can be successively reduced to 2 (see Tab. 10.1). Due to the conjugate-complex localization of the poles for the real-valued lowest subband, an even number of the poles has to be chosen for which four is sufficient.

Table 10.1. Chosen model orders for speech in dependence of the subbands. For the chosen conjugate-complex filterbank, the model orders for subband 10-16 can be derived based on the given figures.

Sub-band	Frequency frame	Model order
1	0 – 250 Hz	4
2	250 – 750 Hz	5
3	750 – 1250 Hz	5
4	1250 – 1750 Hz	5
5	1750 – 2250 Hz	4
6	2250 – 2750 Hz	4
7	2750 – 3250 Hz	3
8	3250 – 3750 Hz	3
9	3750 – 4000 Hz	2

The next step is then the actual estimation of the clean speech models for each subband. This is rather challenging since only disturbed speech signals are available. The goal of the model estimation is to determine the AR models coefficients $a_i(n)$ as well as the speech excitation power $\sigma_w^2(n)$ under these adverse conditions. In the following, first, methods for the AR model estimation are presented and compared. Then, possibilities for determining the speech signal excitation power are considered.

In this contribution, we do not further investigate methods that

- can only be applied for estimating models of very low order [26] or
- exhibit too strong requirements on the SNR of the input signal [7,8].

One possibility for estimating the AR parameters of speech is simply to utilize the noisy speech as input. Since usually the power spectral density of the superimposing car noise – except for the lowest subband – is smooth with respect to the frequency, with this approach, the AR models of speech can be well modelled for a high SNR down to a medium SNR. For this estimation procedure the Burg method is first choice, since this method models the spectral maxima (pitch components) better compared to the alternative autocorrelation method. Results of this estimation procedure are depicted in Fig. 10.20 in comparison with results of the competing EM method which is presented next.

Iterative EM Algorithm

The iterative EM (estimate maximize) method according to [11] is based on the EM method [5, 6] and enhances the AR model estimation iteratively, starting with an initial estimate. For determining this initial estimate, the before explained Burg estimation method can be utilized.

The iterative EM method applies the Kalman filter equations several times, iteratively. The goal is to utilize the enhanced model estimates for iterative Kalman filtering of the same signal frame. The enhanced speech and noise model estimates are determined based on the matrix:

$$\boldsymbol{Q}(n) = \sum_{n_0=n-N_{\text{model}}/2+1}^{n+N_{\text{model}}/2} \boldsymbol{P}(n_0|n_0) + \hat{\boldsymbol{x}}(n_0|n_0)\,\hat{\boldsymbol{x}}^{\text{T}}(n_0|n_0) \qquad (10.52)$$

determined as the sum of the covariance matrix $\boldsymbol{P}(n|n)$ and the state vector $\hat{\boldsymbol{x}}(n|n)$ of the Kalman equations (see Eqs. 10.30-10.34) over the current signal frame for the previous iteration step.

The upper left and the lower right quarter of the matrix $\boldsymbol{Q}(n)$ are estimates for the autocorrelation matrices of speech and noise and can be utilized for determining enhanced estimates for the speech and noise models with the help of the Yule-Walker equations.

The iterative algorithm converges after 10 – 15 iterations. The results show – compared to the initial estimate – higher maxima at multiples of the pitch frequency. For this reason, the pitch components are better resolved and the enhanced speech signal sounds more rich.

However, the algorithm also tends to increase local maxima in speech sections without speech excitation which often causes musical tones. Another disadvantage of the EM method is the strongly increased complexity due to the iterative procedure. In Fig. 10.20 the described properties are clearly shown: The depicted speech models on the right are modified with the EM method. Especially, in the lowest subband, the results of the iterative EM method are better compared to the initial Burg estimate because the pitch components are better resolved. Nevertheless, the model estimates of the iterative EM method vary stronger, especially during speech pauses which causes *musical tones*.

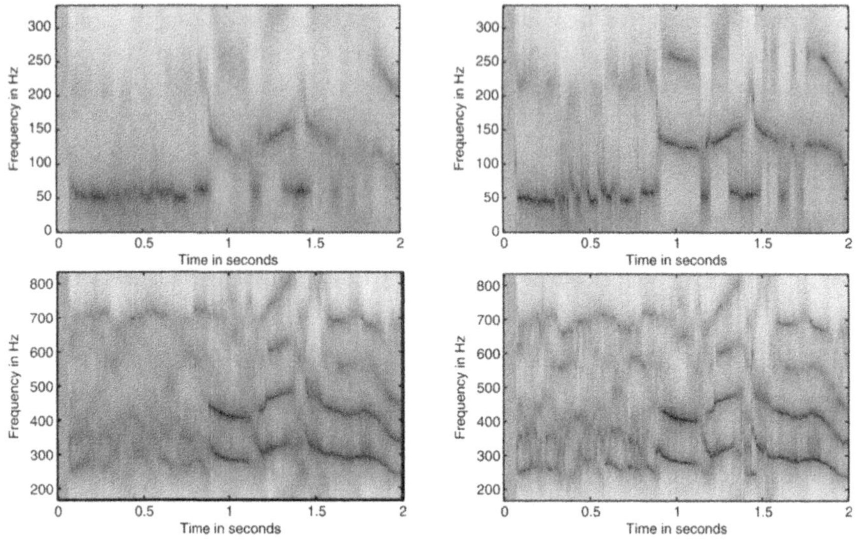

Fig. 10.20. Comparison of the model estimates of the Burg algorithm (left) and the iterative EM method (right). The upper graphs show the estimation results for the lowest subband and below the results for the second subband are depicted, respectively.

Autocorrelation Method Based on the Estimated Autocorrelation Function of Speech

An alternative procedure for an enhanced model estimation of speech is to estimate the autocorrelation of the noisy speech signal and noise to subtract these estimates.

$$\hat{s}_{ss,n}(i) = \hat{s}_{xx,n}(i) - \hat{s}_{bb,n}(i). \tag{10.53}$$

Using the estimates of the clean speech, in the next step, the speech model coefficients may be estimated with the autocorrelation method (see Sec. 10.3.2.1). However, this procedure can not be applied. The reason is that only strongly smoothed estimates of the noise autocorrelation function $\hat{s}_{bb,n}(i)$ can be estimated in speech pauses whereas $\hat{s}_{xx,n}(i)$ has to be estimated with low smoothing in order to follow fast changing speech properties. For this reason, the estimation variance of $\hat{s}_{xx,n}(i)$ is higher than for $\hat{s}_{bb,n}(i)$. The difference then may lead to estimates for $\hat{s}_{ss,n}(i)$ for which the corresponding autocorrelation matrix is not positive definite. Determining AR models based on these estimates, they are not guaranteed to be stable.

The problem can be avoided by calculating the difference in the frequency domain when assuring that these values are larger than zero:

$$\hat{S}_{ss}(k,n) = \max\left\{\hat{S}_{xx}(k,n) - \hat{S}_{bb}(k,n), 0\right\}. \tag{10.54}$$

Thus, it is guaranteed that the corresponding autocorrelation function

$$\hat{s}_{ss,n}(i) = \mathrm{IDFT}\Big\{\hat{S}_{ss}(k,n)\Big\}, \tag{10.55}$$

determined as the Fourier inverse of the speech power spectral density exhibits a corresponding autocorrelation matrix which is positive-definite. The procedure thus allows to estimate stable AR speech models. The required estimates of the power spectral densities of speech and noise are determined via recursively smoothed periodograms [15].

The following steps have to be performed for estimating the power spectral densities of speech and noise:

1. Transformation of the n-th signal frame of length N_{model} into the frequency domain with *zero-padding* for avoiding cyclic convolution distortion:
$$\begin{aligned} X(k,n) = \mathrm{DFT}\Big\{ \big[x(n - N_{\text{model}}/2+1), \\ \ldots, x(n + N_{\text{model}}/2),\, 0,\, \ldots,\, 0 \big] \Big\}. \end{aligned} \tag{10.56}$$
2. Estimation of the power spectral density of the noisy input signal and noise:
$$\hat{S}_{xx}(k,n) = \big|X(k,n)\big|^2, \tag{10.57}$$
$$\overline{B}(k,n) = \begin{cases} \alpha_b\,\overline{B}(k,n-1) + (1-\alpha_b)\,\big|X(k,n)\big| & : \big[VAD(n)=0\big] \vee \big[|X(k,n)| < \beta_B\,\overline{B}(k,n-1)\big], \\ \overline{B}(k,n-1) & : \text{else}, \end{cases}$$
$$\hat{S}_{bb}(k,n) = \overline{B}^2(k,n). \tag{10.58}$$

Here, the noise estimation is enabled during speech pauses and for time frames where the current spectral amplitude is not larger the β_B times the current mean noise amplitude.

Since the spectral components $X(k,n)$ are determined for a signal frame of length 48 msec, no further smoothing has to be applied for estimating $\hat{S}_{xx}(k,n)$.

The advantage of this procedure compared to the model estimates based on the disturbed speech signal is that speech components are better modeled. However, this procedure tends to generate – such as the iterative EM method – local maxima of the estimated speech model which cause musical tones. Results are depicted in Fig. 10.22.

Combined and Optimized Estimation Procedure

In this section, so far, three different methods for the speech model estimation have been presented with specific advantages and disadvantages:

- The estimation with the Burg algorithm, on the basis of the disturbed speech signal leads to models which, in general, do not provoke musical tones. However, this procedure only models those spectral components with an appropriate accuracy which exhibit a sufficient SNR.
- The iterative EM method shows a high computational complexity and is, mainly for this reason, no further considered in this chapter. Additionally, the obtained results are not remarkably better compared to other methods that were examined since the method shows an increased tendency to musical tones.
- The autocorrelation method, i.e. determining the speech models with the help of the difference of the power spectral densities of the input signal and noise, finally allows to obtain better speech model estimates. The accuracy of this method is approximately comparable to the iterative EM method, showing a comparable tendency to musical tones, nevertheless, with a reduced computational complexity. This method is called differential autocorrelation (DACF) method in the following.

Comparing the estimates of the Burg algorithm and the DACF method, one observes that both show different advantages which can be usefully combined: The DACF method models speech components better, whereas the Burg method shows a smaller tendency to musical tones.

Examinations showed that both methods can be advantageously combined as follows:

- The musical tones, which are mainly provoked by the DACF method, can be reduced so far such that the remaining ones can be masked by a residual noise (see Sec. 10.4.2.3).
- Additionally, the combined methods allow a better modeling of speech components compared to the Burg algorithm especially for the subbands with a low SNR.

The combined estimation procedure consists of the following steps:

1. Estimation of the reflection coefficients $\Gamma_j^{\mathrm{B}}(n)$ of the speech model with the Burg algorithm based on the noisy speech signal according to Eq. 10.50.
2. Calculation of the reflection coefficients $\Gamma_j^{\mathrm{DACF}}(n)$ with the Schur recursion [12] based on the estimated autocorrelation function $\hat{s}_{ss,n}(i)$ according to Eqs. 10.55 - 10.58.
3. Calculation of the combined reflection coefficients by the weighted sum of the two estimates: $\Gamma_j(n) = g_{\mathrm{B}}\Gamma_j^{\mathrm{B}}(n) + g_{\mathrm{DACF}}\Gamma_j^{\mathrm{DACF}}(n)$. By choosing the weights such that the condition $g_{\mathrm{B}} + g_{\mathrm{DACF}} = 1$ is fulfilled, one assures that the combined model is also stable.
4. Determining the combined AR parameters of the speech model with the Step-up recursion [12].

In the latter procedure, Schur recursion and Step-up recursion describe the first and the second part of the Levinson-Durbin recursion, respectively.

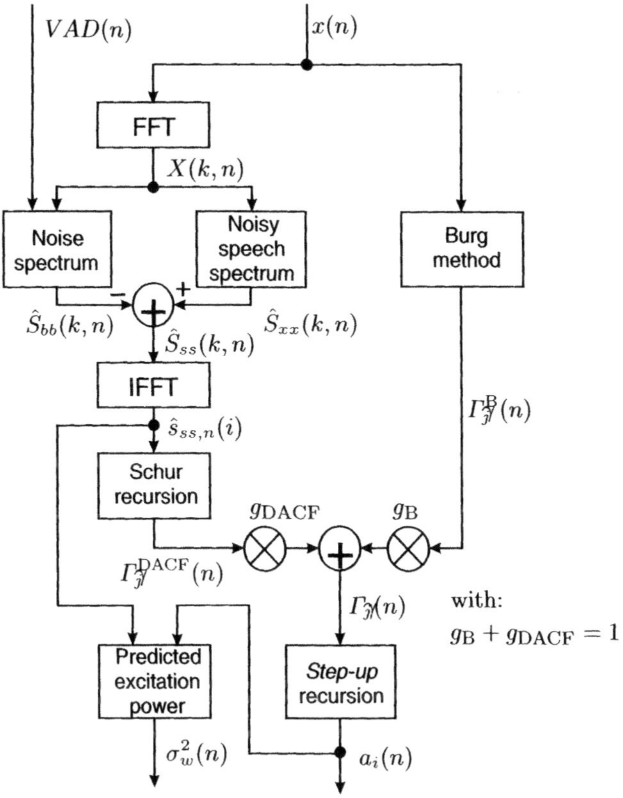

Fig. 10.21. Diagram for estimating the speech model $a_i(n)$ and the power $\sigma_w^2(n)$ of the white speech excitation signal.

The weights g_{B} and g_{DACF} can be chosen differently for each subband. Since for speech disturbed by car noise, the SNR for the lowest subband is usually low, estimates with the Burg method are much worse than with the DACF method. The reason is that the Burg method models the disturbed speech. Additionally, the human hearing is less sensitive to musical tones for low frequencies. For these reasons, the model estimation for the lowest subband is performed only based on the DACF method, i.e. $g_{\mathrm{DACF}} = 1$ and $g_{\mathrm{B}} = 0$. For the other subbands, a weighting accorings to $g_{\mathrm{DACF}} = 0.7$ and $g_{\mathrm{B}} = 0.3$ showed good results. This procedure is shown in Fig. 10.21.

Estimation results according to this procedure are depicted in Fig. 10.22. The two upper graphs show the model estimate for the lowest subband. One clearly observes that the DACF method (right) is able to model the pitch components better than the Burg method (left).

The problems occurring when utilizing only the DACF method for model estimates are depicted below, for the second subband, as an example: strong fluctuations in speech pauses (0 - 0.8 sec) can be observed. Considering both

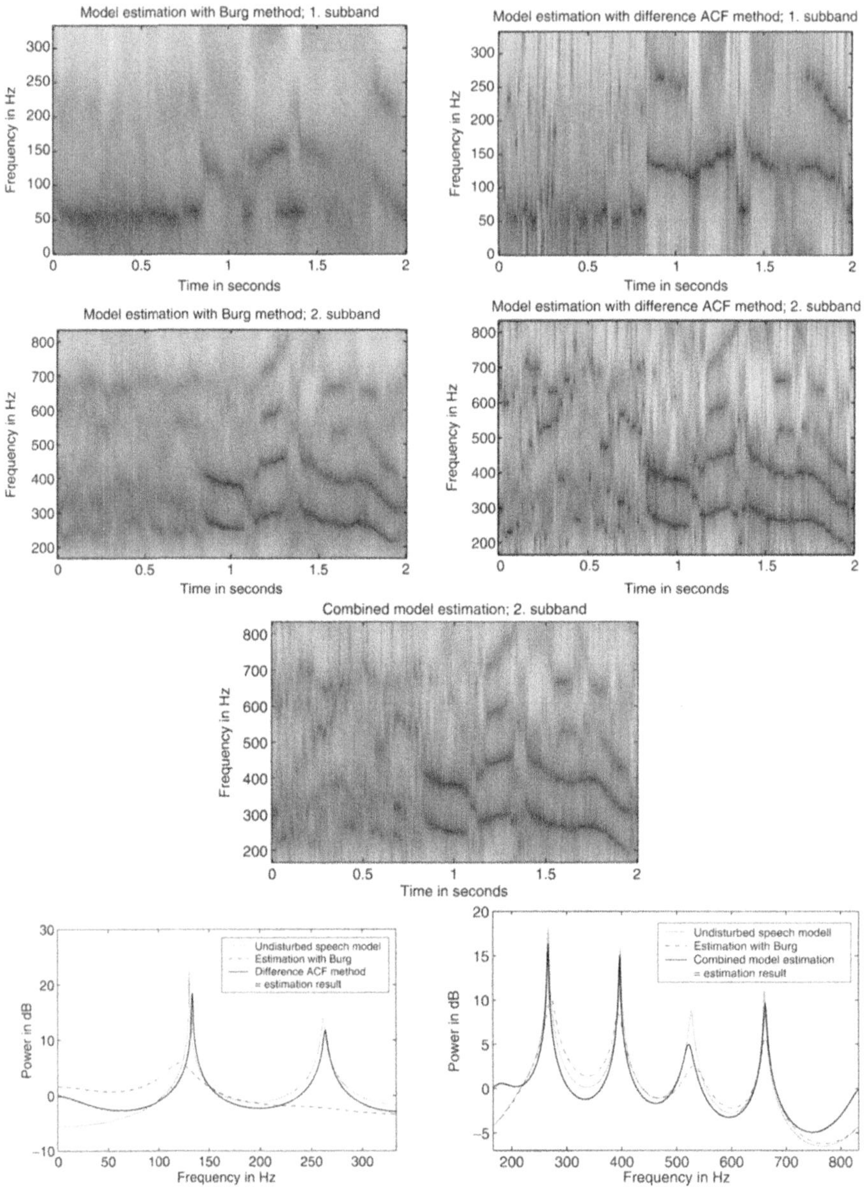

Fig. 10.22. Results of the different speech model estimation methods. For the lowest subband, the DACF method shows the best results, whereas for the second and the higher subbands, the combination of the Burg and DACF methods shows the best results. In the lower two graphs, estimation results for one voiced speech frame are depicted. Here, it is getting obvious that the proposed estimation method gives results which are close to the optimum estimate based on the clean speech signal.

estimation methods, according to the procedure of Fig. 10.21 results in the combined estimate. The deficits of both estimation methods can be well compensated: The fluctuations are well reduced, but the pitch components are still well resolved. This is especially shown on the two lower graphs of Fig. 10.22.

Estimation of the White Speech Model Excitation Signal Power

The excitation signal power can be estimated according to

$$\begin{aligned}\sigma_w^2(n) &= \mathrm{E}\left\{ \left| s(n) - \sum_{i=1}^{p} a_i^*(n)\, s(n-i) \right|^2 \right\}, \\ &= \hat{s}_{ss,n}(0) - 2\,\mathrm{Re}\left\{ \sum_{i=1}^{p} a_i^*(n) \hat{s}_{ss,n}(i) \right\} \\ &\quad + \sum_{i=1}^{p} \sum_{j=1}^{p} a_i^*(n) a_j(n)\; \hat{s}_{ss,n}(i-j),\end{aligned} \tag{10.59}$$

where $\hat{s}_{ss,n}(i)$ is the estimate of the speech autocorrelation function according to Eq. 10.55. The model parameters are determined according to the before mentioned procedure. The block for determining the excitation signal power is also shown in Fig. 10.21. To avoid misunderstandings: The excitation signal power estimate which is implicitly obtained when determining the Burg model estimates cannot be utilized since the noise excitation provokes an estimation bias. An example comparing results of the proposed method and the estimate obtained with the Burg method is shown in Fig. 10.23. In contrast to the Burg estimate, the proposed method shows the desired low estimation power during speech pauses. And, as desired, during speech activity, speech components are not under-estimated. The low fluctuations during speech pauses correspond to the low sensitivity of this estimation procedure towards musical tones.

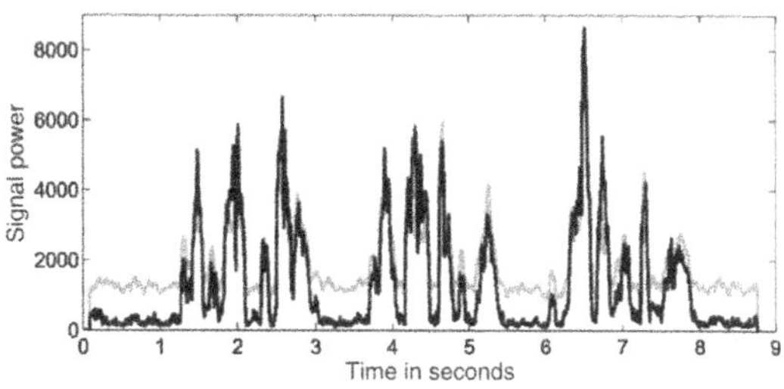

Fig. 10.23. Estimated excitation power of the speech model $\sigma_w^2(n)$ determined with the Burg method (gray) and according to Eq. 10.59 (black), respectively.

10.4.2.2 Estimation of Noise Models

Since car noise spectra, especially after suppressing harmonic engine components [21, 23], are typically smooth and do not exhibit strong local maxima at certain frequency components, model order of $p = 2$ are sufficient for each subband. Utilizing the estimation of the power spectral density according to Eq. 10.58, the first three values of the autocorrelation function $\hat{s}_{bb,n}(i)$ can be determined via the Inverse Fourier Transform. Then, utilizing the autocorrelation method, the two model coefficients $c_1(n)$ and $c_2(n)$ of the noise model are calculated via:

$$\begin{bmatrix} c_1(n) \\ c_2(n) \end{bmatrix} = \begin{bmatrix} \hat{s}_{bb,n}(0) & \hat{s}^*_{bb,n}(1) \\ \hat{s}_{bb,n}(1) & \hat{s}_{bb,n}(0) \end{bmatrix}^{-1} \begin{bmatrix} \hat{s}_{bb,n}(1) \\ \hat{s}_{bb,n}(2) \end{bmatrix}. \tag{10.60}$$

The noise model excitation power $\sigma^2_\eta(n)$ can be computed comparably to Eq. 10.59:

$$\begin{aligned} \sigma^2_\eta(n) = \hat{s}_{bb,n}(0) - 2\,\mathrm{Re}\left\{ \sum_{i=1}^{q} c_i^*(n)\hat{s}_{bb,n}(i) \right\} \\ + \sum_{i=1}^{q}\sum_{j=1}^{q} c_i^*(n)c_j(n)\; \hat{s}_{bb,n}(i-j). \end{aligned} \tag{10.61}$$

10.4.2.3 Overestimation of the Noise Model's Excitation Power

When applying the Kalman filtering for real signals, one also observes musical tones which, however, are less powerful compared to the classical Wiener filters. For suppressing these musical tones, the Wiener filter typically utilizes an overestimation of the noise and a limit of the noise suppression. This is also a good measure for Kalman filtering.

The overestimation of noise should be done in two steps for the Kalman filtering:

1. When calculating the speech power spectral density $\hat{S}_{ss}(k,n)$ according to Eq. 10.54, $\hat{S}_{bb}(k,n)$ should be overestimated by a factor of approximately 1.5:

$$\hat{S}_{ss}(k,n) = \max\left\{ \hat{S}_{xx}(k,n) - 1.5\,\hat{S}_{bb}(k,n), 0 \right\}. \tag{10.62}$$

2. When applying the Kalman equations, the estimated excitation power of the noise $\sigma^2_\eta(n)$ should also be raised by a factor $\beta_{\mathrm{Kalman}}(n)$:

$$\tilde{\sigma}^2_\eta(n) = \beta_{\mathrm{Kalman}}(n)\,\sigma^2_\eta(n). \tag{10.63}$$

When determining $\beta_{\mathrm{Kalman}}(n)$, speech pauses and speech activity should be treated differently. For the differentiation, the same speech activity detector can be utilized as for the noise power spectral density estimate (see Eq. 10.58).

In speech pauses, an overestimation factor around $\beta_{\text{Kalman}}(n) = 3$ and during speech activity a factor of 1.5 allows to obtain good results:

$$\beta_{\text{Kalman}}(n) = \begin{cases} 3 & : VAD(n) = 0, \\ 1.5 & : \text{else}. \end{cases} \tag{10.64}$$

A limitation with a *spectral floor* such as it is usually applied for Wiener filtering cannot directly be incorporated in the Kalman filter. A possible alternative it to add a portion of the noisy input signal to the output according to Fig. 10.24.

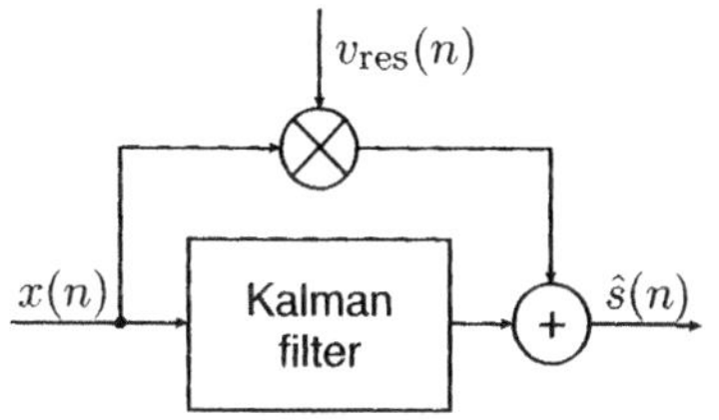

Fig. 10.24. Summation of a portion of the noisy input signal to the output of the Kalman filter. This measure is equivalent to the Spectral Floor of the Wiener filter and preserves the natural sound of speech.

For the factor $v_{\text{res}}(n)$, we obtained good results with 0.08 for the two lowest subbands and 0.15 for the other subbands. The lower value for the low frequencies makes sense since here the car noise exhibits especially large components. Thus, subjectively a higher noise suppression can be obtained. Additionally, mostly the higher subbands are related to a subjectively high speech quality impression.

10.4.3 Pitch-Adaptive Enhanced Speech Model Estimation

The proposed combined speech model estimation method already allows to obtain good noise reduction results with the Kalman filter approach. Comparing, however, the estimated speech models with the true values (see Fig. 10.22) one observes that especially the pitch components for the low-frequency subbands with a low SNR are partly insufficiently resolved.

Considering speech models for the lower subbands, one observes that the pitch components are modeled with poles of the AR models close to the unit circle. Additionally, methods for the pitch frequency estimation [1, 14], also based on disturbed speech signals [25], are known.

Utilizing these relations, methods can be designed which allow the AR model reconstruction or enhancement for voiced speech frames based on the pitch frequency.

The presentation of possibilities for such a pitch-adaptive speech model enhancement is the goal of this section. However, pitch frequency estimation methods are not analyzed here. For them, we refer to the above cited literature references.

One observes that especially in frequency ranges up to 750 Hz one can obtain enhanced model estimates when incorporating the knowledge of the pitch frequency. This for the following reasons:

- Here, pitch components are particularly dominant,
- pitch components are located exactly at multiples of the pitch frequency (in contrast to higher frequencies), and
- the estimation of speech models is especially disturbed for these frequency components.

The presented relations describe the potential for an improvement of the model estimation for the two lower subbands of the proposed subband Kalman noise reduction procedure. In the following, for these two subbands, two different procedures will be described.

10.4.3.1 Speech Model Enhancement for the Lowest Subband

The initial estimate for the lowest subband can be so bad that it may be appropriate to severely modify this estimate. The basis for this modification is the estimated pitch frequency. The enhanced model estimate is then obtained by first determining the poles of the initial AR model and then shifting these poles via the unit circle.

The task of determining the poles, which is equivalent to calculating the zeros of a polynomial, can be written as an Eigenvalue problem for which solution approaches are known [19].

For the AR model modification, we obtained the best results when performing the following two steps:

1. Poles which are located at angles corresponding to frequency components below the current estimate of the pitch frequency are suppressed by reducing their radius, i.e. shifting them far into the unit circle.
2. For voiced frequency frames with pitch components, poles are shifted towards the unit circle.

These two steps are further explained in the following.

Suppression of Poles Below the Detected Pitch Frequency

Poles below the pitch frequency model only noise components and should be suppressed. This can be obtained by multiplying these poles with a factor of 0.1 which corresponds to a shift into the unit circle:

$$\tilde{p}_i = p_i \cdot 0.1, \qquad \text{if } \arg(p_i) < \frac{f_{\text{pitch}}}{f_s} \cdot 2\pi \cdot 0.8 \,. \tag{10.65}$$

Here, f_{pitch} denotes the latest detected pitch frequency and f_s the sampling frequency of the subband signals. With the factor 0.8, one assures a low risk to attenuate true pitch components close to the estimated pitch frequency. It is important that these pole shifts are performed continuously in order to guarantee a good suppression of the low-frequency noise components also during non-voiced speech frames.

Shifting Poles Towards Multiples of the Pitch Frequency

Here, the goal is to place conjugate-complex pole pairs at multiples of the pitch frequency according to the following equation:

$$\tilde{p}_{\pm i} = \exp\left(\pm j \frac{i\, f_{\text{pitch}}}{f_s} 2\pi\right) \frac{1 + r_{\text{pitch}}}{2}. \tag{10.66}$$

The pole radius has been chosen as $(1 + r_{\text{pitch}})/2$, where r_{pitch} is the maximum radius of the poles of the second subband.

The used procedure is the following: First, poles in proximity of the pitch frequency (± 20 Hz) are localized, with a pole radius larger than 0.6. These poles are then replaced by the poles according to Eq. 10.66. Thus, one avoids double pole pairs of the AR model at multiples of the pitch frequency.

If necessary, then, the missing pitch components, according to Eq. 10.66, are generated by poles which have not yet been shifted. This procedure makes sense, since their original location far away from pitch components does not contribute to an appropriate speech model. A result obtained with this procedure is depicted in Fig. 10.25 on the left side. The improvement of the model estimation in comparison with the original estimate in Fig. 10.22 is obvious.

10.4.3.2 Speech Model Enhancement for the Second Subband

The speech model estimated for the second subband usually better match the correct models. When comparing, the estimates only exhibit smaller maxima at the pitch frequency. Model enhancements are possible by shifting the poles at multiples of the estimated pitch frequency closer to the unit circle, according to the following formula:

$$\tilde{p}_i = p_i \left(\frac{1}{2\,|p_i|} + \frac{1}{2}\right). \tag{10.67}$$

The shift is only performed when the difference of the frequency of poles and the corresponding multiple of the pitch frequency is not larger than 10 Hz and the original magnitude of the poles is larger than 0.8. The angles of the poles are not modified, since the pitch components are not exactly periodic. An artificially generated periodic relation could provoke signal distortion.

In Fig. 10.25, on the right, the enhanced model estimation for the second subband is depicted. In comparison to the original estimate, especially the strong pitch components can be well observed.

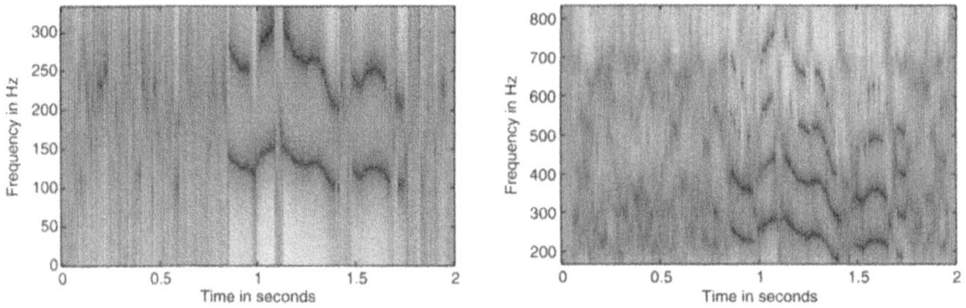

Fig. 10.25. Results of the enhanced model estimates for the first (left) and second (right) subband. The pitch frequency has been estimated with a procedure comparable the method proposed in [1].

When applying this kind of pitch-frequency based model enhancement also to the higher subbands, however, only small model enhancements are possible. Weighing the computational complexity against, in particular provoked by the determination of the poles, it makes no sense to utilize the pitch-adaptive enhancement procedure also for the third and higher subbands.

10.5 Comparison of the Results with Classical Frequency Domain Noise Reduction Approaches

The goal of this section is to compare the proposed Kalman filter noise reduction approach against classical frequency domain Wiener and Ephraim-Malah methods with respect to speech quality and the sound properties of residual noise.

For both frequency domain methods, a filter bank analysis is performed utilizing a prototype low-pass filter with 512 taps in order to decompose the input signal into 256 spectral components. The sub-sampling rate has been chosen to 64. For the estimation of the noise power spectral density the calculation is performed according to Eq. 10.58

$$\overline{B}(k,n) = \begin{cases} \begin{array}{l} \alpha_b\,\overline{B}(k,n-1) \\ +(1-\alpha_b)\,|X(k,n)| \end{array} : & \begin{array}{l} \Big[VAD(n)=0\Big] \vee \\ \Big[|X(k,n)| < \beta_B\,\overline{B}(k,n-1)\Big], \end{array} \\ \overline{B}(k,n-1) & : \text{else}, \end{cases}$$

$$\hat{S}_{bb}(k,n) = \overline{B}^2(k,n), \tag{10.68}$$

with $\alpha_b = 0.985$ and $\beta_B = 2.15$.

Wiener Filter Approach

For the Wiener filter approach, the spectral weighting coefficients are chosen according to the modified Wiener formula:

$$G_{\text{Wiener,para}}(k,n) = \max\left\{1 - \beta_{\text{OV}} \frac{S_{bb}(k,n)}{S_{xx}(k,n)}, spfl\right\}, \tag{10.69}$$

where β_{OV} denotes the overestimation factor and $spfl$ the spectral floor. Here, the overestimation factor has been chosen to $\beta_{\text{OV}} = 5$ and spectral floor to $spfl = 0.1$.

Ephraim-Malah Approach

The utilized formula for the Ephraim-Malah approach is based on the minimization of the log-spectral amplitudes:

$$\begin{aligned}\hat{S}(k,n) = {} & \frac{\xi(k,n)}{1+\xi(k,n)} \exp\left[\frac{1}{2}\int\limits_{v(k,n)}^{\infty} \frac{\exp(-z)}{z}\, dz\right] \\ & \cdot p\Big(H_1(k,n)\big|X(k,n)\Big)\, X(k,n),\end{aligned} \tag{10.70}$$

with

$$\xi(k,n) = \frac{S_{ss}(k,n)}{S_{bb}(k,n)}, \tag{10.71}$$

$$\gamma(k,n) = \frac{\big|X(k,n)\big|^2}{S_{bb}(k,n)}, \tag{10.72}$$

$$v(k,n) = \frac{\xi(k,n)}{1+\xi(k,n)}\, \gamma(k,n), \tag{10.73}$$

where $\xi(k,n)$ is estimated with the decision directed approach. Here, the smoothing constant has been chosen to $\alpha_{\text{DDA}} = 0.98$. In order to avoid musical tones, the estimated a-priori SNR is limited to $\xi_{\text{min}} = -13$ dB:

$$\begin{aligned}\hat{\xi}(k,n) = {} & \alpha_{\text{DDA}} \frac{|\hat{S}(k,n-1)|^2}{S_{bb}(k,n-1)} \\ & + \big(1 - \alpha_{\text{DDA}}\big) \max\Big\{\gamma_k(k,n) - 1, \xi_{\text{min}}\Big\}.\end{aligned} \tag{10.74}$$

The term $p(H_1(k,n)|X(k,n))$ which considers the conditional probability for speech activity is determined with the following formula:

$$p\Big(H_1(k,n)\big|X(k,n)\Big) = \frac{\dfrac{1-p\Big(H_0(k,n)\Big)}{p\Big(H_0(k,n)\Big)\Big(1-\xi(k,n)\Big)}\exp\big[v(k,n)\big]}{1+\dfrac{1-p\Big(H_0(k,n)\Big)}{p\Big(H_0(k,n)\Big)\Big(1-\xi(k,n)\Big)}\exp\big[v(k,n)\big]}. \qquad (10.75)$$

The required value $p(H_0(k,n))$, i.e. the probability for a speech pause of the k-th spectral component is based on $\gamma(k,n)$ [17], where $\gamma(k,n)-1$ denotes the a-posteriori SNR. First, $\gamma(k,n)$ is compared with a fixed threshold. Based on this comparisons, one obtains a binary, time-frequency dependent decision $p_{\mathrm{B}}(k,n)$, which can be utilized to determine, with a recursive smoothing, the required frequency dependent probability for speech pauses:

$$p_{\mathrm{B}}(k,n) = \begin{cases} 1 & : \ \gamma(k,n) < 0.8\,, \\ 0 & : \ \text{else}\,, \end{cases} \qquad (10.76)$$

$$p\Big(H_0(k,n)\Big) = 0.95\,p\Big(H_0(k,n-1)\Big) + 0.05\,p_{\mathrm{B}}(k,n). \qquad (10.77)$$

In the following, the noise reduction approaches are compared with respect to speech quality and the natural sound of the residual noise. The comparison is done by a description of the subjective impression and objective spectral analyses.

For the spectral analysis, in particular, pitch components and the formant structure are analyzed. Pitch components are required for a full, rich, and natural sound. The formant structure, especially the attenuation between the formants is an indication for speech distortion [24].

The algorithms were evaluated with different speech and noise signals. For the detailed objective spectral comparison, a male speech signal has been chosen.

Residual Noise

The parameters of the algorithms were chosen such that the residual noise sounds natural and the musical tones are suppressed. Only for the Wiener filter approach, sometimes slight musical tones are audible. With a higher overestimation factor, they could also be completely suppressed, however, with the disadvantage to further reduce the natural speech sound.

The residual noise of the Ephraim-Malah algorithm is very smooth and natural, which is also the case for the Kalman filter. Additionally, the residual noise for the Kalman filter can be chosen a little lower without risking speech distortion or musical tones.

Speech Quality

The Wiener filter shows the undesirable property to attenuate speech components rather strongly which provokes audible signal distortion and a dull speech sound.

The output of the Ephraim-Malah filter sounds natural, in speech frames with low speech excitation, however, speech attenuation is audible.

The Kalman filter provides a speech output with the most natural sound. A slight reverberant sound can be observed by trained listeners which is, however, hardly disturbing. This is a tribute to the speech model estimation based on frames of the length 48 ms which is necessary for a reliable estimation. A very small speech distortion can be observed rather for female speech signals. A possible reason is the good modeling of pitch components. Especially male speech with a high number of pitch components may profit from this.

In Fig. 10.26 spectrograms of the three compared methods are depicted which show the described properties, although acoustic differences are usually only partly represented by spectrograms.

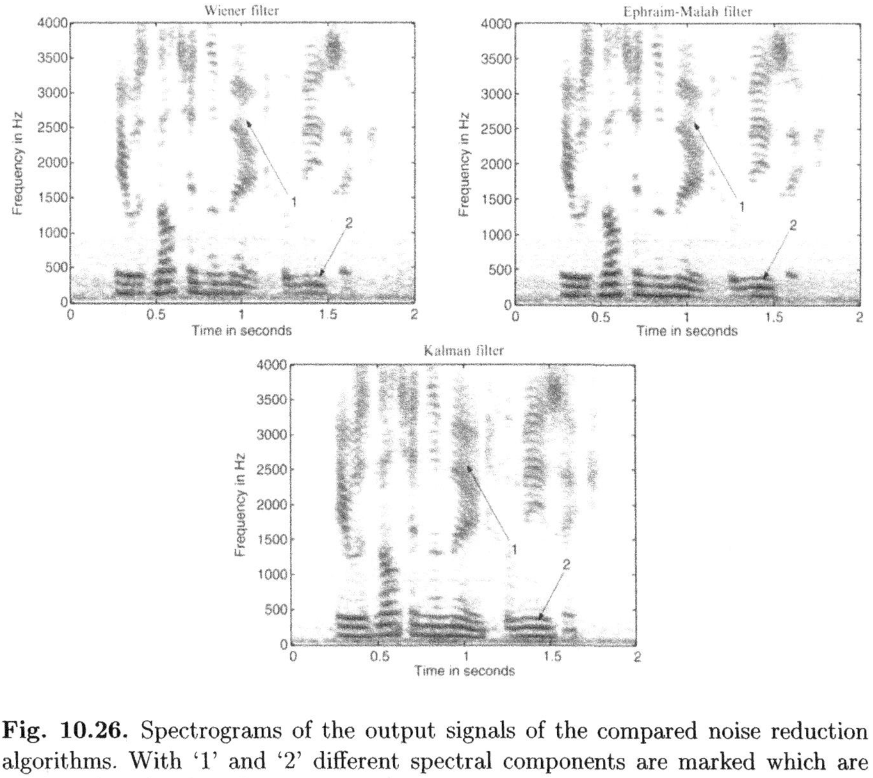

Fig. 10.26. Spectrograms of the output signals of the compared noise reduction algorithms. With '1' and '2' different spectral components are marked which are compared in detail in Figs. 10.27 und 10.28.

Comparing the pitch structure of the signals at 1.4 s (see Fig. 10.27) marked with '1' in Fig. 10.26, one observes that the Kalman filter resolves best the pitch structure and is able to selectively attenuate the signals in between.

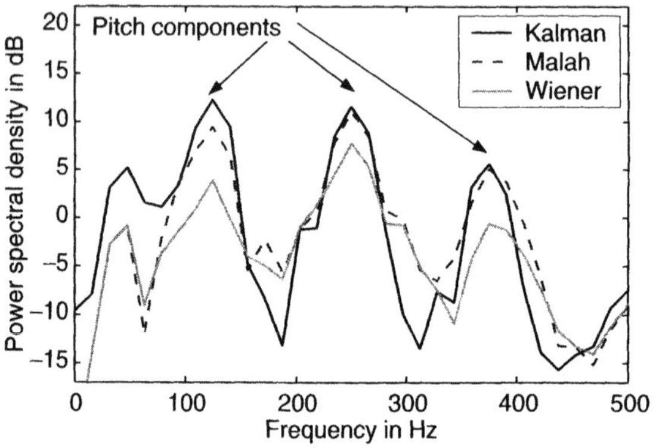

Fig. 10.27. Comparison of the pitch components of the enhanced speech signals at 1.4 s.

The Wiener filter shows a strong attenuation of the pitch components which provokes signal distortion. Much better is the Ephraim-Malah filter output which also resolves well the pitch components. However, the attenuation in between is not as large.

A second indication for speech distortion is the attenuation between formants [24] which is equivalent to a modification of the signal's spectral envelope. In Fig. 10.28 the signal frame is depicted which is marked with '2' in Fig. 10.26. One observes that the attenuation between the formants is lowest for the Kalman filter and strongest for the Ephraim-Malah algorithm. This result coincides with the impression that the Kalman filter provokes the lowest signal distortion.

So far, the comparisons were done with stationary interference noise. For non-stationary noise suppression, especially, the noise power spectral density estimation is the crucial point. Supposing good noise power spectral density estimates are available, the compared noise reduction methods show comparable performance differences as for stationary noise. In the other case, the non-reliable spectral density estimation dominates the signal distortion.

In order to reduce the non-stationary properties of car noise, as mentioned, an interesting possibility for a noise reduction preprocessing unit is to perform a suppression of engine noise harmonics [21, 23] to pre-enhance the SNR and reduce to non-stationary components.

A final remark is dedicated to the performance limit of these kinds of single noise reduction methods. All methods show a reduced performance

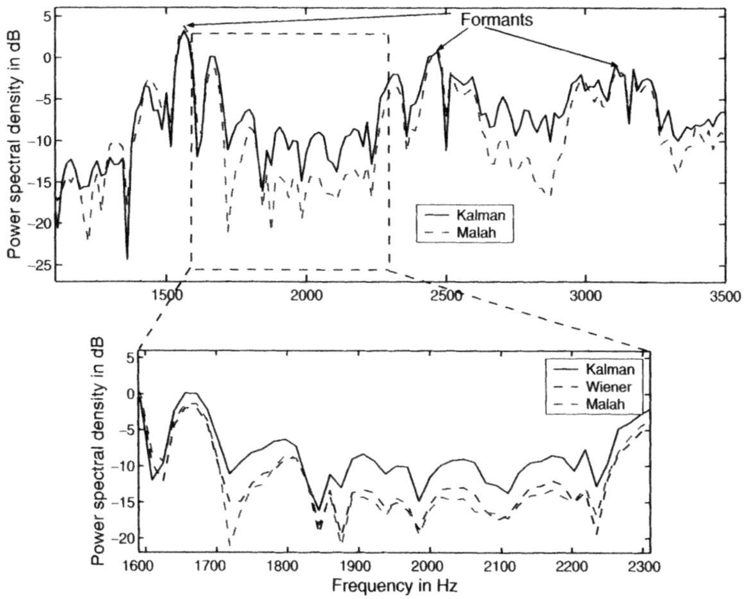

Fig. 10.28. Comparison of the formant structure of the enhanced speech signals at 1.0 s.

for SNRs below approximately 0 dB. In this case, the Kalman filter exhibits increased problems to perform the parametric spectral estimation based on the disturbed speech signal. Also the pitch-adaptive model enhancement shows problems since the pitch frequency estimation is less reliable for a low SNR. In case of a low SNR, best choice is to increase the portion of residual noise in order to avoid non-acceptable speech distortion.

10.6 Conclusions

In this chapter, a subband Kalman filter method was presented in order to enhance speech signals disturbed by car noise, with the specific application focus on hands-free car phones.

After an introduction, this chapter started with a detailed speech and car noise analysis. The important properties of both signals were described which are required for an optimum Kalman filter design. These main important properties are that car noise is a signal with slowly changing signal characteristics, car noise exhibits dominant low-frequency components and shows a rather smooth spectrum. Speech, in contrast, exhibits fast time-varying signal characteristics and a typical spectral structure with pitch components and formants.

In the third section, the theoretical basis of Kalman filters and the required parametric spectral estimation procedures are presented.

The fourth section, then, is dedicated to the presentation of a Kalman filter noise reduction method tailored specifically to the application for speech signals disturbed by car noise. The number of subbands, the orders of the parametric speech and noise models and the methods that are utilized for the estimation of these parametric models were chosen based on sophisticated analyses and with respect to the basis of the speech signal and car noise properties analyzed in the second section.

Finally, in the fifth section, the proposed subband Kalman filter noise reduction procedure was compared to other well-known frequency domain noise reduction procedures, such as Wiener-filter and Ephraim-Malah noise reduction procedures. Here, it could be shown that down to a SNR of approximately 0 dB, the proposed Kalman filter method outperforms the other methods especially with respect to the natural sound. Objective comparisons with respect to the pitch and formant structure confirm these subjective observations.

References

[1] L. Arevalo: *Beiträge zur Schätzung der Frequenzen gestörter Schwingungen kurzer Dauer und eine Anwendung auf die Analyse von Sprachsignalen,* Dissertation, Bochum, Germany, 1991 (in German).

[2] M. Berouti, R. Schwarz, J. Makhoul: Enhancement of speech corrupted by acoustic noise, *Proc. ICASSP '79*, 208-211, Washington, DC, USA, 1979.

[3] S.F. Boll: Suppression of acoustic noise in speech using spectral subtractions, *IEEE Trans. on Acoustics, Speech, and Signal Processing*, **27**(2), 113-120, 1979.

[4] C. Breining, P. Dreiseitel, E. Hänsler, A. Mader, B. Nitsch, H. Puder, T. Schertler, G. Schmidt, and J. Tilp: Acoustic echo control, *IEEE Signal Processing Magazine*, **16**(4), 42-69, 1999.

[5] A.P. Dempster, N.M. Laird, D.B. Rubin: Maximum likelihood from incomplete data via the EM algorithm, *Journal of the Royal Statistical Society*, **39**, 1-8, 1977.

[6] M. Deriche: AR parameter estimation from noisy data Uusing the EM algorithm, *Proc. ICASSP '94*, **4**, 69-72, Adelaine, Australia, 1994.

[7] G. Doblinger: An adaptive Kalman filter for the enhancement of noisy AR signals, *Proc. ISCAS-98*, **5**, 305-308, Monterey, CA, USA, 1998.

[8] G. Doblinger: Adaptive Kalman smoothing of AR signals disturbed by impulses and colored noise, *Proc. of IEEE Symposium on Advances in Digital Filtering and Signal Processing*, 72-76, Victoria, BC, Canada, 1998.

[9] Y. Ephraim, D. Malah: Speech enhancement using a minimum mean-square error short-time spectral amplitude estimator, *IEEE Trans. on Acoustics, Speech, and Signal Processing*, **32**(6), 1109-1121, 1984.

[10] Y. Ephraim, D. Malah: Speech enhancement using a minimum mean-square error log-spectral amplitude estimator, *IEEE Trans. on Acoustics, Speech, and Signal Processing*, **33**(2), 443-445, 1985.

[11] S. Gannot, D. Burshtein, E. Weinstein: Iterative and sequential Kalman filter-based speech enhancement algorithm, *IEEE Trans. on Speech and Audio Processing*, **6**(4), 373-385, 1998.

[12] M.H. Hayes: *Statistical digital signal processing and modelling,* New York, NY, USA: Wiley, 1996.

[13] S. Haykin: *Adaptive Filter Theory,* 4th ed., Englewood Cliffs, NJ, USA: Prentice Hall, 2002.

[14] W. Hess: *Pitch Detection of Speech Signals,* Springer, 1983.

[15] K.D. Kammeyer, K. Kroschel: *Digitale Signalverarbeitung,* 4th ed., Stuttgart, Germany: Teubner, 1998 (in German).

[16] A. Mader, H. Puder, and G.U. Schmidt: Step-size control for acoustic echo cancellation filters - An overview, *Signal Processing*, **80**(9), 1697-1719, 2000.

[17] D. Malah, R.V. Cox, A.J. Accardi: Tracking speech-presence uncertainty to improve speech enhancement in non-stationary noise environments, *Proc. ICASSP '99*, **2**, 1761-1764, Phoenix, AR, USA, 1999.

[18] D.C. Popescu, I. Zeljkovic: Kalman filtering of colored noise for speech enhancement, *Proc. ICASSP '98*, **2**, 997-1000, Seattle, WA, USA, 1998.

[19] W.H. Press, S.A. Teukolsky, W.T. Vetterling, B.P. Flannery: *Numerical Recipes in C, The Art of Scientific Computing*, 2nd ed., chapter 15: Modeling of Data, 656-699, West Nyack, NY, USA: Cambridge University Press, 1992.

[20] J. Proakis, C.M. Radar,F. Lung, C.L. Nikias: *Advanced Digital Signal Processing,* New York, NY, USA: Maxwell Macmillan Div., 1992.

[21] H. Puder and F. Steffens: Improved noise reduction for hands-free car-phones utilizing information on vehicle and engine speeds, *Proc. EUSIPCO '00*, **3**, 1851-1854, Tampere, Finland, 2000.

[22] H. Puder: Kalman-filters in subbands for noise reduction with enhanced pitch-adaptive speech model estimation, *European Trans. on Telecommunication,* **13**(2), 139-148, 2002.

[23] H. Puder: *Geräuschreduktionsverfahren mit modellbasierten Ansätzen für Freisprecheinrichtungen in Kraftfahrzeugen,* Düsseldorf, Germany, Fortschr.-Ber. VDI-Reihe 10(721), VDI Verlag, 2003 (in German).

[24] J. Tilp: Formant-based detection of speech distortions for a single-channel spectral-subtraction scheme, *Proc. IWAENC '99*, 72-75, Pocono Manor, NJ, USA, 1999.

[25] J. Tilp: *Verfahren zur Verbesserung gestörter Sprachsignale unter Berücksichtigung der Grundfrequenz stimmhafter Sprachlaute,* Düsseldorf, Germany, Fortschr.-Ber. VDI-Reihe 10(703), VDI Verlag, 2002 (in German).

[26] W.R. Wu, P.C. Chen: Subband Kalman filtering for speech enhancement, *IEEE Trans. on Circuits and Systems-II,* **45**(8), 1072-1083, 1998.

Part V

Selected Applications

11

Evaluation of Algorithms for Speech Enhancement

Pia Dreiseitel[1] and Gerhard Schmidt[2]

[1] Smiths Heimann, Wiesbaden, Germany
[2] Harman/Becker Automotive Systems, Ulm, Germany

In many situations and particularly in the context of developing and improving speech enhancement algorithms, performance measures are needed to evaluate whether one algorithm or one algorithmic version is in some sense superior to another algorithm or to its preceding version. Speech enhancement systems are often evaluated utilizing objective measures, such as distance measures [13] between the clean and the enhanced signal or segmental signal-to-noise ratios. These measures can be quantified in straightforward objective terms. During the optimization of existing algorithms and for parameter optimization this kind of evaluation is certainly appropriate. We will therefore present a few objective measures and how they are designed and trained for specific applications. However, due to the complexity of speech quality these methods usually do not deliver an estimate for the quality that comprises all the various aspects of a high quality speech signal when a new application has to be tested.

A more general – but also more expensive and time consuming – way to evaluate the speech intelligibility and quality are subjective listening tests. These tests are the most reliable tool available for the evaluation of speech enhancement algorithms [39]. The challenge of these tests is to design them in such a way that the quality of the enhancement system can be measured in a reliable and reproducable manner.

11.1 The Focus of this Chapter

For several well-established applications such as hands-free telephony a variety of measurement standards [22] have been established. Within these standards subjective [25] and objective [12,25] evaluation tools such as *TOSQA*, *PAMS*, or *PESQ* [11] are specified.[3] Describing the details of only the most important

[3] TOSQA abbreviates Telecommunication Objective Speech Quality Assessment, PAMS is short for Perceptual Analysis/Measurement System, and PESQ stands for Perceptual Evaluation of Speech Quality.

ones would go far beyond the scope of this chapter. The focus of this contribution is on the evaluation of new speech enhancement algorithms – especially on the evaluation of the (speech) quality improvement during the design and optimization of an algorithm.

In the first part of this chapter we will introduce objective tests that can be applied for algorithms that suppress background noise. Since we focus on the design and optimization stage we assume that we have in addition to the noisy input signal also the clean speech signal as a reference. In this case we can apply several distance measures between the clean speech signal and the output of the noise suppression system.

Afterwards we will concentrate on the subjective evaluation of algorithms for speech enhancement. A variety of listening tests have been published, each of them optimized for a special purpose. We will focus here only on two different listening tests:

- comparison mean opinion scores (CMOS), and
- rhyme tests (diagnostic and modified rhyme tests).

For these two kinds of subjective tests standards have been published by the International Telecommunication Union (ITU) [24] and by the American National Standards Association (ANSI) [1]. We will apply CMOS tests for evaluating the speech quality and rhyme tests for investigating the speech intelligibility. For this reason, we will use common phrases such as popular song refrains or well-known proverbs as audio examples for CMOS tests. In this case it is sufficient to understand only a part of the utterance to get the meaning of it and one can concentrate on the quality of the presented sounds. On the other hand we will use word groups which differ only in one vowel or consonant, such as *meat*, *need*, *feed*, and *heat*, when evaluating the speech intelligibility with diagnostic or modified rhyme tests.

The subjective tests are described in the second part of this chapter. We will utilize rather new applications such as bandwidth extension and in-car communication systems as examples and show how subjective tests are designed, applied, and analyzed.

11.2 Objective Tests for Noise Suppression

In most communication applications the recording of a speech signal takes place in a noisy environment. In case of hands-free telephony, e.g., the local speech signal $s(n)$ is corrupted by background noise $b(n)$ and echo components $d(n)$ (see Fig. 11.1). The level of the background noise depends on the area of application. While a moderate noise level can be assumed in quiet offices, a signal-to-noise ratio (SNR) up to 0 dB might be expected if a phone call is made out of a car, a train, or an airplane.

The aim of noise suppression systems is to reduce the distorting background noise component while keeping the local speech signal as natural as

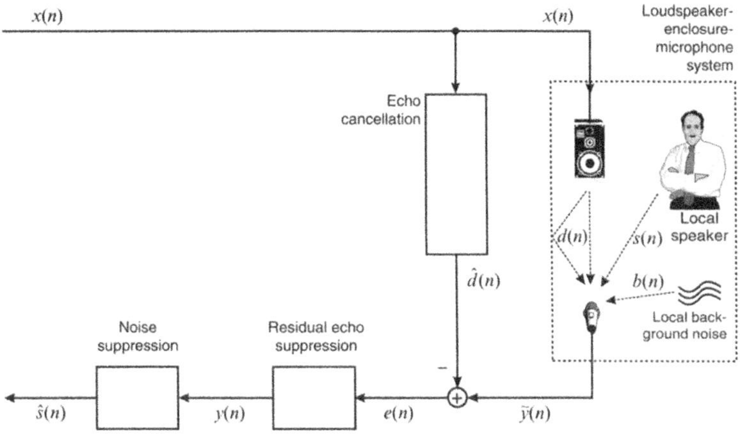

Fig. 11.1. Basic structure of the processing units within a hands-free telephone.

possible. Thus, the desired and the distorting components of the microphone signal $\tilde{y}(n)$ need to be separated. To achieve this, the input signal is split into overlapping blocks of appropriate size (e.g. 32 ms). Within such a block all signals are assumed to be stationary. Each block is transformed via a filter bank or a DFT into the frequency domain. In order to remove the distorting echo and noise components each subband or frequency bin is weighted with an attenuation factor $G(e^{j\Omega_\mu}, n)$, that depends on the current signal-to-noise ratio. Additionally, postprocessing such as pitch-adaptive filtering [43] or automatic gain control can be applied. The resulting representation of the enhanced signal spectrum is transformed back into the time domain (see Fig. 11.2). This

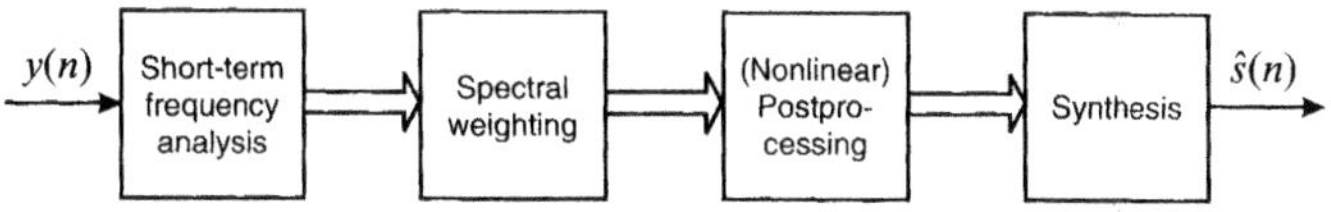

Fig. 11.2. Basic units of a noise suppression system.

basic principle is common to most systems for noise suppression. A detailed description of noise suppression systems can be found in Chapters 9 and 10 of this book.

Since the spectral power subtraction rule is the most straight forward and easiest noise reduction method and it is well-known, it is used in this chapter as a reference for speech enhancement. The transfer function of the spectral power subtraction rule follows:

$$G\left(e^{j\Omega}, n\right) = \max\left\{\beta, 1 - \alpha\frac{\widehat{S}_{bb}(\Omega, n)}{\widehat{S}_{yy}(\Omega, n)}\right\}, \tag{11.1}$$

where $\widehat{S}_{yy}(\Omega, n)$ and $\widehat{S}_{bb}(\Omega, n)$ are the estimated short-term power spectral densities of the distorted input signal and the background noise, respectively. The parameters α and β denote an overestimation factor for the power spectrum of the background noise and a limitation of the attenuation, called *spectral floor*. The algorithm shows a well-known behavior concerning these two parameters which will be evaluated later. The typical performance of a spectral subtraction rule is depicted in Figs. 11.3 (progression of the input power and the estimated noise power) and 11.4 (time-frequency analyses of the input and output signals).

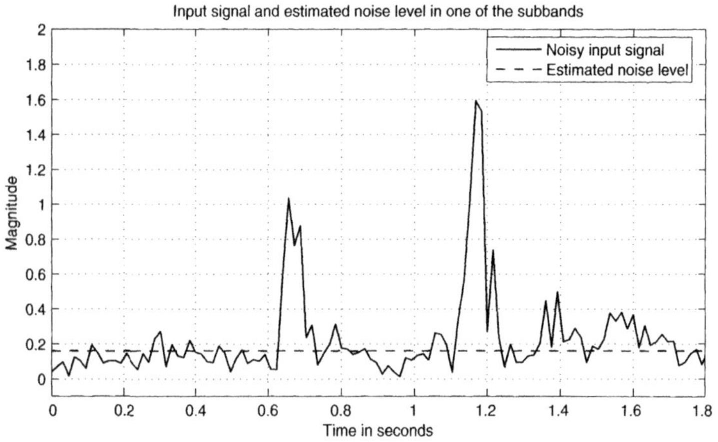

Fig. 11.3. Spectral subtraction for one spectral (subband) magnitude with a steady background noise level, indicated with the straight line.

It can be seen that without the introduction of a limitation of the transfer function towards zero or another small value (parameter β) the result of a spectral subtraction could be negative and that without an overestimation of the background noise level (parameter α) the spectral estimation leaves noise parts in the remaining signal which cause unnatural sounding tonal distortions. In a time-frequency plot the effect of a spectral subtraction rule on heavily distorted speech is clearly visible (see Fig. 11.4). Here the clean speech signal, the distorted input signal, and the output signal after spectral subtraction can be compared. A clear attenuation of the background noise is visible even though the quality of the noise reduction algorithm depicted here does not show satisfying results due to the tonal distortions. The speech signal, however, shows a good quality.

11.2.1 Measuring the Quality of Noise Suppression Systems

As far as a quality measure for noise reduction algorithms is concerned, very often only the attenuation of the background noise is taken as a measure.

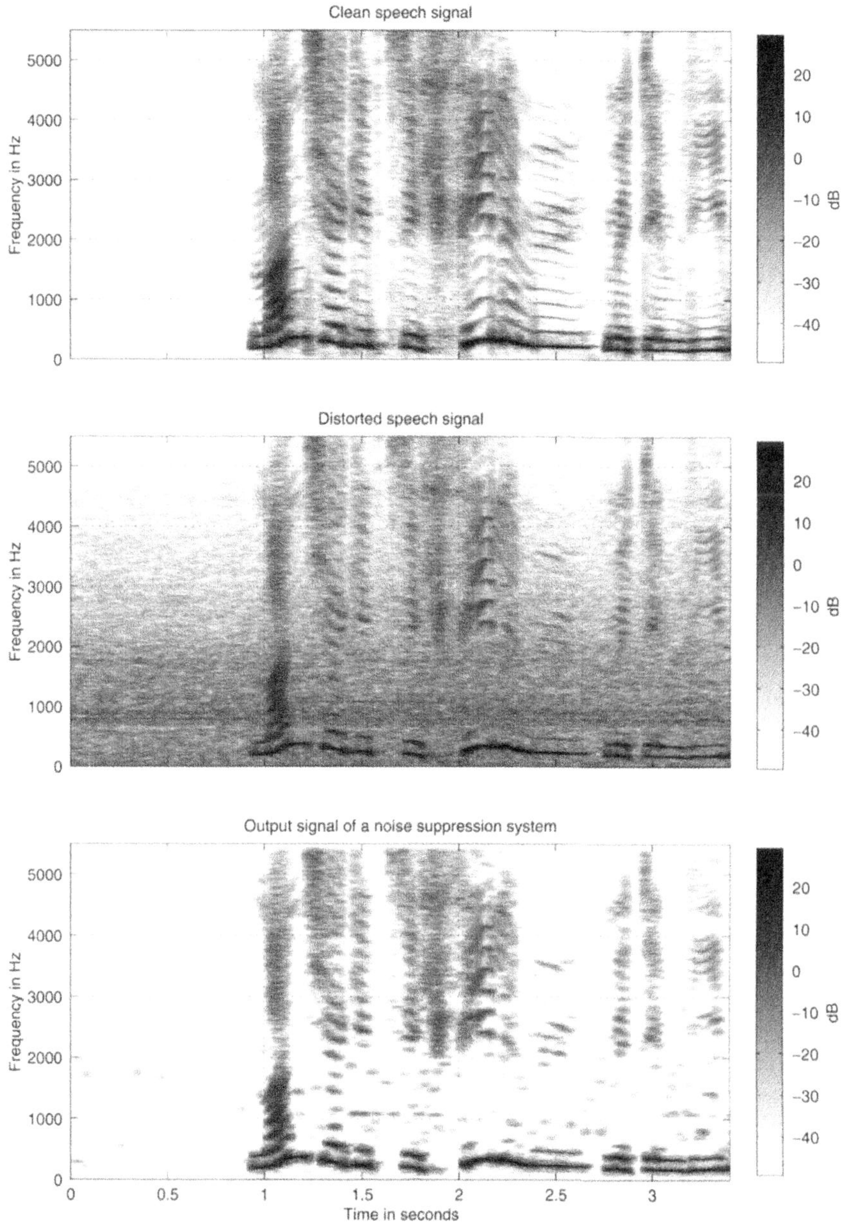

Fig. 11.4. Sample time-frequency analyses of a clean speech signal (top), of a distorted speech signal (center), and of an output signal after noise reduction (bottom). Tonal distortions are clearly visible as dots in the time-frequency domain within the lowest diagram (noise reduction was performed without spectral floor).

Sometimes two properties of the signal are under investigation: the degradation of the speech signal and the attenuation of the background noise [3]. However, this appears not to be sufficient. We propose three different signal properties instead:

- variations in the pure speech signal (negative),
- variations in the noise characteristics (negative), and
- attenuation of the background noise (positive).[4]

All of these classes can be examined by a large number of distance measures. Examples for the respective measures will be given later in this chapter. It seems to be obvious that different noise reduction systems affect the symptoms in different ways leading to a different performance. We want to find out which of the symptoms above are relevant for a general quality measure.

After extensive listening tests for a subjective quality measure (thirty test persons had to listen to sixty sequences each), an opinion poll about speech enhancement systems was carried out. The listeners were asked about their impression of what was most important for a speech enhancement system. The poll results are depicted in Fig. 11.5. It may be a bit astonishing that

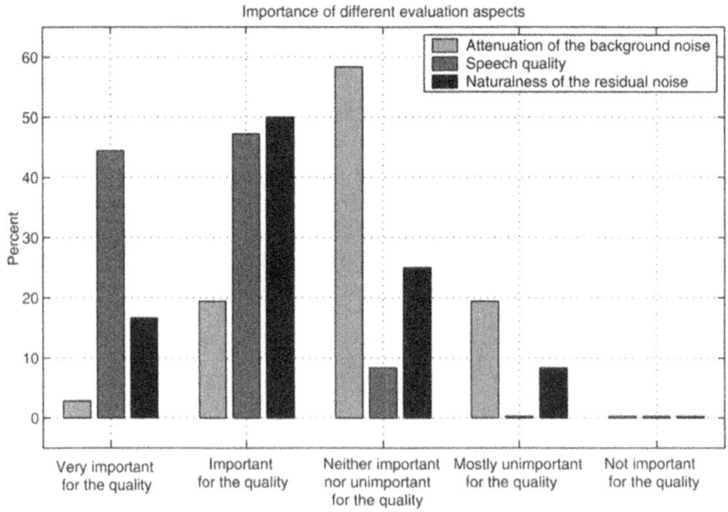

Fig. 11.5. Results of an opinion poll after an extensive listening test. Speech quality appears to be the most crucial property of speech quality concerning noise suppression algorithms.

the actual noise attenuation is the least crucial point of the noise reduction

[4] The terms *positive* or *negative* should indicate whether a better or a worse signal quality is expected when increasing the amount of variation or attenuation, respectively.

system. Speech degradation or an unnatural characteristic of the remaining noise is far more important to the overall judgement. Because of the outcome of this opinion poll the attempt of measuring the different effects separately becomes more and more promising.

11.2.2 Distance Measures

As we learnt from Fig. 11.5, the speech quality of the output signal is the most crucial part of a noise reduction system. Therefore we start with investigating the speech quality separately. Speech quality degradation again can occur in different ways. While phase distortions are usually neglected, nonlinear distortions or attenuation of parts of the speech signal change the speech quality significantly. Typical measures for speech quality, known from speech coding, were used for the speech quality evaluation [40].

Before details on objective measurements are presented in the next few sections we will introduce the notation for a few artificially generated signals that are required for the following investigations. In a real noise suppression system one can usually monitor only the microphone signal[5]

$$y(n) = s(n) + b(n) \tag{11.2}$$

(but not the input speech signal $s(n)$ or the background noise $b(n)$) and the output signal of the system $\hat{s}(n)$. During the development stage of a noise suppression system, however, one has access to all internal parameters as well as to individual components of the microphone signal. This allows to split the output signal $\hat{s}(n)$ into speech $\tilde{s}(n)$ and noise $\tilde{b}(n)$ components:

$$\hat{s}(n) = \tilde{s}(n) + \tilde{b}(n)\,. \tag{11.3}$$

This can be achieved by applying the spectral weights (see Eq. 11.1) that are computed with the noisy input signal $y(n)$ separately to the noise components $b(n)$ and to the speech components $s(n)$ (see Fig. 11.6).

11.2.2.1 Cepstral Distance

The cepstral distance is based on the logarithmic separation of sinusoids in the frequency range due to the pitch frequency and its harmonics from the spectral envelope of the speech signal. The cepstrum of a speech signal $s(n)$ is defined as

$$c_i = \frac{1}{2\pi} \int\limits_{\Omega=-\pi}^{\pi} \log\left|S\left(e^{j\Omega}\right)\right| e^{j\Omega i}\, d\Omega. \tag{11.4}$$

[5] For the sake of simplicity we assume here and in the following that we have no echo components. Thus, we can omit any echo suppression device. In this case the signal $y(n)$ is equal to the microphone signal $\tilde{y}(n)$ (see Fig. 11.1).

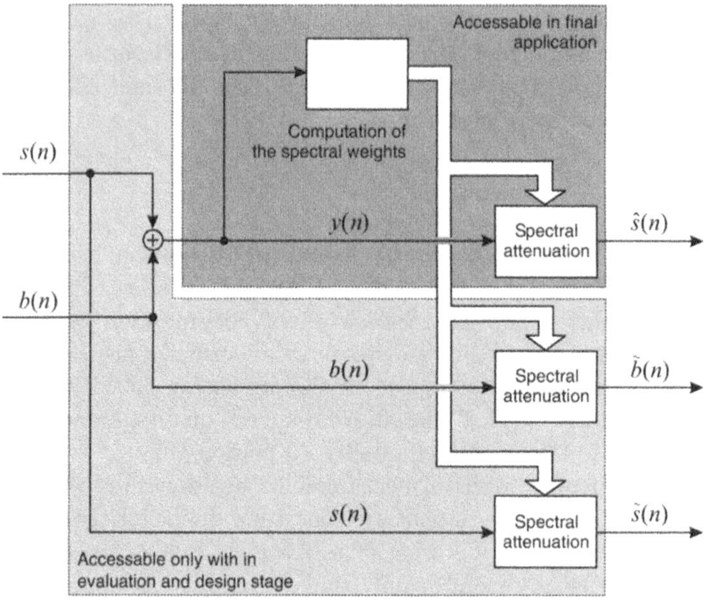

Fig. 11.6. Generation of the output signal of a noise suppression system $\hat{s}(n)$ as well as the individual signal components: output noise $\tilde{b}(n)$ and output speech $\tilde{s}(n)$.

The comparison between two cepstra is a widely used tool for the investigation of speech signals. For a short frame of a speech signal and its estimated counterpart the cepstral distance is computed according to

$$d_{\text{cep}}(n) = \frac{\sum\limits_{i=1}^{C} \Big[c_{s,i}(n) - c_{\hat{s},i}(n)\Big]^2}{\sum\limits_{i=1}^{C} c_{s,i}^2(n)} . \tag{11.5}$$

The cepstral coefficients $c_{s,i}(n)$ and $c_{\hat{s},i}(n)$ can be also obtained from predictor coefficients. In general, a finite number of predictor coefficients is transformed into an infinite number of cepstral parameters. However, the resulting cepstral series has only very limited energy at coefficients with large indices. Thus, the series is often truncated after computing $C = 1.5\,P$ coefficients, whereas P is denoting the predictor order. A computationally effective, order recursive method for transforming predictor coefficients $a_i(n)$ into cepstral coefficients $c_i(n)$ is given by[6]

[6] Note that Eqs. 11.6 and 11.7 are denoted for the reason of brevity only for the speech components within the microphone signal $s(n)$. The predictor coefficients

$$c_{s,i}(n) = \begin{cases} a_{s,i}(n) + \frac{1}{i} \sum\limits_{k=1}^{i-1} k\, c_{s,k}(n)\, a_{s,i-k}(n), & \text{for } i = 1 \dots P, \\ \frac{1}{i} \cdot \sum\limits_{k=1}^{i-1} k\, c_{s,k}(n)\, a_{s,i-k}(n), & \text{else.} \end{cases} \tag{11.6}$$

The predictor coefficients $a_i(n)$ can be computed by solving the so-called *Yule-Walker* equation system [17]

$$\underbrace{\begin{bmatrix} r_{ss,0}(n) & r_{ss,1}(n) & \dots & r_{ss,P-1}(n) \\ r_{ss,1}(n) & r_{ss,0}(n) & \dots & r_{ss,P-2}(n) \\ \vdots & \vdots & \ddots & \vdots \\ r_{ss,P-1}(n) & r_{ss,P-2}(n) & \dots & r_{ss,0}(0) \end{bmatrix}}_{\boldsymbol{R}_{ss}(n)} \underbrace{\begin{bmatrix} a_1(n) \\ a_2(n) \\ \vdots \\ a_P(n) \end{bmatrix}}_{\boldsymbol{a}_s(n)} = \underbrace{\begin{bmatrix} r_{ss,1}(n) \\ r_{ss,2}(n) \\ \vdots \\ r_{ss,P}(n) \end{bmatrix}}_{\boldsymbol{r}_{ss}(n)} . \tag{11.7}$$

The coefficients $r_{ss,i}(n)$ represent the short-term autocorrelation at lag i estimated around the time index n. Due to the special character of the matrix $\boldsymbol{R}_{ss}(n)$ Eq. 11.7 can be solved in an order-recursive manner by using, e.g., the *Levinson-Durbin* recursion [10, 32]. Since speech can be assumed to have stationary character only for short periods of time, the parameters of the model need to be estimated periodically every 5 to 10 ms. However, by utilizing about $K = 10 \dots 20$ coefficients $a_i(n)$ (resulting in about $C = 15 \dots 30$ cepstral coefficients) one is able to estimate the spectral envelope of a speech signal in a reliable manner.

Finally, all distances resulting from each frame are averaged

$$\overline{d}_{\text{cep}} = \frac{1}{N_{\text{F}}} \sum_{n=0}^{N_{\text{F}}-1} d_{\text{cep}}(n\,L)\,, \tag{11.8}$$

with L being the frameshift in samples between two adjacent frames. The parameter N_{F} denotes the total number of frames within the analysis.

The objective quality measures were tested with a reference noise reduction algorithm where well-known parameter modifications were performed. It is known that increasing the overestimation factor lowers the amount of musical noise, leads to a better noise attenuation, but also distorts the speech signal [6]. Increasing the input signal-to-noise ratio also delivers a known behavior: the speech signal becomes less distorted and also the background noise characteristic becomes more natural. In the lower plot of Fig. 11.7 we see that the speech quality is related to the overestimation factor. The dashed line shows the cepstral distance between the original speech signal and the estimated speech signal at the output of the system. The two signals are most similar for an overestimation factor $\alpha = 1.25$ (see Eq. 11.1). If only the changes in the speech signal are of interest, the overestimation factor α should be as

---

$a_{\hat{s},i}(n)$ and the cepstral coefficients $c_{\hat{s},i}(n)$ for the output signal of the noise suppression system $\hat{s}(n)$ are computed accordingly.

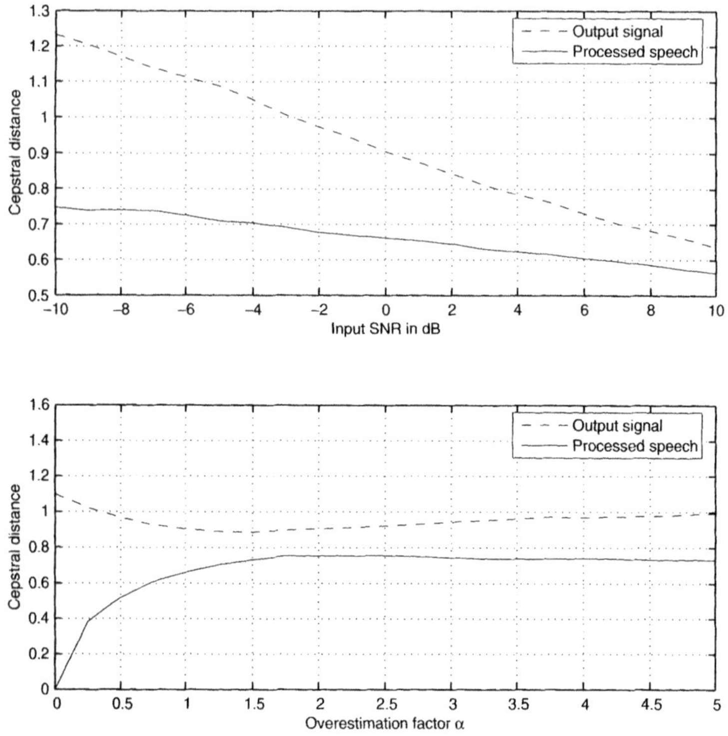

Fig. 11.7. Cepstral distance versus overestimation factor or input signal-to-noise ratio. The solid line compares the clean speech signal $s(n)$ with the processed speech signal $\tilde{s}(n)$. The dashed line compares the clean speech signal $s(n)$ to the output including the remaining noise $\hat{s}(n)$. For the definitions of the signals $s(n)$, $\tilde{s}(n)$, and $\hat{s}(n)$ see Fig. 11.6.

small as necessary since this means little processing. As expected we see in the upper plot of Fig. 11.7 that the distance between the clean speech signal and the processed speech or the output signal becomes smaller for a better input SNR.

11.2.2.2 Itakura Measure

Another distance measure for speech signals is the so-called *Itakura* measure which is also a measure looking for differences in the spectral envelope by linear prediction:

$$d_{\mathrm{Ita}}(n) = \ln\left(\frac{E_{\hat{s}}(n)}{E_s(n)}\right) = \ln\left(\frac{\boldsymbol{a}_{\hat{s}}^{\mathrm{T}}(n)\boldsymbol{R}_{ss}(n)\boldsymbol{a}_{\hat{s}}(n)}{\boldsymbol{a}_s^{\mathrm{T}}(n)\boldsymbol{R}_{ss}(n)\boldsymbol{a}_s(n)}\right). \tag{11.9}$$

The quantities $\boldsymbol{a}_s(n)$ and $\boldsymbol{a}_{\hat{s}}(n)$ are denoting the coefficients of a predictor error filter trained with the two signals to be compared:

$$\boldsymbol{a}_s(n) = \boldsymbol{R}_{ss}^{-1}(n)\, \boldsymbol{r}_{ss}(n)\,, \tag{11.10}$$

$$\boldsymbol{a}_{\hat{s}}(n) = \boldsymbol{R}_{\hat{s}\hat{s}}^{-1}(n)\, \boldsymbol{r}_{\hat{s}\hat{s}}(n)\,. \tag{11.11}$$

Both predictors, however, are excited with the same input signal. Thus, $E_{\hat{s}}(n)$ is the output power of a predictor that is excited with the clean speech signal $s(n)$ but the coefficients are adjusted according to the short-term autocorrelation of the output signal of the noise suppression system $\hat{s}(n)$. Fig. 11.8 shows the signal flow graph for the computation of the distance measure according to Itakura.

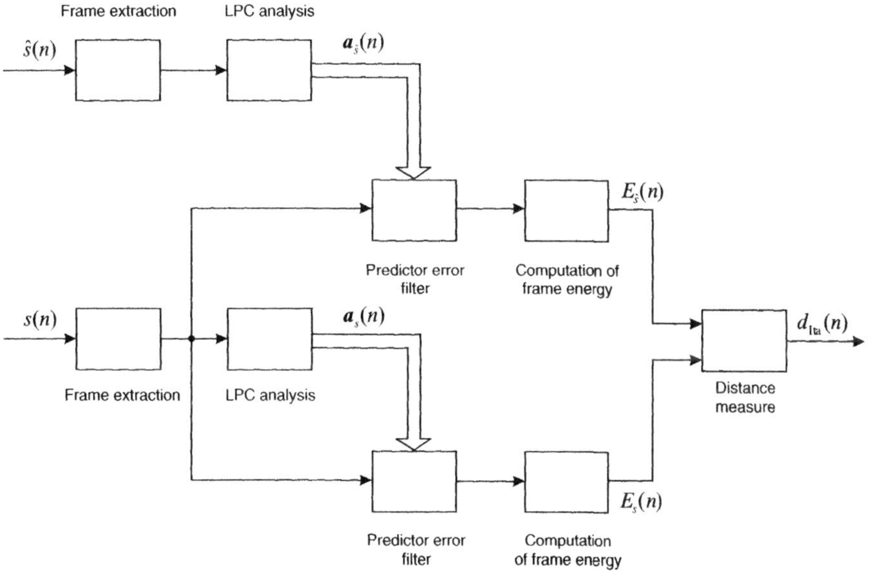

Fig. 11.8. Computation of the distance measure according to Itakura.

It is important to note that the output of the Itakura measure is not symmetric because there is a difference if the linear predictors are used with the clean signal or with the output signal.

As in Sec. 11.2.2.1 all distances $d_{\rm Ita}(n)$ that are computed for individual frames are averaged over the entire test sequence:

$$\bar{d}_{\rm Ita} = \frac{1}{N_{\rm F}} \sum_{n=0}^{N_{\rm F}-1} d_{\rm Ita}(n\,L)\,. \tag{11.12}$$

Again, the parameter L describes the frameshift in samples between two adjacent frames and $N_{\rm F}$ denotes the total number of frames within the analysis.

It can be seen in Fig. 11.9 that the outcome of the Itakura measure over the overestimation factor is similar to that of the cepstral distance. For a change in the input SNR, however, the Itakura measure sees hardly any differences in the speech signals.

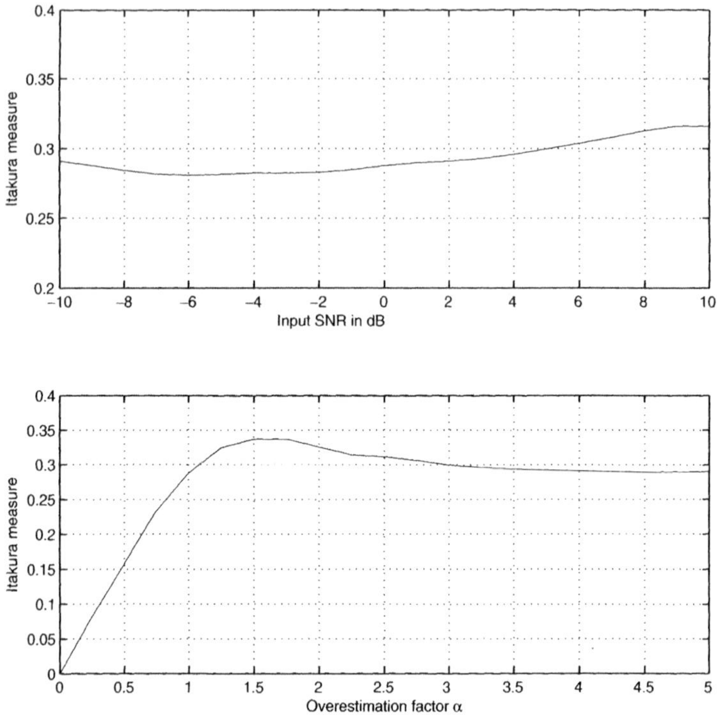

Fig. 11.9. Itakura measure versus overestimation factor or input signal to noise ratio. While for the change of the input SNR almost no change in the Itakura measure appears, the signal differences over the overestimation factor are clearly visible.

11.2.2.3 Itakura-Saito Measure

A similar measure is the *Itakura-Saito* measure which uses the difference between the two vectors of prediction coefficients instead of the single vector in the numerator leading to the following measure rule

$$d_{\mathrm{IS}}(n) = \ln\left(\frac{\Big[\boldsymbol{a}_s(n) - \boldsymbol{a}_{\hat{s}}(n)\Big]^{\mathrm{T}} \boldsymbol{R}_{ss}(n) \Big[\boldsymbol{a}_s(n) - \boldsymbol{a}_{\hat{s}}(n)\Big]}{\boldsymbol{a}_s^{\mathrm{T}}(n)\, \boldsymbol{R}_{ss}(n)\, \boldsymbol{a}_s(n)}\right), \tag{11.13}$$

and its time-averaged version

$$\bar{d}_{\mathrm{IS}} = \frac{1}{N_{\mathrm{F}}} \sum_{n=0}^{N_{\mathrm{F}}-1} d_{\mathrm{IS}}(n\,L), \tag{11.14}$$

where $\boldsymbol{a}_s(n)$ and $\boldsymbol{a}_{\hat{s}}(n)$ are again the coefficients of a predictor error filter trained with the two signals to be compared. More speech quality measures are also discussed in [15, 36, 45].

11.2.3 Noise Characteristics

Early listening tests and also various publications of speech enhancement algorithms point out a typical deficiency of noise reduction or speech enhancement algorithms. They tend to change the characteristics of the background noise. If human listeners are asked for their preferences they want the remaining noise to sound natural. For a hands-free telephone installed in a car, the remaining noise should sound like car noise.

11.2.3.1 Noise Attenuation

The actual goal of a noise reduction algorithm is the attenuation of the background noise without attenuation of the speech signal. A simple measure for the performance of a speech enhancement system is therefore the average attenuation of the background noise or in other words the enhancement of the signal-to-noise ratio which can be defined by the following equation:

$$\overline{d}_{\text{att}} = \frac{\overline{P}_{\tilde{b}}}{\overline{P}_{b}} . \tag{11.15}$$

Both the noise power before ($\overline{P}_b$) and after processing ($\overline{P}_{\tilde{b}}$) are averaged over time:

$$\overline{P}_b = \frac{1}{N_{\mathrm{S}}} \sum_{n=0}^{N_{\mathrm{S}}-1} b^2(n) , \tag{11.16}$$

$$\overline{P}_{\tilde{b}} = \frac{1}{N_{\mathrm{S}}} \sum_{n=0}^{N_{\mathrm{S}}-1} \tilde{b}^2(n) , \tag{11.17}$$

with N_{S} denoting the length of the sequence that is tested.

When applying this measure one should be aware that non-stationary noise components usually are misinterpreted by most noise suppression systems as desired signal components. This leads to small attenuation values of the noise suppression characteristics. If such a noise component has also more power than the residual noise the non-stationary noise burst will dominate within the sums of Eqs. 11.16 and 11.17. As a result the measure $\overline{d}_{\text{att}}$ will show values close to one, no matter how much noise attenuation is achieved in all other situations. For this reason, noise bursts should be excluded before applying this measure.

11.2.3.2 Musical Noise

Especially the tonal parts in the remaining noise disqualify a noise reduction system. The measure denoted in the following equations gives a hint about the tonal distortions present in the outgoing signal. Since tonal distortions are visible as short-term variations in the periodogram, we compare the short-term spectrum to a model-based spectrum estimation

$$d_{\text{tonal}}(n) = \frac{\int\limits_{\Omega=0}^{2\pi} \left| \widehat{S}_{\tilde{b}\tilde{b},\,\text{LPC}}(\Omega,n) - \left|\tilde{B}\left(e^{j\Omega},n\right)\right|^2 \right| d\Omega}{\int\limits_{\Omega=0}^{2\pi} \left| \widehat{S}_{bb,\,\text{LPC}}(\Omega,n) - \left|B\left(e^{j\Omega},n\right)\right|^2 \right| d\Omega}. \tag{11.18}$$

The integration in Eq. 11.18 is usually approximated by a sum over discrete frequency supporting points. For the model based spectrum estimations $S_{\tilde{b}\tilde{b},\,\text{LPC}}(\Omega,n)$ and $S_{bb,\,\text{LPC}}(\Omega,n)$ linear prediction based models of order 10 to 20 are usually applied. The quatities $\tilde{B}(e^{j\Omega},n)$ and $B(e^{j\Omega},n)$ represent the current short-term spectra of the residual noise $\tilde{b}(n)$ and the original noise $b(n)$. As in the computation of the other quality measures $d_{\text{tonal}}(n)$ is averaged on a frame by frame basis over time:

$$\overline{d}_{\text{tonal}} = \frac{1}{N_{\text{F}}} \sum_{n=0}^{N_{\text{F}}-1} d_{\text{tonal}}(n\,L)\,. \tag{11.19}$$

The performance follows the expected behavior. For an increasing overestimation factor (lower plot Fig. 11.10) and for an increasing input SNR (upper plot Fig. 11.10) the outcome of the distortion measure decreases.

Note that musical noise is one of the most annoying artifacts within noise-only periods for a human listener. A lot of the speech recognition systems are, however, quite insensitive to this kind of distortion.

11.2.3.3 Difference in Power Level

Not really surprisingly, the difference in the noise power compared in sequences with or without speech activity also gives a hint of the noise reduction quality. A high power level in speech sequences and at the same time a very low power level in speech pauses leads to a low subjective mark.[7] Therefore we evaluate the signal with the following rule:

$$\overline{d}_{\text{pow}} = \frac{\left|\overline{P}_{\tilde{b},\text{speech}} - \overline{P}_{\tilde{b}}\right|}{\overline{P}_{\tilde{b}}} + \frac{\left|\overline{P}_{\tilde{b},\text{pause}} - \overline{P}_{\tilde{b}}\right|}{\overline{P}_{\tilde{b}}}, \tag{11.20}$$

with $\overline{P}_{\tilde{b},\text{pause}}$ and $\overline{P}_{\tilde{b},\text{speech}}$ being the current values of background noise in pauses

[7] The term *low subjective mark* should indicate a bad signal quality. A detailed mapping between marks and quality descriptions can be found in Tab. 11.3.

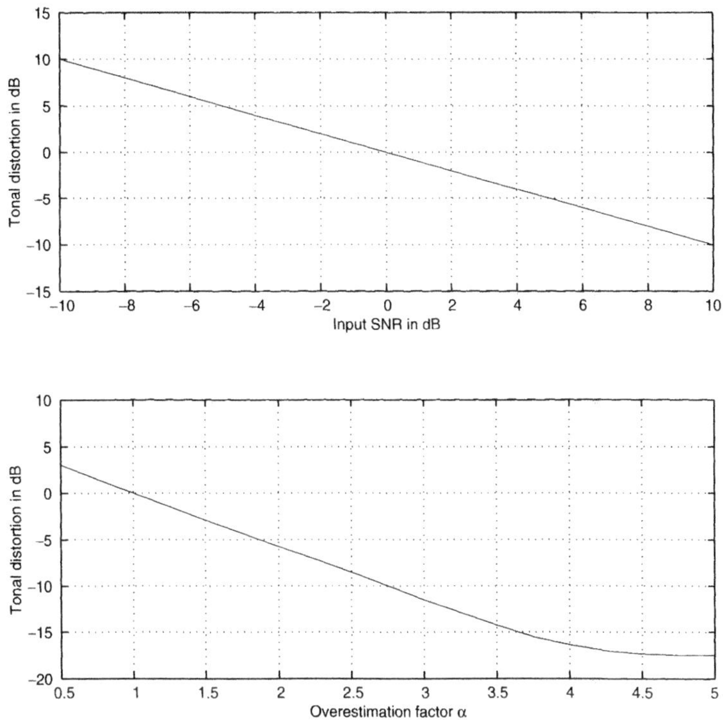

Fig. 11.10. Simulation results for the tonal distortion measure with input signals of well-known behavior. Note that for the change in the overestimation factor the amount of input tonal distortions is equal for all overestimation factors.

$$\overline{P}_{\tilde{b},\text{pause}} = \frac{1}{N_{\text{S,pause}}} \sum_{\text{Pauses}} \tilde{b}^2(n) \tag{11.21}$$

or speech activity,

$$\overline{P}_{\tilde{b},\text{speech}} = \frac{1}{N_{\text{S,speech}}} \sum_{\text{Speech}} \tilde{b}^2(n) \tag{11.22}$$

respectively and $\overline{P}_{\tilde{b}}$ the average value of the residual background noise over all samples (see Eq. 11.17). The parameters $N_{\text{S,pause}}$ and $N_{\text{S,speech}}$ are denoting the amount of samples within pauses and during speech activity, respectively. It is recommended to label the periods of speech and pause manually, especially for low SNR conditions.

11.2.4 Psycho-Acoustic Methods

There are many other quality measures available. A very common approach is to evaluate the psycho-acoustical masking properties of the human hearing system, which is also tested here for improving the signal-to-noise ratio enhancement. Only audible parts of the speech and noise signals are compared [46]. Typically psychoacoustic methods outperform simple signal-to-noise ratio enhancement measures. But since the discussion of psychoacoustic methods opens a completely new field, we restrict ourselves to the ideas above.

11.2.5 Coherence Between Instrumental Measures and Listening Tests

If the coherence between the subjective and the objective quality measures is required, usually the correlation coefficient is used. Since objective quality measures are typically not on the same scale as subjective measures, usually a nonlinear fitting is performed beforehand. However, the amount of training data is usually not very large, due to the enormous effort that listening tests require. This causes the results to be heavily related with the fitting curves. One can avoid this problem by using the so-called *rank correlation.*

11.2.5.1 Rank Correlation

For the rank correlation [19, 30] the data under investigation (subjective as well as objective) is put in a rising order. Assuming that we have made four simulations leading to four different output samples, the quality of which we have compared both subjectively and objectively getting the results $V_{\Sigma,1}$ to $V_{\Sigma,4}$ for the subjective and $\overline{d}_{t,1}$ to $\overline{d}_{t,4}$ for the objective evaluation.[8] Writing both results in increasing order leads to

$$\begin{aligned} & \overline{d}_{t,3} < \overline{d}_{t,2} < \overline{d}_{t,1} < \overline{d}_{t,4} \quad \text{(objective)} \\ \text{and} \quad & V_{\Sigma,1} < V_{\Sigma,2} < V_{\Sigma,4} < V_{\Sigma,3} \quad \text{(subjective)} \end{aligned}$$

and we get the following ranks:

$$\begin{aligned} \mathcal{R}(\overline{d}_{t,3}) = \mathcal{R}(V_{\Sigma,1}) = 1, \quad \mathcal{R}(\overline{d}_{t,2}) = \mathcal{R}(V_{\Sigma,2}) = 2, \\ \mathcal{R}(\overline{d}_{t,1}) = \mathcal{R}(V_{\Sigma,4}) = 3, \quad \mathcal{R}(\overline{d}_{t,4}) = \mathcal{R}(V_{\Sigma,3}) = 4, \end{aligned}$$

where $\mathcal{R}(V_{\Sigma,n})$ denotes the rank of a subjective quality measure for algorithm n, $\mathcal{R}(\overline{d}_{t,n})$ that of an objective quality measure, respectively. Analogous to the correlation coefficient, we calculate the Spearman rank correlation coefficient ρ_t:

$$\rho_t = \frac{\sum\limits_{n=1}^{N} \left[\mathcal{R}(V_{\Sigma,n}) - \overline{\mathcal{R}}(V_\Sigma)\right]\left[\mathcal{R}(\overline{d}_{t,n}) - \overline{\mathcal{R}}(\overline{d}_t)\right]}{\sqrt{\sum\limits_{n=1}^{N} \left[\mathcal{R}(V_{\Sigma,n}) - \overline{\mathcal{R}}(V_\Sigma)\right]^2 \sum\limits_{n=1}^{N} \left[\mathcal{R}(\overline{d}_{t,n}) - \overline{\mathcal{R}}(\overline{d}_t)\right]^2}}, \tag{11.23}$$

[8] The quantities $\overline{d}_{t,n}$ are abbreviating one of the objective distance measures presented before. Thus, the variable t could be one of the subscripts *cep*, *Ita*, *IS*, *att*, *tonal*, or *P*.

where $\overline{\mathcal{R}}(\bar{d}_t)$ and $\overline{\mathcal{R}}(V_\Sigma)$ denote the respective average rank and have the same value, namely $(N+1)/2$. If the two input sequences are both strictly increasing or both decreasing the rank correlation delivers one as output. The Spearman correlation coefficient can therefore be used like its traditional counterpart.

11.2.5.2 Judging Quotient

In order to compute rank correlation coefficients according to Eq. 11.23 subjective tests have to be performed. Therefore, a listening test with 35 persons was carried out. The test persons were asked to mark the speech signal and the background noise separately, and to say if they have the impression that the speech enhancement is a general improvement. Various speech enhancement schemes were used as test sequences. For the speech and noise signal, respectively marks[9] between one and five had to be given (very poor to very good) and for the general opinion a mark between minus two and plus two. A smaller value always represents a lower quality. We will not describe the details about the subjective tests right now[10] since we will focus on the question about the quality of the objective measures in this and in the next section. For details about the boundary conditions of the noise suppression tests the interested reader is referred to [8].

In contrast to the widely used mean opinion scores, we do not average directly the outcomings of an opinion poll. To achieve a scaled outcome of the listening tests we use a judgement quotient similar to quantiles as known in statistics:

$$Q_i = \frac{\text{No. of judgements} \geq \text{level } i}{\text{Total number of judgements}} \,. \tag{11.24}$$

This gives the highest results to the best sequences. However, the question arises which of the possible quotients gives the best result. We average over all levels for the evaluation:

$$Q_\Sigma = \frac{1}{N_{\text{max}} - N_{\text{min}}} \sum_{i=N_{\text{min}}+1}^{N_{\text{max}}} Q_i \,. \tag{11.25}$$

The summation starts at $i = N_{\text{min}} + 1$ because for $i = N_{\text{min}}$ we receive $Q(N_{\text{min}}) = 1$ which only leads to a global offset. Due to the normalization we ensure that the average judgement quotient stays within the interval

$$0 \leq Q_\Sigma \leq 1 \,. \tag{11.26}$$

[9] In the following the terms *mark* and *level* are used in an equivalent manner. The lowest possible level is denoted by N_{min}, the highest one by N_{max}.

[10] Two types of subjective tests will be introduced in detail in Sec. 11.3 (CMOS) and Sec. 11.4 (DRT), respectively.

If all listeners vote in all tests with the worst judgement all judgement quotients will be $Q_i = 0$, except for the first quotient which will be $Q_{N_{\min}} = 1$. As a result the average judgement quotient will be

$$Q_\Sigma \Big|_{\text{All listeners vote with most negative judgement.}} = 0\,. \tag{11.27}$$

On the other hand, if all listeners give in all tests best marks all judgement quotients will be $Q_i = 1$. The average judgement quotient in this case will be

$$Q_\Sigma \Big|_{\text{All listeners vote with most positive judgement.}} = 1\,. \tag{11.28}$$

11.2.5.3 Results

Tab. 11.1 shows the rank correlation between the distance measures for the speech signal and the subjective quality of the speech signal. If the quality of the speech signal alone is required, the cepstral distance shows the best correlation.[11] However, all distance measures show only a small rank correlation with the subjective evaluation. By applying a linear combination of all presented distance measures a rank correlation coefficient of about $\rho_{\text{combined}}|_{Q_\Sigma} = 0.81$ can be achieved [9].

Table 11.1. Rank correlation coefficients of subjective judgement and speech quality measures.

Judging Quotients	Rank correlation coefficients		
	Cepstral dist.	Itakura	Itakura-Saito
Q_2	$\rho_{\text{ceps}}\big\|_{Q_2} = 0.61$	$\rho_{\text{Ita}}\big\|_{Q_2} = 0.50$	$\rho_{\text{IS}}\big\|_{Q_2} = 0.50$
Q_3	$\rho_{\text{ceps}}\big\|_{Q_3} = 0.62$	$\rho_{\text{Ita}}\big\|_{Q_3} = 0.47$	$\rho_{\text{IS}}\big\|_{Q_3} = 0.47$
Q_4	$\rho_{\text{ceps}}\big\|_{Q_4} = 0.60$	$\rho_{\text{Ita}}\big\|_{Q_4} = 0.47$	$\rho_{\text{IS}}\big\|_{Q_4} = 0.47$
Q_5	$\rho_{\text{ceps}}\big\|_{Q_5} = 0.60$	$\rho_{\text{Ita}}\big\|_{Q_5} = 0.51$	$\rho_{\text{IS}}\big\|_{Q_5} = 0.52$
Q_Σ	$\rho_{\text{ceps}}\big\|_{Q_\Sigma} = 0.62$	$\rho_{\text{Ita}}\big\|_{Q_\Sigma} = 0.48$	$\rho_{\text{IS}}\big\|_{Q_\Sigma} = 0.48$

For the analysis of the change in the noise characteristics, we also compare the objective measures for the background noise with the results of the

[11] Note that negative values for the rank correlation coefficients would be obtained for most distance measures since on one hand larger subjective ranks indicate better quality but smaller distance measures on the other hand. For this reason the rank correlations were computed using the negative distance measures.

listening test (see Tab. 11.2). The measure for the tonal distortions delivers the highest agreement with the human listeners, while the actual noise attenuation is not suitable for a prediction of the noise quality.

Table 11.2. Rank correlation coefficients of subjective judgement and noise quality measures.

Judging Quotients	Rank correlation coefficients		
	Noise att.	Tonal noise	Power diff.
Q_2	$\rho_{\text{att}}\big\|_{Q_2} = 0.20$	$\rho_{\text{tonal}}\big\|_{Q_2} = 0.40$	$\rho_{\text{pow}}\big\|_{Q_2} = 0.49$
Q_3	$\rho_{\text{att}}\big\|_{Q_3} = 0.34$	$\rho_{\text{tonal}}\big\|_{Q_3} = 0.67$	$\rho_{\text{pow}}\big\|_{Q_3} = 0.60$
Q_4	$\rho_{\text{att}}\big\|_{Q_4} = 0.36$	$\rho_{\text{tonal}}\big\|_{Q_4} = 0.66$	$\rho_{\text{pow}}\big\|_{Q_4} = 0.64$
Q_5	$\rho_{\text{att}}\big\|_{Q_5} = 0.32$	$\rho_{\text{tonal}}\big\|_{Q_5} = 0.45$	$\rho_{\text{pow}}\big\|_{Q_5} = 0.53$
Q_Σ	$\rho_{\text{att}}\big\|_{Q_\Sigma} = 0.35$	$\rho_{\text{tonal}}\big\|_{Q_\Sigma} = 0.67$	$\rho_{\text{pow}}\big\|_{Q_\Sigma} = 0.60$

As we see from the results a perfect correspondence between subjective and objective quality measures is still not possible. However, for the design of a new noise reduction algorithm the proposed quality measures help to yield information for tuning and comparing noise reduction systems. Human listeners tend to put their main emphasis on the speech quality and only a minor emphasis on the actual noise attenuation.

11.3 Comparison Mean Opinion Scores (CMOS)

When evaluating the quality of speech coding and decoding systems mean opinion scores (MOS) are often applied. In these tests audio examples which have been coded and decoded with a specific algorithm are presented to listeners. They have to evaluate the quality of each signal in terms of marks from 5 (excellent) down to 1 (bad). Tab. 11.3 gives an overview of the entire listening scale.

The disadvantages of MOS tests are the limited repeatability and the variable interpretation of scores such as excellent, good, fair, etc. For this reason MOS tests should be performed only with trained listeners. The quality or reliability of the listeners can be measured using quantities called *intra-person* and *inter-person deviation*. For the first quantity a few audio examples are presented twice within a test, randomly distributed within the sequences. If the listeners vote in a reliable manner the difference between the two ratings

Table 11.3. Listening scale of MOS tests.

Speech quality	Score
Excellent	5
Good	4
Fair	3
Poor	2
Bad	1

should be very small or even zero. If we denote the number of equal audio examples by N_{eq} and the voting result according to Tab. 11.3 of the first example with $V_m(k,1)$ and the one of the second example with $V_m(k,2)$ the intra-person deviation is defined as

$$U_m = \sqrt{\frac{1}{N_{\text{eq}}} \sum_{k=0}^{N_{\text{eq}}-1} \Big[V_m(k,1) - V_m(k,2)\Big]^2} \tag{11.29}$$

This quantity is evaluated for each listener (denoted by the subscript m) and only those individuals with a small intra-person deviation should be selected for performing a MOS test. To test how precise the adjectives *excellent*, *good*, etc. (see Tab. 11.3) describe a certain type of quality to different listeners[12] the inter-person deviation

$$T(k) = \sqrt{\frac{1}{N_{\text{lis}}-1} \sum_{m=0}^{N_{\text{lis}}-1} \Big[V_m(k) - \overline{V}(k)\Big]^2} \tag{11.30}$$

can be evaluated. Again $V_m(k)$ denotes the vote of listener m for the audio example k. The quantity

$$\overline{V}(k) = \frac{1}{N_{\text{lis}}} \sum_{m=0}^{N_{\text{lis}}-1} V_m(k) \tag{11.31}$$

estimates the average voting for audio example k. In [8] a test with $N_{\text{lis}} = 36$ untrained listeners has been performed for the purpose of evaluating the quality of different noise suppression schemes. During each test $N_{\text{eq}} = 12$ equal audio examples have been presented twice. The listeners were mostly students of an electrical engineering faculty. An average intra-person deviation (mean over all listeners) of about

[12] We assume that a pre-selection of the listeners in order to achieve a small average intra-person deviation was already made.

$$\overline{U} = \sum_{m=0}^{N_{\text{lis}}-1} U_m \approx 1.3 \tag{11.32}$$

was measured. For these reasons it is more reliable to compare between two signals and evaluate the difference of both. These tests are called comparison mean opinion scores (CMOS). In its simplest form the listeners evaluate only which of the two examples sounds better (see Tab. 11.4). The speech samples

Table 11.4. Listening scale of simple CMOS tests.

Voting	Score
A is better than B	1
A is worse than B	−1

are presented to the listeners by pairs (version A – version B) or by repeated pairs (version A – version B – version A – version B). In half of the trials the order of presentation should be reversed. Between the audio examples a pause of about 0.5 to 1 second should be inserted. After each trial the listeners have to evaluate which algorithmic version produces the better result. Note that in this test with only two choices it is not possible to rate the quality of the two approaches under test as equal. For some tests it is very desirable to force the listeners to decide for one version. However, if a more detailed analysis is preferred a seven score CMOS test with scores according to Tab. 11.5 can be performed. The intra-person and inter-person deviation in

Table 11.5. Listening scale of extended CMOS tests.

Voting	Score
A is much worse than B	−3
A is worse than B	−2
A is slightly worse than B	−1
A and B are about the same	0
A is slightly better than B	1
A is better than B	2
A is much better than B	3

CMOS tests is usually smaller than in MOS tests, especially with untrained listeners. However, to test the intra-person deviation a few of the trials in seven score CMOS tests can be performed with equal audio examples (version A = version B). These pairs of audio examples are called *null pairs.*

Furthermore, the selection of the listeners has influence on the result of the test. On one hand trained listeners, such as professional audio reviewers, have a much better reliability than untrained listeners [2]. In [37] it was reported that for loudspeaker evaluation the number of untrained listeners that is required to achieve a certain reliability was about 7 times higher than the number of trained listeners. On the other hand trained listeners tend to rate with a lower average quality level [37]. However, this is important only for absolute category ratings as they appear in MOS tests. For comparison ratings the effect is not very critical. As a consequence it is suggested that untrained listeners should be trained using a short supervised training phase before the actual subjective test is started.

When CMOS tests are performed to investigate (and to improve) the quality of speech enhancement algorithms they are usually accomplished more than once, e.g., once to investigate the influence on the overall quality of an enhanced background noise estimation, the second time to find out the best attenuation characteristic, and so forth. In this case one has to ensure that the boundary conditions, such as the audio presentation equipment, the characteristics (reverberation time) of the listening room, or the environmental background noise in the listening room, are equal for each test. Furthermore, the boundary conditions should be as close as possible to the final application. If, for example, a noise reduction is developed for the receiving path of a mobile phone, then the audio examples should be presented via this mobile phone.

11.3.1 Example

To show how a CMOS test is realized in detail we will present the following example where two versions of a bandwidth extension algorithm were evaluated. This was part of a pilot study for investigating the potential of both algorithmic approaches. For those readers who are not familiar with bandwidth extension algorithms a brief introduction in this topic is given in the next section. The others may continue with Sec. 11.3.1.2.

11.3.1.1 Basics of Bandwidth Extension Algorithms

Speech signals that are transmitted over current public telephone networks exhibit only a very limited bandwidth, e.g. 300 Hz up to 3400 Hz for analog lines. When comparing those speech signals to other audio sources such as radio or CD the quality difference is obvious and bothersome. Thus, great efforts have been made to increase the quality of telephone speech signals in recent years. Wideband codecs are able to increase the bandwidth up to 7 kHz or even higher at only moderate complexity. Nevertheless, applying these codecs would mean to exchange current networks. Another (cheaper) possibility is to extend the bandwidth after transmission over the unchanged network [27,28]. The basic idea of these enhancements is to estimate the speech signal components above 3400 Hz and below 300 Hz and to complement the signal in the new frequency bands with this estimate. In Fig. 11.11 the basic structure of a system for bandwidth extension of telephony speech is depicted.

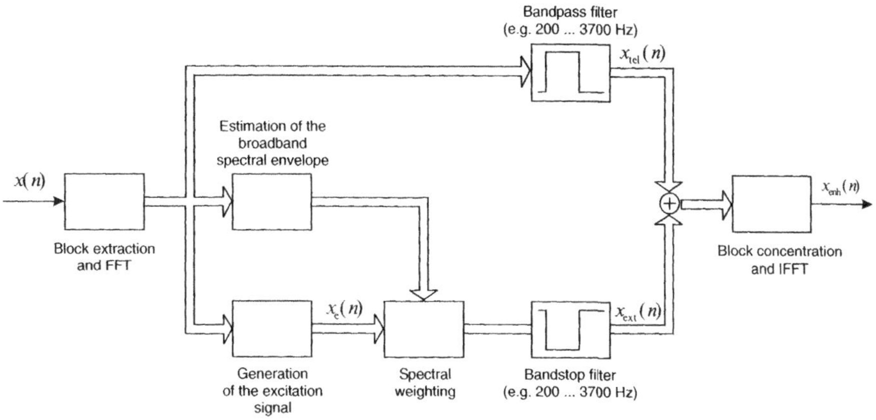

Fig. 11.11. Structure of a system for bandwidth extension of speech signals, which have been transmitted over a public telephone network.

The generation of this estimate can be divided into two separate tasks assuming that the well-known source-filter model of speech generation [4] is applied. First, a so-called *excitation* signal is required. This excitation signal corresponds to the signal that can be observed directly behind the vocal chords, which means that this signal contains information about voicing and pitch but not about formant structures or the spectral shaping in general. Consequently, this excitation signal has to be weighted with the spectral envelope of the speech signal. Thus, one key element in bandwidth extension of speech signals is the estimation of the spectral envelope. Two methods for this estimation were investigated with a CMOS test: mapping the narrow-band envelope to a broad-band envelope by either using a codebook or a neural network.

After generating the excitation signal $x_e(n)$ and weighting $x_e(n)$ with the spectral envelope, power adjustment of the synthesized signal to the input signal $x(n)$ is necessary. Before adding the complementary signals the phase of the extended frequency bands can be manipulated. Both, power adjustment and phase manipulation, are not depicted in Fig. 11.11. For computing the bandwidth extension block processing in the frequency domain is applied. The input signal is divided into overlapping blocks of length $N_B = 256$ (sampling rate $=$ 11025 Hz). The blocks are overlapping by 75 percent resulting in a frameshift of 64 samples. Further details about the specific algorithms can be found in [20].

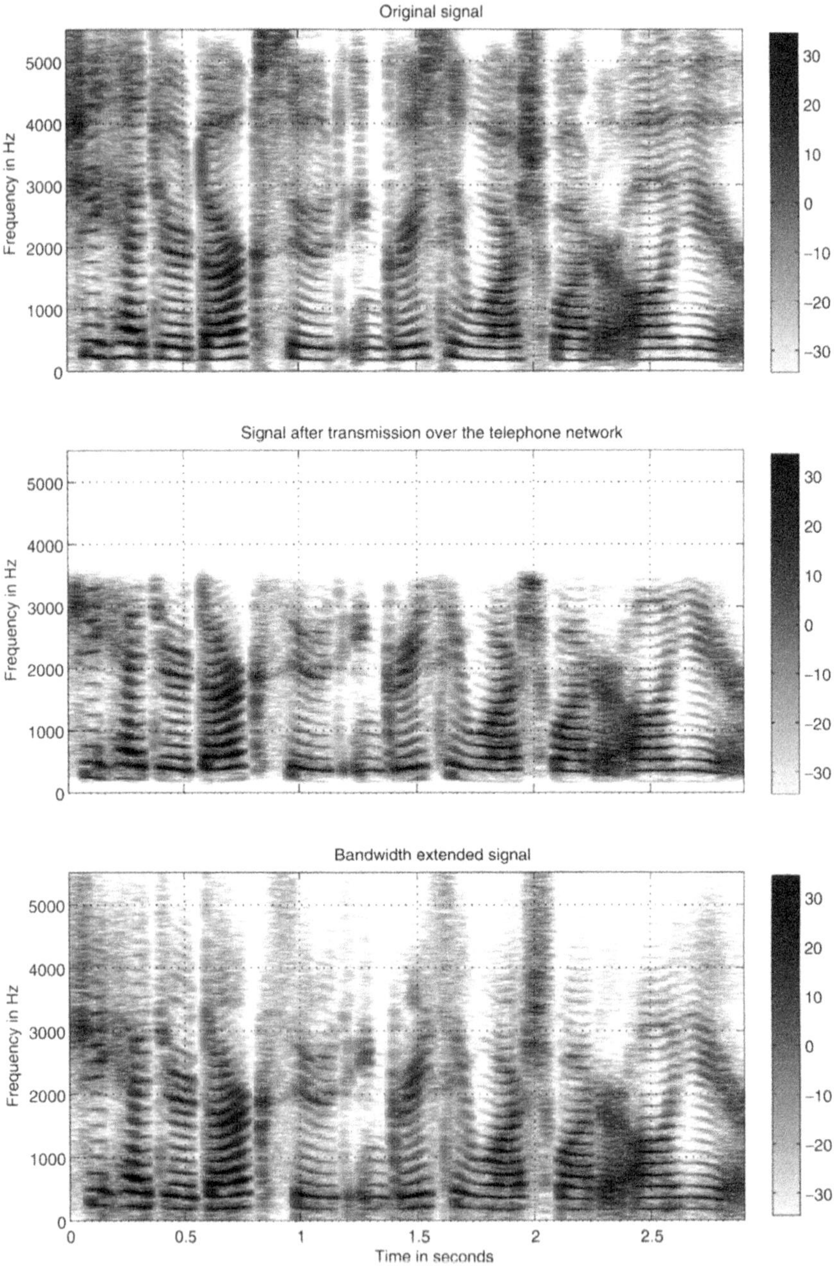

Fig. 11.12. Time-frequency analysis of wideband speech (top), bandlimited speech (middle), and reconstructed speech (bottom).

In Fig. 11.12 three time-frequency analyses are presented. The upper most analysis depicts a wideband speech signal $x_{\mathrm{orig}}(n)$ as it would be recorded close to the mouth of the communication partner on the remote side. If we assume not to have any kind of errors or distortions on the transmission a bandlimited signal $x(n)$ as depicted in the center diagram would be received at the local side. The truncation of the frequency range is clearly visible. Without any additional processing the local communication partner would be listening to this signal. If bandwidth extension is applied a signal $x_{\mathrm{enh}}(n)$ as depicted in the lowest part of Fig. 11.12 would be reconstructed. Even if the signal is not exactly the same as the original one, it sounds more natural and - as a variety of listening test indicate - the speech quality in general is increased as well [20].

11.3.1.2 Performing the CMOS Test

When designing a system for bandwidth extension it is quite interesting which method for generating the broadband envelope should be chosen. Codebooks are quite easy to train, e.g., using the LBG algorithm [33]. However, if a large codebook is required the search for the optimal entry is quite expensive in terms of computational complexity. On the other hand, the utilization of a codebook means some kind of discretization and algorithms which take past input vectors into account are quite simple to realize. Furthermore, stability can be guaranteed if the training of the codebook is performed carefully. A neural network on the other hand can generate the broadband spectral envelope in a computationally more effective manner. However, if the desired outputs are cepstral coefficients or predictor error filter coefficients, the resulting filters have to be checked for stability and a correction might be necessary.

In order to make this design decision the subjective quality of the differently extended signals have been evaluated by a CMOS test. About 20 people of different age and gender have participated in the test. At the beginning of each test the listeners were asked to complete the fields of a first program window. Within a comments field the age and the gender of the subjects were noted. Furthermore, it was made sure that none of the listeners had any hearing impairment.

After completing the fields of the first window the listeners were asked to compare the quality of two signals (pairs of bandlimited and extended signals) by choosing one of the statements listed in Tab. 11.5. This decision (extended signal versus original telephone signal) was requested for the extended signal using a codebook and for the neural network extension. Finally, the listeners were asked whether they preferred the signal which was extended by the neural network or the one which was extended with the codebook (choice according to Tab. 11.4). During a test 20 groups of audio signals (two extended signals and the bandlimited reference) were presented to each listener. The program window of this part of the CMOS test is depicted in Fig. 11.13.

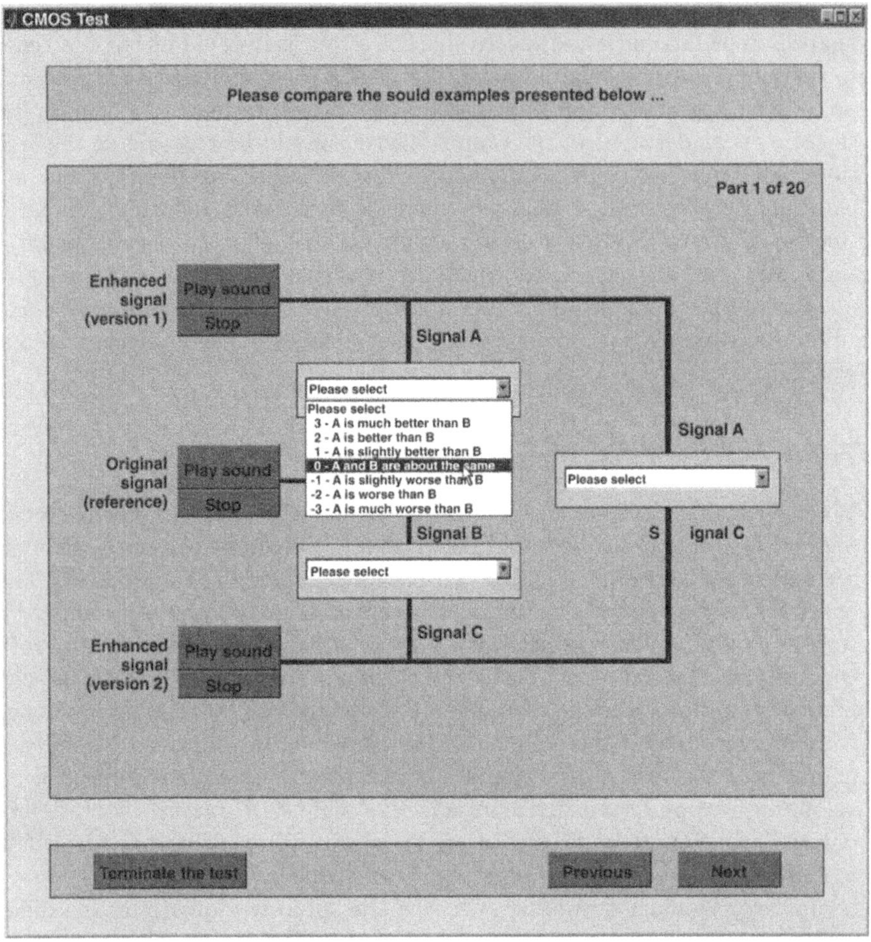

Fig. 11.13. Program windows of a CMOS test – part II.

The listeners were allowed to play the signals as often as they wanted until they were able to make their decisions. This way of audio presentation is not common in CMOS tests. However, because of the larger number of comparisons (two times reference against extended signal and once a comparison between the two extended signals) the possibility of repeated audio presentation was given to the listeners. During all tests the same audio equipment (computer, soundcard and loudspeakers) was utilized. Also the individuals did not know which path in Fig. 11.13 belonged to the codebook version and which to the neural network version. The connection of both approaches to pathes A and C was chosen randomly for each triple of audio signals. Only version B corresponds always to the telephone bandlimited (not extended) signal. Note furthermore, that the extended signals have not been used for network training or codebook generation, respectively.

11.3.1.3 Evaluation of the Test

Tab. 11.6 shows the absolute results of the seven score parts of the CMOS test. As already described the tests were performed by $N_{\text{lis}} = 20$ individuals, each voted $N_{\text{t}} = 20$ times during a test. This means that the sum of each column in Tab. 11.6 is 400. When choosing which approach produces better results 80 percent (320 of 400 trials) voted for the codebook based scheme. Fig. 11.14 shows the relative results of the CMOS test.

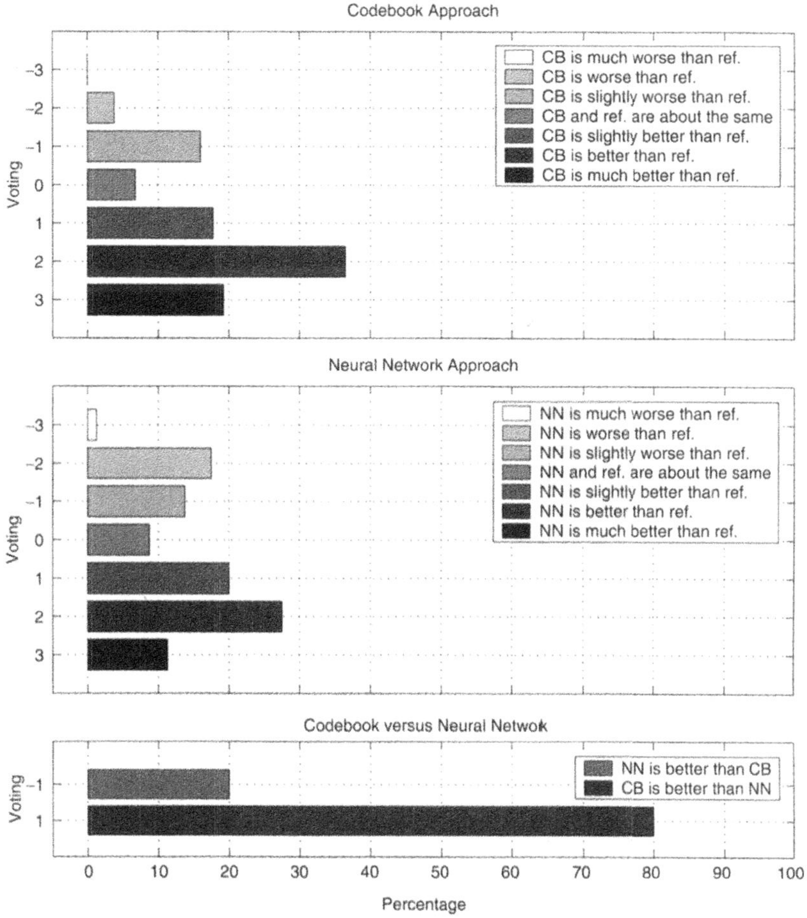

Fig. 11.14. Results of the CMOS test. The abbreviations CB, NN, and ref. stand for codebook, neural network, and reference, respectively.

Table 11.6. Results of the bandwidth extension CMOS tests (CB abbreviates codebook, NN stands for neural network).

Statement (... than reference)	Amount of results	Amount of results	Statement (... than reference)
CB is much worse ...	0	5	NN is much worse ...
CB is worse ...	15	70	NN is worse ...
CB is slightly worse ...	64	55	NN is slightly worse ...
CB is about the same ...	27	35	NN is about the same ...
CB is slightly better ...	71	80	NN is slightly better ...
CB is better ...	146	110	NN is better ...
CB is much better ...	77	45	NN is much better ...

When analyzing the results of CMOS tests often the average rating V_Σ in terms of summing all scores (see Tabs. 11.4 and 11.5) and dividing it by the number of votings

$$V_\Sigma = \frac{1}{N_\mathrm{t}\, N_\mathrm{lis}} \sum_{m=0}^{N_\mathrm{lis}-1} \sum_{k=0}^{N_\mathrm{t}-1} V_m(k) \tag{11.33}$$

is computed. $V_m(k)$ denotes the vote of listener m for audio example k (see Sec. 11.3). In our example this would lead to an average mark of

$$V_{\Sigma,\mathrm{NN}} \approx 0.56$$

(between equal and slightly better than the bandlimited signals) for the network approach and to

$$V_{\Sigma,\mathrm{CB}} = 1.25$$

for the codebook scheme (between slightly better and better than the bandlimited signals). This evaluation method has the drawback that statements and especially the differences between statements are mapped onto a linear scale. To avoid this linearization we prefer the quantile based evaluation that was presented in Sec. 11.2.5.2. The so-called *judging quotients* are listed in Tab. 11.7.

Before we start comparing both methods for envelope estimation an important question is whether each of the methods is able to improve the speech quality (compared to the pure telephone signal). To answer this question the judging quotient Q_0 should be analyzed. A large value indicates that most of the listeners are of the opinion that the extended signals sound at least as good as the original signals. The codebook approach achieves

$$Q_{0,\mathrm{CB}} \approx 0.80\,.$$

Table 11.7. Judgement quotients obtained with the CMOS test (for the definitions of Q_i and Q_Σ see Sec. 11.2.5.2).

Voting	Judging Quotients	
	Codebook	Neural network
CB/NN is much worse than ref.	$Q_{-3} = 1.000$	$Q_{-3} = 1.000$
CB/NN is worse than ref.	$Q_{-2} = 1.000$	$Q_{-2} = 0.988$
CB/NN is slightly worse than ref.	$Q_{-1} = 0.963$	$Q_{-1} = 0.813$
CB/NN and ref. are about the same	$Q_0 = 0.803$	$Q_0 = 0.675$
CB/NN is slightly better than ref.	$Q_1 = 0.735$	$Q_1 = 0.588$
CB/NN is better than ref.	$Q_2 = 0.556$	$Q_2 = 0.388$
CB/NN is much better than ref.	$Q_3 = 0.193$	$Q_3 = 0.113$
Average judging quotient	$Q_\Sigma = 0.708$	$Q_\Sigma = 0.594$

This means that 80 percent of the tests result in equal quality or in quality improvement. Taking the judgement quotient

$$Q_{-1,\mathrm{CB}} \approx 0.96$$

into account shows that for most of the residual sound examples (20 percent) only a slight degradation can be observed (only 4 percent of the examples sound worse or much worse).

Even if the neural network approach performs slightly worse the method is also able to improve most of the sound examples considerably, since judgement quotients

$$Q_{0,\mathrm{NN}} \approx 0.68 \quad \text{and} \quad Q_{-1,\mathrm{NN}} \approx 0.81$$

have been achieved. However, a judgement ratio $Q_{-1,\mathrm{NN}} = 0.81$ indicates that a non-negligible amount of the presented examples (19 percent) were degraded in quality. Fig. 11.15 shows the judgement quotients for both 7-level CMOS tests. When computing the average judgement quotient for the codebook approach according to Eq. 11.25 we obtain

$$Q_{\Sigma,\mathrm{CB}} \approx 0.71\,.$$

The neural network performs slightly worse with an average judgement quotient of about

$$Q_{\Sigma,\mathrm{NN}} \approx 0.59\,.$$

Note, that the average judgement quotient Q_Σ can be easily transformed for a 7-level CMOS test into the average voting V_Σ (and vice versa):

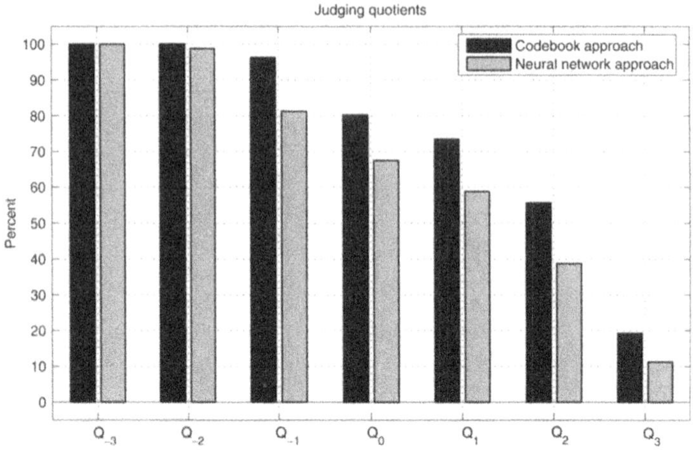

Fig. 11.15. Judgement quotients of both 7-level CMOS tests.

$$V_\Sigma = 6\,Q_\Sigma - 3\,, \tag{11.34}$$

where $N = N_{\text{lis}}\,N_{\text{t}}$ denotes the total number of votings (in our example we have $N = 400$). Since both average judgement ratios are larger than 0.5 both methods seem to enhance the speech quality.[13]

As a second result, the test indicates that the codebook approach outperforms the neural network scheme:

- On one hand the average judgement ratio of the codebook scheme is larger than that of the neural network approach

$$Q_{\Sigma,\text{CB}} > Q_{\Sigma,\text{NN}}\,.$$

- On the other hand and even more important, in 80 percent of the tests the listeners voted for the codebook approach.

After receiving these absolute and relative results that indicate which of the two versions is the better one, the question about statistical significance of the test comes up. Before this subject is treated in Sec. 11.3.2 allow us a few final remarks about the comparison between the two schemes for spectral envelope estimation in bandwidth extension systems in the next section.

11.3.1.4 Remark

Note that this test does not mean that codebook approaches are in general the better choice for bandwidth extension systems. The results depend crucially

[13] If all listeners had voted with "*CB/NN and ref. are about the same*" the first four judgement quotients would have been $Q_{-3} = Q_{-2} = Q_{-1} = Q_0 = 1$, and the last three $Q_1 = Q_2 = Q_3 = 0$. This results in an average judging ratio of $Q_\Sigma = 0.5$.

on the stabilization method that is applied as a necessary postprocessing unit to the output of the neural network.[14] The main problem of the neural network approach was that it has produced quite rarely audible artifacts. This resulted in lower votings. However, the examples without artifacts were scored rather high. As a result the neural network approach was evaluated undecidedly.

11.3.2 Statistical Analysis

When investigating statistical significance of the results of the subjective tests we will start for simplicity reasons with the analysis of the question which of the two extension schemes produces the better result. The analysis of this kind of test will be done in Sec. 11.3.2.2.

11.3.2.1 Analysis of the Two-Level Test

For the following statistical analysis we assume that one of the schemes (or algorithmic versions) was rated better than the other. This scheme, in our example the codebook approach, is denoted with version 1, the one which had produced a lower result is called version 2. The results of the CMOS test were grouped into two categories:

a_-	Number of results in which version 1 was rated worse than version 2.
a_+	Number of results in which version 1 was rated better than version 2.

The total number of tests is denoted by

$$N = a_+ + a_- \,. \tag{11.35}$$

We further assume all tests to be mutually independent and that results indicating that version 1 is better than version 2 will be produced with probability p_+. Worse quality is voted with probability p_-.

Under the assumptions and definitions given above the probability of getting $\bar{a}_+$ positive and $\bar{a}_-$ negative results is given by[15]

$$p\Big((a_+ = \bar{a}_+) \wedge (a_- = \bar{a}_-)\Big) = \begin{pmatrix} N \\ \bar{a}_+ \end{pmatrix} p_+^{\bar{a}_+} \, p_-^{\bar{a}_-} \,. \tag{11.36}$$

[14] The spectral envelope is coded in terms of an inverse predictor error filter. Since the network approach does not necessarily generate a minimum phase predictor error filter, stability can not be guaranteed for the inverse all-pole filter.

[15] The term *positive* indicates here that version 1 is rated better than version 2. Analogously, *negative* means that version 1 is rated worse than version 2. Furthermore, a_+ and a_- are denoting the possible results of the test. The quantities $\bar{a}_+$ and $\bar{a}_-$ are describing the actually achieved results of the test.

Both probabilities, p_+ and p_- sum up to 1

$$p_+ + p_- = 1\,. \tag{11.37}$$

Thus, one of the parameters in Eq. 11.36, e.g. a_-, can be omitted:

$$p(a_+ = \bar{a}_+) = \binom{N}{\bar{a}_+}\, p_+^{\bar{a}_+}\,(1-p_+)^{(N-\bar{a}_+)}\,. \tag{11.38}$$

For the following derivation we will make use of the fact that all possible probabilities $p(a_+ = 0)$, ..., $p(a_+ = N)$ also sum up to one:

$$\sum_{k=0}^{N} p(a_+ = k) = \sum_{k=0}^{N} \binom{N}{k}\, p_+^{k}\,(1-p_+)^{(N-k)} = 1\,. \tag{11.39}$$

Thus, we can write the probability $p(a_+ = \bar{a}_+)$ also as

$$p(a_+ = \bar{a}_+) = \frac{\binom{N}{\bar{a}_+}\, p_+^{\bar{a}_+}\,(1-p_+)^{(N-\bar{a}_+)}}{\sum\limits_{k=0}^{N} \binom{N}{k}\, p_+^{k}\,(1-p_+)^{(N-k)}}\,. \tag{11.40}$$

The probability for achieving $\bar{a}_+$ or even more positive results can be computed as

$$\begin{aligned} p(a_+ \geq \bar{a}_+) &= \sum_{k=\bar{a}_+}^{N} p(a_+ = k) \\ &= \frac{\sum\limits_{k=\bar{a}_+}^{N} \binom{N}{k}\, p_+^{k}\,(1-p_+)^{(N-k)}}{\sum\limits_{k=0}^{N} \binom{N}{k}\, p_+^{k}\,(1-p_+)^{(N-k)}} \\ &= \frac{\sum\limits_{k=\bar{a}_+}^{N} \binom{N}{k} \left(\frac{p_+}{1-p_+}\right)^{k}}{\sum\limits_{k=0}^{N} \binom{N}{k} \left(\frac{p_+}{1-p_+}\right)^{k}}\,. \end{aligned} \tag{11.41}$$

Writing

$$r = \frac{p_+}{p_-} = \frac{p_+}{1-p_+}\,, \tag{11.42}$$

and inserting the definition of the binomial coefficient

$$\binom{N}{k} = \frac{N!}{k!\,(N-k)!}\,, \tag{11.43}$$

we can write Eq. 11.41 shortly as

$$p(a_+ \geq \bar{a}_+) = \frac{\sum\limits_{k=\bar{a}_+}^{N} \frac{r^k}{k!\,(N-k)!}}{\sum\limits_{k=0}^{N} \frac{r^k}{k!\,(N-k)!}} \,. \tag{11.44}$$

In the following we will utilize Eq. 11.44 to compute an upper limit for the probability to obtain $\bar{a}_+$ or more positive results under certain assumptions. In particular, these assumptions are:

Hypothesis H_0 : We assume that $p_+ \leq p_-$, meaning that version 2 produces better or at least equal results as version 1.

Hypothesis H_1 : We assume that $p_+ > p_-$, meaning that version 1 produces better results as version 2.

If H_0 is true our convention at the beginning of this section for version 1 has proved wrong. We compute an upper limit for the conditional probability

$$p(a_+ \geq \bar{a}_+)\big|_{H_0} \leq p_0 \,. \tag{11.45}$$

If this upper limit is sufficiently small we can discard hypothesis H_0 and use H_1 instead.

In Fig. 11.16 four density examples are depicted. We assume to have performed $N = 400$ tests. The upper left diagram shows the density according to Eq. 11.44 for equal probabilities $p_+ = p_- = 0.5$, resulting in a ratio $r = 1$. In this case and under the restriction H_0, meaning that $p_- \geq p_+$, the conditional probability $p(a_+ \geq 210)$ to obtain 210 positive results or even more reaches its maximum value.[16] In the upper right diagram of Fig. 11.16 the density obtained for $p_+ = 0.49$ and $p_- = 0.51$ is depicted. The highlighted area representing $p(a_+ \geq 210)$ is clearly smaller than the one for $p_+ = p_- = 0.5$. This trend holds for even smaller ratios r (depicted in the lower two diagrams).

The trend can be explained analytically by analyzing Eq. 11.41. The sum in the denominator of Eq. 11.41 originates from adding all probabilities according to the distribution given in Eq. 11.36 with respect to all restrictions given until now. Furthermore, the terms

$$a_k = \frac{1}{k!\,(N-k)!} \tag{11.46}$$

[16] We have chosen $\bar{a}_+ = 210$ (instead of $\bar{a}_+ = 320$ as achieved with the test) in order to obtain results that are clearly visible. The shaded areas in Fig. 11.16 representing the probabilities $p(a_+ \geq \bar{a}_+)$ would be hardly visible for the case $\bar{a}_+ = 320$.

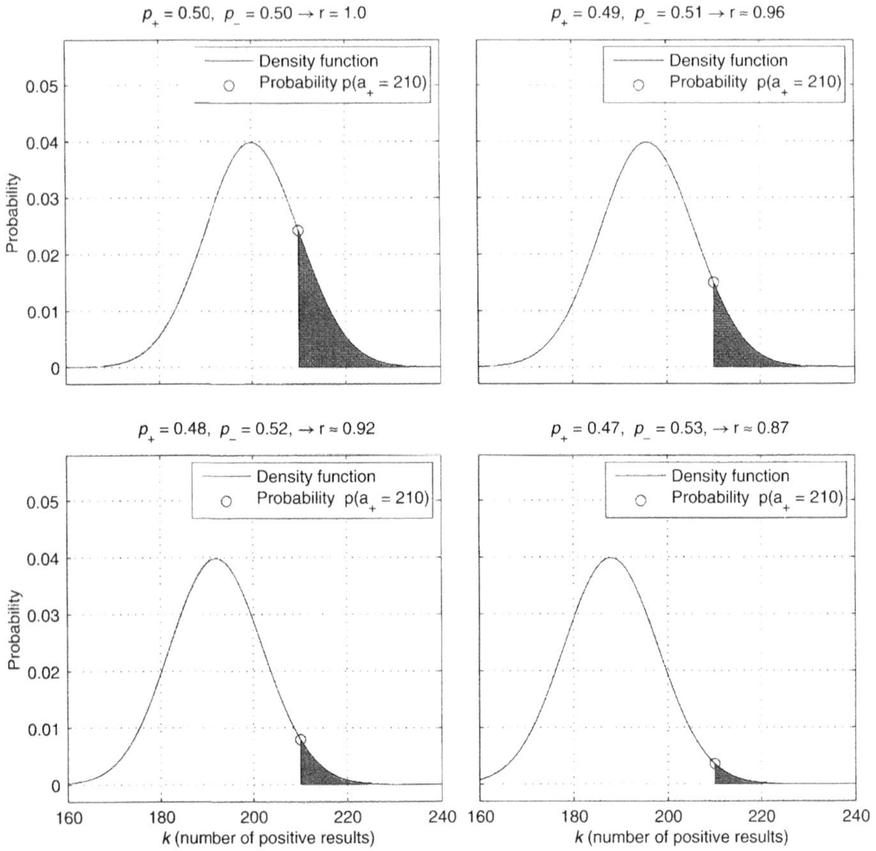

Fig. 11.16. Binomial density functions for different values of r. For the sake of better visibility, the discrete density functions have been plotted using continuous lines.

are symmetric with respect to the center of the summation interval if the term r^k is neglected:

$$a_k = a_{N-k} \,. \tag{11.47}$$

If the ratio r is smaller than 1 the summands a_k of the higher order half of the summation interval are attenuated more than the lower order half. Fig. 11.17 shows in the upper diagram the addends for the ratios $r = 1$ (depicted by circles) and $r = 0.8$ (depicted by squares) for a survey with just $N = 10$ examples. The attenuation of the higher order terms (compared to their lower order counterpart) is clearly visible. The attenuation itself (r^k) is depicted in the lower diagram (the curve has been interpolated for better visibility for non-integer values of k).

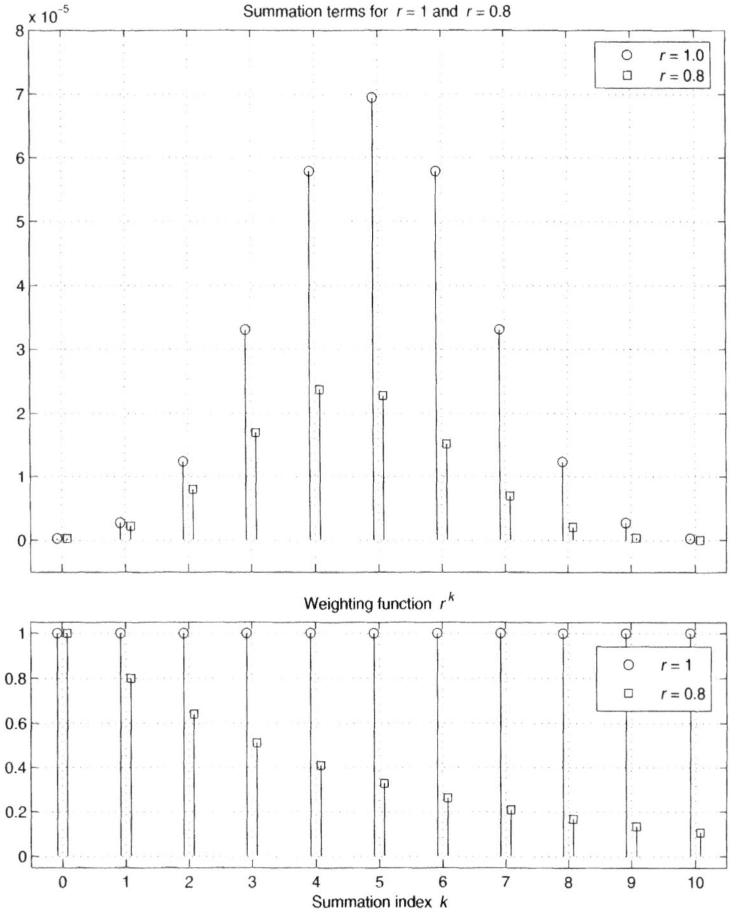

Fig. 11.17. Weighted (with r^k) and non-weighted summation terms.

In the numerator of Eq. 11.44 only a subset of these terms is added. This subset contains – according to the start index $k = \bar{a}_+$ – only the higher order values. For this reason the conditional probability $p(a_+ \geq \bar{a}_+)$ is bounded if r is restricted to $r \geq 1$:

$$p(a_+ \geq \bar{a}_+)\Big|_{p_+ \leq p_-} \leq p(a_+ \geq \bar{a}_+)\Big|_{p_+ = p_-} = \frac{\sum\limits_{k=\bar{a}_+}^{N} \dfrac{1}{k!\,(N-k)!}}{\sum\limits_{k=0}^{N} \dfrac{1}{k!\,(N-k)!}} = p_0\,. \quad (11.48)$$

This result is important since we are able now to compute an upper limit for the probability p_0 that hypothesis H_0 is true. In our case ($N = 400$, $\bar{a}_+ = 320$) we obtain an upper limit for

$$p_0\Big|_{N=400,\bar{a}_+=320} < 10^{-10}\,. \tag{11.49}$$

This means that we can discard hypothesis H_0 – meaning that the neural network scheme produces better results than the codebook approach – since H_0 is true only with a probability of at most 10^{-10}. As a result we can conclude that codebooks produce better results than neural networks (at least with the algorithmic setup utilized within our tests).

The computation of Eq. 11.48 is often difficult due to numerical inaccuracies the binomial distribution can be approximated for large N by a normal distribution of appropriate mean and variance. In this case the probability can be approximated by [31]

$$p_0(\bar{a}_+, N)\Big|_{N\gg 1} \approx 1 - \Phi\left(\frac{2\bar{a}_+ - N}{\sqrt{N}}\right) \tag{11.50}$$

with

$$\Phi(x) = \frac{1}{\sqrt{2\pi}} \int\limits_{-\infty}^{x} e^{-\frac{1}{2}t^2}\, dt\,. \tag{11.51}$$

11.3.2.2 Analysis of the Seven-Level Test

The analysis of the seven-level CMOS test is very much related to the analysis of the two-level test. For this reason, the derivation is rather brief and we will focus on the differences compared to the previous section. We will group the results of the seven-level CMOS test into three (respectively two) categories:

a_-	Number of results in which version 1 was rated slightly worse, worse, or much worse than version 2.
a_0	Number of results in which version 1 was rated about the same as version 2.
a_+	Number of results in which version 1 was rated slightly better, better, or much better than version 2.

This kind of grouping takes the difficulty of mapping adjectives such as slightly better, better, and much better onto numbers such as 1, 2, and 3 into account. Furthermore, we will combine equal and worse results within the category

$$a_{0,-} = a_0 + a_-\,.$$

This means that we distinguish in the following only between the categories "better" and "equal or worse". As in the last analysis we will denote the amount of votings that have been achieved in the test as $\bar{a}_+$ for positive results, respectively $\bar{a}_{0,-}$ for equal or negative results.[17] In the test we obtained for the codebook approach

$$\bar{a}_{+,\mathrm{CB}} = 294\,, \tag{11.52}$$
$$\bar{a}_{0,\,-,\mathrm{CB}} = 106\,, \tag{11.53}$$

and for the network scheme

$$\bar{a}_{+,\mathrm{NN}} = 235\,, \tag{11.54}$$
$$\bar{a}_{0,\,-,\mathrm{NN}} = 165\,. \tag{11.55}$$

As a next step we postulate – as in the previous section – two hypotheses:

Hypothesis H_0 : We assume that $p_+ \leq p_{0,-}$, meaning that the reference (the non-processed signal) produces better or at least equal results as the codebook / neural network approach.

Hypothesis H_1 : We assume that $p_+ > p_{0,-}$, meaning that the codebook / neural network approach produces better results than the reference.

Assuming that H_0 is true we can compute an upper limit for the conditional probability to obtain more than $\bar{a}_+$ positive results:

$$\begin{aligned} p(a_+ \geq \bar{a}_+)\Big|_{p_+ \leq p_{0,-}} &\leq p(a_+ \geq \bar{a}_+)\Big|_{p_+ = p_{0,-}} \\ &= \frac{\sum\limits_{k=\bar{a}_+}^{N} \dfrac{1}{k!\,(N-k)!}}{\sum\limits_{k=0}^{N} \dfrac{1}{k!\,(N-k)!}} = p_0\,. \end{aligned} \tag{11.56}$$

Again, we have abbreviated the upper limit of the conditional probability defined in Eq. 11.56 with p_0. In our test we obtained for the codebook approach

$$p_{0,\mathrm{CB}}\Big|_{N=400,\bar{a}_{+,CB}=294} < 10^{-10} \tag{11.57}$$

and

$$p_{0,\mathrm{NN}}\Big|_{N=400,\bar{a}_{+,NN}=235} < 0.00027 \tag{11.58}$$

for the neural network scheme. Both conditional probabilities are sufficiently small to discard hypothesis H_0 and assume instead that the extended signals have better sound quality on average.

[17] Again, the term *positive* indicates votings such as "much better", "better", or "slightly better" – *negative* summarizes "much worse", "worse", and "slightly worse".

11.4 Rhyme Tests

In the previous tests well-known phrases such as popular song texts or proverbs were used for evaluating the speech quality. The speech intelligibility, also an important part of the speech quality, is usually underweighted.

If the speech intelligibility of a speech enhancement system is examined segmental evaluation methods are the better choice. In these tests, the so-called *rhyme tests*, several lists with homophone sounding (rhyming) words, such as "west", "test", and "best", are presented to the listeners. *Presenting* means here that all words are displayed, e.g., on a computer monitor or on a sheet of paper. One of the words is presented also acoustically and the listeners have to decide which item on the list was actually played.

The stimulus word lists have to be designed individually for each language. In Tab. 11.8 and Tab. 11.9 an English stimulus list as published in the American National Standard ANSI S3.2-1989 [1] is presented. The list consists of 50 sets of six monosyllable words, resulting in a total set of 300 words. Half of the sets can be used to evaluate the intelligibility of the initial consonants, the other half was designed to test the final consonant. Besides this list, several institutes and vendors such as Bellcore and AT&T have also designed word lists for their individual applications, e.g. for the evaluation of speech synthesis systems [5].

Such a test is called a *modified rhyme test* (MRT). When the listeners had made their choices which of the six words was the one that was acoustically presented, average error rates can be computed. Usually, the total error rate is of main interest but also single consonants and how they are confused with each other can be investigated. However, usually the test material is rather limited and not all possible confusion cases might appear. Thus, confusions presented in terms of matrices are not easy to evaluate.

The one-out-of-six choice per stimulus yields to an amount of information of about $\log_2(6) = 2.59$ bits per selection.[18] On average it takes a human listener about 4 seconds to make such a decision [44]. Thus, one can obtain about 0.65 bit/s of information using a modified rhyme test. Another well-known rhyme test is the so-called *diagnostic rhyme test* (DRT). In this test the listeners are provided with lists that contain only two items. Thus, an information of only one bit per selection is obtained. However, a human listener needs on average only 1.33 seconds for such a one-out-of-two choice. As a result the average information rate of a diagnostic rhyme test is about 0.75 bit/s. As for the modified rhyme test, stimulus lists for the DRT are published in the American National Standard ANSI S3.2-1989 [1] (see Tab. 11.10).

Even if the diagnostic rhyme test is in some ways only a subset of the modified rhyme test (word pairs of the MRT could be taken as a stimulus basis for the DRT), it is also an extension of the MRT in terms of its evaluation

[18] Here and in the following example it is assumed that all stimulus words appear with the same a priori probability.

Table 11.8. Examples for stimulus words used in a modified rhyme test [1] – part 1.

List with 50 groups of homophone sounding (rhyming) words (part 1)										
went	–	sent	–	bent	–	dent	–	tent	–	rent
hold	–	cold	–	told	–	fold	–	sold	–	gold
pat	–	pad	–	pan	–	path	–	pack	–	pass
lane	–	lay	–	late	–	lake	–	lace	–	lame
kit	–	bit	–	fit	–	hit	–	wit	–	sit
must	–	bust	–	gust	–	rust	–	dust	–	just
teak	–	team	–	teal	–	teach	–	tear	–	tease
din	–	dill	–	dim	–	dig	–	dip	–	did
bed	–	led	–	fed	–	red	–	wed	–	shed
pin	–	sin	–	tin	–	fin	–	din	–	win
dug	–	dung	–	duck	–	dud	–	dub	–	dun
sum	–	sun	–	sung	–	sup	–	sub	–	sud
seep	–	seen	–	seethe	–	seek	–	seem	–	seed
not	–	tot	–	got	–	pot	–	hot	–	lot
vest	–	test	–	rest	–	best	–	west	–	nest
pig	–	pill	–	pin	–	pip	–	pit	–	pick
back	–	bath	–	bad	–	bass	–	bat	–	ban
way	–	may	–	say	–	pay	–	day	–	gay
pig	–	big	–	dig	–	wig	–	rig	–	fig
pale	–	pace	–	page	–	pane	–	pay	–	pave
cane	–	case	–	cape	–	cake	–	came	–	cave
shop	–	mop	–	cop	–	top	–	hop	–	pop
coil	–	oil	–	soil	–	toil	–	boil	–	foil
tan	–	tang	–	tap	–	tack	–	tam	–	tab
fit	–	fib	–	fizz	–	fill	–	fig	–	fin

depth. As one can see in Tab. 11.10 the pairs of stimulus words in a diagnostic rhyme test as proposed by W. Voiers in 1965 [44] are grouped into six categories: voicing, nasality, sustention, sibilation, graveness, and compactness. If a speech transmission or speech enhancement system has, e.g., problems in terms of maintaining the periodicity of a signal the amount of errors within the voicing category will be higher than in the other categories. Thus, analyzing the error rates individually for each category of a diagnostic rhyme test does not only give information about the speech intelligibility in general but also about the specific weaknesses of the system or algorithm under test.

Table 11.9. Examples for stimulus words used in a modified rhyme test [1] – part 2.

List with 50 groups of homophone sounding (rhyming) words (part 2)										
same	–	name	–	game	–	tame	–	came	–	fame
peel	–	reel	–	feel	–	eel	–	keel	–	heel
hark	–	dark	–	mark	–	bark	–	park	–	lark
heave	–	hear	–	heat	–	heal	–	heap	–	heath
cup	–	cut	–	cud	–	cuff	–	cuss	–	cub
thaw	–	law	–	raw	–	paw	–	jaw	–	saw
pen	–	hen	–	men	–	then	–	den	–	ten
puff	–	puck	–	pub	–	pus	–	pup	–	pun
bean	–	beach	–	beat	–	beak	–	bead	–	beam
heat	–	neat	–	feat	–	seat	–	meat	–	beat
dip	–	sip	–	hip	–	tip	–	lip	–	rip
kill	–	kin	–	kit	–	kick	–	king	–	kid
hang	–	sang	–	bang	–	rang	–	fang	–	gang
took	–	cook	–	look	–	hook	–	shook	–	book
mass	–	math	–	map	–	mat	–	man	–	mad
ray	–	raze	–	rate	–	rave	–	rake	–	race
save	–	same	–	sale	–	sane	–	sake	–	safe
fill	–	kill	–	will	–	hill	–	till	–	bill
sill	–	sick	–	sip	–	sing	–	sit	–	sin
bale	–	gale	–	sale	–	tale	–	pale	–	male
wick	–	sick	–	kick	–	lick	–	pick	–	tick
peace	–	peas	–	peak	–	peach	–	peat	–	peal
bun	–	bus	–	but	–	bug	–	buck	–	buff
sag	–	sat	–	sass	–	sack	–	sad	–	sap
fun	–	sun	–	bun	–	gun	–	run	–	nun

11.4.1 Performing a Rhyme Test

When performing a rhyme test – either a DRT or an MRT – several boundary conditions should be considered:

- The recording conditions of the stimulus words should be as close as possible to the final application. If, for example, a speech enhancement system should be tested under several noise conditions the Lombard effect[19] [16, 35] should be taken into account within the recording session. Also the electro-acoustic transducers (microphones, AD and DA convert-

[19] People alter their way of speaking according to the level and type of background noise. This behavior is called *Lombard* effect.

Table 11.10. Stimulus words used in a diagnostic rhyme test [1,44].

Voicing			Nasality			Sustention		
veal	–	feel	meat	–	beat	vee	–	bee
bean	–	peen	need	–	deed	sheet	–	cheat
gin	–	chin	mitt	–	bit	vill	–	bill
dint	–	tint	nip	–	dip	thick	–	tick
zoo	–	Sue	moot	–	boot	foo	–	pooh
dune	–	tune	news	–	dues	shoes	–	choose
voal	–	foal	moan	–	bone	those	–	doze
goat	–	coat	note	–	dote	though	–	dough
zed	–	said	mend	–	bend	then	–	den
dense	–	tense	neck	–	deck	fence	–	pence
vast	–	fast	mad	–	bad	than	–	Dan
gaff	–	calf	nab	–	dab	shad	–	chad
vault	–	fault	moss	–	boss	thong	–	tong
daunt	–	taunt	gnaw	–	daw	shaw	–	chaw
jock	–	chock	mom	–	bomb	von	–	bon
bond	–	pond	knock	–	dock	vox	–	box
Sibilation			**Graveness**			**Compactness**		
zee	–	thee	weed	–	reed	yield	–	wield
cheep	–	keep	peak	–	teak	key	–	tea
jilt	–	gilt	bid	–	did	hit	–	fit
sing	–	thing	fin	–	thin	gill	–	dill
juice	–	goose	moon	–	noon	coop	–	poop
chew	–	coo	pool	–	tool	you	–	rue
Joe	–	go	bowl	–	dole	ghost	–	boast
sole	–	thole	fore	–	thor	show	–	so
jest	–	guest	met	–	net	keg	–	peg
chair	–	care	pent	–	tent	yen	–	wren
jab	–	dab	bank	–	dank	gat	–	bat
sank	–	thank	fad	–	thad	shag	–	sag
jaws	–	gauze	fought	–	thought	yawl	–	wall
saw	–	thaw	bond	–	dong	caught	–	taught
jot	–	got	wad	–	rod	hop	–	fop
chop	–	cop	pot	–	tot	got	–	dot

ers, loudspeakers) should possibly be the same as in the final application. This could be achieved either by using, e.g., the same microphone or a calibrated microphone and applying a correction filter. The latter approach has the advantage that the recorded data base can be used for several applications.

- The presentation of the stimulus words to the listeners should also be as close to the final application as possible. If, for example, a noise reduction scheme for a hands-free telephone is tested, the listeners should listen to the audio examples after appropriate coding and decoding as well as via an appropriate amount of different mobile phones, hand-sets, or hands-free telephones.
  If, for example, a noise suppression scheme for hands-free telephones has to be tested, the input signals should be connected to the system with a high sampling rate even if the output sampling rate is only 8 kHz. A few noise suppression or speech coding systems include so-called *fricative spreading* [14, 18] that is able to increase the speech intelligibility by downmixing frequency components above 4 kHz.[20]
- The order of the visual and acoustical presentation of the word alternatives should be randomized. Without randomization the listeners quickly get a preference depending on the order of the visual presentation (learning effect). It is also important to note that at least two systems or algorithmic approaches should be tested. Also the amount of examples that have been processed by one version should be spread randomly over the whole test. If only one system is tested it should be compared with a nearly ideal system that is free from noise or distortion. The intelligibility of such a system serves as a reference. Note, that the error rate of such an ideal system is usually not zero!
- After performing the test the results should be analyzed. According to, e.g., Voiers [44] the estimated probability of correct answers, adjusted for the effects of contents, is given by

$$\hat{p}_{\rm c} = \frac{N_{\rm correct} - \dfrac{N_{\rm wrong}}{M-1}}{N}\,, \tag{11.59}$$

  with the following abbreviations:

[20] In most current voice transmission systems, the bandwidth is still limited to 3.4 or 4 kHz. This bandwidth is sufficiently large for vowels as spoken by a majority of speakers. However, for consonants, especially for fricatives like /s/ or /f/, this is not always true because their spectral energy is often located above 4 kHz. To improve the quality of the narrowband signal, techniques to shift the spectrum of fricatives below the cutoff frequency of the transmission system can be applied. To achieve this, the signal is recorded first at a high sample rate (e.g., 11.025 or 16 kHz) and noise suppression is applied to the wideband signal. Afterwards frequencies above 4 kHz are shifted down according to various rules [18]. An increase of the speech intelligibility from 90.2% (without fricative spreading) to 94.3% (with fricative spreading) was measured.

N_{correct} : total number of correct answers,
N_{wrong} : total number of wrong answers,
M : number of alternatives (6 for an MRT, 2 for a DRT),
N : total number of answers.

The right side of Eq. 11.59 leads to a good approximation of the true probability p_c only if the entire set of permissible responses are equally attractive. A set consisting of the stimulus words *went*, *sent*, *bent*, and *subjective* would obviously not fulfill the assumption. Furthermore, it is assumed that the listeners respond to all lists, meaning that

$$N_{\text{correct}} + N_{\text{wrong}} = N . \tag{11.60}$$

Additionally, a statistical analysis about the significance of the test should be made. Details about such an analysis will be given in Sec. 11.4.3.

11.4.2 Example

As in Sec. 11.3.1 we will show further details about the realization of a rhyme test for evaluating the speech intelligibility by using an example. In contrast to Sec. 11.3.1 where two bandwidth extension algorithms have been compared we will evaluate the quality of in-car communication systems (intercom) here. For those readers who are not familiar with in-car communication systems a brief introduction is given in the next section. Further details can be found in Chapter 14 of this book. Those readers that have already basic knowledge about intercom systems may continue with Sec. 11.4.2.2.

11.4.2.1 Basics of In-Car Communication Systems

In limousines and vans communication between passengers in the front and in the rear may be difficult. Driver and front passengers speak towards the windshield. Thus, they are hardly intelligible for those sitting behind them. In the directions rear-to-front and left-to-right the acoustic loss is smaller. This can be measured by placing a so-called artificial mouth loudspeaker[21] at the driver's seat and torsos with earmicrophones [23] at the passenger's seat and at the backseats, respectively. On average the acoustic loss is 5 to 15 dB larger to the backseat passenger (as compared to the front passenger).

Fig. 11.18 sketches the structure of a simple car interior communication system [34,38] aimed to support only front-to-rear conversations with one microphone and one loudspeaker. Since driver and front passenger are located at well defined positions, fixed microphone arrays (not depicted in Fig. 11.18) can point towards each of them. Feedback suppression by means of an adaptive notch filter can improve the system. Thus, the howling margin is improved. A

[21] This is a loudspeaker which has (nearly) the same radiation pattern as the human speech apparatus.

device with nonlinear characteristic attenuates large signal amplitudes. The output gain of a car interior communication system needs to be adjusted continuously according to the current driving condition. While only a moderate gain is required whenever the car does not drive a large gain is required and more artifacts will be tolerated at high speed.

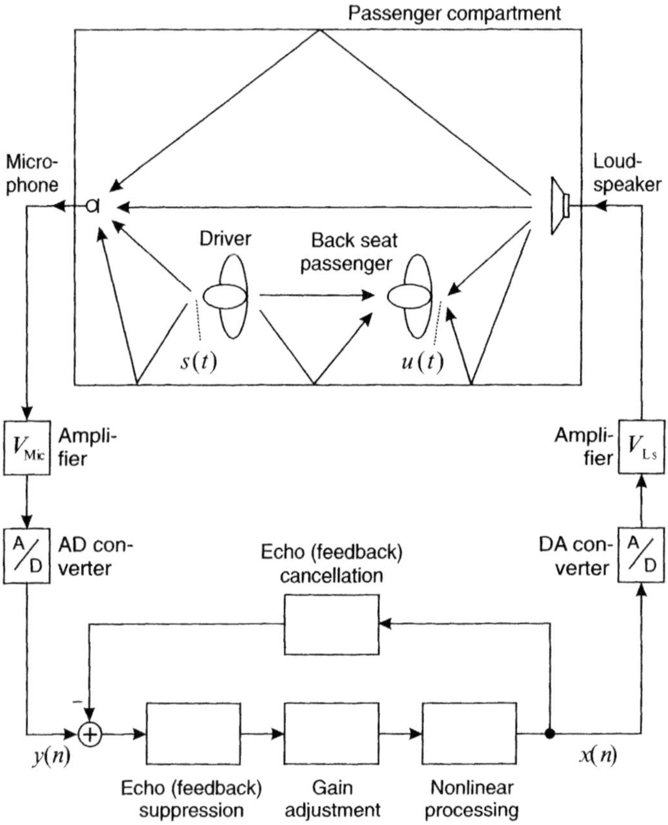

Fig. 11.18. Structure of a car interior communication system. Further details about such systems can be found in Chapter 14 of this book.

Fig. 11.19 shows the results of a car interior communication system. The system utilizes 8 microphones (2 per passenger) and 6 loudspeaker channels (standard car loudspeakers). To obtain high speech intelligibility beamforming, feedback cancellation, loss control and dynamic processing are applied. Especially at high speeds (90 km/h or more) a clear improvement of the communication quality can be achieved. To visualize this gain a binaural recording was made with a torso located on the seat behind the front passenger. Fig. 11.19 shows a time-frequency analysis of the output signal of the microphone located in the torso's left ear. The car was driving at about 160 km/h.

The driver was talking with the same loudness during the entire recording. After 16 seconds the system was deactivated for about 7 seconds to demonstrate the system performance. Within the time-frequency analysis the speech components of the driver are recognizable whenever the system is activated. During deactivation, however, the driver's speech is mostly masked by the driving noise.

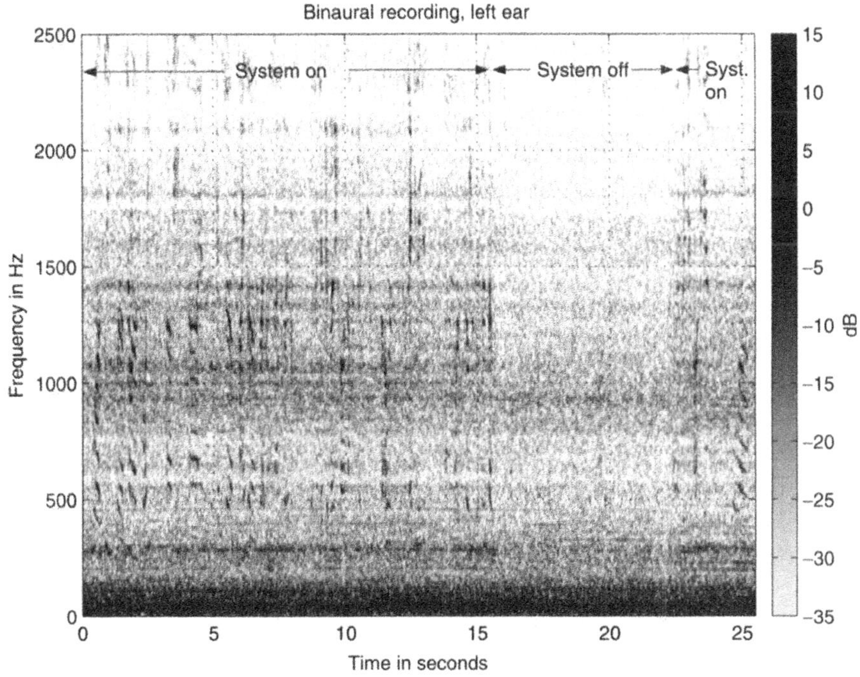

Fig. 11.19. Time-frequency analysis of the left channel of a binaural recording made on the right backseat.

As mentioned before, further details about in-car communication systems can be found in Chapter 14 of this book.

11.4.2.2 Rhyme Test for In-Car Communication Systems

For evaluating the improvement in terms of speech intelligibility of the intercom system two pairs of rhyme tests were performed – each pair consisted of one test with an activated system and another one without the system. In order to have best reliability of the tests prerecorded speech instead of spontaneous was utilized within the tests.

For the recordings a list containing 100 groups (six words per group) of monosyllable, rhyming words [41, 42] was used – resulting in 600 stimulus words. Since the way of speaking is changing with the amount of background noise (Lombard effect [16, 35]) the recordings were made in different simulated noise conditions.[22] To achieve this, first several binaural recordings of real background noise were made within a car driving at different speeds. By using a calibrated combined recording and playback device we were able to achieve a high fidelity when presenting the recorded background noises to the speakers via headphones (see Fig. 11.20). After a short period of getting accustomed to each of the noise scenarios the speakers read the list containing 600 stimulus words. For recording the stimulus words two microphones were used – a reference microphone (omnidirectional, located at a distance of about 15 cm in front of the mouth reference point [21]) and a close talking microphone (located about 5 cm left of the mouth reference point in order to get rid of the effects of breathing). In Fig. 11.20 two photos depict the setup of the microphones and the headphones.

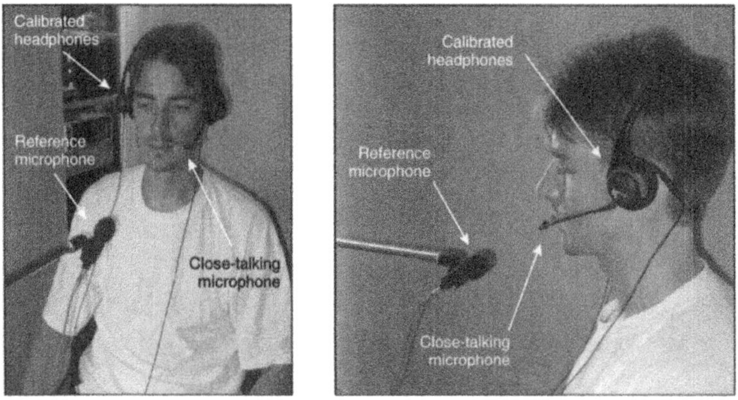

Fig. 11.20. Setup for recording of the stimulus words for the rhyme tests.

The recordings were made in an acoustically *dry* environment, meaning that the walls and the ceiling of the recording room were covered with sound absorbing panels. The signals recorded by the close talking microphone were used – after calibrating it according to the mouth reference point – for finally playing the stimulus words within the car. Since the test was performed with German listeners also the stimulus words were German. In contrast to the word lists presented before we have used stimulus words that differ either within their initial consonant, their center vowel, or their final consonant.

For obtaining the audio examples that were presented to the listeners finally a so-called *artificial mouth loudspeaker* was placed on the front passenger's seat (see upper two photos in Fig. 11.21). This loudspeaker had (nearly)

[22] Note that the recordings differ significantly – both in terms of power as well as in the way how the speakers articulate vowels, etc. – between the different noise conditions.

the same radiation pattern as a human head. Rhyme tests were performed in two scenarios:

- during stand-still at a parking area and
- at a speed of about 130 km/h (about 80 miles/h) on a motorway.

For each scenario all utterances that belong – in terms of their Lombard level – to the appropriate noise conditions were selected and played via the artificial mouth loudspeaker. The playback was calibrated such that a reference microphone located about 15 cm in front of the loudspeaker[23] records the same power as during the recording of the stimulus words.

Due to changing environmental conditions like weather and traffic, the background noise would have changed too from listener to listener. Thus, the rhyme test was carried out in a laboratory instead of the actual motorway. For this reason, we have used again the binaural recording device worn by one of the backseat passengers (see lower two photos in Fig. 11.21). With this device all stimulus words that were played via the artificial mouth loudspeaker were recorded binaurally and the resulting stereo signals were utilized as sound examples for the rhyme test that was performed in the laboratory.

After performing the playback of the stimulus words for one speaker with an activated intercom system the car was driven back to the entry point of the motorway and the same part of the road was driven again. All recordings

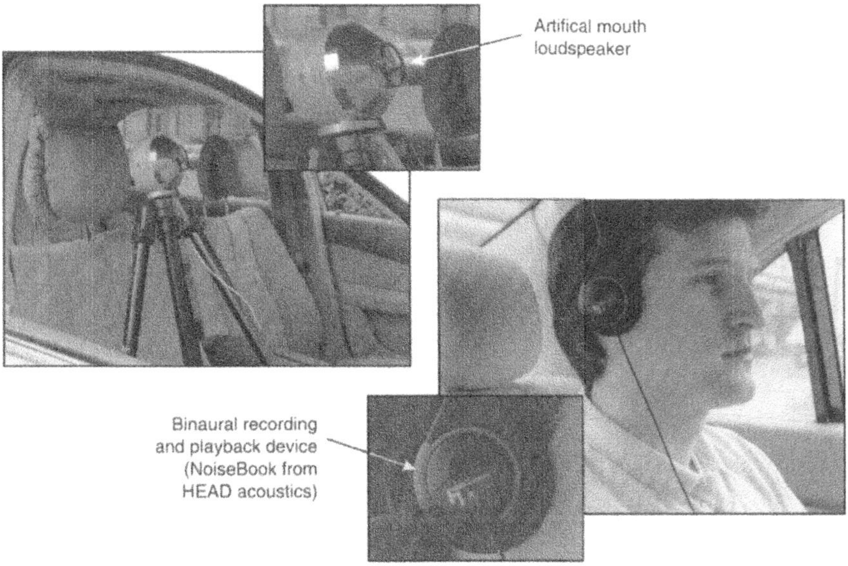

Fig. 11.21. Setup for recording the sound examples for the rhyme test.

[23] To be precise: in front of the mouth reference point of the loudspeaker.

were repeated but then with a deactivated system.[24] Even though this way of obtaining the audio examples is more reliable and less time-consuming compared to putting each listener into the driving car one major drawback still exists: when comparing the binaural recordings for the individual stimulus word with and without the intercom system the background noise level varies slightly. It might happen that, e.g., during the recording with the intercom system another car overtook but not when the recording without the system was carried out. However, we assume that the amount of audio examples was large enough in order to average out those effects.

After having finished the recordings the actual rhyme tests were performed in the laboratory. For each of the four conditions (intercom system off at 0 km/h, intercom on at 0 km/h, intercom off at 130 km/h, and intercom on at 130 km/h) 10 to 15 listeners of different ages and genders participated in the tests. For each listener 40 pairs of rhyming words were selected randomly from the recorded data bases. Both words were presented visually first. After that, one of the examples was selected (again randomly) and played via headphones. Afterwards the listeners had to decide which of the two stimulus words was acoustically presented. In Fig. 11.22 the amount of correct answers for each rhyme test is depicted.

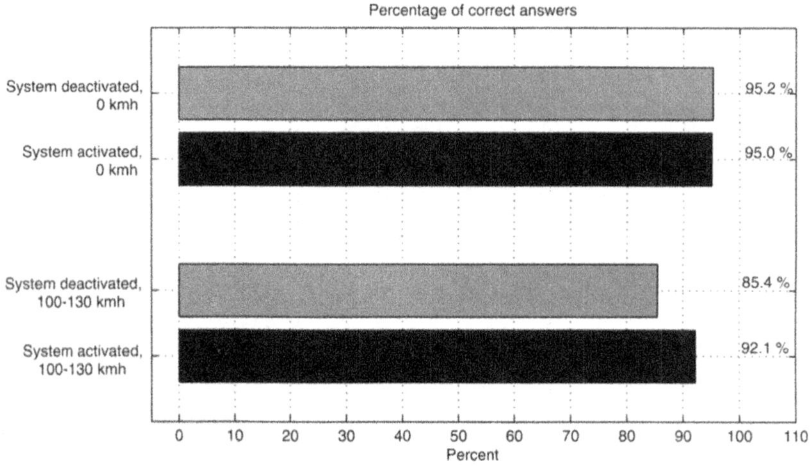

Fig. 11.22. Results of the rhyme tests.

Since the intercom system adjusts its gain automatically according to the background noise it is not surprising that no or nearly no difference was measured at 0 km/h (95.0 % for the activated system and 95.2 % for the deacti-

[24] Since more than one speaker was used for speaking the 600 rhyming words under different simulated noise condition, the process of doing the binaural recordings within the car took about 2 days.

vated system). The conditions of those two tests were more or less optimum – meaning that all stimulus words were clearly understandable. Most of the errors were made such that the word that was presented on the left of the computer monitor (that was read first) was also selected by the listeners even if the second was actually presented acoustically. As a result of the first two rhyme tests one can conclude the following:

- The tested intercom system did not improve the speech intelligibility when the car is in stand-still.
- Even under optimum conditions the listeners do not achieve a correctness of 100 %.

As a result of the last point one can conclude further that an intercom system that is able to improve the word correctness under high noise conditions such that the same correctness as during stand still is achieved works perfectly.

Even though the system under test did not achieve such a high rate of correct results, the amount of correct results could still be increased impressively: from 85.4 % without the intercom system to 92.1 % with an activated system.

In Table 11.11 the detailed results in terms of absolute numbers are presented. Furthermore, separate analysis of the correctness concerning stimulus words that differ within the initial or final consonant or within their center vowel are presented. In all cases the rates do not change much at a speed of 0 km/h, but large improvements can be observed at a higher speed.

11.4.3 Statistical Analysis

As in the case of CMOS tests we will end this section with answering the question about the statistical significance of the obtained DRT results. A detailed analysis – comparable with the analysis performed in Sec. 11.3.2 – would lead in this case to a so-called *exact Fisher test* [29]. With a few approximations, however, a simpler hypothesis test can be performed. If we assume the results of each rhyme test to be Gaussian distributed and to be statistically independent of the other tests, then a single sided so-called *t-test* [31] can be performed. Due to the Gaussian assumption the probability density function is fully described by the mean μ and the standard deviation σ. Within a t-test both quantities are assumed to be unknown.

When comparing the DRT results with and without the intercom system the main question is whether the mean μ_{on} obtained with the activated system is larger than the mean μ_{off} without the system. The corresponding standard deviations σ_{on} and σ_{off} need not necessarily to be equal.

11.4.3.1 Hypotheses

Under these assumptions the following hypotheses are set up:

Table 11.11. Results of the rhyme tests.

	Driving speed:	0 km/h		130 km/h	
	Intercom system:	on	off	on	off
All utterances	Correct	456 (95.0 %)	457 (95.2 %)	479 (92.1 %)	410 (85.4 %)
	Wrong	24 (5.0 %)	23 (4.8 %)	41 (7.9 %)	70 (14.6 %)
Difference within the first consonant	Correct	96 (92.3 %)	97 (91.5 %)	104 (85.2 %)	79 (73.8 %)
	Wrong	8 (7.7 %)	9 (8.5 %)	18 (14.8 %)	28 (26.2 %)
Difference within the center vowel	Correct	190 (98.4 %)	188 (97.4 %)	198 (99.0 %)	178 (92.7 %)
	Wrong	3 (1.6 %)	5 (2.6 %)	2 (1.0 %)	14 (7.3 %)
Difference within the final consonant	Correct	170 (92.9 %)	172 (95.0 %)	177 (89.4 %)	153 (84.5 %)
	Wrong	13 (7.1 %)	9 (5.0 %)	21 (10.6 %)	28 (15.5 %)

Hypothesis H_0 : We assume that $\mu_{\text{off}} > \mu_{\text{on}}$, meaning that better DRT results are obtained without the intercom system.

Hypothesis H_1 : We assume that $\mu_{\text{off}} \leq \mu_{\text{on}}$, meaning that better or equal DRT results are obtained when the intercom system is activated.

11.4.3.2 Results

When testing the Hypothesis H_0 with a t-test we obtain the results that are depicted in Tab. 11.12. Since the intercom system contained an automatic gain unit that adjusted the output gain according to the background noise level it is not surprising that no or nearly no difference can be detected during stand still. Also the probability for rejecting H_0 is only 0.01 (if all utterances are taken into account).

At high speed (130 km/h), however, the improvement was clearly significant. Taking all tests into account leads to a probability smaller than 10^{-10} that H_0 is true. Since a lower number of tests has been performed for the individual differences at the beginning, in the middle, or at the end of the stimulus words the probabilities of H_0 are larger too. Nevertheless, a maximum probability of 0.02 is obtained for the differences in the first consonant.

During stand still the communication within a car is possible without any support by in-car communication systems since the speech intelligibility is sufficiently high. This result is also obtained with the statistical analysis – the probability $p(H_0)$ is bounded by $p(H_0) \leq 0.99$.

Table 11.12. Results of the rhyme tests.

	Driving speed:	0 km/h		130 km/h	
	Intercom system:	on	off	on	off
All utterances	Correct	456	457	479	410
	Wrong	24	23	41	70
	$p(H_0$ is true)	≤ 0.99		$\leq 10^{-10}$	
Difference within the first consonant	Correct	96	97	104	79
	Wrong	8	9	18	28
	$p(H_0$ is true)	≤ 0.41		≤ 0.02	
Difference within the center vowel	Correct	190	188	198	178
	Wrong	3	5	2	14
	$p(H_0$ is true)	≤ 0.23		$\leq 10^{-3}$	
Difference within the final consonant	Correct	170	172	177	153
	Wrong	13	9	21	28
	$p(H_0$ is true)	≤ 0.80		$\leq 10^{-3}$	

11.5 Outlook

In this chapter we have presented some details about objective tests for noise reduction systems and subjective tests for new applications. It was not our aim to cover all current algorithms and tests in this topic. We have focussed on how these tests can be applied for new applications where no standard evaluation methods have been successfully introduced, yet. We hope that the readers are encouraged now to modify current objective and subjective tests in order to evaluate algorithms and systems for new speech and audio applications.

References

[1] ANSI S3.2-1989: *Method for Measuring the Intelligibility of Speech over Communication Systems,* American National Standard, 1989.

[2] S. Bech: Selection and training of subjects for listening tests on sound-reproducing equipment, *J. Audio Eng. Soc.,* **40**(7/8), 590–610, July/August 1992.

[3] R. Le Bouquin, G. Faucon, A. Akbari Azirani: Proposal of a composite measure for the evaluation of noisecancelling methods in speech processing, *Proc. EUROSPEECH '93,* **1**, 227–230, Berlin, Germany, 1993.

[4] J. R. Deller Jr., J. H. L. Hansen, J. G. Proakis: *Discrete-Time Processing of Speech Signals,* New York, NY, USA: IEEE Press, 2000.

[5] C. Delogu, A. Paolini, P. Ridolfi, K. Vagges: Intelligiblity of speech produced by text-to-speech systems in good and telephonic conditions, *Acta Acoustica,* **3**, 89–96, 1995.

[6] P. Dreiseitel, E. Hänsler, H. Puder: Acoustic echo and noise control - a long lasting challenge, *Proc. EUSIPCO '98,* **2**, 945–952, Island of Rhodes, Greece, 1998.

[7] P. Dreiseitel: Quality evaluation of noise reduction algorithms, *Proc. IWAENC '99*, 88–91, Pocono Manor, NJ, USA, 1999.

[8] P. Dreiseitel: *Untersuchung und Bewertung von Geräuschreduktionsverfahren zur Verbesserung gestörter Sprache*, Aachen, Germany: Shaker, 2001 (in German).

[9] P. Dreiseitel: Hybrid quality measures for single-channel speech enhancement algorithms, *European Transactions on Telecommunication,* **13**(2), 159–166, 2002.

[10] J. Durbin: The fitting of time series models, *Rev. Int. Stat. Inst.,* **28**, 233–244, 1960.

[11] ETSI recommendation EG 201 377-1: *Speech processing transmission and quality aspects,* ETSI, France, 2002.

[12] H. W. Gierlich: The auditory perceived quality of hands-free telephones: auditory judgements, instrumental measurements and their relationship, *Speech Communication,* **20**(3-4), 241–254, 1996.

[13] A. H. Gray, J. D. Markel: Distance measures for speech processing, *IEEE Trans. Acoust. Speech Signal Process.,* **ASSP-24**(5), 380–391, 1976.

[14] E. Hänsler, G. Schmidt: *Acoustic Echo and Noise Control - A Practical Approach,* New York, NY, USA: Wiley, 2004.

[15] J. H. L. Hansen, M. A. Clements: Use of objective speech quality measures in selecting effective spectral estimation techniques for speech enhancement, *Proc. of the 32nd Midwest Symposium on Circuits and Systems,* Urbana, IL, USA, 105–108, 1990.

[16] J. H. L. Hansen: Morphological constrained feature enhancement with adaptive cepstral compensation (MCE-ACC) for speech recognition in noise and Lombard effect, *IEEE Trans. Speech Audio Process.,* **T-SA-2**(4), 598–614, 1994.

[17] M. H. Hayes: *Statistical Digital Signal Processing and Modelling,* New York, NY, USA: Wiley, 1996.

[18] D. A. Heide, G. S. Kang: Speech enhancement for bandlimited speech, *Proc. ICASSP '98*, **1**, 393–396, Washington, DC, USA, 1998.

[19] R. V. Hogg, A. T. Craig: *Introduction to Mathematical Statistics,* 5th ed., New York, NY, USA: Macmillan, 1995.

[20] B. Iser, G. Schmidt: Neural networks versus codebooks in an application for bandwidth extension of speech signals, *Proc. EUROSPEECH '03,* **1**, 237–240, Geneva, Switzerland, 2003.

[21] ITU-T Recommendation P.64: *Determination of sensivity/frequency characteristics of local telephone systems,* Geneva, Switzerland, 1999.

[22] ITU-T recommendation P.340: *Transmission characteristics and speech quality parameters of hands-free terminals,* Geneva, Switzerland, 2000.

[23] ITU-T recommendation P.581: *Use of head and torso simulator (HATS) for hands-free terminal testing,* Geneva, Switzerland, 2000.

[24] ITU-T recommendation P.800: *Methods for subjective determination of transmission quality,* Geneva, Switzerland, 1996.

[25] ITU-T recommendation P.831: *Perceptual evaluation of speech quality (PESQ), an objective method for end-to-end speech quality assessment of narrowband telephone networks and speech codecs,* Geneva, Switzerland, 2001.

[26] ITU-T recommendation P.862: *Subjective performance evaluation of network echo cancellers,* Geneva, Switzerland, 1998.

[27] P. Jax, P. Vary: On artificial bandwidth extension of telephone speech, *Signal Processing,* **83**(8), 1707–1719, August 2003.

[28] U. Kornagel: Spectral widening of telephone speech using an extended classification approach, *Proc. EUSIPCO '02,* **2**, 339–342, Toulouse, France, 2002.

[29] E. L. Lehmann: *Testing Statistical Hypothesis,* 2nd ed., Berlin, Germany: Springer, 1997.

[30] E. L. Lehmann, H. J. M. D'Abrera: *Statistical Methods Based on Ranks,* rev. ed. Englewood Cliffs, NJ, USA: Prentice Hall, 1998.

[31] J. Lehn, H. Wegmann: *Einführung in die Statistik,* Stuttgart, Germany: Teubner, 2004 (in German).

[32] N. Levinson: The Wiener RMS error criterion in filter design and prediction, *J. Math. Phys.*, **25**, 261–268, 1947.

[33] Y. Linde, A. Buzo, R. M. Gray: An algorithm for vector quantizer design, *IEEE Trans. Comm.*, **COM-28**(1), 84–95, Jan. 1980.

[34] E. Lleida, E. Masgrau, A. Ortega: Acoustic echo and noise reduction for car cabin communication, *Proc. EUROSPEECH '01,* **3**, 1585–1588, Aalborg, Denmark, 2001.

[35] E. Lombard: Le signe de l'elevation de la voix, *Ann. Maladies Oreille, Larynx, Nez. Pharynx,* **37**, 101–119, 1911 (in French).

[36] N. Magotra, M. Kirstein, S. Sirivara, T. Hamill: Quantitative and qualitative (subjective) perceptual measures for speech processing applica-

tions, *Conference Record of the Thirtieth Asilomar Conference on Signals, Systems and Computers, 1996,* 766–769, Asilomar, CA, USA, 1997.
[37] S. Olive: Differences in performance and preference of trained versus untrained listeners in loudspeaker tests: a case study, *Proc. 114th AES Convention,* Amsterdam, Netherlands, 2003.
[38] A. Ortega, E. Lleida, E. Masgrau, F. Gallego: Cabin car communication system to improve communication inside a car, *Proc. ICASSP '02,* **4**, 3836–3839, Orlando, FL, USA, 2002.
[39] T. Painter, A. Spanias: Perceptual coding of digital audio, *Proc. of the IEEE,* **88**(4), 451–513, April 2000.
[40] S. R. Quackenbusch, T. P. Barnwell, M. A. Clements: *Objective measures of speech quality,* Englewood Cliffs, NJ, USA: Prentice Hall, 1988.
[41] J. Sotscheck: Ein Reimtest für Verständlichkeitsmessungen mit deutscher Sprache als ein verbessertes Verfahren zur Bestimmung der Sprachübertragungsgüte, *Der Fernmeldeingenieur,* **36**, 1–84, 1982 (in German).
[42] J. Sotscheck: Sprachverständlichkeit bei additiven Störungen, *Acoustica,* **57**(4/5), 257–267, 1989 (in German).
[43] J. Tilp: Single-channel noise reduction with pitch-adaptive postfiltering, *Proc. EUSIPCO '00,* 101–104, Tampere, Finland, 2000.
[44] W. Voiers: Evaluating processed speech using the diagnostic rhyme test, *Speech Technology,* 30–39, Jan./Feb. 1983.
[45] S. Wang, A. Sekey, A. Gersho: An objective measure for predicting subjective quality of speech coders, *IEEE Journal on Selected Areas in Communications,* 819–829, June 1992.
[46] E. Zwicker, H. Fastl: *Psychoacoustics – Facts and Models,* Berlin, Germany: Springer, 1990.

12

An Auditory Scene Analysis Approach to Monaural Speech Segregation

Guoning Hu[1] and DeLiang Wang[2]

[1] The Ohio State University, Biophysics Program
Columbus, OH 43210
[2] The Ohio State University, Department of Computer Science & Engineering and Center for Cognitive Science
Columbus, OH 43210

A human listener has the remarkable ability to segregate an acoustic mixture and attend to a target sound. This perceptual process is called auditory scene analysis (ASA). Moreover, the listener can accomplish much of auditory scene analysis with only one ear. Research in ASA has inspired many studies in computational auditory scene analysis (CASA) for sound segregation. In this chapter we introduce a CASA approach to monaural speech segregation. After a brief overview of CASA, we present in detail a CASA system that segregates both voiced and unvoiced speech. Our description covers the major stages of CASA, including feature extraction, auditory segmentation, and grouping.

12.1 Introduction

We live in an environment rich in sound from many sources. The presence of multiple sound sources complicates the processing of the target sound we are interested in, and often causes serious problems for many applications, such as automatic speech recognition and voice communication. There has been extensive effort to develop computational systems that automatically separate target sound or attenuate background interference. When target and interference come from different directions and multiple microphones are available, one may remove interference using spatial filtering that extracts the signal from the target direction or cancels the signals from the interfering directions [29], or independent component analysis [26]. These approaches do not apply to the situations when target and interference originate from the same direction or only mono-recordings are available. In the monaural (one microphone) situation, one must consider the intrinsic properties of target or interference to distinguish and separate them.

As a special case of monaural separation, monaural speech segregation is of particular importance. Here a major challenge is the variety of interference;

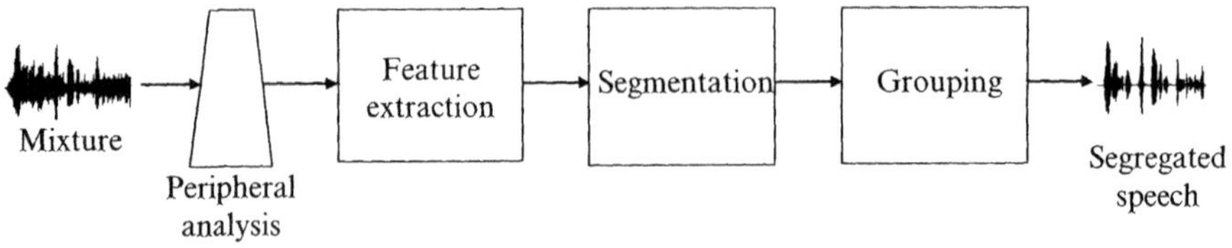

Fig. 12.1. Schematic diagram of a typical CASA system.

the interference can change in time and space in an unpredictable manner. For decades, various methods have been proposed for monaural speech enhancement, such as spectral subtraction [5], subspace analysis [17], hidden Markov modeling [46], and sinusoidal modeling [28]. These methods usually assume certain properties (or models) of interference and then enhance speech or attenuate interference based on these assumptions. Their capacity for dealing with the variability of interference is much limited in comparison with human speech segregation. This contrast has motivated a different approach to monaural speech segregation – mimicking the auditory process of source separation.

The auditory segregation process is termed by Bregman as *auditory scene analysis (ASA)* [6], which is considered to take place in two main stages: Segmentation and grouping. In segmentation, the acoustic input is decomposed into segments or sensory elements, each of which should originate from a single source. In grouping, the segments that likely arise from the same source are grouped together. Segmentation and grouping are guided by perceptual principles that determine how the auditory scene is organized according to ASA cues. These cues characterize intrinsic sound properties, including harmonicity, onset and offset, location, and prior knowledge of specific sounds.

Research in ASA has inspired considerable work to build CASA (computational auditory scene analysis) systems for sound segregation (for reviews see [8, 44]). A main advantage is that CASA does not make strong assumptions about interference. A typical CASA system is shown in Fig. 12.1. It contains four stages: Peripheral analysis, feature extraction, segmentation, and grouping. The peripheral processing decomposes the auditory scene into a time-frequency (T-F) representation via bandpass filtering and time windowing. The second stage extracts auditory features corresponding to ASA cues, which will be used in subsequent segmentation and grouping. In segmentation and grouping, the system generates segments for both target and interference and groups the segments originating from the target into a target stream. A stream corresponds to a sound source. The waveform of segregated target can then be resynthesized from the target stream [7, 52, 53].

As an illustration, Figs. 12.2(a) and 12.2(b) show a T-F decomposition and the waveform of a male utterance, "Her right hand aches whenever the barometric pressure changes," from the TIMIT database [18]. Figs. 12.2(c) and 12.2(d) show a T-F decomposition and the waveform of this utterance

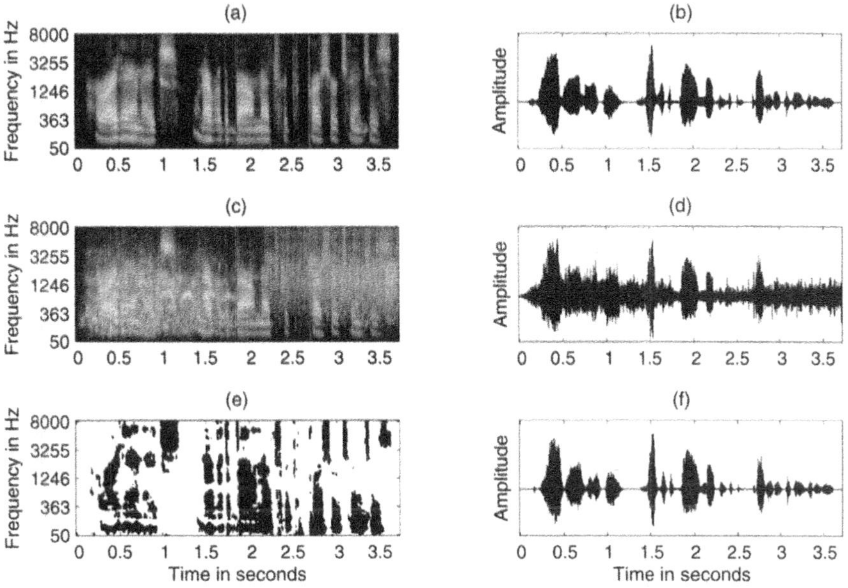

Fig. 12.2. Signal representation. (a) T-F decomposion of a male utterance, "Her right hand aches whenever the barometric pressure changes." (b) Waveform of the utterance. (c) T-F decomposition of the utterance mixed with a crowd noise in playground. (d) Waveform of the mixture. (e) Target stream composed of all the T-F units (black regions) dominated by the target (ideal binary mask). (f) The waveform resynthesized from the target stream.

mixed with a crowd noise in playground, at the overall SNR of 0 dB. Here the input is decomposed using a filterbank with 128 gammatone filters [36] and 20-ms rectangular time windows with 10-ms window shift (see Sec. 12.3 for implementation details). The small T-F area within each filter channel and time window is referred to as a T-F unit. Figs. 12.2(a) and 12.2(c) show the energy within each T-F unit, where a brighter pixel indicates stronger energy. Fig. 12.2(e) shows the target stream we aim to segregate, which contains all the T-F units dominated by the target. To obtain this stream, a typical CASA system first merges neighboring T-F units dominated by target speech into segments, shown as the contiguous black regions in the figure, in the stage of segmentation. In this stage, the system may also generate segments for interference. Then in the stage of grouping, the system determines for each segment whether it belongs to the target and groups them accordingly. Fig. 12.2(f) shows the waveform resynthesized from the target stream in Fig. 12.2(e).

Brown and Wang have recently written a review chapter on CASA for speech segregation, also included in a Springer volume [8]. Instead of another review, this chapter mainly describes our systematic effort on monaural speech segregation. The chapter is organized as follows. In Sec. 12.2, we give a brief overview of other CASA studies on monaural speech segregation. We then describe in depth the major stages of our CASA system in the subsequent four sections. Sec. 12.7 concludes the chapter.

12.2 Computational Auditory Scene Analysis

Natural speech contains both voiced and unvoiced portions. Voiced speech is periodic or quasi-periodic. Periodicity and temporal continuity are two major ASA cues for voiced speech. A well-established representation for periodicity and pitch perception is a correlogram - a running autocorrelation of each filter response across an auditory filterbank [31, 48]. The correlogram has been adopted by many CASA systems for monaural segregation of voiced speech [7, 13, 16, 23, 52, 53]. In what is regarded as the first CASA model, Weintraub used a coincidence function, a version of autocorrelation, to capture periodicity as well as amplitude modulation (AM) [53]. He then used the coincidence function to track pitch contours of multiple utterances. Sounds from different speakers are separated by using iterative spectral estimation according to pitch and temporal continuity. Cooke proposed a model that first generates local segments based on filter response frequencies and temporal continuity [13]. These segments are merged into groups based on common harmonicity and common AM. A pitch contour is then obtained for each group, and groups with similar pitch contours are put into the same stream. Brown and Cooke proposed to form segments based on correlation of filter responses across frequency and frequency transition across time [7]. These segments are grouped by common periodicity and common onset and offset. Wang and Brown used a two-layer oscillator network for speech segregation [52]. In the first layer, segments are formed based on cross-channel correlation and temporal continuity. In the second layer, segments are grouped into two streams, one for the target and the other its background on the basis of dominant pitch in each time frame. The above systems are mainly data-driven approaches. Ellis developed a prediction-driven system which generates predictions using a world model and compares the predictions against the input [16]. The world model includes three types of sound elements: Noise cloud, transient click, and harmonic sound.

12.2.1 Computational Goal of CASA

A critical issue in developing a CASA system is to determine its computational goal [32]. With the initial analysis into T-F units described in Sec. 12.1, we have suggested that the computational goal of CASA should be to retain the T-F units where target speech is more intense than interference and remove others [21, 23]. In other words, the goal is to identify a binary T-F mask, referred to as the *ideal binary mask*, where 1 indicates that target is stronger than interference in the corresponding T-F unit and 0 otherwise. Target speech can then be resynthesized with the ideal mask by retaining the acoustic energy from T-F regions corresponding to 1's and rejecting other energy. This computational goal is supported by the auditory masking phenomenon: Within a critical band, a weaker signal tends to be masked by a stronger one [35]. In addition, there is considerable evidence supporting the ideal binary mask as the CASA objective from both human speech intelligibility [9,12,42] and automatic speech recognition [14,42] studies (for an extensive discussion see [51]). What Fig. 12.2(e) shows, in fact, is an ideal binary mask for the mixture in Fig. 12.2(c). As shown in Fig. 12.2(f), the speech resynthesized from the ideal binary mask is close to the original clean utterance in Fig. 12.2(b).

12.2.2 Motivation and Overview of the Approach

A common problem in earlier CASA systems is that they do not separate voiced speech well in the high-frequency range from broadband interference. This problem is closely related to the peripheral analysis of the input scene. Most CASA systems perform initial frequency analysis with an auditory filterbank, where the bandwidth of a filter increases quasi-logarithmically with its center frequency. These filters are usually derived from psychophysical data and mimic cochlear filtering. An important observation is that the structure of cochlear filtering limits the ability of human listeners to resolve harmonics [38,40]. In the low-frequency range, harmonics are resolved since the corresponding auditory filters have narrow passbands including only one harmonic. In the high-frequency range, harmonics are generally unresolved since the corresponding auditory filters have wide passbands including multiple harmonics. In addition, psychophysical evidence suggests that the human auditory system processes resolved and unresolved harmonics differently [3, 11]. Hence, one should carefully consider the distinctions between resolved and unresolved harmonics. The earlier CASA systems employ the same strategy to segregate all the harmonics, which works reasonably well for resolved harmonics but poorly for unresolved ones.

A basic fact of acoustic interaction is that the filter responses to multiple harmonics are amplitude-modulated and the response envelopes fluctuate at the fundamental frequency (f_0) of target speech [19]. Fig. 12.3 shows the response and its envelope of a gammatone filter centered at 2.5 kHz within a time frame (from 0.7 s to 0.72 s). The input is the clean utterance in Fig. 12.2(b). The response in Fig. 12.3 is strongly amplitude-modulated, and its envelope fluctuates at the f_0 rate in this frame.

Motivated by the above considerations, we have proposed to employ different methods to segregate resolved and unresolved harmonics of target speech [23]. For resolved harmonics, we generate segments based on temporal continuity and cross-channel correlation, and these segments are grouped according to common periodicity, similar to [52]. For unresolved harmonics, we generate segments based on common AM in addition to temporal continuity.

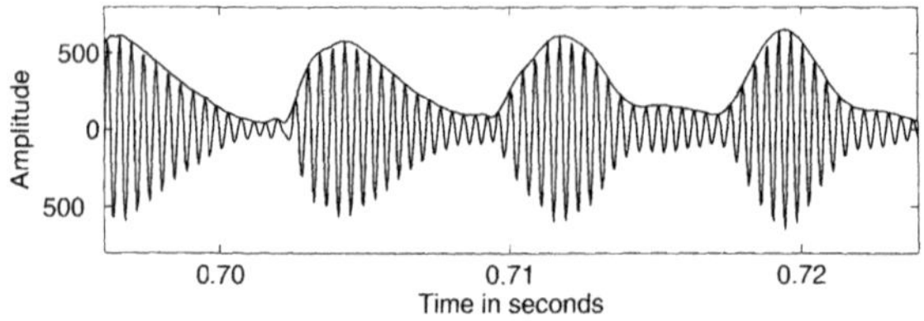

Fig. 12.3. AM effects for filter responses to multiple harmonics. The input is the utterance in Fig. 12.2(b). The filter is centered at 2.5 kHz.

These segments are further grouped based on AM rates, which are obtained from the temporal fluctuations of the corresponding response envelopes.

So far the discussion is focused on voiced speech. Compared with voiced speech, unvoiced speech is generally much weaker and more susceptible to interfering sounds. In addition, unvoiced speech lacks harmonic structure and is noise-like itself. As a result, segregating unvoiced speech is significantly more challenging and little previous work has addressed this problem.

We have proposed to segment unvoiced speech based on onset and offset analysis [24]. Onsets and offsets are important ASA cues [6] because different sound sources in an environment seldom start and end simultaneously. In addition, there is strong evidence for onset detection by auditory neurons [37]. In the time domain, onsets and offsets likely form boundaries between sounds from different sources. Common onsets and offsets also provide natural cues to integrate sounds from the same source across frequency. In addition, onset/offset based segmentation is applicable to both voiced and unvoiced speech.

Given segments, the next task is to group segments of unvoiced speech. When interference is non-speech, we may formulate this as a classification task, i.e., to classify segments as unvoiced speech or interference. Since each segment should belong to one source, segments dominated by unvoiced speech are likely to have similar acoustic-phonetic characteristics as those of clean speech, whereas segments dominated by interference are likely to be different. Therefore, we can group segments for unvoiced speech by analyzing their acoustic-phonetic features [25].

In the following sections, we describe our systematic investigation into segregation of both voiced and unvoiced speech. Our model includes all the major stages of a typical CASA system shown in Fig. 12.1.

12.3 Peripheral Analysis and Feature Extraction

We describe below early auditory processing that first decomposes the input in the T-F domain, and then extracts auditory features corresponding to ASA cues.

12.3.1 Auditory Periphery

Cochlear filtering is commonly modeled by a gammatone filterbank that decomposes the input in the frequency domain [36]. The impulse response of a gammatone filter centered at frequency f is:

$$g(f,t) = \begin{cases} b^a t^{a-1} e^{-2\pi bt} \cos(2\pi f t), & t \geq 0, \\ 0, & \text{else}, \end{cases} \tag{12.1}$$

where $a = 4$ is the order of the filter. b is the equivalent rectangular bandwidth, which increases as the center frequency f increases. For a filter channel c, let f_c be the center frequency. Let $x(t)$ be the input signal, the response from channel c, $x(c,t)$, is then

$$x(c,t) = x(t) * g(f_c, t), \tag{12.2}$$

where "$*$" denotes convolution. The response is shifted backwards by $(a-1)/(2\pi b)$ to compensate for the filter delay [20]. We find that this delay compensation gives a small but consistent performance improvement. In addition, the gain of each filter is adjusted according to equal loudness contours [27] in order to simulate the pressure gains of the outer and middle ears.

The response of a filter channel can be further processed by the Meddis model of auditory nerve transduction [33]. This model simulates the nonlinear processes of the auditory nerve, such as rectification, saturation, and phase locking. Its output represents the firing rate of an auditory nerve fiber, denoted by $h(c,t)$.

In each filter channel, the output is divided into 20-ms time frames with 10-ms overlapping between consecutive frames. This frame size is commonly used for speech analysis. Examples of this T-F decomposition are shown in Figs. 12.2(a) and 12.2(c). The resulting time-frequency representation is called a *cochleagram*.

12.3.2 Correlogram and Cross-Channel Correlation

As discussed in Sec. 12.2, a correlogram is a commonly used periodicity representation, which consists of autocorrelations of filter responses across all the filter channels. Let u_{cm} denote a T-F unit for frequency channel c and time frame m, the corresponding normalized autocorrelation of the filter response is given by

$$A_{\mathrm{H}}(c,m,\tau) = \frac{\sum\limits_n h\big(c, mT_{\mathrm{f}} - nT_{\mathrm{s}}\big)\, h\big(c, mT_{\mathrm{f}} - nT_{\mathrm{s}} - \tau T_{\mathrm{s}}\big)}{\sum\limits_n h^2\big(c, mT_{\mathrm{f}} - nT_{\mathrm{s}}\big)}. \tag{12.3}$$

Here, τ is the delay and n denotes digitized time. $T_{\mathrm{f}} = 10$ ms, the time shift from one frame to the next and T_{s} is denoting the sampling time. The above summation is over the period of a time frame.

As shown in [7, 52], cross-channel correlation measures the similarity between the responses of two adjacent filter channels and indicates whether the filters respond to the same sound component. For T-F unit u_{cm}, its cross-channel correlation with $u_{c+1,m}$ is given by

$$C_{\mathrm{H}}(c,m) = \sum_{\tau=0}^{L} \widetilde{A}_{\mathrm{H}}(c,m,\tau)\, \widetilde{A}_{\mathrm{H}}(c+1,m,\tau)\,, \tag{12.4}$$

where $\widetilde{A}_{\mathrm{H}}(c,m,\tau)$ denotes $A_{\mathrm{H}}(c,m,\tau)$ normalized to 0 mean and unity variance and $LT_{\mathrm{s}} = 12.5$ ms - the maximum delay for A_{H}.

The AM information is carried by the response envelope. A general way to obtain response envelope is to perform half-wave rectification followed by low-pass filtering. Since we are interested in the envelope fluctuations corresponding to target pitch, here we perform a bandpass filtering instead, where the passband corresponds to the plausible f_0 range of target speech. Let $h_{\mathrm{E}}(c,t)$ denote the resulting envelope.

Similar to Eqs. 12.3 and 12.4, we can compute a normalized envelope autocorrelation to represent AM rates:

$$A_{\mathrm{E}}(c,m,\tau) = \frac{\sum\limits_{n} h_{\mathrm{E}}\big(c, mT_{\mathrm{f}} - nT_{\mathrm{s}}\big)\, h_{\mathrm{E}}\big(c, mT_{\mathrm{f}} - nT_{\mathrm{s}} - \tau T_{\mathrm{s}}\big)}{\sum\limits_{n} h_{\mathrm{E}}^{2}\big(c, mT_{\mathrm{f}} - nT_{\mathrm{s}}\big)} \tag{12.5}$$

and cross-channel correlation of response envelopes,

$$C_{\mathrm{E}}(c,m) = \sum_{\tau=0}^{L} \widetilde{A}_{\mathrm{E}}(c,m,\tau)\, \widetilde{A}_{\mathrm{E}}(c+1,m,\tau)\,. \tag{12.6}$$

Figs. 12.4(a) and 12.4(b) illustrate the correlogram and the envelope correlogram as well as the cross-channel correlation at time frame 70 (i.e., 0.7 s from the beginning of the stimulus) for the utterance in Fig. 12.2(b), and Figs. 12.4(c) and 12.4(d) the corresponding responses to the mixture in Fig. 12.2(d). As shown in the figure, the autocorrelation of filter response generally reflects the periodicity of a single harmonic for a channel in the low-frequency range where harmonics are resolved. The autocorrelation is amplitude-modulated in high-frequency channels where harmonics are unresolved. As a result, these autocorrelations are not as highly correlated between adjacent channels. On the other hand, the corresponding autocorrelations of response envelopes are more correlated, as shown in the cross-channel correlations of response envelopes, since they have similar fluctuation patterns.

12.3.3 Onset and Offset

Onsets and offsets correspond to sudden amplitude increases and decreases. A standard way to identify such intensity changes is to take the first-order

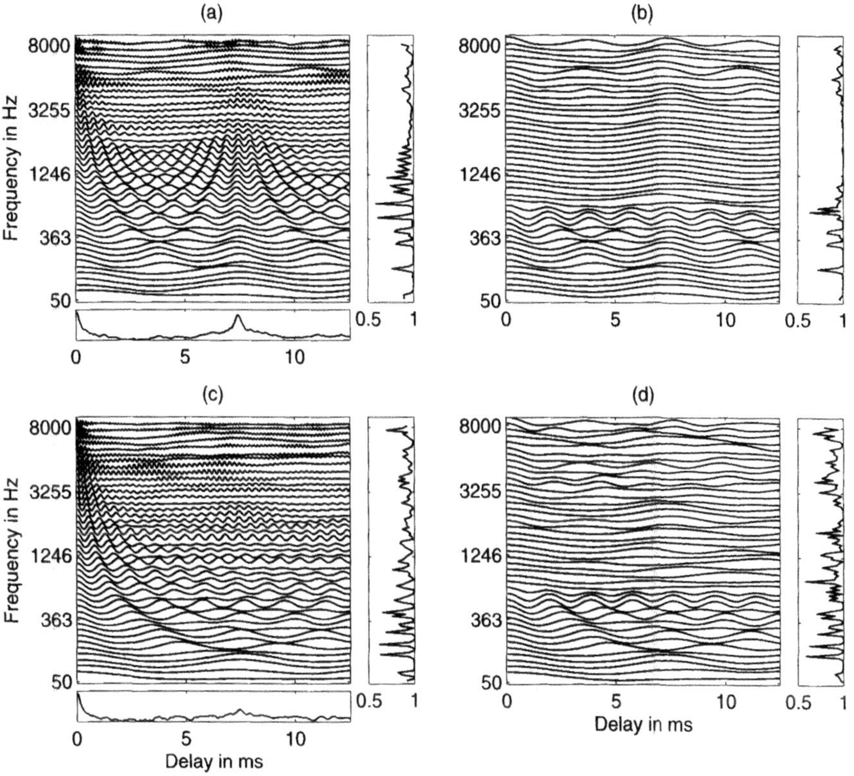

Fig. 12.4. Auditory features. (a) Correlogram at frame 70 (i.e. 0.7 second after the onset) for the utterance in Fig. 12.2(b). For clarity, every third channel is displayed. The corresponding cross-channel correlation is given in the right panel, and the summary correlogram in the bottom panel. (b) Envelope correlogram for the utterance. The corresponding cross-channel envelope correlation is shown in the right panel. (c) Correlogram and cross-channel correlation for the mixture in Fig. 12.2(d). (d) Envelope correlogram and cross-channel envelope correlation for the mixture.

derivative of intensity with respect to time and then find the peaks and valleys of the derivative. Because of intrinsic intensity fluctuations, many peaks and valleys of the derivative do not correspond to actual onsets and offsets. To reduce such fluctuations, we smooth the intensity over time, as is commonly done in edge detection for image analysis. The intensity is basically the square of the envelope of filter response. Smoothing can be performed through either a diffusion process [43] or lowpass filtering. Here we consider a special case of Gaussian smoothing. First we calculate the response envelope with half-wave rectification and lowpass filtering. Since here we are interested in low-rate fluctuations of envelope, the cutoff frequency of the lowpass filter should be set smaller than 30 Hz. The obtained low-rate envelope is denoted by $x_{\mathrm{E}}(c,t)$. The smoothed intensity is obtained by the convolution of the intensity (in decibels) and a Gaussian kernel with mean 0 and variance σ^2. The derivative

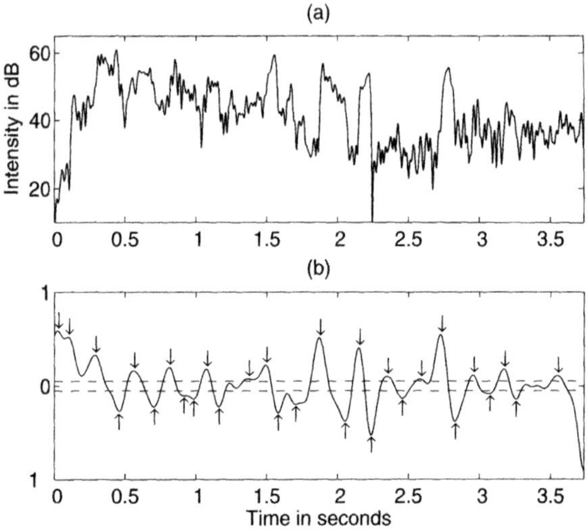

Fig. 12.5. Onset and offset detection. The input is the response of a gammatone filter to the mixture in Fig. 12.2(d). The upper panel shows the response intensity, and the lower panel shows the results of onset and offset detection using Gaussian smoothing ($\sigma = 16$). The threshold for onset detection is 0.05 and for offset detection is -0.05, indicated by the dash lines. Detected onsets are marked by downward arrows and offsets by upward arrows.

of the smoothed response is:

$$\frac{d}{dt}\left\{10\left[\log_{10} x_{\mathrm{E}}^2(c,t)\right] * \left[\frac{1}{\sqrt{2\pi}\,\sigma}\exp\left(-\frac{t^2}{2\sigma^2}\right)\right]\right\}$$
$$= -20\log_{10}\left|x_{\mathrm{E}}(c,t)\right| * \left[\frac{t}{\sqrt{2\pi}\,\sigma^3}\exp\left(-\frac{t^2}{2\sigma^2}\right)\right].$$

Onsets correspond to the peaks of the derivative above a certain threshold, and offsets the valleys below a certain threshold. The purpose of thresholding is to remove peaks and valleys corresponding to insignificant intensity fluctuations. The above procedure is very similar to the standard Canny edge detector in image processing [10]. An example of the above procedure is shown in Fig. 12.5.

12.3.4 Pitch Determination

A periodic sound consists of a harmonic series, each harmonic having a frequency that equals or is a multiple of f_0. A frequently-used method for pitch determination is to simply pool autocorrelations across all the channels and then identify a global peak in the summary correlogram [34]. When a harmonic sound is presented, the autocorrelations of the activated filters in a correlogram all exhibit a peak at the delay corresponding to the pitch period. Let $A_{\mathrm{H}}(m,\tau)$ be the summary correlogram at frame m, that is,

$$A_{\mathrm{H}}(m,\tau) = \sum_{c} A_{\mathrm{H}}(c,m,\tau)\,. \tag{12.7}$$

The estimated pitch period at frame m, $\tau_{\mathrm{S}}(m)$, is the lag corresponding to the maximum of $A_{\mathrm{H}}(m,\tau)$ in the plausible pitch range of target speech. The bottom panels of Figs. 12.4(a) and 12.4(c) shows examples of summary correlogram. The peak at 7.21 ms in Fig. 12.4(c), representing the estimated pitch period, turns out to equal that of target speech (indicated by the peak in Fig. 12.4(a)).

There are several problems with the above method. First, it gives a pitch value at each frame no matter whether the signal at a particular frame is periodic or not. Second, detected pitches in neighboring frames may correspond to different sound sources. Third, it may not give a reliable estimate of target pitch even if it exists, when the signal-to-noise ratio (SNR) is low. This is because the autocorrelations in many channels exhibit peaks not corresponding to the periodicity of the target. To address these problems, we apply the Wang and Brown algorithm [52] in an initial grouping stage. The grouping in their algorithm is based on the dominant pitch of each time frame, and can eliminate many T-F units that unlikely belong to the target. With this initial grouping, we track a *target pitch contour* by pooling autocorrelations from the remaining T-F units. The initial grouping is not accurate in the high-frequency range; however, this stage is employed only for the purpose of pitch tracking. Note that pitch detection requires only a portion of harmonics; the fact that the Wang and Brown algorithm works reasonably well in the low-frequency range accords well with the perceptual evidence that human pitch detection primarily relies on lower harmonics [39]. To deal with the third problem, we take advantage of the pitch continuity to enhance the reliability of target pitch tracking [23]. Specifically, we first determine the reliability of an estimated pitch based on its coherence with the periodicity patterns of the retained T-F units in initial grouping, and then use pitch continuity to interpolate for unreliable pitch points on the basis of reliable ones.

The algorithm given in [23] assumes that the target has a continuous pitch contour throughout the whole utterance. We note that it can be applied iteratively to handle the general situation when the target utterance contains multiple pitch contours separated by unvoiced speech or silence. This is because the initial grouping by the Wang-Brown algorithm is based on the longest segment. Specifically, after extracting the first pitch contour based on the longest segment, the algorithm can then be applied to extract the next longest pitch contour from remaining time frames where no target pitch has been detected. This process can repeat until no more significant pitch contour is detected. However, when interference also contains periodic signal, the above procedure may generate pitch contours for interference as well. To determine the

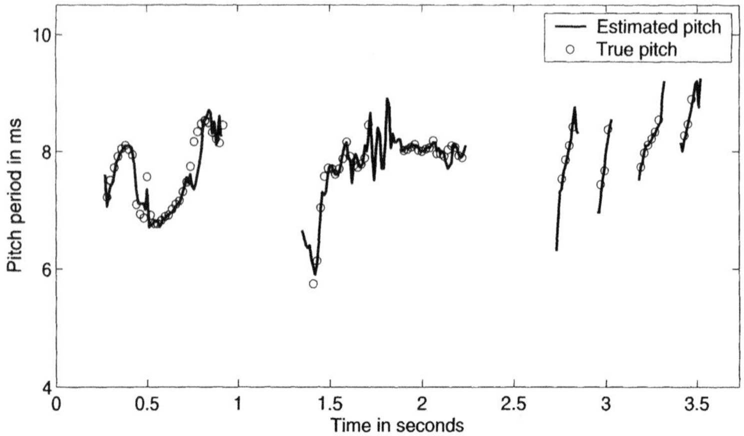

Fig. 12.6. Results of pitch tracking for the mixture in Fig. 12.2(d). Solid lines indicate estimated target pitch contours. True pitch points are marked by circles. For clarity, every other frame is displayed.

source for each pitch contour is the task of sequential grouping, which is not addressed by this algorithm.

Fig. 12.6 shows several estimated pitch contours from the mixture in Fig. 12.2(d) obtained iteratively as described above. For most time frames, the detected contours well match the reference pitch contours generated from the clean utterance using *Praat* - a standard pitch determination algorithm for clean speech [4].

The above algorithm only tracks one pitch at a frame. When interference also contains a harmonic component, e.g., another utterance, it is probably more helpful to track multiple pitch contours from different sources simultaneously. Wu et al. [54] proposed a robust multipitch tracking algorithm, which works as follows. After a T-F analysis and computing the correlogram, their algorithm selects channels that likely contain signals dominated by harmonic sources. The other channels mostly contain aperiodic sounds and therefore are ignored in subsequent processing. Within each channel, the algorithm treats a peak in the auto-correlation as a pitch hypothesis. Then it integrates periodicity information across the selected channels in order to formulate the conditional probabilities of multiple pitch hypotheses given the periodicity information in these channels. Finally, a continuous hidden Markov model (HMM) is used to model pitch dynamics across successive time frames and the Viterbi algorithm is then used to find optimal pitch contours. The Wu et al. algorithm is illustrated in Fig. 12.7 for pitch tracking of two simultaneous utterances. The algorithm successfully tracks the pitch contours of both utterances at most time frames.

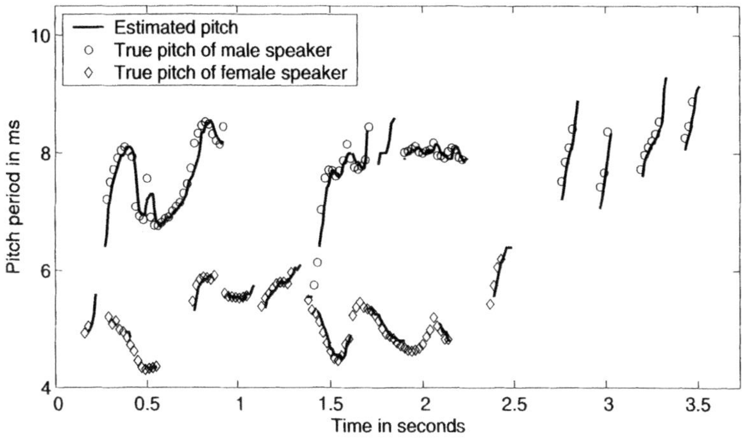

Fig. 12.7. Results of multipitch tracking by the Wu et al. algorithm. The input is the mixture of the utterance in Fig. 12.2(b) and a female utterance: "That noise problem grows more annoying each day." Solid lines indicate estimated target pitch contours. True pitch points of the male utterance are marked by circles, and those of the female utterance are marked by diamonds. For clarity, every other frame is displayed.

12.4 Auditory Segmentation

In addition to the conceptual importance of segmentation in ASA, a segment as a region of T-F units contains more global information of the source that is missing from individual T-F units, such as spectral and temporal envelope. This information could be key for distinguishing sounds from different sources. One may skip the stage of segmentation by grouping individual T-F units directly. However, such grouping based on local information will not be very robust. In our view, auditory segmentation provides a foundation for grouping and is essential for successful CASA.

12.4.1 Segmentation for Voiced Speech

Speech signal lasts for a period of time, within which it has good temporal continuity. Therefore, T-F units neighboring in time tend to originate from the same source. In addition, because the passbands of adjacent channels have significant overlap, a harmonic usually activates a number of adjacent channels, which leads to high cross-channel correlation. Therefore, we perform segmentation by merging T-F units based on temporal continuity and cross-channel correlation [52]. More specifically, only units with sufficiently high cross-channel correlation of correlogram responses are marked, and neighboring marked units are iteratively merged into segments. To account for AM effects of unresolved harmonics, we separately mark and merge high-frequency units on the basis of cross-channel correlation of response envelopes.

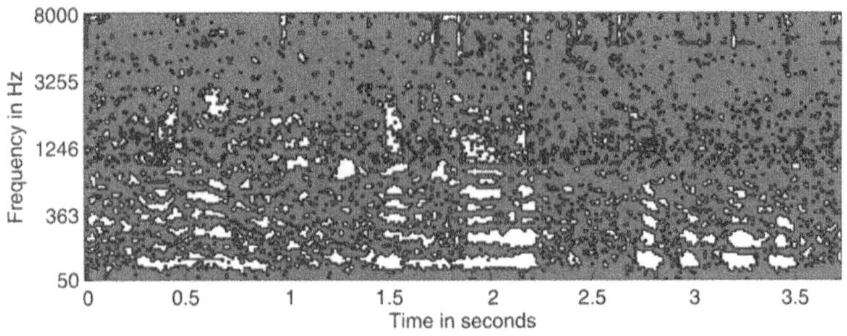

Fig. 12.8. The bounding contours of estimated segments based on cross-channel correlation and temporal continuity. The background is represented by gray.

Fig. 12.8 shows the segments generated in this process for the mixture in Fig. 12.2(d). Compared with Fig. 12.2(e), computed segments cover most T-F regions dominated by voiced speech. In addition, T-F regions dominated by target and interference are well separated into different segments. If desired, very small segments can be easily removed [23]. Note that the correlogram is a periodicity representation, and correlogram-based segmentation therefore is not expected to work well for aperiodic signal, such as unvoiced speech.

12.4.2 Segmentation Based on Onset/Offset Analysis

Unvoiced speech lacks the harmonic structure, and as a result is more difficult to segment. We have proposed a general method for segmentation based on analysis of event onset and offset. This method has three stages: Smoothing, onset/offset detection, and multiscale integration [24], and it works for both voiced and unvoiced speech since onsets and offsets are generic sound properties.

As discussed in Sec. 12.3.3, onsets and offsets correspond to sudden intensity increases and decreases, or the peaks and valleys of the time derivative of the intensity. In smoothing, the intensity is first smoothed over time in order to reduce insignificant fluctuations. We then perform smoothing over frequency to enhance synchronized onsets and offsets across frequency. The degree of smoothing is referred to as the scale [43]. A larger scale leads to smoother intensity.

In the stage of onset/offset detection and matching, our system detects onsets and offsets in each filter channel and merges them into onset and offset fronts if they are sufficiently close. A front corresponds to a boundary along the frequency (vertical) axis in a 2-D cochleagram representation. Individual onset and offset fronts are matched, and a matching pair encloses a segment.

Smoothing with a large scale may blur onsets and offsets of a short acoustic event. Consequently, segmentation may miss short events or combine different events into one segment. On the other hand, smoothing with a small (fine)

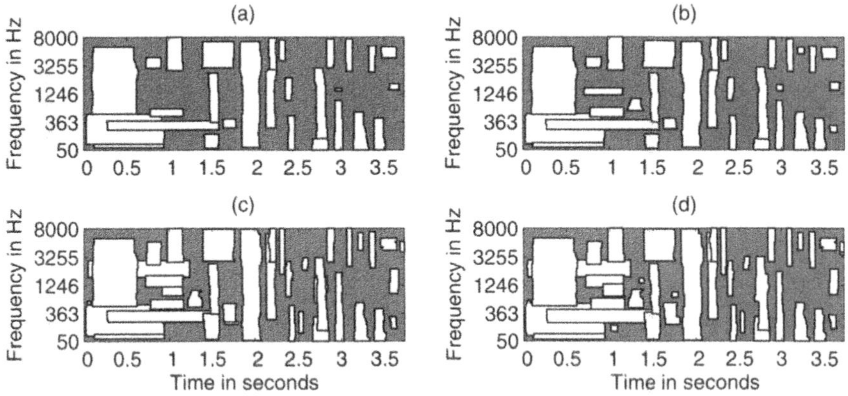

Fig. 12.9. Bounding contours of estimated segments from multiscale analysis of onset and offset. (a) One scale analysis. (b) Two-scale analysis. (c) Three-scale analysis. (d) Four-scale analysis. The input is the mixture shown in Fig. 12.2(d). The background is represented by gray.

scale may not adequately remove insignificant intensity fluctuations. Consequently, segmentation may separate a continuous event into several segments. In general, it is difficult to obtain satisfactory segmentation with a single scale. The multiscale analysis stage is designed to detect and localize different events at appropriate scales. In this stage, we start at a large scale and then gradually move to the finest scale. At each scale, the system generates new segments from within the current background and locates more accurate onset and offset positions for existing segments.

Figs. 12.9(a), 12.9(b), 12.9(c), and 12.9(d) show the bounding contours of obtained segments by integrating 1, 2, 3, and, 4 scales, respectively (see [24] for implementation details). The input is the mixture in Fig. 12.2(d). Comparing it with Fig. 12.2(e), we can see that at the largest scale, the system captures most of the speech events, but misses some small segments. As the system integrates more fine scales, more segments for speech as well as for interference appear.

12.5 Voiced Speech Grouping

To group voiced speech, we use the segments obtained by the simple algorithm described in Sec. 12.4.1. Given pitch contours from the target pitch tracking described in Sec. 12.3.4, we label each T-F unit as target dominant or interference dominant according to target pitch. To label a T-F unit, we first compare the periodicity of its response with the estimated pitch. Specifically, a T-F unit u_{cm} is labeled as target if the correlogram response at the estimated pitch period $\tau_S(m)$ is close to the maximum of the autocorrelation

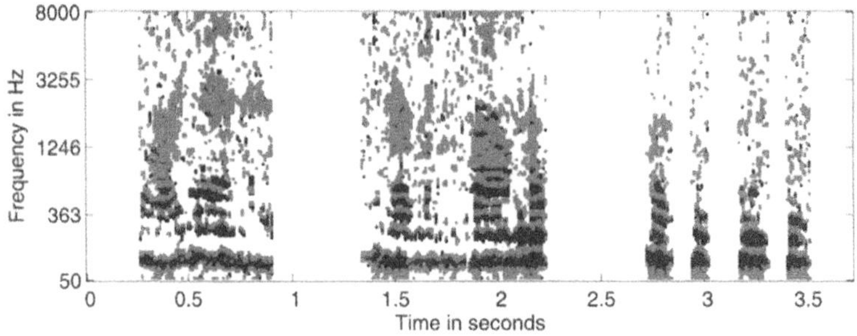

Fig. 12.10. Results of T-F unit labeling for the mixture in Fig. 12.2(d). Black regions: units labeled as target by the periodicity criterion; gray regions: units labeled as target by the AM criterion.

within the plausible pitch range, Γ:

$$\frac{A_{\mathrm{H}}\big(c, m, \tau_{\mathrm{S}}(m)\big)}{\max\limits_{\tau \in \Gamma} A_{\mathrm{H}}\big(c, m, \tau\big)} > \theta_{\mathrm{T}} \,. \tag{12.8}$$

The above criterion, referred to as the *periodicity criterion*, works well for resolved harmonics.

For units responding to multiple harmonics, their responses are amplitude-modulated. We have found that the periodicity criterion does not work well for such units. Observe that the envelope of such a response fluctuates at the f_0 rate of the source. Therefore, we label these T-F units by comparing their AM rates with the estimated pitch. A straightforward way is to check the autocorrelation of response envelopes:

$$\frac{A_{\mathrm{E}}\big(c, m, \tau_{\mathrm{S}}(m)\big)}{\max\limits_{\tau \in \Gamma} A_{\mathrm{E}}\big(c, m, \tau\big)} > \theta_{\mathrm{A}} \,. \tag{12.9}$$

This criterion is referred to as the *AM criterion.*

In practice, we use the periodicity criterion to label T-F units that belong to segments formed on the basis of high cross-channel correlation of filter responses. Such units correspond to resolved harmonics. The remaining units are labeled by the AM criterion.

Fig. 12.10 shows the T-F units labeled as target for the mixture in Fig. 12.2(d). Compared with Fig. 12.2(e), one can see that most units dominated by target voiced speech are correctly labeled. However, some units containing stronger intrusion are also labeled as target speech, especially in the high-frequency range.

With unit labels, we group a segment into the target stream if the acoustic energy corresponding to its T-F units labeled as target exceeds half of the total energy of the segment. Furthermore, significant T-F regions labeled as inference are removed from the target stream. Finally, to group more target

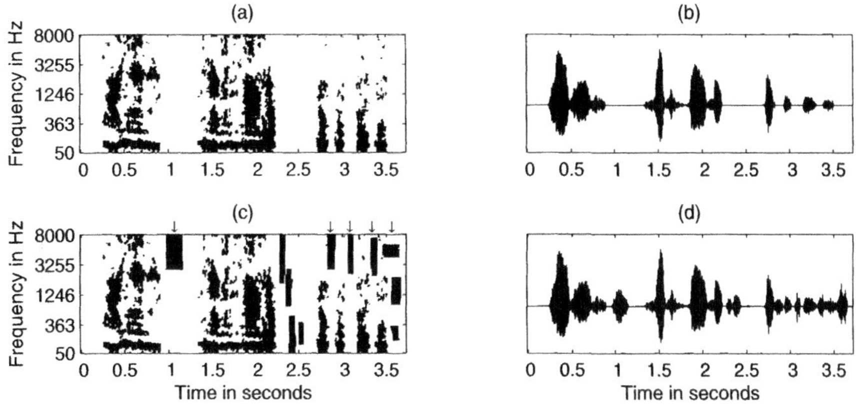

Fig. 12.11. Results of segregation for the mixture in Fig. 12.2(d). (a) Segregated voiced target. (b) The corresponding resynthesized voiced target. (c) Segregated final target. The arrows indicate the segregated fricatives and affricates. (d) Corresponding resynthesized final target.

energy we expand each target segment by iteratively grouping its neighboring units that are labeled as target and do not belong to any segment. When this expansion ends, the system yields a target stream and its background that consists of the remaining T-F units.

Figs. 12.11(a) and 12.11(b) shows the final target stream and the corresponding resynthesized speech for the mixture in Fig. 12.2(d). Compared with Fig. 12.2(e), this stream contains a majority of the T-F units where voiced target speech dominates. In addition, only a small number of units where intrusion dominates are incorrectly included. The segregated speech waveform in Fig. 12.11(b) within voiced speech sections is much more similar to that of the clean speech in Fig. 12.2(b) than the mixture waveform in Fig. 12.2(d).

The performance of the system on voiced speech segregation has been evaluated using a corpus of 100 mixtures composed of 10 voiced utterances mixed with 10 intrusions collected by Cooke [13]. This corpus has been used to test previous CASA systems [7, 13, 15, 16, 52]. The intrusions have a considerable variety; specifically they are described in Tab. 12.1.

As discussed in Sec. 12.2, our computational goal is to estimate the ideal binary mask. Therefore, our evaluation compares the segregated speech, $\hat{s}(n)$, against the speech waveform resynthesized from the ideal binary mask, $s(n)$. Let $e_1(n)$ denote the signal present in $s(n)$ but missing from $\hat{s}(n)$, and $e_2(n)$ the signal present in $\hat{s}(n)$ but missing from $s(n)$. Then, we measure the percentage of energy loss, P_{EL}, and the percentage of noise residue, P_{NR}:

Table 12.1. Types of intrusions.

Intrusion	Description
N0	1kHz pure tone
N1	white noise
N2	noise bursts
N3	"cocktail party" noise
N4	rock music
N5	siren
N6	trill telephone
N7	female speech
N8	male speech
N9	female speech

$$P_{\text{EL}} = \sum_n e_1^2(n) \Big/ \sum_n s^2(n)\,, \tag{12.10}$$

$$P_{\text{NR}} = \sum_n e_2^2(n) \Big/ \sum_n \hat{s}^2(n)\,. \tag{12.11}$$

P_{EL} indicates the percentage of target speech excluded from segregated speech, and P_{NR} the percentage of intrusion included. They provide complementary error measures of a segregation system and a successful system needs to achieve low errors in both measures.

The results from our model are shown in Tab. 12.2. Each value in the table represents the average result of one intrusion with 10 voiced utterances, and a further average across all intrusions is also shown. On average, our system retains 96.28% of target speech energy, and the percentage of noise residue is kept at 2.81%. The percentage of noise residue for the original mixtures is 36.05%, also shown in the table; energy loss is obviously zero for the original mixtures. As indicated by the table, our model achieves very good performance across the noise types. In particular, the errors measured by P_{EL} and P_{NR} are balanced in our system.

Since our model applies different mechanisms to segregate resolved and unresolved harmonics, it is instructive to present the performance in the high-frequency range separately. For this purpose, we calculate the percentages of energy loss and noise residue for only the filter channels with center frequencies greater than 1 kHz, denoted by P_{EL}^{H} and P_{NR}^{H}, respectively. Note that for the evaluation corpus, target harmonics in the frequency range above 1 kHz are generally unresolved. The corresponding results are shown in Tab. 12.2. Most of the voiced energy in the high-frequency range is recovered and not much interference is included. The performance in high-frequency range is not as good as that in the low-frequency range since intrusions are relatively much stronger in the high-frequency range, which is clear from the average noise residue of the original mixtures and that in the high-frequency range.

Table 12.2. P_{EL} and P_{NR} for segregation of voiced speech.

Intrusion	Segregated target				Mixture	
	$P_{EL}(\%)$	$P_{EL}^{H}(\%)$	$P_{NR}(\%)$	$P_{NR}^{H}(\%)$	$P_{NR}(\%)$	$P_{NR}^{H}(\%)$
N0	1.47	14.97	0.05	0.52	67.76	96.82
N1	4.61	32.48	3.78	61.00	57.16	96.00
N2	1.01	8.18	0.42	7.98	5.04	44.02
N3	4.04	12.90	2.14	6.44	18.15	42.57
N4	2.81	21.42	3.58	43.28	27.17	81.31
N5	1.32	7.47	0.06	0.46	78.84	97.90
N6	0.95	8.99	0.94	16.27	39.24	91.26
N7	2.01	9.76	2.25	8.68	16.68	43.49
N8	1.16	8.59	0.65	4.32	7.37	31.07
N9	17.80	19.25	14.22	5.47	43.09	27.72
Average	3.72	14.40	2.81	15.44	36.05	65.22

To compare waveforms directly we can measure SNR in decibels:

$$SNR = 10 \log_{10} \frac{\sum_n s^2(n)}{\sum_n \left[s(n) - \hat{s}(n)\right]^2} . \qquad (12.12)$$

The SNR for each intrusion averaged across 10 target utterances is shown in Fig. 12.12, together with the SNR of the original mixtures and the results from the Wang-Brown system [52], whose performance is representative of previous CASA systems, and a spectral subtraction method [5], a standard method for speech enhancement. Our system shows substantial improvements. In particular, it yields a 12.1 dB gain on average over the original mixtures, a 5.8 dB gain over the Wang-Brown model, and a 7.0 dB gain over spectral subtraction.

We point out that, although the above algorithm for voiced speech segregation is similar to that presented in [23], it is simplified a good deal. The guiding principle for the algorithm presented in this chapter is to simplify that in [23] as much as possible without sacrificing the segregation performance. Also the delay compensation for gammatone filters discussed in Sec. 12.3.1 is not implemented in [23]. Indeed, the SNR performance for the simplified version is even slightly better than that in [23]. For completeness, we give the entire algorithm in the Appendix along with a few further notes.

12.6 Unvoiced Speech Grouping

Unvoiced speech lacks the periodicity feature, which plays the primary role in voiced speech segregation, and segregation of unvoiced speech is particularly

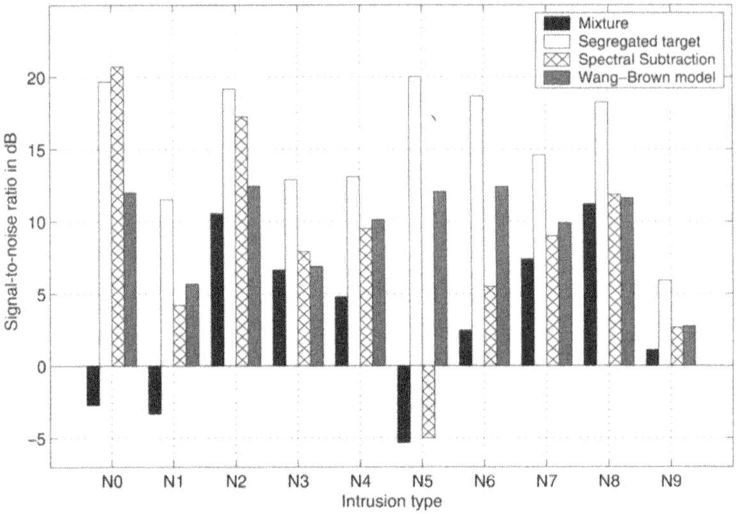

Fig. 12.12. Signal-to-noise ratio (SNR) results against the ideal binary mask for segregated speech and original mixtures. White bars show the results from our system, gray bars those from the Wang-Brown system, cross bars those from a spectral subtraction method, and black bars those of original mixtures.

challenging. Unvoiced speech in English contains three categories of consonants: Stops, fricatives, and affricates [30]. Stops consist of /t/, /d/, /p/, /b/, /k/, and /g/, and fricatives consist of /s/, /z/, /f/, /v/, /θ/, /ð/, /ʃ/, /ʒ/, and /h/. There are two affricates, /tʃ/ and /dʒ/, each of which is a stop followed by a fricative. Although about half of these consonants are phonetically voiced, their acoustic realizations often contain weak voicing [50], and they cannot be reliably segregated with pitch-based analysis. Hence all these consonants are treated in this section. As stated in Sec. 12.2, here we only deal with non-speech interference. Because of the similarity between fricatives and affricates, we consider them together. In this section, we first describe segregation of stop consonants and then segregation of fricatives and affricates.

12.6.1 Segregation of Stop Consonants

A stop consonant starts with a closure corresponding to the stop of airflow in the vocal tract, followed by a burst corresponding to a sudden release of airflow. The closure contains little energy and is usually masked by interference. The focus here is to segregate stop bursts.

In a previous study, we have proposed to segregate stop consonants in two steps: Stop detection and stop grouping [22]. In the first step, onset detection is performed in each frequency channel, and onset fronts are formed by connecting close onsets at neighboring channels. We distinguish onset fronts belonging to stop consonants from others via featured-based classification.

Table 12.3. P_{EL} and P_{NR} for stop consonants.

Overall SNR (dB)	$P_{\mathrm{EL}}(\%)$	$P_{\mathrm{NR}}(\%)$
0	84.79	9.62
10	70.68	2.81
20	41.56	0.81
30	28.01	0.04

Stop bursts are characterized by the following features: Spectral envelope, intensity, duration, and formant transition (see [1] for example). However, the formant transition from a stop to its neighboring voiced phoneme is very difficult to obtain; moreover, it is closely related to the spectrum. Therefore we use the following features for classification: Spectral envelope, intensity, and duration.

Stop consonants are grouped based on onset synchrony. Specifically, for each detected stop, the frequency channels that contain onsets synchronous with the onset of the stop burst are grouped together. The temporal boundary within each such channel is determined as from the minimum filter response immediately before the burst duration to the minimum point immediately after the burst. This pair of minima approximately marks the onset and the offset of the stop for the filter channel. The T-F units within this interval are hence labeled as belonging to the stop consonant.

The above method has been tested with 10 utterances from the TIMIT database mixed with the following 10 interference: White noise, pink noise, airplane noise, car noise, factory noise, noise burst, clicks, bar noise, fireworks, and rain. Average P_{EL} and P_{NR} for stop consonants at different SNR levels are shown in Tab. 12.3. The system performs well when SNR is relatively high. As SNR decreases, P_{EL} increases significantly while P_{NR} remains relatively low.

12.6.2 Grouping of Fricatives and Affricates

We group fricatives and affricatives with the segments obtained by the segmentation algorithm described in Sec. 12.4.2. Because fricatives and affricates are relatively steady acoustically [50], most T-F units dominated by these consonants are well organized into obtained segments. The task here is to distinguish these segments from those corresponding to interference. This is performed in two steps [25]. First, we remove those segments dominated by non-fricative and non-affricate sounds within voiced sections. Then we apply a Bayesian classifier to determine whether each remaining segment belongs to a fricative, an affricate, or interference.

The motivation of the first step is to take advantage of segregated voiced speech. In the segmentation stage described in Sec. 12.4.2, obtained segments containing significant portions of fricatives and affricates tend to contain little signal from other phonemes or interference. Therefore, segments overlapping significantly with non-fricative and non-affricate sounds are removed. To identify these segments, our system first uses the segregated voiced speech to determine time frames containing phonemes other than fricatives and affricates as follows.

Let H_0 be the hypothesis that a T-F region is dominated by interference, $H_{1,k}$ a T-F region dominated by a fricative or an affricate, indexed by k, and $H_{2,l}$ a T-F region dominated by another phoneme, indexed by l. Let $X(m)$ be the power spectrum of the input mixture at frame m, and $X_S(m)$ be the corresponding power spectrum within segregated target stream. Frame m is labeled as non-fricative and non-affricate if

$$\max_k P(H_{1,k}|X_S(m)) < \max_l P(H_{2,l}|X_S(m)) . \tag{12.13}$$

By applying the Bayesian rule, we have

$$\max_k \left[p(X_S(m)|H_{1,k})\, P(H_{1,k})\right] < \max_l \left[p(X_S(m)|H_{2,l})\, P(H_{2,l})\right] . \tag{12.14}$$

Note that frames not occupied by the segregated target are not considered. The segments whose energy is dominated by such frames are removed.

For each remaining segment, which lasts from frame m_1 to m_2, let $Y(m)$ be the power spectrum within the segment at frame m, and

$$\boldsymbol{Y} = \left[Y(m_1), Y(m_1+1), \ldots, Y(m_2)\right] . \tag{12.15}$$

This segment is classified as dominated by a fricative or an affricate if:

$$\max_k \left[p(\boldsymbol{Y}|H_{1,k})\, P(H_{1,k})\right] > p(\boldsymbol{Y}|H_0)\, P(H_0) . \tag{12.16}$$

Because segments have varied sizes, the complexity for computing $p(\boldsymbol{Y}|H_{1,k})$ and $p(\boldsymbol{Y}|H_0)$ directly is very high. Fortunately, we find that, by considering only the dependence between two consecutive frames, a good estimate of $p(\boldsymbol{Y}|H_0)$ can be obtained,

$$p(\boldsymbol{Y}|H_0) = p(Y(m_1)|H_0) \prod_{m=m_1}^{m_2-1} p(Y(m+1)|Y(m),\, H_0) . \tag{12.17}$$

This observation holds for $p(\boldsymbol{Y}|H_{1,k})$ also. Then Eq. 12.16 becomes

$$\begin{aligned} &\max_k \left[p(Y(m_1)|H_{1,k})\, P(H_{1,k}) \prod_{m=m_1}^{m_2-1} p(Y(m+1)|Y(m),\, H_{1,k})\right] \\ &> p(Y(m_1)|H_0)\, P(H_0) \prod_{m=m_1}^{m_2-1} p(Y(m+1)|Y(m),\, H_0) . \end{aligned} \tag{12.18}$$

In Eq. 12.18, segment duration is implicitly given. To emphasize the contribution of duration in classification, we insert duration D as an additional feature into Eq. 12.18:

$$\max_k \left[p\big(Y(m_1),\, D\big|H_{1,k}\big)\, P\big(H_{1,k}\big) \prod_{m=m_1}^{m_2-1} p\big(Y(m+1),\, D\big|Y(m),\, H_{1,k}\big)\right]$$
$$> p\big(Y(m_1),\, D\big|H_0\big)\, P\big(H_0\big) \prod_{m=m_1}^{m_2-1} p\big(Y(m+1),\, D\big|Y(m),\, H_0\big)\,, \tag{12.19}$$

so that the contributions from spectrum and duration are well balanced.

We use the two features of spectrum (including the spectral envelope and intensity) and duration for the classification task in both of the steps. The formant transition is another feature for identifying fricatives and affricates. As discussed in Sec. 12.6.1, the formant transition is partly captured by the spectrum. In addition, it is very difficult to extract. Therefore, it is not utilized here.

The prior distributions and probabilities required for calculating Eq. 12.14 and Eq. 12.19 are obtained from training using the training part of the TIMIT database and 90 environmental intrusions, including crowd noise, traffic noise, and wind, etc. A Gaussian mixture model with 8 components and a full covariance matrix for each mixture is used for training the probability density function for all the spectral features and duration. Then in calculating Eq. 12.14 and Eq. 12.19, we use marginal distribution since only a subset of spectral features is included in the formula.

All the segments identified as dominated by fricatives or affricates are added to the segregated voiced target. As an illustration, Figs. 12.11(c) and 12.11(d) show the final target stream and the corresponding resynthesized speech for the mixture in Fig. 12.2(d). The target utterance, "*H*er right *h*and ache*s* whene*v*er *th*e barometric pre*ss*ure *ch*an*g*e*s*" contains 7 fricatives and 2 affricates, italicized in the sentence. Among them, /h/ in "hand", /v/ in "whenever", and /ð/ in "the" are mainly voiced and portions of their energy are recovered in voiced speech segregation (see Fig. 12.11(a)). /h/ in "her" is mostly masked by the intrusion, hence not recoverable. The remaining 5 are successfully segregated by the system, as indicated by the arrows in Fig. 12.11(c). At the same time, some intrusion-dominated T-F regions are also included in the segregated target.

The performance of fricative and affricate segregation is systematically evaluated with 20 utterances from the testing part of the TIMIT database, mixed with 10 intrusions at different SNR levels. The intrusions are white noise, electrical fan, rooster crowing and clock alarm, traffic noise, crowd noise in playground, crowd noise with music, crowd noise with clapping, bird chirping and water flow, wind, and rain.

Tab. 12.4 shows the average P_{EL} and P_{NR} for segregation of fricatives and affricates. As shown in the table, our system extracts about 70% of the fricative and affricate energy from the mixture under different SNR situations. On the other hand, it retains certain amounts of interference, which are much less than those included in the original mixture. Our system performs significantly better than a spectral subtraction method, especially in low SNR situations [25].

Table 12.4. P_{EL} and P_{NR} for fricatives and affricates.

Overall SNR (dB)	Segregated target		Mixture
	$P_{\mathrm{EL}}(\%)$	$P_{\mathrm{NR}}(\%)$	$P_{\mathrm{NR}}(\%)$
0	33.48	35.11	82.17
5	32.39	21.19	61.38
10	29.39	8.47	36.05
15	29.60	5.34	16.39
20	29.88	3.30	6.21

12.7 Concluding Remarks

We should point out that our approach is primarily feature-based. The features used by the system, such as periodicity, AM, and onset, are general properties. Our system does not employ specific prior knowledge of target or interference, except in unvoiced speech grouping where we perform phonetic classification. Prior knowledge helps human ASA in the form of schema-based grouping [6]. Schema-based organization has been emphasized by Ellis [16], and is a subject of several recent studies. Roweis trained HMMs to separate mixtures from two speakers [45]. Barker et al. coupled segmentation with explicit speech models [2]. Srinivasan and Wang used word models to restore phonemes that are masked by interference [49]. These model-based approaches should help to improve the performance of a feature-based system.

A natural speech utterance contains silent gaps and other sections masked by interference. In practice, one needs to group the utterance across such time intervals. This is the problem of sequential grouping, which is not addressed in this chapter. One way of grouping segments across time uses speech recognition in a top-down manner [2]. Recently, Shao and Wang proposed to perform sequential grouping [47] using trained speaker models. Such methods can be integrated with simultaneous grouping addressed in this chapter. Room reverberation is another important issue that must be addressed before speech segregation systems can be deployed in real world environments (see [41] for a recent study on pitch-based segregation of reverberant speech).

To conclude, we have described a CASA approach to monaural speech segregation. Our system segregates voiced speech based on periodicity and AM as well as temporal continuity. Unvoiced speech is segregated via onset/offset analysis and feature-based classification. Evaluation results show that the system performs well on both voiced and unvoiced speech. Note that unvoiced speech is particularly challenging for monaural speech segregation, and our research is the first systematic study on separating unvoiced speech.

Appendix: Voiced Speech Segregation Algorithm

In this appendix, we provide the complete algorithm for voiced speech segregation along with several notes. To facilitate the reader's use of this algorithm, we also post the C++ code for the algorithm on the website (http://www.cse.ohio-state.edu/pnl/software.html). See text for notations. The parameter values used in our implementation are: $\theta_{\mathrm{C}} = 0.99$, $\theta_{\mathrm{P}} = 0.95$, $\theta_{\mathrm{T}} = 0.85$, and $\theta_{\mathrm{A}} = 0.7$. The plausible pitch period range, Γ, is [2 ms, 12.5 ms]. The algorithm is given below.

1. **Cochlear filtering.** A bank of 128 gammatone filters centered from 80 Hz to 5000 Hz is used.

2. **Auditory nerve transduction.** The Meddis model is used.

3. **Feature extraction.** The following features are extracted: Correlogram, envelope correlogram, cross-channel correlation, and dominant pitch. The envelope is obtained through half-wave rectification and bandpass filtering with the passband from 50 Hz to 550 Hz.

4. **Segmentation**
   4.1. Mark two adjacent T-F units, u_{cm} and $u_{c+1,m}$, according to their cross-channel correlation:
      4.1.1. If $C_{\mathrm{H}}(c, m) > \theta_{\mathrm{C}}$, both units are marked as 1.
      4.1.2. Else if $C_{\mathrm{E}}(c, m) > \theta_{\mathrm{C}}$ and the center frequency of channel c is above 1 kHz, both units are marked as 2.
   4.2. Neighboring T-F units with the same mark are merged into segments. Two types of segments are obtained, type 1 and type 2, according to their marks. Two units are considered neighbors if they share the same channel and appear in consecutive time frames, or if they share the same frame and appear in adjacent filter channels. Note that there are unmarked units.

5. **Target pitch tracking**
   5.1. Initial grouping. Only type-1 segments are considered.
      5.1.1. u_{cm} is labeled as the dominant source if

$$\frac{A_{\mathrm{H}}\big(c, m, \tau_S(m)\big)}{\max\limits_{\tau \in \Gamma} A_{\mathrm{H}}\big(c, m, \tau\big)} > \theta_{\mathrm{P}} .$$

      $\tau_S(m)$ initially indicates the dominant pitch period at frame m.
      5.1.2. At a frame of a segment, the segment is labeled as the dominant source if its T-F units labeled as the dominant source contain

more than half of the total energy of the segment at the frame; otherwise, it is labeled as the background.

5.1.3. Find a seed segment that has the largest number of frames labeled as the dominant source.

5.1.4. Determine whether a segment agrees with the seed segment. A segment agrees with the seed segment if they share the same label (either dominant source or background) for more than 2/3 of their overlapping frames. All the segments agreeing with the seed segment form an initial estimate of the target stream, S_0.

5.2. Estimate the target pitch contour from S_0 for every frame of the seed segment. For each such frame, m, the estimated target pitch period, $\tau_S(m)$, is the lag corresponding to the maximum of $\sum_{c, u_{cm} \in S_0} A_{\mathrm{H}}(c, m, \tau)$ in Γ.

5.3. Label individual T-F units and check the reliability of the estimated pitch against the *consistency constraint*: A reliable target pitch is consistent with the periodicity of S_0.

5.3.1. Label a T-F unit at frame m with an estimated pitch as target if

$$\frac{A_{\mathrm{H}}\big(c, m, \tau_S(m)\big)}{\max\limits_{\tau \in \Gamma} A_{\mathrm{H}}\big(c, m, \tau\big)} > \theta_{\mathrm{P}} .$$

Otherwise, label it interference.

5.3.2. If less than half of the T-F units of S_0 at frame m are labeled as target, the estimated pitch, $\tau_S(m)$, is considered inconsistent and all the T-F units of frame m are labeled as interference.

5.4. Re-estimate target stream with labeled T-F units. A segment is labeled as target if its T-F units labeled as target contain more than half of its total energy. All the segments labeled as target form a new estimate of target, S_1.

5.5. Estimate target pitch for all the frames of S_1 as done in Step 5.2. Label individual T-F units and check the consistency of the estimated pitch as done in Step 5.3.

5.6. Pitch interpolation for frames with unreliable pitch:

5.6.1. Consistent pitch points in consecutive frames are connected to form a set of smooth contours. A smooth contour is the one where consecutive frames on the contour satisfy the *smoothness constraint*: The pitch contour of speech changes slowly. Specifically, the change from a pitch period to the one at the next frame is considered smooth if the change is less than 20% of both pitch periods.

5.6.2. Find the longest smooth contour and denote it the seed contour.

5.6.3. Re-estimate the pitch periods for the frames before the seed contour. Set m to the first frame of the seed contour. Iterate until m is the first frame of S_1:
   i. Denote the current frame, m, as a reliable frame (i.e. it has a reliable pitch estimate) and denote c as a selected channel if $u_{cm} \in S_1$ and is labeled as target.
   ii. Decrease m by 1.
   iii. Check if $\tau_S(m)$ satisfies both the consistency and the smoothness constraints. If yes, go directly to Step 5.6.3.i.
   iv. Summate the autocorrelations of u_{cm}'s at frame m where $u_{cm} \in S_1$ and c is a selected channel of the nearest reliable frame. Replace $\tau_S(m)$ by the lag corresponding to the maximum of the summation in the range $[0.65\tau_R, 1.55\tau_R]$, where τ_R indicates the estimated pitch period at the nearest reliable frame.
   v. Check if the new $\tau_S(m)$ satisfies the smoothness constraint. If not, $\tau_S(m)$ is considered unreliable, and then go directly to Step 5.6.3.ii.

5.6.4. Re-estimate the pitch periods for the frames after the seed contour in a symmetric way, until the last frame of S_1.

5.6.5. For any interval of unreliable pitch estimates between two intervals of reliable estimates, the pitch periods within this interval are obtained by linear interpolation from the last frame of the preceding reliable interval and the first frame of the succeeding one.

6. **T-F unit labeling**

6.1. For unit u_{cm} belonging to a type-1 segment, label it as target if

$$\frac{A_H(c, m, \tau_S(m))}{\max_{\tau \in \Gamma} A_H(c, m, \tau)} > \theta_T .$$

Otherwise, label it as interference.

6.2. For a remaining unit, u_{cm}, label it as target if

$$\frac{A_E(c, m, \tau_S(m))}{\max_{\tau \in \Gamma} A_E(c, m, \tau)} > \theta_A .$$

Otherwise, label it as interference.

7. **Grouping**

7.1. A segment is labeled as target if its T-F units labeled as target contain more than half of its total energy. These segments form S_2.

7.2. In S_2, find all the contiguous T-F regions that are all labeled as interference, and remove those regions longer than 40 ms.

7.3. Expand S_2 by iteratively grouping neighboring unmarked T-F units that are labeled as target.

The resulting S_2 represents the segregated target speech by the algorithm. A few further notes are in order. Regarding Step 2 – the modeling of the auditory nerve transduction – we find that the performance without the step is similar for all intrusions except N9, a female utterance. Step 2 helps the system to obtain a better target pitch estimate with the N9 intrusion.

Note also that the algorithm segregates only one continuous section of voiced speech since the pitch determination algorithm provides one pitch contour. If multiple pitch contours are given, one can easily use the given contours instead of Step 5. As discussed in Sec. 12.3.4, we can also apply Step 5 iteratively to estimate multiple pitch contours. However, there is no guarantee that a pitch contour generated this way corresponds to target speech. As mentioned in Sec. 12.7, to determine whether a pitch contour is a target contour is the task of sequential grouping, not addressed here. Step 5.6 in the above algorithm performs pitch interpolation and is relatively complicated. A simpler way is to perform linear interpolation between smooth contours obtained in Step 5.6.1. However, we find this simple method does not work as well for two reasons. First, our tracking algorithm attempts to re-estimate unreliable pitch points from selected frequency channels at the nearest reliable frame, an instance of applying temporal continuity. Second, some smooth contours are inaccurate – e.g. reflecting doubles of pitch frequencies – and when this happens, the smoothness of the overall pitch contour tends to be violated. The tracking algorithm from a seed contour guarantees the smoothness of an overall pitch contour.

Acknowledgments

The preparation of this chapter was supported in part by an AFOSR grant (FA9550-04-01-0117) and an AFRL grant (FA8750-04-1-0093).

References

[1] A.M.A. Ali, J. Van der Spiegel: Acoustic-phonetic features for the automatic classification of stop consonants, *IEEE Trans. Speech Audio Process.*, **9**, 833–841, 2001.
[2] J.P. Barker, M.P. Cooke, D.P.W. Ellis: Decoding speech in the presence of other sources, *Speech Comm.*, **45**, 5–25, 2005.
[3] J. Bird, C.J. Darwin: Effects of a difference in fundamental frequency inseparating two sentences, in A.R. Palmer, A. Rees, A.Q. Summerfield, R. Meddis (eds.), *Psychophysical and Physiological Advances in Hearing,* London, UK: Whurr, 263–269, 1998.

[4] P. Boersma, D. Weenink: *Praat: Doing Phonetics by Computer,* Version 4.2.31, http://www.fon.hum.uva.nl/praat/, 2004.
[5] S.F. Boll: Suppression of acoustic noise in speech using spectral subtraction, *IEEE Trans. Acoust. Speech Signal Process.*, **27**, 113–120, 1979.
[6] A.S. Bregman: *Auditory Scene Analysis,* Cambridge, MA, USA: MIT Press, 1990.
[7] G.J. Brown, M.P. Cooke: Computational auditory scene analysis, *Comput. Speech and Language,* **8**, 297–336, 1994.
[8] G.J. Brown, D.L. Wang: Separation of speech by computational auditory scene analysis, J. Benesty, S. Makino, J. Chen (eds.), *Speech Enhancement,* Berlin, Germany: Springer, 371–402, 2005.
[9] D.S. Brungart, P.S. Chang, B.D. Simpson, D.L. Wang: Isolating the energetic component of speech-on-speech masking with an ideal binary mask, *Submitted for journal publication,* 2005.
[10] J. Canny: A computational approach to edge detection, *IEEE Trans. Pattern Analysis and Machine Intelligence,* **8,** 679–698, 1986.
[11] R.P. Carlyon, T.M. Shackleton: Comparing the fundamental frequencies of resolved and unresolved harmonics: evidence for two pitch mechanisms? *J. Acoust. Soc. Am.*, **95**, 3541–3554, 1994.
[12] P.S. Chang: *Exploration of Behavioral, Physiological, and Computational Approaches to Auditory Scene Analysis,* M.S. Thesis, The Ohio State University Dept. Comput. Sci. & Eng., 2004 (available at http://www.cse.ohio-state.edu/pnl/theses).
[13] M.P. Cooke: *Modelling Auditory Processing and Organisation,* Cambridge, UK: Cambridge University Press, 1993.
[14] M.P. Cooke, P. Green, L. Josifovski, A. Vizinho: Robust automatic speech recognition with missing and unreliable acoustic data, *Speech Comm.*, **34**, 267–285, 2001.
[15] L.A. Drake: *Sound Source Separation via Computational Auditory Scene Analysis (CASA) - Enhanced Beamforming,* Ph.D. Dissertation, Northwestern University Dept. Elec. Eng., 2001.
[16] D.P.W. Ellis: *Prediction-driven Computational Auditory Scene Analysis,* Ph.D. Dissertation, MIT Dept. Elec. Eng. & Comput. Sci., 1996.
[17] Y. Ephraim, H.L. van Trees: A signal subspace approach for speech enhancement, *IEEE Trans. Speech Audio Process.*, **3**, 251–266, 1995.
[18] J. Garofolo, L. Lamel, et al.: Darpa TIMIT acoustic-phonetic continuous speech corpus, *NISTIR 4930,* 1993.
[19] H. Helmholtz: *On the Sensation of Tone,* 2nd English ed., New York, NY, USA: Dover Publishers, 1863.
[20] J. Holdsworth, I. Nimmo-Smith, R.D. Patterson, P. Rice: Implementing a gammatone filter bank, *MRC Applied Psych. Unit,* 1988.
[21] G. Hu, D.L. Wang: Speech segregation based on pitch tracking and amplitude modulation, *Proc. WASPAA '01*, 79–82, New Paltz, New York, USA, 2001.

[22] G. Hu, D.L. Wang: Separation of stop consonants, *Proc. ICASSP '03,* **2**, 749–752, 2003.

[23] G. Hu, D.L. Wang: Monaural speech segregation based on pitch tracking and amplitude modulation, *IEEE Trans. Neural Net.,* **15**, 1135–1150, 2004.

[24] G. Hu, D.L. Wang: Auditory segmentation based on event detection, *Proc. ISCA Tutorial and Research Workshop on Stat. & Percept. Audio Process.,* 2004.

[25] G. Hu, D.L. Wang: Separation of fricatives and affricates, *Proc. ICASSP '05,* **1**, 1101–1104, Philadelphia, PA, USA, 2005.

[26] A. Hyvärinen, J. Karhunen, E. Oja: *Independent Component Analysis,* New York, NY, USA: Wiley, 2001.

[27] ISO: *Normal Equal-loudness Level Contours for Pure Tones under Free-field Listening Conditions (ISO 226),* International standards organization.

[28] J. Jensen, J.H.L. Hansen: Speech enhancement using a constrained iterative sinusoidal model, *IEEE Trans. Speech Audio Process.,* **9**, 731–740, 2001.

[29] H. Krim, M. Viberg: Two decades of array signal processing research: The parametric approach, *IEEE Signal Process. Mag.,* **13**, 67–94, 1996.

[30] P. Ladefoged: *Vowels and Consonants,* Oxford, UK: Blackwell, 2001.

[31] J.C.R. Licklider: A duplex theory of pitch perception, *Experientia,* **7**, 128–134, 1951.

[32] D. Marr: *Vision,* New York, NY, USA: Freeman, 1982.

[33] R. Meddis: Simulation of auditory-neural transduction: Further studies, *J. Acoust. Soc. Am.,* **83**, 1056–1063, 1988.

[34] R. Meddis, M. Hewitt: Modelling the identification of concurrent vowels with different fundamental frequencies, *J. Acoust. Soc. Am.,* **91**, 233–245, 1992.

[35] B.C.J. Moore: *An Introduction to the Psychology of Hearing,* 5th ed., San Diego, CA, USA: Academic Press, 2003.

[36] R.D. Patterson, I. Nimmo-Smith, J. Holdsworth, P. Rice: An efficient auditory filterbank based on the gammatone function, *MRC Applied Psych. Unit. 2341,* 1988.

[37] J.O. Pickles: *An Introduction to the Physiology of Hearing,* 2nd ed., London, UK: Academic Press, 1988.

[38] R. Plomp: The Ear as a Frequency Analyzer, *J. Acoust. Soc. Am.,* **36**, 1628–1636, 1964.

[39] R. Plomp: *The Intelligent Ear,* Mahwah, NJ, USA: Lawrence Erlbaum Associates, 2002.

[40] R. Plomp, A.M. Mimpen: The ear as a frequency analyzer II, *J. Acoust. Soc. Am.,* **43**, 764–767, 1968.

[41] N. Roman, D.L. Wang: A pitch-based model for separation of reverberant speech, *Proc. INTERSPEECH '05,* 2109–2112, Lisbon, Portugal, 2005.

[42] N. Roman, D.L. Wang, G.J. Brown: Speech segregation based on sound localization, *J. Acoust. Soc. Am.*, **114**, 2236–2252, 2003.
[43] B.H. Romeny, L. Florack, J. Koenderink, M. Viergever (eds.): *Scale-space Theory in Computer Vision,* Berlin, Germany: Springer, 1997.
[44] D.F. Rosenthald, H.G. Okuno (eds.): *Computational Auditory Scene Analysis,* Mahwah, NJ: Lawrence Erlbaum Associates, 1998.
[45] S.T. Roweis: One microphone source separation, *Proceedings of the Annual Neural Information Processing Systems (NIPS 2000) Conference,* 2001.
[46] H. Sameti, H. Sheikhzadeh, L. Deng, R.L. Brennan: HMM-based strategies for enhancement of speech signals embedded in nonstationary noise, *IEEE Trans. Speech Audio Process.*, **6**, 445–455, 1998.
[47] Y. Shao, D.L. Wang: Model-based sequential organization in cochannel speech, *IEEE Trans. Speech Audio Process.*, in press, 2005.
[48] M. Slaney, R.F. Lyons: A perceptual pitch detector, *Proc. ICASSP '90*, **1**, 357–360, Albuquerque, NM, USA, 1990.
[49] S. Srinivasan, D.L. Wang: A schema-based model for phonemic restoration, *Speech Comm.*, **45**, 63–87, 2005.
[50] K.N. Stevens: *Acoustic Phonetics,* Cambridge, MA, USA: MIT Press, 1998.
[51] D.L. Wang: On ideal binary mask as the computational goal of auditory scene analysis, P. Divenyi (ed.), *Speech Separation by Humans and Machines,* Norwell, MA, USA: Kluwer, 181–197, 2005.
[52] D.L. Wang, G.J. Brown: Separation of speech from interfering sounds based on oscillatory correlation, *IEEE Trans. Neural Net.*, **10**, 684–697, 1999.
[53] M. Weintraub: *A Theory and Computational Model of Auditory Monaural Sound Separation,* Ph.D. Dissertation, Stanford University Dept. Elec. Eng., 1985.
[54] M. Wu, D.L. Wang, G.J. Brown: A multipitch tracking algorithm for noisy speech, *IEEE Trans. Speech Audio Process.*, **11**, 229–241, 2003.

13

Wave Field Synthesis Techniques for Spatial Sound Reproduction

Rudolf Rabenstein, Sascha Spors, and Peter Steffen

Telecommunications Laboratory, University Erlangen-Nuremberg, Germany

13.1 Introduction

Wave field synthesis (WFS) is a sound reproduction technique which overcomes certain limitations of conventional surround sound methods. It is based on a physical description of the propagation of acoustic waves. Wave field synthesis uses loudspeaker array technology to correctly reproduce sound fields without the "sweet spot" limitation well-known from stereophonic surround sound methods.

The main applications of wave field synthesis are in the areas of entertainment and the performing arts. Due to its rigorous physical foundations, wave field synthesis is also used for reproduction of sound fields caused by room reverberation or for the creation of virtual noise fields. It may not only recreate sound fields of virtual theaters and concert halls, but also acoustic environments for human communication. This way, wave field synthesis provides acoustical testbeds for echo and noise control solutions.

Wave field synthesis techniques are formulated in terms of the acoustic wave equation and the description of its solutions by Green's functions. These foundations have been initially developed by the Technical University of Delft [3,6,11,18,22–25] and were later extended within the European project CARROUSO [5].

This chapter discusses the signal processing aspects of state-of-the-art wave field synthesis systems. The most important of these aspects is the generation of the correct driving signals for each loudspeaker by suitable digital signal processing. Sec. 13.2 presents the notation and some elements from the foundations of acoustics. They are required for the presentation of the concept of wave field synthesis and the resulting signal processing structure in Sec. 13.3. Finally, an implementation example is given in Sec. 13.4.

13.2 Elements from the Foundations of Acoustics

This section starts with a review of some elements from the foundations of acoustics. At first the notation of the required coordinate systems is presented. Then follows a short discussion of the acoustical wave equation and the representation of its solutions in terms of plane waves and Green's functions. Finally the Kirchhoff-Helmholtz integral is introduced for later reference. These foundations of acoustics and wave physics are found in more detail e.g. in [4, 8, 14, 16, 26].

13.2.1 Coordinate Systems

The correct description of sound propagation in space requires a three-dimensional (3D) formulation of the respective acoustical processes. On the other hand, in many applications the source and receiver positions are located in a plane, e.g. a horizontal plane at the height of the listeners' ears. In these cases, a two-dimensional (2D) description is appropriate. The notation for 2D and 3D coordinates is shown in Fig. 13.1 and is introduced below.

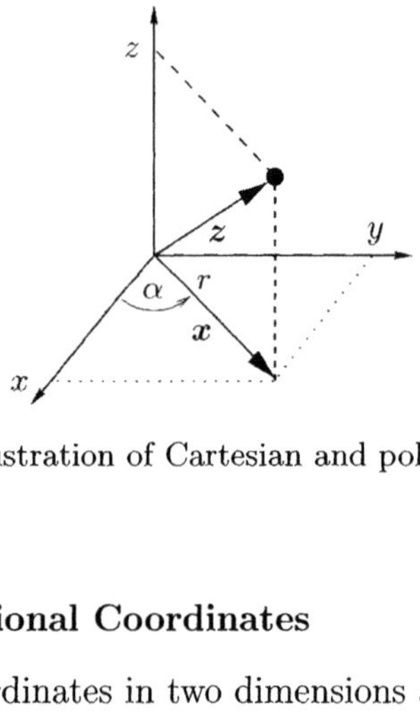

Fig. 13.1. Illustration of Cartesian and polar coordinates.

13.2.1.1 Two-Dimensional Coordinates

Cartesian and polar coordinates in two dimensions are denoted by

$$\boldsymbol{x} = \begin{bmatrix} x \\ y \end{bmatrix}, \qquad \boldsymbol{r} = \begin{bmatrix} r \\ \alpha \end{bmatrix}. \tag{13.1}$$

Their components are related by

$$\begin{bmatrix} x \\ y \end{bmatrix} = r \begin{bmatrix} \cos\alpha \\ \sin\alpha \end{bmatrix}, \qquad \begin{bmatrix} r \\ \alpha \end{bmatrix} = \begin{bmatrix} \sqrt{x^2 + y^2} \\ \tan^{-1}\left(\dfrac{x}{y}\right) \end{bmatrix}. \tag{13.2}$$

The 2D volume elements used for integration are

$$d\boldsymbol{x} = dx\,dy\,, \qquad d\boldsymbol{r} = r\,dr\,d\alpha\;. \tag{13.3}$$

13.2.1.2 Three-Dimensional Coordinates

The position vector $\boldsymbol{z}$ in Cartesian coordinates is defined as

$$\boldsymbol{z} = \begin{bmatrix} \boldsymbol{x} \\ z \end{bmatrix} = \begin{bmatrix} x \\ y \\ z \end{bmatrix} . \tag{13.4}$$

The volume element for 3D integration is

$$d\boldsymbol{z} = dx\, dy\, dz . \tag{13.5}$$

The assignment $p_0(\boldsymbol{z}) = p_1(\boldsymbol{x})$ means that $p_0(\boldsymbol{z})$ is independent of the z-coordinate.

13.2.2 The Wave Equation

The *wave equation* is given by [14, 16, 26]

$$\Delta p(t, \boldsymbol{z}) - \frac{1}{c^2} \frac{\partial^2}{\partial t^2} p(t, \boldsymbol{z}) = 0 . \tag{13.6}$$

$p(t, \boldsymbol{z})$ is the sound pressure at time t and at the location $\boldsymbol{z}$. $\Delta = \nabla^2$ denotes the Laplace operator [2, 9, 10], i.e. second order spatial derivation and c is the propagation speed. Possible solutions of the wave equation are constrained to signals with equal second order partial derivatives in time and space. Solutions of the wave equation are also called *wave fields* or *sound fields*.

Application of the Fourier transform with respect to time turns the wave equation into the *Helmholtz equation*

$$\Delta P(\omega, \boldsymbol{z}) + \left(\frac{\omega}{c}\right)^2 P(\omega, \boldsymbol{z}) = 0 . \tag{13.7}$$

Here the differentiation theorem of the Fourier transform has been applied to substitute the second order time derivative in (13.6) by $(j\omega)^2$ in (13.7). The validity of the conditions for the application of the differentiation theorem have been tacitly assumed. The relation between the temporal frequency variable ω and the propagation speed c is frequently called the wave number $k = \omega/c$.

13.2.2.1 Plane Wave Solution

A *plane wave* is a special solution of the wave equation, which has a very simple form for Cartesian coordinates. It is determined by its wave form and by the direction from which the wave form emanates. The wave form is given by a time function $f(t, \theta)$ and the direction is given by the unit vector $\boldsymbol{n}_\theta$. Later, only plane waves with a zero component in the z-direction are considered. Then $\boldsymbol{n}_\theta$ is a vector in the xy-plane and it is uniquely determined by its x- and y-components

$$\boldsymbol{n}_\theta = \begin{bmatrix} \cos\theta \\ \sin\theta \end{bmatrix} . \tag{13.8}$$

The plane wave solution of Eq. 13.6 is the 3D signal

$$p(t, \boldsymbol{x}) = f\left(t + \frac{1}{c}\langle \boldsymbol{x}, \boldsymbol{n}_\theta \rangle, \theta\right) , \tag{13.9}$$

where $\langle \boldsymbol{x}, \boldsymbol{n}_\theta \rangle$ is the scalar product between $\boldsymbol{x}$ and $\boldsymbol{n}_\theta$. It describes a planar wave front which propagates through space from the direction of $\boldsymbol{n}_\theta$ with speed c. In the origin $\boldsymbol{x} = \boldsymbol{0}$, the wave form $f(t, \theta)$ is observed directly as $p(t, \boldsymbol{0}) = f(t, \theta)$. The Fourier transform with respect to time gives

$$P(\omega, \boldsymbol{x}) = F(\omega, \theta)\, e^{j\frac{\omega}{c}\langle \boldsymbol{x}, \boldsymbol{n}_\theta \rangle} \tag{13.10}$$

as the plane wave solution of Eq. 13.7.

A more general wave field is obtained by superposition of plane wave solutions from all possible directions θ

$$P(\omega, \boldsymbol{r}) = \int_0^{2\pi} F(\omega, \theta)\, e^{j\frac{\omega}{c} r \cos(\theta - \alpha)}\, d\theta , \tag{13.11}$$

where the scalar product $\langle \boldsymbol{x}, \boldsymbol{n}_\theta \rangle$ has been expressed in polar coordinates

$$\langle \boldsymbol{x}, \boldsymbol{n}_\theta \rangle = r \cos(\theta - \alpha) . \tag{13.12}$$

The relation 13.11 is closely related to the plane wave decomposition of a wave field [12].

13.2.2.2 Green's Functions

Arbitrary solutions of the wave equation with homogeneous boundary conditions are described in terms of Green's functions. They can be regarded as the response of a sound field to an impulse in time and space. Since there are various kinds of impulse functions in 3D space, also the possible forms of the corresponding Green's functions differ. Here, the Green's functions of point sources and line sources are considered.

Point Source

A point source in 3D space is defined by the 3D Dirac-impulse function in Cartesian coordinates [4,8]

$$\delta_{3\mathrm{D}}(\boldsymbol{z}) = \delta(x)\, \delta(y)\, \delta(z) , \tag{13.13}$$

where $\delta(x)$ denotes the 1D Dirac-impulse. The 3D Dirac-impulse can also be defined for cylindrical and other 3D coordinates. A point source at the location $\boldsymbol{z}'$ with time varying source strength is described by

$$q_0(t, \boldsymbol{z}) = q_0(t, \boldsymbol{z}')\, \delta_{3\mathrm{D}}(\boldsymbol{z} - \boldsymbol{z}') \tag{13.14}$$

or after Fourier transform with respect to time by

$$Q_0(\omega, \boldsymbol{z}) = Q_0(\omega, \boldsymbol{z}')\, \delta_{3\mathrm{D}}(\boldsymbol{z} - \boldsymbol{z}')\,. \tag{13.15}$$

The index zero indicates a point sources with dimension zero. An arbitrary spatial distribution of point sources is given by

$$Q_0(\omega, \boldsymbol{z}) = \iiint\limits_V Q_0(\omega, \boldsymbol{z}')\, \delta_{3\mathrm{D}}(\boldsymbol{z} - \boldsymbol{z}')\, d\boldsymbol{z}'\,. \tag{13.16}$$

The spatial sound field $P_0(\omega, \boldsymbol{z})$ caused by a spatially distributed source $Q_0(\omega, \boldsymbol{z})$ is given by

$$P_0(\omega, \boldsymbol{z}) = \iiint\limits_V G_0(\omega, \boldsymbol{z}|\boldsymbol{z}')\, Q_0(\omega, \boldsymbol{z}')\, d\boldsymbol{z}'\,. \tag{13.17}$$

The Green's function $G_0(\omega, \boldsymbol{z}|\boldsymbol{z}')$ describes the contribution of a point source at position $\boldsymbol{z}'$ to the sound field at position $\boldsymbol{z}$. The integration is carried out on the volume V where the solution of the wave equation is considered. The position $\boldsymbol{z}$ is also referred to as the listener position.

The form of the Green's function $G_0(\omega, \boldsymbol{z}|\boldsymbol{z}')$ depends on the shape of the volume V and on the kind of the boundary condition on its surface. In the free-field, i. e. $V = \mathbb{R}^3$ the Green's function for all kinds of boundary conditions is given by [14]

$$G_0^{\mathrm{f}}(\omega, \boldsymbol{z}|\boldsymbol{z}') = \frac{1}{4\pi} \frac{e^{-j\frac{\omega}{c}\|\boldsymbol{z} - \boldsymbol{z}'\|}}{\|\boldsymbol{z} - \boldsymbol{z}'\|}\,. \tag{13.18}$$

It describes a spherical wave and is also called the free-field Green's function. The denominator accounts for the decay of the amplitude over distance and the exponential term accounts for the time delay of the propagating spherical wave.

Line Source

A line source consists of a superposition of equal point sources along a line. When the line is oriented in parallel to the z-axis then all point sources with the same xy-coordinates have equal source strength $Q_0(\omega, \boldsymbol{z})$. Consequently, $Q_0(\omega, \boldsymbol{z})$ does not depend on z and the integration in Eq. 13.17 degenerates to an integration in the xy-plane

$$\begin{aligned} Q_0(\omega, \boldsymbol{z}) &= \iiint\limits_V Q_0(\omega, \boldsymbol{z}')\, \delta_{3\mathrm{D}}(\boldsymbol{z} - \boldsymbol{z}')\, d\boldsymbol{z}' \\ &= \iint\limits_L Q_1(\omega, \boldsymbol{x}')\, \delta_{2\mathrm{D}}(\boldsymbol{x} - \boldsymbol{x}') \underbrace{\int_{-\infty}^{\infty} \delta(z - z')\, dz'}_{=1}\, d\boldsymbol{x}' \\ &= Q_1(\omega, \boldsymbol{x})\,, \end{aligned} \tag{13.19}$$

where

$$\delta_{2D}(\boldsymbol{x}) = \delta(x)\,\delta(y) \tag{13.20}$$

denotes a 2D Dirac-impulse in Cartesian coordinates and L a horizontal cut through the volume V. This result may be interpreted in two ways:

- As before, $Q_0(\omega, \boldsymbol{z})$ describes a collection of point sources (index zero). The three components of $\boldsymbol{z}$ denote the location of each point source in 3D space. However, the source strength is constant in z-direction and therefore the result does not depend on the z-coordinate.
- $Q_1(\omega, \boldsymbol{x})$ describes a collection of line sources parallel to the z-axis. The index one denotes the one-dimensional character of the line sources. The two components of $\boldsymbol{x}$ denote the coordinates of the root points of each line in the xy-plane.

Note that $Q_0(\omega, \boldsymbol{z})$ is a function of three spatial variables (x, y, z) which denote the location of a zero-dimensional entity (a point source) in 3D space. On the other hand, $Q_1(\omega, \boldsymbol{x})$ is a function of two variables (x, y) which denote the location of a one-dimensional entity (a line source parallel to the z-axis). So both $Q_0(\omega, \boldsymbol{z})$ and $Q_1(\omega, \boldsymbol{x})$ describe a 3D sound field with a special structure, i.e. without dependence on the z-coordinate.

The sound field caused by a collection of line sources $Q_1(\omega, \boldsymbol{x})$ can be obtained from Eq.13.17 when $Q_0(\omega, \boldsymbol{z})$ does not depend on z. Then the integration with respect to z' is only performed for the Green's function $G_0(\omega, \boldsymbol{z}|\boldsymbol{z}')$ and yields the Green's function $G_1(\omega, \boldsymbol{x}|\boldsymbol{x}')$ of a line source in parallel to the z-axis

$$G_1(\omega, \boldsymbol{x}|\boldsymbol{x}') = \int_{-\infty}^{\infty} G_0(\omega, \boldsymbol{x}|\boldsymbol{z}')\,dz'\,. \tag{13.21}$$

The resulting sound field is also constant in z-direction and is described by a function $P_1(\omega, \boldsymbol{x})$ depending only on x and y

$$P_1(\omega, \boldsymbol{x}) = \iint_L G_1(\omega, \boldsymbol{x}|\boldsymbol{x}')Q_1(\omega, \boldsymbol{x}')d\boldsymbol{x}'\,. \tag{13.22}$$

The interpretation of $Q_0(\omega, \boldsymbol{z})$ and $Q_1(\omega, \boldsymbol{x})$ in Eq. 13.19 applies in a similar fashion also to $P_0(\omega, \boldsymbol{z})$ and $P_1(\omega, \boldsymbol{x})$.

Evaluating the integral in Eq. 13.21 for the integrand from Eq. 13.18 gives [9, 3.876]

$$G_1^{\mathrm{f}}(\omega, \boldsymbol{x}|\boldsymbol{x}') = -\frac{j}{4}\,H_0^{(2)}\left(\frac{\omega}{c}\rho\right)\,. \tag{13.23}$$

where $H_0^{(2)}(\frac{\omega}{c}\rho)$ is the Hankel function of the second kind and order zero. Due to the circular symmetry, $G_1^{\mathrm{f}}(\omega, \boldsymbol{x}|\boldsymbol{x}')$ depends only on the distance ρ between the listener position $\boldsymbol{x}$ and the line source at $\boldsymbol{x}'$. It is given by (see Fig. 13.2)

$$\rho = \|\boldsymbol{x} - \boldsymbol{x}'\| = \sqrt{(x-x')^2 + (y-y')^2}\,. \tag{13.24}$$

Thus the notation for the Green's function of the line source may be shortened to

$$G_1^{\mathrm{f}}(\omega, \boldsymbol{x}|\boldsymbol{x}') = \tilde{G}_1^{f}(\omega, \rho) = -\frac{j}{4}\, H_0^{(2)}\left(\frac{\omega}{c}\,\rho\right) . \qquad (13.25)$$

In this notation, the Hankel function $H_0^{(2)}$ is given by [9]

$$H_0^{(2)}\left(\frac{\omega}{c}\rho\right) = J_0\left(\frac{\omega}{c}\rho\right) - jN_0\left(\frac{\omega}{c}\rho\right) , \qquad (13.26)$$

where $J_0(\frac{\omega}{c}\rho)$ and $N_0(\frac{\omega}{c}\rho)$ are the Bessel and Neumann functions of the first kind and order zero.

13.2.2.3 Relation Between the Green's Functions for Line and Point Sources for the Free Field Case

Now the relation between a line source parallel to the z-axis and a point source at the root point of the line source is investigated for the free-field case. The orientation of the line source and the position of the point source are shown in Fig. 13.2. The effect of both sources on the sound field in the xy-plane is now compared. The effect of the line source is described by $\tilde{G}_1^{\mathrm{f}}(\omega, \rho)$ introduced in Eq. 13.25. The effect of the point source is described by its Green's function (Eq. 13.18) for $z = 0$ and $z' = 0$, i.e. by

$$G_0^{\mathrm{f}}(\omega, \boldsymbol{z}|\boldsymbol{z}')\bigg|_{\substack{z=0\\ z'=0}} = \tilde{G}_0^{\mathrm{f}}(\omega, \rho) = \frac{1}{4\pi\,\rho}\, e^{-j\left(\frac{\omega}{c}\rho\right)} . \qquad (13.27)$$

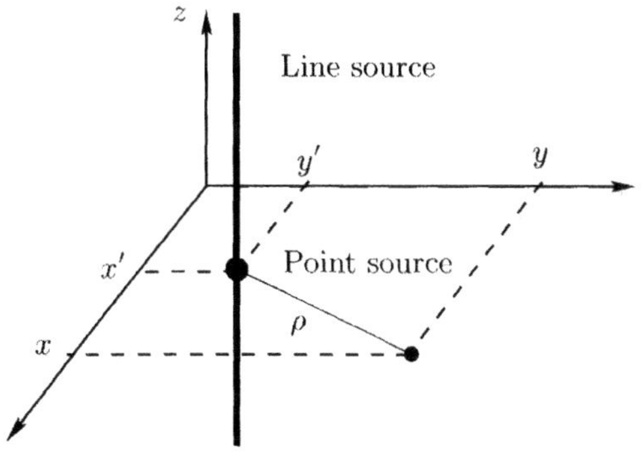

Fig. 13.2. Line source parallel to the z-axis and point source at the root point of the line source in the xy-plane.

The relation between the sound field of a point source in the xy-plane $G_0^{\mathrm{f}}(\omega, \rho)$ and a line source parallel to the z-axis $G_1^{\mathrm{f}}(\omega, \rho)$ is established by an approximation. This approximation can be derived in two different ways, i.e.

through the *method of stationary phase* [26] for the integral in Eq. 13.21 and by the *far-field approximation* [1] for the Green's function $G_l^{\mathrm{f}}(\omega, \rho)$.

The method of stationary phase allows to express certain integrals by the value of their integrand at a fixed argument, the so-called stationary phase point zs through

$$\int_{-\infty}^{\infty} f(\zeta)\, e^{j\phi(\zeta)} d\zeta \approx \sqrt{\frac{2\pi j}{\phi''(z_{\mathrm{s}})}}\, f(z_{\mathrm{s}})\, e^{j\phi(z_{\mathrm{s}})} \tag{13.28}$$

where $\phi''(\zeta)$ denotes the second derivative of $\phi(\zeta)$. This approximation holds for $\phi(\zeta) \gg 1$. The stationary phase point is found by setting the first derivative $\phi'(\zeta)$ of $\phi(\zeta)$ to zero, i.e. $\phi'(z_{\mathrm{s}}) = 0$.

Applying the method of stationary phase to $G_0^{\mathrm{f}}(\omega, \boldsymbol{z}|\boldsymbol{z}')$ with $z = 0$, $z' = \zeta$, and

$$\phi_\rho(\zeta) = -\frac{\omega}{c}\sqrt{\rho^2 + \zeta^2}\,, \qquad f_\rho(\zeta) = \frac{1}{4\pi\,\sqrt{\rho^2 + \zeta^2}} \tag{13.29}$$

leads to $z_{\mathrm{s}} = 0$ and

$$\tilde{G}_1^{\mathrm{f}}(\omega, \rho) = \int_{\infty}^{\infty} f_\rho(\zeta)\, e^{j\phi_\rho(\zeta)}\, d\zeta \approx \frac{1}{\sqrt{j8\pi\left(\frac{\omega}{c}\rho\right)}}\, e^{-j\left(\frac{\omega}{c}\rho\right)}\,. \tag{13.30}$$

The same approximation can also be obtained from the asymptotic expansion of the Hankel function for large $|\rho|$ [1]

$$H_0^{(2)}\left(\frac{\omega}{c}\rho\right) \approx \sqrt{\frac{2}{\pi\left(\frac{\omega}{c}\rho\right)}}\; e^{-j\left(\left(\frac{\omega}{c}\rho\right) - \frac{\pi}{4}\right)}\,. \tag{13.31}$$

In acoustics, this expansion is called the far-field approximation of the Hankel function.

Comparing Eqs. 13.25 and 13.30 shows that the effect of a line source on the sound field in the xy-plane may be approximated by the effect of a point source. In particular, $G_1^{\mathrm{f}}(\omega, \rho)$ may be approximated by

$$\tilde{G}_1^{\mathrm{f}}(\omega, \rho) = H(\omega)\; A(\rho)\; \tilde{G}_0^{\mathrm{f}}(\omega, \rho)\,, \tag{13.32}$$

where

$$H(\omega) = \sqrt{\frac{c}{j\omega}} \tag{13.33}$$

causes a spectral shaping and

$$A(\rho) = \sqrt{2\pi\,\rho} \tag{13.34}$$

causes an amplitude modification of $\tilde{G}_0^{\mathrm{f}}(\omega, \rho)$.

The derivation from Eq. 13.23 to Eq. 13.34 on the relation between line sources and point sources may be summarized as follows:

- The evaluation of the integral in Eq. 13.21 with the stationary phase method leads to the far-field approximation of the Green's function of a line source.
- The effect of a line source on the xy-plane can be approximated by a point source at the root point of the line source. The sound field of the point source has to be corrected by spectral shaping and an amplitude modification. This approximation is valid in the far-field of the line source.

13.2.3 Kirchhoff-Helmholtz Integral

The Kirchhoff-Helmholtz integral is the key element of the wave field synthesis principle. It provides the relation between the sound field inside a volume of arbitrary shape and on the enclosing boundary. The Kirchhoff-Helmholtz-Integral is presented first for a general 3D volume and then specialized to a 3D prism.

13.2.3.1 Kirchhoff-Helmholtz Integral for a General 3D Volume

The *Kirchhoff-Helmholtz integral* or *Helmholtz integral equation* expresses the values $P_0(\omega, \boldsymbol{z})$ inside a volume V by an integral on the surface ∂V [14,16,26]

$$-\oint\!\!\!\oint_{\partial V} \left(G_0(\omega, \boldsymbol{z}|\boldsymbol{z}') \frac{\partial}{\partial \boldsymbol{n}} P_0(\omega, \boldsymbol{z}') - P_0(\omega, \boldsymbol{z}') \frac{\partial}{\partial \boldsymbol{n}} G_0(\omega, \boldsymbol{z}|\boldsymbol{z}') \right) d\boldsymbol{z}' =$$

$$= \begin{cases} P_0(\omega, \boldsymbol{z}), & \boldsymbol{z} \in V \\ 0, & \boldsymbol{z} \notin V \end{cases} . \quad (13.35)$$

$G_0(\omega, \boldsymbol{z}|\boldsymbol{z}')$ is a Green's function which satisfies suitable boundary conditions on ∂V.

The Kirchhoff-Helmholtz integral states that at any point within the source-free region V the sound pressure $P_0(\omega, \mathbf{z})$ can be calculated if both the sound pressure $P_0(\omega, \boldsymbol{z}')$ and its directional gradient $\frac{\partial}{\partial \boldsymbol{n}} P_0(\omega, \boldsymbol{z}')$ are known on the boundary ∂V enclosing the volume. The boundary ∂V does not necessary have to be a real physical existing surface. The Kirchhoff-Helmholtz integral is typically used in three areas: (1) the calculation of a sound field emitted by a vibrating surface into a region, (2) the calculation of a sound field inside a finite region produced by a source outside the volume from measurements on the surface and (3) the acoustic control over the sound field within a volume. The third application area leads to sound reproduction according to the principle of wave field synthesis [22,24].

13.2.3.2 Kirchhoff-Helmholtz Integral for a Prism

The Kirchhoff-Helmholtz integral is now specialized to sound fields which do not depend on the z-coordinate. The shape of the volume for the integration in Eq. 13.35 turns into a prism oriented in parallel to the z-axis (see

Fig. 13.3). This rather special spatial arrangement allows the transition to a 2D description of the Kirchhoff-Helmholtz integral. The stages of this transition are shown for the first term in Eq. 13.35. The presented procedure applies equally to the second term.

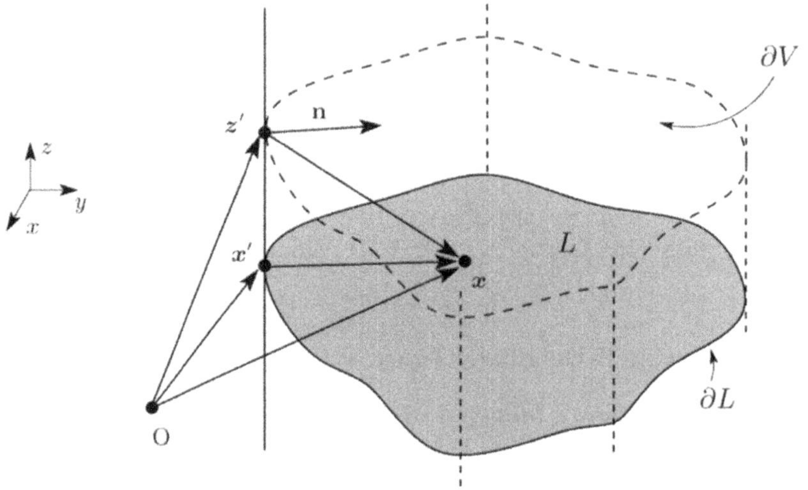

Fig. 13.3. Illustration of a sound field with does not depend on the z-coordinate.

Since the sound field is assumed to be independent of z, $P_0(\omega, \boldsymbol{x})$ depends only on x and y. Furthermore, any vector normal to the surface ∂V has no component in the z-direction and also the normal derivative of $P_0(\omega, \boldsymbol{x})$ is independent of z. Thus the surface integration with respect to $\boldsymbol{z}'$ in Eq. 13.35 can be split into a contour integration with respect to $\boldsymbol{x}'$ and an integration with respect to z'. The contour ∂L is defined by the intersection of the prism with the xy-plane. This procedure turns the first term of Eq. 13.35 into

$$\begin{aligned}
&\oiint_{\partial V} G_0(\omega, \boldsymbol{x}|\boldsymbol{z}')\, \frac{\partial}{\partial \boldsymbol{n}} P_0(\omega, \boldsymbol{z}')\, d\boldsymbol{z}' \\
&= \oint_{\partial L} \int_{-\infty}^{\infty} G_0(\omega, \boldsymbol{x}|\boldsymbol{z}')\, \frac{\partial}{\partial \boldsymbol{n}} P_0(\omega, \boldsymbol{z}')\, dz'\, d\boldsymbol{x}' \\
&= \oint_{\partial L} \left[\int_{-\infty}^{\infty} G_0(\omega, \boldsymbol{x}|\boldsymbol{z}')\, dz' \right] \frac{\partial}{\partial \boldsymbol{n}} P_0(\omega, \boldsymbol{x}')\, d\boldsymbol{x}' \,. \qquad (13.36)
\end{aligned}$$

With Eq. 13.21 follows a 2D version of the Kirchhoff-Helmholtz-Integral

$$P_1(\omega, \boldsymbol{z}) = - \oint_{\partial L} \left(G_1(\omega, \boldsymbol{x}|\boldsymbol{x}') \frac{\partial}{\partial \boldsymbol{n}} P_1(\omega, \boldsymbol{x}') - P_1(\omega, \boldsymbol{x}') \frac{\partial}{\partial \boldsymbol{n}} G_1(\omega, \boldsymbol{x}|\boldsymbol{x}') \right) d\boldsymbol{x}'. \tag{13.37}$$

13.3 Wave Field Synthesis

13.3.1 Introduction

In the following the sound reproduction scenario depicted in Fig. 13.4 will be considered. The wave field emitted by an arbitrary virtual source $Q_0(\omega, \boldsymbol{z})$

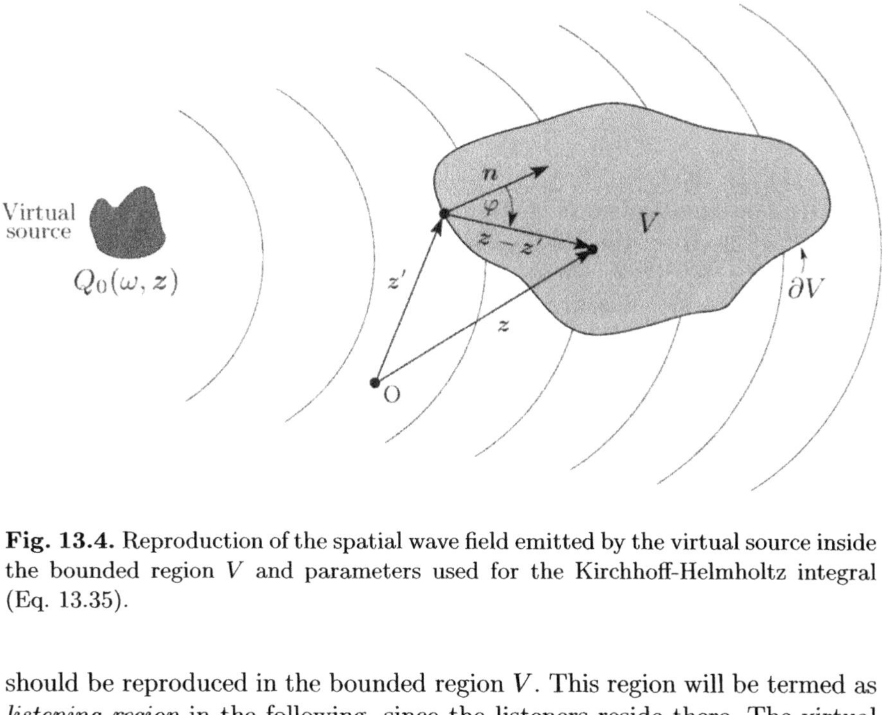

Fig. 13.4. Reproduction of the spatial wave field emitted by the virtual source inside the bounded region V and parameters used for the Kirchhoff-Helmholtz integral (Eq. 13.35).

should be reproduced in the bounded region V. This region will be termed as *listening region* in the following, since the listeners reside there. The virtual source $Q_0(\omega, \boldsymbol{z})$ may not have contributions in V. The limitation to one virtual source poses no constraints on the wave field to be reproduced, since this source may have arbitrary shape and frequency characteristics. Additionally, multiple sources can be reproduced by the principle of superposition.

The basic principle of sound reproduction can be illustrated with the principle of Huygens [14]. Huygens stated that any point of a propagating wave front at any time-instant conforms to the envelope of spherical waves emanating from every point on the wavefront at the prior instant. This principle can be used to synthesize acoustic wavefronts of arbitrary shape. Of course, it is not very practical to position the acoustic sources on the wavefronts for synthesis. By placing the loudspeakers on an arbitrary fixed curve and by weighting and delaying the driving signals, an acoustic wavefront can be synthesized with a loudspeaker array. Fig. 13.5 illustrates this principle. The

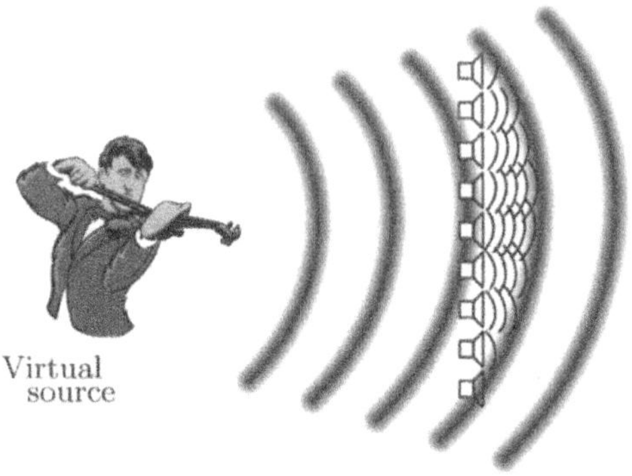

Fig. 13.5. Application of Huygens principle to perform sound reproduction.

mathematical foundation of this more illustrative description to sound reproduction is given by the Kirchhoff-Helmholtz integral. It was introduced in Sec. 13.2.3 and will be utilized in the following to derive a generic theory of sound reproduction systems.

13.3.2 Kirchhoff-Helmholtz Integral based Sound Reproduction

The Kirchhoff-Helmholtz integral (Eq. 13.35) comprises a number of different problems as already addressed in Sec. 13.2.3.1. Each of these issues is characterized by its specific type of boundary conditions and thus by the corresponding Green's function.

For the sound reproduction scenario according to Fig. 13.4 the Green's function $G_0(\omega, \boldsymbol{z}|\boldsymbol{z}')$ and its directional gradient can be understood as the field emitted by sources placed on ∂V. These sources will be termed as *secondary sources* in the following. The strength of these sources is determined by the pressure $P_0(\omega, \boldsymbol{z}')$ and the directional pressure gradient $\frac{\partial}{\partial n} P_0(\omega, \boldsymbol{z}')$ of the virtual source field $Q_0(\omega, \boldsymbol{x}')$ on ∂V.

Thus, this specialized Kirchhoff-Helmholtz integral can be interpreted as follows: If appropriately chosen secondary sources are driven by the sound pressure and the directional pressure gradient of the wave field emitted by the virtual source $Q_0(\omega, \boldsymbol{x}')$ on the boundary ∂V, then the wave field within the region V is equivalent to the wave field which would have been produced by the virtual source inside V. Thus, the theoretical basis of sound reproduction is described by the Kirchhoff-Helmholtz integral (Eq. 13.35).

The three-dimensional free-field Green's function is given by Eq. 13.18. In the context of sound reproduction it can be interpreted as the field of a monopole point source distribution on the surface ∂V. The Kirchhoff-Helmholtz integral (Eq. 13.35) also involves the directional gradient of the Green's function. The directional gradient of the three-dimensional free-field Green's function can be interpreted as the field of a dipole source whose main axis lies in direction of the normal vector $\boldsymbol{n}$. Thus, the Kirchhoff-Helmholtz integral states in this case, that the acoustic pressure inside the volume V can be controlled by a monopole and a dipole point source distribution on the surface ∂V enclosing the volume V.

This interpretation of the Kirchhoff-Helmholtz integral sketches a first draft of a technical system for spatial sound reproduction. In rough terms, such a system would consist of technical approximations of acoustical monopoles and dipoles by appropriate loudspeakers. These loudspeakers cover the surface of a suitably chosen volume around the possible listener positions. They are excited by appropriate driving functions to reproduce the desired sound field inside the volume.

However, there remain a number of fundamental questions to be resolved on the way to a technical realization. Four major areas can be identified. They are listed below and are discussed in detail in the following sections.

Monopole and Dipole Sources.

Technical approximations of acoustical monopoles and dipoles consist of loudspeakers with different types of enclosures. A restriction to only one type of sources would be of advantage for a technical realization. For example the use of monopole sources only facilitates a technical solution with small loudspeakers in closed cabinets.

Reduction to Two Spatial Dimensions.

The volume V certainly has to be large enough to enclose at least a small audience or to give a single listener room to move within the sound field. Covering the whole surface with suitable sound sources appears to be a technological and economical challenge. Furthermore, it may not be required to reproduce the sound field within the entire volume. A correct reproduction in a horizontal plane at the level of the listeners' ears may be sufficient. Such a simplification requires to reduce the 3D problem to two spatial dimensions.

Spatial Sampling.

The Kirchhoff-Helmholtz integral prescribes a continuous source distribution over the surface ∂V. However, an approximation of sound sources by loudspeakers results in a spatially discrete source distribution. The resulting discretization effects may be described in terms of spatial sampling.

Driving Signals.

Once the source distribution is approximated by a sufficiently dense grid of loudspeakers, their driving signals have to be generated by signal processing hardware and digital-to-analog converters.

13.3.3 Monopole and Dipole Sources

The use of monopole and dipole sources in Eq. 13.35 allows a very precise reproduction of the desired wave field: It is recreated as $P_0(\omega, z)$ for all positions inside of V and it is zero outside. Such a restriction is usually not required for spatial sound reproduction. As long as the reproduction is correct inside of V, almost arbitrary sound fields outside may be tolerated, as long as their reproduction volume is moderate. This situation suggests the following trade-off: Use one type of sound sources only and tolerate some sound pressure outside of V.

To realize this trade-off, a Green's function $G_0(\omega, \boldsymbol{z}|\boldsymbol{z}')$ is constructed which satisfies boundary conditions of the first or second kind on the surface ∂V. Since it is desirable to drop the dipoles and to keep the monopole sources, the Green's function of a point source $G_0^{\mathrm{f}}(\omega, \boldsymbol{z}|\boldsymbol{z}')$ according to Eq. 13.18 is chosen as the basic building block. Then for each position z' on the boundary, the Green's function $G_0^{\mathrm{f}}(\omega, \boldsymbol{z}|\boldsymbol{z}')$ for a position z inside of V and the Green's function $G_0^{\mathrm{f}}(\omega, \bar{\boldsymbol{z}}(\boldsymbol{z})|\boldsymbol{z}')$ for a position $\bar{\boldsymbol{z}}(\boldsymbol{z})$ outside of V are superposed

$$G_0(\omega, \boldsymbol{z}|\boldsymbol{z}') = G_0^{\mathrm{f}}(\omega, \boldsymbol{z}|\boldsymbol{z}') + G_0^{\mathrm{f}}(\omega, \bar{\boldsymbol{z}}(\boldsymbol{z})|\boldsymbol{z}') . \tag{13.38}$$

The position $\bar{\boldsymbol{z}}(\boldsymbol{z})$ is chosen as the mirror image of $\boldsymbol{z}$ with respect to the tangent plane in $\boldsymbol{z}'$ on the surface ∂V (see Fig. 13.6). The tangent plane is characterized by the unit vector $\boldsymbol{n}$. The notation $\bar{\boldsymbol{z}}(\boldsymbol{z})$ indicates that $\bar{\boldsymbol{z}}$ depends on $\boldsymbol{z}$.

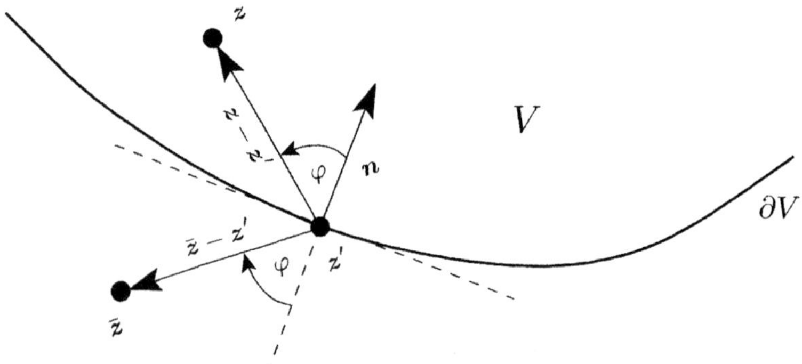

Fig. 13.6. Illustration of the geometry used for the derivation of the modified Green's function for a sound reproduction system using monopole secondary sources only.

From the symmetry of the mirror images follows

$$\left\|\boldsymbol{z} - \boldsymbol{z}'\right\| = \left\|\bar{\boldsymbol{z}}(\boldsymbol{z}) - \boldsymbol{z}'\right\| = \rho_z , \tag{13.39}$$

and thus the functional dependence of $G_0^{\mathrm{f}}(\omega, \boldsymbol{z}|\boldsymbol{z}')$ and $G_0^{\mathrm{f}}(\omega, \bar{\boldsymbol{z}}(\boldsymbol{z})|\boldsymbol{z}')$ on $\boldsymbol{z}$ and $\boldsymbol{z}'$ is identical

$$G_0^{\mathrm{f}}(\omega, \boldsymbol{z}|\boldsymbol{z}') = G_0^{\mathrm{f}}(\omega, \bar{\boldsymbol{z}}(\boldsymbol{z})|\boldsymbol{z}') \,. \tag{13.40}$$

However, $G_0^{\mathrm{f}}(\omega, \boldsymbol{z}|\boldsymbol{z}')$ and $G_0^{\mathrm{f}}(\omega, \bar{\boldsymbol{z}}(\boldsymbol{z})|\boldsymbol{z}')$ have to be carefully distinguished as Green's functions for positions inside of V and outside of V, respectively. This difference becomes apparent for the gradients

$$\begin{aligned} \nabla G_0^{\mathrm{f}}(\omega, \boldsymbol{z}|\boldsymbol{z}') &= -\frac{1+jk\rho_z}{\rho_z}\, G_0^{\mathrm{f}}(\omega, \boldsymbol{z}|\boldsymbol{z}')\, \boldsymbol{n}_{\boldsymbol{z}} \\ \nabla G_0^{\mathrm{f}}(\omega, \bar{\boldsymbol{z}}|\boldsymbol{z}') &= -\frac{1+jk\rho_z}{\rho_z}\, G_0^{\mathrm{f}}(\omega, \bar{\boldsymbol{z}}|\boldsymbol{z}')\, \boldsymbol{n}_{\bar{\boldsymbol{z}}} \,, \end{aligned} \tag{13.41}$$

with the unit vectors

$$\boldsymbol{n}_{\boldsymbol{z}} = \frac{\boldsymbol{z}-\boldsymbol{z}'}{\|\boldsymbol{z}-\boldsymbol{z}'\|} \,, \qquad \boldsymbol{n}_{\bar{\boldsymbol{z}}} = \frac{\bar{\boldsymbol{z}}-\boldsymbol{z}'}{\|\bar{\boldsymbol{z}}-\boldsymbol{z}'\|} \,. \tag{13.42}$$

The derivative with respect to $\boldsymbol{n}$ is given by

$$\frac{\partial}{\partial \boldsymbol{n}} G_0^{\mathrm{f}}(\omega, \boldsymbol{z}|\boldsymbol{z}') = \left\langle \nabla G_0^{\mathrm{f}}(\omega, \boldsymbol{z}|\boldsymbol{z}'), \boldsymbol{n} \right\rangle = -\frac{1+jk\rho_z}{\rho_z}\, G_0^{\mathrm{f}}(\omega, \boldsymbol{z}|\boldsymbol{z}') \left\langle \boldsymbol{n}_{\boldsymbol{z}}, \boldsymbol{n} \right\rangle \tag{13.43}$$

and similarly for the derivative of $G_0^{\mathrm{f}}(\omega, \bar{\boldsymbol{z}}|\boldsymbol{z}')$. With Eq. 13.40 and (see Fig. 13.6)

$$\left\langle \boldsymbol{n}_{\boldsymbol{z}}, \boldsymbol{n} \right\rangle + \left\langle \boldsymbol{n}_{\bar{\boldsymbol{z}}}, \boldsymbol{n} \right\rangle = 0 \tag{13.44}$$

follows

$$\frac{\partial}{\partial \boldsymbol{n}} G_0(\omega, \boldsymbol{z}|\boldsymbol{z}') = \frac{\partial}{\partial \boldsymbol{n}} G_0^{\mathrm{f}}(\omega, \boldsymbol{z}|\boldsymbol{z}') + \frac{\partial}{\partial \boldsymbol{n}} G_0^{\mathrm{f}}(\omega, \bar{\boldsymbol{z}}|\boldsymbol{z}') = 0 \,, \quad \boldsymbol{z}' \in \partial V \,. \tag{13.45}$$

In summary, the Green's function $G_0(\omega, \boldsymbol{z}|\boldsymbol{z}')$ according to Eq. 13.38 induces a sound field not only inside of V but also on the outside. On the other hand, the normal derivative of $G_0(\omega, \boldsymbol{z}|\boldsymbol{z}')$ is zero for all positions $\boldsymbol{z}'$ on the boundary ∂V. Thus inserting $G_0(\omega, \boldsymbol{z}|\boldsymbol{z}')$ as Green's function into the Kirchhoff-Helmholtz integral (Eq. 13.35) leads to

$$P_0(\omega, \boldsymbol{z}) = -\iint_{\partial V} G_0(\omega, \boldsymbol{z}|\boldsymbol{z}')\, \frac{\partial}{\partial \boldsymbol{n}} P_0(\omega, \boldsymbol{z}')\, d\boldsymbol{z}' \,, \qquad \boldsymbol{z} \in V \,. \tag{13.46}$$

Since $G_0(\omega, \boldsymbol{z}|\boldsymbol{z}')$ is equal to $G_0^{\mathrm{f}}(\omega, \boldsymbol{z}|\boldsymbol{z}')$ for sources inside of V, the result of Eq. 13.46 is equal to the desired wave field $P_0(\omega, \boldsymbol{z})$ inside of V. Outside of V the wave field consists of a mirrored version of the wave field inside of V. An example is shown in Fig. 13.8. The result (Eq. 13.46) is also known as the type-I Raleigh integral [22].

It states that the sound field inside of a volume V may be reproduced by a distribution of point sources if a mirrored version of this sound field is tolerated outside of V. If only the monopole properties of this Green's function are or interest, then Eq. 13.21 may be expressed with Eq. 13.40 as

$$G_0(\omega, \boldsymbol{z}|\boldsymbol{z}') = 2\, G_0^{\mathrm{f}}(\omega, \boldsymbol{z}|\boldsymbol{z}')\,, \tag{13.47}$$

i.e. the boundary sources as free-field point sources with double strength. The factor of two follows from the fact that these point sources describe the contribution inside and outside of V in equal terms.

These considerations are summarized by Eq. 13.48 below. The sound field inside $P_0(\omega, \boldsymbol{z})$ of a volume V can be generated by a distribution of monopole sources on the surface ∂V. The field outside of V will not vanish as in Eq. 13.35, since no more dipole sources are involved. Furthermore, the construction of the Green's function according to Eq. 13.38 induces boundary conditions of the second kind (Neumann) on the surface ∂V

$$P_0(\omega, \boldsymbol{z}) = -\iint\limits_{\partial V} 2\, G_0^{\mathrm{f}}(\omega, \boldsymbol{z}|\boldsymbol{z}')\, \frac{\partial}{\partial \boldsymbol{n}} P_0(\omega, \boldsymbol{z}')\, d\boldsymbol{z}'\,, \qquad \boldsymbol{z} \in V\,. \tag{13.48}$$

The effects of these boundary conditions are considered for the determination of the driving signals in Sec. 13.3.6.

13.3.4 Reduction to Two Spatial Dimensions

The requirement of creating a distribution of sources on a whole surface around a listening space may be impractical for many sound reproduction systems. This section shows how to reduce the source distribution from a surface around the listeners to a closed curve in a horizontal plane preferably in the height of the listeners' ears. For convenience this height is denoted by $z = 0$.

The mathematical tools for the reduction of the source distribution have already been presented by considering the Kirchhoff-Helmholtz integral for a prism in Sec. 13.2.3.2 and the relations between line and point sources in Sec. 13.2.2.3. These considerations are now applied in two steps to the representation of a spatial sound field in Eq. 13.48.

The first step is the conversion of the general surface ∂V to a prism. Performing the mathematical operations in Eq. 13.36 on Eq. 13.48 results in a representation of the 3D sound field in a prism which is independent of z (compare Eq. 13.37)

$$P_1(\omega, \boldsymbol{x}) = -\oint\limits_{\partial L} 2\, G_1^{\mathrm{f}}(\omega, (\boldsymbol{x}|\boldsymbol{x}')\, \frac{\partial}{\partial \boldsymbol{n}} P_1(\omega, \boldsymbol{x}')\, d\boldsymbol{x}'\,, \qquad \boldsymbol{x} \in L\,. \tag{13.49}$$

This relation describes a distribution of line sources parallel to the z-axis. However, a technical realization would not be easy to implement.

To arrive at a model for a practical solution, a second step replaces the line sources by point sources according to Sec. 13.2.2.3. With Eq. 13.32 follows from Eq. 13.49

$$P_1(\omega, \boldsymbol{x}) = -\oint_{\partial L} G_0^{\mathrm{f}}(\omega, \boldsymbol{x}|\boldsymbol{x}')\, D(\omega, \boldsymbol{x}|\boldsymbol{x}')\, d\boldsymbol{x}' , \qquad \boldsymbol{x} \in L , \tag{13.50}$$

with

$$D(\omega, \boldsymbol{x}|\boldsymbol{x}') = 2A\big(\|\boldsymbol{x} - \boldsymbol{x}'\|\big)\, H(\omega)\, \frac{\partial}{\partial \boldsymbol{n}} P_1(\omega, \boldsymbol{x}') . \tag{13.51}$$

To show the dependence on $\boldsymbol{x}$ and $\boldsymbol{x}'$, ρ in Eq. 13.32 has been expanded by Eq. 13.24.

In Eq. 13.50, $G_0^{\mathrm{f}}(\omega, \boldsymbol{x}|\boldsymbol{x}')$ denotes the Green's function of the monopole sources on the contour ∂L in the xy-plane. It describes the wave propagation in 3D space, however, the receiver locations $\boldsymbol{x}$ (the listeners' ears) are assumed to reside in the xy-plane as well. $D(\omega, \boldsymbol{x}|\boldsymbol{x}')$ denotes the source signal of the monopoles.

13.3.5 Spatial Sampling

The previous sections showed how the rather general statement of the Kirchhoff-Helmholtz integral can be narrowed down to a model for a spatial reproduction system. A hypothetical distribution of monopole and dipole sources on a 2D surface around the listener has been replaced by a distribution of monopoles on a 1D contour in a horizontal plane in the height of the listeners' ears.

For a technical solution, this spatially continuous source distribution has to be replaced by an arrangement of a finite number of loudspeakers with a monopole-like source directivity. The resulting wave field is given by a modification of Eq. 13.50, where the integral is substituted by a sum over the discrete loudspeaker positions $\boldsymbol{x}'_n$

$$P_1(\omega, \boldsymbol{x}) \approx -\sum_n G_0^{\mathrm{f}}(\omega, \boldsymbol{x}|\boldsymbol{x}'_n)\, D(\omega, \boldsymbol{x}|\boldsymbol{x}'_n)\, \Delta x'_n . \tag{13.52}$$

$\Delta x'_n$ is the length of the spatial increment $\Delta \boldsymbol{x}'_n$ between the samples. It is not required to be equidistant.

The representation of a continuous function by a finite number of spatially discrete sources is known as spatial sampling in terms of signal theory. Unfortunately, spatial sampling of wave fields is not yet satisfactorily described in the technical literature. Therefore, some general remarks have to suffice at this point.

Sampling of multidimensional functions is well understood, e.g. in image or video processing. However, deriving a suitable loudspeaker spacing from the requirement of the sampling theorem demands to place two loudspeakers per shortest permissible wavelength. For the usual audio range up to 20 kHz, loudspeakers would have to be placed at a distance of less than 1 cm. Such a loudspeaker array is not technically feasible, considering both the size of available loudspeakers as well es their total number.

It appears that there are two factors of influence, which allow to reduce the number of loudspeakers significantly. At first, a wave field is not an arbitrary signal restricted only by an upper bound of its frequency range. Instead, all signals in acoustics are solutions of the wave equation. This special property restricts the frequency domain representations of wave fields significantly [17]. The consequences for sampling of wave fields have been described e.g. in [7].

The second factor is the human perception of spatial aliasing effects. Experience with existing wave field synthesis implementations with loudspeaker spacing between 10 cm and 20 cm suggests that aliasing terms in sound fields are subject to effective masking by other sound components. However, working knowledge in human perception of spatial aliasing seems to be still rather restricted. For some further comments on spatial sampling see e.g. [21]. In short, spatial sampling seems to be a useful approach for spatial reproduction, although the human perception of its effects is largely unexplored.

13.3.6 Driving Signals

It remains to determine the driving signals of the loudspeakers. They follow from an analysis of $D(\omega, \boldsymbol{x}|\boldsymbol{x}')$ according to Eq. 13.51. This analysis has to take the nature of the desired wave field into account. Wave fields may be modeled by arrangements of different types of sources, e.g. monopoles and dipoles, and by plane waves. The determination of the driving signals from a model of the wave field is called *model based rendering.* On the other hand, a wave field can be recorded in a natural environment like a concert hall or a church. Obtaining the driving signals from a recorded wave field is called *data based rendering.*

The determination of the driving signals is shown here for a rather general case of model based rendering, where the desired wave field is given by a decomposition into plane waves according to Eq. 13.11. The discussion focusses on three major points:

1. The correct consideration of the boundary conditions (Eq. 13.45) induced by the choice of the Green's function for the elimination of the dipole sources in Sec. 13.3.3.
2. The determination of the normal derivative in Eq. 13.51.
3. The independence of the driving signals from the listener position.

13.3.6.1 Boundary Conditions

The elimination of the dipole sources in Sec. 13.3.3 was based on the choice of a Green's function with homogeneous boundary conditions of the second kind (Neumann), i.e. a vanishing normal derivative on the boundary (see Eq. 13.45). Boundary conditions of this kind are known to produce reflections on the boundary. At first sight it is not clear, how these reflections should occur, since the boundary ∂L is only an arbitrary contour for the placement of point sources and not a solid wall.

For a better understanding of the situation, consider Fig. 13.7. It shows a contour ∂L in a plane with two different monopole positions. These positions are spatial samples $\boldsymbol{x}'_n$ of the coordinate $\boldsymbol{x}'$ on ∂L. Since these positions are arbitrary, the indication of the sample number n will be suppressed for ease of notation. Note that the monopole positions are not only characterized by their coordinates $\boldsymbol{x}'$ but also by the normal vector $\boldsymbol{n}$ on the contour ∂L at $\boldsymbol{x}'$ and by the angle γ according to

$$\boldsymbol{n} = \begin{bmatrix} \cos\gamma \\ \sin\gamma \end{bmatrix} . \tag{13.53}$$

The monopoles shall be driven such that they produce a plane wave in the direction $\boldsymbol{n}_\theta$. Obviously the contour ∂L exhibits two different sections separated by circles in Fig. 13.7. Monopoles in the left section emanate waves into the domain L with a component in the direction $\boldsymbol{n}_\theta$ of the plane wave. Producing circular waves, they also radiate a component into the direction opposite to $\boldsymbol{n}_\theta$, but it does not effect the domain L. However a monopole in the right section emanates into L a component opposite to $\boldsymbol{n}_\theta$. The superposition of monopoles and dipoles required by the Kirchhoff-Helmholtz integral creates a directivity which prevents source components in the wrong direction. But with monopoles only, there is no other way than accepting equal sound radiation in all directions.

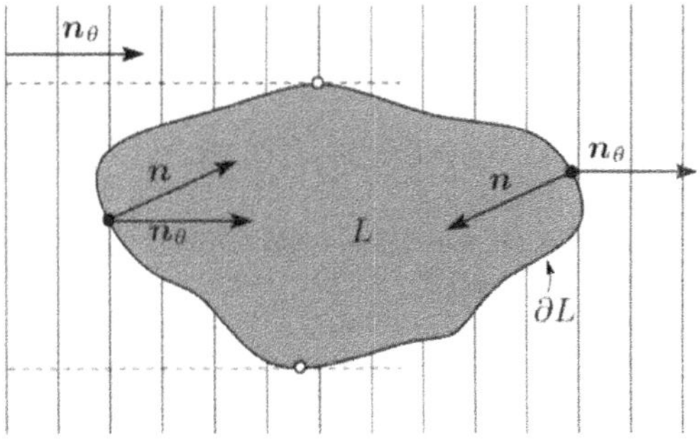

Fig. 13.7. Reproduction of a plane wave by a monopoles on a contour ∂L.

On the other hand, the radiation of the monopoles in the right section of Fig. 13.7 into the domain L equals the reflections that would have been produced by a hard surface at ∂L. Although such a reflecting wall is not physically present, its reflections are produced by the monopoles in the right section. However these components travelling in the direction opposite to $\boldsymbol{n}_\theta$ do not belong to the desired wave field. Since dipoles are no more available to solve the problem in the domain of acoustics, the reflections in the right section have to be counteracted by suitable processing of the driving signals.

A simple but effective way to prevent sound radiation in the wrong direction is to cancel the driving signals of the loudspeakers in the right section. This measure is described by a rectangular window function $v(\boldsymbol{x}', \theta)$ which determines the source activity for each location $\boldsymbol{x}'$ on the contour ∂L and for each possible direction θ of a plane wave. The values of the window function depend on the scalar product $\langle \boldsymbol{n}, \boldsymbol{n}_\theta \rangle$ between the direction $\boldsymbol{n}_\theta$ of plane wave propagation and the normal vector $\boldsymbol{n}$ in each position $\boldsymbol{x}'$ on ∂L. Since $\langle \boldsymbol{n}, \boldsymbol{n}_\theta \rangle = \cos(\theta - \gamma)$, the window function is defined as

$$v(\boldsymbol{x}', \theta) = \begin{cases} 1, & \text{if} \quad \langle \boldsymbol{n}, \boldsymbol{n}_\theta \rangle > 0 \quad \text{or} \quad |\theta - \gamma| < \dfrac{\pi}{2}, \\ 0, & \text{else.} \end{cases} \tag{13.54}$$

With this window function, the source signal for the monopoles from Eq. 13.51 can be written as

$$D(\omega, \boldsymbol{x}|\boldsymbol{x}') = 2v(\boldsymbol{x}', \theta)\, A\big(\|\boldsymbol{x} - \boldsymbol{x}'\|\big)\, H(\omega)\, \frac{\partial}{\partial \boldsymbol{n}} P_1(\omega, \boldsymbol{x}')\,. \tag{13.55}$$

As an example, the wave field reproduced by a circular distribution of monopole sources is shown in Fig. 13.8. The radius of the circular region was chosen as $R = 1.50$ m. Two cases where evaluated: (1) the left row shows the results when the window function $v(\boldsymbol{x}', \theta)$ is discarded, (2) the right row shows the results when incorporating the window function.

From the results it can be clearly seen that the window function eliminates the reflections introduced by the Neumann boundary conditions. The wave field is reproduced correctly within the circular distribution of secondary monopole sources. The wave field outside of that region does not vanish, as this would be the case when using both monopole and dipole sources. Instead it is a mirrored version of the plane wave within the circle (see Fig. 13.6).

13.3.6.2 Determination of the Normal Derivative

To calculate the driving signal from Eq. 13.55 requires to express the normal derivative of $P_1(\omega, \boldsymbol{x}')$ by a suitable characterization of the wave field. Here, wave fields with a representation as a plane wave decomposition are considered. Then the normal derivative has to be expressed by the plane wave coefficients (see Sec. 13.2.2.1), i.e. by the wave forms of the individual plane wave components.

First, only a single plane wave is investigated. According to Eq. 13.10 its influence at the loudspeaker position $\boldsymbol{x}'$ is given by

$$P_1(\omega, \boldsymbol{x}') = F(\omega, \theta)\; e^{\,j\frac{\omega}{c}\langle \boldsymbol{x}', \boldsymbol{n}_\theta \rangle}\,, \tag{13.56}$$

For the gradient $\nabla P_1(\omega, \boldsymbol{x}')$ follows

$$\nabla P_1(\omega, \boldsymbol{x}') = j\frac{\omega}{c}\, P_1(\omega, \boldsymbol{x}')\, \boldsymbol{n}_\theta \tag{13.57}$$

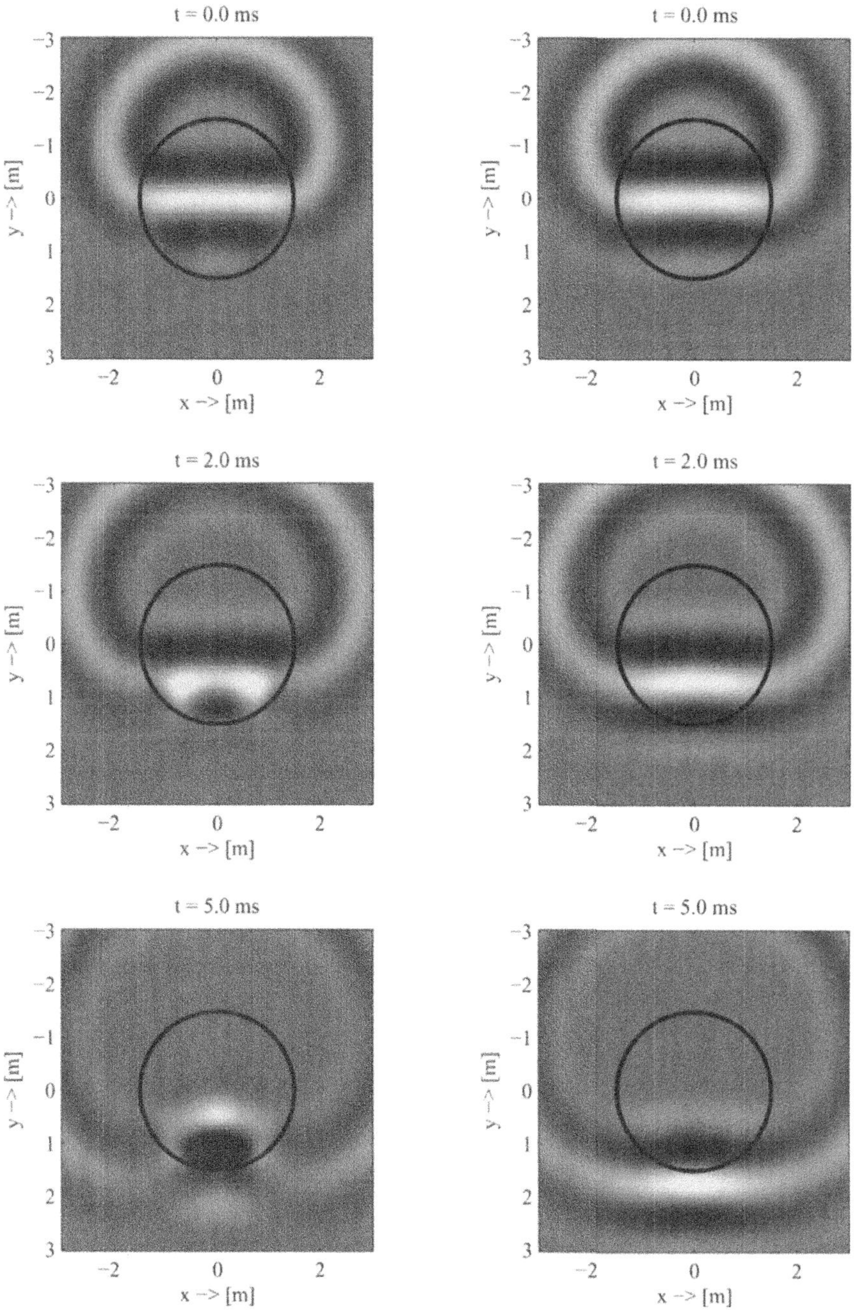

Fig. 13.8. Reproduction of a plane wave with a two-dimensional circular distribution of monopole sources. The left row shows the wave field when discarding the window function $v(\boldsymbol{x}', \theta)$, the right row when taking it into account.

and for the normal derivative

$$\begin{aligned}\frac{\partial}{\partial \boldsymbol{n}} P_1(\omega, \boldsymbol{x}') &= \left\langle \nabla P_1(\omega, \boldsymbol{x}'), \boldsymbol{n} \right\rangle \\ &= j\frac{\omega}{c}\, P_1(\omega, \boldsymbol{x}')\, \langle \boldsymbol{n}_\theta, \boldsymbol{n} \rangle \\ &= j\frac{\omega}{c}\, P_1(\omega, \boldsymbol{x}')\, \cos(\theta - \gamma). \end{aligned} \tag{13.58}$$

Inserting the normal derivative (Eq. 13.58) into Eq. 13.55 gives the driving signal $D_\theta(\omega, \boldsymbol{x}|\boldsymbol{x}')$ for a plane wave with the angle of incidence θ. After some manipulations, the result may be written as

$$D_\theta(\omega, \boldsymbol{x}|\boldsymbol{x}') = 2w(\boldsymbol{x}', \theta)\, A(\|\boldsymbol{x} - \boldsymbol{x}'\|)\, K(\omega)\, e^{j\frac{\omega}{c}\langle \boldsymbol{x}', \boldsymbol{n}_\theta \rangle}\, F(\omega, \theta)\,. \tag{13.59}$$

The various terms of $D_\theta(\omega, \boldsymbol{x}|\boldsymbol{x}')$ are now discussed in detail.

- The window function $w(\boldsymbol{x}', \theta)$ combines the effects of the rectangular window function $v(\boldsymbol{x}', \theta)$ from Eq. 13.54 and the cos-term from the normal derivative in Eq. 13.58

$$w(\boldsymbol{x}', \theta) = \begin{cases} \cos(\theta - \gamma), & \text{if } \ |\theta - \gamma| < \frac{\pi}{2}, \\ 0, & \text{else.} \end{cases} \tag{13.60}$$

- The spectrum $F(\omega, \theta)$ is the Fourier transform of the wave form $f(t, \theta)$. The wave form is observed directly at the origin of the coordinate system (see Sec. 13.2.2.1).
- The exponential term describes the delay of the plane wave from the origin to the position $\boldsymbol{x}'$ of the monopole. It can be realized by a time delay of the wave form $f(t, \theta)$.
- The high-pass frequency response $K(\omega)$ combines the term $H(\omega)$ from the approximation of a line source by a point source in Eq. 13.33 and the effect of the differentiation in Eq. 13.58

$$K(\omega) = \sqrt{\frac{\omega}{c}}\, e^{j3\pi/4}\,. \tag{13.61}$$

  It can be realized by filtering the wave form $f(t, \theta)$.
- The amplitude modification $A(\|\boldsymbol{x} - \boldsymbol{x}'\|)$ comes from the approximation of a line source by a point source in Eq. 13.34. It is discussed in detail in Sec. 13.3.6.3.

Finally, the driving signals for a wave field that is composed of various plane waves are obtained by superposition of the individual plane wave driving signals D_θ

$$D(\omega, \boldsymbol{x}|\boldsymbol{x}') = \int_0^{2\pi} D_\theta(\omega, \boldsymbol{x}|\boldsymbol{x}')\, d\theta\,. \tag{13.62}$$

13.3.6.3 Independence of the Driving Signals from the Listener Position

The driving signals for a plane wave in Eq. 13.59 or for a wave field composed of plane waves in Eq. 13.62 formally depend on time t or frequency ω, the loudspeaker position $\boldsymbol{x}'$ on the contour ∂L and on the listener position $\boldsymbol{x}$ within the area L. Time dependence is a necessary feature, dependence on the loudspeaker position is manageable by multichannel filtering, but dependence on the position of a listener is highly undesirable. Even if the position of a listener was known at all times, e.g. by a tracking system, this would not solve the problem of sound reproduction for multiple listeners in a larger audience.

Fortunately, the dependence on the listener position is rather mild and is not at all comparable with the well-known sweet spot limitation of stereophonic systems. The only component of $D_\theta(\omega, \boldsymbol{x}|\boldsymbol{x}')$ which depends on $\boldsymbol{x}$ is the amplitude modification $A(\|\boldsymbol{x} - \boldsymbol{x}'\|)$ due to the approximation of the line sources. A less drastic approximation, e.g. by multiple point sources, would result in less amplitude modification. However, most important, all the other components of $D_\theta(\omega, \boldsymbol{x}|\boldsymbol{x}')$ do not depend on the listener position at all. Delay and filtering of the wave form are correct for the total area L.

For practical realizations, the amplitude correction is set to a fixed listener position $A(\|\boldsymbol{x}_0 - \boldsymbol{x}'\|)$. Then the resulting modified driving signals $D_{0,\theta}$ are independent of the listener position $\boldsymbol{x}$

$$D_{0,\theta}(\omega, \boldsymbol{x}') = D_\theta(\omega, \boldsymbol{x}_0|\boldsymbol{x}') \,. \tag{13.63}$$

They cause a slight deviation in the reproduction volume, but delay and filtering operations remain unaffected. The amount of this deviation and its spatial distribution can be controlled by suitable choice of $\boldsymbol{x}_0$ [18].

13.3.7 Signal Processing Structure

Now that the driving signals for the loudspeakers are determined, the signal processing structure for their production is investigated. From now on, only the listener independent driving signals $D_{0,\theta}(\omega, \boldsymbol{x}')$ are used. They are the output of a signal processing chain with the wave forms $F(\omega, \theta)$ of the plain wave components as input. In short form this processing chain is written as

$$D_{0,\theta}(\omega, \boldsymbol{x}') = M(\omega, \boldsymbol{x}', \theta)\, F(\omega, \theta) \tag{13.64}$$

with

$$M(\omega, \boldsymbol{x}', \theta) = 2w(\boldsymbol{x}', \theta)\, A\big(\|\boldsymbol{x}_0 - \boldsymbol{x}'\|\big)\, K(\omega)\, e^{j\frac{\omega}{c}\langle \boldsymbol{x}', \boldsymbol{n}_\theta \rangle} \,. \tag{13.65}$$

Complex wave fields can be represented by a superposition of plane waves as shown in Eq. 13.62. In practical spatial reproduction systems the number of plane wave components is limited. Then the driving signals contain contributions from a finite number of plane waves from a discrete set of angles denoted by θ_m with $m = 1, 2, \ldots$

$$D_0(\omega, \boldsymbol{x}') = \sum_m D_{0,\theta_m}(\omega, \boldsymbol{x}') = \sum_m M(\omega, \boldsymbol{x}', \theta_m)\, F(\omega, \theta_m)\,. \tag{13.66}$$

Furthermore, from this finite number of plane wave components, driving signals for each discrete loudspeaker position $\boldsymbol{x}'_n$ have to be generated. The resulting structure is best formulated in vector notation with

$$\boldsymbol{D}_0(\omega) = \begin{bmatrix} \vdots \\ D_0(\omega, \boldsymbol{x}'_n) \\ \vdots \end{bmatrix}, \qquad \boldsymbol{F}(\omega) = \begin{bmatrix} \vdots \\ F(\omega, \theta_m) \\ \vdots \end{bmatrix}, \tag{13.67}$$

$$\boldsymbol{M}(\omega) = \begin{bmatrix} & \vdots & \\ \ldots & M(\omega, \boldsymbol{x}'_n, \theta_m) & \ldots \\ & \vdots & \end{bmatrix}. \tag{13.68}$$

Then the vector of driving signals for each loudspeaker is calculated from the vector of wave forms for each plane wave component by

$$\boldsymbol{D}_0(\omega) = \boldsymbol{M}(\omega)\, \boldsymbol{F}(\omega)\,. \tag{13.69}$$

In the time domain the driving signals are the result of a multichannel convolution

$$\boldsymbol{d}_0(t) = \boldsymbol{m}(t) * \boldsymbol{f}(t)\,. \tag{13.70}$$

In short, the signal processing structure for the calculation of the loudspeaker driving signals is a multiple-input, multiple-output (MIMO) system which performs a multichannel convolution with the wave forms of the plane wave components. The convolution filters comprise various operations in time and space as discussed in Sec. 13.3.6.2 and 13.3.6.3.

13.4 Implementation of a Wave Field Synthesis System

Finally, the implementation of a wave field synthesis system is shown by an example. Fig. 13.9 sketches a typical configuration of a virtual acoustical scene. The grey background shows the floor plan of a church with the apsis on the right. Here a singer or a musician is placed as a primary sound source. Its sound waves propagate through the church via a direct path and multiple reflections (indicated by dashed lines). The intention is now to reproduce the sound field in the center part of the church by a wave field synthesis system installed at a remote location. To this end, the sound waves arriving at the

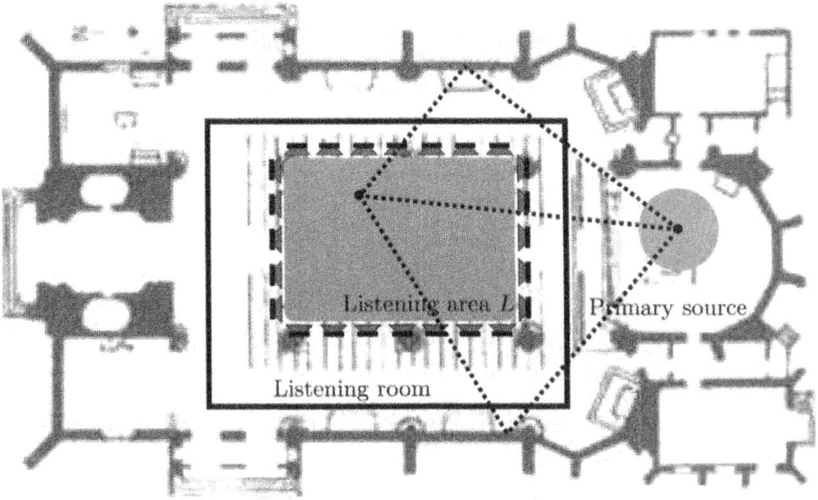

Fig. 13.9. Reproduction of a virtual environment with a wave field system.

boundary of the listening area serve as source signals for the loudspeakers of the wave field synthesis system.

There are different ways to obtain these loudspeaker signals corresponding to model based and data based rendering techniques (see Sec. 13.3.6). In this example, a virtual scene model based on a decomposition into plane waves is used for simplicity. For more advanced approaches using data based rendering see e.g. [12, 21].

The acoustical scene inside the church can be simplified e.g. by an image source model. It starts from a point source model for the primary source in the apsis and approximates the reflections by sources mirrored on the reflecting surfaces. The total sound field is then represented by a multitude of point sources. Each of these point sources can be decomposed into a superposition of plane waves. The directions for two selected reflections to an arbitrary listener position are indicated by the dashed lines in Fig. 13.9. Varying the secondary source position along the boundary ∂L of the listening area gives the normal directions $\boldsymbol{n}_\theta$ for the determination of the driving signals e.g. in Eqs. 13.59 and 13.60. Playing back the driving signals with a wave field installation in a remote listening room (indicated by the solid line in Fig. 13.9) then reproduces the sound field within the listening area.

An implementation of a wave field synthesis system with a circular loudspeaker array is shown in Fig. 13.10. Here the planar area L is a disc with a radius of 1.5 m. A total of 48 two-way loudspeakers are mounted on the circumference ∂L with a spacing of about 20 cm. The analog driving signals are delivered by three 16-channel audio amplifiers with digital inputs shown in Fig. 13.11. The digital input signals are the result of the multichannel convolution (Eq. 13.70). It is realized by fast convolution techniques in real-time on a personal computer. The system described here is located at the Telecommunications Laboratory (Multimedia Communications and Signal Processing) of the University of Erlangen-Nuremberg in Germany [13].

Fig. 13.10. Circular loudspeaker array with a radius of 1.5 m and 48 channels.

Fig. 13.11. Three 16-channel audio amplifiers for the array in Fig. 13.10 mounted in a 19 inch rack.

13.5 Conclusions

This chapter has given an introduction to wave field synthesis. It is based on the acoustic wave equation and the representation of its solutions by plane waves and Green's functions. Starting from these physical foundations, it has been shown how to derive the driving signals for the loudspeaker array of the resulting wave field synthesis systems.

The derivation is valid for rather general geometries and sizes of loudspeaker arrays. Furthermore, no assumption on the position of the listener is required. Then the reproduced sound field is physically correct within the limitations imposed by amplitude deviation and by spatial discretization effects. The computation of the loudspeaker driving signals is conceptually simple and is performed by a multichannel convolution. In short, the realizing technique of wave field synthesis systems is obtained by mapping the acoustic wave equation to a multiple-input, multiple-output system (MIMO) system.

However, the practical realization of wave field synthesis has some pitfalls, which can be avoided by further signal processing techniques. These pertain the simplified monopole model of the loudspeakers and the acoustical reflections of the loudspeakers within the listening room.

So far it has been assumed that acoustical monopoles can be approximated well by small loudspeakers with closed enclosures. If required, this approximation can be improved with digital compensation of non-ideal loudspeaker properties [20]. The second pitfall consists of the reflections of the loudspeaker array signals in the listening room. They may degrade the performance level predicted from theory. Countermeasures are passive or active cancellation of these reflections. Especially, active cancellation seems promising by using the loudspeaker arrays for reproduction also for the cancellation of room reflections [19].

The presentation here has been focussed on the determination of the driving signals, once a representation of an existing or virtual sound field is given in terms of plane waves. Such a representation is always possible through the so-called plane wave decomposition [12]. How to obtain this decomposition from microphone array measurements in a real room belongs to the area of wave field analysis. This area has not been discussed here in detail. For more information see e.g. [12, 21]. Also not discussed here were the relations of wave field synthesis to other spatial reproduction techniques. For a comparison with Ambisonics see [15].

References

[1] M. Abramowitz, I.A. Stegun: *Handbook of Mathematical Functions,* New York, NY, USA: Dover Publications, 1972.

[2] G.B. Arfken, H.J. Weber: *Mathematical Methods for Physicists,* 5th ed., San Diego, CA, USA: Academic Press, 2001.

[3] A.J. Berkhout: A holographic approach to acoustic control, *Journal of the Audio Engineering Society,* **36**, 977–995, December 1988.

[4] R.N. Bracewell: *Fourier Analysis and Imaging,* Boston, MA, USA: Kluwer, 2003.

[5] S. Brix, T. Sporer, J. Plogsties: CARROUSO - An European approach to 3D-audio, *110th AES Convention,* Audio Engineering Society (AES), Amsterdam, Netherlands, May 2001.

[6] W. de Bruin: *Application of wave field synthesis in videoconferencing,* PhD thesis, Delft University of Technology, 2004.

[7] J. Coleman: Ping-pong sample times on a linear array halve the Nyquist rate, *Proc. ICASSP '04*, **4**, 925–928, Montreal, Canada, 2004.

[8] J.D. Gaskill: *Linear Systems, Fourier Transforms, and Optics,* New York, NY, USA: Wiley, 1978.

[9] I.S. Gradshteyn, I.M. Ryzhik: *Tables of Integrals, Series, and Products,* San Diego, CA, USA: Academic Press, 1965.

[10] S. Hassani: *Mathematical Physics, A Modern Introduction to its Foundations,* Berlin, Germany: Springer, 1999.

[11] E. Hulsebos: *Auralization using Wave Field Synthesis,* PhD thesis, Delft University of Technology, 2004.

[12] E. Hulsebos, D. de Vries, E. Bourdillat: Improved microphone array configurations for auralization of sound fields by wave field synthesis, *Journal of the Audio Engineering Society (AES),* **50**(10), 779–790, Oct. 2002.

[13] Chair of Multimedia Communications and Signal Processing, University Erlangen-Nuremberg, Germany, `http://www.lnt.de/LMS`.

[14] P.M. Morse, H. Feshbach: *Methods of Theoretical Physics – Part I,* New York, NY, USA: McGraw-Hill, 1953.

[15] R. Nicol, M. Emerit: Reproducing 3D-sound for videoconferencing: A comparison between holophony and ambisonic, *Proc. DAFX '98,* Barcelona, Spain, Nov. 1998.

[16] A.D. Pierce: *Acoustics – An Introduction to its Physical Principles and Applications,* Acoustical Society of America, 1991.

[17] R. Rabenstein, P. Steffen, S. Spors: Representation of two-dimensional wave fields by multidimensional signals, *Signal Processing,* to be published.

[18] J.-J. Sonke, D. de Vries, J. Labeeuw: Variable acoustics by wave field synthesis: A closer look at amplitude effects, *Proc. 104th AES Convention,* prepr. 4712, Audio Engineering Society (AES), Amsterdam, Netherlands, May 1998.

[19] S. Spors, H. Buchner, R. Rabenstein: Adaptive listening room compensation for spatial audio systems, *Proc. EUSIPCO '04,* 1381–1385, Vienna, Austria, 2004.

[20] S. Spors, D. Seuberth, R. Rabenstein: Multiexciter panel compensation for wave field synthesis, *Proc. DAGA '05,* Munich, Germany, 2005.

[21] S. Spors, H. Teutsch, A. Kuntz, R. Rabenstein: Sound field synthesis, in Y.Huang and J.Benesty (eds.), *Audio Signal Processing for Next-Generation Multimedia Communication Systems,* Boston, MA, USA: Kluwer, 2004.

[22] E.W. Start: *Direct Sound Enhancement by Wave Field Synthesis,* PhD thesis, Delft University of Technology, 1997.

[23] E.N.G. Verheijen: *Sound Reproduction by Wave Field Synthesis,* PhD thesis, Delft University of Technology, 1997.

[24] P. Vogel: *Application of Wave Field Synthesis in Room Acoustics,* PhD thesis, Delft University of Technology, 1993.
[25] D. de Vries, E.W. Start, V.G. Valstar: The wave field synthesis concept applied to sound reinforcement: Restrictions and solutions, *96th AES Convention,* Audio Engineering Society (AES), Amsterdam, Netherlands, February 1994.
[26] E.G. Williams: *Fourier Acoustics: Sound Radiation and Nearfield Acoustical Holography,* San Diego, CA, USA: Academic Press, 1999.

14

Signal Processing for In-Car Communication Systems

Gerhard Schmidt and Tim Haulick

Harman/Becker Automotive Systems, Ulm, Germany

In limousines and vans communication between passengers in the front and in the rear may be difficult – especially if the car is driven at medium or high speed, resulting in a large background noise level. Furthermore, driver and front passenger speak towards the windshield. Thus, they are hardly intelligible for those sitting behind them. To improve the speech intelligibility the passengers start speaking louder and lean or turn towards their communication partners (see Fig. 14.1). For longer conversations this is usually tiring and uncomfortable.

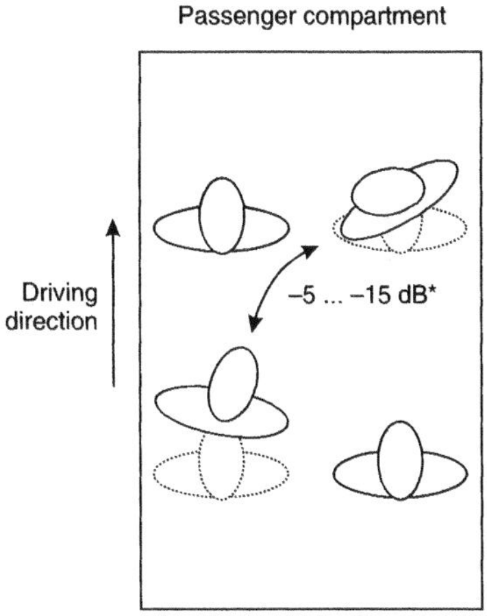

Fig. 14.1. Communication between passengers in a car (*acoustic loss, referred to the right ear of the driver).

A way to improve the speech intelligibility within a passenger compartment is to use an in-car communication system [31, 35] – often shortly called *intercom system.* These systems record the speech of the speaking passengers by means of microphones and improve the communication by playing the recorded signals via those loudspeakers located close to the listening passengers. Fig. 14.2 sketches the structure of a simple car interior communication system aimed to support front-to-rear conversations with one microphone and one loudspeaker.

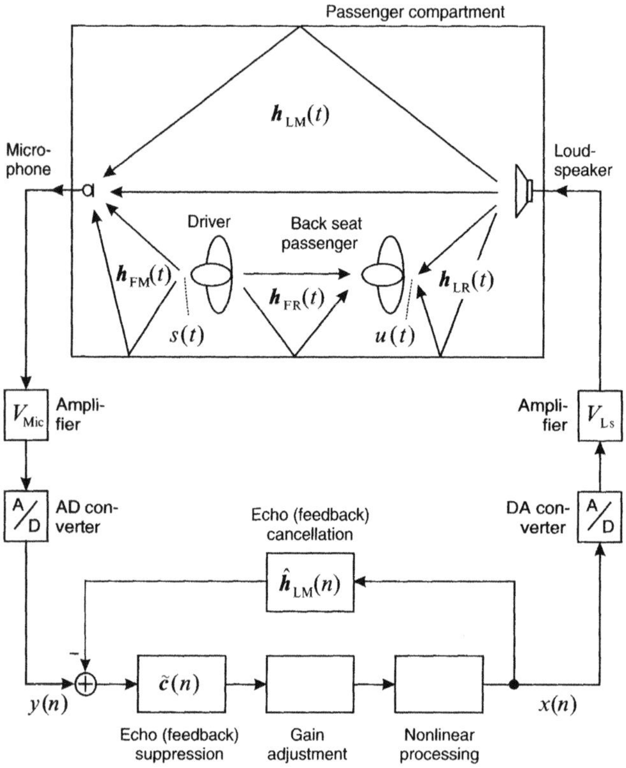

Fig. 14.2. Structure of a basic car interior communication system.

As it is clearly visible in Fig. 14.2, intercom systems operate in a closed electro-acoustic loop. The microphone picks up at least a portion of the loudspeaker signal. If this portion is not sufficiently small sustained oscillations appear – which can be heard as howling or whistling. The howling threshold depends on the output gain of the intercom system as well as on the gains of the analog amplifiers V_{Mic} and V_{Ls}. For this reason all gain settings within the system need to be adjusted carefully.

To improve the stability margin signal processing, such as beamforming, feedback and echo cancellation, adaptive notch filtering, adaptive gain adjustment, and nonlinear processing can be applied. A few basic processing units are already depicted in Fig. 14.2.

Before we will describe the signal processing units in more detail in Sec. 14.2, we will discuss the boundary conditions we have to fulfill when designing communication systems for passenger compartments in the next section. In contrast to hands-free telephones or speech recognition engines no methods for evaluating the quality of intercom systems have been standardized or even published yet.[1] Thus, evaluation is not as easy as in other speech and audio applications. However, a few measurements (binaural recordings) as well as subjective tests (performed in a car equipped with an intercom system) are presented at the end of this contribution.

14.1 Basics

When designing an intercom system a variety of limiting conditions and system demands will appear. In order to understand the origin of these demands a few – mostly physical or psychoacoustic – phenomena will be described within this section. Furthermore, models for all important transmission paths are introduced. This allows to give a first motivation for some of the signal processing units, such as feedback cancellation and beamforming.

14.1.1 Communication without Intercom Systems

We will start with the analysis of the acoustic scenario within a passenger compartment without an intercom system. Comparisons of communication channels between standing vehicle and high speed as well as between passengers sitting side by side and behind each other allow to get an impression of the required gain or SNR improvement of intercom systems.

Because of the directionality of a human head[2] – depicted for two frequency ranges in Fig. 14.3 – it is harder to understand someone from behind than it is during an eye-to-eye conversation. In contrast to the rear passengers, driver and front passenger do not speak towards their communication partners. Thus, they are less intelligible. The frequency range from 1400 to 2000 Hz, for example, is attenuated by more than 10 dB when listening to someone from behind ($\phi = 180^o$) compared to an eye-to-eye communication. For this reason, it might be sufficient to enhance only the communication from front to rear within a passenger compartment. However, this is only true for cars with only two rows of seats. In minibusses or very large limousines an intercom system should support both directions.

Another important question is how much "enhancement" (in terms of amplification) is required. In most cars the speech intelligibility is good or at least sufficient if the car is not driving. In such a scenario an intercom system

[1] For hands-free telephones, e.g., a variety of measurements including test signals [24], measurement procedures, as well as system requirements, such as minimal echo attenuation or maximum delay [21] have been standardized.

[2] Here the directionality of the human speaking apparatus (mouth, head) – not of the listening system – is meant.

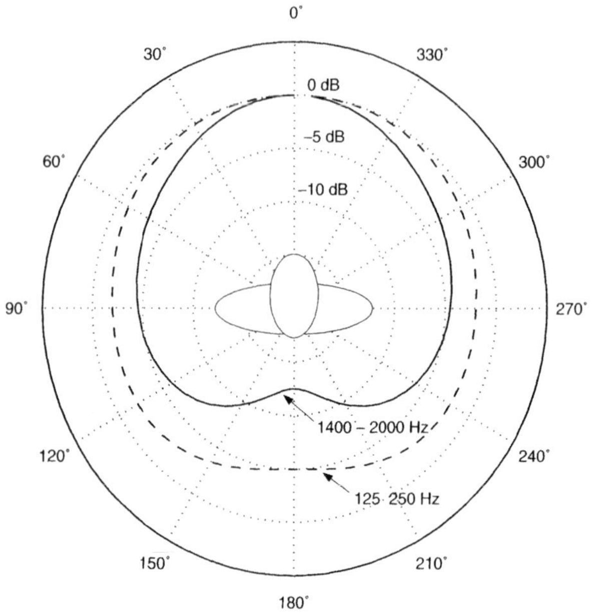

Fig. 14.3. Average directionality of a human mouth and head (according to [29]).

would make the car sound more reverberant and, thus, reduce the communication quality. However, at medium or high speed things are changing and an intercom system is able to enhance the speech intelligibility considerably.

Fortunately, an in-car communication system does not have to compensate – in terms of loudspeaker gain – the full amount of the noise power difference between standing car and high speed. Car noise results from a large number of sources. The main components are engine noise, wind noise, tire noise, and noise from devices (e.g., fans) inside the passenger compartment. Fig. 14.4 shows estimates of power spectral densities of noises measured in a car at various speeds.

As one can see in Fig. 14.4 the power of the background noise increases by more than 30 dB at nearly all frequencies when the car accelerates from 0 to 150 km/h. However, because of the so-called *Lombard effect* [33] it is not necessary to increase the amplification of an intercom system by the same amount. Any person who speaks in a noisy environment will automatically alter the speech characteristics in order to increase the efficiency of communication over the noisy channel. This effect can be described in more detail by the following statements [14]:

- The average bandwidth of most phonemes decreases.
- The formant frequencies of vowels are increased.
- The first formant frequency of most phonemes increases.
- The formant amplitudes increase, leading to an increased spectral tilt.

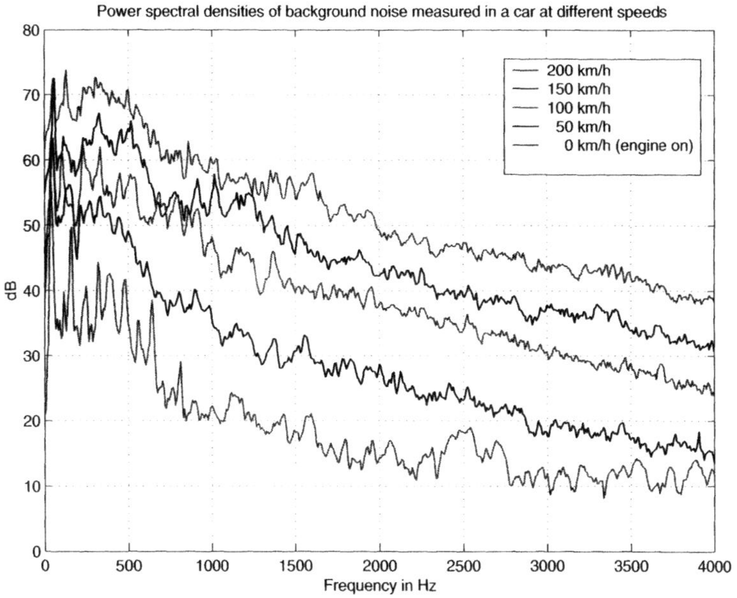

Fig. 14.4. Estimates of the power spectral densities of noise measured in a car. For the analysis the same part of motorway was driven at various speeds (with permission from [13]).

As a result of the last point, the overall speech level rises with increasing background noise power. Rates of about 0.3–1 dB of speech power increment per decibel of background noise increment have been reported [14]. Thus, the power of the same speech sequence recorded in a car can vary by about 10–30 dB, depending on whether the car is parked or moving at high speed.

In Fig. 14.5 the results of an analysis of the speech power in dependence of the noise power (A-weighted) are depicted. About 2000 utterances were recorded in a car driven at different speeds including standstill. For the recordings a close-talking microphone has been utilized. By using a correction filter the microphone was calibrated such that the same power was measured as in the *mouth reference point* (MRP).[3] The speakers were sitting on the front passengers seat and the car was driven at a speed of either 0 km/h, 90 km/h, 130 km/h, or 160 km/h. Furthermore, the speakers were asked to spell city names and to command a speech recognition system in terms of speaking a phone number. In order to get also natural utterances the speakers were asked what they had eaten for breakfast and what kind of electronic devices they own. Each of these answers results in one of the entries in terms of pairs of speech power and A-weighted noise power in Fig. 14.5.

[3] The mouth reference point (MRP) is defined as the point that is about 25 mm in front of the center of the lips. Further details about the definition of this point can be found in the ITU-T recommendation P.64 [22].

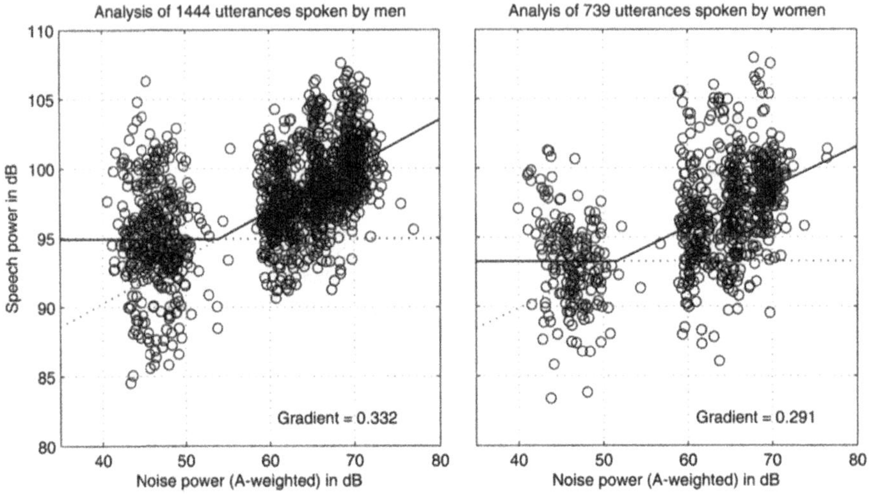

Fig. 14.5. Analysis of the Lombard effect. About 2000 utterances were recorded in a car driven at different speeds. At higher speed (noise power > 55 dBA) an increase of the average speech power of about 0.84 dB per dB noise power increment was measured.

In order to get a model for the dependency between the speech power and the background noise level the data set was split in two categories. It has been assumed that the speech power is not dependent on the noise level if the latter is below a certain threshold. For this reason, we put each speech-noise-power pair with an A-weighted noise level below 55 dBA into the first category and the others into a second. For the first category we modelled the speech power simply by the average over all measurement points (horizontal lines in Fig. 14.5). For the second category we computed a regression model of first order (lines in the right part of the diagrams in Fig. 14.5). The gradient of this regression line was about 0.3 for both male and female. That means that an increase of the noise level (A-weighted) by one decibel results in an increase of the speech power by about 0.3 dB. However, also the large variations of the speech power within this analysis should be mentioned. As a result one can conclude that the gain of an intercom system should be computed adaptively from both the measured noise power as well as the average speech power.

For further answering the question about the required gain of intercom systems an analysis of the *mouth-to-ear* transfer functions within a car without such a system is helpful. These frequency responses can be measured by placing a so-called artificial mouth loudspeaker[4] at the speaker's seat and torsos with earmicrophones [25] at the listeners seats. In Fig. 14.6 the frequency responses measured between the driver's mouth and the left ear of the second

[4] This is a loudspeaker which has (nearly) the same radiation pattern as the human speech apparatus.

front passenger, respectively the left ear of the rear passenger behind the front passenger are depicted. On average the acoustic loss to the rear passenger is 5 to 15 dB larger (compared to the front passenger).

If we assume that even at high speed the communication quality between two passengers sitting in the same row of seats within a car is at least sufficient, such measurements give a first hint about the required gain of intercom systems enhancing front-to-rear communications. Of course, whenever the power of the speech components is increased the noise level should be kept constant – otherwise the signal-to-noise ratio (SNR) will not be increased and the speech quality will not be enhanced.

Finally, two further aspects should be mentioned. An SNR improvement of about 10 dB is only an average value. The amount of required gain varies in relation to the distance of the front and rear seat rows and is dependent on the materials which cover the passenger compartment. Diffuse field distances measured in various cars indicate that up to a distance of 1.5 m the radiated acoustic power decreases with $1/r^2$, if r describes the distance from the sound source. Thus, the larger the distance between speaking and listening passenger is, the more gain is required. Furthermore, most materials used for lining passenger compartments absorb high frequency sound energy better than low frequency energy. As a consequence it is more important to enhance medium and high frequencies than low ones if the speech intelligibility should be increased.

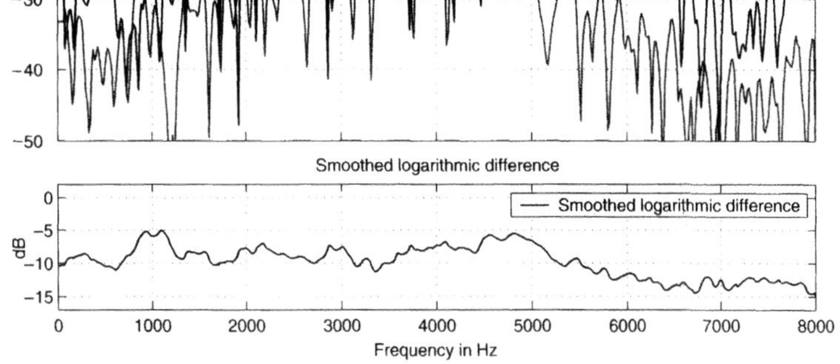

Fig. 14.6. Frequency responses of different (driver to right front passenger and driver to right rear passenger) communication directions within a passenger compartment.

14.1.2 Communication with Intercom Systems

In Sec. 14.1.1 we have analyzed the communication within a passenger compartment without an intercom system. In this section we will analyze the effects that appear if a communication system is activated. For reason of simplicity we will model all impulse responses that appear within the passenger compartment as time discrete systems. Furthermore, all systems are considered time-invariant. Thus, all time indices, denoted by (n), can be omitted:[5]

$$h_{\mathrm{XY},i}(n) = h_{\mathrm{XY},i}. \tag{14.1}$$

In this notation the subscript i denotes the coefficient index of the impulse response. The terms X and Y describe the source and the sink of the transmission. Table 14.1 shows an overview about the utilized abbreviations.

Table 14.1. Abbreviations utilized for description of impulse responses.

Abbreviation	Meaning
F	Front
L	Loudspeaker
M	Microphone
R	Rear
S	Speaking person
I	Intercom system

Whenever necessary we will assume further that the impulse responses are causal and can be modeled with finite memory. Thus, vector notation is utilized:

$$\boldsymbol{h}_{\mathrm{XY}} = \left[h_{\mathrm{XY},0},\, h_{\mathrm{XY},1},\, \ldots,\, h_{\mathrm{XY},N_{\mathrm{XY}}-1}\right]^{\mathrm{T}}. \tag{14.2}$$

Finally, we will assume that the intercom system with input $y(n)$ and output $x(n)$ (see Fig. 14.7) can be modeled as a linear, time-invariant system with frequency response

$$H_{\mathrm{I}}\big(e^{j\Omega}\big) = \frac{X\big(e^{j\Omega}\big)}{Y\big(e^{j\Omega}\big)}. \tag{14.3}$$

This last assumption is surely not true in reality. However, it allows to apply the theory of linear, time-invariant (LTI) systems, that gives us a deeper insight into a few basic problems of intercom systems.

[5] Within the figures of this section the time dependency—denoted by (n)—has not been dropped.

14.1.2.1 Frequency Response of the Closed-Loop System

An important issue for intercom systems is to guarantee stability. Due to the closed-loop operation an upper bound for the gain $|H_\mathrm{I}(e^{j\Omega})|$ of the stable system exists. If we define the signal of the speaking passenger $s(n)$ received at the microphone

$$\tilde{s}(n) = \sum_{i=0}^{N_\mathrm{FM}-1} h_{\mathrm{FM},i}\, s(n-i) \tag{14.4}$$

as the input (see Fig. 14.7) and the signal of the loudspeaker $x(n)$ as the

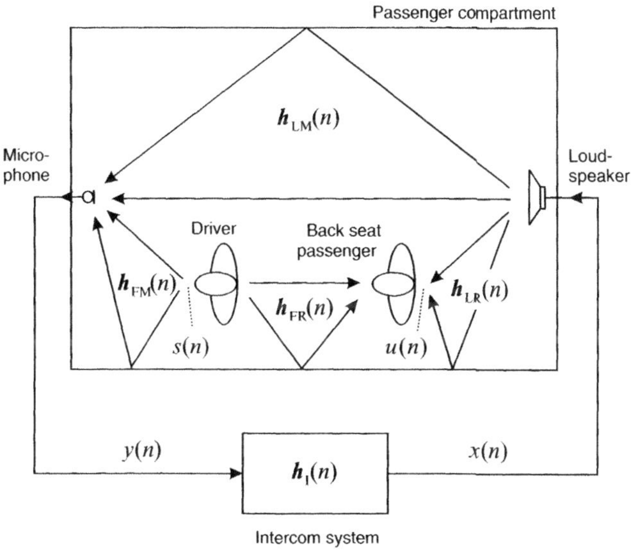

Fig. 14.7. Impulse responses within a passenger compartment equipped with an intercom system.

output of the closed-loop system, we can write the frequency response as

$$H_\mathrm{ML}\left(e^{j\Omega}\right) = \frac{H_\mathrm{I}\left(e^{j\Omega}\right)}{1 - H_\mathrm{I}(e^{j\Omega})\, H_\mathrm{LM}(e^{j\Omega})}. \tag{14.5}$$

The term $H_\mathrm{I}(e^{j\Omega})\,H_\mathrm{LM}(e^{j\Omega})$ is often called the *open loop gain* [29] of the system. In order to assure a stable system, the open loop gain has to be smaller than unity at all frequencies:

$$\left| H_\mathrm{I}\left(e^{j\Omega}\right)\, H_\mathrm{LM}\left(e^{j\Omega}\right) \right| < 1. \tag{14.6}$$

By rearranging Eq. 14.6 we can specify an upper bound for the system gain

$$\left|H_{\mathrm{I}}\left(e^{j\Omega}\right)\right| \leq \left|H_{\mathrm{LM}}\left(e^{j\Omega}\right)\right|^{-1} \tag{14.7}$$

As we can see the upper limit for $|H_{\mathrm{I}}(e^{j\Omega})|$ is directly bounded by the inverse of $|H_{\mathrm{LM}}(e^{j\Omega})|$. This motivates the application of a feedback cancellation filter. An adaptive filter $\hat{\boldsymbol{h}}_{\mathrm{LM}}(n)$ is placed in parallel to the loudspeaker-enclosure-microphone (LEM) system $\boldsymbol{h}_{\mathrm{LM}}$ and its output $\hat{d}(n)$ is subtracted from the microphone signal. Fig. 14.8 shows the structure of an intercom system consisting of an adaptive feedback cancellation filter and residual processing.

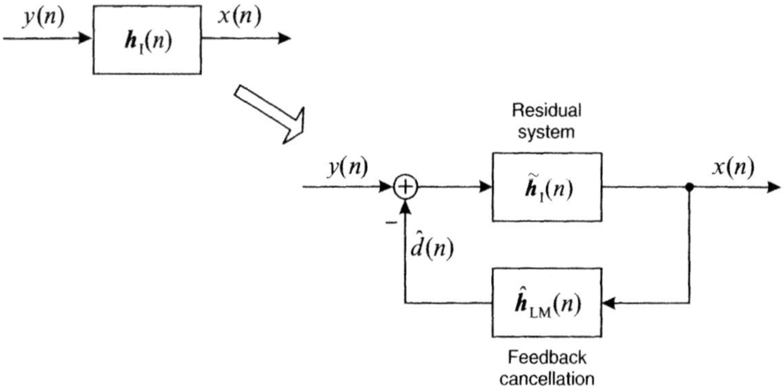

Fig. 14.8. Structure of a basic intercom system consisting of a feedback cancellation filter $\hat{\boldsymbol{h}}_{\mathrm{LM}}(n)$ and a residual system $\tilde{\boldsymbol{h}}_{\mathrm{I}}(n)$.

The frequency response of an intercom system as depicted in Fig. 14.8 can be described as:

$$H_{\mathrm{I}}\left(e^{j\Omega}\right) = \frac{\tilde{H}_{\mathrm{I}}\left(e^{j\Omega}\right)}{1 + \tilde{H}_{\mathrm{I}}(e^{j\Omega})\,\hat{H}_{\mathrm{LM}}(e^{j\Omega})}. \tag{14.8}$$

Inserting this result into Eq. 14.5 leads to

$$H_{\mathrm{ML}}\left(e^{j\Omega}\right) = \frac{\tilde{H}_{\mathrm{I}}\left(e^{j\Omega}\right)}{1 - \tilde{H}_{\mathrm{I}}(e^{j\Omega})\left[H_{\mathrm{LM}}(e^{j\Omega}) - \hat{H}_{\mathrm{LM}}(e^{j\Omega})\right]}. \tag{14.9}$$

As we can see the maximum gain of the stable intercom system is bounded now by the difference of the real and the estimated transmission from the loudspeaker to the microphone. Whenever an optimal match between both frequency responses is achieved,

$$\hat{H}_{\mathrm{LM,\,opt}}\left(e^{j\Omega}\right) = H_{\mathrm{LM}}\left(e^{j\Omega}\right), \tag{14.10}$$

the stability problem does not exist any more and the frequency response of the residual system

$$H_{\mathrm{ML}}\left(e^{j\Omega}\right)\Big|_{\mathrm{opt}} = \tilde{H}_{\mathrm{I}}\left(e^{j\Omega}\right) \tag{14.11}$$

can be designed arbitrarily. However, a perfect match of the feedback cancellation filter according to Eq. 14.10 cannot be guaranteed in reality for all situations, since the LEM system is strongly time-variant (e.g., the system changes whenever one of the passengers moves) and the convergence speed of every adaptation algorithm is limited. Furthermore, most adaptive algorithms converge towards the Wiener solution [42]. Unfortunately, this is not the desired solution here, as we will see in Sec. 14.2.5.

14.1.2.2 Transmission from Speaking to Listening Passenger

Besides analyzing the stability and the closed-loop gain of an intercom system it is also important to set up a model for the transmission from the speaking to the listening person. According to the notation of the impulse responses depicted in Fig. 14.7 we can describe this transmission in terms of a frequency response as (see Fig. 14.9)

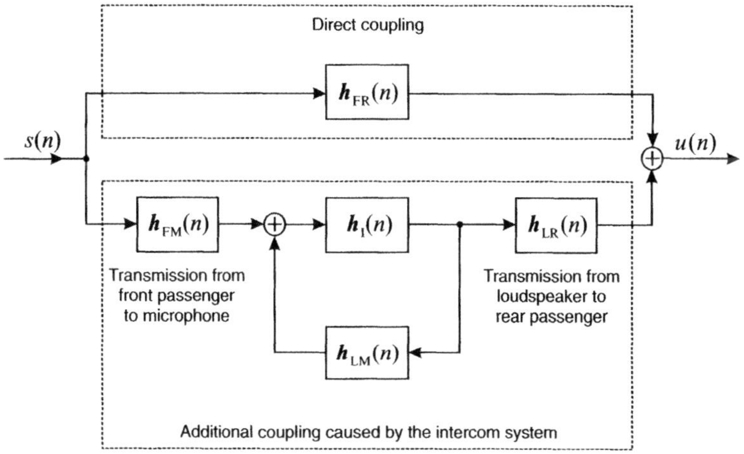

Fig. 14.9. Transmission from the speaking to the listening passenger.

$$H_{\mathrm{SR}}\left(e^{j\Omega}\right) = \underbrace{H_{\mathrm{FR}}\left(e^{j\Omega}\right)}_{\text{Direct coupling}} + \underbrace{H_{\mathrm{FM}}\left(e^{j\Omega}\right) H_{\mathrm{ML}}\left(e^{j\Omega}\right) H_{\mathrm{LR}}\left(e^{j\Omega}\right)}_{\text{Coupling caused by the intercom system}} . \tag{14.12}$$

The first term on the right side of Eq. 14.12 $H_{\mathrm{FR}}\left(e^{j\Omega}\right)$ describes the direct coupling from the speaking to the listening passenger. The second term, in the following abbreviated by

$$\tilde{H}_{\mathrm{FR}}\left(e^{j\Omega}\right) = H_{\mathrm{FM}}\left(e^{j\Omega}\right) H_{\mathrm{ML}}\left(e^{j\Omega}\right) H_{\mathrm{LR}}\left(e^{j\Omega}\right) , \tag{14.13}$$

describes the additional coupling caused by the intercom system. With these definitions we can set up a basic rule for designing intercom systems:

- Maximize the ratio

$$R_{\mathrm{FR}}(\Omega)=\frac{\left|H_{\mathrm{FR}}(e^{j\Omega})+\tilde{H}_{\mathrm{FR}}(e^{j\Omega})\right|}{\left|H_{\mathrm{FR}}(e^{j\Omega})\right|}\rightarrow \max, \tag{14.14}$$

  with respect to Eq. 14.7.

It is important to note that besides the stability margin (Eq. 14.7) a few other limiting conditions have to be considered. One of these is that visual and acoustic source localization should match. This is especially a problem for the rear passengers as they see the front passengers in front of them. However, if the rear loudspeakers are installed behind the rear seats and the gain of these loudspeakers is too high, the acoustic localization indicates that the speaking person is behind the listening one. This mismatch of different senses causes a very unnatural impression of the communication. To overcome this problem the gain of the rear loudspeakers has to be limited according to the delay between the primary source (e.g., the driver) and the secondary source (e.g., loudspeaker in the rear). The amount of amplification until the localization mismatch effect appears is given by the so-called *precedence effect* also called Haas effect or law of the first wave front [11].

In Fig. 14.10 the results of a psycho-acoustic experiment [34] are depicted. Two loudspeakers were placed at angles of 40° and −40° in front of a listener. Both loudspeakers emit a prerecorded speech signal but one of the loudspeakers was delayed. About 20 subjects were asked to adjust the gain of the delayed loudspeaker until they have the impression that a) the loudness of both loudspeakers is about the same, b) the signal of the earlier loudspeaker is not audible any more, and c) the delayed loudspeaker is not audible any more. As one can see in Fig. 14.10, a second loudspeaker, which emits a 15 ms delayed signal, can be amplified by 10 to 12 dB until the equal loudness impression from both directions is achieved. The overall loudness, however, could be enlarged by 10 to 12 dB. These results correspond very well with experiments made within cars. Rear loudspeakers in an intercom system can significantly improve the loudness without changing the acoustically perceived localization of the source. At a delay of 10 to 20 ms best results were achieved. However, the maximum gain has to be adjusted carefully and individually for each type of car.

14.1.2.3 Transmission from Speaking to Speaking Passenger

In the preceding section the restriction of the system gain with respect to localization mismatch was discussed. However, there are further and sometimes even stricter constraints to the system gain. If the loudspeaker signals are amplified too much the speaking passenger becomes aware of his or her own echo. This is very annoying for the speaking person and usually leads to non-acceptance of the intercom system. The coupling from the speaking

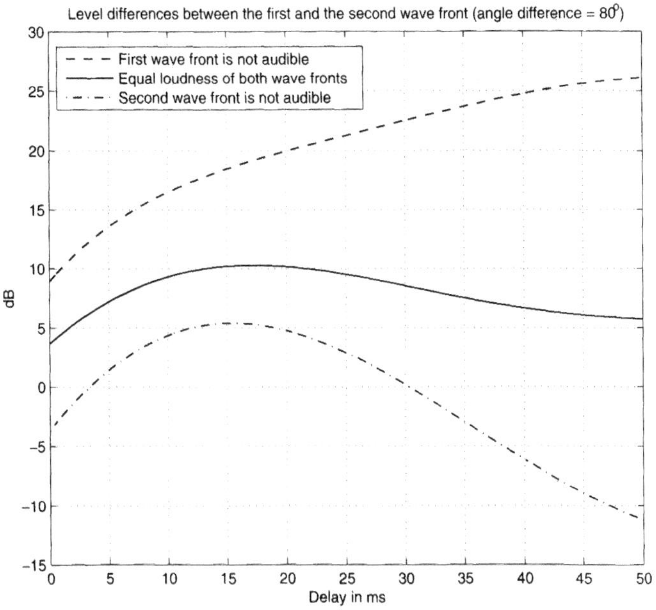

Fig. 14.10. Results of a psycho-acoustic experiment (according to [34]) for determining the relationship between the loudness of different loudspeakers and human auditory source localization.

person back to him or herself can be described in a similar manner as the coupling from the speaking to the listening passenger. We will assume for simplicity that all coupling components are linear and time-invariant – thus we can describe the coupling in terms of the frequency response

$$H_{\mathrm{SS}}\big(e^{j\Omega}\big) = \underbrace{H_{\mathrm{FF}}\big(e^{j\Omega}\big)}_{\text{Direct coupling}} + \underbrace{H_{\mathrm{FM}}\big(e^{j\Omega}\big)\, H_{\mathrm{ML}}\big(e^{j\Omega}\big)\, H_{\mathrm{LF}}\big(e^{j\Omega}\big)}_{\text{Coupling caused by the intercom system}} . \quad (14.15)$$

Fig. 14.11 shows the underlying structure. The term $H_{\mathrm{FF}}(e^{j\Omega})$ denotes the natural coupling from the mouth over the air into the ears when someone is speaking (nonlinear characteristics such as coupling over the bones, etc. are not treated here). The second term on the right side of Eq. 14.15 describes the additional coupling caused by the intercom system. This coupling component should be kept below a threshold which originates in self-masking effects of the human auditory system. At a delay of about 10 ms the ratio

$$R_{\mathrm{SS}}(\Omega) = \frac{\left|H_{\mathrm{FM}}\big(e^{j\Omega}\big)\, H_{\mathrm{ML}}\big(e^{j\Omega}\big)\, H_{\mathrm{LF}}\big(e^{j\Omega}\big)\right|}{\left|H_{\mathrm{FF}}\big(e^{j\Omega}\big)\right|} \quad (14.16)$$

should be smaller than -15 dB. If the intercom system introduces more delay an even smaller ratio is required. According to a study of AT&T [37] each

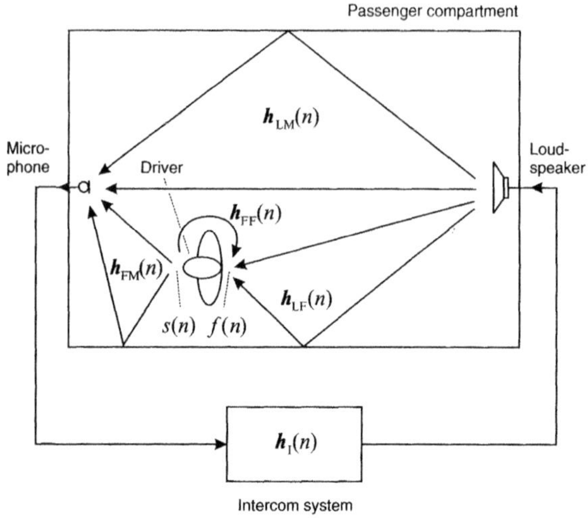

Fig. 14.11. Transmission from the speaking to the speaking passenger.

doubling of the delay results in a 6 to 8 dB lower maximum playback volume. Combining these results with the other boundary conditions, it turns out that the overall delay introduced by the system (AD and DA converters, processing delay, and delay caused by the acoustic paths) should not exceed 10 ms.

When comparing the last terms on the right sides of Eqs. 14.12 and 14.15 one realizes that the only difference is the acoustic transmission from the loudspeaker to either the listening or the speaking person. In order to get best intelligibility for the listening passenger one would like to make $|H_{LR}(e^{j\Omega})|$ as large as possible. On the other hand $|H_{LF}(e^{j\Omega})|$ should be as small as possible in order not to disturb the speaking passenger. One way of achieving both is to place the loudspeaker as close as possible to the listeners ears and as far as possible from the ears of the speaking person. Another possibility is the usage of a loudspeaker array which should be designed for achieving maximum output into the direction of the listening passengers while blocking the direction to the speaking passengers. However, due to the size of typical loudspeakers compared to the involved wave lengths beamforming with loudspeakers is not as simple and effective as using an array of microphones.

14.2 Signal Processing for Intercom Systems

Fig. 14.12 sketches the structure of an intercom system aimed to support front-to-rear conversations (for the opposite direction a similar structure is applied). Compared to the basic system depicted in Fig. 14.2 now much more details are shown. Since driver and front passenger are located at well defined positions, specially designed microphone arrays can point towards each of them, which allows to use fixed beamformers. This allows to start with the echo and

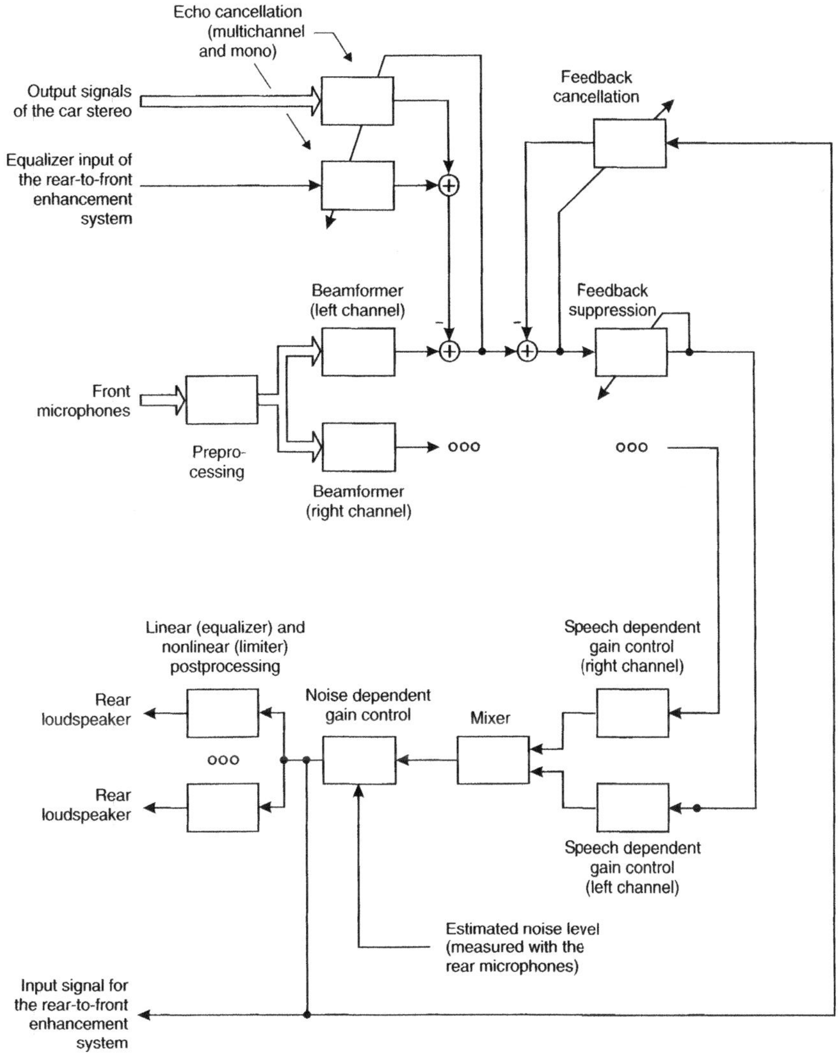

Fig. 14.12. Structure of a car interior communication system aimed to support front-to-rear conversations.

feedback cancellation after the beamformer (and to reduce the computational complexity because only one echo cancellation filter per reference channel is required). Feedback suppression by means of an adaptive notch filter can improve the system stability by rising the howling margin. A mixer combines the signals of driver and second front passenger according to the detected speech activity. A device with nonlinear characteristic attenuates large signal amplitudes before the signals are played back via the loudspeakers. The output gain of a car intercom system needs to be adjusted continuously according to the current driving situation. While only a moderate gain is required whenever the car is in low noise conditions, a large gain is required and more artifacts will be tolerated at high speed. Finally, loudspeaker equalization (either adaptive or fixed) can be applied.

In this section we will introduce the basic building blocks of an in-car communication system as described above (and depicted in Fig. 14.12). Going into the details of each block would go far beyond the scope of this chapter. For this reason, the interested reader is referred to the references cited within the corresponding sections.

14.2.1 Processing Structures

Beside selecting adaptive algorithms [9] like NLMS, affine projection, RLS, etc., the system designer also has the freedom to choose between different processing structures. The most popular ones are broadband processing, block processing[6], and subband processing. The special challenge in in-car communication systems consists in designing a system with an overall delay of not more than 10 ms. Signals from the loudspeakers delayed for more than that will be perceived as echoes and reduce the subjective quality of the system. For this reason, only broadband processing or block processing with very small block sizes can be applied if a high system quality should be achieved.

14.2.2 Preprocessing

The signals picked up by the front microphones are highpass filtered first as the energy of the background noise in a car is typically concentrated in the low frequency range (see Fig. 14.4). Furthermore, the attenuation of the speech signal resulting from the directionality of the human head at low frequencies is not as high as at medium or high frequencies (see Fig. 14.3). For this reason low order Butterworth highpass filters with a 3 dB cut-off frequency of about 300 Hz have been applied.

Additionally it is checked whether the signals at the AD converters are clipped – in this case the assumption of a linear relationship between loudspeaker and microphone signals is no longer valid. As a result, any adaptive algorithm (echo and feedback cancellation) which relies on this assumption is paused. Clipping of the AD converters can appear if, for example, a loudspeaker located close to the microphones emits loud radio signals.

[6] By block processing we mean performing the convolution and/or the adaptation in the frequency domain and using overlap-add or overlap-save techniques.

14.2.3 Beamforming

If more than one microphone is installed array processing – in terms of beamforming – can be utilized for enhancing the incoming signals.[7] The output signal $u(n)$ of a beamformer is given by the addition of the (FIR) filtered microphone signals (see Fig. 14.13):

$$u(n) = \sum_{i=0}^{L-1} \boldsymbol{y}_i^{\mathrm{T}}(n)\, \boldsymbol{g}_i(n), \tag{14.17}$$

where L and M denote the number of microphones and the length of the FIR filters, respectively. The vectors $\boldsymbol{y}_i(n)$ and $\boldsymbol{g}_i(n)$ are defined as

$$\boldsymbol{y}_i(n) = \Big[y_i(n),\, y_i(n-1),\, ...,\, y_i(n-M+1)\Big]^{\mathrm{T}}, \tag{14.18}$$

$$\boldsymbol{g}_i(n) = \Big[g_{i,0}(n),\, g_{i,1}(n),\, ...,\, g_{i,M-1}(n)\Big]^{\mathrm{T}}. \tag{14.19}$$

The simplest type of beamformer is the delay-and-sum beamformer. In this case the filters are time-invariant ($\boldsymbol{g}_i(n) = \boldsymbol{g}_i$) and designed such that

$$G_i\big(e^{j\Omega}\big) \approx \frac{1}{L}\, e^{-j\Omega\tau_i f_s}, \tag{14.20}$$

where τ_i represents the delay of the ith microphone channel for time-aligning the signals from a predefined source direction [30]. If this direction is not known a priori it has to be estimated [20, 28]. The so-called *beampattern* for linear arrays (all microphones are located within one line and are equally spaced) is defined as the squared magnitude of

$$B(\Omega, \phi) = \sum_{i=0}^{L-1} G_i\big(e^{j\Omega}\big)\, e^{-j\Omega i \frac{d\, f_s}{c} \sin\phi}. \tag{14.21}$$

It is depicted in the right part of Fig. 14.13 for an array consisting of $L = 4$ cardioid sensors with a distance of $d = 5$ cm between two adjacent microphones. The quantity c is denoting the speed of sound ($c \approx 340$ m/s), the angle ϕ describes the angle of incidence, and f_s is denoting the sampling rate. If a better directivity at low frequencies should be achieved the delay-and-sum principle can be extended to a filter-and-sum approach. In this case the filters $\boldsymbol{g}_i$ are designed such that the output power of the beamformer is minimized while keeping the receiving characteristic of the desired direction ϕ_0 as a pure delay of K_0 samples:

$$\mathrm{E}\big\{u^2(n)\big\} \to \min\,, \quad \text{with} \quad B(\Omega, \phi_0) = e^{-j\Omega K_0}. \tag{14.22}$$

[7] In Fig. 14.12 the same microphones are used to record the speech of the driver and the second front passenger.

When minimizing the output power according to Eq. 14.22 the temporal and spatial correlation properties of the noise need to be known. A superposition of the correlation properties of a diffuse noise field, of one or more directional noise sources (those loudspeakers which emit the signals for the rear seat passengers) and of sensor noise is often utilized. For the design depicted in Fig. 14.13 a desired source was assumed at 22° and an undesired source (a loudspeaker for the rear passengers) at an angle of $-22°$. Each filter $\boldsymbol{g}_i$ consists of $M = 48$ coefficients. Additionally more constraints can be added to the design process, making the result more robust against positioning tolerances and sensor imperfections [17].

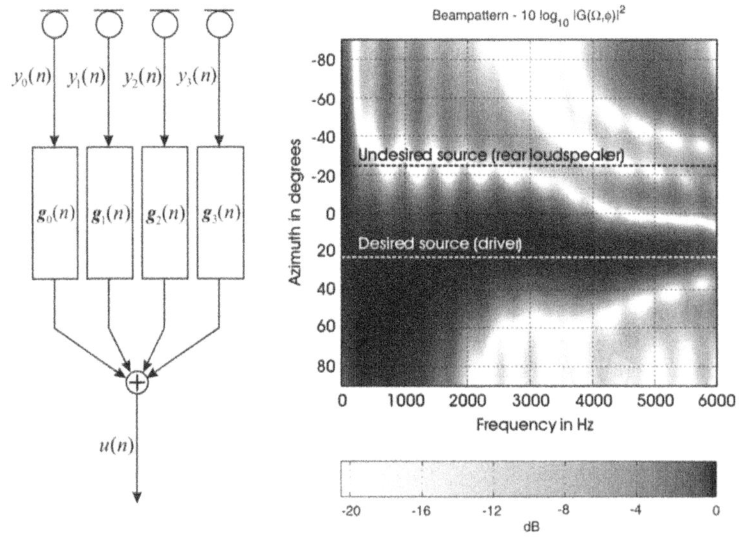

Fig. 14.13. Characteristics of a microphone array.

If the temporal and spatial correlation properties are not known a priori adaptive beamforming is usually applied. The most popular approach is the so-called generalized sidelobe cancellation (GSC) according to [10]. In order to make the adaptive approach more robust against several kinds of distortions, a variety of extensions such as an adaptive blocking matrix [19] or adaptive microphone calibration have been proposed. Because of reverberation effects it is very important to adapt a beamformer in GSC structure only during speech pauses of the desired speaker [18] – otherwise signal cancellation may occur. The signal cancellation is caused by delayed versions of the desired signal. This restriction in particular makes it very difficult to use adaptive beamformers for in-car communication systems. For this reason it is assumed that only non-adaptive beamformers are utilized here. In this case echo cancellation can be applied after the beamformer (see Fig. 14.12). This reduces the computational complexity since only one echo cancellation filter has to be computed for each beamformer output (instead of one echo canceller per microphone).

If we recall a few of the definitions that were introduced in Secs. 14.1.2.1 – 14.1.2.3 one can see the influence of beamforming on the system parameters. In Sec. 14.1.2.1 the stability of the entire system was analyzed. As long as the gain of the entire intercom system $|H_{\mathrm{I}}(e^{j\Omega})|$ is smaller than the inverse gain of the transmission from the loudspeaker to the microphone

$$\left|H_{\mathrm{I}}\left(e^{j\Omega}\right)\right| \leq \left|H_{\mathrm{LM}}\left(e^{j\Omega}\right)\right|^{-1} \tag{14.23}$$

the system is stable. By applying beamforming we have now several LEM systems – one for each microphone.[8] The stability condition changes to

$$\left|H_{\mathrm{I}}\left(e^{j\Omega}\right)\right| \leq \left|\sum_{i=0}^{L-1} G_i\left(e^{j\Omega}\right)\, H_{\mathrm{LM},i}\left(e^{j\Omega}\right)\right|^{-1}. \tag{14.24}$$

As long as we make sure that the weighted sum of the individual LEM frequency responses is smaller than the frequency response that we get with just one microphone

$$\left|\sum_{i=0}^{L-1} G_i\left(e^{j\Omega}\right)\, H_{\mathrm{LM},i}\left(e^{j\Omega}\right)\right| \leq \left|H_{\mathrm{LM}}\left(e^{j\Omega}\right)\right|, \tag{14.25}$$

the maximal gain of the intercom system can be increased. As a result a larger ratio $R_{\mathrm{FR}}(\Omega)$ can be achieved.

Finally, an important feature of multi-microphone processing should be considered. By using more than one microphone it becomes possible to estimate the direction of a sound source. With this information it is possible to distinguish between different sound sources (e.g. front passengers and loudspeakers which emit the signals for the rear passengers) which are very similar in their statistical properties. This spatial information can be exploited for enhanced system control [7].

14.2.4 Echo Cancellation

As depicted in Fig. 14.12 echo cancellation might be performed for two kinds of echoes: on one hand the car radio (or the CD player, etc.) might be activated and on the other hand the enhanced signals from the rear passengers, emitted via the front loudspeakers, are coupling into the front microphones. For both kinds of sources, echo cancellation can be applied in order to remove those signal components from the microphone signals. We will start with a description of the cancellation of the radio signals in the next section. Cancellation of the output signals of the rear-to-front enhancement system is principally a very basic approach. This will be described briefly in Sec. 14.2.4.2.

[8] For the reason of simplicity we assume here again that we have only one loudspeaker.

14.2.4.1 Cancellation of the Output Signal of the Car Radio

If the car radio as well as the intercom system are activated at the same time the front microphones do not pick up only the speech of speaking passengers but also the radio signals (convolved with the corresponding loudspeaker-enclosure-microphone impulse responses). Those latter components are also processed by the intercom system and played back via the output loudspeakers. Due to the delay of the system the sound impression of the radio signals becomes reverberant. The degree of reverberation depends on the gain and the delay of the in-car communication system. Usually such a behavior is very undesired. For this reason multichannel echo cancellation is applied to remove the reverberant impression. In the following we will assume to have a stereo car radio. However, if more than two output signals are produced, e.g., by a DVD player, the number of channels need to be increased.

The radio signals of the left loudspeaker $x_{\mathrm{L}}(n)$ and of the right loudspeaker $x_{\mathrm{R}}(n)$ are transmitted via two different impulse responses, $\boldsymbol{h}_{\mathrm{L}}(n)$ and $\boldsymbol{h}_{\mathrm{R}}(n)$, to the microphones (see Fig. 14.14).[9] Two adaptive filters with impulse responses $\hat{\boldsymbol{h}}_{\mathrm{L}}(n)$ and $\hat{\boldsymbol{h}}_{\mathrm{R}}(n)$, respectively, try to replicate the echo signal $u(n)$ by appropriate outputs $\hat{d}_{\mathrm{L}}(n)$ and $\hat{d}_{\mathrm{R}}(n)$, in order to cancel the echo of the car radio.

A dual-channel structure as depicted in Fig. 14.14 poses the following problems with regard to the adaptation of the echo cancellation filters [2]:

- For certain types of radio signals, such as news presentations or interviews without background music, the signals $x_{\mathrm{L}}(n)$ and $x_{\mathrm{R}}(n)$ are strongly cross-correlated since they are filtered versions of a common source signal. For this reason, the correlation matrix

$$\boldsymbol{R}_{\boldsymbol{xx}}(n) = \mathrm{E}\left\{ \begin{bmatrix} \boldsymbol{x}_{\mathrm{L}}(n) \\ \boldsymbol{x}_{\mathrm{R}}(n) \end{bmatrix} \left[\boldsymbol{x}_{\mathrm{L}}^T(n) \; \boldsymbol{x}_{\mathrm{R}}^T(n) \right] \right\} \tag{14.26}$$

  is ill-conditioned and consequently, the performance of the adaptation algorithm is degraded. However, noise components in the excitation signals $x_{\mathrm{L}}(n)$ and $x_{\mathrm{R}}(n)$ help reduce this effect to a certain but limited amount. The signal vectors $\boldsymbol{x}_{\mathrm{L}}(n)$ and $\boldsymbol{x}_{\mathrm{R}}(n)$ contain the last N input signals of the left and right radio signal, respectively. The quantity N is denoting the filter length.
- With an ill-conditioned correlation matrix $\boldsymbol{R}_{\boldsymbol{xx}}(n)$, the misalignment of the echo cancellation filters is much worse for the dual-channel case than for the single-channel case.

---

[9] Note that for the reason of simplicity the notation of this section is not consistent with the rest of the paper. The subscripts L and R abbreviate *left* and *right* here (instead of *loudspeaker* and *rear* as introduced in Sec. 14.1.1). Furthermore, the impulse responses describe the transmission between the loudspeakers and the output of the beamformer.

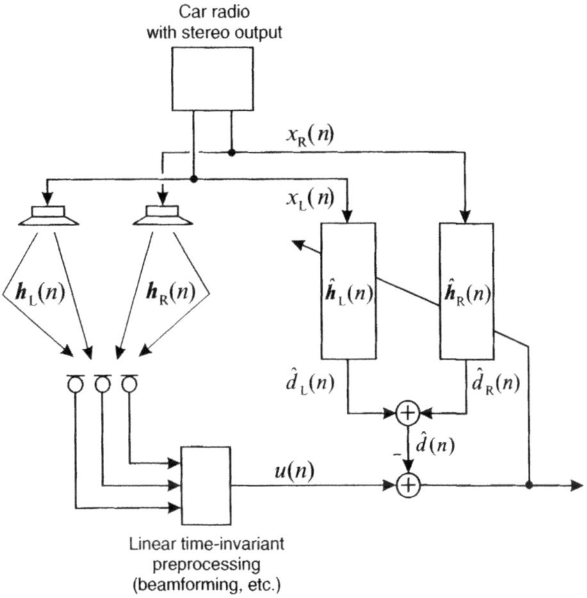

Fig. 14.14. Basic scheme for stereophonic acoustic echo cancellation of the radio signals. The subscripts L and R abbreviate *left* and *right* here.

- Since $\boldsymbol{x}_L(n)$ and $\boldsymbol{x}_R(n)$ might be fully correlated, the optimal impulse responses $\hat{\boldsymbol{h}}_L(n)$ and $\hat{\boldsymbol{h}}_R(n)$ are not uniquely defined [38] (in terms of minimizing the power of the output signal).

In general, any imaginable method should avoid affecting stereophonic perception. Several proposals for a solution of the problems listed above have been made. According to Benesty et al. [1], any adaptive algorithm should be backed up by a decorrelating component, in order to enhance the conditioning of the correlation matrix $\boldsymbol{R}_{\boldsymbol{xx}}(n)$ and to get a robust solution that is no longer dependent on cross correlation properties. Recently a variety of methods for reducing the correlation between left and right channel have been proposed:

- The *addition of independent random noise* to each channel may help reduce the correlation of the stereo signals and therefore enhance the performance of the echo cancellation filters. Unfortunately, an improvement of the conditioning of the correlation matrix requires rather high and thus disturbing noise levels [38]. A possible solution to this dilemma may be to use the auditory masking properties of the human ear. The idea consists of adding spectrally shaped random noise that is masked by the loudspeaker input signals to each channel [8]. The additional costs for this method can be kept relatively low when frequency-domain adaptive algorithms for the echo cancellation filters are applied.
- Further research has been directed towards the application of *nonlinear transformations* [1]. In principle, a small signal is added to the excitation

signals. The added signal is derived by a nonlinear function, such as half-wave rectification, from the excitation signal itself:

$$\tilde{x}_{\mathrm{L}}(n) = x_{\mathrm{L}}(n)\,, \tag{14.27}$$

$$\tilde{x}_{\mathrm{R}}(n) = \begin{cases} x_{\mathrm{R}}(n), & \text{if } x_{\mathrm{R}}(n) > 0, \\ (1+\alpha)\, x_{\mathrm{R}}(n), & \text{else.} \end{cases} \tag{14.28}$$

For small amplifications ($0 \geq \alpha > 0.3$), the distortion introduced by the use of such nonlinearities is hardly perceptible.

- Another method for decorrelating stereo signals proposes a periodically changing delay of the excitation signal in one channel by a simple filter with time-variable coefficients [26]. The *time-variant filter* consists of a 2-tap FIR filter whose coefficients $\tilde{a}_i(n)$, $i \in \{0, 1\}$, are controlled by a periodic function $t(n)$, $\tilde{a}_0(n) = t(n)$, $\tilde{a}_1(n) = 1 - t(n)$, with period Q. For the first $Q/2$ iterations $t(n) = 1$, meaning that the filter output $\tilde{x}_{\mathrm{R}}(n)$ is equal to the input signal $x_{\mathrm{R}}(n)$. When $t(n) = 0$ for the following $Q/2$ iterations, the filter output is a one-sample delayed version $\tilde{x}_{\mathrm{R}}(n-1)$ of the input signal. This alternating procedure is repeated every Q samples. Unfortunately, the sudden change of the coefficient $t(n)$ every $Q/2$ iterations leads to audible distortions, clicks, in the processed signal $\tilde{x}_{\mathrm{R}}(n)$. To avoid these clicks, $t(n)$ can be varied smoothly between zero and one over L samples. For a sampling rate of 16 kHz, the parameters Q and L were suggested to be $Q = 4000$ and $L = 400$ [39].

Note that in practice the short-term coherence between the left and right channel is not close to one for most types of radio signals. This means that the above mentioned decorrelation methods should only be applied for certain types of signals, such as newscasts. For all other signal types, such as music or advertisements, preprocessing should not be applied in order to avoid any distortion of the audio presentation. For not completely correlated signals selective coefficient update schemes can be applied [27]. On one hand, these schemes decrease the computational complexity and on the other hand, the convergence speed is improved by further decorrelation of the input signals.

Fig. 14.15 shows the results of a radio echo canceller within a car. The driver's speech was recorded with a 4 element microphone array integrated in the rear view mirror. Besides the front door loudspeakers the radio signal was also emitted via a center speaker integrated within the dashboard of the car. Due to the closeness to the microphone array the coupling of the radio signals into the microphone array of the in-car communication system was very high. For this reason speech activity of the driver is hard to detect. By applying stereo echo cancellation the short-term power of the received signal can be reduced by about 20 dB (compare output power of the beamformer and short-term power after subtracting the estimated radio signal in Fig. 14.15). A detection of the periods containing speech activity of the driver is now easily

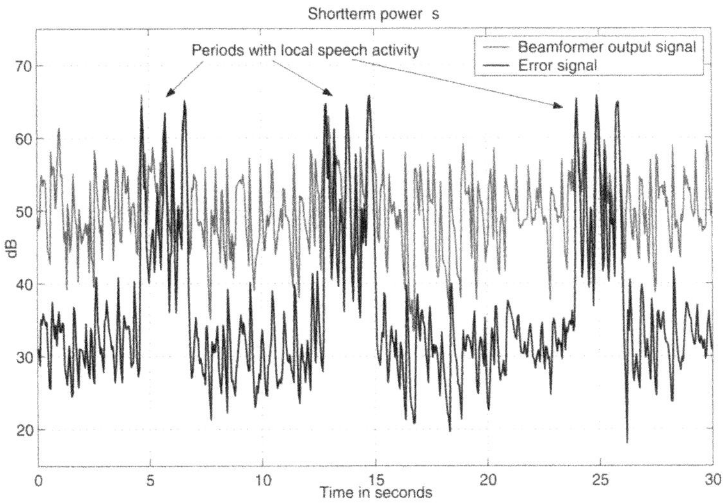

Fig. 14.15. Results of radio echo cancellation within a car.

possible. Furthermore, the reverberant impression of the music presentation can be reduced to a non-audible level.

14.2.4.2 Cancellation of the Output Signal of the Intercom System

In contrast to the cancellation of the radio signal the elimination of echoes $d_F(n)$ resulting from the output of the intercom system is rather simple. Because the echo components $d_F(n)$ within the front microphone signals are caused by the output signals of the enhancement branch rear-to-front (and vice versa), this kind of echo cancellation is only necessary if both directions are supported by the intercom system. Fig. 14.16 shows the basic structure of the front echo cancellation (driver's side) of a two-way system. Note that for the reason of clarity not all signal processing units of each branch are depicted in Fig. 14.16. For the adaptation of the filter $\hat{\boldsymbol{h}}_F(n)$ standard algorithms such as the NLMS or low order affine projection can be utilized. Due to the low signal-to-noise ratio (caused by a large amount of background noise as well as by speech activity of the front passengers) a reliable adaptation control scheme is required. An overview about those schemes can be found in [13, Chapter 13].

Comparable to the cancellation of the radio signals, cancellation of the output signals of the reverse branches of an intercom system is rather important, since the error rate of speech activity detection would be much higher without it. Usually each intercom system has a loss control which opens only those microphone channels where speech activity was detected (see Sec. 14.2.8). This improves not only the system stability, but also reduces the correlation between the excitation signal $x_F(n)$ of the adaptive filter $\boldsymbol{h}_F(n)$ and those signal

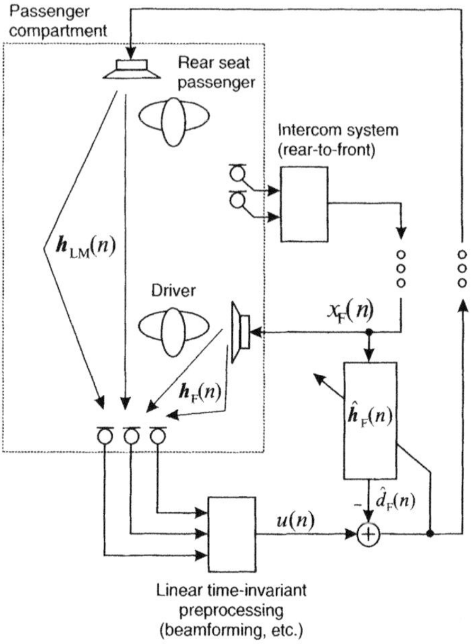

Fig. 14.16. Basic scheme for echo cancellation of output of the rear-to-front intercom branch.

components which disturb the adaptation process (here the output signals of the rear loudspeakers). This correlation, however, is a severe problem for the adaptation of feedback cancellation filters as described in the next section.

14.2.5 Feedback Cancellation

The feedback cancellation turns out to be extremely difficult since the adaptation of the filter $\hat{\boldsymbol{h}}_{\text{LM}}(n)$ is disturbed by the strong correlation between the excitation signal of the adaptive filter $x(n)$ and the speech signals of the driver and second front passenger $s(n)$:[10]

$$\mathrm{E}\{\, x(n)s(n+l) \,\} \neq 0\,. \tag{14.29}$$

Algorithms which are converging towards the Wiener solution [16] will converge towards

$$\hat{H}_{\text{LM,opt}}\left(e^{j\Omega}\right) = \frac{S_{xy}(\Omega)}{S_{xx}(\Omega)} = H_{\text{LM}}\left(e^{j\Omega}\right) + \frac{S_{xs}(\Omega)}{S_{xx}(\Omega)}\, H_{\text{FM}}\left(e^{j\Omega}\right)\,, \tag{14.30}$$

which is not the desired solution $(\hat{H}_{\text{LM,opt}}(e^{j\Omega}) = H_{\text{LM}}(e^{j\Omega}))$. For this reason, the adaptation is usually carried out only at falling signal edges of the excitation signal (whenever the speaking person stops talking for a short moment).

[10] For the definitions of the signals, impulse and frequency responses see Fig. 14.2.

During such periods the correlation between the excitation signal and the echo component is much stronger than the correlation between $x(n)$ and $s(n)$. Furthermore, during noise only periods (none of the passengers is speaking) the output signal consists only of background noise. By replacing the output signal with artificially generated noise the undesired cross-correlations can be forced to become zero. However, in this case only a low excitation-to-noise ratio can be expected and the convergence speed is slowed down. Another approach uses nonlinear or time-variant procedures – as described in Sec. 14.2.4.1 – applied on the system output signal $x(n)$ in order to reduce the correlation with the signal $s(n)$. In current systems for feedback cancellation a combined approach consisting of nonlinear preprocessing, comfort noise injection, and time-variant filtering is often applied.

Even if the adaptation of the feedback cancellation filters turns out to be rather difficult, the resulting enhancements on one hand, in terms of dereverberation of the system, especially in case of large gain values and on the other hand, regarding the improvement of the system stability are of major importance. Note the close relationship with the feedback problem in hearing aids and public address systems.

14.2.6 Feedback Suppression

Whenever closed-loop acoustic echo control systems are operating close to the stability margin, some sort of "emergency brake" should be implemented. One possibility to realize this is to implement a feedback suppression filter according to Fig. 14.17. For $\alpha = 0$ the structure resembles a predictor error filter. The FIR filter $\mathbf{c}(n)$ is an adaptive filter which is adjusted such that the power of the output signal $e(n)$ is minimized – e.g., by using the NLMS algorithm. If howling occurs at a certain frequency the feedback suppression filter tries to attenuate this frequency. According to the filter structure this is possible as long as the inverse of the howling frequency is larger than N_D and smaller than $N_\mathrm{D} + N_\mathrm{C}$ sample intervals, where N_C is denoting the length of the filter $\mathbf{c}(n)$.

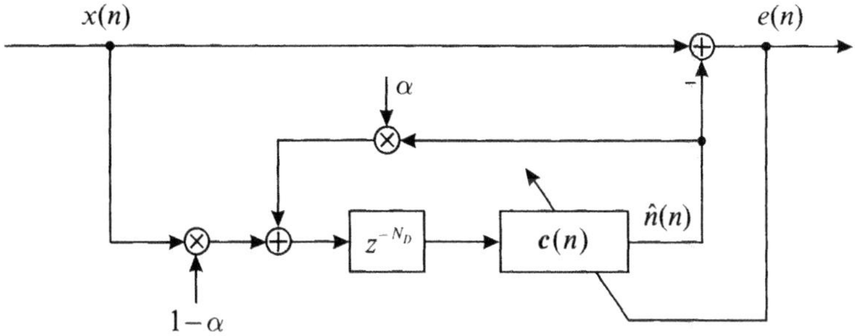

Fig. 14.17. Structure of a basic feedback suppression system.

Delaying the incoming signal before filtering is necessary because otherwise the short-term correlation within the speech signal would be removed. In this case the spectral envelope would be flattened. A delay of about 2 ms is sufficient to avoid this. Due to periodic components of speech signals the memory of the adaptive filter should not exceed a time interval equivalent to the pitch period[11]. For this reason the filter should not contain more than 80 to 120 coefficients (at 16 kHz sampling rate). Otherwise, the periodic components of the speech signal would be suppressed as well.

Due to short-term correlated speech components the filter also tries to suppress parts of the speech signal. By using a small step size μ this behavior can be avoided and only periodic distortions which are present for a longer time interval are cancelled. A small step size, on the other hand, leads to a slow convergence. Sudden (periodic) distortions would be attenuated only after a non-negligible period of time. For this reason a compromise for the step size has to be found. Usually, the NLMS algorithm with a fixed, but small step size $\mu \in \{0.01, 0.00001\}$ is utilized. It is needless to say that other adaptation algorithms can be used as well. Besides the ones suitable for echo cancellation, algorithms that use higher order statistics have also been applied to this problem [5].

Fig. 14.18 shows time-frequency analyses of output signals of an intercom system consisting only of an amplifier and a feedback suppression filter. In a first stage the feedback suppression unit was switched off and the gain of the system was adjusted such that it operates close to the stability margin. The slow decay of several formerly excited frequencies is clearly visible in the upper diagram. The lower analysis depicts the output of the intercom system with activated feedback suppression. All boundary conditions of the measurement were exactly the same as in the previous experiment. Even if the decay rate is still not as good as in the case without the additional feedback path a significant improvement is visible (and audible). The application of feedback suppression can improve the maximum system gain by a few decibels [32].

The basic FIR structure of a feedback suppression filter can be extended by a weighted feedback path [3] as depicted in Fig. 14.17. By varying the feedback gain α it is possible to modify the filter from an FIR structure ($\alpha = 0$) to an adaptive oscillator ($\alpha = 1$). The motivation behind this IIR configuration is to achieve some of the benefits of a noise canceller with a separate pure periodic reference. With the extended structure it is possible to achieve very narrow notches. Nevertheless, due to the IIR structure the filter might become instable if the adaptation process forces the poles to move out of the unit circle within the z-domain. For this reason, the stability of the structure needs to be checked periodically.

[11] The specification is only true for FIR structures. In case of IIR schemes the group delay should not exceed the specified range.

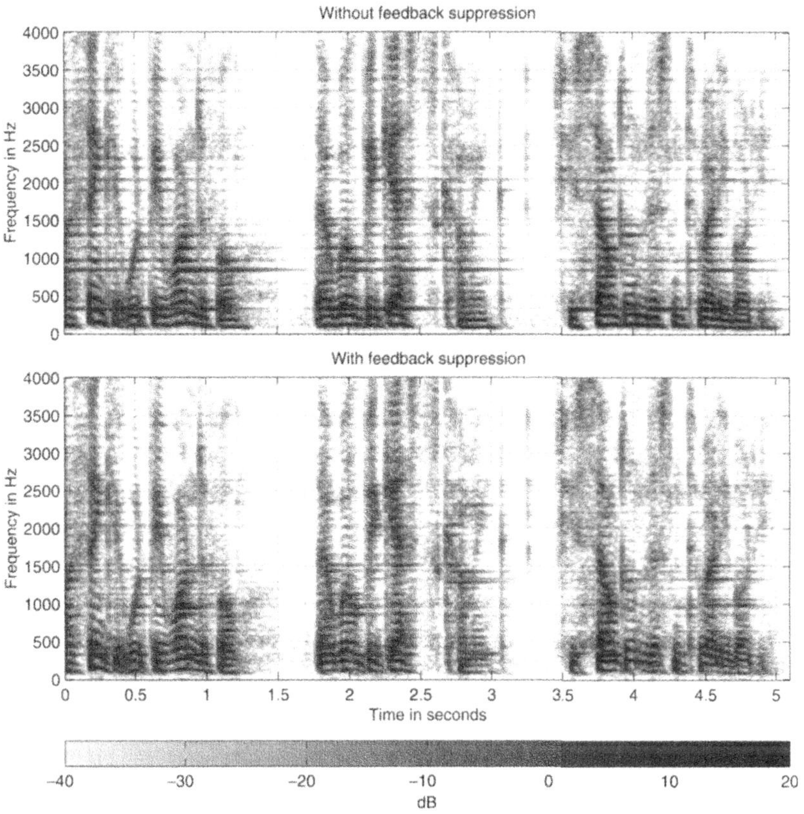

Fig. 14.18. Time-frequency analysis of the output signals of an intercom system (without and with feedback suppression).

14.2.7 Combining Feedback Cancellation and Feedback Suppression

Even though both algorithms that have been described in the last two sections are able to reduce the feedback problem still several drawbacks exist:

- The feedback cancellation on one hand can adapt only very slowly due to the correlation between the loudspeaker signal $x(n)$ and the speech signals of the driver and front passenger $s(n)$. For this reason, the achievable feedback reduction as well as the required adaptation time are much worse compared to standard echo cancellation approaches known from hands-free telephone systems.
- The feedback suppression as described in Sec. 14.2.6, on the other hand, is able to attenuate feedback components much faster. This approach, however, suffers from the fact that each time a certain feedback component

is suppressed it does no longer contribute to the cost function.[12] This leads to a release of the once achieved attenuation and the component (respectively its center frequency) can pass the feedback suppression filter. After a short period of time the whole system might start howling or whistling again. As soon as this happens the filter is able to attenuate this component again.

The drawbacks of both approaches can be avoided if feedback cancellation and feedback suppression are combined (see Fig. 14.19). The basic building block of such a combined system is again a predictor like feedback suppression filter $\boldsymbol{c}(n)$ without a feedback path ($\alpha = 0$). The adaptation of the filter, however, is not performed with the input signal $x(n)$ – as before – but with an artificially generated signal $\tilde{x}(n)$. This signal consists not only of the input components but also of those components that the basic structure according to Fig. 14.17 subtracts from the input signals. With this mechanism the problem with the cost function of the pure suppression approach can be avoided. Since the artificially added components should not be audible the structure of the feedback suppression is split into an adaptation branch (with backward addition of the feedback components) and a filtering branch (without these signal components).

Before adding the estimated feedback components $\hat{n}(n)$ to the input signal they are filtered such that the amplitude and phase modifications of the loudspeaker-enclosure-microphone system are also considered. This is achieved by applying a copy of the feedback cancellation filter $\hat{\boldsymbol{h}}_{\mathrm{LM}}(n)$ before adding the signal $\hat{n}(n)$ to the input signal.

A slow adaptation of the feedback cancellation filter is not as critical as in the direct approach (described in Sec. 14.2.5) since the filter is only used to modify the estimated feedback components. It is only important that the filter adapts at those frequencies where large feedback components appear – and that is exactly the behavior of the filter. Mismatch at the other frequencies is not as critical as before because the filter is not used within a signal path that is connected to a loudspeaker.

To show the advantages of the combined structure versus the basic structure that has been described in Sec. 14.2.6 three time-frequency analyses are depicted in Fig. 14.20. The lowest diagram shows an analysis of the output signal of the intercom system without any processing at all ($u(n) = x(n)$). However, due to the digital-to-analog conversion the signal is limited to the maximum range of the converter. Only a very low signal quality can be achieved and howling starts at several frequencies, e.g. at 800 Hz or 2500 Hz.

If a feedback suppression scheme according to Sec. 14.2.6 is applied it is possible to avoid instabilities (see center diagram of Fig. 14.20). However, due to the problems with the cost function of the basic structure the output signal contains several artifacts. A few of these artifacts – slowly decaying

[12] Remember that for adjusting the coefficients $c_i(n)$ of the feedback suppression filter the output power $\mathrm{E}\{e^2(n)\}$ was minimized.

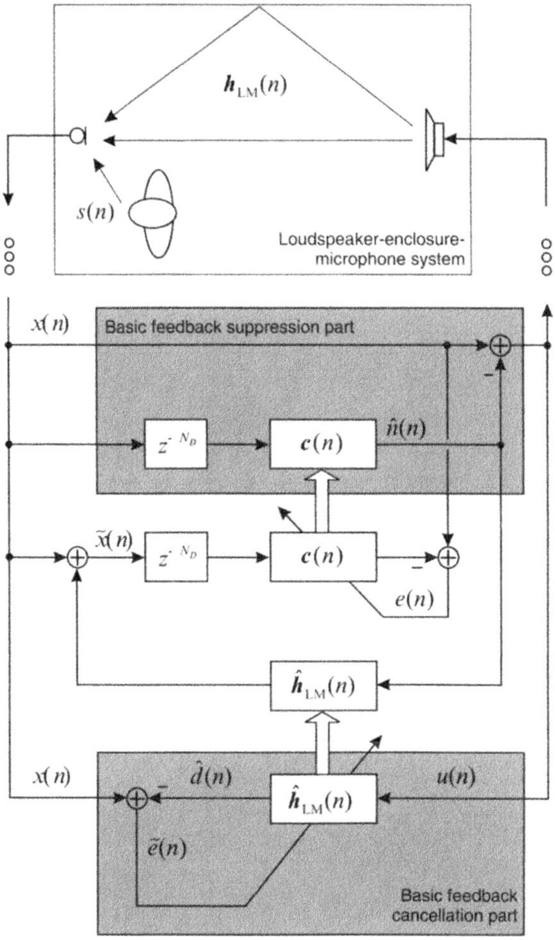

Fig. 14.19. Structure of a system that combines feedback suppression and cancellation.

oscillations – are marked by circles in the center diagram. If the combined scheme is applied it is possible to reduce the amount of artifacts significantly, especially after convergence of the feedback suppression filter $\hat{\boldsymbol{h}}_{\mathrm{LM}}(n)$. After a convergence time of about one second the oscillations disappear mostly (see upper diagram of Fig. 14.20).

Beside the improvements of the new scheme also the drawbacks should be mentioned. When comparing the basic and the extended structure (see Figs. 14.17 and 14.20) it is obvious that the new scheme requires much more computational power as well as memory. Furthermore, it is much more complicated to analyze the convergence and stability of the new scheme since two adaptive filters operate in an interlocked manner.

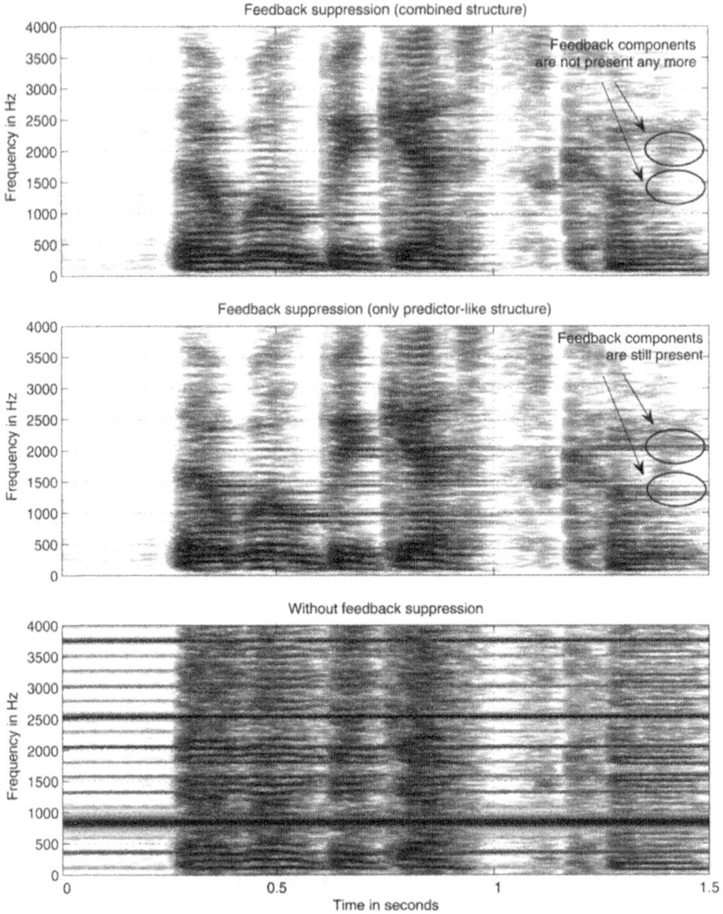

Fig. 14.20. Time-frequency analyses of the output signals of three intercom systems. Bottom: system without feedback suppression, center: system with feedback suppression according to Sec. 14.2.6, top: system with feedback suppression and cancellation as presented in this section.

14.2.7.1 Extensions

Beside the basic combination of feedback cancellation and feedback suppression that was presented in the last section several extensions are possible:

- The feedback suppression part can be extended from FIR ($\alpha = 0$) to IIR structure (as presented in Sec. 14.2.6). The advantages but also the disadvantages of the additional feedback path are still the same as in the basic scheme.
- To improve the adaptation speed fixed decorrelation filters (also called prewhitening filters) can be inserted at the inputs of the system and in-

verse counterparts at the outputs [13]. Due to the foreground-background structure of both algorithmic parts it is very simple to realize the decorrelation filters in an adaptive manner. This leads to an even higher convergence speed.

- If the feedback cancellation filter performs very well[13] it is also possible to place it directly into the signal path (see Fig. 14.21). In this case most of the feedback components would be cancelled and only the residual signals would be suppressed. However, this structure requires a highly sophisticated and reliable control of the feedback cancellation part.

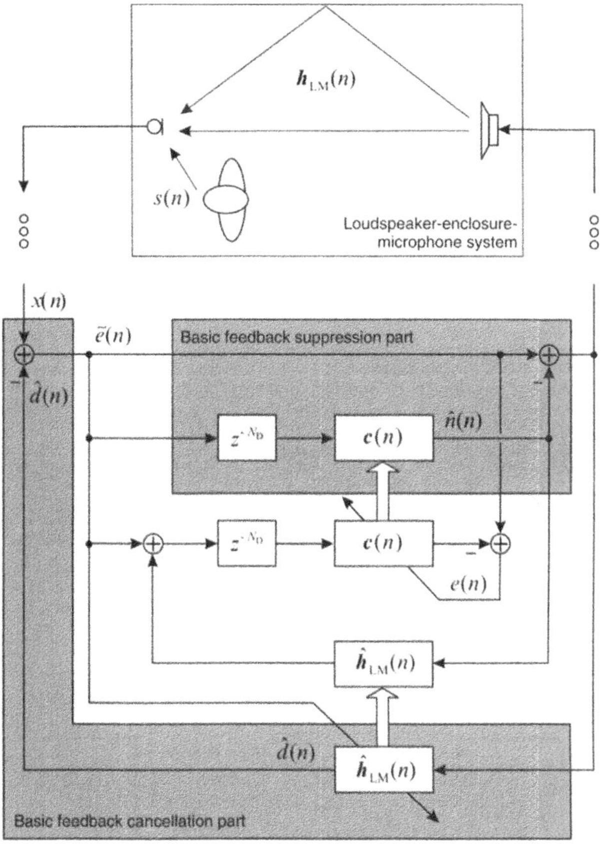

Fig. 14.21. Structure of the enhanced combined feedback cancellation and suppression system. This time the echo cancellation is performed within the signal path of the intercom system.

[13] This depends most crucially on the residual parts of the intercom system.

14.2.8 Gain Control

An adaptive gain control is the central element of an in-car communication system. Its task can be split into three subunits:

- At first, a speech activity detection has to be performed for each seat position. Only if the passenger on a specific seat really speaks, then his or her signals are played back via those loudspeakers which are close to the other communication partners.
- Since the exact seat position and thus also the exact distance to the microphones is a priori known only approximately, a gain control is computed adaptively for each beamformer output signal. This gain compensates not only for gain variation according to the mouth-microphone distance, but also for different speech levels.
- Finally, an individual playback volume for each seat position is computed. This loudspeaker gain depends on the individual background noise level and varies according to the driving situation (standstill, city traffic, or motorway driving). If, for example, one of the passengers opens a window, the playback volume of the loudspeakers close to this seat will be increased more than that of the other loudspeakers.

Fig. 14.22 depicts an overview of the loss control unit. Details about the three sub-units introduced above will be given in the next sections.

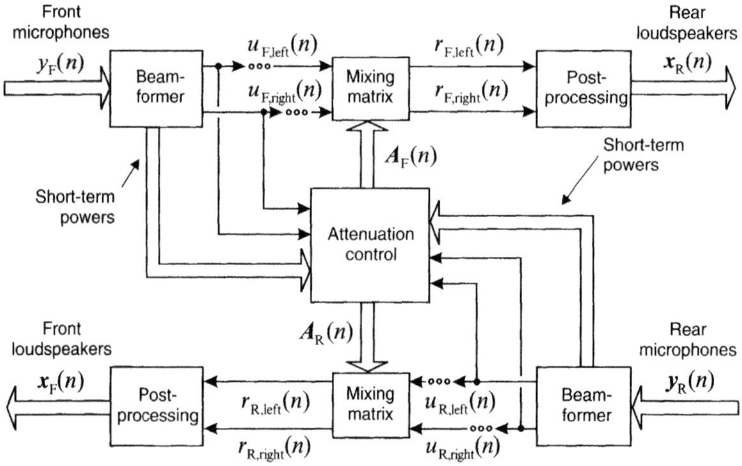

Fig. 14.22. Basic structure of a loss control unit.

14.2.8.1 Basic Control Structure

The front and rear mixing matrices $\boldsymbol{A}_{\mathrm{F}}(n)$ and $\boldsymbol{A}_{\mathrm{R}}(n)$ contain the weights that are applied to the beamformer output signals. These signals are mixed, resulting in the output signals $\boldsymbol{r}_{\mathrm{F}}(n)$ and $\boldsymbol{r}_{\mathrm{R}}(n)$ for postprocessing (as depicted in Fig. 14.22). The mixing process of each channel can be described as

$$\underbrace{\begin{bmatrix} r_{\text{left}}(n) \\ r_{\text{right}}(n) \end{bmatrix}}_{\boldsymbol{r}(n)} = \underbrace{\begin{bmatrix} a_{11}(n)\, a_{12}(n) \\ a_{21}(n)\, a_{22}(n) \end{bmatrix}}_{\boldsymbol{A}(n)} \underbrace{\begin{bmatrix} u_{\text{left}}(n) \\ u_{\text{right}}(n) \end{bmatrix}}_{\boldsymbol{u}(n)} . \tag{14.31}$$

For better readability the subscripts R and F, indicating whether the vectors are denoting the front (F) or rear (R) mixing process, have been omitted. For the determination of the mixing weights $a_{ij}(n)$ short-term powers of highpass filtered versions of the output signals $\tilde{u}(n) = h_{\mathrm{HP}}(n) * u(n)$ of each beamformer are computed:[14]

$$\hat{\sigma}(n) = \beta\, \hat{\sigma}(n-1) + (1-\beta)\, |\tilde{u}(n)| . \tag{14.32}$$

The time constant β of this first order IIR filter is usually chosen from the interval $[0.98, 0.998]$. As we will see later each of the before mentioned subtasks contributes its own part to the entire weighting factor:

- During speech activity the peak power is estimated and compared with a predefined reference level. If the current peak power (corrected by its peak power correction gain) is smaller than the reference value the gain value responsible for the peak power adjustment is increased slowly. In the other case a slow decrease is applied.
- A second stage detects which passenger currently speaks. Only those beamformer output signals are passed to the loudspeakers without attenuation, where speech activity was detected. In this stage only front-to-rear and rear-to-front enhancement is supported. Recording the left passenger and playing the recorded signals via the right loudspeaker in the same seat row (and vice versa) is not necessary, since in these directions communication is usually possible with sufficient quality.
- In a third stage the playback volume of each loudspeaker group is adjusted according to individually estimated noise levels. A group of loudspeakers may consist of only one transducer but usually all loudspeakers that are located close to one seat position are grouped together.

If we denote the gains resulting from peak power adjustment by $a_{\mathrm{p},i}(n)$, the attenuations caused by the speech activity detection by $a_{\mathrm{a},i}(n)$, and the loudness settings by $a_{\mathrm{l},i}(n)$ the computation of the mixing weights can be described by[15]

[14] Again for the reason of simplicity we have omitted here all subscripts that would indicate whether front or rear as well as left or right is addressed.

[15] The subscript i stands for either left or right.

$$a_{11}(n) = a_{\mathrm{p,left}}(n)\, a_{\mathrm{a,left}}(n)\, a_{\mathrm{l,left}}(n)\,, \tag{14.33}$$
$$a_{12}(n) = a_{\mathrm{p,right}}(n)\, a_{\mathrm{a,right}}(n)\, a_{\mathrm{l,left}}(n)\,, \tag{14.34}$$
$$a_{21}(n) = a_{\mathrm{p,left}}(n)\, a_{\mathrm{a,left}}(n)\, a_{\mathrm{l,right}}(n)\,, \tag{14.35}$$
$$a_{22}(n) = a_{\mathrm{p,right}}(n)\, a_{\mathrm{a,right}}(n)\, a_{\mathrm{l,right}}(n)\,. \tag{14.36}$$

Details about the computation of the coefficients $a_{\mathrm{p},i}(n)$, $a_{\mathrm{a},i}(n)$, and $a_{\mathrm{l},i}(n)$ are presented in the next three sections.

14.2.8.2 Automatic Gain Control

For determining the coefficient $a_{\mathrm{p},i}(n)$ the average peak power $\hat{\sigma}_{\mathrm{p}}(n)$ of each recorded signal is estimated during speech activity using a multiplicative correction approach:

$$\hat{\sigma}_{\mathrm{p}}(n) = \begin{cases} \hat{\sigma}_{\mathrm{p}}(n-1)\, K(n), & \text{during speech activity,} \\ \hat{\sigma}_{\mathrm{p}}(n-1), & \text{else.} \end{cases} \tag{14.37}$$

The correcting factor is computed as

$$K(n) = \begin{cases} K_{\mathrm{r}}, & \text{if}\, \hat{\sigma}_{\mathrm{p}}(n-1) < \hat{\sigma}(n), \\ K_{\mathrm{f}}, & \text{else,} \end{cases} \tag{14.38}$$

with $0 < K_{\mathrm{f}} < 1 < K_{\mathrm{r}}$. If the current peak power (corrected by its peak power correction gain) is smaller than the reference value $\sigma_{\mathrm{p,ref}}$ the current gain correction is increased slowly. In the other case a slow decrease is applied:

$$a_{\mathrm{p},i}(n) = \begin{cases} \min\{a_{\mathrm{p},i}(n-1)\,(1+\Delta), a_{\mathrm{p,max}}\}\,, & \text{if}\ \hat{\sigma}_{\mathrm{p}}(n)\, a_{\mathrm{p},i}(n-1) < \sigma_{\mathrm{p,ref}}, \\ \max\{a_{\mathrm{p},i}(n-1)\,(1-\Delta), a_{\mathrm{p,min}}\}\,, & \text{else.} \end{cases} \tag{14.39}$$

The quantity Δ is adjusted such that gain modifications of about 2 to 4 dB per second are possible. Additionally, the gain values are limited to a maximum and a minimum threshold. To determine periods with speech activity the background noise level for each highpass filtered beamformer output signal is computed according to

$$\hat{\sigma}_{\mathrm{n}}(n) = \min\left\{\hat{\sigma}_{\mathrm{n}}(n-1),\, \hat{\sigma}(n)\right\}\, (1+\epsilon)\,, \tag{14.40}$$

with ϵ being a small positive value. Fig. 14.23 shows an example of a background noise estimation according to Eq. 14.40. Besides the estimated noise level $\hat{\sigma}_{\mathrm{n}}(n)$ the input signal is depicted as well. To detect speech activity (Eq. 14.37) the following condition has to be fulfilled:

$$\hat{\sigma}(n) > K_{\mathrm{n}}\, \hat{\sigma}_{\mathrm{n}}(n)\,, \tag{14.41}$$

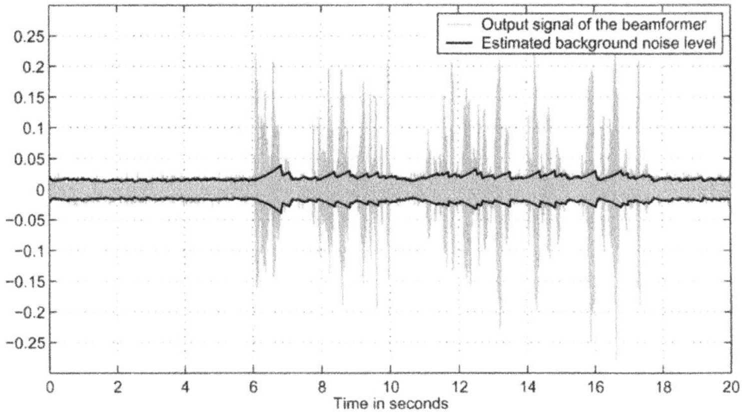

Fig. 14.23. Output signal of a beamformer and estimated background noise level.

with K_{n} being an appropriately chosen constant. Furthermore, spatial conditions – in terms of the ratio between short-term powers of a summing and a blocking beamformer [7] – can be utilized. However, especially at high background noise levels spatial criteria turn out to be very robust. Spatial speech activity detection is only possible if more than one microphone is used per passenger.

14.2.8.3 Speech Activity Controlled Attenuation

For the adjustment of the attenuation factors $a_{\mathrm{a},i}(n)$ first a so-called target state is determined. This target state defines which of the passengers are assumed to speak. Determining the state is done as in the last section: condition 14.41 is checked and spatial criteria are evaluated. If both criteria indicate speech activity for more than one seat position the loudest passenger is detected by comparing the individual beamformer output powers. For this seat position the target attenuation is set to 0 dB, all other target attenuations are set to a certain attenuation level, e.g., -10 dB. The current attenuation values $a_{\mathrm{a},i}(n)$ are computed by IIR smoothing of the target values. If two or more people speak simultaneously the detection of the loudest speaker will vary over time (according to the current speaking loudness). Due to the recursive smoothing the beamformer output signals of the active passengers will be attenuated only slightly, while the beamformer output signals where speech activity was not detected are attenuated strongly. Thus, it is possible to support more than one communication direction at the same time.

14.2.8.4 Adjustment of the Playback Volume

Finally, an individual playback volume correction $a_{\mathrm{l},i}(n)$ is computed for each seat position. The coefficients $a_{\mathrm{l},i}(n)$ are normalized to 0 dB for the non-driving and engine-off scenario. With increasing speed of the car noise resulting from the engine and from wind as well as tire hiss is also increasing. To take this into account the playback volume is increased accordingly. To determine the amount of gain increment, the estimated background noise levels according to Eq. 14.40 are compared with several thresholds. As soon as one of the thresholds is exceeded a slow increment of the playback volume is applied. The gain corrections should be varied within an interval of 0 to 10 dB. Since the background noise estimation is performed for each beamformer output an individual volume adjustment can be computed for each seat position. Note that the estimated background noise levels of the rear beamformer outputs are mapped on the gains of the front beamformer outputs and vice versa.

14.2.9 Loudspeaker Equalization

Before the signals are radiated via the loudspeakers, equalization filters should be applied. The objective of these filters is twofold: on one hand the output sound should be optimized from a subjective perspective. If certain frequency ranges are attenuated during the transmission from the mouth of the speaking passenger to the ears of the listening passengers, e.g. by superposition of several paths within the passenger compartment, an amplification of these frequencies should be applied. On the other hand the stability threshold of the entire system can be increased if those frequencies that have the highest coupling (from the loudspeakers to the microphones) are attenuated.

A simple and delay optimized approach for equalization are cascades of so-called *peak filters* [6]. Peak filters are able to attenuate or amplify certain frequency bands while keeping the residual frequency range unaffected. These filters are second order IIR filters with a transfer function

$$H_{\mathrm{Eq},i}(z) = 1 + \gamma_i \left(1 - A_i(z)\right) . \tag{14.42}$$

The term $A_i(z)$ is denoting a second order allpass filter with transfer function

$$A_i(z) = \frac{\alpha_i + \beta_i\, z^{-1} + z^{-2}}{1 + \beta_i\, z^{-1} + \alpha_i\, z^{-2}} . \tag{14.43}$$

The structure of one stage of the equalizer cascade is depicted in Fig. 14.24. It is important to note that due to the special structure of the IIR filters it is possible to adjust the center frequencies, the bandwidths, and the gain factors independently.

If we define the following parameters

f_s : sampling frequency,
f_c : center frequency of the notch or peak filter,
f_w : 3-dB-width of the filter,
V : gain or attenuation at the frequency f_c,

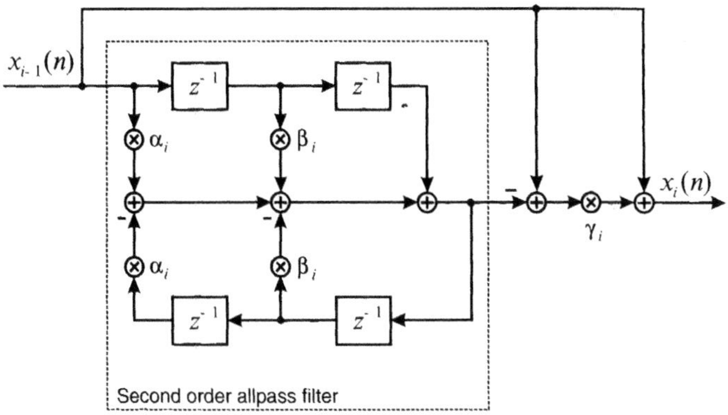

Fig. 14.24. Structure of one stage of the equalizer cascade.

the parameter α_i is adjusted according to

$$\alpha_i = \begin{cases} \dfrac{\tan\left(\pi \frac{f_w}{f_s}\right) - 1}{\tan\left(\pi \frac{f_w}{f_s}\right) + 1}, & \text{if } V > 1, \\[2ex] \dfrac{\tan\left(\pi \frac{f_w}{f_s}\right) - V}{\tan\left(\pi \frac{f_w}{f_s}\right) + V}, & \text{else.} \end{cases} \tag{14.44}$$

With this parameter the 3-dB-bandwidth of the filters is mainly controlled. With the second parameter of the allpass filter β_i the center frequency can be set:

$$\beta_i = -\cos\left(2\pi \frac{f_c}{f_s}\right) (1 + \alpha_i)\,. \tag{14.45}$$

Finally, the depth of the notch or the peak is adjusted mainly via the last parameter

$$\gamma_i = \frac{V - 1}{2}\,. \tag{14.46}$$

In Fig. 14.25 the frequency response of a single equalization unit is depicted for several variation of either the depth of the notch (left diagram) or of the 3-dB-bandwidth (right diagram).

To show an example for an entire loudspeaker equalization the squared magnitude of a frequency response measured between the loudspeaker located on the left side of the hat rack and the output of the driver's beamformer is depicted in the upper diagram of Fig. 14.26 (solid gray line). If the system gain is increased close to the stability margin howling starts first at frequencies around 3 kHz and 4.7 kHz. At those frequencies the frequency response also

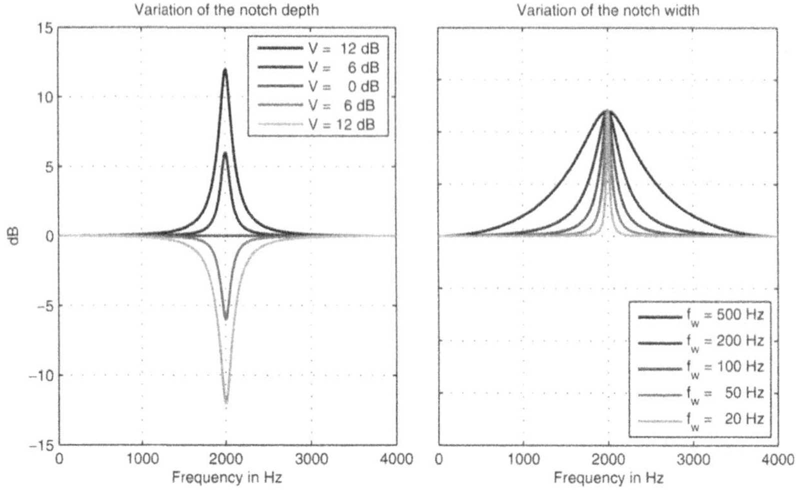

Fig. 14.25. Frequency responses of several notch and peak filters. Left: variations of the depth of the notch via the parameter V, right: variation of the 3-dB-bandwidth via the parameter f_W.

has local maxima. If the transmission is equalized using a five-stage equalization filter with two attenuation stages (around 3 kHz and 4.7 kHz) and three amplification stages (around 1.2 kHz, 1.7 kHz, and 2.3 kHz) the stability margin can be increased by about 4 dB. Additionally due to the amplification stages the system sounds broader and slightly louder. The frequency response of the entire equalization filter is depicted in the lower diagram of Fig. 14.26. The frequency response of the equalized system is plotted in the upper diagram (solid black line).

Adjustment of the equalization parameters can be done off-line using adequate impulse response measurements. However, in this case only those frequencies which show a certain behavior (attenuation or gain) under various conditions (all or just two seats are occupied, different seat adjustments, different temperatures, etc.) should be equalized. Further improvement can be achieved by adaptive equalization. In this case the coefficients should be changed only very slowly and the gain and attenuation should be limited carefully.

14.2.10 Further Signal Processing Units

Besides the schemes described above there is a variety of additional processing components. Not all of these processing units have been described so far. A description of a noise suppression, for example, was omitted. The reason for this omission was that traditional noise suppression schemes are usually introducing a small delay due to a spectral analysis-synthesis stage. As described in

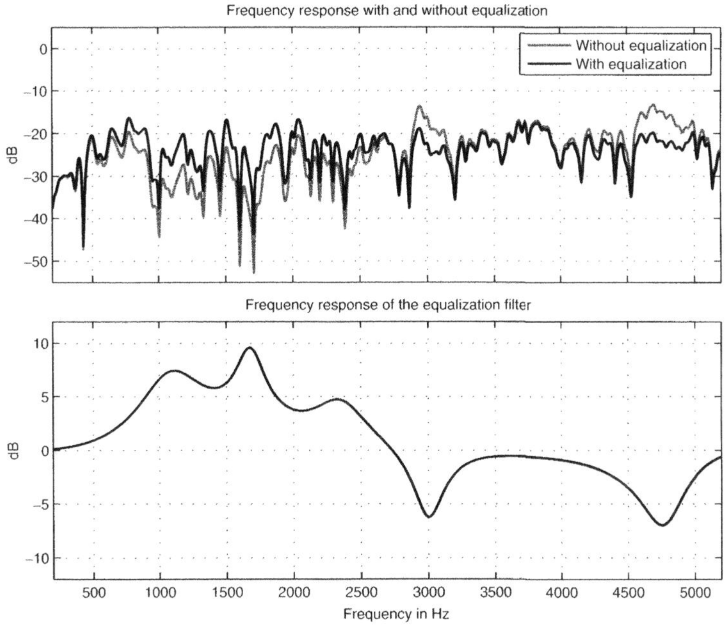

Fig. 14.26. Equalization of the loudspeakers.

Sec. 14.1.2.3 intercom systems are very restrictive concerning the delay introduced by signal processing. Thus, only noise suppression approaches with no or nearly no delay should be applied. Within the system presented in Sec. 14.4 a noise reduction was not implemented, since the noise played via the loudspeakers is typically masked by the environmental noise within the passenger compartment even at maximum output gain. However, if the microphones cannot be placed close to the speaking passengers and the loudspeakers are not located close to the listening passenger some kind of noise reduction is required.

Beside noise reduction several other algorithmic parts have not been described in this section. This should not indicate that additional processing units might not enhance the overall system quality.

14.3 Evaluation of Intercom Systems

After presenting algorithms or algorithmic parts of intercom systems, the question arises how to evaluate the quality of these algorithms and how to compare two competing approaches.[16] The fairest way of answering this question are, of course, subjective tests. For this reason we will present two subjective tests

[16] By *algorithm* or *approach* usually the entire intercom system is meant here.

in the next section: a rhyme test and a comparison mean opinion score. These tests are, however, quite expensive and time consuming since a lot of test subjects need to be involved. Objective tests, on the other hand, are much simpler and – if well designed – can give also a good indication about the quality of intercom systems. These kind of tests are presented in Sec. 14.3.4. All tests presented in the following are based on a real intercom system. Details about this system are described in Sec. 14.4.

14.3.1 Subjective Methods

For evaluating the improvements concerning speech quality and speech intelligibility two subjective tests can be utilized. Improvements or degradations of the speech intelligibility can be measured with a so-called *diagnostic* or *modified rhyme test* [41]. In that test pairs or even larger groups of rhyming words are used to focus on the intelligibility of each syllable or even on each phoneme of a word. A good speech intelligibility is one of the basic requirements of any communication system. If this quantity has reached a certain level communication is possible and people focus also on other aspects such as the naturalness of speech or the amount of reverberation. These aspects can be analyzed with a so-called *comparison mean opinion score* (CMOS) [23]. In order not to focus again on the speech intelligibility often very well known phrases such as popular song texts or proverbs are utilized within a CMOS test. In that case the listeners are able to get the full meaning of an utterance even if they understand just a few parts of it. Both subjective tests are used here to compare between two different situations: communication within a car that has an activated intercom system and one that is not equipped with such a system.

The next two sections give only a brief overview about both subjective tests. We are more focussed on the results here. If the reader is interested in more details about the tests, Chapter 11 (especially Sec. 11.4 and 11.3) of this book as well as the references cited herein are recommended for further reading.

14.3.2 Rhyme Tests

Let us describe first the boundary conditions of the rhyme test. We performed four tests at the following conditions:

- intercom system off at 0 km/h,
- intercom system on at 0 km/h,
- intercom system off at 130 km/h,
- intercom system on at 130 km/h.

10 to 15 listeners of different age and gender participated in each test. For each listener 40 pairs of rhyming words, such as *game* and *name* or *peace* and *peach*, were selected randomly from a prerecorded database. Both words were

presented visually first. Secondly, one of the examples was selected (again randomly) and played via a headset.[17] Afterwards the listeners had to decide which of the two stimulus words was acoustically presented. In Fig. 14.27 the amount of correct results for each rhyme test is depicted.

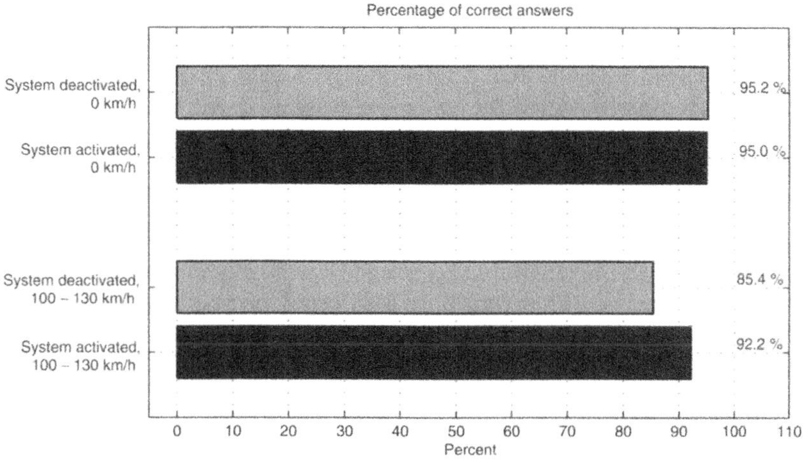

Fig. 14.27. Results of the rhyme tests.

Since the intercom system adjusts its gain automatically according to the background noise it is not surprising that no or nearly no difference (95.0 % for the activated system and 95.2 % for the deactivated system) was measured at 0 km/h. The conditions of those two tests were more or less optimal – meaning that all stimulus words were clearly understandable. Most of the errors were made such that the word that was presented on the left of the computer monitor (that was read first) was also selected by the listeners even if the second was actually presented acoustically.

As a result one can conclude that the intercom system that was tested does not improve the speech intelligibility when the car is in stand-still. This was not surprising since the intelligibility of the speech was already quite high. Under noisy conditions (130 km/h on a German autobahn), however, the amount of correct results could be increased impressively: from 85.4 % without the intercom system to 92.1 % with an activated system. The relative error rate has been reduced by about 50 %. Having in mind that rhyme tests do not achieve a rate of correct answers of 100 % even under ideal conditions the error rate improvement would be even larger.

[17] The recording and playback devices were calibrated in such a way that a true binaural impression with calibrated output levels could be achieved.

14.3.3 Comparison Mean Opinion Scores

In contrast to the rhyme test in which short and very similar stimulus words have been used longer and well known phrases are utilized for a CMOS test. For evaluation of the intercom system a list with 50 German phrases consisting of popular song texts, proverbs, and advertisements was used. Each sentence was played back using an artificial mouth loudspeaker (loudspeaker with approximately the same radiation pattern as a human mouth) at the passenger's seat in two different noise environments (0 km/h and 130 km/h) with activated and deactivated intercom system. As in the previous section binaural recordings were made on one of the back seats. These recordings were used as audio examples for all participants of the CMOS test.

Per scenario 10 to 15 subjects were asked about the speech quality of the presented signals. Per subject 25 pairs of audio examples were presented. Each pair consists of the same stimulus sentence – once recorded with an activated intercom system and once without. The order of presentation was chosen randomly. After listening to each pair of signals the subjects were asked to rate the differences between both signals on a scale consisting of seven levels:

- A is much better than B,
- A is better than B,
- A is slightly better than B,
- A and B are about the same,
- A is slightly worse than B,
- A is worse than B,
- A is much worse than B.

Before starting a short introduction was given by a supervisor to the listeners and a few examples were presented. Each test lasts about 4 to 8 minutes.

In Fig. 14.28 the detailed results of the CMOS test are depicted. As in the rhyme test no significant difference in the speech quality was observed for the low noise condition (0 km/h, within a parking area). In this scenario

- 19.7 % of the subjects preferred the system to be switched off,
- 29.7 % of the subjects had no preference, and
- 50.6 % of the subjects preferred an activated system.

Even though more than 50 % prefer the intercom system to be activated this result is far away from having statistical significance (see Chapter 11). However, in noisy driving conditions (130 km/h, on a German autobahn)

- only 4.3 % of the subjects preferred the system to be switched off,
- 7.1 % of the subjects had no preference, and
- 88.6 % of the subjects preferred an activated intercom system.

This shows a clear (and significant) preference for the intercom system.

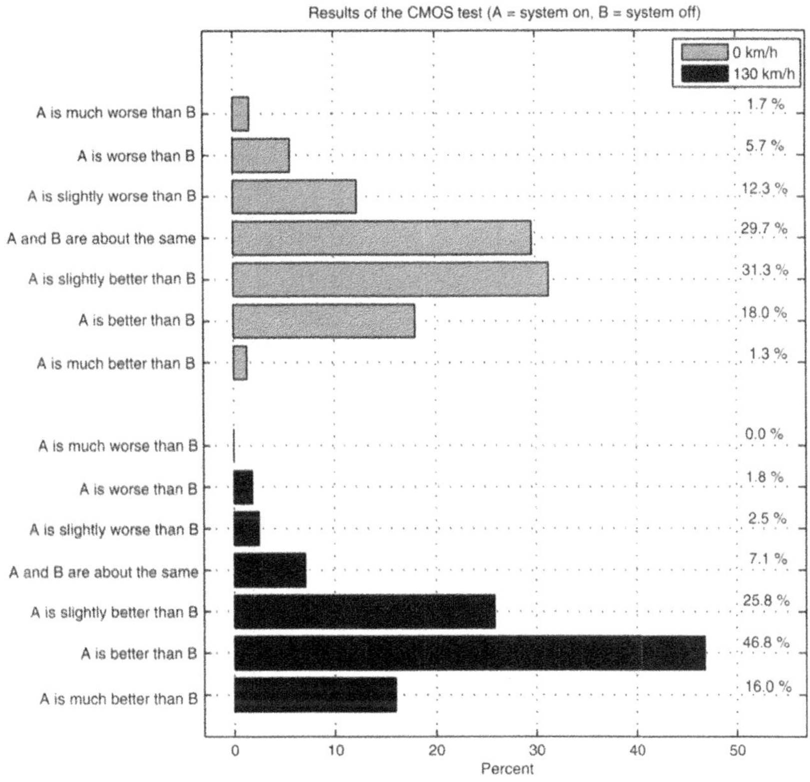

Fig. 14.28. Results of the comparison mean opinion score.

14.3.4 Objective Methods

Subjective tests have two main drawbacks: on one hand they are quite time consuming – and thus expensive – and on the other hand small differences between different systems or algorithmic versions are quite hard to evaluate with a small group of listeners. For this reason, objective evaluation methods should be applied not for replacing but for extending subjective tests. In this section we will focus on two measurements: the frequency response (respectively its absolute value) of the transmission from the speaking to the listening passenger and the impulse response from the mouth of the speaking passenger to his or her ears.

14.3.4.1 Improvements for the Listening Passengers

One way of measuring the improvement of the speech quality due to an intercom system is to measure the impulse or frequency responses from the mouth of the speaking passenger, say the driver, to the ears of the listening passenger, e.g. the left rear passenger. Such measurements should be performed with and without the intercom system. If the background noise is not increased due to the intercom system the ratio between the absolute values of the frequency responses is a good indicator for the signal-to-noise ratio improvement. If we denote – according to the definitions in Sec. 14.1.2.2 – the frequency responses with $H_{\mathrm{SR},i}(e^{j\Omega})$ we can compute for each measurement pair i the ratio

$$R_{\mathrm{SR},i}\big(e^{j\Omega}\big) = \frac{\Big|H_{\mathrm{SR},i}\big(e^{j\Omega}\big)\Big|_{\mathrm{on}}}{\Big|H_{\mathrm{SR},i}\big(e^{j\Omega}\big)\Big|_{\mathrm{off}} + \epsilon} \,. \tag{14.47}$$

The subscripts *on* and *off* indicate whether the intercom system should be activated or not. The constant ϵ avoids division by zero. In Fig. 14.29 two frequency responses are depicted. Both have been measured at a speed of about 70 km/h[18] between the front passenger's mouth reference point and the right ear of the right rear passenger. The upper curve shows the measurement with an activated intercom system, the lower curve shows the frequency response without an intercom system. The index i is used for distinguishing between

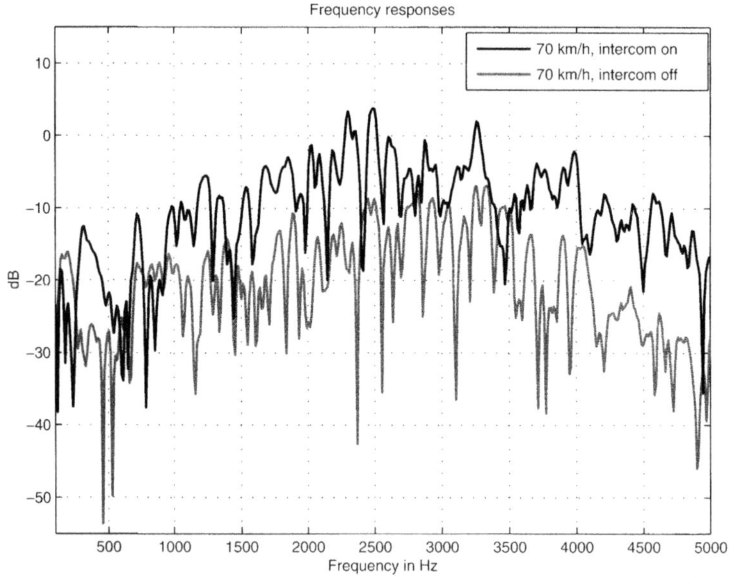

Fig. 14.29. Frequency responses that were measured at a speed of about 70 km/h. The measurement was made once with an activated intercom system (upper curve) and once without the system (lower curve).

[18] The speed, respectively the corresponding background noise level, has influence on the output gain of the system and thus on the frequency response.

different measurement points such as the left or right ear, and left or right passenger, respectively:

$$\begin{aligned} i = 1 &\quad : \quad \text{left passenger, left ear,} \\ i = 2 &\quad : \quad \text{left passenger, right ear,} \\ i = 3 &\quad : \quad \text{right passenger, left ear,} \\ i = 4 &\quad : \quad \text{right passenger, right ear.} \end{aligned}$$

Usually four measurements are made for each frequency response and the corresponding ratios are averaged afterwards:

$$R_{\mathrm{SR}}\left(e^{j\Omega}\right) = \frac{1}{4} \sum_{i=1}^{4} R_{\mathrm{SR},i}\left(e^{j\Omega}\right) \tag{14.48}$$

Additionally the average ratio can be smoothed along the frequency axis. For the measurements an artificial mouth loudspeaker and a head-and-torso simulator with ear-microphones should be used. Both is depicted in Fig. 14.30. As a measurement signal artificial voice [24] should be utilized. This is mainly because the intercom system should operate as in normal conditions during the measurement.

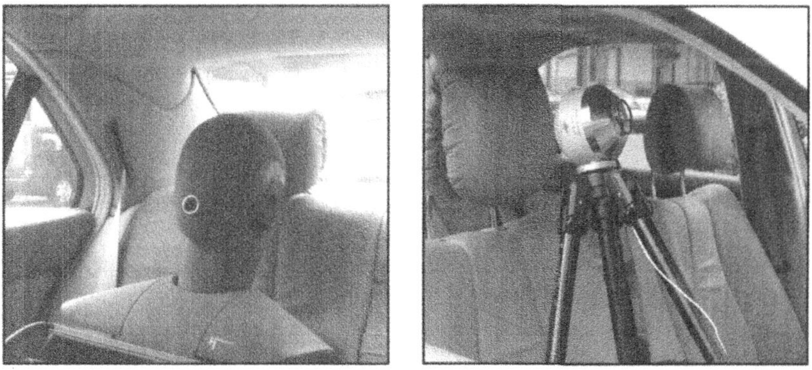

Fig. 14.30. Head-and-torso simulator (left) and artificial mouth loudspeaker (right).

With a real system installed in a limousine-type car (for further details see Sec. 14.4) ratios of about 5 to 15 dB were measured. In all cases the background noise level was not affected by the intercom system. The amount of improvement depends on many influences such as the size of the car, the position of the microphones and loudspeakers and of course on the adjusted output gain of the system. Usually more improvement is achieved at higher frequencies since several of the processing units, such as beamforming, are not very effective at low frequencies and also the directivity pattern of a human mouth is not as distinctive at low frequencies as it is at high ones.

14.3.4.2 Distortions for the Speaking Passenger

A large ratio $R_{SL}(e^{j\Omega})$ indicates an improvement of the communication quality for the listening passenger. The communication quality from the point-of-view of the speaking passenger, on the other hand, cannot be measured by $R_{SL}(e^{j\Omega})$. If the gain of the intercom system is rather large the speaking passenger might be disturbed by getting aware of his own echo. This might happen even before reaching the stability margin.[19] The longer the delay of the system is the more disturbing is a certain amount of echo.

At a system delay of about 5 ms and a coupling of about −10 dB from the mouth of a speaker to his ears, for example, most people do not realize any echo due to the self masking effect of the human auditory system [43]. If the delay exceeds 30 ms (again with a coupling of about −10 dB) nearly everyone gets aware of the echo and is disturbed by it.

These disturbing echo effects can be detected – at least approximately – by measuring the impulse response between the mouth of a speaker and his ears. Again a head-and-torso simulator (see left part of Fig. 14.30) can be utilized for this purpose. By comparing the absolute value of the impulse response with a masking envelope curve audible echoes can be detected. In the upper part of Fig. 14.31 an impulse response that was measured in a car with a well-adjusted intercom system was measured. The first coefficients represent the natural coupling from the mouth to the ears. Coefficients with an index larger than 50 are also influenced by the feedback caused by the intercom system. All coefficients stay, however, within the threshold (dotted line).

If the gain of the intercom system was increased by a few decibels an impulse response as depicted in the lower diagram of Fig. 14.31 was measured.[20] In this setup most of the speaking passengers reported that they get aware of their own echo. In the lower diagram one can see that the impulse response is above the threshold at several time lags.

The problem with this kind of analysis is the determination of the margin envelopes. These envelopes depend on many boundary conditions. For example, different slopes are necessary for different types and levels of the background noise. For this reason, a large amount of research is required until an appropriate echo masking model will be set up for this special purpose.

14.4 A Real System

In this last section the intercom system that has been used for most of the measurements and analyses of the previous sections will be described. The system was realized on a floating-point digital signal processor. It supports both directions (front-to-rear and rear-to-front) und makes use of the standard

[19] This means that the listening passenger might have a good impression of the system, while the speaking passenger is really disturbed by the intercom system.

[20] Note that only the coefficients with an index larger than 50 should have changed.

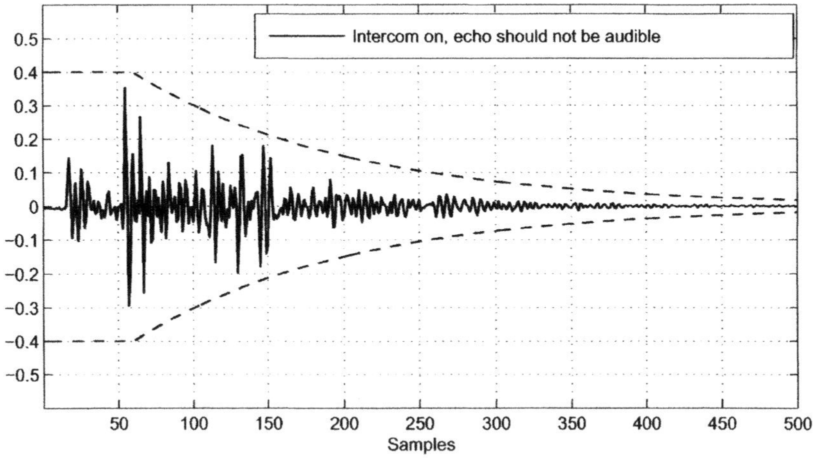

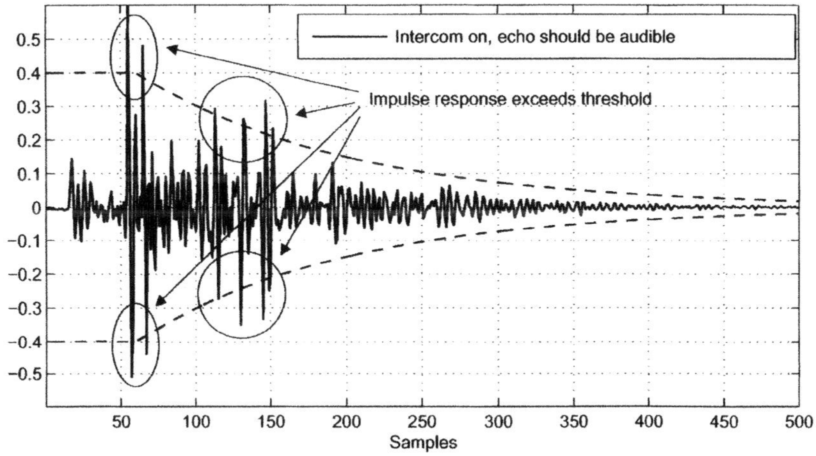

Fig. 14.31. Impulse responses measured between the mouth and the ears of a head-and-torso simulator for two different gain adjustments of an intercom system (solid lines). The dotted lines depict masking envelopes. If the impulse responses stay within the masking envelopes the speaking passengers should not be disturbed by their own echoes.

car loudspeakers (see Fig. 14.32). The loudspeaker setup consists of midrange-tweeter combinations within each door and two tweeters on the hat rack. The subwoofer, also integrated within the hat rack, was not used for the intercom system (but for playing the signals of the radio or CD player).

The intercom system can be integrated in several cars, but most of the tests were performed in a Mercedes S-Class (see again Fig. 14.32). Several microphone positions have been evaluated. Finally, four microphones, integrated within the front top control unit (two for each front passenger) were used. For the rear seat passengers also two microphones per passenger were utilized.

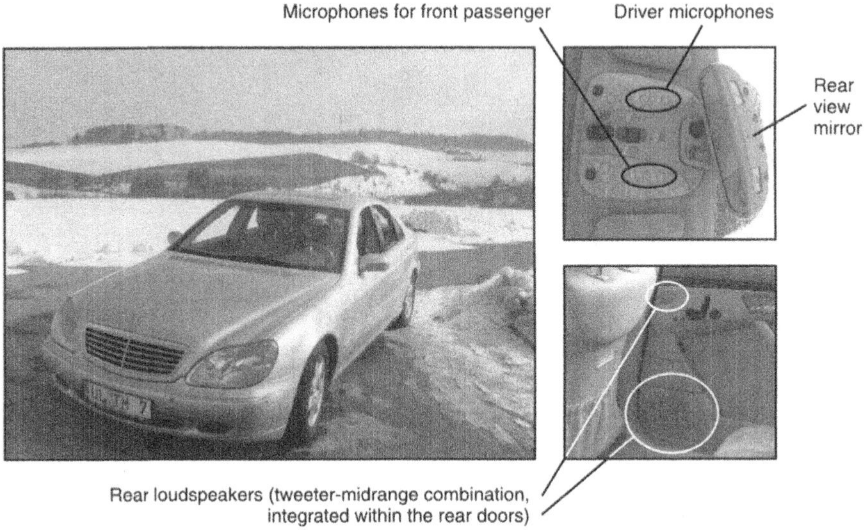

Fig. 14.32. Some pictures of the intercom system. Left: car in which the system was installed, top right: front microphones, bottom right: rear loudspeakers.

The microphones were integrated within the grab handles above the rear side windows. Standard cardioid microphones with a maximum sensitivity of 102 dB SPL were chosen for the system.

Enhancement from the left to right seat of the same seat row was not supported. All algorithmic parts that were described in Sec. 14.2 were implemented. It was possible to activate and deactivate each part. This made it possible to operate the system from pure amplification to highly sophisticated signal processing.

During the last three years the intercom system has been used for extensive testing, demonstrations, subjective evaluations, and measurements. Virtually all people who have tested the system on one of the back seats can attest a clear improvement of the communication quality. On the front seats usually the impression of a *broader* sound was reported but the improvements in terms of speech quality and speech intelligibility were not as large as for the rear seat passengers.

To visualize the improvement three binaural recordings that were made with a torso located on the seat behind the driver are depicted in Fig. 14.33. At the beginning of each recording the intercom system was activated. After a few seconds the system was switched off for a couple of seconds to show the difference of the speech quality. Finally the system was activated again. In all diagrams time-frequency analysis of the right ear are depicted.

To emphasize on the differences between the periods with enhancement and periods without any enhancement, all signals were normalized concerning their average power. Additionally, for better visualization, all signals were

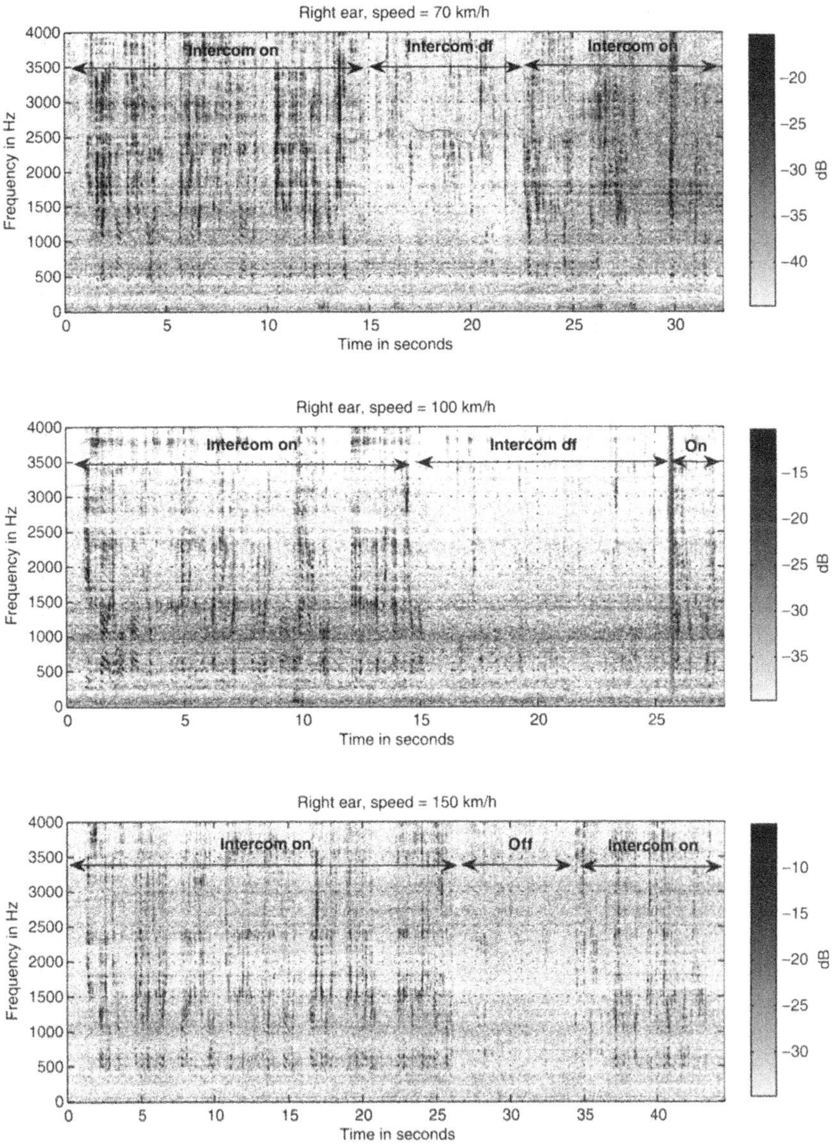

Fig. 14.33. Time-frequency analysis of the right channels of three binaural recordings made on the left back seat.

predictor error filtered. The adjustment of the prediction filters of order 8 were performed such that the background noise of each recording was whitened. In all diagrams only the lower frequency range (up to 4 kHz) is depicted.

Within the time-frequency analyses the speech components of the driver are recognizable in a much better way whenever the system is activated. During deactivation, however, the driver's speech is mostly masked by the driving noise.[21] Furthermore, the characteristics of the background noise do not change during the deactivation period. This means that the noise impression of the rear passenger is not affected – in terms of a higher noise level – when the intercom system is activated.

14.5 Conclusions and Outlook

In this chapter the basic signal processing components of an in-car communication system have been described. Even if most algorithms are already known for other applications such as hands-free telephones, public address system, or hearing aids, the specific conditions in which intercom systems have to operate require several modifications of the standard algorithms. For this reason the boundary conditions have been described in detail at the beginning of this contribution.

Undoubtedly, in-car communication systems are able to significantly enhance the quality of a conversation in a car driven at moderate or even high speed. In current automobiles all hardware components that are necessary to build such systems – microphones, loudspeakers, and powerful signal processing devices – are often already installed. Thus, the step for building such systems is really small. For this reason, we believe that intercom systems will soon satisfy the needs of customers for enhanced communication quality within a passenger compartment on a broad range.

References

[1] J. Benesty, D. R. Morgan, M. M. Sondhi: A better understanding and an improved solution to the problems of stereophonic acoustic echo cancellation, *Proc. ICASSP '97*, **1**, Munich, Germany, 303–306, 1997.

[2] J. Benesty, D. R. Morgan, M. M. Sondhi: A better understanding and an improved solution to the specific problems of stereophonic acoustic echo cancellation, *IEEE Trans. Acoust. Speech Signal Process.*, **6**(2), 156–165, 1998.

[3] J. Chang, J.R. Glover: The feedback adaptive line enhancer: a constrained IIR adaptive filter, *IEEE Trans. Signal Process.*, **41**(11), 3161–3166, 1993.

[4] W. F. Clemency, F. F. Romanow, A. F. Rose: The Bell system speakerphone, *AIEE. Trans.*, **76**(1), 148–153, 1957.

[5] A. G. Constantinides, K.M. Knill, J.A. Chambers: A novel orthogonal set adaptive line enhancer tuned with fourth-order cumulants, *Proc. ICASSP '92*, **4**, 241–244, San Francisco, CA, USA, 1992.

[21] When analyzing Fig. 14.32 the reader should not expect a very clean signal as known from publications on speech analysis or noise reduction. The earmicrophones of the torso do record a lot of noise, since the recording is made in a car driven at high speed. The focus should be on trying to see the typical time-frequency patterns of speech. When the system was switched off this is nearly impossible.

[6] P. Dutilleux, U. Zölzer: Filters, in U. Zölzer (ed.), *DAFX – Digital Audio Effects,* New York, NY: Wiley, 2002.
[7] M. Fuchs, T. Haulick, G. Schmidt: Noise suppression for automotive applications based on directional information, *Proc. ICASSP '04,* **1**, 237–240, Montreal, Canada, 2004.
[8] A. Gilloire, V. Turbin: Using auditory properties to improve the behaviour of stereophonic acoustic echo cancellation, *Proc. ICASSP '98,* **6**, 3681–3684, Washington, DC, USA, 1998.
[9] G. Glentis, K. Berberidis, S. Theodoridis: Efficient least squares adaptive algorithms for FIR transversal filtering: a unified view, *IEEE Signal Process. Mag.*, **16**(4), 13–41, 1999.
[10] L. J. Griffiths, C. W. Jim: An alternative approach to linearly constrained adaptive beamforming, *IEEE Trans. Antennas and Propagation,* **AP-30**(1), 24–34, 1982.
[11] H. Haas: The influence of a single echo on the audibility of speech, *Journal of the Audio Engineering Society,* **20**, 145–159, March 1972.
[12] V. Hamacher: Comparison of advanced monaural and binaural noise reduction algorithms for hearing aids, *Proc. ICASSP '02,* **4**, 4008–4011, Orlando, FL, USA, 2002.
[13] E. Hänsler, G. Schmidt: *Acoustic Echo and Noise Control – A Practical Approach,* New York, NY: Wiley, 2004.
[14] J. H. L. Hanson: Morphological constrained feature enhancement with adaptive cepstral compensation (MCE-ACC) for speech recognition in noise and Lombard effect, *IEEE Trans. Speech Audio Process.,* **T-SA-2**(4), 598–614, 1994.
[15] T. Haulick, G. Schmidt: Signalverarbeitungskomponenten zur Verbesserung der Kommunikation in Fahrzeuginnenräumen, *Proc. ESSV '03,* 130–137, Karlsruhe, Germany, 2003 (in German).
[16] S. Haykin: *Adaptive Filter Theory,* 4th ed., Englewood Cliffs, NJ: Prentice Hall, 2002.
[17] W. Herbordt, W. Kellermann: Adaptive Beamforming for audio signal acquisition, in J. Benesty, Y. Huang (eds.), *Adaptive Signal Processing,* Berlin, Germany: Springer, 2003, 155–194.
[18] O. Hoshuyama, B. Begasse, A. Sugiyama, A. Hirano: A realtime robust adaptive microphone array, *Proc. ICASSP '98,* **6**, 3605–3608, Washington, DC, USA, 1998.
[19] O. Hoshuyama, A. Sugiyama, A. Hirano: A robust adaptive beamformer for microphone arrays with a blocking matrix using constrained adaptive filters, *IEEE Trans. Signal Process.,* **47**(10), 2677–2684, 1999.
[20] Y. Huang, J. Benesty, G. Elko: Microphone arrays for video camera steering, in S. Gay, J. Benesty, (eds.), *Acoustic Signal Processing for Telekommunication,* Boston, MA: Kluwer, 239–259, 2000.
[21] ITU-T Recommendation G.167: *General Characteristics of International Telephone Connections and International Telephone Circuits – Acoustic echo Controllers,* Helsinki, Finland, 1993.
[22] ITU-T Recommendation P.64: *Determination of sensivity/frequency characteristics of local telephone systems,* Geneva, Switzerland, 1999.
[23] ITU-T recommendation P.800: *Methods for subjective determination of transmission quality,* Geneva, Switzerland, August 1996.

[24] ITU-T Recommendation P.501: *Test Signals for Use in Telephonometry,* Geneva, Switzerland, Geneva, Switzerland, 2000.

[25] ITU-T Recommendation P.581: *Use of Head And Torso Simulator (HATS) for Hands-Free Terminal Testing,* Geneva, Switzerland, 2000.

[26] Y. Joncour, A. Sugiyama, A. Hirano: DSP implementations and performance evaluation of a stereo echo canceller with pre-processing, *Proc. EUSIPCO '98,* **2**, 981–984, Rhodos, Greece, 1998.

[27] A. Khong, P. Naylor: Reducing inter-channel coherence in stereophonic acoustic echo cancellation using partial update adaptive filters, *Proc. EUSIPCO '04,* 405–408, Vienna, Austria, 2004.

[28] C. H. Knapp, G. C. Carter: The generalized correlation method for estimation of time delay, *IEEE Trans. Acoust. Speech Signal Process.,* **ASSP-24**(4), 320–327, 1976.

[29] H. Kuttruff: *Room Acoustics,* London, GB: Spon Press, 2000.

[30] T. I. Laakso, V. Välimäki, M. Karjalainen, U. K. Laine: Splitting the unit delay – tools for fractional delay filter design, *IEEE Signal Processing Magazine,* **13**(1), 30–60, 1996.

[31] E. Lleida, E. Masgrau, A. Ortega: Acoustic echo and noise reduction for car cabin communication, *Proc. EUROSPEECH '01,* **3**, 1585–1588, Aalborg, Denmark, 2001.

[32] K. Linhard, J. Freudenberger: Passenger in-car communication enhancement, *Proc. EUSIPCO '04,* **1**, 21–24, Vienna, Austria, 2004.

[33] E. Lombard: Le signe de l'elevation de la voix, *Ann. Maladies Oreille, Larynx, Nez. Pharynx,* **37**, 101–119, 1911 (in French).

[34] E. Meyer, G. R. Schodder: Über den Einfluss von Schallrückwürfen auf Richtungslokalisation und Lautstärke bei Sprache, *Nachrichten der Akademie der Wissenschaften in Göttingen,* **Math-phys. Kl. 6**, 31–42, 1952 (in German).

[35] A. Ortega, E. Lleida, E. Masgrau, F. Gallego: Cabin car communication system to improve communication inside a car, *Proc. ICASSP '02,* **4**, 3836–3839, Orlando, FL, USA, 2002.

[36] M. R. Schroeder: Improvement of acustic-feedback stability by frequency shifting, *J. Acoust. Soc. Am.,* **36**(9), 1718–1724, 1964.

[37] K. Shenoi: *Digital Signal Processing in Telecommunications,* Englewood Cliffs, NJ: Prentice Hall, 1995.

[38] M. M. Sondhi, D. R. Morgan, J. L. Hall: Stereophonic acoustic echo cancellation – an overview of the fundamental problem, *IEEE Signal Process. Letters,* **2**(8), 148–151, 1995.

[39] A. Sugiyama, Y. Joncour, A. Hirano: A stereo echo canceller with correct echo-path identification based on an input-sliding technique, *IEEE Trans. Signal Process.,* **49**(1), 2577–2587, 2001.

[40] B. D. Van Veen, K. M. Buckley: Beamforming: a versatile approach to spatial filtering, *IEEE Signal Process. Mag.,* **5**(2), 4–24, 1988.

[41] W. Voiers: Evaluating processed speech using the diagnostic rhyme test, *Speech Technology,* 30–39, Jan./Feb. 1983.

[42] N. Wiener: *Extrapolation, Interpolation, and Smoothing of Stationary Time Series, with Engineering Applications,* Cambridge, MA: MIT Press, 1949 (originally published as confidential report in 1942).

[43] E. Zwicker, H. Fastl: *Psychoacoustics – Facts and Models,* 2nd ed., Berlin, Germany: Springer, 1999.

15

Applications of Adaptive Signal Processing Methods in High-End Hearing Aids

Volkmar Hamacher, Eghart Fischer, Ulrich Kornagel, Henning Puder

Siemens Audiological Engineering Group, Erlangen, Germany

15.1 Introduction

In the past ten years the technical capabilities of hearing aids have considerably increased. One important mile stone was the changeover from analog to digital technology enabled by the continuous progress in semi-conductor technology. This chapter focuses on the powerful digital signal processing of modern high-end hearing aids.

In principal, the development of hearing aids incorporates two aspects, namely the audiological and the technical point of view. The former focuses on items like the recruitment phenomenon, the speech intelligibility of hearing impaired persons or just on the question of hearing comfort. Concerning these topics different algorithms intending to improve the hearing ability are presented in this chapter. These are automatic gain controls, directional microphones and noise reduction algorithms. Besides the audiological point of view there are several purely technical problems which have to be solved. An important one is the acoustic feedback. Another instance is the proper automatic control of all hearing aid components by means of a classification unit.

Fig. 15.1 schematically shows the main signal processing units of a high-end hearing aid [23, 24]. We will follow the depicted signal flow and discuss the state-of-the-art and the challenges for the different components. A coarse overview is given below.

First, the acoustic signal is captured by up to three microphones. The microphone signals are processed into a single signal within the directional microphone unit which will be discussed in Sec. 15.2.

The obtained mono-signal is further processed separately for different frequency ranges. In general this requires an analysis filterbank and a corresponding signal synthesis. The main frequency band dependent processing steps are noise reduction as detailed in Sec. 15.3 and signal amplification combined with dynamic compression as discussed in Sec. 15.4.

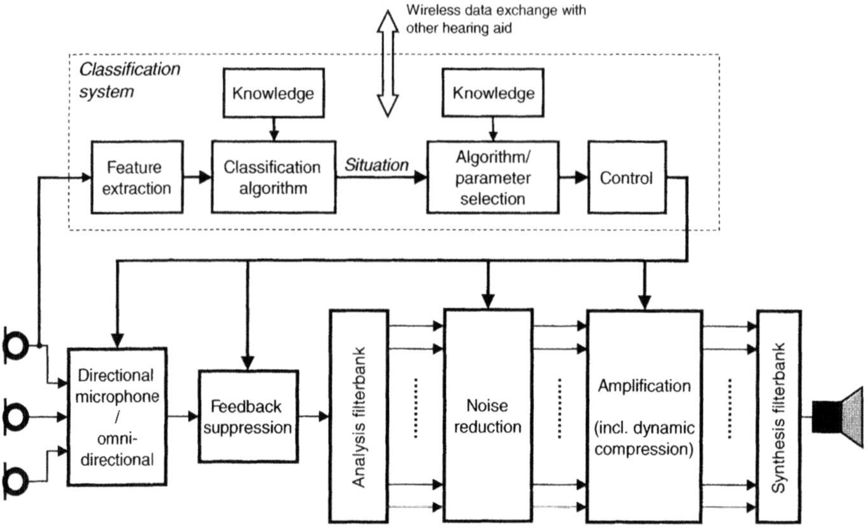

Fig. 15.1. Processing stages of a high-end hearing aid.

A technically challenging problem of hearing aids is the risk of acoustic feedback that is provoked by strong signal amplification in combination with microphones and receiver being close to each other. The two major alternatives to remedy feedback are the feedback suppression approach and the feedback compensation approach. Details regarding the feedback problem and possible solutions are discussed in Sec. 15.5. Note that feedback suppression can be applied at different stages of the signal flow dependent on the chosen strategy. One reasonable solution is shown in Fig. 15.1, where feedback suppression is applied right after the (directional) microphone unit.

Almost all mentioned hearing aid components can be tuned differently for optimal behavior in various listening situations. Providing different "programs" that can be selected by the hearing impaired is a simple means to account for this difficulty. However, the usability of the hearing aid can be significantly improved if control of the signal processing algorithms can be handled by the hearing aid itself. Thus, a classification and control unit, as shown in the upper part of Fig. 15.1 and described in Sec. 15.6, is required and offered by advanced hearing aids. In binaural use, the effectiveness of this unit can be significantly improved by means of wireless coupling of both hearing aids.

15.2 Directional Microphones

One of the main problems for the hearing impaired is the reduction of speech intelligibility in noisy environments, which is mainly caused by the loss of temporal and spectral resolution in the auditory processing of the impaired ear. The loss in signal-to-noise ratio (SNR) is estimated to be about 4-10 dB [12]. Additionally, the natural directivity of the outer ear is not effective when BTE (behind-the-ear) instruments are used. To compensate for these disadvantages, directional microphones have been used in hearing aids for several years and have proved to significantly increase speech intelligibility in various noisy environments [61].

15.2.1 First-Order Differential Arrays

In advanced hearing aids, directivity is achieved by differential processing of two nearby omni-directional microphones in endfire geometry (first-order differential array) to create a direction-dependent sensitivity. The directivity pattern of the system is defined by the ratio r of the internal delay T_i and the external delay due to the microphone spacing d (typically 7-16 mm). In this example the ratio was set to $r = 0.57$ resulting in a super-cardioid pattern also shown in Fig. 15.2. To compensate for the high-pass characteristic introduced by the differential processing, an appropriate low-pass filter (LPF) is usually added to the system.

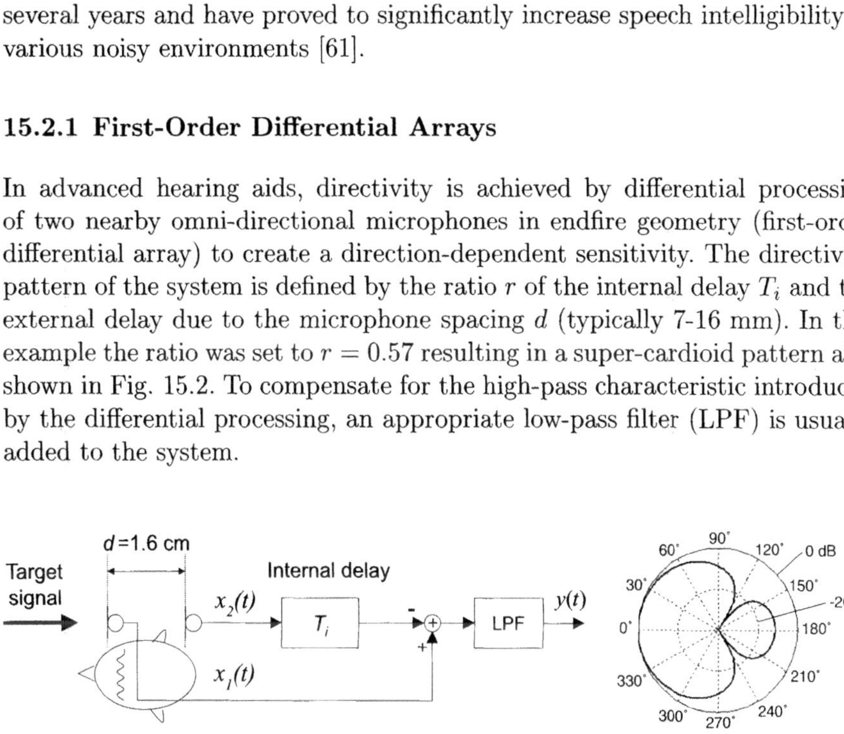

Fig. 15.2. Signal processing of a first order differential microphone.

The performance of a directional microphone is quantified by the directivity index (DI),

$$DI\left(e^{j\Omega}\right) = 10\log_{10}\left(\frac{\left|H\left(e^{j\Omega},\varphi_0,\theta_0\right)\right|^2}{\frac{1}{4\pi}\int\limits_{-\pi/2}^{\pi/2}\int\limits_{0}^{2\pi}\left|H\left(e^{j\Omega},\varphi,\theta\right)\right|^2\sin\left(\theta\right)d\varphi d\theta}\right) \tag{15.1}$$

where $H\left(e^{j\Omega},\varphi,\theta\right)$ denotes the spatial-temporal transfer function of the array depending on azimuth φ and elevation θ in a spherical coordinate system.

The DI is defined by the power ratio of the output signal (in dB) between sound incidence only from the front and the diffuse case, i.e. sound coming equally from all directions. Consequently, the DI can be interpreted as the improvement in SNR that can be achieved for frontal target sources in a diffuse noise field. The hyper-cardioid pattern ($r = 0.34$) provides the best directivity with a DI of 6 dB, which is the theoretical limit for any two-microphone array processing [4]. However, in practical use these DI values cannot be reached due to shading and diffraction effects caused by the human head. Fig. 15.3 illustrates the impact of the human head on the directivity of a BTE hearing aid with a two-microphone array. The most remarkable point is that the direction of maximum sensitivity is shifted aside by approximately 40°, if the device is mounted behind the ear of a KEMAR (Knowles Electronic Manikin for Acoustic Research). Consequently, the DI which is related to the 0° front direction, decreases typically by 1.5 dB compared to the free-field condition.

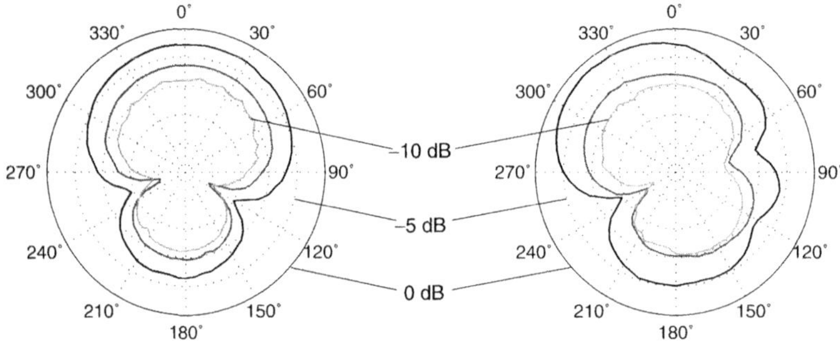

Fig. 15.3. Impact of head shadow and diffraction on the directivity pattern of a BTE hearing aid with a two-microphone differential array in free field (left plot) and mounted behind the left ear of a KEMAR (right plot). The black, dark gray and light gray curves show the directivity pattern for 2 kHz, 1 kHz, and 500 Hz, respectively (10 dB grid).

The performance related to speech intelligibility is quantified by a weighted average of the DI across frequency, commonly referred to as the AI-DI. The weighting function is the importance function used in the Articulation Index (AI) method [46] and takes into account that SNR improvements in different frequency bands contribute differently to the speech intelligibility. As shown in Fig. 15.4 for a hyper-cardioid pattern, the AI-DI (as measured on KEMAR) of two-microphone arrays in BTE instruments ranges from 3.5 to 4.5 dB. For speech intelligibility tests in mainly diffuse noise the effect of directional microphones typically leads to improvements of the Speech-Reception-Threshold (SRT) in the range from 2 to 4 dB (e.g. [53]).

In high-end hearing aids, the directivity is normally adaptive in order to achieve a higher noise suppression effect in coherent noise, i.e. in situations with one dominant noise source [12,50]. As depicted in Fig. 15.6, the primary direction from which the noise arrives is continually estimated and the direc-

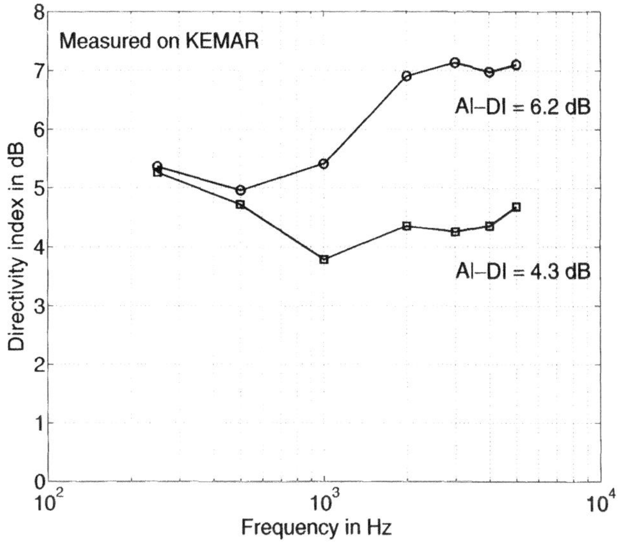

Fig. 15.4. DI and AI-DI for a fist-order array (Siemens Triano S) and the combination with a second-order array (see Sec. 15.2.2) in the upper frequency range (Siemens Triano 3).

tivity pattern is automatically adjusted so that the directivity notch matches the main direction of noise arrival. Instead of implementing computationally expensive fractional delay filters, the efficient method proposed by Elko and Pong [15] can be used. In this approach, the shape of the directivity pattern is steered by a weighted sum of the output signals of two cardioid patterns, one facing to the front (0°), the other one facing to the back (180°). The position of the directivity notch is monotonically related to the weighting factor. Great demands are made on the adaptation algorithm. The steering of the directional notch has to be reliable and accurate and should not introduce artifacts or perceivable changes in the frequency response for the 0°-target direction, which would be annoying for the user. The adaptation process must be fast enough (< 100 ms) to compensate for head movements and to track moving sources in common listening situations, such as conversation in a street-cafe with interfering traffic noise. To ensure that no target sources from the front hemisphere are suppressed, the directivity notches are limited to the back hemisphere (90° – 270°). Finally, the depth of the notches is limited to prevent hazardous situations for the user, e.g. when crossing the street while a car is approaching.

Fig. 15.5 shows a measurement in an anechoic test chamber with an adaptive directional microphone BTE instrument mounted on the left KEMAR ear. A noise source was moved around the head and the output level of the hearing aid was recorded (dashed line). Compared to the same measurement for a non-adaptive super-cardioid directional microphone (solid line), the higher

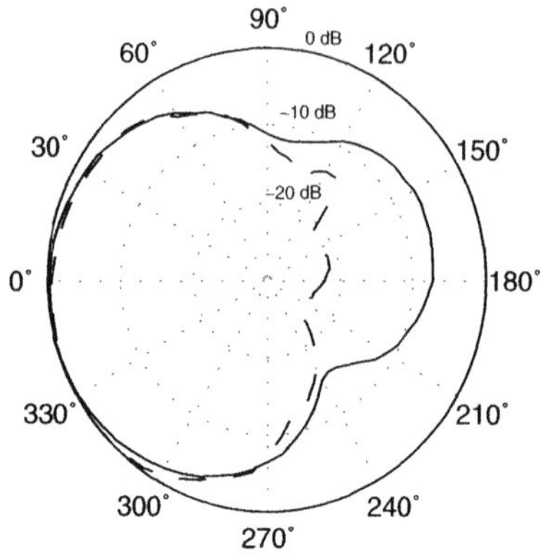

Fig. 15.5. Suppression of a noise source moving around the KEMAR for a BTE instrument (mounted on left ear) with directional microphone in adaptive mode (dashed line) and non-adaptive mode (solid line).

suppression effect for noise incidence from the back hemisphere is clearly visible.

In order to achieve optimum performance also for natural sound fields with non-diffuse spatial and non-white frequency distribution, it is advantageous to use a frequency specific implementation of the adaptive directional microphone principle. Fig. 15.6 shows the principle of a four-channel adaptive differential microphone with four directional characteristics that can independently adapt to the main direction of incidence of the interferer within the corresponding frequency band.

Studies have shown the advantage of using adaptive directional processing instead of static directivity. For a situation with three interferers from 90, 180 and 270°, as shown in Fig. 15.7 an improvement of 1.5 dB for the SRT could be achieved, see Fig. 15.8. The speech reception threshold SRT is a measure from speech audiometry, which determines the lowest intensity level of speech, presented in noise, at which the patient can correctly identify 50% of the words.

15.2.2 Second-Order Differential Arrays

Using second order arrays that are using three omnidirectional microphones [4] instead of two, generally a significantly higher DI can be achieved. (6-8 dB instead of 4-5 dB for hearing aids worn on the head.)

Unfortunately, this increase in DI has to be payed by a higher self induced noise with second order differential processing compared to first order processing. Fig. 15.9 and Fig. 15.10 compare directivity and the self induced noise for first vs. second order processing.

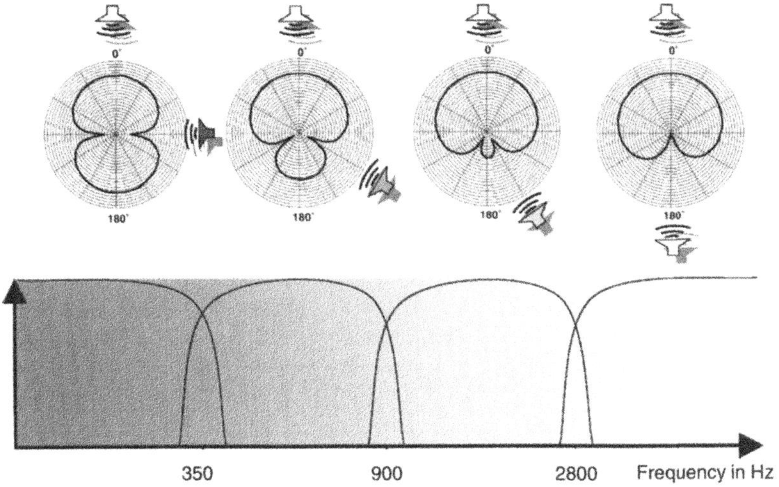

Fig. 15.6. Principle of a 4-channel adaptive directional microphone.

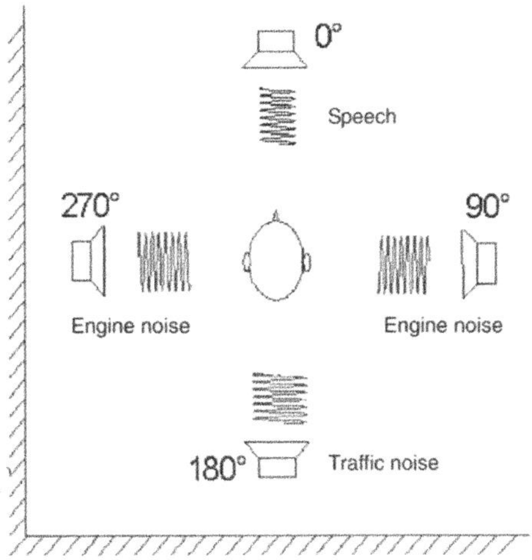

Fig. 15.7. Setup with one speaker and three interferers.

There are different possibilities to deal with the noise problem for higher order differential microphones.

One is the realization of a combined first- and second-order directional processing in a hearing aid with three microphones [50], which is shown in Fig. 15.11. Due to the high sensitivity to microphone noise especially in the low frequency range the second-order processing is limited to the frequencies above approx. 1 kHz which are most important for speech intelligibility.

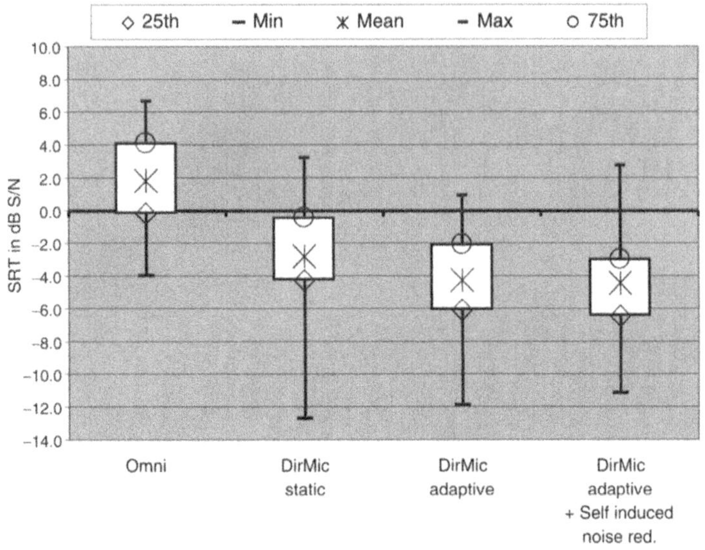

Fig. 15.8. Adaptive directional processing is showing an improvement in SRT of about 1.5 dB compared to static directional processing.

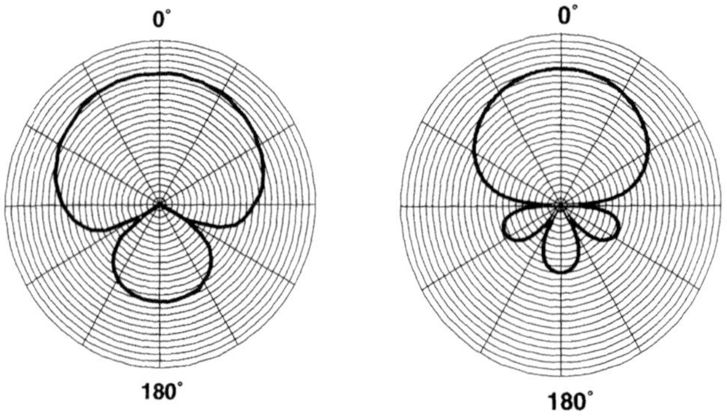

Fig. 15.9. Polar plots of first (left) and second (right) order differential microphones.

As shown in Fig. 15.4 calculation of the AI-DI leads to values of 6.2 dB, i.e. an improvement in AI-DI of about 2 dB compared to a first-order system worn at the head. It should be noted that for many listening situations, improvements of 2 dB in the AI-DI can have a significant impact on speech intelligibility [54].

Another possibility to handle the noise problem is to make the directional microphone system not only adaptive in terms of spatial shape of the directivity but also adaptive in terms of adaptation to the level of the target input

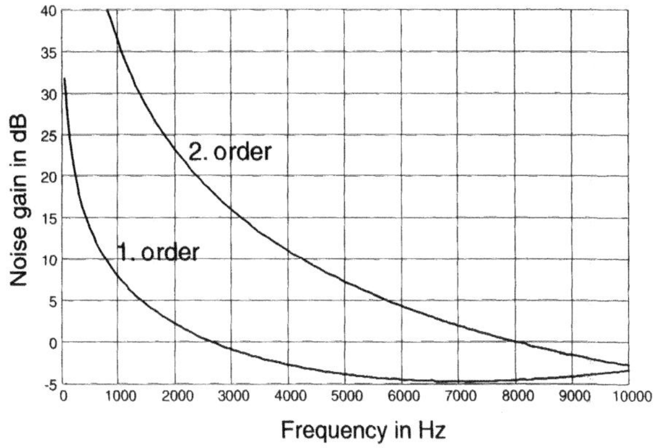

Fig. 15.10. Self induced microphone noise gain for first and second order processing with flat frequency response for the zero degree direction.

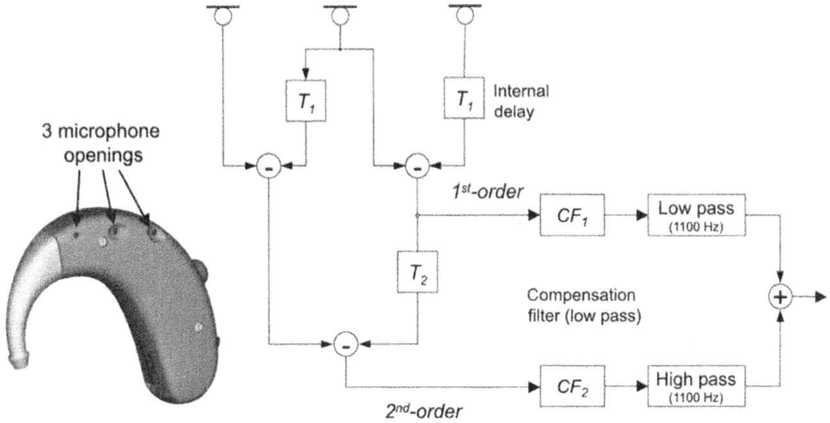

Fig. 15.11. Combined first- and second-order processing in a BTE hearing aid with three microphones.

signal: If the input level is high enough, more directivity with more self induced noise may be applied in the considered frequency band as the noise is to a large extent masked by the target signal. If the input level is low, less directivity with less self induced noise would be used in order to avoid audible noise. Fig. 15.12 illustrates the principle of this approach.

The differential approach for directional microphones as described above is of course just one - though very effective - method of generating directivity. There are several other ways with their specific advantages and disadvantages to build directional systems in hearing aids, e.g. adaptive beamformers (e.g. [13, 29, 30, 33, 34, 62]), beamformers, taking head shadow effects into account [44] and blind source separation techniques (e.g. [1, 14]).

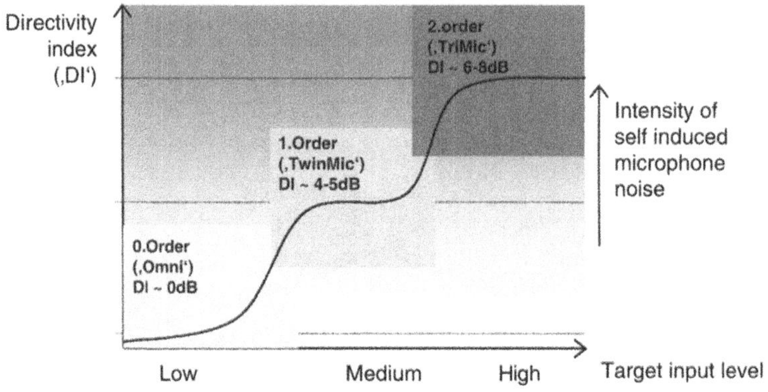

Fig. 15.12. Principle of the input level dependent directivity: Order of directivity is increasing with increasing input level.

15.3 Noise Reduction

Directional microphones, as described in the preceding section are usually not applicable to small ear canal instruments for reasons of size constraints and the assumption of a free sound field which is not met inside the ear canal. Consequently, one-microphone noise reduction algorithms became an essential signal processing stage of today's high-end hearing aids. Due to the lack of spatial information, these approaches are based on the different signal characteristics of speech and noise. Usually, despite of the fact that these methods may improve the SNR, they could yet not prove to enhance the speech intelligibility.

In the following, we will focus on two noise reduction methods which both showed their suitability for hearing aids. The first method is also one of the early ones in the field. It decomposes the noisy signal into many subbands and applies a long-term smoothed attenuation to those subbands for which the average SNR is very low. The second, a Wiener-filter based method applies a short-term attenuation to the subband signals and is thus able to enhance the SNR even for those signals for which the desired signal and the noise cover the same frequency range. Both methods can also be well combined: During speech activity, the Wiener filter approach exhibits the stronger impact whereas during speech pauses or for frequency bands with a very low long-term SNR the long-term noise reduction shows the stronger impact. Both effects are desired and by choosing the maximum noise reduction of both methods they can be both achieved.

At the end of this section, we will have an outlook to the application of Ephraim-Malah based short-term noise reduction approaches for hearing aids.

15.3.1 Long-Term Smoothed, Modulation Frequency Based Noise Reduction

Supposing a noisy input signal which has been decomposed into several frequency bands, the task of this noise reduction unit is to apply a long-term smoothed attenuation to frequency bands with a very low SNR which do not contain any remarkable speech components.

15.3.1.1 Theoretical Basis

The theoretical basis for distinguishing speech components from others is that speech signals exhibit a characteristic modulation frequency at 4 Hz [47].

For calculating this characteristic modulation frequency,

- first the envelope of a speech signal is determined according to Fig. 15.13,

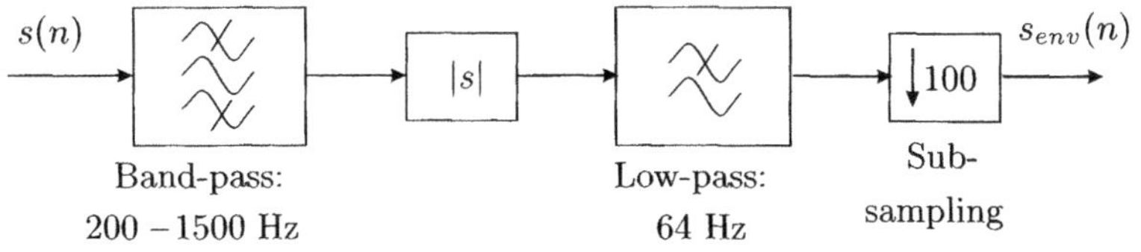

Fig. 15.13. Processing for determining the signal envelope.

- then DC component is removed by an IIR filter of first order:

$$s_{\text{env,AC}}(n) = \frac{1+\beta}{2}\Big[s_{\text{env}}(n) - s_{\text{env}}(n-1)\Big] + \beta\, s_{\text{env,AC}}(n-1) \quad (15.2)$$

  with a typical value for $\beta = 0.995$.
- The power spectral density (PSD) of the envelope has to be calculated, normalized by the mean power $m_s^{(2)} = \mathrm{E}\{{s_{\text{env}}}^2(n)\}$:

$$S_{ss}(\Omega) = PSD\big\{s_{\text{env,AC}}(n)\big\}. \quad (15.3)$$

$$S_{ss,\text{norm}}(\Omega) = \frac{S_{ss}(\Omega)}{m_s^{(2)}}, \quad (15.4)$$

- and finally the PSD is summed over Terz-bands for determining the modulation spectrum at a logarithmic scale:

$$S_{\text{mod\_spec}}(i) = \frac{1}{2\pi} \int\limits_{\Omega=\Omega_i}^{\Omega_{i+1}} S_{ss,\text{norm}}(\Omega)\, d\Omega, \quad (15.5)$$

  where Ω_i are the limits of the Terz-bands.

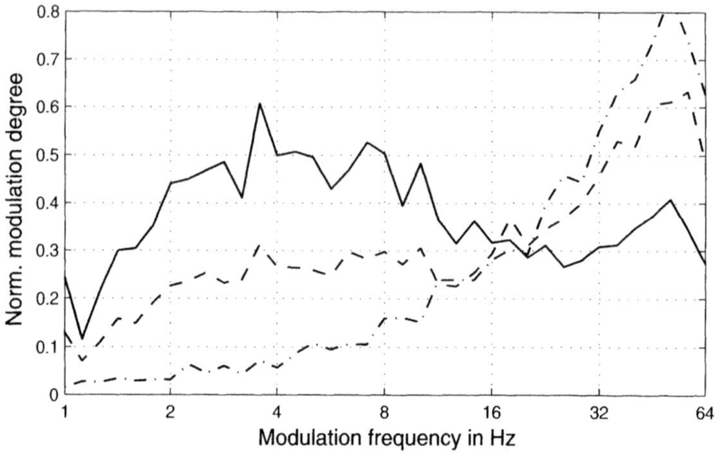

Fig. 15.14. Modulation spectrum of clean speech (solid), noisy speech (dash) and noise (dash-dot).

Such a spectrum is depicted in Fig. 15.14 for three types of signals: clean speech, noisy speech and noise.

One clearly observes that the quantity of the modulation spectrum at 4 Hz is directly related to the SNR of the corresponding signal: For the given example this values decreases form 0.6 for clean speech to 0.3 for noisy speech and to nearly zero for pure noise.

Based on the discussed properties of the modulation spectrum, a long-term noise reduction method can be designed: After the decomposition of the noisy input signal into several frequency subbands, the modulation spectrum at 4 Hz is determined for each subband. Then, this value has to be mapped to a noise reduction gain value, e.g. by

$$g = \max\left[\min\left\{v \cdot \left[S_{\text{mod\_spec}}(4\text{ Hz}) - b\right], 1\right\}, spfl\right]. \tag{15.6}$$

Here, the additive constant b and the gain v map the time-frequency dependent 4 Hz-modulation spectrum to the range of the noise-reduction gains, limited between the Spectral Floor ($spfl$) and 1. The Spectral Floor assures that the attenuation does not exceed an adjustable maximum attenuation of approximately 10 to 15 dB, i.e.

$$max\_atten = -20\,\log_{10}(spfl). \tag{15.7}$$

15.3.1.2 Computational Efficient Realization

Since the procedure for determining the modulation spectrum around 4 Hz is computationally expensive, it would be advantageous to provide an alternative calculation method which also shows the desired relation without explicitly determining the modulation spectrum.

A very simple method which fulfills these requirements mainly consists of two short-term average magnitude (SAM) units, which perform a calculation according to:

$$s_{\mathrm{SAM}}(n) = \begin{cases} \alpha_r \, s_{\mathrm{SAM}}(n-1) + (1-\alpha_r)\,|s(n)| \, : \, |s(n)| > s_{\mathrm{SAM}}(n-1)\,, \\ \alpha_f \, s_{\mathrm{SAM}}(n-1) + (1-\alpha_f)\,|s(n)| : \, |s(n)| \le s_{\mathrm{SAM}}(n-1)\,. \end{cases} \tag{15.8}$$

For the two SAM units different settings α_r and α_f are chosen. One unit estimates the long-term smoothed average magnitude by setting $\alpha_r = \alpha_f$, whereas the other estimator is parametrized by $\alpha_r < \alpha_f$, i.e. the output follows a raising signal power faster than a falling signal power.

With an appropriate choice of the smoothing parameters α_r and α_f for both units, the ratio of these two SAM units is equivalent to the quantity of the modulation spectrum around 4 Hz, but computationally clearly less consuming. The equivalence of the approach utilizing SAM units and the modulation spectrum around 4 Hz is shown in Fig. 15.15. Here the ratio of these two SAM units is depicted in dependence of the modulation frequency of the input signal. It can be well observed that this ratio reaches its maximum around 4 Hz.

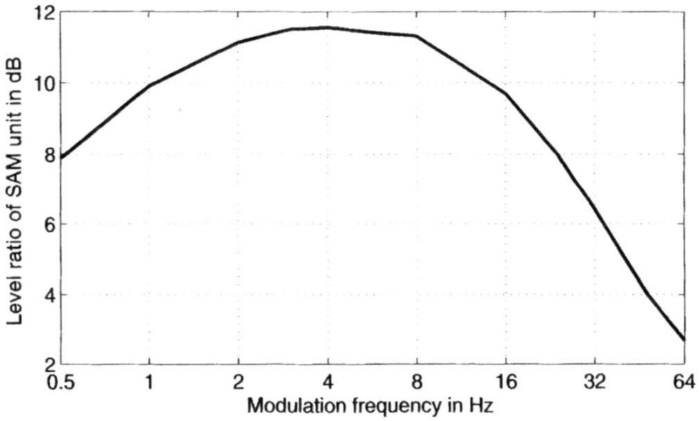

Fig. 15.15. Ratio of two SAM units with different parameter settings.

That the computational efficient approach can be well utilized for determining the long-term modulation-based noise reduction is also obvious by the results depicted in Fig. 15.16. Here, a clean and noisy speech, as well a pure noise signal are depicted in the top. Below, the corresponding modulation spectra and the applied attenuation is depicted, determined based on the

SAM-unit approach. The desired dependence of the applied attenuation on the the modulation spectrum around 4 Hz is obvious.

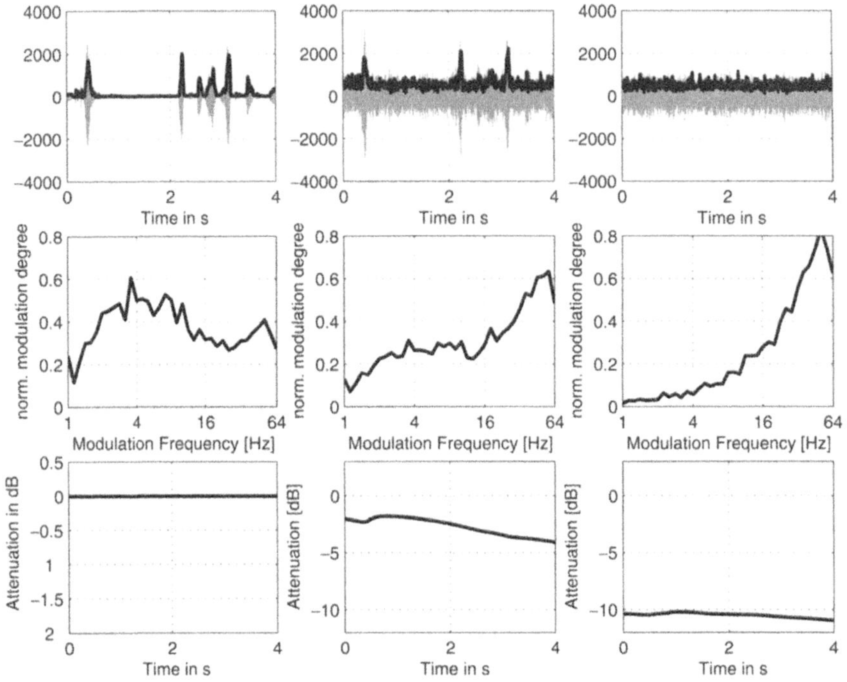

Fig. 15.16. Above: Clean speech (left), noisy speech (mid) and noise (right); Mid: Corresponding modulation spectrum; Below: Long-term noise reduction gain.

15.3.2 Wiener-Filter Based, Short-Term Smoothed Noise Reduction Methods

The aim of these noise reduction procedures is to obtain significant noise reduction performance even for signals whose desired signal and noise components are located in the same frequency range.

Applying the Wiener-filter attenuation:

$$H(\Omega,n) = \frac{S_{ss}(\Omega,n)}{S_{ss}(\Omega,n)+S_{bb}(\Omega,n)} = 1 - \frac{S_{bb}(\Omega,n)}{S_{yy}(\Omega,n)} \tag{15.9}$$

where n denotes the time indices and Ω the normalized frequency. Utilizing short-term estimates for the required power spectral densities $S_{ss}(\Omega,n)$,

$S_{bb}(\Omega, n)$ and $S_{yy}(\Omega, n)$ of speech, noise, and noisy speech, respectively, noticeable noise reduction can be obtained. In these cases, the filter coefficients $H(\Omega, n)$ directly follow short-term fluctuations of the desired signal.

However, a high audio quality noise-reduced signal cannot be easily obtained with this method. The main reason is the non-optimal estimation of power spectral densities which are required in Eq. 15.9. Here, especially the estimation of the noise power spectral density poses problems since the noise signal alone is not available.

In order to obtain reliable estimates, despite of these problems, well-known methods can be utilized. These are:

- Estimating the noise power spectral density in pauses of the desired signal which requires an algorithm to detect these pauses.
- Estimating the noise power spectral density with the Minimum Statistics Method [38] or its modifications [39].

Both methods, however, exhibit a major disadvantage: They only provide long-term smoothed noise power estimates.

However, for power spectral density estimation of the noisy signal, $S_{yy}(\Omega, n)$, which can easily be obtained by smoothing the subband input signal power, short-term smoothing has to be applied in order that the Wiener-filter gains can follow short-term fluctuations of the desired signal.

Calculating the Wiener-filter gain with differently smoothed power spectral density estimates causes the well-known Musical Tones phenomenon [5].

To avoid this unpleasant noise, a large number of procedures have been investigated of which the most widely used are

- Overestimating the noise power spectral density and
- Lower-limiting the Wiener-filter values to a minimum, the so-called Spectral Floor.

With the overestimation of the noise power spectral density, short-time fluctuations of the noise no more provoke a random "opening" of the Wiener-filter coefficients – the cause of Musical Tones.

However, this overestimation reduces the audio quality of the desired signal since especially low power signal components are more strongly attenuated or vanish due to the overestimation. Limiting the noise reduction to the Spectral Floor reduces this problem but, unfortunately, also reduces the overall noise reduction performance. Nevertheless, this reduced noise reduction performance is generally preferred against strong audio quality distortion. More sophisticated methods utilize, e.g., speech characteristics [51] or masking properties [21] of the ear to limit the Wiener attenuation and thus reduce the signal distortion without compromising the noise reduction effect too much.

The noise reduction gain one obtains with the Wiener-filter approach are depicted in Fig. 15.17 for the same signal section which had been chosen to show the long-term noise reduction in Fig. 15.16. One clearly observes that the signal attenuation follows the short-term signal power variations of the input signal: The attenuation is only reduced when short-term speech signal components are present.

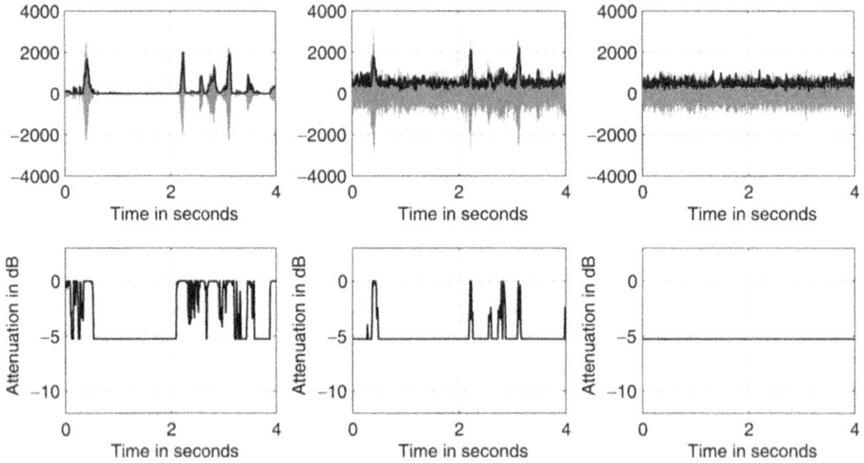

Fig. 15.17. Above: Clean speech (left), noisy speech (mid) and noise (right); Below: Short-term Wiener-filter based noise reduction gain.

However the noise attenuation has to be limited to a smaller value than the long-term noise reduction in order not to reduce speech quality. By combining the noise reduction methods one can profit by the advantages of both: the short-term selective noise reduction during speech presence of the Wiener-filter approach and the stronger noise reduction of the modulation frequency based approach during speech pauses and for frequency bands with negligible SNR. The combination is simply possible by choosing the minimum noise gain which, for the selected signal samples, is shown in Fig. 15.18.

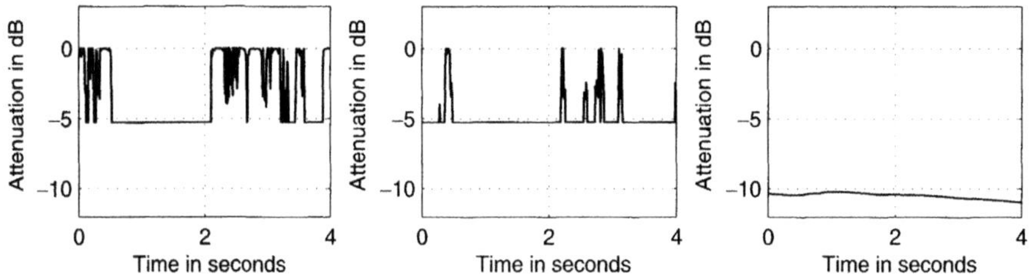

Fig. 15.18. Combined short and long-term noise reduction gain.

15.3.3 Ephraim-Malah Based, Short-Term Smoothed Noise Reduction Methods

An alternative approach to the above outlined Wiener-based noise reduction procedures is the MMSE (Minimum Mean Square Estimation) spectrum amplitude estimator which was initially proposed by Ephraim and Malah [16].

This single channel noise reduction framework is depicted in Fig. 15.19.

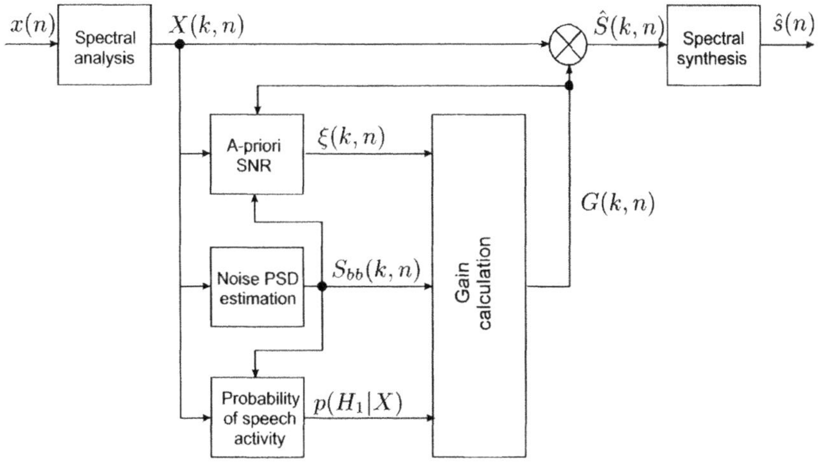

Fig. 15.19. Structure of an Ephraim-Malah based noise reduction method. After the spectral analysis, first the noise power spectral density $S_{bb}(k,n)$ has to be estimated. Then, the a-priori SNR $\xi(k,n)$ is estimated. Optionally, also the probability of speech activity $p(H_1|X)$ may be considered.

First the power spectral density $S_{bb}(k,n)$ of the background noise has to be estimated, e.g., by the Minimum Statistics approach. Then the a-priori SNR is estimated, e.g. by the *Decision directed* approach. Additionally, according [36,43,59] the probability for speech activity may be incorporated, by the additional factor $p(H_1(k,n)|X(k,n))$.

Based on these three estimates the noise reduction gain $G(k,n)$ is determined according to

$$\begin{aligned} G(k,n) &= \frac{\xi(k,n)}{1+\xi(k,n)} \exp\left[\frac{1}{2}\int\limits_{v(k,n)}^{\infty} \frac{\exp(-z)}{z}\,dz\right] \\ &\quad \cdot p\big(H_1(k,n)\big|X(k,n)\big), \\ \text{with: } v(k,n) &= \frac{\xi(k,n)}{1+\xi(k,n)}\gamma(k,n);\ \ \gamma(k,n) = \frac{|X(k,n)|^2}{S_{bb}(k,n)}. \end{aligned} \tag{15.10}$$

For the deriving the calculation formula for the filter weights $G(k, n)$ according to Eqn. 15.10, the knowledge of the distribution of the real and imaginary parts of the speech and noise components is required. They are often assumed as Gaussian [16].

This assumption holds for many noise signals in everyday acoustic environments, but it is not exactly true for speech. A performance investigation for the application in hearing aids can be found, e.g., in [42]. More appropriate models for speech are mentioned in the next section.

15.3.4 Future Trends

So far, the application of well-known noise reduction methods for hearing aids has been explained. Now, we want to outline some methods and ideas for further enhancing the quality of noise reduction.

A big problem of noise reduction procedures is addressed by the first proposal: The estimation of the noise PSD. The basis of this proposal is to utilize both hearing aids on each side of the head for obtaining a more reliable noise PSD estimate, in particular during speech activity. The theoretical basis is the cross-correlation property of the signals of both hearing aids [14]. It is different for speech and noise components. Due to the diffuse character of noise, its components are less correlated than speech components, especially for high frequencies.

The calculation of the cross-correlation requires a full rate audio signal transmission between both hearing aids. When only lower data rate transmission is possible, also some binaural enhancements are possible: Supposing a voice activity detector is utilized for determining the time instances when the noise PSD is preferably estimated, a combined and more reliable activity detector can be obtained by logically combining the detection results of both sides.

As mentioned before, another possibility for a better noise reduction methods is to further advance the Ephraim-Malah noise reduction method by utilizing more appropriate models for the probability density of speech than the Gaussian model. One possibility is to utilize supergaussian statistical modelling for the speech DFT coefficients [32, 40, 41]. Noise reduction algorithms based on this modified estimator outperform the classical approaches using the Gaussian assumption. The noise reduction effect can be increased at an equal target signal distortion level. A computationally efficient realization has been published [32, 33] which allows a parametrization of the probability density function for speech spectral amplitudes so that an implementation in hearing aids is feasible in the near future.

Also model based noise reduction methods such as proposed in Chapter 10 are a promising idea in particular for the enhancement of speech. Since the proposed approach is optimized for car noise as disturbing signal, it has to be further generalized for other kinds of noise signals.

However, independent of the different rules for calculating the filter weights, the estimation quality of the power spectral density shows the strongest impact on the noise reduction quality. Since hearing impaired people wear their hearing aids during the whole day they are very sensitive to signal distortion which is therefore a more critical issue compared to noise reduction for hands-free telephones. For strictly avoiding desired signal distortion, for all noise reduction methods the noise attenuation has to be strongly limited. Unfortunately, for most short-term noise reduction approaches, alternative to the Wiener-filter, the gain of the acceptable noise reduction limit is not very high but has to be paid by a strongly increased computational complexity.

15.4 Multi-Band Compression

Whereas most signal processing algorithms in hearing aids can also be useful for normal hearing (e.g. noise reduction in telecommunications), multi-band compression directly addresses the individual hearing loss. A phenomenon typically observed in sensorineaural hearing loss is "recruitment" [60], which can be measured by categorical loudness scaling procedures (e.g. "Würzburger Hörfeld" [25]) and also could be demonstrated in physiological measurements of basilar membrane velocity [55]. Fig. 15.20 shows the growth of loudness as a function of level for a typical hearing impaired listener in comparison to the normal hearing reference.

With increasing frequency the level difference between normal and hearing-impaired listeners for soft sounds ($<$ 10 CU; CU = Categorical Loudness Unit) increases, whereas curves cross at high levels. The arrows in the right bottom graph indicate the necessary level dependent gain to achieve the same loudness perception at 4 kHz for normal and hearing-impaired listeners. Thus, this measurement directly calls for the need of a frequency specific and level dependent gain - if loudness shall be restored to normal. Since more gain is needed for low input levels than for high input levels, the resulting input-output curves of an appropriate automatic gain control (AGC) system have a compressive characteristic.

Restoration of loudness - often also called "loudness normalization" - has been shown, both theoretically [10] and empirically [56], to be capable of also restoring temporal and spectral resolution (as measured by masking patterns) to normal. However, despite many years of research related to loudness normalization [31, 60], the benefits of this approach are difficult to prove [45]. Thus, over the years, many alternative rationales and design goals have been developed resulting in a large variety of AGC systems.

15.4.1 State-of-the-Art

Practically every modern hearing aid employs some form of AGC. The first stage of a multi-band AGC is a spectral analysis. In order to restore loudness,

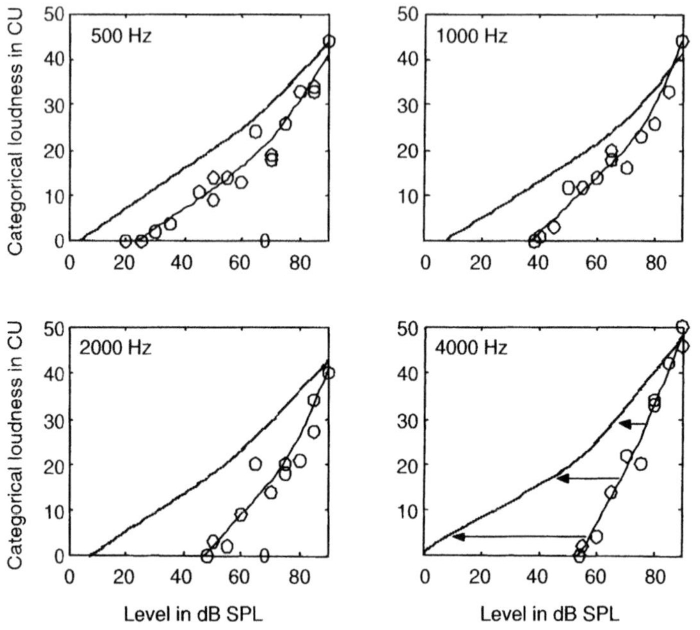

Fig. 15.20. Loudness as a function of level for a hearing-impaired listener (right curve, surrounded by circles) and normal listeners (left curve).

this spectral analysis should be similar to the human auditory system (for details see [65]). Therefore, often non-uniform filterbanks are used: constant bandwidth of about 100 Hz up to 500 Hz and approximately 1/3-octave filters above 500 Hz. In each channel the envelope is extracted as input to the nonlinear input-output function.

Depending on the time constants used for envelope extraction, different rationales can be realized. With very slow attack and release times (several seconds) the gain is adjusted to varying listening environments. These systems are often referred to as *automatic volume control* (AVC), whereas systems with fast time constants (several milliseconds) are called "syllabic compression" as they are able to adjust the gain for vowels and consonants within a syllable. For loudness normalization (also of time varying sounds) gains must be adjusted quasi-instantaneously, i.e., the gains follow the magnitude of the complex band pass signals. Moreover, combinations of both slow and fast time constants ("dual compression") have been developed [57].

To avoid a flattening of the spectral structure of speech signals - which is regarded to be important for speech intelligibility - neighboring channels are coupled or the control signal is calculated as a weighted sum of narrowband and broadband level [57]. The input-output function (see component in Fig. 15.21) calculates a time-varying gain which is multiplied by the band pass signal or the magnitude of the complex bandpass signal prior to the spec-

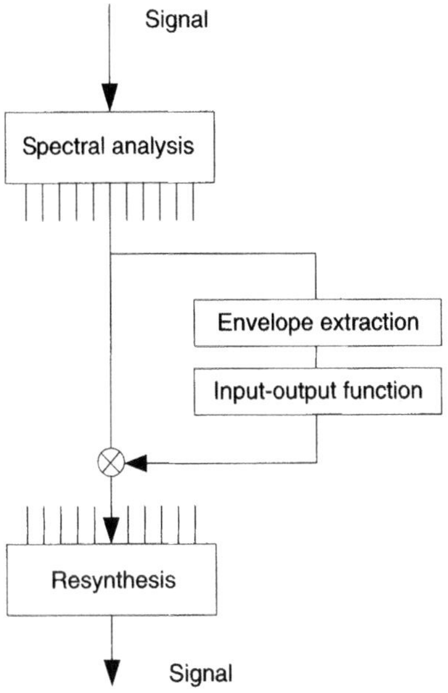

Fig. 15.21. Signal-flow for multi-band AGC processing.

tral resynthesis stage. There are many rationales to determine the frequency specific input-output functions from an individual audiogram, e.g. loudness restoration (see above), restoration of audibility (DSL i/o [11]) or optimization of speech intelligibility without exceeding normal loudness (NAL-NL1 [9]). The optimum rationale usually depends on many variables like hearing loss, age, hearing aid experience and actual acoustical situation.

Whereas the above mentioned AGC systems branch off the control signal before the multiplication of bandpass signal by nonlinear gain ("AGC-i"), output controlled systems ("AGC-o") get the control signal afterwards. AGC-o is often used to ensure that the maximum comfortable level is not exceeded and is thus typically implemented subsequent to an AGC-i. Recently, an AGC-o system has been proposed which is based on percentile levels and keeps the output not only below a maximum level but also above a minimum level in order to optimize audibility [37].

15.4.2 Future Trends

A possibility to cope with situation dependent fitting rationales is to control the AGC parameters (e.g. attack and release time, input-output function) by the classifier. In a situation where speech intelligibility is most important, e.g. a conversation in a crowded restaurant, the appropriate parameters for realizing NAL-NL1 are loaded, whereas when listening to music a setting with optimized sound quality is activated. A wireless link between hearing aids might be beneficial to synchronize the settings on both sides in order to avoid localization problems.

Another promising scenario is to implement psychoacoustic models (e.g. speech intelligibility, loudness, pleasantness) and use them for a continuous and situation dependent constrained optimization of the AGC parameters or directly of the time-varying gain. The latter can be realized by estimating the spectra of noise, speech and the composite signal block by block, similar to the Wiener-filter approach. The speech and noise spectra are used to calculate speech intelligibility (e.g. according to the SII [2]), whereas the overall spectrum is used to determine the current loudness (e.g. according to [10]). Then the channel gains are optimized for each block with the goal to maximize speech intelligibility and the constrained that the aided loudness for the individual hearing impaired listener does not exceed the unaided loudness for a normal listener. In this case, the hearing aid setting is not optimized for the average male speaker in a quiet surrounding (as is done with NAL-NL1), but for the individual speaker in the given acoustical situation.

15.5 Feedback Cancellation

Acoustic feedback ("whistling") is a major problem when fitting hearing aids because it limits the maximum amplification. Feedback describes the situation when output signal components are fed back to the hearing aid microphone and are again amplified. In cases where the hearing aid amplification is larger than the attenuation of the feedback path, and the feedback signal is in phase, instabilities occur and whistling is provoked. The feedback path describes the frequency response of the acoustic coupling between the receiver and the microphones as depicted in Fig. 15.22.

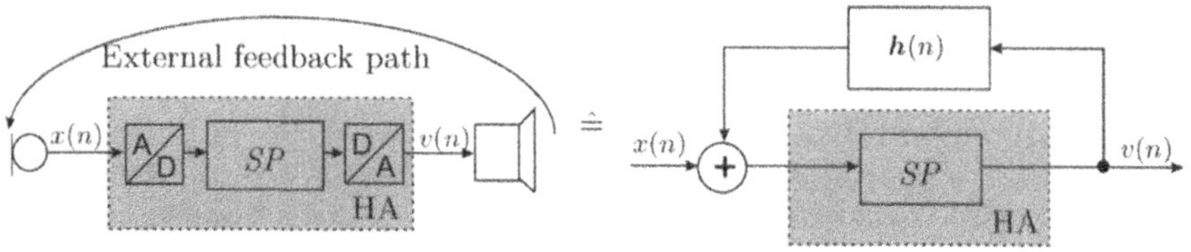

Fig. 15.22. On the left, the acoustic coupling between the hearing aid output and its microphone is shown and on the right the corresponding signal model where the acoustic path is modelled as a FIR filter with impulse response $\boldsymbol{h}(n)$. (HA: hearing aid).

Increasing the ear mold venting or even using open-fitting hearing aids, is more and more preferred by hearing aid users. The reason is that the occlusion effect [12] is usually reduced and the open fitting hearing aids are very comfortable to wear. However, increasing the vent diameter or even using open fitting hearing aids automatically increases the feedback risk and lowers the achievable amplification of the hearing aid. Therefore, well-performing feedback cancellation systems are becoming more and more important.

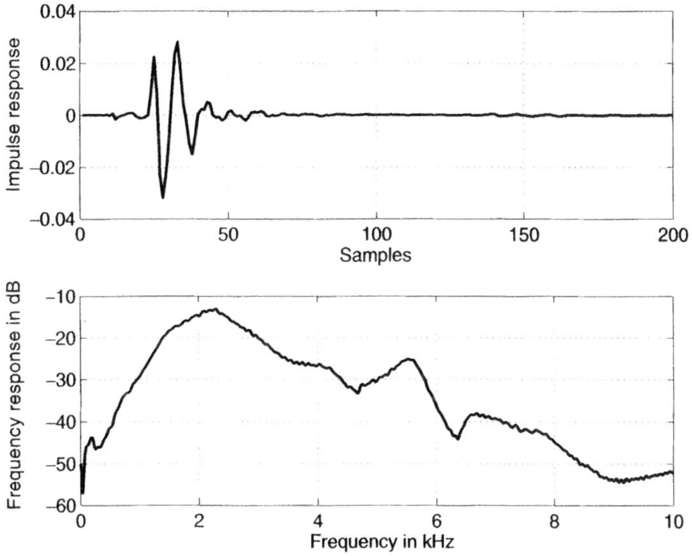

Fig. 15.23. Impulse (top) and frequency (bottom) responses of a typical hearing aid feedback path sampled at 20 kHz.

A typical hearing aid feedback path is depicted in Fig. 15.23. Here, one can observe that generally the paths exhibit a band-pass characteristic with the highest amount of coupling at frequency components between 1 and 5 kHz. The typical length of feedback paths which has to be modelled be a feedback cancellation system is approximately 64 coefficients long for a sampling rate of 20 kHz. Additionally, the current feedback path is highly dependent on many parameters of which the three most important are:

- the type of the hearing aid: BTE (behind-the-ear) or ITE (in-the-ear),
- the vent size,
- obstacles around the hearing aid (hands, hats, telephone receivers),
- the physical fit in the ear canal and leaks from jaw movements.

The first two parameters are static whereas the third is highly time-varying during the operation of the hearing aid. In Fig. 15.24 the variance of the feedback paths can be observed in dependence for the above given parameters.

Corresponding to the time-dependent or static parameters, fixed and dynamic measures are utilized in today's hearing aids to avoid feedback.

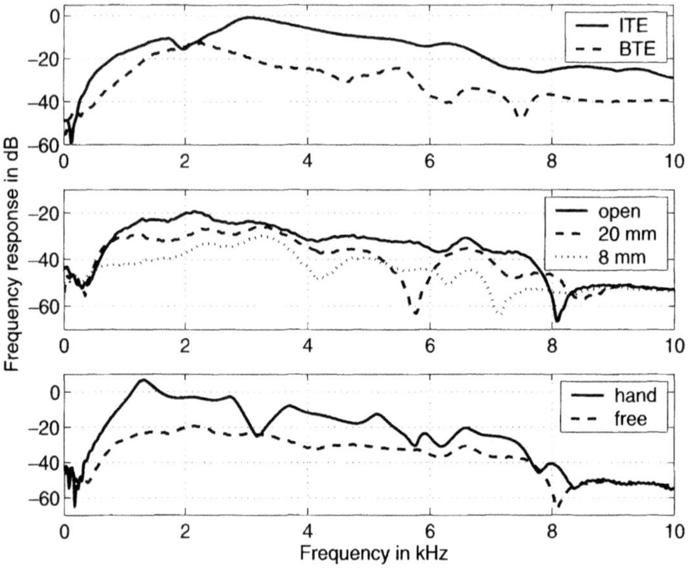

Fig. 15.24. Typical feedback paths for different types of hearing aids (top), different vent sizes (middle), and obstacles, i.e. a hand near the hearing aid compared to the normal situation (bottom).

A static method is to measure the normal feedback path (without obstacles) once after the hearing aid has been fitted. Limiting the gain of the hearing aid so that the closed loop gain is smaller than one for all frequency components, generally can prevent feedback.

Nevertheless, a totally feedback-free performance of the hearing aid can usually not be obtained without additional measures, especially when the closed-loop gain of the hearing aid in normal situations is close to one. Reflection obstacles such as a hand may then provoke feedback. To avoid this, dynamic methods are necessary for cancelling feedback adaptively when it appears.

For these dynamic measures, two methods are widely spread:

1. Selectively attenuating the frequency components for which feedback occurs is utilized in today's hearing aids. This method is normally efficient to avoid feedback. However, it is equivalent to a narrow-band hearing aid gain reduction.
2. Another method is the feedback compensation method where the feedback path is modelled with an internal filter in parallel to the feedback path and which subtracts the feedback signal. Thus, the hearing aid gain is not affected by this method. Additionally, it even allows hearing aid gain settings with closed-loop gains larger than one. This method is currently becoming state-of-the-art for hearing aids.

15.5.1 Feedback Suppression: Dynamic and Selective Attenuation of Feedback Components

An effective and selective attenuation of feedback components can be reached by notch filters. These notch filters are generally characterized by three parameters: the notch frequency, the notch width and the notch depth. It is most important to choose the appropriate notch frequency, i.e. when feedback occurs, the feedback frequency has to be determined fast and precisely.

Different methods, in the time and frequency domains, are applicable for the estimation of the feedback frequency. These are comparable to methods which can also be found for pitch frequency estimation [63]. These methods are, e.g., the zero crossing rate, the autocorrelation function and the linear predictive analysis. Most important is the fast reaction to feedback but also to apply the notch filters only where and as long as necessary in order to minimize the negative effect of the reduced hearing aid gain.

15.5.2 Feedback Compensation

The reduced hearing aid gain can be totally avoided by the compensation approach. Here, a filter is internally put in parallel to the external acoustic feedback path, as shown in Fig. 15.25. The output of the filter models the feedback signal.

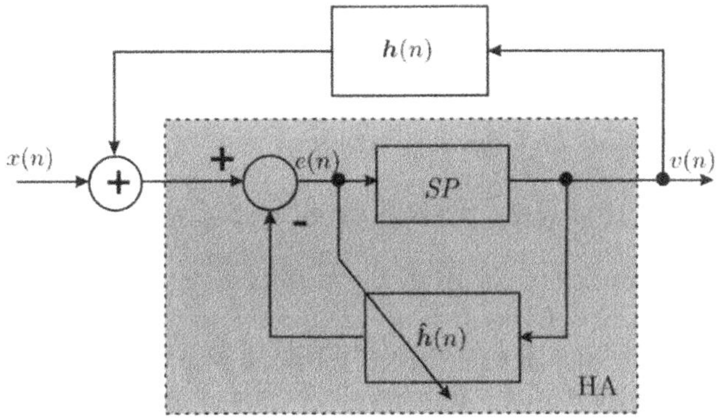

Fig. 15.25. General setup of a feedback cancellation system with SP modeling the hearing aid signal processing, $\boldsymbol{h}(n)$ the external feedback path, $\hat{\boldsymbol{h}}(n)$ the adaptive filter.

The challenge of this approach is to properly estimate the external feedback path with an adaptive filter. This is hard to realize due to the correlation of the input signal and the signal which is acoustically fed back to the microphones. For reliable estimates of the feedback path, the adaptation has to be controlled by sophisticated methods.

Adaptive algorithms generally estimate the filter coefficients, based on an optimization criterion. The criterion which is very often utilized is the minimization of the mean square error signal, i.e., the signal $e(n)$ after the subtraction of the adaptive filter's output signal. Writing $e(n)$ as

$$e(n) = x(n) + \sum_{l=0}^{N-1} \left[h_l(n) - \hat{h}_l(n)\right] v(n-l), \tag{15.11}$$

where the adaptive filter is assumed to model the complete feedback path of length N, and deriving the mean square error $\mathrm{E}\{e^2(n)\}$ with respect to $\hat{h}_l(n)$, one obtains the following relation:

$$\mathrm{E}\left\{e(n)\left[v(n-\nu) - \sum_{l=0}^{N-1}\left[h_l(n) - \hat{h}_l(n)\right]\frac{\partial v(n-l)}{\partial \hat{h}_\nu(n)}\right]\right\} \stackrel{!}{=} 0 \tag{15.12}$$

$$\forall\, \nu \in [0, N-1] \tag{15.13}$$

Under the assumption that the adaptive filter is nearly converged to the feedback path, one obtains the well-known orthogonality theorem:

$$\mathrm{E}\{v(n-l)\, e(n)\} \stackrel{!}{=} 0 \qquad \forall\, l \in [0, N-1]. \tag{15.14}$$

Writing Eqn. 15.11 in vector notation as

$$e(n) = x(n) + \left[\boldsymbol{h}(n) - \hat{\boldsymbol{h}}(n)\right]^T \boldsymbol{v}(n) \tag{15.15}$$

with $\boldsymbol{v}(n) = [v(n), \ldots, v(n-N+1)]^T$, $\hat{\boldsymbol{h}}(n) = [\hat{h}_0(n), \ldots, \hat{h}_{N-1}(n)]^T$ and $\boldsymbol{h}(n) = [h_0(n), \ldots, h_{N-1}(n)]^T$ and deriving the mean square error with respect to $\hat{\boldsymbol{h}}(n)$, one obtains the following equation:

$$\boldsymbol{r}_{x\boldsymbol{v}}(n) + \boldsymbol{R}_{\boldsymbol{vv}}(n)\left[\boldsymbol{h}(n) - \hat{\boldsymbol{h}}(n)\right] = [0, \ldots, 0]^T, \tag{15.16}$$

with the cross-correlation vector $\boldsymbol{r}_{x\boldsymbol{v}}(n) = \mathrm{E}\{x(n)\,\boldsymbol{v}(n)\}$ and the autocorrelation matrix $\boldsymbol{R}_{\boldsymbol{vv}}(n) = \mathrm{E}\{\boldsymbol{v}(n)\boldsymbol{v}^T(n)\}$, respectively.

Resolving this equation with respect to $\hat{\boldsymbol{h}}(n)$, it becomes obvious that the optimum solution which minimizes the mean square error shows a bias compared to the true feedback path:

$$\hat{\boldsymbol{h}}_{opt}(n) = \boldsymbol{h}(n) + \boldsymbol{R}_{\boldsymbol{vv}}^{-1}(n)\,\boldsymbol{r}_{x\boldsymbol{v}}(n). \tag{15.17}$$

The second term $\boldsymbol{R}_{\boldsymbol{vv}}^{-1}(n)\,\boldsymbol{r}_{x\boldsymbol{v}}(n)$ distorts the input signal $x(n)$ as the signal $v(n)$ is filtered such that all predictable components of $x(n)$ are subtracted, i.e. $x(n)$ is whitened. For an alternative derivation of correlation effects, see e.g. [58].

To demonstrate the relations, simulations were performed where the *SP* block of Fig. 15.25 was simply set to a gain g. The filter $\hat{\boldsymbol{h}}(n)$ was adapted under three different conditions:

1. for a white input signal,
2. for a colored input signal with the external feedback path turned to zero: $\boldsymbol{h} = [0, \ldots, 0]^T$, and
3. for a colored input signal with an activated model of the external feedback path.

For the feedback path, a very simple model was used with $\boldsymbol{h} = [\;0\;0\;0\;1\;-0.6\;0.1\;-0.3\;-0.2]^T$. The colored input signal was generated by a MA (moving average) process: $x(n) = u(n) + \sum_{l=0}^{L} a(l)\,u(n-l-1)$, with a white signal $u(l)$ and $L = 20$.

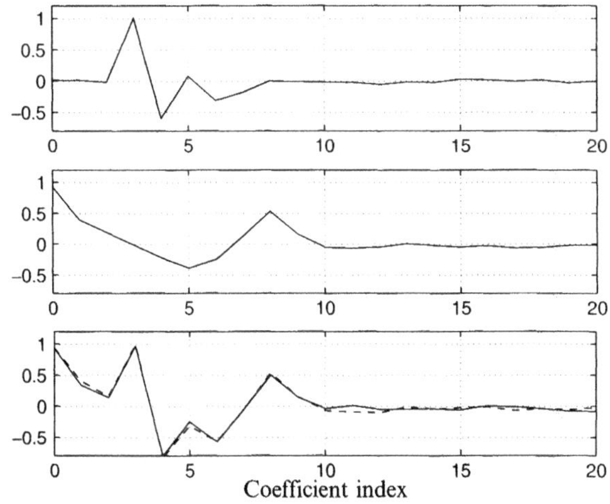

Fig. 15.26. Results for $\hat{\boldsymbol{h}}(n)$ for white (above) and colored excitation with external feedback path off (middle) and on (below). In the lower graph it is shown that the filter (solid line) nearly converges to the sum of the upper and middle graph (dashed line).

The results are depicted in Fig. 15.26. For the white input signal, the filter $\hat{\boldsymbol{h}}(n)$ adapts – as desired – to the feedback path (upper graph). When the feedback path is turned off and the colored signal is used as input, however, the filter acts as a decorrelation filter: If the *SP* block simply is a gain $g = 1$ the filter coefficients model the coefficients $a(l)$ of the input signal's model (middle graph). The result, which is obtained for the case when a colored signal is used to identify the feedback path, shows the superposition of both, the true feedback path and the FIR model of the input signal (lower graph).

Unfortunately, the last case corresponds to the general application for which the decorrelating effect of the feedback cancellation filter can hardly be avoided. This bias causes a distortion of the hearing aid output and has to be reduced as much as possible.

Thus, the main objective for enhancing the adaptation should be to reduce this correlation. Here, different methods exist [52]:

- Decorrelating the input signal with fast-adaptive decorrelation filters,
- delaying the output signal, or
- putting a nonlinear processing unit before the output stage of the hearing aid.

However, none of these methods is a straight-forward solution to the given problem, since many problems occur while implementing the proposals.

We made good experiences with three main settings:

- We reduce the step size, when music is detected as excitation signal,
- we utilize an internal feedback detector which allows a fast feedback reduction when suddenly the external feedback signal decreases, and
- we avoid gain settings of the hearing aid which provoke a closed loop gain setting strongly larger the critical gain.

The music detection is based on the decision of the classificator (see Sec. 15.6). In case when music is present as excitation signal, the risk of a correlated input and thus a biased adaptation of the filter is high. Therefore, to avoid this the step size is reduced. The drawback that the tracking of the filter is reduced can be accepted since when listening to music people usually move less and thus the risk of feedback provoked by feedback path changes is not very high.

The internal feedback detector steadily compares the input signal of the feedback cancellation system and the output signal of the adaptive filter which is the estimated feedback signal. In case the estimated feedback signal is larger than the input signal, this is a clear indication of a mis-adjustment of the adaptive filter. Either an increased step size or a complete reset of the filter coefficients can assure a fast readaptation of the filter coefficients. Usually this case occurs when a obstacle near the hearing aid (hand, telephone receiver, hat, etc.) which provokes a larger feedback path is suddenly removed.

Finally, one has to be aware of the limits of a feedback compensation system: The larger the hearing aid gain exceeds the critical gain, i.e. the gain when feedback occurs without feedback cancellation, the higher are the demands for the feedback compensator, and the more accurately the feedback path has to be estimated to avoid feedback. In other words, only slight mis-adjustment of the feedback path may already provoke strong feedback. This also has a direct impact to the adaption control: Only weak correlations of the input signal and thus a small bias of the estimated filter coefficients may provoke feedback in case of hearing aid gains that exceed the critical gain strongly, i.e. more than 10-15 dB.

15.6 Classification

Hearing aid users encounter a lot of different hearing situations in every day life, e.g. conversation in quiet or in noise, telephone calls, being in a theater or in road traffic noise. They expect real benefits from a hearing aid in each of the mentioned situations. As was shown in the previous part of this paper, modern digital hearing aids provide multiple signal processing algorithms and possible parameter settings, e.g. concerning directivity, noise reduction and dynamic compression. This portfolio of algorithms is expected to still grow with increasing IC computational power. Single algorithms and their multitude of possible parameter settings are mostly working in a situation specific way, i.e. these algorithms are beneficial in certain hearing situations whereas they have no or even negative impact in other situations. For example noise reduction algorithms as described in Sec. 15.3 reduce stationary background noise efficiently, whereas they may have some negative influence on the sound of music and should therefore be disabled in such situations. Even if the optimal signal processing algorithm for any relevant situation would be available, the problem to activate it reliably in the current specific hearing situation remains. A promising solution for this problem is to use a classification system, which can be understood as a superordinate, intelligent algorithm that continuously analyzes the hearing situation and automatically enables the optimal hearing aid setting. The alternative would be a great number of situation specific hearing aid programs, which have to be chosen manually. However, this approach would certainly overextend the mental and motor abilities of many hearing aid users, especially for the small ITE (in-the-ear) devices, and therefore, seems not to be a very attractive alternative [22].

15.6.1 Basic Structure of Monaural Classification

Fig. 15.1 shows the basic structure of a digital hearing aid with a superordinated classification system controlling the different signal processing blocks like directional microphone, noise reduction, shaping of the frequency response and dynamic compression. Classification systems consist of different functional stages:

As a first step, "features" are extracted from the microphone signal. "Features" are certain properties of the signals, whose magnitude is as different as possible for selected situation classes like "speech in quiet", "(speech in) noise" or "music" and can therefore be used to distinguish between situation classes. In literature several spectral and temporal features have been proposed, mostly in the context of separation of "speech in quiet" and "speech in noise": profile and temporal changes of the frequency spectrum [7, 17, 27], statistical distribution of signal amplitudes [35] or analysis of modulation frequencies [48].

To illustrate the principle of feature extraction, Fig. 15.27 shows the extraction of a modulation feature from three different signals belonging to the classes "speech in quiet", "speech in noise" and "music".

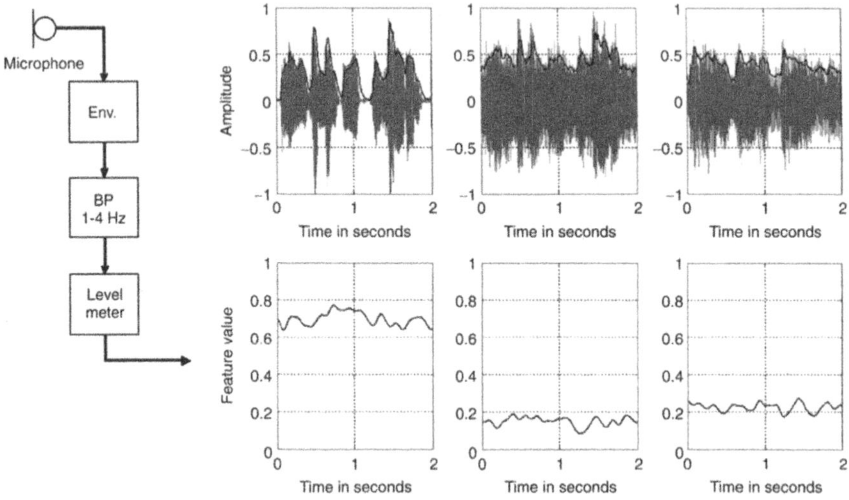

Fig. 15.27. Example for the calculation of an envelope-modulation feature.

The fluctuations of the signal envelope which are calculated by taking the absolute value and lowpass filtering are called "modulation". Typical for speech are strong modulations in the range of 1-4 Hz. The magnitude curves of this feature for the three examples as depicted in Fig. 15.27, show that values of this feature are obviously higher for "speech in quiet" than for the other signals. Consequently, the modulation feature allows to separate "speech in quiet" from "speech in noise" and "music", whereas separation of "speech in noise" and "music" is not possible due to similar feature values. Therefore, most applications of classification techniques require the simultaneous evaluation of a larger number of features to ensure sufficient decision reliability. The assignment of feature values and their combinations to the different classes can be achieved with standard approaches like the Bayes Classifier [48] or Neural Networks [17]. These algorithms learn the necessary a-priori knowledge about the relationship between feature values and situation classes in appropriate training procedures, which have to be based on large and representative databases of every-day life signals.

Fig. 15.28 demonstrates the performance of a classification system in a commercially available high-end hearing aid, which uses a Bayes classification system based on two envelope-modulation and a rhythm features to detect the four classes "speech", "speech in noise", "noise" and "music".

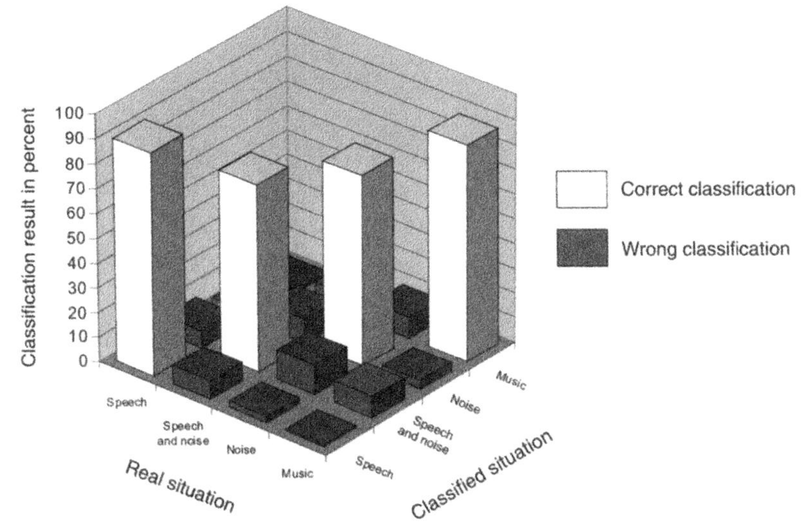

Fig. 15.28. Performance evaluation of a classification system of a modern hearing aid, based on classification of 15 hours of recorded hearing aid microphone signals comprising 500 hearing-situations in total.

Following the Bayes approach, for each detected value of the three features the probability of the four different classes is calculated. After summing up the probabilites across the three different features, the decision is made for the class with the highest cumulated probability. The underlying probability density functions, which are shown in Fig. 15.29, were derived in training with a large training database.

They can be implemented in hearing aids as look-up table or more efficiently in terms of hearing aid memory as polynomial approximations.

Every second a classification decision was made finally leading to the detection and error rates calculated for each of the four classes. Obviously, detection rates between 75 and 90 percent can be achieved, which have shown to be sufficient for a robust and beneficial control of the hearing aid signal processing. The perceptual influence of the misdetections can be reduced to a negligible level by nonlinear temporal averaging of the classification results and, as described in the next section, by smooth transitions between different operation states.

The adaption of the hearing aid signal processing to the detected listening situation is divided into two parts as shown in Fig. 15.1. The block "selection of algorithm and parameters" contains an "action matrix" describing which of the settings for the algorithms and parameters are optimal in each situation. The definition of the action matrix is based on detailed knowledge of the properties of the particular algorithms in the different situations. Extensive investigations and tests are the base for this knowledge. Every time the

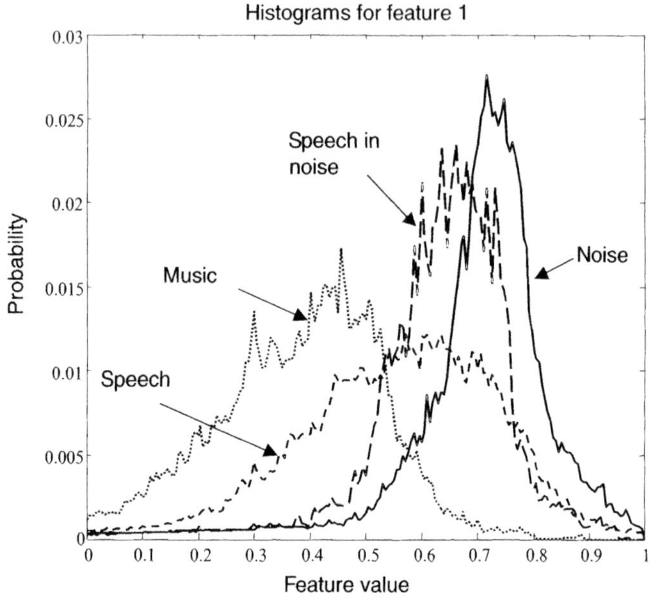

Fig. 15.29. Calculated probability density functions of an envelope-modulation feature in four different classes of hearing situations.

detected situation class is updated, the next block generates "on/off"-control signals for each hearing aid algorithm. Sudden "off/on"-switching of signal processing components like the directional microphone are considered as irritating and unpleasant. Thus, appropriate fading mechanisms which realize a gliding smooth transition from one state of operation to another are advantageous. In many cases, this can easily be achieved by low pass filtering of the control signals. Fig. 15.30 illustrates the fading from omnidirectional to directional microphone mode.

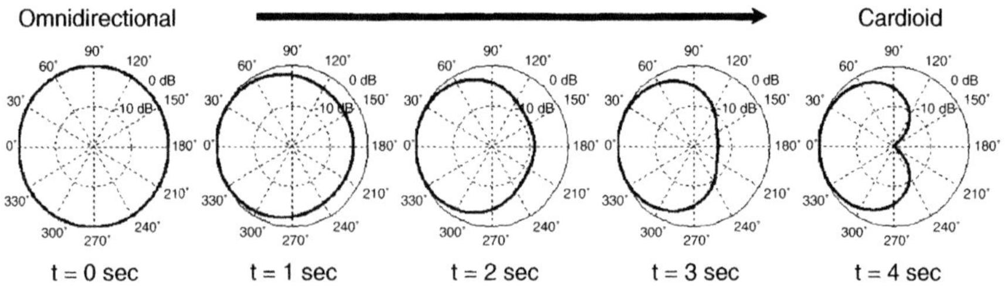

Fig. 15.30. Fading from omnidirectional to directional microphone mode

15.6.2 Binaural Classification

A problem in bilateral fittings of hearing aids with classification systems is that different classes can be detected in the left and right hearing aid resulting in different processing schemes. These differences, e.g. if the directional microphone is activated only on one side, can temporarily reduce the sound quality as well as the speech intelligibility and in addition to that, introduce artificial interaural time and level differences reducing the localization ability of the hearing impaired, which is mainly based on analyzing these signal cues [28].

Differences in classification results are mainly caused by head shading effects in asymmetrical hearing situation, e.g. a hearing situation with a music source on one side of the head and a talker on the other side, would lead to local classification decisions dominated by the ispilateral source, since the contralateral source is shaded, i.e. attenuated, by the head. Real-life evaluations with BTE (behind-the-ear) hearing aids showed that the percentage of asymmetrical classification results can reach up to 20 %. To overcome the problems described above, a binaural synchronization of the classification systems based on a bidirectional low-power wireless link between both hearing aids was introduced recently. In this realization both hearing aids first analyze the sound field independently, then exchange information of the local classification results and then follow exactly the same procedure in parallel to determine the global "binaural" class, see Fig. 15.1. Finally, both hearing aids are adapted synchronously to the signal processing and parameter settings prescribed for the common class. Doing so, the above mentioned disadvantages in unsymmetrical hearing situations can be avoided.

15.6.3 Future Trends

Using multi-microphone signals is the most important step from classification based on the statistical information of one microphone signal towards a future sound scene classification [49]. Typical situations where single-signal based classification systems fail are, for example, listening to music from the car radio while driving or conversing in a cafe with background music. To classify these situations correctly so that the algorithms can take advantage of the result requires information about the sound incidence direction, and the number, distance and type of sound sources in the room. This information can be derived from future multi-microphone localization and classification algorithms. Methods known from the Computational Auditory Scene Analysis (CASA) [6] can be used to further develop today's classification systems. For example, simultaneous speech sources in noisy environment can be recognized by pitch tracking [64].

15.7 Summary

The development of hearing aids covers a wide range of different signal processing components. They are mainly motivated by audiological questions. This chapter focuses on algorithms dealing with the compensation of the recruitment phenomenon, the improvement of speech intelligibility and the enhancement of comfort while using the hearing aid in everyday life.

As one important component of hearing aids, the directional microphone and its effect on the improvement of speech intelligibility is discussed. Directional microphones of different complexities are investigated starting with simple methods like first-order and second-order differential arrays. A description of a four-channel adaptive beamformer closes this topic.

One component which mainly focuses on the improvement of comfort is the noise reduction unit. Algorithms of different complexities, with different amounts of statistical a priori knowledge concerning the computed signal and different speeds of reaction are described. Noise reduction algorithms which exploit the binaural wireless link of future high-end digital hearing aids are discussed as well.

A significant unit in hearing aids is the AGC which compensates the recruitment phenomenon. This chapter discusses state-of-the-art systems and future trends.

Another important aspect is the feedback phenomenon which may occur at high levels of amplification in the hearing aid. This chapter presents two concepts to reduce feedback, namely the feedback compensation approach and the feedback suppression approach.

Finally, the ability of modern hearing aids to detect different hearing situations on the basis of binaurally coupled classification algorithms using a wireless link and to properly adapt to the optimal processing for the specific situation is discussed.

References

[1] J. Anemueller: *Across-Frequency Processing in Convolutive Blind Source Separation*, Univ. Oldenburg, 2001.

[2] *ANSI S3.5-1997*, Methods for calculation of the speech intelligibility index, 1997.

[3] J. Benesty, D. R. Morgan: A better understanding and an improved solution to the specific problems of stereophonic acoustic echo cancellation, *Transactions on Speech and Audio Processing*, **6**(2), 156-165, March, 1998.

[4] J. Benesty, S. Gay (Eds.): *Acoustic Signal Processing for Telecommunication*, Boston/Dordrecht/London: Kluwer Academic Publishers, 2000.

[5] M. Berouti, R. Schwarz, J. Makhoul: Enhancement of speech corrupted by acoustic noise, *Proc. ICASSP '79*, Tulsa, USA, 1979.

[6] A. S. Bregman: *Auditory Scene Analysis*, Cambridge, Massachusetts: MIT Press, 1990.

[7] M. Büchler, N. Dillier, S. Allegro, S. Launer: Klassifizierung der akustischen Umgebung für Hörgeräte-Anwendungen, *Proc. DAGA*, 282-283, Oldenburg, Germany, 2000 (in German).

[8] H. Buchner, R. Aichner, W. Kellermann, Y. Huang, J. Benesty: *Audio Signal Processing for Next-Generation Multimedia Communication Systems*, Blind Source Separation for Convolutive Mixtures: A Unified Treatment, Boston: Kluwer Academic Publishers, Feb., 2004.

[9] D. Byrne, H. Dillon, T. Ching, R. Katsch, G. Keidser: NAL-NL1 Procedure for fitting nonlinear hearing aids: characteristics and comparisons with other procedures, *J. Am. Acad. Audiol.*, **12**, 37-51, 2001.

[10] J. Chalupper, H. Fastl: Dynamic loudness model (DLM) for normal and hearing-impaired listeners, *Acta Acustica united with Acustica*, **88**, 378-386, 2002.

[11] L. E. Cornelisse, R. C. Seewald, D. G. Jamieson: The input/output formula: A theoretical approach to the fitting of personal amplification devices, *J. Acoust. Soc. Am.*, **97**, 1854-1864, 1995.

[12] H. Dillon: *Hearing aids*, New York, Stuttgart: Boomerang Press, 2001.

[13] S. Doclo: *Multi-Microphone Noise Reduction and Dereverberation Techniques for Speech Applications*, Univ. of Leuven, ISBN 90-6582-409-0, 2003.

[14] M. Doerbecker, S. Ernst: Combination of two-channel spectral subtraction and adaptive wiener post-filtering for noise reduction and dereverberation, *Proc. EUSIPCO*, 995-998, Triest, Italy, 1996.

[15] G. W. Elko, A. N. Pong: A simple first order directional microphone, *Proc. IEEE Workshop Appl. Signal Process. Audio Acoust.*, 169-172, Mohonk, New Paltz, NY, 1995.

[16] Y. Ephraim, D. Malah: Speech enhancement using a minimum mean-square error short-time spectral amplitude estimator, *IEEE Trans. on Acoustics, Speech, and Signal Processing*, **ASSP-32**(6), 1109-1121, 1984.

[17] F. Feldbusch: Geräuscherkennung mittels Neuronaler Netze, *Zeitschrift für Audiologie*, **1**, 30-36, 1998 (in German).

[18] S. Gannot, D. Burshtein, E. Weinstein: Signal enhancement using beamforming and non-stationarity with application to speech, *IEEE Trans. on Sig. Proc.*, **49**(8), 1614-1626, Aug., 2001.

[19] J. E. Greenberg, P. M. Zurek: Evaluation of an adaptive beamforming method for hearing aids, *JASA*, **91**(3), 1662-1676, 1992.

[20] L. J. Griffiths, C. W. Jim: An alternative approach to linearly constrained adaptive beamforming, *IEEE Trans. Antennas Propagat.*, **AP-30**, 2734, Jan., 1982.

[21] S. Gustafsson: *Enhancement of Audio Signals by Combined Acoustic Echo Cancellation and Noise Reduction*, RWTH Aachen, ABDN Band 11, Aachen, P. Vary (Hrsg. Verlag der Augustinus Buchhandlung), 1999.

[22] V. Hamacher, E. Fischer, I. Holube: Methods to classify listening situations, *UHA-Tagungsband*, 167-180, Nürnberg, Median Verlag, 2001.

[23] V. Hamacher: Algorithms for future commercial hearing aids, *Proc. ICA '04*, **2**, 1385-1388, Kyoto, Japan, April, 2004.

[24] V. Hamacher, J. Chalupper, J. Eggers, E. Fischer, U. Kornagel, H. Puder, U. Rass: Signal Processing in High-End Hearing Aids: State of the Art, Challenges, and Future Trends, *EURASIP Journal on Applied Signal Processing*, **2005**(18), 2915-2929, 2005.

[25] O. Heller: Hörfeldaudiometrie mit dem Verfahren der Kategorienunterteilung, *Psychologische Beiträge*, **27**, 478-493, 1985 (in German).

[26] I. Holube, V. Hamacher, M. Wesselkamp: Hearing instruments: Noise reduction strategies, *Proc. 18th Danavox Symposium: Auditory Models and Non-linear Hearing Instruments*, Kolding, Denmark, Danavox, Copenhagen, 1999.

[27] J. M. Kates: Classification of background noises for hearing-aid applications, *JASA*, 461-470, 1997.

[28] G. Keidser, K. Rohrseitz, H. Dillon, V. Hamacher, U. Rass, E. Convery, L. Carter, J. Mejia: The effect of multi-channel wide dynamic range compression, noise reduction, and directionality on horizontal localisation performance, *submitted to Ear & Hearing*, 2005.

[29] B. Kollmeier, J. Peissig, V. Hohmann: Binaural noise reduction hearing aid scheme with realtime processing in the frequency domain, *Scand. Audiol. Suppl.*, **38**, 28-38, 1993.

[30] M. Kompis, N. Dillier: Performance of an adaptive beamforming noise reduction scheme for hearing aid applications, *JASA*, **109**(3), 1123-1143, 2001.

[31] A. Leijon: Hearing aid gain for loudness-density normalization in cochlear hearing losses with impaired frequency resolution, *Ear & Hearing*, **12**, 242-250, 1990.

[32] T. Lotter, P. Vary: Noise reduction by maximum a posteriori spectral amplitude estimation with supergaussian speech modeling, *Proc. IWAENC '03*, 83-86, Kyoto, Japan, 2003.

[33] T. Lotter: *Single and Multimicrophone Speech Enhancement for Hearing Aids*, Aachener Beiträge zu Digitalen Nachrichtensystemen, **18**, ISBN 3-86130-645-X, 2004.

[34] C. Liu et al.: A two-microphone dual delay-line approach for extraction of speech sound in the presence of multiple interferers, *JASA*, **110**(6), 3218-3231, 2001.

[35] C. Ludvigsen: Schaltungsanordnung für die automatische Regelung von Hörhilfsgeräten, *Europäische Patentschrift, EP 0 732 036 B1* (in German).

[36] D. Malah, R.V. Cox, A.J. Accardi: Tracking speech-presence uncertainty to improve speech enhancement in non-stationary noise environments, *Proc. ICASSP '99*, **2**, 1761-1764, Phoenix, USA, 1999.

[37] L. F. A. Martin, P. J. Blamey, C. J. James, K. L. Galvin, D. Macfarlane: Adaptive dynamic range optimization for hearing aids, *Acoustics Australia*, **29**, 21-24, 2001.

[38] R. Martin: Spectral subtraction based on minimum statistics, *Proc. EUSIPCO '94*, 1182-1185, Edinburgh, Scotland, 1994.

[39] R. Martin: Noise power spectral density estimation based on optimal smoothing and minimum statistics, *IEEE Trans. on Speech and Audio Processing*, **9**(5), July, 2001.

[40] R. Martin: Speech enhancement using MMSE short time spectral estimation with gamma distributed priors, *Proc. ICASSP '02*, 504-512, Orlando, FL, USA, May, 2002.

[41] R. Martin, C. Breithaupt: Speech Enhancement in the DFT domain using Laplacian speech priors, *Proc. IWAENC '03*, 87-90, Kyoto, Japan, 2003.

[42] M. Marzinzik: *Noise Reduction Schemes for Digital Hearing Aids and Their Use for the Hearing Impaired*, Univ. Oldenburg, Medical Physics, Oldenburg, Germany, 2000.

[43] R.J. McAulay, M.L. Malpass: Speech enhancement using a soft-decision noise suppression filter, *IEEE Trans. on Acoustics and Signal Processing*, ASSP-28,**2**, 1980.

[44] J. Meyer: *Beamforming für Mikrofonarrays unter Berücksichtigung von akustischen Streukörpern*, TU Darmstadt, 2001 (in German).

[45] B. C. J. Moore: *Perceptual Consequences of Cochlear Damage*, London: Oxford University Press, 1995.

[46] H. G. Mueller, M. Killion: An easy method for calculating the articulation index, *The Hearing Journal*, **43**(9), 14-17, 1990.

[47] M. Ostendorf, V. Hohmann, B. Kollmeier: Empirische Klassifizierung verschiedener akustischer Signale und Sprache mittels einer Modulationsfrequenzanalyse, *Proc. DAGA '97*, 608-609, Kiel, Germany, 1997 (in German).

[48] M. Ostendorf, V. Hohmann, B. Kollmeier: Klassifikation von akustischen Signalen basierend auf der Analyse von Modulationsspektren zur Anwendung in digitalen Hörgeräten, *Proc. DAGA*, 402-403, Zürich, Switzerland, 1998 (in German).

[49] V. Peltonen, J. Tuomi, A. Klapuri, J. Huopaniemi, T. Sorsa: Computational auditory scene recognition, *Proc. ICASSP '02*, **2**, 1941-1944, Orlando, FL, USA, 2002.

[50] T. A. Powers, V. Hamacher: Three microphone instrument is designed to extend benefits of directionality, *The Hearing Journal*, **55**(10), 38-45, 2002.

[51] H. Puder: *Geräuschreduktionsverfahren mit modellbasierten Ansätzen für Freisprecheinrichtungen in Kraftfahrzeugen*, TU Darmstadt, Darmstädter Dissertation D17, Fortschritt-Berichte VDI-Reihe 10(721), 2003 (in German).

[52] H. Puder, B. Beimel: Controlling the adaptation of feedback cancellation filters – problem analysis and solution approaches, *Proc. EUSIPCO '04*, Vienna, Austria, Sept., 2004.
[53] T. A. Ricketts: Impact of noise source configuration on directional hearing aid benefit and performance, *Ear and Hearing*, **21**(3), 194-205, 2000.
[54] T. A. Ricketts: Directional hearing aids, *Trends in Amplification*, **5**(4), 139-176, 2001.
[55] M. A. Ruggero, N. C. Rich: Furosemide alters organ of corti mechanics: Evidence for feedback of outer hair cells upon the basilar membrane, *J. Neurosci.*, **11**, 1057-1067, 1991.
[56] N. Sasaki, T. Kawase, H. Hidaka, M. Ogura, T. Takasaka, K. Ozawa, Y. Suzuki, T. Sone: Apparent change of masking functions with compression-type digital hearing aid, *Scand. Audiol.*, **29**, 159-169, 2000.
[57] T. Schneider, R. Brennan: A Multichannel compression strategy for a digital hearing aid, *Proc. ICASSP '97*, *1*, 411-414, Munich, Germany, 1997.
[58] M. G. Siqueira, A. Alwan: Steady-state analysis of continuous adaptation in acoustic feedback reduction systems for hearing-aids, *Transactions on Speech and Audio Processing*, **8**(4), 443-453, July, 2000.
[59] I.Y. Soon, N.K. Soo, C.K. Yeo: Improved noise suppression filter using self-adaptive estimator of probability of speech absence, *Signal Processing*, **75**, 151-159, 1999.
[60] J. Steinberg, M. Gardner: Dependence of hearing impairment on sound intensity, *Journal of the Acoustical Society of America*, **9**, 11-23, 1937.
[61] M. Valente, R. E. Sandlin: *The Textbook of Hearing Aid Amplification*, Use of Microphone Technology to Improve User Performance in Noise, 2nd edition, San Diego, California: Singular Publishing Group, 2000.
[62] J. Vanden Berghe, J. Wouters: An adaptive noise canceller for hearing aids using two nearby microphones, *JASA*, **103**, 3621-3626, 1998.
[63] H. L. Van Trees: *Detection, Estimation, and Modulation Theory*, part 1, New York: John Wiley and Sons, Inc., 1968.
[64] M. Wu, D. Wang, G. J. Brown: A multi-pitch tracking algorithm for noisy speech, *Proc. ICASSP '02*, **1**, 369-372, Orlando, FL, USA, 2002.
[65] E. Zwicker, H. Fastl: *Psychoacoustics: Facts and Models*, 2nd edition, Berlin, Heidelberg, New York: Springer, 1999.

Index

GPSR Compliance

The European Union's (EU) General Product Safety Regulation (GPSR) is a set of rules that requires consumer products to be safe and our obligations to ensure this.

If you have any concerns about our products, you can contact us on ProductSafety@springernature.com

In case Publisher is established outside the EU, the EU authorized representative is:

Springer Nature Customer Service Center GmbH
Europaplatz 3
69115 Heidelberg, Germany

Batch number: 10755248

Printed by Printforce, the Netherlands